Design of Unmanned Aerial Systems

Aerospace Series

Helicopter Flight Dynamics: Including a Treatment of Tiltrotor Aircraft, 3rd Edition
Gareth D. Padfield, CEng, PhD, FRAeS

Space Flight Dynamics, 2nd Edition
Craig A. Kluever

Performance of the Jet Transport Airplane: Analysis Methods, Flight Operations, and Regulations
Trevor M. Young

Small Unmanned Fixed-wing Aircraft Design: A Practical Approach
Andrew J. Keane, András Sóbester, James P. Scanlan

Advanced UAV Aerodynamics, Flight Stability and Control: Novel Concepts, Theory and Applications
Pascual Marqués, Andrea Da Ronch

Differential Game Theory with Applications to Missiles and Autonomous Systems Guidance
Farhan A. Faruqi

Introduction to Nonlinear Aeroelasticity
Grigorios Dimitriadis

Introduction to Aerospace Engineering with a Flight Test Perspective
Stephen Corda

Aircraft Control Allocation
Wayne Durham, Kenneth A. Bordignon, Roger Beck

Remotely Piloted Aircraft Systems: A Human Systems Integration Perspective
Nancy J. Cooke, Leah J. Rowe, Winston Bennett Jr., DeForest Q. Joralmon

Theory and Practice of Aircraft Performance
Ajoy Kumar Kundu, Mark A. Price, David Riordan

Adaptive Aeroservoelastic Control
Ashish Tewari

The Global Airline Industry, 2nd Edition
Peter Belobaba, Amedeo Odoni, Cynthia Barnhart

Modeling the Effect of Damage in Composite Structures: Simplified Approaches
Christos Kassapoglou

Introduction to Aircraft Aeroelasticity and Loads, 2nd Edition
Jan R. Wright, Jonathan Edward Cooper

Theoretical and Computational Aerodynamics
Tapan K. Sengupta

Aircraft Aerodynamic Design: Geometry and Optimization
András Sóbester, Alexander I. J. Forrester

Stability and Control of Aircraft Systems: Introduction to Classical Feedback Control
Roy Langton

Aerospace Propulsion
T. W. Lee

Civil Avionics Systems, 2nd Edition
Ian Moir, Allan Seabridge, Malcolm Jukes

Aircraft Flight Dynamics and Control
Wayne Durham

Modelling and Managing Airport Performance
Konstantinos Zografos, Giovanni Andreatta, Amedeo Odoni

Advanced Aircraft Design: Conceptual Design, Analysis and Optimization of Subsonic Civil Airplanes
Egbert Torenbeek

Design and Analysis of Composite Structures: With Applications to Aerospace Structures, 2nd Edition
Christos Kassapoglou

Aircraft Systems Integration of Air-Launched Weapons
Keith A. Rigby

Understanding Aerodynamics: Arguing from the Real Physics
Doug McLean

Design and Development of Aircraft Systems, 2nd Edition
Ian Moir, Allan Seabridge

Aircraft Design: A Systems Engineering Approach
Mohammad H. Sadraey

Introduction to UAV Systems, 4th Edition
Paul Fahlstrom, Thomas Gleason

Theory of Lift: Introductory Computational Aerodynamics in MATLAB/Octave
G. D. McBain

Sense and Avoid in UAS: Research and Applications
Plamen Angelov

Morphing Aerospace Vehicles and Structures
John Valasek

Spacecraft Systems Engineering, 4th Edition
Peter Fortescue, Graham Swinerd, John Stark

Unmanned Aircraft Systems: UAVS Design, Development and Deployment
Reg Austin

Gas Turbine Propulsion Systems
Bernie MacIsaac, Roy Langton

Aircraft Systems: Mechanical, Electrical, and Avionics Subsystems Integration, 3rd Edition
Ian Moir, Allan Seabridge

Basic Helicopter Aerodynamics, 3rd Edition
John M. Seddon, Simon Newman

System Health Management: with Aerospace Applications
Stephen B. Johnson, Thomas Gormley, Seth Kessler, Charles Mott, Ann Patterson-Hine, Karl Reichard, Philip Scandura Jr.

Advanced Control of Aircraft, Spacecraft and Rockets
Ashish Tewari

Air Travel and Health: A Systems Perspective
Allan Seabridge, Shirley Morgan

Principles of Flight for Pilots
Peter J. Swatton

Handbook of Space Technology
Wilfried Ley, Klaus Wittmann, Willi Hallmann

Cooperative Path Planning of Unmanned Aerial Vehicles
Antonios Tsourdos, Brian White, Madhavan Shanmugavel

Design and Analysis of Composite Structures: With Applications to Aerospace Structures
Christos Kassapoglou

Introduction to Antenna Placement and Installation
Thereza Macnamara

Principles of Flight Simulation
David Allerton

Aircraft Fuel Systems
Roy Langton, Chuck Clark, Martin Hewitt, Lonnie Richards

Computational Modelling and Simulation of Aircraft and the Environment, Volume 1: Platform Kinematics and Synthetic Environment
Dominic J. Diston

Aircraft Performance Theory and Practice for Pilots, 2nd Edition
Peter J. Swatton

Military Avionics Systems
Ian Moir, Allan Seabridge, Malcolm Jukes

Aircraft Conceptual Design Synthesis
Denis Howe

Design of Unmanned Aerial Systems

Dr. Mohammad H. Sadraey
Southern New Hampshire University
Manchester, NH, USA

WILEY

This edition first published 2020

Registered Offices
John Wiley & Sons, Inc., 111 River Street, Hoboken, NJ 07030, USA
John Wiley & Sons Ltd, The Atrium, Southern Gate, Chichester, West Sussex, PO19 8SQ, UK

Editorial Office
The Atrium, Southern Gate, Chichester, West Sussex, PO19 8SQ, UK

For details of our global editorial offices, customer services, and more information about Wiley products visit us at www.wiley.com.

Wiley also publishes its books in a variety of electronic formats and by print-on-demand. Some content that appears in standard print versions of this book may not be available in other formats.

Library of Congress Cataloging-in-Publication Data

Names: Sadraey, Mohammad H., author.
Title: Design of unmanned aerial systems / Dr. Mohammad H. Sadraey.
Description: First edition. | Hoboken, NJ: John Wiley & Sons, 2020. |
Series: Aerospace series | Includes bibliographical references and index.
Identifiers: LCCN 2019024537 (print) | LCCN 2019024538 (ebook) | ISBN 9781119508700 (hardback) |
ISBN 9781119508694 (adobe pdf) | ISBN 9781119508625 (epub)
Subjects: LCSH: Drone aircraft–Design and construction.
Classification: LCC TL685.35 .S235 2019 (print) | LCC TL685.35 (ebook) | DDC 629.133/39–dc23
LC record available at https://lccn.loc.gov/2019024537
LC ebook record available at https://lccn.loc.gov/2019024538

Cover image: © NASA, © NASA/Tony Landis
Cover design by Wiley

Set in 10/12pt Warnock by SPi Global, Pondicherry, India

10 9 8 7 6 5 4 3 2 1

To Fatemeh Zafarani, Ahmad, and Atieh, for all their love and understanding

Contents

Preface

Definitions

An Unmanned Aerial System (UAS) is a group of coordinated multidisciplinary elements for an aerial mission by employing various payloads in flying vehicle(s). In contrast, an Unmanned Aerial Vehicle (UAV) is a remotely piloted or self-piloted aircraft that can carry payloads such as camera, radar, sensor, and communications equipment. All flight operations (including takeoff and landing) are performed without on-board human pilot. In news and media reports, the expression "drone" – as a short term – is preferred.

A UAS basically includes five main elements: 1. Air vehicle; 2. Control station; 3. Payload; 4. Launch and recovery system, 5. Maintenance and support system. Moreover, the environment in which the UAV(s) or the systems elements operate (e.g., the airspace, the data links, relay aircraft, etc.) may be assumed as the sixth (6) inevitable element.

A UAV is much more than a reusable air vehicle. UAVs are to perform critical missions without risk to personnel and more cost effectively than comparable manned system. UAVs are air vehicles; they fly like airplanes and operate in an airplane environment. They are designed like air vehicles; they have to meet flight critical air vehicle requirements. A designer needs to know how to integrate complex, multi-disciplinary systems, and to understand the environment, the requirements and the design challenges.

UAVs are employed in numerous flight missions; in scientific projects and research studies such as hurricane tracking, volcano monitoring, and remote sensing; and in commercial applications such as tall building and bridge observation, traffic control, tower maintenance, and fire monitoring. UAVs also present very unique opportunities for filmmakers in aerial filming/photography.

The UAVs are about to change how directors make movies in capturing the perfect aerial shot. In military arenas, UAVs may be utilized in flight missions such as surveillance, reconnaissance, intelligent routing, offensive operations, and combat. A UAV must typically be flexible, adaptable, capable of performing reconnaissance work, geo-mapping ready, able to collect samples of various pollutants, ready to conduct "search and destroy" missions, and prepared to research in general.

There is no consensus for the definition of autonomy in UAV community. The main systems drivers for autonomy are that it should provide more flexible operation, in that the operator tells the system what is wanted from the mission (not how to do it) with the flexibility of dynamic changes to the mission goals being possible in flight with minimal operation re-planning. Autonomy is classified in 10 levels, from remotely piloted, to fully autonomous swarm. Autonomy includes a level of artificial intelligence. An autopilot is the main element by which the level of autonomy is determined. For instance, stabilization of an unstable UAV is a function for autopilot.

In 2018, at least 122000 people in the U.S. are certified to fly UAVs professionally, according to the Federal Aviation Administration (FAA), which sparked the UAVs explosion in 2016 when it simplified its process for allowing their commercial use. FAA has ruled that commercial UAV flight outside a pilot's line of sight is not allowed. About three million UAVs were sold [1] worldwide in 2017, according to Time Magazine, and more than one million UAVs are registered for US use with the FAA.

By January 2019, at least 62 countries are developing or using over 1300 various UAVs. The contributions of unmanned UAV in sorties, hours, and expanded roles continue to increase. These diverse systems range in cost from a few hundred dollars (Amazon sells varieties) to tens of millions of dollars. Range in capability from Micro Air Vehicles (MAV) weighing less than 1 lb to aircraft weighing over 40000 lbs. UAVs will have to fit into a pilot based airspace system. Airspace rules are based on manned aircraft experience.

Objectives

The objective of this book is to provide a basic text for courses in the design of UASs and UAVs at both the upper division undergraduate and beginning graduate levels. Special effort has been made to provide knowledge, lessons, and insights into UAS technologies and associated design techniques across various engineering disciplines. The author has attempted to comprehensively cover all the main design disciplines that are needed for a successful UAS design project. To cover such a broad scope in a single book, depths in many areas have to be sacrificed.

UAVs share much in common with manned aircraft. The design of manned aircraft and the design of UAVs have many similarities; and some differences. The similarities include: 1. Design process; 2. Constraints (e.g., g-load, pressurization); and 3. UAV main components (e.g., wing, tail, fuselage, propulsion system, structure, control surfaces, and landing gear). The differences include: 1. Autopilot, 2. Communication system, 3. Sensors, 4. Payload, 5. Launch and recovery system, and 6. Ground control station.

The book is primarily written with the objective to be a main source for a UAS chief designer. The techniques presented in this book are suitable for academic study, and teaching students. The book can be adopted as the main text for a single elective course in UAS and UAV design for engineering programs. This text is also suitable for professional continuing education for individuals who are interested in UASs. Industries engineers with various backgrounds can learn about UAS and prepare themselves for new roles in UAS design project.

Approach

The process of UAS design is a complex combination of numerous disciplines which have to be blended together to yield the optimum design to meet a given set of requirements. This is a true statement "the design techniques are not understood unless practiced." Therefore, the reader is highly encouraged to experience the design techniques and concepts through application projects. The instructors are also encouraged to define an open-ended semester-/year-long UAS design project to help the students to practice and learn through the application and experiencing the iterative nature of the design technique. It is my sincere wish that this book will help aspiring students and design engineers to learn and create more efficient and safer UASs, and UAVs.

In this text, the coverage of the topics which are similar to that of a manned aircraft is reviewed. However, the topics which are not covered in a typical manned aircraft design book, are presented in detail. The author has written a book on manned aircraft design – Aircraft Design, a Systems Engineering Approach – published by Wiley. In several topics, the reader recommends the reader to study that text for the complete details. Some techniques (e.g., matching plot) deviate from traditional aircraft design. Throughout the text, the systems engineering approach is examined and implemented.

A UAV designer must: (a) be knowledgeable on the various related engineering topics; (b) be aware of the latest UAV developments; (c) be informed of the current technologies; (d) employ lessons learned from past failures; and (e) appreciate breadth of UAV design options.

A design process requires both integration and iteration. A design process includes: 1. Synthesis: the creative process of putting known things together into new and more useful combinations. 2. Analysis: the process of predicting the performance or behavior of a design candidate. 3. Evaluation: the process of performance calculation and comparing the predicted performance of each feasible design candidate to determine the deficiencies.

UAVs are typically smaller than manned aircraft, have a reduced radar signature, and an increased range and endurance. A UAV designer is also involved in mission planning. Payload type has a direct effect of mission planning. For any mission, the commander seeks to establish criteria that maximize his probability of success. Planning considerations are cost dependent. A UAV can be designed for both scientific purposes and for the military. Their once reconnaissance only role is now shared with strike, force protection, and signals collection.

Beyond traditional aircraft design topics, this text presents detail design of launchers, recovery systems, communication systems, electro-optic/infrared cameras, ground control station, autopilot, radars, scientific sensors, flight control system, navigation system, guidance system, and microcontrollers.

Outline

The objective of the book is to review the design fundamentals of UAVs, as well as the coverage of the design techniques of the UASs. The book is organized into 14 Chapters. Chapter 1 is devoted to design fundamentals including design process, and three design phases (i.e., conceptual, preliminary, and detail). The preliminary design phase is presented in Chapter 2 to determine maximum takeoff weight, wing reference planform area, and engine thrust/power. Various design disciplines including propulsion system, electric system, landing gear, and safety analysis are covered in Chapter 3. The aerodynamic design of wing, horizontal tail, vertical tail, and fuselage is provided in Chapter 4.

Fundamentals of autopilot design including UAV dynamic modeling, autopilot categories, flight simulation, flying qualities for UAVs, and autopilot design process is discussed in Chapter 5. The detail design of control system, guidance system, and navigation system are covered in Chapters 6, 7, and 8 respectively. As the heart of autopilot, the design and application of microcontrollers are explained in Chapter 9. In this Chapter, topics such as microcontroller circuitry, microcontroller elements, embedded systems, and programming are described. Moreover, features of a number of open-source commercial microcontrollers and autopilots (e.g., Arduino and Ardupilot) are introduced. Chapters 10 and 11 are dedicated to two subsystems of a UAS; namely launch and recovery systems, and ground control station. In both chapters, fundamentals, equipment, types, governing equations, ergonomics, technologies, and design techniques are presented.

The payload selection and design is provided in Chapter 12. Various types of payloads including cargo, electro-optic cameras, infrared sensors, range finders, radars, lidars, scientific payloads, military payloads, and electronic counter measure equipment are considered in this chapter. The communications system (including transmitter, receiver, antenna, datalink, frequencies, and encryption) design is discussed in Chapter 13. Finally, in Chapter 14, various design analysis and evaluation techniques; mainly weight and balance, stability analysis, control analysis, performance analysis, and cost analysis techniques are discussed.

Special effort has been made to provide example problems so that the reader will have a clear understanding of the topic discussed. The book contains many fully solved examples in various chapters to exhibit the applications of the design techniques presented. Each chapter concludes with questions and problems; and some chapters with design problems and lab experiments. A solutions manual and figures library are available for instructors who adopt this book.

Quadcopters

Due to the popularity and uniqueness of quadcopters in aeronautics/aviation and commercial applications, this type of UAV is specially treated in this book. A number of sections in various chapters are dedicated to the configuration design, aerodynamic design, and control of quadcopters as follows: Section 2.10. Quadcopter configuration, Section 4.8. Aerodynamic design of quadcopters, and Section 5.7. Quadcopter dynamic model.

Unit System

In this text, the emphasize is on the SI units or metric system; which employs the meter (m) as the unit of length, the kilogram (kg) as the unit of mass, and the second (s) as the unit of time. The metric unit system is taken as fundamental, this being the educational basis in the most parts of the world. It is true that metric units are more universal and technically consistent than British units. However, currently, many Federal Aviation Regulations (FARs) are published in British Units; where the foot (ft) is the unit of length/altitude, the slug is the unit of mass, pound (lb) is the unit of force (weight), and the second (s) as the unit of time. British/imperial units are still used extensively, particularly in the USA, and by industries and other federal agencies and organizations in aviation, such as FAA and NASA.

In FARs, the unit of pound (lb) is used as the unit for force and weight, knot for airspeed, and foot for altitude. Thus, in various locations, the knot is mainly used as the unit of airspeed, lb for weight and force and, ft as the unit of altitude. Therefore, in this text, a combination of SI unit and British unit systems is utilized. For dimensional examples in the text and diagrams, both units are used which it is felt have stood the test of time and may well continue to do so.

In many cases, units in both systems are used, in other cases reference may need to be made to the conversion tables. In either system, units other than the basic one are sometimes used, depending on the context; this is particularly so for weight/mass and airspeed. For instance, the UAV airspeed is more conveniently expressed in kilometers/hour or in knots than in meters/second or in feet/second. For the case of weight/mass, the unit of kg is employed for maximum takeoff mass, while the unit of pound (lb) is utilized for the maximum takeoff weight.

Acknowledgment

Putting a book together requires the talents of many people, and talented individuals abound at Wiley Publishers. My sincere gratitude goes to Eric Willner and Steven Fassioms, executive editors of engineering, Thilagavathy Mounisamy, production editor, and Sashi Samuthiram for composition. My special thanks go to Mary Malin, as outstanding copy editor and proof-reader that are essential in creating an error-free text. I especially owe a large debt of gratitude to my students and the reviewers of this text. Their questions, suggestions, and criticisms have helped me to write more clearly and accurately and have influenced markedly the evolution of this book.

January 2019 *Mohammad H. Sadraey*

Acronyms

2d	Two dimensional
3d	Three dimensional
AC	Alternating Current, aerodynamic center
ADF	Automatic direction finder
AI	Artificial intelligence
AIA	Aerospace Industries Association
AFCS	Automatic flight control systems
APU	Auxiliary power unit
ATC	Air Traffic Control
C2	Command and Control
C3	Command, Control, and Communications
C4ISR	Command, Control, Communications, Computer, Intelligence, Surveillance, and Reconnaissance
CFD	Computational Fluid Dynamics
cg	Center of gravity
CMOS	Complementary metal oxide semiconductor; sensors
COTS	Commercial off-the-shelf
DARPA	Defense Advanced Research Projects Agency
DC	Direct Current
DOD	Department of Defense
DOF	Degree of freedom
DoS	Denial of Service
EO/IR	Electro-Optic/Infra-Red
ECM	Electronic Counter Measures
EM	Electro Magnetic
FAA	Federal Aviation Administration
FAR	Federal Aviation Regulations
FBW	Fly-by-wire
FLIR	Forward looking infrared
FOV	Field of view
fps	ft/sec, frame/sec
GA	General aviation
GCS	Ground control station
GIS	Geographic Information System
GNC	Guidance-Navigation-Control

GPS	Global Positioning System
GUI	Graphical user interface
HALE	High altitude long endurance
HLD	High Lift Device
HTOL	Horizontal takeoff and landing
HVAC	Heating, Ventilation, and Air Conditioning
IC	Integrated Circuit
I2C	Inter-Integrated Circuit
ILS	Instrument landing system
IMU	Inertial measurement unit
INS	Inertial navigation system
IR	Infra-Red
ISA	International Standard Atmosphere
JATO	Jet assisted takeoff
KEAS	Knot Equivalent Air Speed
KTAS	Knot True Air Speed
LED	Light emitting diode
LIDAR	Light detection and ranging
LOS	Line-of-sight
LQR	Linear Quadratic Regulator
MAC	Mean Aerodynamic Chord
mAh	mili Ampere hour
MAV	Micro Air Vehicle
MCE	Mission control element
MDO	Multidisciplinary design optimization
MEMS	Microelectromechanical system
MIL-STD	Military Standards
MIMO	Multiple-input multiple-output
MTBF	Mean time between failures
MTI	Moving Target Indicator
MTOW	Maximum takeoff weight
NACA	National Advisory Committee for Aeronautics
NASA	National Administration for Aeronautics and Astronautics
NTSB	National Transportation Safety Board
OS	Operating System
PIC	Pilot-in-Command
Pot	Potentiometer
PRF	Pulse-repetition frequency
PWM	Pulse Width Modulation
rad	Radian
RC	Remote control, Radio control
RCS	Radar Cross Section
rpm	Revolution per minute
RPV	Remotely piloted vehicle
SAR	Synthetic aperture radar
SAS	Stability augmentation system
Satcom	Satellite Communication

SDRAM	Synchronous dynamic random access memory
SFC	Specific fuel consumption
SIGINT	Signals Intelligence
SISO	Single-Input Single-Output
sUAS	small unmanned aircraft system
sUAV	small unmanned aerial vehicle
TCA	Traffic collision avoidance
TCAS	Traffic Alert and Collision Avoidance System
TE	Trailing Edge
UAS	Unmanned aerial system
UAV	Unmanned Aerial Vehicle
UCAV	Unmanned combat air vehicle
USB	Universal Serial Bus
VHF	Very High Frequency
UHF	Ultra High frequency
VOR	Very High Frequency Omni-Directional Range
VTOL	Vertical takeoff and landing
WGS	World Geodetic System

Nomenclature

Symbol	Name	Unit
a	Speed of sound	m/s, ft/s
a	Acceleration	m/s^2, ft/s^2
a_C	Commanded normal acceleration, Coriolis acceleration	m/s^2, ft/s^2
A	Area	m^2, ft^2
A_r	Effective aperture	m^2, ft^2
AR	Aspect ratio	—
b	Lifting surface (wing, tail) span	m, ft
C	Specific fuel consumption	N/h·kW, lb/h·hp
c	Wave/light velocity	m/s, km/h
C, c	Local chord, moment arm for an accelerometer	m, ft
$\bar{C}$	Mean aerodynamic chord	m, ft
C_D, C_L, C_y	Drag, lift, and side force coefficients	—
C_l, C_m, C_n	Rolling, pitching, and yawing moment coefficients	—
C_{LR}	Rotation lift coefficient	—
$C_{m_{acwf}}$	Wing–fuselage pitching moment coefficient (about wing–fuselage aerodynamic center)	—
C_{Lmax}	Maximum lift coefficient	—
Cm_α	Rate of change of pitching moment coefficient w.r.t. angle of attack	1/rad
Cm_q	Rate of change of pitching moment coefficient w.r.t. pitch rate, $\partial C_m/\partial q$	1/rad
C_{l_β}	Rate of change of rolling moment coefficient w.r.t. sideslip angle, $\partial C_l/\partial \beta$	1/rad
Cn_β	Rate of change of yawing moment coefficient w.r.t. sideslip angle, $\partial C_n/\partial \beta$	1/rad
Cn_r	Rate of change of yawing moment coefficient w.r.t. yaw rate	1/rad
C_{Do}	Zero-lift drag coefficient	—
C_D	Drag coefficient	—
C_{DG}	Ground drag coefficient	—
C_{DTO}	Takeoff drag coefficient	—

Symbol	Name	Unit
$C_{L\alpha}$	Wing/tail/aircraft (3D) lift curve slope	1/rad
$C_{l\alpha}$	Airfoil (2D) lift curve slope	1/rad
C_{Lmax}	Maximum lift coefficient	—
C_p	Pressure coefficient	—
D	Drag force, drag	N, lb
D, d	Distance	m, ft
E	Endurance	h, s
E	Energy	J, ft·lb
E_D	Energy density	Wh/kg
e	Oswald span efficiency factor, natural logarithm base (i.e., 2.72), error, Earth eccentricity	—
f	Wave frequency; number of pulses per second	Hz
F	Force, friction force	N, lb
F_C	Centrifugal force, Coriolis force	N, lb
g	Gravity constant	9.81 m/s^2, 32.17 ft/s^2
G	Fuel weight fraction	—
G_t	Gain of transmitting antenna	—
h	Altitude	m, ft
h, h_o	Non-dimensional distance from cg (h) or ac (h_o) to a reference line	—
H	Angular momentum	kg m^2/s, slug ft^2/s
i_h	Tail incidence	deg, rad
i_w	Wing incidence	deg, rad
l	Length, tail arm	m, ft
I	Mass moment of inertia	kg·m^2, slug·ft^2
I	Current	A, mA
J_{TP}	Rotor inertia	kg·m^2, slug·ft^2
K	Induced drag factor, gain in transfer function, gain in a controller	—
k	Cord spring constant	N/m
L, L_A	Rolling moment	Nm, lb·ft
l	Screw lead	m, in
L	Length	m, ft
L	Lift force, lift	N, lb
$(L/D)_{max}$	Maximum lift-to-drag ratio	—
M	Mach number	—
M, M_A	Pitching moment	Nm, lb·ft
m	Mass	kg, slug
m_B	Battery mass	kg, slug
$\dot{m}$	Engine air mass flow rate	kg/s, lb/s
MTOW	Maximum takeoff weight	N, lb
MAC	Mean aerodynamic chord	m, ft
n	Load factor	—

Symbol	Name	Unit
n	Rotational speed	rpm, rad/s
n_C	Commanded acceleration	—
N	Normal force	N, lb
N'	Guidance gain	—
N, N_A	Yawing moment	N·m, lb·ft
P	Pressure	N/m^2, Pa, lb/in^2, psi
P	Power	W, kW, hp, lb·ft/s
p	Screw pitch	m, in
P_{req}	Required power	W, kW, hp, lb·ft/s
P_{av}	Available power	W, kW, hp, lb·ft/s
P_{exc}	Excess power	W, kW, hp, lb·ft/s
P, p	Roll rate	rad/s, deg/s
q, $\bar{q}$	Dynamic pressure	N/m^2, Pa, lb/in^2, psi
Q, q	Pitch rate	rad/s, deg/s
R	Range	m, km, ft, mile, mi, nmi
R	Air gas constant	287.26 J/kg·K
R	Radius, turn radius	m, ft
Re	Reynolds number	—
ROC	Rate of climb	m/s, ft/min, fpm
R, r	Yaw rate	rad/s, deg/s
s	Laplace transform variable	—
S	Planform area of a lifting/control surface	m^2, ft^2
S_A	Airborne section of the takeoff run	m, ft
S_G	Ground roll	m, ft
S_{TO}	Takeoff run	m, ft
SFC	Specific fuel consumption	N/h/kW, lb/h/hp, 1/s, 1/ft
t	Time	S, min, h
T	Engine thrust	N, lb
T	Temperature	°C, °R, K, °F
T	Torque	Nm, lb·ft
T, t	Thickness	m, ft
t/c	Airfoil thickness-to-chord ratio	—
T/W	Thrust-to-weight ratio	—
U	Forward airspeed	m/s, ft/min, km/h, mi/h, knot
u	Control input in state space	—
V	Velocity, speed, airspeed	m/s, ft/min, km/h, mi/h, knot
V_C	Cruising velocity, closing velocity	m/s, ft/min, km/h, mi/h, knot

Symbol	Name	Unit
V	Volume	m^3, ft^3
V	Voltage	V
V_n	Normal velocity	m/s, knot
V_{max}	Maximum speed	m/s, ft/min, km/h, mi/h, knot
V_{Emax}	Maximum endurance speed	m/s, ft/min, km/h, mi/h, knot
$Vmin_D$	Minimum drag speed	m/s, ft/min, km/h, mi/h, knot
V_{Pmin}	Minimum power speed	m/s, ft/min, km/h, mi/h, knot
V_R	Rotation speed	m/s, ft/min, km/h, mi/h, knot
V_{ROCmax}	Maximum rate of climb speed	m/s, ft/min, km/h, mi/h, knot
V_s	Stall speed	m/s, ft/min, km/h, mi/h, knot
V_T	True airspeed	m/s, ft/min, km/h, mi/h, knot
V_t	Terminal velocity	m/s, ft/min, km/h, mi/h, knot
V_{TO}	Takeoff speed	m/s, ft/min, km/h, mi/h, knot
V_W	Wind speed	m/s, ft/min, km/h, mi/h, knot
V^*	Corner speed	m/s, knot
$\bar{V}_H$, $\bar{V}_V$	Horizontal/vertical tail volume coefficient	—
W	Weight	N, lb
W_A	Autopilot weight	N, lb
WB	Battery weight	N, lb
W_E	Empty weight	N, lb
W_f	Fuel weight	N, lb
W_L	Landing weight	N, lb
W_{PL}	Payload weight	N, lb
W_{TO}	Maximum takeoff weight	N, lb
W/P	Power loading	N/W, lb/hp
W/S	Wing loading	N/m^2, lb/ft^2
x, y, z	Displacement in x-, y-, and z-direction	m, ft
x	State variable in state-space equation	—
Y	Side force	N, lb
y	Output variable in state space	—
z	Variable in transfer function for digital form	—

Greek Symbols

Symbol	Name	Unit
α	Angle of attack	deg, rad
β	Sideslip angle	deg, rad
ε	Downwash angle	deg, rad
ε	Cross-track error	m, ft
γ	Climb angle	deg, rad
θ	Pitch angle, angular displacement, launch angle	deg, rad
λ	Taper ratio, roots of characteristic equation	—
λ	Wavelength	m, in
λ	Localizer error angle, line of sight angle	deg
λ	Longitude	deg
ϕ	Bank angle, latitude	deg, rad
δ	Control surface deflection	deg, rad
σ	Air density ratio	—
σ	Sidewash angle	deg, rad
σ	Radar cross section	m^2, ft^2
σ_{max}	Maximum actuation stress	N/m^2, psi
ρ	Air density, materials density	kg/m^3, $slug/ft^3$
μ	Dynamic viscosity	kg/m·s, $lb{\cdot}s/ft^2$
μ	Friction coefficient	—
η_P	Prop efficiency	—
Λ	Sweep angle	deg, rad
ω	UAV angular velocity, engine rotational speed, turn rate	rad/s, deg/s, rpm
ω	Frequency	Hz
ω_n	Natural frequency	rad/s, deg/s
ω_C	Angular velocity of the line of sight	rad/s, deg/s
Ω	Overall propeller speed, angular velocity of the Earth	rad/s, deg/s
ψ, φ	Yaw angle, heading angle	deg, rad
π	3.14	—
τ	Torque in quadcopter	Nm, lb·ft
ξ	Damping ratio	—
τ	Time constant	s
Γ	Dihedral angle	deg, rad
$d\varepsilon/d\alpha$	Downwash slope	—
$d\sigma/d\beta$	Sidewash slope	—
$\ddot{\theta}$	Takeoff rotation angular acceleration	deg/s^2, rad/s^2

Subscripts

Note: AR, S, b, λ, Λ, Γ, and C without a subscript indicate a wing property

0, o	Zero-lift, sea level, about aerodynamic center, total
0.25	Quarter chord
1	Steady-state value
a, A	Aileron
aft	The most aft location
A	Aerodynamic
AC	Absolute ceiling
ac	Aerodynamic center
avg	Average
a	Aircraft
c/4	Relative to the quarter chord
cg	Center of gravity
C	Ceiling, cruise
cp	Center of pressure
d	Design, dynamic
D	Drag
e, E	Elevator, equivalent, Empty
E	Engine
f	Fuel, fuselage, flap, friction
for	Most forward location
G	Ground
g	Gust
h	Horizontal tail
i	Incidence angle, item number, inboard, ideal, initial
ISA	International Standard Atmosphere
L	Lift, left, landing
LE	Leading edge
LG	Landing gear
max	Maximum
min	Minimum
m	Pitching moment
np	Neutral point
opt	Optimum
p	Propeller
PL	Payload
r, R	Rudder
R	Rotation
r	Root
ref	Reference
s	Stall, stick
SL	Sea level
t	Tip, twist, horizontal tail, transmitted

ft	Fastest turn
tt	Tightest turn
T	True
TO	Takeoff
ult	Ultimate
v, V	Vertical tail
VT	Vertical tail
w, W	Wing, wind
wf	Wing–fuselage
x, y, or z	In the x-, y-, or z-direction
xx, yy, or zz	About the x-, y-, or z-axis

About the Companion Website

This book is accompanied by a companion website:

www.wiley.com/go/sadraey/unmanned-aerial-systems

The website includes:

- Solutions manual.

Scan this QR code to visit the companion website.

1

Design Fundamentals

CONTENTS

Educational Outcomes

After study of this chapter, the reader will be able to:

1) Discuss features of current and modern Unmanned Aerial Vehicles UAVs
2) Manage a Unmanned Aerial System (UAS) design project
3) Develop UAV design requirements

Design of Unmanned Aerial Systems, First Edition. Mohammad H. Sadraey.

Companion website: www.wiley.com/go/sadraey/unmanned-aerial-systems

4) Be familiar with the current UAS technologies
5) Be familiar with the current UAS design techniques
6) Describe UAV classifications
7) Explain design process
8) Conduct feasibility analysis
9) Organize a UAV design team
10) Describe systems engineering approach in designing UAVs
11) Conduct UAV conceptual, preliminary, and detail design phases
12) Know alternatives in a UAV configuration design
13) Analyze design reviews, evaluations, and feedbacks
14) Implement safety analysis
15) Demonstrate familiarity with contemporary issues and challenges related to UAVs
16) Learn from past UAV design lessons
17) Prepare and present a design report

1.1 Introduction

Unmanned Aerial Vehicles (UAVs) are remotely-piloted or self-piloted air vehicles that can carry payloads such as cameras, sensors, and communications equipment. All flight operations (including take-off and landing) are performed without an on-board human pilot. In some reports of Department of Defense (DOD), Unmanned Aerial System (UAS) is preferred. In media reports, the term "drone" is utilized. The UAV mission is to perform critical flight operations without risk to personnel and more cost effectively than comparable manned system. A civilian UAV is designed to perform a particular mission at a lower cost or impact than a manned aircraft equivalent.

UAV design is essentially a branch of engineering design. Design is primarily an analytical process which is usually accompanied by drawing/drafting. Design contains its own body of knowledge that is independent of the science-based analysis tools that is usually coupled with it. Design is a more advanced version of a problem-solving technique that many people use routinely.

Research in UAVs has grown in interest over the past couple decades. There has been tremendous emphasis in UAVs, both of fixed and rotary wing types over the past decades. Historically, UAVs were designed to maximize endurance and range, but demands for UAV designs have changed in recent years. Applications span both civilian and military domains, the latter being the more important at this stage. Early statements about performance, operation cost, and manufacturability are highly desirable already early during the design process. Individual technical requirements have been satisfied in various prototype, demonstrator and initial production programs like Predator, Global Hawk and other international programs. The possible break-through of UAV technology requires support from the aforementioned awareness of general UAV design requirements and their consequences on cost, operation, and performance of UAV systems.

In June 2016, the Department of Transportation's Federal Aviation Administration finalized the first operational rules [2] for commercial use of small unmanned aircraft systems (sUAS), opening pathways toward fully integrating UAS into the USA's airspace. These new regulations aim to harness new innovations safely, to spur job growth, advance critical scientific research and save lives. From FAR (Federal Aviation Regulations) Part 107, a small UAV is defined as the one with a weight of less than 55 lb (i.e., its mass is less than 25 kg). Please note that, Part 107 does not cover recreational applications or hobby use of small UAVs. Moreover, as given in Part 107.51(a)–(d), an sUAS can be operated only below 400 ft. Above Ground Level (i.e., local altitude), 500 ft below clouds, and must have at least three statute mile visibility.

The design principles for UAV's are similar to the principles developed over the years and used successfully for the design of manned UAV. The size of the UAV varies according to the purpose of the utility. In many cases the design and construction of UAVs face new challenges and, as a result of these new requirements, several recent works are concerned with the design of innovative UAVs. Autonomous vehicle technologies for small and large fixed-wing UAVs are being developed by various startups and established corporations such as Lockheed Martin. A number of conceptual design techniques, preliminary design methodologies, and optimization have been applied to the design of various UAVs including Medium-Altitude, Long-Endurance (MALE) UAV using a multi-objective genetic algorithm.

The first UAV designs that appeared in the early 1990s were based on the general design principles for full UAV and findings of experimental investigations. The main limitation of civil UAVs is often low cost. An important area of UAV technology is the design of autonomous systems. The tremendous increase of computing power in the last two decades and developments of general-purpose reliable software packages made possible the use of full configuration design software packages for the design, evaluation, and optimization of modern UAV.

UAVs are air vehicles, they fly like airplanes and operate in an airplane environment. They are designed like air vehicles. They have to meet flight critical air vehicle requirements. You need to know how to integrate complex, multi-disciplinary systems. You need to understand the environment, the requirements and the design challenges.

A UAV system is much more than a reusable air vehicle or vehicles. The UAV system includes five basic elements: 1. The Environment in which the UAV(s) or the Systems Element operates (e.g., the airspace, the data links, relay UAV, etc.). 2. The air vehicle(s) or the Air Vehicle Element. 3. The control station(s) or the Mission Control Element. 4. The payload(s) or the Payload Element. 5. The maintenance and support system or the Support Element.

The design of manned UAV and the design of UAVs have some similarities; and some differences such as: design process; constraints (e.g., g-load, pressurization); and UAV main components (Autopilot, ground station, communication system, sensors, payload). A UAV designer must be aware of: (a) the latest UAV developments; (b) current technologies; and (c) lessons learned from past failures. The designer should appreciate breadth of UAV design options.

The number of small RC model (hobbyist) airplanes registered in the US is projected to grow from 1.1 million units in 2017 to 2.4 million in 2022. Moreover, commercial (small non-model UAVs) registrations are totaled 110 604 in 2017, and are projected to grow to 451 800 in 2022. FAA (Federal Aviation Administration) has required owners of model and commercial UAVs weighing more than 0.55 and less than 55 pounds to register them, beginning December 2015. Large UAVs – those weighing over 55 pounds – must register with the FAA as traditional aircraft.

There are five primary military UAVs in service currently: The US Army's Hunter and Shadow, the US Navy's Pioneer, and the US Air Force's Predator and Global Hawk. The features and characteristics of these UAVs are presented in the next section.

By January 2019, at least 62 countries are using or developing over 1300 various UAVs. The contributions of unmanned UAV in sorties, hours, and expanded roles continue to increase. As of September 2004, some 20 types of coalition UAVs, large and small, have flown over 100 000 total flight hours in support of Operation Enduring Freedom and Operation Iraqi Freedom. Large numbers of UAVs presently exist, both domestically and internationally. Their payload weight carrying capability, their accommodations (volume, environment), their mission profile (altitude, range, duration) and their command, control and data acquisition capabilities vary significantly.

As of January 2019, the following countries developed and employed UAVs with civil payloads: Algeria, Argentina, Armenia, Australia, Austria, Azerbaijan, Belarus, Belgium, Brazil, Bulgaria, Canada, Chile, China, Colombia, Costa Rica, Croatia, Czech Republic, Estonia, Finland, France, Georgia, Germany, Greece, Hungary, India, Indonesia, International, Iran, Israel, Italy, Japan, Jordan, Latvia, Malaysia, Mexico, Netherlands, New Zealand, Nigeria, Norway, North Korea, Pakistan, Peru, Philippines, Poland, Portugal, Romania, Russia, Saudi Arabia, Serbia, Singapore, Slovenia, South Africa, South Korea, Spain, Switzerland, Sweden, Taiwan, Thailand, Tunisia, Turkey, United Arab Emirates, United Kingdom, United States, Vietnam.

Their once reconnaissance only role is now shared with strike, force protection, and signals collection. As of 2018, the following countries developed and employed UAVs with military payloads: Argentina, Australia, Brazil, Canada, China, France, Germany, Israel, Iran, Iraq, India, Italy, Nigeria, Pakistan, Russia, Somalia, Spain, South Africa, Turkey, United Kingdom, and the USA.

General Atomics Aeronautical Systems (as the General Atomic division of General Dynamics), on 9 April 2018 announced that its Predator-series family of Remotely Piloted Aircraft, encompassing MQ-1 Predator, Predator B, Gray Eagle, MQ-9 Reaper, MQ-9B SkyGuardian, and Predator C Avenger, has achieved a historic industry milestone: five million flight hours. The milestone was achieved on 4 April, with 360 311 total missions completed and more than 90% of all missions flown in combat.

UAVs can help companies in fields like energy, telecommunications, videography, sport, and construction to get new perspectives on their works. According to a 2013 report from the Aerospace Industries Association (AIA), the integration of UAS into civil airspace is projected to generate $89 billion through 2023. In 2013, the US has more than one million unmanned flight hours annually, and the Department of Defense operates more than 7000 UAV operations.

Recently, United States Marine Corps procures the new InstantEye Mk-3 GEN5-D1 sUAS system to expand mission scope and to provide additional reconnaissance, surveillance, and target acquisition capability to the individual marine or sailor.

In this chapter, definitions, design process, UAV classifications, current UAVs, and some design challenges will be covered. In addition, conceptual design, preliminary design, and detailed design of a UAV based on systems engineering approach are introduced. At each stage, application of this approach is described by presenting the design flow chart and practical steps of the design.

1.2 UAV Classifications

It is a must for a UAV designer to be aware of classifications and applications of UAVs, which are based on various parameters such as cost, size, weight, mission, and the user. For instance, UAVs range in weight from Micro Air Vehicles (MAV) weighing less than one pound to UAVs weighing over 40 000 pounds. Moreover, these diverse systems range in cost from less than one hundred dollars (Amazon sells varieties) to tens of millions of dollars (e.g., Global Hawk). In addition, UAV missions range from: reconnaissance; combat; target acquisition; electronic warfare; surveillance; special purpose UAV; target and decoy; communication relay; logistics [3]; research and development; civil and commercial UAVs; to Environmental application (e.g., University of Kansas North Pole UAV for measuring ice thickness).

For instance, in humanitarian aid/disaster response, a communication relay UAV can be utilized to fill the gaps in the communications grid when cellphone towers are damaged, or to enhance existing communications past the normal quality, range, or security. Moreover, in a search and rescue mission, swarm technology allows multiple sUASs to work in tandem to cover larger areas and communicate back to a single ground control station.

UAVs are considered as a great force multiplier within military use, as they offer many advantages. Commonly these advantages are attained at a lower risk and a lower cost than if a corresponding manned aircraft would do the same mission. Typical applications for the Navy include [4, 5]: (a) shadowing enemy fleets; (b) decoying missiles by the emission of artificial signatures; (c) electronic intelligence; (d) relaying radio signals; (e) protection of ports from offshore attack; (f) placement and monitoring of sonar buoys and possibly other forms of anti-submarine warfare; (g) optical surveillance and reconnaissance; and (h) Command, Control, Communications, Computer, Intelligence, Surveillance, and Reconnaissance (C4ISR).

For instance, sUASs can be considered as an effective way [5] to conduct aerial swarm raids to neutralize radar defenses by saturating the target acquisition area with a swarm of UAVs that either operate as decoys, or have weapons payloads to destroy designated targets. This tactic has the potential to be more cost effective than traditional stealth technology.

The early classification includes target drones, and remotely piloted vehicles (RPVs). The current classification ranges from Micro UAVs (less than 15 cm long, or 1 lb), to High-Altitude Long-Endurance (HALE), to tactical and combat UAVs. In this section, characteristics of various classifications are briefly presented.

The Micro Unmanned Aerial Vehicles (MUAV) was originally a DARPA (Defense Advanced Research Projects Agency) program to explore the military relevance of Micro Air Vehicles for future military operations, and to develop and demonstrate flight enabling technologies for very small UAV (less than 15 cm/6 in. in any dimension). The Tactical UAV (e.g., Outrider) is designed to support tactical commanders with near-real-time imagery intelligence at ranges up to 200 km. The Joint Tactical UAV (Hunter) were developed to provide ground and maritime forces with near-real-time imagery intelligence at ranges up to 200 km.

The MALE UAV (Predator) provides imagery intelligence to satisfy Joint Task Force and Theater Commanders at ranges out to 500 nautical miles. The High-Altitude Long-Endurance UAV (Global Hawk) are intended for missions requiring long-range deployment and wide-area surveillance or long sensor dwell over the target area.

Table 1.1 shows the UAV classifications from a few aspects including: (a) size (e.g., micro, mini, and small); (b) mass; and (c) mission (e.g., HALE, UCAV). In Table 1.1, the term "size" for fixed-wing UAVs, refers to the largest of the wing span and the fuselage length. However, for quadcopters, it refers to the outer distance between the tip of one propeller to the neighboring one. Moreover, the term weight refers to the maximum takeoff weight (MTOW) of the UAV.

The Boeing-Insitu Scan Eagle (Figure 3.3) is classified as a small UAV, RQ-4 Global Hawk is a HALE UAV, and MQ-1 Predator and MQ-9 Reaper are medium UAVs. X-45 is a tactical/ combat UAV or UCAV (Unmanned Combat Aerial Vehicle). Quadcopters may be in various classes including micro through medium.

In US military, the classification is mainly based on a tier system. For instance, in US Air Force the Tier I is for low altitude, long endurance missions, while Tier II is for MALE missions (e.g., Predator). Moreover, Tier II+ is for HALE missions, and Tier III- denotes HALE low observable. MALE UAVs usually have a continental operating scenario, while HALE UAVs usually have an intercontinental operating scenario.

For other military forces, the following is the classification: Marine Corp: Tier I: Mini UAV; (e.g., Wasp); Tier II: (e.g., Pioneer); and Tier III: Medium range, (e.g., Shadow). Army: Tier I: Small UAV, (e.g., Raven); Tier II: Short range, tactical UAV, (e.g., Shadow 200); and Tier III: Medium range, tactical UAV.

Micro, mini, very small, small, and HALE UAVs, and quadcopters are typically allowed to fly within the category G of international airspace. However, MALE UAVs are allowed to fly within the category A of airspace, and tactical UAVs can fly within the categories B, C, D, and E of airspace. It should be clear that flight within all categories must have a permission prior to the flight operation from the respective authorities. Large UAVs must always send radio signals (via equipment such as transponder) within the labeled airspace to declare their flight characteristics.

Table 1.1 Unmanned Aerial Vehicles (UAVs) classification.

No	UAV Class	Weight (lb)	Size	Normal operating altitude	Range (km)	Endurance
1	Micro	<0.55	≤10 cm	<100 ft	0.1–0.5	≤1 hr
2	Mini	0.55–2	10–30 cm	<500 ft	0.5–1	≤1 hr
3	Very small	2–5	30–50 cm	<1000 ft	1–5	1–3 hr
4	Small	5–55	0.5–2 m	1000–5000 ft	10–100	0.5–2 hr
5	Medium	55–1000	5–10 m	10 000–15 000 ft	500–2000	3–10 hr
6	Large	10 000–30 000	20–50 m	20 000–40 000 ft	1000–5000	10–20 hr
7	Tactical/ combat	1000–20 000	10–30 m	10 000–30 000 ft	500–2000	5–12 hr
8	MALE	1000–10 000	15–40 m	15 000–30 000 ft	20 000–40 000	20–40 hr
9	HALE	>5000	20–50 m	50 000–70 000 ft	20 000–40 000	30–50 hr
10	Quadcopter	0.5–100	0.1–1 m	<500 ft	0.1–2	20 min–1 hr
11	Helicopter	0.001–200	13 mm–2 m	<500 ft	0.2–5	10 min–2 hr

Another basis for classification for UAVs in the military is echelon: (a) Class 1 supports platoon echelon, (e.g., Raven), MUAV, and small UAV; (b) Class 2 supports company echelon, (e.g., Interim Class 1 and 2 UAV); (c) Class 3 supports battalion echelon, (e.g., Shadow 200 Tactical UAV); and (d) Class 4 supports unit of action (brigade), (e.g., Hunter), Extended Range/Multipurpose (ER/MP) UAV. In naming military UAVs in the USA, "R" stands for reconnaissance, and "Q" for unmanned aircraft system. As an example, Predator was the first UAV for a reconnaissance mission (Thus RQ-1).

In terms of wing, there are two groups of UAVs: (a) fixed-wing; and (b) rotary wing. A fixed-wing UAV often needs a runway or a launcher to take-off, while a rotary-wing UAV can take off and land vertically. Two popular groups of rotary-wing UAVs are: (a) unmanned helicopter; and (b) quadcopter. Figure 1.1 shows a Yamaha RMAX unmanned helicopter with a maximum mass of 94 kg that can have an endurance of one hour. This UAV has a main rotor diameter of 3.115 m, and a length of 3.63 m.

Some Current UAVs are listed here:

- **US Army UAV Systems**: RQ-1L I-GNAT Organization; RQ-5/MQ-5 Hunter Aerial Reconnaissance Company; RQ-7 Shadow Aerial Reconnaissance Platoon; RQ-11 Raven Team.
- **Air Force UAV Systems**: RQ-4 Global Hawk; RQ/MQ-1 Predator; MQ-9 Predator B; Force Protection Aerial Surveillance System–Desert Hawk.
- **Navy UAV Systems**: RQ-2 Pioneer; RQ-8B Fire Scout.
- **Marine Corps UAV Systems**: FQM-151 Pointer; Dragon Eye; Silver Fox; Scan Eagle.
- **Coast Guard UAV Systems**: Eagle Eye.
- **Special Operations Command UAV Systems**: CQ-10 SnowGoose; FQM-151 Pointer; RQ-11 Raven; Dragon Eye; RQ-170 Sentinel.

Any UAV must be registered under a particular part of FAR. Table 1.2 presents UAV registration coverage by FAA under three parts: Part 48, Part 47, and Part 107. The FAA has approved [6] the use of a new UAV weighing over 55 pounds for commercial crop-spraying operations.

Figure 1.1 Yamaha RMAX unmanned helicopter.

Table 1.2 Unmanned Aerial Vehicles (UAV) registration coverage.

No	UAV weight (mass)	FAR
1	> or = 0.55 lbs. (250 g)	Part 48
2	> or = 55 lbs. (25 kg)	Part 47
3	< 55 lbs. (25 kg)	Part 107

1.3 Review of a Few Successful UAVs

It is very helpful for a UAV designer to be familiar with the features of some old and current UAVs. The features of the ground control stations of these UAVs are presented in Chapter 11. Moreover, the payloads of these air vehicles are provided in Chapter 12. Furthermore, lessons learned in designing other UAVs are provided in Chapter 2. This covers success stories for some UAV design projects, as well as some lessons from crashes and cancelations of a few discontinued projects.

1.3.1 Global Hawk

The Global Hawk (Figure 1.2) is an advanced intelligence, surveillance, and reconnaissance unmanned air system composed of a HALE vehicle and a ground control segment for command, control, and data collection. Its primary mission is to provide overt, continuous, all-weather, day/night, and near-real-time, large geographic area reconnaissance and surveillance. The Global Hawk, designed and manufactured by Northrop Grumman, is the largest of the military unmanned aircraft systems.

This flew for the first time at Edwards Air Force Base, California, on Saturday, 28 February 1998. The first flight of the Global Hawk signified the first UAV to cross the Pacific Ocean in April 2001, when it flew from the United States to Australia. The entire mission, including the take-off and landing, was performed autonomously by the UAV based on its mission plan. The launch and recovery element of the system's ground segment continuously monitored the status of the flight.

Figure 1.2 Global Hawk.

Guinness World Records has recognized the flight as the longest (13 840 km) by a full-scale unmanned aircraft. The Global Hawks, monitored by shifts of pilots in a ground control station in California, fly 24-hour missions, and they are cheaper to operate than spy plane U-2.

Global Hawk is the largest active current UAVs with successful flights, and with a high altitude and long endurance. Some mass and geometry features are: wingspan: 39.9 m, length: 14.5 m, maximum takeoff mass: 14 628 kg. This HALE UAV has a single Rolls Royce F137-RR-100 turbofan engine with a maximum thrust of 34 kN. The performance statistics of the UAV are: maximum speed: 340 knot, cruise speed: 310 knot, range: 22 779 km, endurance: 32+ hours, and service ceiling: 60 000 ft. Global Hawk has an aluminum fuselage and contains pressurized payload and avionics compartments.

1.3.2 RQ-1A Predator

The Predator UAV (Figure 8.5) – developed by General Atomics – is a relatively small aircraft (with a wing span of 48.4 ft) constructed of high strength composite materials. The Predator made its first flight in June 1994, and is considered as one of the most successful unmanned machines in the US Military.

Predator requires a full-time personnel of about 55 individuals to fly and operate. Predator can conduct an automatic flight along a path that has been programmed in advance. By using satellite communication and an encrypted data link, it can fly remotely anywhere in the world. The wings and tail surfaces are removable, and the entire vehicle fits into a purpose-built container, several of which can be transported in a C-130 along with the ground control trailer.

Predator has a relatively conventional configuration, a high aspect ratio (19) wing, Y-tail, single pusher piston engine, and retractable tricycle landing gear. Due to its MTOW of 2100 lb and a low engine power (115 hp), it has a relatively low performance (Cruise speed of 90 knot, and ceiling of 26 000 ft). However, with a payload of 450 lb, its endurance of 25 hour is relatively long.

Since its first operational deployment in the mid-1990s, the Predator has evolved from a pure surveillance UAV to an armed flying weapon that has become a symbol of American military presence in Middle East countries (e.g., Afghanistan, Iraq, Pakistan, and Syria). The Predator can carry two Hellfire missiles, at a maximum speed of 135 mi/hr and has a range of 770 miles. The Hellfire missiles organizationally do not belong to Airforce; they are utilized, since they are light (14.2 kg), small (53 cm) enough to be attached and carried by Predator. The air force missiles are relatively heavy (86 kg) and large (2.89 m). In 2013, the cost per hour of Predator flight was at $3679. The Predator fleet reached 500 000 flight hours on 18 February 2009. The Predator is planned to retire from US Air Force service soon, its retirement has been pushed to 2018.

1.3.3 MQ-9 Predator B Reaper

There were originally two Predator variants, currently designated as MQ-1 Predator and MQ-9 Predator B Reaper. The MQ-9 Reaper is a larger, heavier, and more capable aircraft than the earlier version (i.e., Predator A). After more than 20 years of service, the US Air Force announced on 27 February 2017 that the MQ-1 Predator UAV would be phased out within a year. The retirement of the Predator does not mean that they will not be present in the skies. Instead General Atomics Aeronautical Systems, the manufacturer, has redeveloped the UAV for civilian use, renamed it the Predator B Reaper, and is trying to sell it to police forces and firefighters.

General Atomics has examined four variants for the Predator B:

- Naval version, named Mariner
- US Customs and Border Protection variants, known as Guardians
- International version, Sky Guardian
- United Kingdom version Protector RG.1.

The Predator B (which first flew on 2001) is equipped with a Honeywell TPE-331-10T turboprop engine, flat-rated to 750 hp and 1360 kg of fuel. The greater engine power allows the Reaper to carry 15 times more ordnance payload and cruise at about three times the speed of the MQ-1 Predator A.

The UAV, with a take-off gross mass of 4500 kg and a wingspan of 20 m, has a maximum speed 210 knot, and ceiling of 50 000 ft. The Predator B has two types of payloads: (a) 340 kg of internal payload for sensors and cameras; and (b) 1360 kg of external payload, for 6 store stations/14 Hellfire missiles. An MQ-9 with two 1000-pound external fuel tanks and 1000 pounds of munitions has an endurance of 42 hours.

Avenger is the combat drone that has been developed by the General Atomics Aeronautical System Company. It is one of the latest unmanned aircrafts of the US Army. In 2012, the company improved the ground control station. A Predator variant, Avenger uses the same ground support infrastructure as the MQ-1 and MQ-9, including the ground control station and existing communications networks. By the end of 2018, variants of General Atomics' Predator (Predator A, Gray Eagle, SkyGuardian, and Reaper) have surpassed five million flight hours.

1.3.4 RQ-5A Hunter

The RQ-5A Hunter UAV is based on the MQ-5 Hunter, the first project that was canceled in January 1996 after some 20 air vehicle crashes. On 13 January 2014, Northrop Grumman's RQ-5A Hunter, in use with the US Army since 1996, surpassed 100 000 combat flight hours in service. This UAV with a wingspan of 10.57 m is powered by two Twin Mercedes HFE Diesel engines, each with a maximum power of 56 hp. Hunter with a maximum TO mass of 885 kg has a maximum speed of 90 knot, a range of 125 km, an endurance of 21 hours and a service ceiling of 18 000 ft.

The UAV has two pilots, one in the Ground Control Station (GCS) to control the flight after takeoff, and one pilot on the runway to conduct takeoff and landing. A hook (a cable system) located below the aircraft is used to snag the aircraft on a set of arresting cables positioned across the runway. As of October 2012, the US Army had 20 MQ-5B Hunters in service, but retirement of the Hunter was expected to be completed in 2013.

1.3.5 RQ-7 Shadow 200

The RQ-7 Shadow 200 unmanned aircraft is of a high- rectangular wing, pusher engine, with a twin tail-boom and an inverted V-tail. The aircraft is powered by a 38 hp piston engine. Unlike the Hunter, the Shadow does not use an external pilot, depending instead on a launcher for takeoffs, and an automated landing system for recovery. The landing system, called the tactical automated landing system controls the aircraft during approach and landing, usually without intervention from the GCS pilot.

A cable system, similar to the one used for the Hunter, is used to stop the aircraft after landing. Aircraft control during flight is accomplished by the GCS pilot through a computer menu interface that allows selection of altitude, heading, and airspeed. During landing, GCS personnel have no visual contact with the aircraft, nor do they have any sensor input from onboard sensors. A command to stop the aircraft engine is given by the GCS pilot, who must rely on an external observer to communicate that the plane has touched down.

1.3.6 RQ-2A Pioneer

Like the US Army's Hunter UAV, the Pioneer (First flight: 1985) requires an external pilot for takeoff and landing. After takeoff, the aircraft can be controlled from a GCS in one of three modes. In the first mode the air vehicle is operated autonomously, and the autopilot uses global positioning system preprogrammed coordinates to fly the air vehicle to each waypoint. In the second mode, the internal pilot commands the autopilot to command airspeed, altitude, and heading or bank angle. In the third mode, the internal pilot flies the aircraft using a joystick. The Pioneer can be landed at a runway using arresting cables, but because it is a US Navy/Marine operated aircraft, it is also landed on board a ship by flying into a net.

This vehicle with a wing span of 16.9 ft, and a length of 14 ft has a Max TO Weight: 450 lbs. The maximum speed is only 110 knots.

Since 1994, it has flown over a number of countries including Bosnia, Haiti, and Somalia. It was also used extensively in Falujeh, Iraq, 2006. During Operation Desert Shield, the US deployed 43 Pioneers that flew 330 sorties, completing over 1000 flight hours. In 10 years, the Pioneer system has flown nearly 14 000 flight hours.

1.3.7 RQ-170 Sentinel

The RQ-170 Sentinel (Figure 1.3) was developed by Lockheed Martin as a stealth UAV, operated by the US Air Force for the Central Intelligence Agency. It has a tail-less

Figure 1.3 Lockheed Martin RQ-170 Sentinel.

configuration, and is fitted with aerial reconnaissance equipment; about 20 of this UAV were built. The wingspan is approximately 20 m, length about 4.5 m, takeoff mass is estimated as being greater than 3900 kg. The UAV is equipped with a single turbofan engine (GE TF34) with a thrust of 40.3 kN. The UAV which was introduced in 2007 has a service ceiling which is estimated to be 50 000 ft. The RQ-170 has three crew members in the GCS to remotely control the UAV; one for launch and recovery, one for mission control, and one as the sensor operator.

1.3.8 X-45A UCAV

The X-45A UCVA (Figure 4.10) – developed by Boeing [7] – is a concept demonstrator for the next generation of completely autonomous military fighter. It is of a swept-wing, stealthy design, no vertical tail, and composite construction. The fuselage is blended into a swept wing, with a small exhaust outlet. The air vehicle can carry advanced precision guided munitions, bombs, or other weapons systems. This vehicle with a wing span of 10.3 m, and a length of 8.08 m has a turbofan engine with a maximum thrust of 28 kN. The maximum speed is Mach 0.75, the range is 2405 km, and the service ceiling is 13 200 m.

In 2003, DARPA announced the cancelation of the X-45B, and the approval for the development of an improved UCAV, called the X-45C. Hence, after the completion of the flight test program, both X-45As were delivered to two national museums.

1.3.9 Epson Micro-flying Robot

In 2004, Epson announced that it has successfully developed one of the world's smallest and lightest micro-flying robots (Figure 1.4) with a mass of (with battery) 12.3 g.

The UAV is equipped with an image sensor unit that can capture and transmit aerial images via a Bluetooth wireless connection to a monitor on ground, and two LED lamps that can be controlled as a means of signaling. General specifications are: Power: 4.2 V, power consumption: 3.5 W, diameter: 136 mm, height: 85 mm, Flight time: about 3 minutes, and mass of the structure: 3 g.

Table 1.3 exhibits characteristics of a number of UAVs, including maximum takeoff mass, wing area, wing span, engine power or thrust. It also demonstrates a few performance features such as service ceiling, maximum speed, range, endurance.

1.4 Design Project Planning

The development and certification of a new UAS project is not cheap and easy, and usually takes several years to conduct. For a design project schedule to be effective, it is necessary to have some procedures for monitoring progress; and in a broader sense for encouraging personnel to progress. An effective general form of project management control device is the Gantt chart. It presents a project overview which is almost immediately understandable to non-systems personnel; hence it has great value as a means of informing management of project status. A Gantt chart has three main features:

1) It informs the manager and chief designer of what tasks are assigned and who has been assigned to them.

Figure 1.4 Epson micro flying robot.

2) It indicates the estimated dates on which tasks are assumed to start and end, and it represents graphically the estimated duration of the task.
3) It indicates the actual dates on which tasks were started and completed and pictures this information.

Like many other planning/management tools, Gantt charts provide the manager/chief designer with an early warning if some jobs will not be completed on schedule and/or if others are ahead of schedule. Gantt charts are also helpful, in that they present graphically immediate feedback regarding estimates of personnel skill and job complexity. A Gantt chart provides the chief designer with a scheduling method and enables him/her to rapidly track and assess the design activities on a weekly/monthly basis. An aircraft project such as Global Hawk (Figure 1.2) will not be successful without a design project planning.

1.5 Decision Making

Not every design parameter is the outcome of a mathematical/technical calculations. There are UAV parameters which are determined through a selection process. In such cases, the designer should be aware of the decision-making procedures. The main challenge in decision making is that there are usually multiple criteria along with a risk associated with each one. Any engineering selection must be supported by logical and scientific reasoning and analysis. The main challenge in decision making is that there are usually multiple criteria along with a risk associated with each one. There are no straightforward governing equations to be solved mathematically.

Table 1.3 Characteristics of a number of Unmanned Aerial Vehicles (UAVs).

No	UAV name	m_{TO} (kg)	m_{PL} (kg)	S (m^2)	b (m)	P or T	Service ceiling	Max speed	Range	Endurance (hr)
1	Predator B Reaper	4760	1700	20.2	20.1	712 kW Turboprop	50 000 ft	260 knot	5926 km	14–28
2	Global Hawk	14 628	2000 lb	69	39.9	31.4 kN Turbofan	65 000 ft	V_C 345 knot	14 000 nm	41
3	Predator A	1020	–	11.5	14.8	86 kW Piston	25 000 ft	117 knot	726 km	24
4	Yamaha R-MAX	94	30 kg	Rotor diameter: 3.115 m	–	15.4 kW Piston	–	–	1 hour	–
5	ScanEagle	18	–	–	3.1	1.5 hp Piston	16 000 ft	70 knot		20+
6	X-45A UCAV	6804	–	–	10.23	31.4 kN Turbofan		Mach 0.75	–	–
7	X-45C UCAV	16 555	–	–	14.9	50.03 kN Turbofan	12.19 km	Mach 1	2220 km	
8	RQ-5A Hunter	885	90 kg	14.28	10.57	2 × 64 hp Piston	18 000 ft	89 knot	125 km	12
9	RQ-7 Shadow 200	170	–	4.5	4.3	28 kW Piston	15 000 ft	110 knot	400 km	6
10	Raven	1.9	0.4	0.32	1.37	250 W Electric	–	30 km/hr	10 km	1

A designer must recognize the importance of making the best decision and the adverse of consequence of making the poorest decision. In majority of the design cases, the best decision is the right decision, and the poorest decision is the wrong one. The right decision implies the design success, and the wrong decision results in a fail in the design. As the level of design problem complexity and sophistication increases in a particular situation, a more sophisticated approach is needed.

1.6 Design Criteria, Objectives, and Priorities

One of the preliminary tasks in UAV configuration design is identifying system design considerations. The definition of a need at the system level is the starting point for determining customer requirements and developing design criteria. The requirements for the system as an entity are established by describing the functions that must be performed. Design criteria constitute a set of "design-to" requirements, which can be expressed in both qualitative and quantitative terms. Design criteria are customer specified or negotiated target values for technical performance measures. These requirements represent the bounds within which the designer must "operate" when engaged in the iterative process of synthesis, analysis, and evaluation. Both operational functions (i.e., those required to accomplish a specific mission scenario, or series of missions) and maintenance and support functions (i.e., those required to ensure that the UAV is operational when required) must be described at the top level.

Various UAV designer have different priorities in their design processes. These priorities are based on different objectives, requirements and mission. There are primarily three groups of UAV designers, namely: (a) military UAV designer; (b) civil UAV designer; and (c) homebuilt UAV designer. These three groups of designers have different interests, priorities, and design criteria. There are mainly 10 figures of merit for every UAV configuration designer. They are: (a) production cost; (b) AV performance; (c) flying qualities; (d) design period; (e) beauty (for civil UAV) or scariness (for military UAV); (f) maintainability; (g) producibility; (h) UAV weight; (i) disposability; and (j) stealth requirement. Table 1.4 demonstrates objectives and priorities of each UAV designer against some figures of merit.

In design evaluation, an early step that fully recognizes design criteria is to establish a baseline against which a given alternative or design configuration may be evaluated. This baseline is determined through the iterative process of requirements analysis (i.e., identification of needs, analysis of feasibility, definition of UAV operational requirements, selection of a maintenance concept, and planning for phase-out and disposal). The mission that the UAV must perform to satisfy a specific customer should be described, along with expectations for cycle time, frequency, speed, cost, effectiveness, and other relevant factors. Functional requirements must be met by incorporating design characteristics within the UAV and its configuration components.

As an example, Table 1.5 illustrates three scenarios of priorities (in percent) for military UAV designers. Among 10 figures of merit (or criteria), grade "1" is the highest priority and grade "10" is the lowest priority. The grade "0" in this table means that, this figure of merit is not a criterion for this designer. Number one priority for a military UAV designer is UAV performance, while for a homebuilt UAV designer cost is the number one priority. It is also interesting that stealth capability is an important priority

Table 1.4 Design objectives.

No	Objective	Basis for measurement	Criterion	Units
1	Inexpensive in market	Unit manufacturing cost	Manufacturing cost	Dollar
2	Inexpensive in operation	Fuel consumption per km	Operating cost	Liter/km
3	Light weight	Total weight	Weight	N
4	Small size	Geometry	Dimensions	m
5	Fast	Speed of operation	Performance	km/hr
6	Maintainable	Man-hour to maintain	Maintainability	Man-hour
7	Producible	Required technology for manufacturing	Manufacturability	–
8	Recyclable	Amount of hazardous or non-recyclable materials	Disposability	kg
9	Maneuverable	Turn radius; turn rate	Maneuverability	m
10	Detect and avoid	Navigation sensors	Guidance and control	–
11	Airworthiness	Safety standards	Safety	–
12	Autonomy	Autopilot complexity	Crashworthiness / formation flight	–

Table 1.5 Three scenarios of priorities (in percent) for a military Unmanned Aerial Vehicle (UAV) designer.

No	Figure of Merit	Priority	Designer # 1	Designer # 2	Designer # 3
1	Cost	4	8	9	9
2	Performance	1	50	40	30
3	Autonomy	2	10	15	20
4	Period of design	5	7	7	8
5	Scariness	10	1	1	2
6	Maintainability	7	4	5	5
7	Producibility	6	6	6	7
8	Weight	8	3	4	4
9	Disposability	9	2	2	3
10	Stealth	3	9	11	12
		Total→	100%	100%	100%

for a military UAV designer, while for three other groups of designers, it is not important at all. These priorities (later called weights) reflect the relative importance of the individual figure of merit in the mind of the designer.

Design criteria may be established for each level in the system hierarchical structure. The optimization objectives must be formulated in order to determine the optimum design. A selected UAV configuration would be optimum based on only one optimization function. Applicable criteria regarding the UAV should be expressed in terms of technical performance measures and should be prioritized at the UAV (system) level. Technical performance measures are measures for characteristics that are, or derive from, attributes inherent in the design itself. It is essential that the development of design criteria be based on an appropriate set of design considerations, considerations that lead to the identification of both design-dependent and design-independent parameters, and that support the derivation of technical performance measures.

1.7 Feasibility Analysis

In the early stages of design and by employing brainstorming, a few promising concepts are suggested which seems consistent with the scheduling and available resources. Prior to committing resources and personnel to the detail design phase, an important design activity; feasibility analysis; must be performed. There are a number of phases, through which the system design and development process must invariably pass. Foremost among them is the identification of the customer-related need and, from that need, the determination of what the system is to do. This is followed by a feasibility study to discover potential technical solution, and the determination of system requirements.

It is at this early stage in the life cycle that major decisions are made relative to adapting a specific design approach and technology application, which has a great impact on the life-cycle cost of a product. In this phase, the designer addresses the fundamental question of whether to proceed with the selected concept. It is evident that there is no benefit or future in spending any more time and resource attempting to achieve an unrealistic objective. Some revolutionary concepts initially seem attractive, but when it comes to the reality, they are found to be too imaginary.

In this phase, the designer addresses the fundamental question of whether to proceed with the selected concept. A feasibility study distinguishes between a creative design concept and an imaginary idea. Feasibility evaluation determines the degree to which each concept alternative satisfies design criteria.

In the feasibility analysis, the answers to the following two questions are sought:

1) Are the goals achievable? or are the objectives realistic? or are the design requirements meetable?
2) Is the current design concept feasible? If the answer to the first question is no, the design goal and objectives, and design requirements must be changed. Hence, no matter where the source of design requirements is, either through direct customer order or market analysis, they must be changed.

1.8 Design Groups

An aircraft chief designer should be capable of covering and handling a broad spectrum of activities. Thus, an aircraft chief designer should have years of experiences, be knowledgeable of management techniques, and preferably have full expertise and background in the area of "flight dynamics." The chief designer has a great responsibility in planning, coordination, and conducting formal design reviews. He/she must also monitor and review aircraft system test and evaluation activities, as well as coordinating all formal design changes and modifications for improvement. The organization must be such that facilitate the flow of information and technical data among various design departments. The design organization must allow the chief designer to initiate and establish the necessary ongoing liaison activities throughout the design cycle.

A primary building block is organizational patterns as the functional approach, which involves the grouping of functional specialties or disciplines into separately identifiable entities. The intention is to perform similar work within one organizational group. Thus, the same organizational group will accomplish the same type of work for all ongoing projects on a concurrent basis. The ultimate objective is to establish a team approach, with the appropriate communications, enabling the application of concurrent engineering methods throughout.

There are mainly two approaches for handling the design activities and establishing design groups: (a) design groups based on aircraft components; and (b) design groups based on expertise. If the approach of groups based on aircraft components is selected, the chief designer must establish the following teams: (a) structural design group; (b) aerodynamic design group; (c) payloads selection; (d) propulsion system design team; (e) landing gear design team; (f) autopilot design team; (g) ground control station design team; (h) launch and recovery design team; (i) communications system group; (j) landing gear; and (k) mission design and trajectory planning. The aerodynamic design group includes a number of teams such as: (a) wing design team; (b) tail design team; and (c) fuselage design team.

The last team is established for documentation, and drafting. There are various advantages and disadvantages for each of the two planning approaches in terms of ease of management, speed of communication, efficiency, and similarity of tasks. However, if the project is large, such as the design of a large HALE aircraft, both groupings (Figure 1.5) could be applied simultaneously. In Chapter 14, design groups based on expertise are presented. This approach is mainly structured and employed for design analysis.

1.9 Design Process

UAV Design is an iterative process which involves synthesis, analysis, and evaluation. Design (i.e., Synthesis) is the creative process of putting known things together into new and more useful combinations. Analysis refers to the process of predicting the

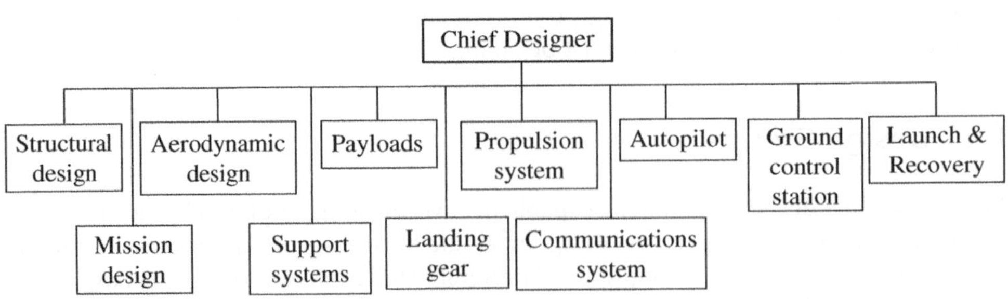

Figure 1.5 Unmanned Aerial Vehicle (UAV) main design groups.

performance or behavior of a design candidate. Evaluation is the process of performance calculation and comparing the predicted performance of each feasible design candidate to determine the deficiencies. A design process requires both integration and iteration. There is an interrelationship between synthesis, analysis, and evaluation. Two main groups of design activities are: (a) problem solving through mathematical calculations; and (b) choosing a preferred one among alternatives.

In general, design considerations are the full range of attributes and characteristics that could be exhibited by an engineered system, product, or structure. These are of interest to both the producer and the customer. Design-dependent parameters are attributes and/or characteristics inherent in the design to be predicted or estimated (e.g., weight, design life, reliability, producibility, maintainability, and disposability). These are a subset of the design considerations for which the producer is primarily responsible. On the other hand, design-independent parameters are factors external to the design that must be estimated and forecasted for use in design evaluation (e.g., fuel cost per gallon, interest rates, labor rates, and material cost per pound). These depend upon the production and operating environment of the UAV.

A goal statement is a brief, general, and ideal response to the need statement. The objectives are quantifiable expectations of performance which identify those performance characteristics of a design that are of most interest to the customer. Restrictions of function of form are called constraints; they limit our freedom to design.

1.10 Systems Engineering Approach

Complex UAV systems, due to the high cost and the risks associated with their development become a prime candidate for the adoption of systems engineering methodologies. The UAV conceptual design process has been documented in many texts, and the interdisciplinary nature of the system is immediately apparent. A successful configuration designer needs not only a good understanding of design, but also systems engineering approach. A competitive configuration design manager must have a clear idea of the concepts, methodologies, models, and tools needed to understand and apply systems engineering to UAV systems.

The design of the configuration for the UAV begins with the requirements definition and extends through functional analysis and allocation, design synthesis and evaluation, and finally validation. An optimized UAV, with a minimum of undesirable side effects, requires the application of an integrated life-cycle oriented "system" approach. Operations and support needs must be accounted for in this process. An optimized UAV, with a minimum of undesirable side effects, requires the application of an integrated life-cycle oriented "system" approach.

The design of the UAV subsystems plays a crucial role in the configuration design and their operation. These subsystems turn an aerodynamically shaped structure into a living, breathing, unmanned flying machine. These subsystems include the: flight control subsystem, power transmission subsystem, fuel subsystem, structures, propulsion, aerodynamics, and landing gear. In the early stages of a conceptual or a preliminary design these subsystems must initially be defined, and their impact must be incorporated into the design layout, weight analysis, performance calculations and cost benefits analysis.

A UAV is a system composed of a set of interrelated components working together toward some common objective or purpose. Primary objectives include safe flight achieved at a low cost. Every system is made up of components or subsystems, and any subsystem can be broken down into smaller components. For example, in an air transportation system, the UAV, terminal, ground support equipment, and controls are all subsystems. The UAV life-cycle is illustrated in Figure 1.6.

A UAV must feature product competitiveness, otherwise the producer and designer may not survive in the world marketplace. Product competitiveness is desired by UAV producers worldwide. Accordingly, the systems engineering challenge is to bring products and systems into being that meet the mission expectations cost-effectively. Because of intensifying international competition, UAV producers are seeking ways to gain sustainable competitive advantages in the marketplace.

It is essential that UAV designers be sensitive to utilization outcomes during the early stages of UAV design and development. They also need to conduct life-cycle engineering as early as possible in the design process. Fundamental to the application of systems engineering is an understanding of the system life-cycle process illustrated in Figure 1.6. It must simultaneously embrace the life cycle of the manufacturing process, the life cycle of the maintenance and support capability, and the life cycle of the phase-out and disposal process.

The requirements need for a specific new UAV first comes into focus during the conceptual design process. It is this recognition that initiates the UAV conceptual design process to meet these needs. Then, during the conceptual design of the UAV, consideration should simultaneously be given to its production and support. This gives rise to a parallel life cycle for bringing a manufacturing capability into being.

Traditional UAV configuration design attempts to achieve improved performance and reduced operating costs by minimizing MTOW. From the point of view of an UAV customer, however, this method does not guarantee the optimality of an UAV program. Multidisciplinary design optimization (MDO) is an important part of the UAV configuration design process. It first discusses the design parameters, constraints, objectives functions and criteria and then UAV configuration classifications. Then the relationship between each major design option and the design requirements are evaluated. Then the systems engineering principals are presented. At the end, the systems engineering approach is applied in the optimization of the UAV configuration design and a new configuration design optimization methodology is introduced.

The design of a UAV within the system life-cycle context is different from the design just to meet a set of performance or stability requirements. Life-cycle focused design is simultaneously responsive to customer needs and to life-cycle outcomes. The design of

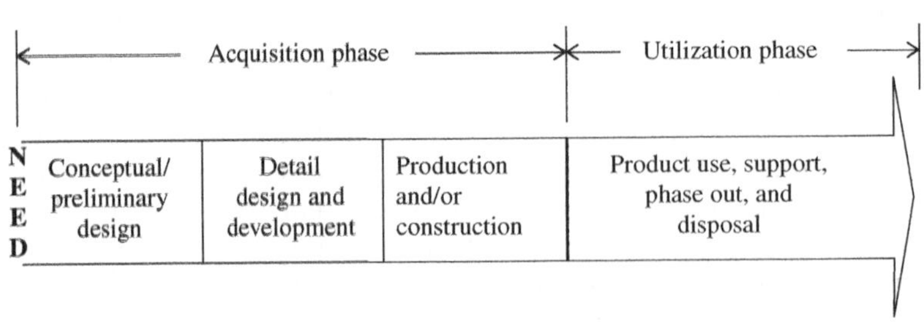

Figure 1.6 The Unmanned Aerial Vehicle (UAV) life-cycle.

the UAV should not only transform a need into a UAV/system configuration, but should also ensure the UAV's compatibility with related physical and functional requirements. Further, it should consider operational outcomes expressed as safety, producibility, affordability, reliability, maintainability, usability, supportability, serviceability, disposability, and others, as well as the requirements on performance, stability, control, and effectiveness.

An essential technical activity within this process is that of evaluation. Evaluation must be inherent within the systems engineering process and must be invoked regularly as the system design activity progresses. However, systems evaluation should not proceed without guidance from customer requirements and specific system design criteria. When conducted with full recognition of design criteria, evaluation is the assurance of continuous design improvement. There are a number of phases through which the system design and development process must invariably pass.

Foremost among them is the identification of the customer related need and, from that need, the determination of what the system is to do. This is followed by a feasibility analysis to discover potential technical solutions, the determination of system requirements, the design and development of system components, the construction of a prototype, and/or engineering model, and the validation of system design through test and evaluation. The system (e.g., UAV) design process includes four major phases: (a) conceptual design; (b) preliminary design; (c) detail design; and (d) test and evaluation. The four phases of the integrated design of an UAV are summarized in Figure 1.7. Sections 1.10 through 1.13 present the details of these design phases.

In the conceptual design phase, the UAV will be designed in concept without precise calculations. In other words, almost all parameters are determined based on a decision-making process and a selection technique. On the other hand, the preliminary design phase tends to employ the outcomes of a calculation procedure. As the name implies, in the preliminary design phase, the parameters that are determined are not final and will be altered later. In addition, in this phase, parameters are essential and will directly influence the entire detail design phase. Therefore, the ultimate care must be taken to ensure the accuracy of the results of the preliminary design phase. In the detail design phase, the technical parameters of all components (e.g., wing, fuselage, tail, landing gear [LG], and engine) including geometry are calculated and finalized.

1.11 UAV Conceptual Design

Throughout the conceptual system design phase (commencing with the need analysis), one of the major objectives is to develop and define the specific design-to requirements for the system as an entry. The results from these activities are combined, integrated, and included in a system specification. This specification constitutes the top "technical-requirements" document that provides overall guidance for system design from the beginning. Conceptual design is the first and most important phase of the UAV system design and development process. It is an early and high-level life cycle activity with potential to establish, commit, and otherwise predetermine the function, form, cost, and development schedule of the desired UAV system. The identification of a problem and associated definition of need provides a valid and appropriate starting point for design at the conceptual level.

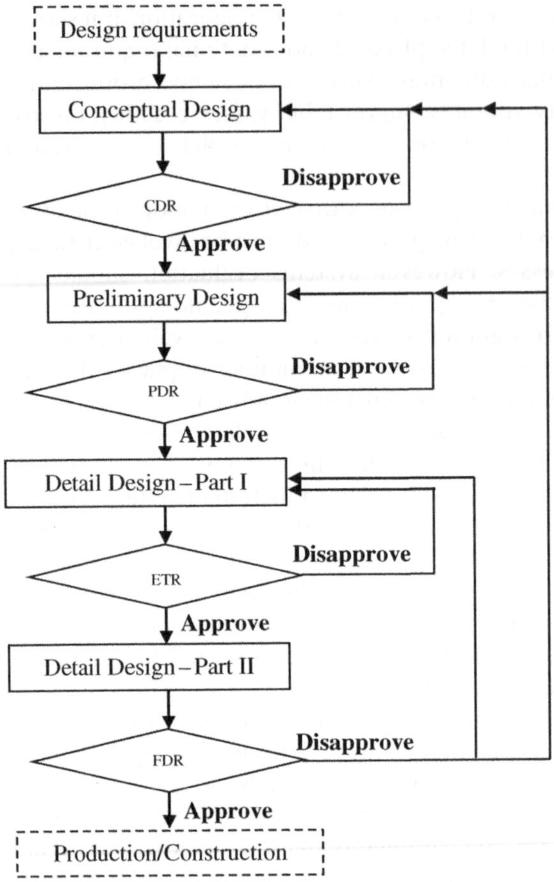

Figure 1.7 Design process and formal design reviews.

Selection of a path forward for the design and development of a preferred system configuration, which will ultimately be responsive to the identified customer requirement, is a major responsibility of conceptual design. Establishing this early foundation, as well as requiring the initial planning and evaluation of a spectrum of technologies, is a critical first step in the implementation of the systems engineering process. Systems engineering, from an organizational perspective, should take the lead in the definition of system requirements from the beginning and address them from a total integrated life-cycle perspective.

As the name implies, the UAV conceptual design phase is the UAV design at the concept level. At this stage, the general design requirements are entered in a process to generate a satisfactory configuration. The primary tool in this stage of design is the "*selection*." Although there is a variety of evaluation and analysis, but there are not many calculations. The past design experience plays a crucial role in the success of this phase. Hence, the members of the conceptual design team must be the most experienced engineers of the corporation. Figure 1.8 illustrates the major activities which are practiced in the UAV conceptual design phase. The fundamental output of this phase is the configuration, with an approximate three-view of the UAV that represents the UAV configuration.

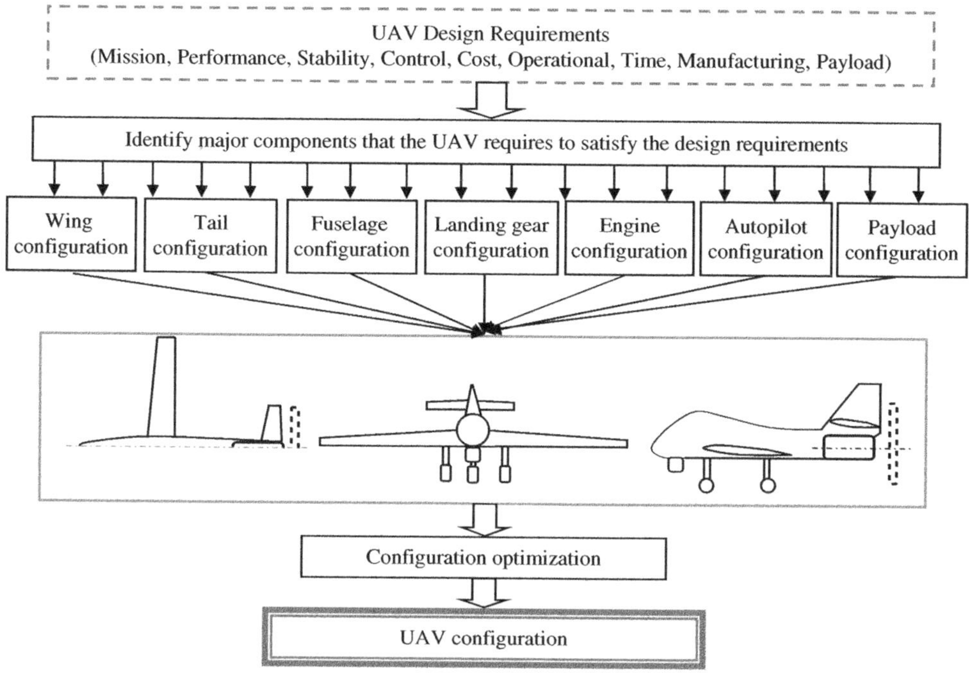

Figure 1.8 Unmanned Aerial Vehicle (UAV) conceptual design.

An UAV comprised of several major components. It mainly includes wing, horizontal tail, vertical tail, fuselage, propulsion system, landing gear, control surfaces, payloads, and autopilot. In order to make a decision about the configuration of each UAV component, the designer must be fully aware of the function of each component. Each UAV component has inter-relationships with other components and interferes with the functions of other components. The above six components are assumed to be the fundamental components of an air vehicle. However, there are other components in an UAV that are not assumed here as a major one. The roles of those components are described in the later sections whenever they are mentioned. Table 1.6 illustrates a summary of UAV major components and their functions. This table also shows the secondary roles and the major areas of influence of each UAV component. The table also specifies the design requirements that are affected by each component.

Table 1.7 illustrates a summary of configuration alternatives for UAV major components. In this table, various alternatives for wing, horizontal tail, vertical tail, fuselage, engine, landing gear, control surfaces, and automatic control system or autopilot are counted. An autopilot tends to function in three areas of guidance, navigation and control. More details are given in the detail design phase Section. For each component, the UAV designer must select one alternative which satisfies the design requirements at an optimal condition. The selection process is based on a trade-off analysis with comparing all pros and cons in conjunction with other components.

In order to facilitate the conceptual design process, Table 1.8 shows the relationship between UAV major components and the design requirements. The third column in this table clarifies the UAV component which affected most; or major design parameter

Table 1.6 Unmanned Aerial Vehicle (UAV) major components and their functions.

No	Component	Primary function	Major areas of influence
1	Fuselage	Payload accommodations	UAV performance, longitudinal stability, lateral stability, cost
2	Wing	Generation of lift	UAV performance, lateral stability
3	Horizontal tail	Longitudinal stability	Longitudinal trim and control
4	Vertical tail	Directional stability	Directional trim and control, stealth,
5	Engine	Generation of thrust	UAV performance, stealth, cost, control
6	Landing gear	Facilitate take-off and landing	UAV performance, stealth, cost
7	Control surfaces	Control	Maneuverability, cost
8	Autopilot	Control, guidance, and navigation	Maneuverability, stability, cost, flight safety
9	Ground station	Control and guide the UAV from the ground	Autonomy, flight safety
10	Launch and recovery	Launching and recovering the UAV	Propulsion, structure, launcher, recovery system

Table 1.7 Unmanned Aerial Vehicle (UAV) major components with design alternatives.

No	Component	Configuration alternatives
1	Fuselage	• Geometry: lofting, cross section • Internal arrangement • What to accommodate (e.g., payload, fuel, engine, and landing gear)?
2	Wing	• Type: Swept, tapered, dihedral • Location: Low-wing, mid-wing, high wing, parasol • High lift device: flap, slot, slat • Attachment: cantilever, strut-braced
3	Horizontal tail	• Type: conventional, T-tail, H-tail, V-tail, inverted V • Installation: fixed, moving, adjustable • Location: aft tail, canard, three surfaces
4	Vertical tail	Single, twin VT, three VT, V-tail
5	Engine	• Type: turbofan, turbojet, turboprop, piston-prop, rocket • Location: (e.g., under fuselage, under wing, beside fuselage) • Number of engines
6	Landing gear	• Type: fixed, retractable, partially retractable • Location: (e.g., nose, tail, multi)
7	Control surfaces	Separate vs. all moving tail, reversible vs. irreversible, conventional vs. non-conventional (e.g., elevon, ruddervator)
8	Autopilot	• UAV: Linear model, nonlinear model • Controller: PID, gain scheduling, optimal, QFT, robust, adaptive, intelligent • Guidance subsystem: Proportional Navigation Guidance, Line Of Sight, Command Guidance, three point, Lead, waypoint • Navigation subsystem: Inertial navigation (Strap down, stable platform), GPS
9	Launch and Recovery	HTOL (Horizontal Take Off and Landing), ground launcher, net recovery, belly landing

Table 1.8 Relationship between Unmanned Aerial Vehicle (UAV) major components and design requirements.

No	Design requirements	UAV component that affected most, or major design parameter
1	Payload (weight) requirements	Maximum take-off weight (MTOW)
	Payload (volume) requirements	Fuselage
2	Performance Requirements (Range and Endurance)	MTOW
3	Performance requirements (maximum speed, Rate of climb, take-off run, stall speed, ceiling, and turn performance)	Engine; landing gear; and wing
4	Stability requirements	Horizontal tail and vertical tail
5	Controllability requirements	Control surfaces (elevator, aileron, rudder), Autopilot
6	Autonomy requirements	Center of gravity, Autopilot, ground station
7	Airworthiness requirements	Minimum requirements, Autopilot
8	Cost requirements	Materials; engine; weight, ...
9	Timing requirements	Configuration optimality
10	Trajectory requirements	Autopilot

by a design requirement. Every design requirement will normally affect more than one component, but we only consider the component that is influenced most. For example, the payload requirement, range and endurance will affect MTOW, engine selection, fuselage design, and flight cost. The influence of payload weight is different than payload volume. Thus, for optimization purpose, the designer must know exactly payload weight and its volume. On the other hand, if the payload can be divided into smaller pieces, the design constraints by the payload are easier to handle. Furthermore, the other performance parameters (e.g., maximum speed, stall speed, rate of climb, take-off run, ceiling) will affect, the wing area and engine power (or thrust).

In order to select the best UAV configuration, a trade-off analysis (Figure 1.9) must be established. Many different trade-offs are possible as the UAV design progresses. Decisions must be made regarding the evaluation and selection of appropriate components, subsystems, possible degree of automation, commercial off-the-shelf parts, various maintenance and support policies, and so on. Later in the design cycle, there may be alternative engineering materials, alternative manufacturing processes, alternative factory maintenance plans, alternative logistic support structures, and alternative methods of material phase-out, recycling, and/or disposal.

The UAV designer must first define the problem statement, identify the design criteria or measures against which the various alternative configurations will be evaluated, the evaluation process, acquire the necessary input data, evaluate each of the candidate under consideration, perform a sensitivity analysis to identify potential areas of risk, and finally recommend a preferred approach. Only the depth of the analysis and evaluation effort will vary, depending on the nature of the component.

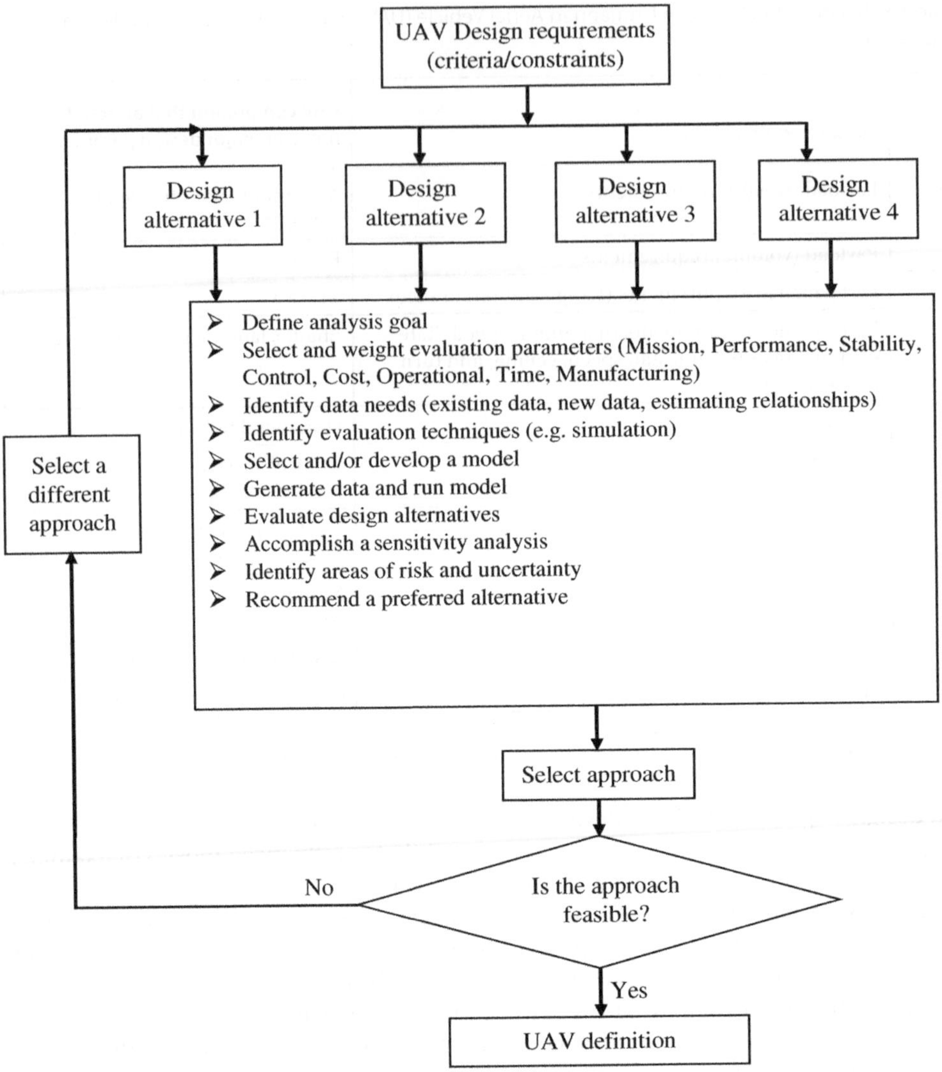

Figure 1.9 Trade-off analysis process.

Trade-off analysis involves synthesis which refers to the combining and structuring of components to create an UAV system configuration. Synthesis is design. Initially, synthesis is used in the development of preliminary concepts and to establish relationships among various components of the UAV. Later, when sufficient functional definition and decomposition have occurred, synthesis is used to further define the "*hows*" at a lower level. Synthesis involves the creation of a configuration that could be representative of the form that the UAV will ultimately take (although a final configuration should not be assumed at this early point in the design process).

One of the most effective techniques in trade-off studies is MDO. Researchers in academia, industry, and government continue to advance MDO and its application to practical problems of industry relevance. MDO is a field of engineering that uses optimization methods to solve design problems incorporating a number of disciplines. MDO allows designers to incorporate all relevant disciplines simultaneously. The optimum solution of a simultaneous problem is superior to the design found by optimizing each discipline sequentially, since it can exploit the interactions between the disciplines. However, including all disciplines simultaneously significantly increases the complexity of the design problem.

Chapter 3 of Reference [8] presents various fixed-wing configuration options, and the technique to conceptually design the air vehicle. Section 2.10 of Chapter 2 of this present book provides the features of the most popular rotary-wing configuration (i.e., quadcopter).

1.12 UAV Preliminary Design

Four fundamental UAV parameters are determined during the preliminary design phase: (a) UAV MTOW (W_{TO}); (b) wing reference area (S); (c) engine thrust (T) or engine power (P); and (d) autopilot preliminary calculations. Hence, four primary UAV parameters of W_{TO}, S, T (or P), and several autopilot data are the output of the preliminary design phase. These four parameters will govern the UAV size, the manufacturing cost, and the complexity of calculation. If during the conceptual design phase, a jet engine is selected, the engine thrust is calculated during this phase. But, if during the conceptual design phase, a prop-driven engine is selected, the engine power is calculated during this phase. A few other non-important UAV parameters such as UAV zero-lift drag coefficient and UAV maximum lift coefficient are also estimated in this phase.

Figure 1.10 illustrates a summary of the preliminary design process. The preliminary design phase is performed in three steps: (a) estimate UAV MTOW; (b) determine wing area and engine thrust (or power) simultaneously; and (c) autopilot preliminary calculations.

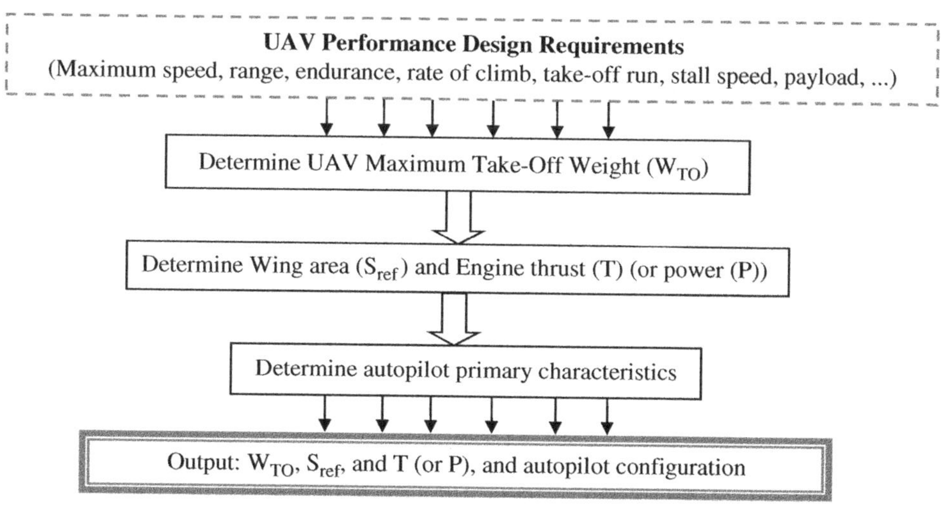

Figure 1.10 Preliminary design procedure.

In this design phase, two design techniques are employed. First a technique based on the statistics is used to determine UAV MTOW. The design requirements which are used in this technique are flight mission, payload weight, range, and endurance.

Second, another technique is employed based on the UAV performance requirements (such as stall speed, maximum speed, range, rate of climb, and take-off run) to determine the wing area and the engine thrust (or engine power). This technique is sometime referred to as the matching plot or matching chart, due to its graphical nature and initial sizing. The principles of the matching plot technique are originally introduced in a NASA technical report and they were later developed by Ref. [8]. The technique is further developed by the author in his new book on UAV design that is under publication.

In general, the first technique is not accurate (in fact, it is an estimation) and the approach may carry some inaccuracies, while the second technique is very accurate, and the results are reliable. Due to the length of the book, the details of these three techniques have not been discussed in detail here in this section. It is assumed that the reader is aware of these techniques which are practiced in many institutions.

1.13 UAV Detail Design

The design of the UAV subsystems and components plays a crucial role in the success of the flight operations. These subsystems turn an aerodynamically shaped structure into a living, breathing, unmanned flying machine. These subsystems include the: wing, tail, fuselage, flight control subsystem, power transmission subsystem, fuel subsystem, structures, propulsion, landing gear, and autopilot. In the early stages of a conceptual or a preliminary design phase, these subsystems must initially be defined, and their impact must be incorporated into the design layout, weight analysis, performance/stability calculations and cost benefits analysis. In this section, the detail design phase of an UAV is presented.

As the name implies, in the detail design phase, the details of parameters of all major components (Figure 1.8) of a UAV is determined. This phase is established based on the results of conceptual design phase and preliminary design phase. Recall that the UAV configuration has been determined in the conceptual design phase and wing area, engine thrust, and autopilot major features have been set in preliminary design phase. The parameters of wing, horizontal tail, vertical tail, fuselage, landing gear, engine, subsystems, and autopilot must be determined in this last design phase. To compare three design phases, the detail design phase contains a huge amount of calculations and a large mathematical operation compared with other two design phases. If the total length of an UAV design is considered to be one year, of that about 10 months is spent on the detail design phase.

This phase is an iterative operation in its nature. In general, there are four design feedbacks in the detail design phase. Figure 1.11 illustrates the relationships between detail design and design feedbacks. Four feedbacks in the detail design phase are: (a) performance evaluation; (b) stability analysis; (c) controllability analysis; and (d) flight simulation. The UAV performance evaluation includes the determination of UAV zero-lift drag coefficient. The stability analysis requires the component weight estimation plus the determination of UAV center of gravity (cg). In the controllability analysis operation, the control surfaces (e.g., elevator, aileron, and rudder) must be designed. When the autopilot is designed, the UAV flight needs to be simulated to assure the success of the flight.

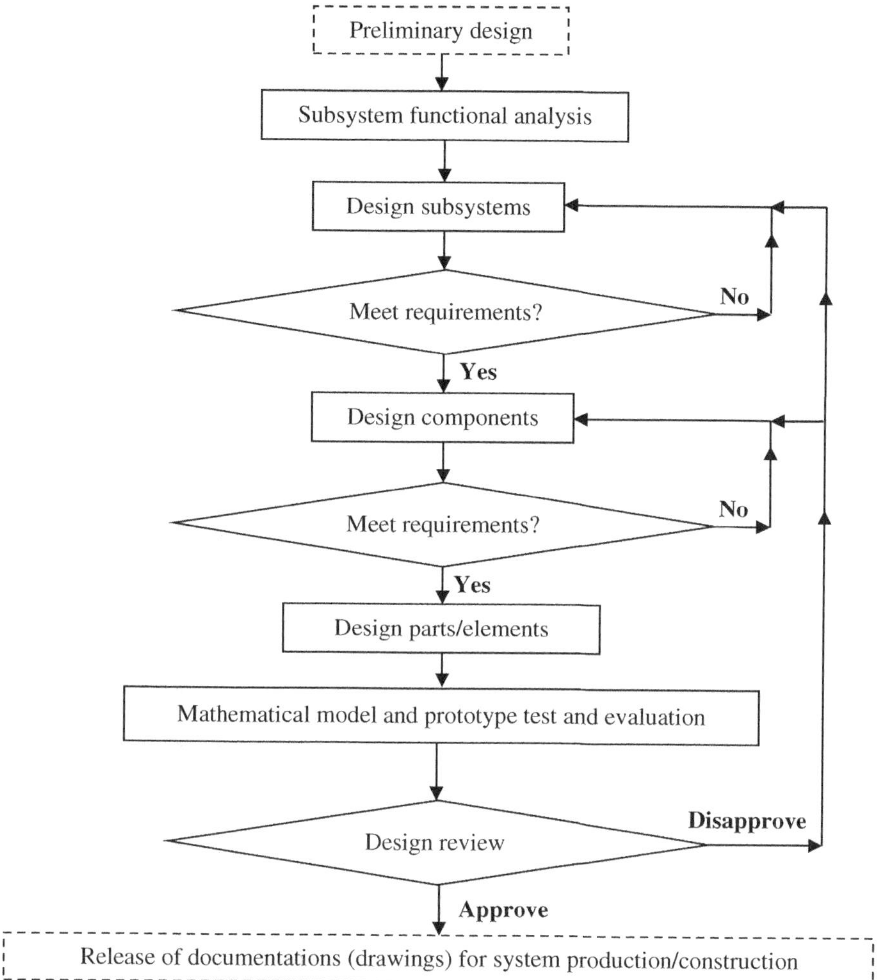

Figure 1.11 Detail design sequence.

As the name implies, each feedback is performed to compare the output with the input and correct the design to reach the design goal. If the performance requirements are not achieved, the design of several components, such as engine and wing, might be changed. If the stability requirements are not met, the design of several components, such as wing, horizontal tail, and vertical tail could be changed. If the controllability evaluation indicates that the UAV does not meet controllability requirements, control surfaces and even the engine must be redesigned. In the case where both stability requirements and controllability requirements were not met, then several components must be moved to change the cg location.

In some instances, this deficiency may lead to a major variation in the UAV configuration, which means the designer needs to return to the conceptual design phase and begin the correction from the beginning. The deviation of the UAV from trajectory during flight simulation necessitates a change in autopilot design.

1.14 Design Review, Evaluation, Feedback

In each major design phase (conceptual, preliminary, and detail), an evaluation should be conducted to review the design and to ensure that the design is acceptable at that point before proceeding with the next stage. There is a series of formal design reviews conducted at specific times in the overall system development process. An essential technical activity within the design process is that of evaluation. Evaluation must be inherent within the systems engineering process and must be invoked regularly as the system design activity progresses. When conducted with full recognition of design criteria, evaluation is the assurance of continuous design improvement. The evaluation process includes both the informal day-to-day project coordination and data review, and the formal design review.

The purpose of conducting any type of review is to assess if (and how well) the design configuration, as envisioned at the time, is in compliance with the initially specified quantitative and qualitative requirements. A design review provides a formalized check of the proposed system design with respect to specification requirements. In principle, the specific types, titles, and scheduling of these formal reviews vary from one design project to the next. The following main four formal design reviews are recommended for a design project:

1) Conceptual Design Review (CDR)
2) Preliminary Design Review (PDR)
3) Evaluation and Test Review (ETR)
4) Critical (Final) Design Review (FDR)

Figure 1.7 shows the position of each design review in the overall design process. Design reviews are usually scheduled before each major design phase. The CDR is usually scheduled toward the end of the conceptual design phase and prior to entering the preliminary design phase of the program. The purpose of CDR is to formally and logically cover the proposed design from the system standpoint. The PDR is usually scheduled toward the end of the preliminary design phase and prior to entering the detail design phase. The FDR is usually scheduled after the completion of the detail design phase and prior to entering the production phase.

The ETR is usually scheduled somewhere in the middle of the detail design phase and prior to the production phase. The ETR accomplishes two major tasks:

1) Finding and fixing any design problems and the subsystem/component level, and then
2) Verifying and documenting the system capabilities for government certification or customer acceptance.

The ETR can range from the test of a single new system for an existing system to the complete development and certification of a new system.

1.15 UAV Design Steps

In a UAV design process, some UAV parameters must be minimized (e.g., weight), while some other variables must be maximized within constraints (e.g., range, endurance, maximum speed, and ceiling), and some others must be evaluated to ensure that they are acceptable. The optimization process must be accomplished through a systems engineering approach. In some cases, the design of the UAV may impose slight to considerable changes to the UAV mission during the conceptual design process. The strong relationship between the analysis and the influencing parameters allow definite, traceable relationships to be constructed. In the case of a UAV design, the major parameters are derived almost completely from operational and performance requirements.

The integration of system engineering principles with the analysis-driven UAV design process demonstrates that a higher level of integrated vehicle can be attained; identifying the requirements/functional/physical interfaces and the complimentary technical interactions which are promoted by this design process. The details of conceptual design phase, preliminary design phase, and detail design phase were introduced earlier.

The following is a suggestion for the UAS major design steps that summarizes the above-mentioned three design phases into steps:

1) Derive UAV design technical requirements, objectives and specifications from the customer order and problem statement.
2) Design program and management planning (e.g., Gantt chart and checklists).
3) Perform feasibility studies.
4) Perform risk analysis.
5) Functional analysis and allocation.
6) Design team allocation.
7) UAV Configuration design.
8) First estimation of UAV MTOW.
9) Estimation of UAV zero-lift drag coefficient (C_{Do}).
10) Calculation of wing reference area (S).
11) Calculation of engine thrust (T) or engine power (P).
12) Wing design.
13) Fuselage design.
14) Horizontal tail design.
15) Vertical tail design.
16) Landing gear design.
17) Propulsion system design.
18) Communication system design.
19) Payload selection/design.
20) First estimate of weight of UAV components.
21) Second estimate of UAV MTOW.
22) First calculation of UAV center of gravity limits.
23) Relocation of components to satisfy stability and controllability requirements.
24) Redesign of horizontal tail and vertical tail design.
25) Design of control surfaces.
26) Autopilot design.
27) Launch/recovery system design.
28) GCS design.
29) Calculation of UAV C_{Do}.
30) Re-selection of engine.
31) Calculation of interferences between UAV components (e.g., wing, fuselage, engine, and tails).

32) Incorporation of design changes.
33) First modifications of UAV components.
34) First calculation of UAV performance.
35) Second modification of UAV to satisfy performance requirements.
36) First stability and control analysis.
37) Third UAV modification to satisfy stability and control requirements.
38) Manufacturing of UAV model.
39) Wind tunnel test.
40) Fourth UAV modification to include aerodynamic considerations.
41) UAV structural design.
42) Design ground station.
43) Design launch and recovery system.
44) Calculation of weight of UAV components.
45) Second calculation of UAV center of gravity limits.
46) Fifth UAV modification to include weight and cg considerations.
47) Second performance, stability and control analysis and design review.
48) Sixth UAV modification.
49) UAV systems design (e.g., electric, mechanical, hydraulic, pressure, and power transmission).
50) Manufacturing of the UAV prototype.
51) Flight tests.
52) Seventh UAV modification to include flight test results.
53) Trade-off studies.
54) Optimization.
55) Certification, validation, or customer approval tests.
56) Eighth modification to satisfy certification requirements.

It is clear that some steps may be moved along with regard to the UAV mission, design team members, past design experiences, design facility, and manufacturing technologies. As it is observed, the design process is truly an iterative process and there are several modification steps to satisfy all design requirements.

Questions

1 What are the five expressions which are currently used for unmanned aircraft?

2 What are the primary design requirements for a UAV?

3 Describe features of a Tier II UAV in the Air Forces.

4 Describe the features of a micro UAV.

5 What is the main objective for the feasibility study?

6 What is the size range for mini UAVs?

7 What do MALE and HALE stand for?

8 What are the operating altitudes for HALE UAVs?

9 What is the endurance range for MALE UAVs?

10 What are the wingspan and MTOW of Global Hawk?

11 What are the cruise speed and endurance for Predator (RQ-1A)?

12 What was the major setback during Phase II flight testing of the Global Hawk on March 29, 1999? What was the reason behind that?

13 Describe the fundamentals of systems engineering approach in UAV design.

14 What are the main four formal design reviews?

15 What are the UAV main design groups based on components?

16 Describe the conceptual design phase.

17 Describe the main outputs of the preliminary design.

18 Describe the process of detail design.

19 Describe the trade-off analysis process.

20 From a systems engineering approach, what are the main design phases?

21 Describe Murphy's Law.

22 Briefly describe the features of Global Hawk.

23 Briefly describe the features of RQ-1A Predator.

24 Briefly describe the features of RQ-9 Predator B Reaper.

25 Briefly describe the features of RQ-5A Hunter.

26 Briefly describe the features of RQ-17 Shadow 200.

27 Briefly describe the features of RQ-170 Sentinel.

28 Briefly describe the features of RQ-2A Pioneer.

29 Briefly describe the features of X-45A UCAV.

30 What is the engine power of Predator A?

31 What is the wing area of Global Hawk?

32 What is the MTOW of Yamaha RMAX?

33 What is the maximum speed of ScanEagle?

34 What is the range of Reaper UAV?

35 What is the maximum speed of X-45A?

36 What is the endurance of RQ-5A Hunter?

37 What is the wingspan of X-45C UCAV?

2

Preliminary Design

CONTENTS

Educational Outcomes

After study of this chapter, the reader will be able to:

1) Conduct Unmanned Aerial Vehicle (UAV) preliminary design
2) Estimate the maximum takeoff weight of a UAV to carry a given payload
3) Determine UAV wing area and engine thrust/power for a fixed-wing UAV for a given flight mission
4) Conceptually design a quadcopter
5) Discuss weight features of current and modern UAVs
6) Determine battery features for a UAV with electric engine(s)

2.1 Introduction

For a fixed-wing UAV, the preliminary design phase is performed in two steps:

1) Estimate aircraft maximum takeoff weight
2) Determine wing area and engine thrust (or power) simultaneously.

Design of Unmanned Aerial Systems, First Edition. Mohammad H. Sadraey.

Companion website: www.wiley.com/go/sadraey/unmanned-aerial-systems

For a rotary-wing UAV (including quadcopters), the preliminary design phase is performed in two steps:

1) Estimate aircraft maximum takeoff weight.
2) Determine engine(s) thrust/power. The estimation of the maximum takeoff weight of both fixed-wing and rotary-wing UAVs are based on the same technique. However, a rotary-wing UAV (e.g., quadcopter) has no wing, so only the calculation technique for its engine power/thrust is presented. In this section, these techniques are presented as part of preliminary design phase.

2.2 Maximum Takeoff Weight Estimation

The purpose of this section is to introduce a technique to obtain the first estimate of the maximum takeoff weight (or all-up weight) for a UAV before it is fully designed and built. The aircraft weight is broken into several parts. Some parts are determined based on statistics, but some are calculated from performance equations. Sections 2.3 through 2.8 will present technique to estimate the UAV maximum take-off weight.

2.3 Weight Buildup

The maximum takeoff weight (W_{TO}) is broken into four elements:

1) Payload weight (W_{PL}).
2) Autopilot weight (W_A).
3) Fuel weight (W_F) if non-electric engine, or Battery weight (W_B), if electric engine.
4) Empty weight (W_E).

$$W_{TO} = W_{PL} + W_A + W_F + W_E \tag{2.1}$$

This equation may be reformatted [8] into the following form:

$$W_{TO} = \frac{W_{PL} + W_A}{1 - \left(\frac{W_F}{W_{TO}}\right) - \left(\frac{W_E}{W_{TO}}\right)} \tag{2.2a}$$

$$W_{TO} = \frac{W_{PL} + W_A}{1 - \left(\frac{W_B}{W_{TO}}\right) - \left(\frac{W_E}{W_{TO}}\right)} \tag{2.2b}$$

Equation 2.2a is used for a fuel-consuming UAV, while equation 2.2b is for an electric UAV. In order to find W_{TO}, one needs to determine four variables of W_{PL}, W_A, W_F/W_{TO}, (or W_B/W_{TO}), and W_E/W_{TO}. The first three parameters, namely payload, autopilot, and fuel (or battery weight) fraction are determined fairly accurately, but the last parameter (i.e., empty weight fraction) is estimated from statistics.

2.4 Payload Weight

The payload (useful load) is the net carrying capacity of an UAV which is originally required and designed to carry. The payload includes sensors (e.g., camera and radar), store, military equipment, and other intended loads. Chapter 11 provides technical features of various payloads including their weights. Table 2.1 presents characteristics of several large and small fixed-wing and rotary-wing UAS along with their payload weight ratios. It is noticed that the payload weight ratio (W_{PL}/W_{TO}) is from 8% (Killer Bee) to 33% (Yamaha RMAX).

The payload weight is a known value, since it is selected by the designer, or given by the customer. The original RQ-7A Shadow air vehicle with a wing span of 3.89 m is carrying a maximum payload of 25.3 kg. Table 2.2 shows the weights for a number of real UAV payloads.

2.5 Autopilot Weight

Another part of the aircraft weight is the weight of the autopilot which is responsible to conduct the flight operations and serving subsystems and payloads. If the UAV is controlled by an operator from the ground control station, a communication system including receiver, transmitter, and data link is required too. In terms of weight buildup, we do

Table 2.1 Payloads weight ratio of several Unmanned Aerial Vehicles (UAVs).

No	Name/Designation	Endurance	Range	TO weight/ mass	Payload weight/ mass	W_{PL}/W_{TO}
1	RQ-4 Global Hawk	32 hours	22 800 km	32 250 lbs	3000 lbs	0.093
2	Raven RQ-11B	90 minutes	10 km	4.2 lbs	6.5 oz	0.097
3	Scan Eagle	23 hours	100 km	48.5 lbs	13 lbs	0.27
4	RQ-7B Shadow	7 hours	110 km	375 lb	100 lbs	0.27
5	MQ-1C Gray Eagle	30 hours	3750 km	3600 lbs	800 lbs	0.22
6	A-160 Hummingbird	24 hours	4023 km	6500 lbs	1000 lbs	0.15
7	MQ-8 Fire Scout	8 hours	203 km	3150 lbs	500 lbs	0.16
8	Black Eagle 50	4 hours	260 km	77 lbs	22 lbs	0.28
9	Silver Fox	12 hours	32 km	28 lbs	4 lbs	0.14
10	Killer Bee	18 hours	100 km	250 lbs	20 lbs	0.08
11	Northrop Grumman Firebird	40 hr	11,200 km	5000 lb	1240 lb	0.25
12	RQ-5 Hunter	21 hr	125 km	885 kg	90 kg	0.102
13	Yamaha RMAX	1 hour	90 km	94 kg	31 kg	0.33
14	Predator B SkyGuardian	30 hours	11,000 km	4500 kg	340 kg internal payload + 1360 kg external payload	0.38

Table 2.2 Weights for a number of Unmanned Aerial Vehicle (UAV) payloads.

No	Payload type	Type/brand	Mass, Weight	Remarks
1	Optical camera	TL300MN FPV	15 g	Size: 21 mm × 26 mm × 30 mm
2	Optical camera	DJI Zenmuse X5	530 g	With gimbal for quadcopters
3	Video camera	CC-1XHRM	0.29 kg	Sony
4	Radar	Garmin GSX 70	12.7 lbs (4.31 kg)	Used in a NASA Global Hawk
5	Radar	Lincoln Lab	175 lb	For research in UAV application
6	Decoy	ALE-50	4.5 kg	Used in ECM
7	ECM system	IDECM	73.4 kg	ITT Electronic Systems
8	Radar warning receiver	AN/ALR-67	45 kg	Raytheon
9	Laser range finder	XP SSXP	2.5 kg	SureShot
10	Laser range finder	BOD 63M-LA01-S115	0.26 kg	Balluaf

not consider the communication system in the autopilot group. The weight of autopilot is a function of the UAV mission.

Figure 2.1 illustrates a commercial flight controller; F4 (V3) Omnibus with the dimension of 30.5 mm × 30.5 mm. This low-cost light-weight autopilot is suitable for a small UAV.

For a regular mission of aerial photography (in a small quadcopter), the autopilot is light weight (as low as 50 g) and is comprised of compact elements for navigation and control systems. However, for a large MALE UAV with a complex and long mission, the autopilot has a number of elements, and is a heavy item (as high as 50 lb).

Figure 2.1 F4 (V3) Omnibus Flight Controller (Dimensions: 30.5 mm × 30.5 mm).

Table 2.3 Mass of a few commercial autopilot and Inertial navigation system (INS) units.

a. Autopilot

No	Autopilot model	Mass	Features
1	APM 2.8 ArduPilot	70 g	Includes 3-axis gyro, accelerometer and magnetometer, 5 Hz GPS module, altimeter, mounting for the vibration damping plate; Dimension: 70.5 × 45 × 13.5 mm
2	Omnibus F4+ (Plus)	24.2 g	Includes flight controller (F4 + OSD + SDCARD + VTX + BUZZER), size: 30.5 mm × 30.5 mm
3	F4 (V3) Omnibus	20 g	With BEC & OSD & Black box (DSHOT1200 / 30.5 mm × 30.5 mm)
4	MP21283X MicroPilot	28 g	integrated with 3-axis gyros/accelerometers, GPS, pressure altimeter, pressure airspeed sensors (4 cm × 10 cm)

b. INS units

No	INS	Weight	Remarks
1	Kearfott KN-4072	11 lb	Used on Global Hawk
2	Kearfott KN-4073	8 lb	Used on Firescot
3	LN-100G	21.6 lb	Northrop Grumman
4	RCCT GS-111	1.7 lb	

If a commercial off-the-shelf (COTS) autopilot is selected, the weight will be a known value. However, if the designer would like to design a custom-made autopilot, the weight will be determined when the design is done. For such conditions, statistical values for the weight of autopilot for a similar UAV could be taken. Table 2.3 demonstrates values of a few commercial autopilot and Inertial navigation system (INS) weights.

2.6 Fuel Weight

A UAV may have a fuel consuming engine (e.g., piston engine), or an electric engine. Thus, there are either of these two calculations:

1) Fuel weight for UAVs with fuel consuming engines.
2) Battery weight for UAVs with electric engines.

For the case of a UAV with fuel consuming engines, the UAV weight at landing is much less than the weight at the takeoff. However, in a UAV with electric engines, the vehicle weight is usually does not change throughout flight, unless a payload (e.g., store) is dropped. This section provides fuel weight calculation for a UAV with fossil fuel consuming engine.

The required amount of the total fuel weight necessary for a complete flight operation depends upon the mission to be followed, the aerodynamic characteristics of the aircraft, and the engine specific fuel consumption. The mission specification is normally given to the designer and must be known.

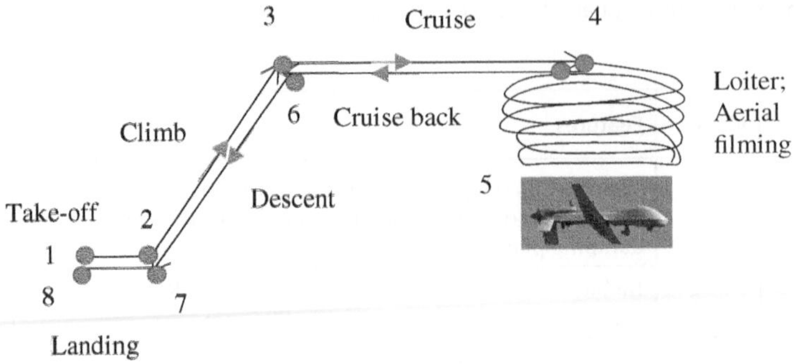

Figure 2.2 Typical mission profile for a remotely controlled Unmanned Aerial Vehicle (UAV).

The first step to determine the total fuel weight is to define the flight mission segments. A typical mission profile is demonstrated in Figure 2.2 for a remotely controlled UAV. This flight operation is composed of taxi/take-off, climb, cruise, loiter (e.g., aerial photography and videography), return back, descent, and landing. Each mission segment is numbered; for instance, 1 denotes the beginning of takeoff and 2 is the end of take-off. The three most fuel/energy consuming are: Cruise, loiter, and cruise back.

In a similar fashion, the aircraft weight at each phase of the flight mission is numbered. Hence, W_1 is the aircraft weight at the beginning of takeoff (i.e., maximum takeoff weight), W_2 is the aircraft weight at the end of takeoff which is the beginning of climb phase, and W_8 is the UAV weight at the end of flight (i.e., landing).

Thus, for any mission segment "*i*," the mission segment weight fraction is expressed as (W_{i+1}/W_i). For instance, W_4/W_3 in the flight mission of Figure 2.2 is the fuel fraction during the cruise segment. During each segment, the fuel is burnt and the aircraft loses weight.

This will make a base for estimating the required fuel weight and fuel fraction during a flight operation. The difference between the aircraft weight at the end of flight (i.e., landing; W_L) and the aircraft weight at the beginning of flight (i.e., take-off) is exactly equal to the fuel weight:

$$W_{TO} - W_L = W_F \tag{2.3}$$

Thus, in a regular flight mission, the ratio between the aircraft weight at the end of flight to the aircraft weight at the beginning of flight is:

$$\frac{W_L}{W_{TO}} = \frac{W_{TO} - W_F}{W_{TO}} \tag{2.4}$$

Therefore, for the case of a mission with seven segments as shown in Figure 2.1, the fuel weight fraction is obtained as follows:

$$\frac{W_F}{W_{TO}} = 1 - \frac{W_8}{W_1} \tag{2.5}$$

where $\frac{W_8}{W_1}$ can be written as:

$$\frac{W_8}{W_1} = \frac{W_2}{W_1}\frac{W_3}{W_2}\frac{W_4}{W_3}\frac{W_5}{W_4}\frac{W_6}{W_5}\frac{W_7}{W_6}\frac{W_8}{W_7} \tag{2.6}$$

For the sake of flight safety, it is recommended to carry a fuel reserve. The extra fuel required for safety purposes is almost 5% of aircraft total weight, so it is applied as follows:

$$\frac{W_F}{W_{TO}} = 1.05\left(1 - \frac{W_8}{W_1}\right) \tag{2.7}$$

Therefore, in order to find the fuel weight fraction, one must first determine these weight fractions for all of the mission segments. There are primarily seven flight segments as take-off, climb, cruise, loiter, cruise back, descent, and landing. These flight phases or segments can be divided into two groups:

Firstly, there are certain segments during which the fuel weight that is burnt is almost nothing and negligible compared with maximum take-of weight. These include, taxi, take-off, climb, descent, approach, and landing. The fuel weight fractions for these mission segments are estimated based on the statistics. Table 2.4 illustrates typical average values for fuel fractions of take-off, climb, descent, and landing.

The segments during which the fuel weight that is burnt is considerable. These segments – 1. cruise to target area, 2. Loiter, and 3. cruise back to the base (or operator) – are determined through mathematical calculations. The fuel weight fraction for cruise (R) segment is determined [9] by employing the Breguet range equations.

$$R_{\max} = \frac{V_{(L/D)_{\max}}}{C}\left(\frac{L}{D}\right)_{\max} \ln\left(\frac{W_i}{W_{i+1}}\right) \quad (UAV \text{ with jet engine}) \tag{2.8}$$

$$R_{\max} = \frac{\eta_P (L/D)_{\max}}{C} \ln\left(\frac{W_i}{W_{i+1}}\right) \quad (UAV \text{ with prop-driven engine}) \tag{2.9}$$

Table 2.4 Typical segment weight fractions.

No	Mission segment	W_{i+1}/W_i
1	Taxi and take-off	0.98
2	Climb	0.97
3	Descent	0.99
4	Approach and landing	0.997

Therefore, the cruise fuel weight ratio is determined as:

$$\frac{W_{i+1}}{W_i} = e^{\frac{-R \cdot C}{0.866V(L/D)_{\max}}} \quad (UAV \text{ with jet engine}) \tag{2.10}$$

$$\frac{W_{i+1}}{W_i} = e^{\frac{-R \cdot C}{\eta_P (L/D)_{\max}}} \quad (UAV \text{ with prop-driven engine}) \tag{2.11}$$

For the flight mission indicated in Figure 2.2, i is equal to 3 and 5. Thus, the fuel weight ratio for two cruise segments; $\frac{W_4}{W_3}$ and $\frac{W_6}{W_5}$ are determined by these two equations. The typical values for the variable specific fuel consumption, C, and the maximum lift-to-drag ratio, $(L/D)_{\max}$ are presented by [8].

Similarly, the fuel weight fraction for endurance (E) segment is determined [8] by employing the Breguet endurance equations.

$$E_{\max} = \frac{(L/D)_{\max}}{C} \ln\left(\frac{W_i}{W_{i+1}}\right) \quad (UAV \text{ with jet engine}) \tag{2.12}$$

$$E_{\max} = \frac{(L/D)_{E\max} \eta_P}{C V_{E_{\max}}} \ln\left(\frac{W_i}{W_{i+1}}\right) \quad (UAV \text{ with prop-driven engine}) \tag{2.13}$$

Therefore, the endurance fuel weight ratio is determined as:

$$\frac{W_{i+1}}{W_i} = e^{\frac{-E \cdot C}{(L/D)_{\max}}} \quad (UAV \text{ with jet engine}) \tag{2.14}$$

$$\frac{W_{i+1}}{W_i} = e^{\frac{-E \cdot C \cdot V_{E_{\max}}}{0.866 \eta_P (L/D)_{\max}}} \quad (UAV \text{ with prop-driven engine}) \tag{2.15}$$

The speed for maximum endurance ($V_{E\max}$) for a prop-driven aircraft [9] happens when the aircraft is flying with the minimum power speed (i.e., $V_{P\min}$). Since the aircraft has not yet been fully designed at preliminary design phase, the calculation of minimum power speed cannot be implemented. Hence, the recommendation is to use a reasonable approximation. The minimum power speed for most prop-driven aircraft is about 20–40% higher than the stall speed. Then:

$$V_{E_{\max}} = V_{P_{\min}} \approx 1.2V_S - 1.4V_S \tag{2.16}$$

The cruising speed for the cruise segment is about 90% of the maximum speed. If cruising speed is not given, you may consider this approximation in the preliminary design phase.

2.7 Battery Weight

An electric propulsion system includes electric motor, battery, and propeller. For a UAV with electric engine(s), a pack of batteries is required to provide electric energy. The battery weight is a function of UAV mission (including cruise and endurance), vehicle aerodynamic features, and electric payload energy consumption. In Section 3.7, the features of electric engines and their applications are provided. Table 3.5 in Chapter 3 provides technical features – including energy density – of some rechargeable batteries. The energy density is defined [10] as:

$$\text{Energy Density} = \frac{\text{battery output power} \times \text{time of energy consumption}}{\text{battery mass}} \tag{2.17}$$

or

$$E_D = \frac{Pt}{m_B} \tag{2.18}$$

So, the battery mass (m_B) will be:

$$m_B = \frac{Pt}{E_D} \tag{2.19}$$

Thus, the battery weight for a number of flight segments is determined from the following equation:

$$W_B = \sum_{i=1}^{n} \frac{Ptg}{E_D} \tag{2.20}$$

where P denotes the battery output power (in Watt, W), t for flight time (in hour, h), E_D for energy density (in Wh/kg), and g for gravitational constant (i.e., 9.81 m/s^2). The parameter stands for the number of flight legs (e.g., leg 1 is the takeoff, leg 2 is the climb, leg 3 is the cruise, etc.). The endurance (i.e., time for the loiter segment) should be given by the customer and is known for the designer. The time for cruising flight segment is a function of range (R) and the cruising velocity (V_C):

$$t_C = \frac{R}{V_C} \tag{2.21}$$

The range may be assumed to be twice the distance to the mission target location, since the UAV has to return to the operator after the mission is done. Other segments of the flight (i.e., takeoff, climb, descent, approach, and landing) does not take significant electric power from the batteries. Table 2.5 provides a typical electric power consumption for these segments.

The power for a moving object is defined as the applied force time the velocity. In addition, the applied force in a range or endurance segment is provided by the engine (i.e., thrust). In a constant velocity level flight (cruising or loitering segment), the engine thrust is equal to the UAV drag.

Table 2.5 Typical electric power consumptions for non-cruise segments.

No	Mission segment	ΔP
1	Taxi and take-off	2%
2	Climb	1%
3	Descent	0.05%
4	Approach and landing	0.05%

$$P = TV = DV \quad \text{(UAV with jet engine)} \tag{2.22a}$$

However, for prop-driven UAV:

$$P = \frac{TV}{\eta_P} = \frac{DV}{\eta_P} \quad \text{(UAV with prop-driven engine)} \tag{2.22b}$$

where η_P is the propeller efficiency. The UAV drag is a function of airspeed (V), wing area (S), air density (ρ), and its configuration (C_D):

$$D = \frac{1}{2}\rho V^2 S C_D \tag{2.23}$$

Therefore, the required power for a UAV with a jet engine is:

$$P_{jet} = \frac{1}{2}\rho V^3 S C_D \tag{2.24}$$

In contrast, the required power for a UAV with a prop engine is:

$$P_{prop} = \frac{1}{2\eta_P}\rho V^3 S C_D \tag{2.25}$$

For a UAV with jet engine, the loiter velocity is recommended to be equal to the minimum drag velocity. For a prop-driven UAV, the loiter velocity is recommended to be equal to the minimum power velocity. The recommended velocity for a loiter mission for both jet and prop-driven is usually 20–40% greater than the stall speed.

$$V_{E_{max}} = V_{P_{min}} \approx 1.2V_S - 1.4V_S \tag{2.26}$$

Moreover, the drag polar for this flight segment is:

$$C_{D-\min P} = C_{D_o} + K\left(\frac{1}{1.3}\right)^2 C_{L\max}^2 \tag{2.27}$$

Using typical values (See Table 2.6) for parameters C_{Do}, K, C_{Lmax}, and η_P, the required power for cruise/loiter flight is estimated.

Table 2.6 Typical values for a number of parameters for a fixed-wing Unmanned Aerial Vehicle (UAV).

No	Parameter	Typical values	Name
1	C_{Do}	0.02–0.03	Zero-lift drag coefficient – retractable landing gear
2	C_{Do}	0.03–0.045	Zero-lift drag coefficient – fixed landing gear
3	AR	5–20	Aspect ratio (AR)
4	C_{Lc}	0.2–0.5	Cruise lift coefficient
5	C_{Lmax}	1.2–1.6	Maximum lift coefficient
6	e	0.6–0.9	Oswald efficiency factor
7	η_P	0.6–0.8	Prop efficiency
8	$(C_L/C_D)_{max}$	6–10	Maximum lift-to-drag ratio – fixed landing gear
9	$(C_L/C_D)_{max}$	8–16	Maximum lift-to-drag ratio – retractable landing gear

To find an expression for the wing area (S), we refer to the level cruising flight, where the lift is equal to UAV weight. Thus, the wing area will be obtained as:

$$S = \frac{2W}{\rho V_C^2 C_{Lc}} \tag{2.28}$$

Inserting this expression into Eq. (2.25), one can obtain:

$$P_{prop} = \frac{1}{2\eta_P}\rho V^3 \frac{2W}{\rho V^2 C_L} C_D \tag{2.29}$$

which can be simplified to:

$$P_{prop} = \frac{WV}{\eta_P}\frac{C_D}{C_L} \tag{2.30}$$

For a pro-driven engine, it is reasonable to assume that, about 5% of the battery energy (See Table 2.5) is consumed in non-cruising flight segments (e.g., takeoff and climb). Plugging "P" from Eq. (2.29), and "t" from Eq. (2.21) into Eq. (2.18), the following expression is obtained:

$$W_B = 1.05\left(\frac{WV}{\eta_P}\frac{C_D}{C_L}\frac{R}{V}\frac{g}{E_D}\right) \tag{2.31}$$

which leads to:

$$W_B = 1.05\left(\frac{W}{\eta_P}\frac{R}{C_L / C_D}\frac{g}{E_D}\right) \tag{2.32}$$

Therefore, the battery weight ratio will be:

$$\frac{W_B}{W} = 1.05\left(\frac{g}{\eta_P E_D}\frac{R}{C_L / C_D}\right) \tag{2.33}$$

Equation (2.33) provides the battery weight ratio for a fixed-wing UAV with prop-driven electric engine(s). Table 3.5 in Chapter 3 provides [11] energy density of some rechargeable batteries. Table 2.6 exhibits typical values for η_P and C_L/C_D of UAVs. Note that lift-to-drag ratio (C_L/C_D) at cruise is about 80–90% of the maximum lift-to-drag ratio, $(C_L/C_D)_{max}$.

Example 2.1
A small fixed-wing RC UAV with a mass of 1 kg is equipped with a pack of Li–Po (Lithium Polymer) batteries for its electric motor. If the energy of batteries is just consumed in a loiter mission (for electric motor), for a maximum endurance of 20 minutes, determine the mass of battery pack. Assume that the electric motor is creating 2 N of thrust, with a prop efficiency of 0.7, and the UAV is flying at a speed of 20 m/s.

Solution
Since only the loiter leg of flight operation is considered, Eq. (2.20) is reduced to:

$$W_B = \frac{Ptg}{E_D} \tag{2.20}$$

Dividing both sides by g results in:

$$m_B = \frac{Pt}{E_D} \tag{2.19}$$

The electric motor is providing the following power:

$$P = TV/\eta = 2 \times 20/0.7 = 57.1\ \text{W} \tag{2.22b}$$

From Table 3.5, the energy density of Lithium Polymer batteries is listed as 100–265 Wh/kg. It is assumed that the battery has the maximum possible energy density (i.e., 265 Wh/kg). Thus,

$$m_B = \frac{Pt}{E_D} = \frac{57.1 \times \frac{20}{60}}{265} = 0.072\,\text{kg} = 72\,\text{gram} \tag{2.18}$$

Since an hour is equivalent to 60 minutes, the term 20/60 is inserted. The battery has about 7% of the UAV mass.

Example 2.2
A fixed-wing RC plane with a prop-driven electric engine (with Lithium-ion batteries) is in the preliminary design phase. You are required to determine its battery weight ratio. The UAV is required to be able to fly to a distance of 10 km and return back.

Solution
From Table 3.5, the energy density of Lithium–ion batteries is listed as 100–250 Wh/kg. It is assumed that the battery has an energy density of 150 Wh/kg or 150 × 3600 W.sec/kg. From Table 2.6, a C_L/C_D of 8, and a prop efficiency of 0.6 are selected. The total range will be 2 × 10 = 20 km = 2 × 10 000 m. Thus,

$$\frac{W_B}{W} = 1.05\left(\frac{g}{\eta_P E_D}\frac{R}{C_L / C_D}\right) = 1.05\left(\frac{9.81}{0.76\times 150\times 3600}\frac{10{,}000\times 2}{8}\right) = 0.079 = 7.9\% \quad (2.33)$$

Thus, the batteries will have 7.9% of the total RC plane weight.

2.8 Empty Weight

The last term in determining maximum takeoff weight in Eq. (2.1) is the empty weight fraction $\left(\frac{W_E}{W_{TO}}\right)$. The empty weight mainly includes weights of structure/airframe, engine(s), landing gear, and systems. At this moment (preliminary design phase), the aircraft has been designed only conceptually, hence, there is no geometry or sizing. Therefore, the empty weight fraction cannot be calculated analytically. The only way is to employ the past history and statistics.

The only known information about the aircraft is the configuration and aircraft type based on the mission. According to this data, the author has developed a series of empirical equations to determine the empty weight fraction. The equations are based on the published data taken from published sources. In general, the empty weight fraction varies from 0.5 to 0.9. The empty weight fraction is mathematically modeled by the following experimental linear equation:

$$\frac{W_E}{W_{TO}} = aW_{TO} + b \quad (2.34)$$

where a and b are presented in Table 2.7. Note that the Eq. (2.34) is curve fitted in the British units system. Thus, the unit for maximum takeoff weight and empty weight is lb. Note that the unit of W_{TO} in Table 2.7 is lb.

The small UAV Arcturus T-20 (Figure 6.37) with a mass of 84 kg, capable of internal and external payloads (e.g., EO/IR camera), has an empty weight of 50 kg. The MALE Reaper UAV with a takeoff gross mass of 4750 kg, a payload mass of 1700 kg, and 1360 kg of fuel, has an empty weight of 2223 kg.

Example 2.3

The design of a small remote controlled (RC) for an aerial filming mission is in the preliminary design phase. A Sony CC-1XHRM camera has been selected. An APM 2.8

Table 2.7 The coefficients "a" and "b" for the empirical equation of (2.34).

No	Unmanned Aerial Vehicle (UAV) type	a	b
1	Small remote controlled (RC)	–0.00296	0.87
2	HALE	1.07×10^{-5}	0.126
3	MALE	5.1×10^{-6}	0.42
4	Quadcopter	-4.6×10^{-5}	0.68

ArduPilot and a single electric engine with Lithium–ion batteries have been chosen. The total desired range is 20 km. Estimate the UAV maximum takeoff weight.

Solution

From Table 12.2, this camera has a weight of 10.2 oz., which is equivalent to a mass of 0.289 kg (i.e., W_{PL} = 0.638 lb). From Table 2.3, the mass of the APM 2.8 ArduPilot is 70 g (i.e., W_A = 0.154 lb). From Table 2.7, values for a and b are −0.002 96 and 0.87 respectively. From Table 3.5, the energy density of Lithium–ion batteries is listed as 100–250 Wh/kg. It is assumed that the battery has an energy density of 150 Wh/kg.

Since, the type of UAV and the range is similar to that for the UAV from example 2.2, we obtain:

$$\frac{W_B}{W} = 0.079$$

The values of a and b are plugged into Eq. (2.34):

$$\frac{W_E}{W_{TO}} = -0.00296 W_{TO} + 0.87 \tag{2.34}$$

The values for W_B, W_{PL}, and W_A and the expression for $\frac{W_E}{W_{TO}}$ are all inserted into Eq. (2.2b):

$$W_{TO} = \frac{W_{PL} + W_A}{1 - \left(\frac{W_B}{W_{TO}}\right) - \left(\frac{W_E}{W_{TO}}\right)} = \frac{0.638 + 0.154}{1 - 0.079 - (-0.00296 W_{TO} + 0.87)} \tag{2.2b}$$

This expression is reduced to:

$$W_{TO} = \frac{0.79}{0.00296 W_{TO} + 0.051}$$

Solution of this nonlinear algebraic equation yields a value of 9.85 for W_{TO}. Thus, the maximum takeoff weight for this UAV is estimated to be 9.85 lb.

An airborne wide-area, nearly all-weather surveillance radar for an UAV – sponsored by the Army and DARPA -was developed by Lincoln Laboratory. Table 2.8 presents [12] the breakdown of the weight for various components in the radar and communications system. As observed, this payload system has a total weight of 175 lb.

2.9 Wing and Engine Sizing

The second crucial step in the aircraft preliminary design phase is to determine wing reference area (S_{ref}) and engine thrust (T), or engine power (P). Unlike the first step in preliminary design phase at which the main reference was statistics, this phase is solely depending upon the aircraft performance requirements and employs flight mechanics theories. The technique – which is referred to as "Matching plot" – initially developed by NASA and later expanded by [8].

Table 2.8 Unmanned Aerial Vehicle (UAV) Radar and support equipment weight breakdown.

No	System	Component	Weight (lb)
1	**Radar**	Transmitter	21
2		Receiver and exciter	10
3		Processor	55
4		Antenna system	14
5		Cables and connectors	10
		Subtotal	**110**
	Support Equipment	Inertial Navigation System (INS) (including heat sink)	17
6		Data link	12
7		Altimeter and GPS receiver	9
8		Support structure	25
9		Cooling fans	2
10		Subtotal	**65**
Total			**175**

Both wing and engine have significant roles in a cruising flight. Four main UAV forces (lift, L; drag, D; weight, W; and thrust, T) are in equilibrium. In a cruising flight, lift is equal to UAV weight (W), and engine thrust (T) is equal to UAV drag (D).

$$L = W \tag{2.35}$$

$$D = T \tag{2.36}$$

In these two equations, the contribution of thrust components due to angle of attack, and engine setting angle are neglected. Equation (2.35) and (2.36) indicate that the weight is balanced by lift, while the drag is balanced by engine thrust. The weight is a function of UAV mass:

$$W = mg \tag{2.37}$$

The aircraft performance requirements that are utilized to size the aircraft in this phase are: (a) stall speed (V_s); (b) maximum speed (V_{max}); (c) maximum rate of climb (ROC_{max}); (d) takeoff run (S_{TO}); and (e) ceiling (h_c). These values should be given to the designer for establishing the preliminary design. For instance, the MQ-1 Predator maximum and stall speeds are 117 and 54 knot respectively. Moreover, the range, endurance, and service ceiling for the RQ-4 Global Hawk are 22 779 km, 32+ hours, and 60 000 ft respectively.

The Matching plot technique is performed in six steps:

1) Derive one equation for each aircraft performance requirement (W/P or T/W as a function of W/S).
2) Sketch all derived equations in one plot.
3) Identify the acceptable region inside the regions that are produced by the axes and the graphs. 4. Determine the design point (i.e., the optimum selection).

4) From the design point, obtain two numbers; corresponding wing loading; $(W/S)_d$ and corresponding power loading; $(W/P)_d$.
5) Calculate the wing area and engine thrust/power from these two values, since the aircraft maximum takeoff weight (W_{TO}) has been previously determined.

A typical graphical technique for a prop-driven UAV is illustrated in Figure 2.3, and for a jet aircraft is illustrated in Figure 2.4.

The following is the list of equations for each performance requirement. For reviewing the details and derivations, refer to [8].

- **Stall**

$$\left(\frac{W}{S}\right)_{V_s} = \frac{1}{2}\rho V_s^2 C_{L_{max}} \tag{2.38}$$

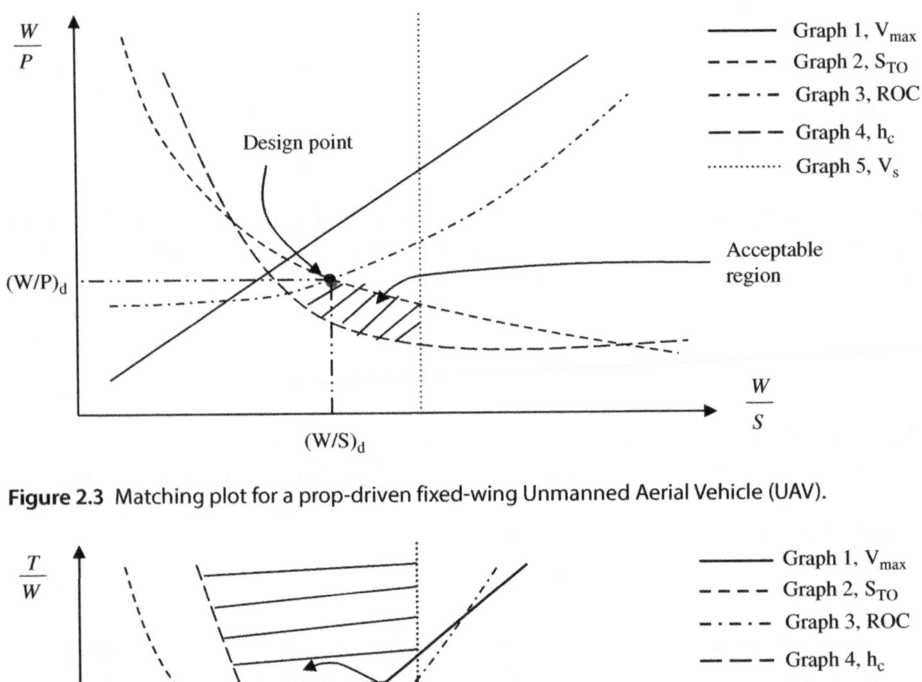

Figure 2.3 Matching plot for a prop-driven fixed-wing Unmanned Aerial Vehicle (UAV).

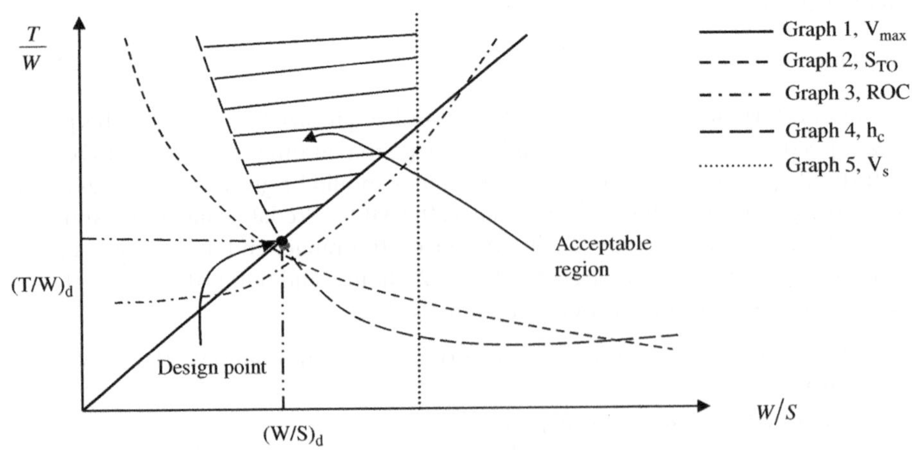

Figure 2.4 Matching plot for a fixed-wing jet Unmanned Aerial Vehicle (UAV).

- **Maximum speed**

$$\left(\frac{T_{SL}}{W}\right)_{V_{max}} = \rho_o V_{max}^2 C_{D_o} \frac{1}{2\left(\frac{W}{S}\right)} + \frac{2K}{\rho\sigma V_{max}^2}\left(\frac{W}{S}\right) \quad \text{(Jet aircraft)} \tag{2.39}$$

$$\left(\frac{W}{P_{SL}}\right)_{V_{max}} = \frac{\eta_P}{\frac{1}{2}\rho_o V_{max}^3 C_{D_o} \frac{1}{\left(\frac{W}{S}\right)} + \frac{2K}{\rho\sigma V_{max}}\left(\frac{W}{S}\right)} \quad \text{(Prop-driven aircraft)} \tag{2.40}$$

- **Takeoff Run**

$$\left(\frac{T}{W}\right)_{S_{TO}} = \frac{\mu - \left(\mu + \frac{C_{D_G}}{C_{L_R}}\right)\left[\exp\left(0.6\rho g C_{D_G} S_{TO} \frac{1}{W/S}\right)\right]}{1 - \exp\left(0.6\rho g C_{D_G} S_{TO} \frac{1}{W/S}\right)} \quad \text{(Jet aircraft)} \tag{2.41}$$

$$\left(\frac{W}{P}\right)_{S_{TO}} = \frac{1 - \exp\left(0.6\rho g C_{D_G} S_{TO} \frac{1}{W/S}\right)}{\mu - \left(\mu + \frac{C_{D_G}}{C_{L_R}}\right)\left[\exp\left(0.6\rho g C_{D_G} S_{TO} \frac{1}{W/S}\right)\right]} \frac{\eta_P}{V_{TO}} \quad \text{(Prop-driven aircraft)} \tag{2.42}$$

where

$$C_{D_G} = \left(C_{D_{TO}} - \mu C_{L_{TO}}\right)$$

- **Rate of Climb**

$$\left(\frac{T}{W}\right)_{ROC} = \frac{ROC}{\sqrt{\frac{2}{\rho\sqrt{\frac{C_{D_o}}{K}}}\left(\frac{W}{S}\right)}} + \frac{1}{(L/D)_{max}} \quad \text{(Jet aircraft)} \tag{2.43}$$

$$\left(\frac{W}{P}\right)_{ROC} = \frac{1}{\frac{ROC}{\eta_P} + \sqrt{\frac{2}{\rho\sqrt{\frac{3C_{D_o}}{K}}}\left(\frac{W}{S}\right)}\left(\frac{1.155}{(L/D)_{max}\,\eta_P}\right)} \quad \text{(Prop-driven aircraft)} \tag{2.44}$$

- **Ceiling**

$$\left(\frac{T}{W}\right)_C = \frac{ROC_C}{\sigma_C\sqrt{\frac{2}{\rho_C\sqrt{\frac{C_{D_o}}{K}}}\left(\frac{W}{S}\right)}} + \frac{1}{\sigma_C (L/D)_{max}} \quad \text{(Jet aircraft)} \tag{2.45}$$

$$\left(\frac{W}{P_{SL}}\right)_C = \frac{\sigma_C}{\frac{ROC_C}{\eta_P} + \sqrt{\frac{2}{\rho_C \sqrt{\frac{3C_{D_o}}{K}}}\left(\frac{W}{S}\right)}\left(\frac{1.155}{(L/D)_{max}\,\eta_P}\right)} \quad \textbf{(Prop-driven aircraft)} \tag{2.46}$$

At the absolute ceiling (h_{AC}), the rate of climb is zero ($ROC_{AC} = 0$). Hence, the corresponding expression for the thrust/power loading are:

$$\left(\frac{T}{W}\right)_{AC} = \frac{1}{\sigma_{AC}(L/D)_{max}} \quad \textbf{(Jet aircraft)} \tag{2.47}$$

$$\left(\frac{W}{P_{SL}}\right)_{AC} = \frac{\sigma_{AC}}{\sqrt{\frac{2}{\rho_{AC}\sqrt{\frac{3C_{D_o}}{K}}}\left(\frac{W}{S}\right)}\left(\frac{1.155}{(L/D)_{max}\,\eta_P}\right)} \quad \textbf{(Prop-driven aircraft)} \tag{2.48}$$

In the Eqs. (2.38) through (2.48), the new parameters are: K, wing induced drag factor, η_P, propeller efficiency, C_{DG}, UAV drag coefficient at the ground, C_{LTO}, UAV drag coefficient at the ground, μ, runway friction coefficient, σ, air density ratio, and $(L/D)_{max}$ denotes the maximum lift to drag ratio.

From the matching plot, the design point is selected, by which two numbers are extracted. The output of the plot would be either 1. $\left(\frac{W}{S}\right)_d$ and $\left(\frac{W}{P}\right)_d$ for a prop-driven UAV, or 2. $\left(\frac{W}{S}\right)_d$ and $\left(\frac{T}{W}\right)_d$ for a UAV with jet engine(s). Then, the wing area and engine power/thrust will be calculated as follows:

$$S = W_{TO} \Big/ \left(\frac{W}{S}\right)_d \tag{2.49}$$

$$P = W_{TO} \Big/ \left(\frac{W}{P}\right)_d \tag{2.50}$$

$$T = \left(\frac{T}{W}\right)_d . W_{TO} \tag{2.51}$$

Therefore, the wing area and engine power/thrust will be readily calculated.

2.10 Quadcopter Configuration

In the past several years, the popularity of quadcopter UAVs has skyrocketed. For many different types of mission, the quadcopter configuration with four electric engines is an optimal choice for small UAVs. The quadcopter is a popular UAV, mainly because of its unique features. There are five main reasons for the popularity of quadcopters: (a) size and portability; (b) low cost; (c) safety; (d) quick in capturing live events; and (e) ability to hover. This section presents technical features, and advantages and disadvantages of the quadcopter configuration.

Quadcopters have proven to be capable of capturing vital information from diverse angles; for example, in filming movies or capturing sport events. They have potential to change the future of journalism in the near future. Quadcopters prices range from as low as \$40–\$10 000 depending on their intended use, battery power and cameras attached to them. The popularity of the quadcopters is increasing as the sensors and control systems are becoming more advanced and less expensive.

The simple control via propeller rotation speed, the possibility to scale down/up, and its low cost are just three advantages of such configuration. The quadcopters are usually equipped with fixed-pitch propellers, so the overall control is implemented by directly altering the rotational speeds of the individual motors via the flight control commands. This in turn generates the rotational torque differences to control the flight path. Engine propellers are responsible for both lift and thrust, hence the vertical component of the engine force is lift, and the horizontal component is the thrust.

There are many similarities and differences between helicopters and quadcopters. A conventional helicopter has one main rotor, a gearbox, hub mechanism, and one tail rotor. However, the quadcopters have no moving parts except for the rotating motors and propellers. The quadcopter is also less prone to vibrations and it is more flexible when it comes to the placement of the center of gravity. Due to smaller size of rotors, they safer to fly indoors.

The motors of the quadcopters are mounted on the frame, and the only way to induce a lateral/directional/longitudinal motion is to tilt the entire body. Unlike a conventional helicopter, the quadcopter does not have a tail rotor to control the yaw motion. The quadcopter has four electric motors – each turns a propeller- (Figure 2.5) where two spins clockwise and two spins counterclockwise. If the pair of clockwise motors are spinning at a different rate than the pair of counterclockwise motors, it will create a torque about an axis. Each motor is placed at 90° from another one. All motors are generating an upward thrust – the total force will be equal to the lift. When the quadcopter is tilted, a horizontal component of the total force will push the vehicle in the horizontal direction.

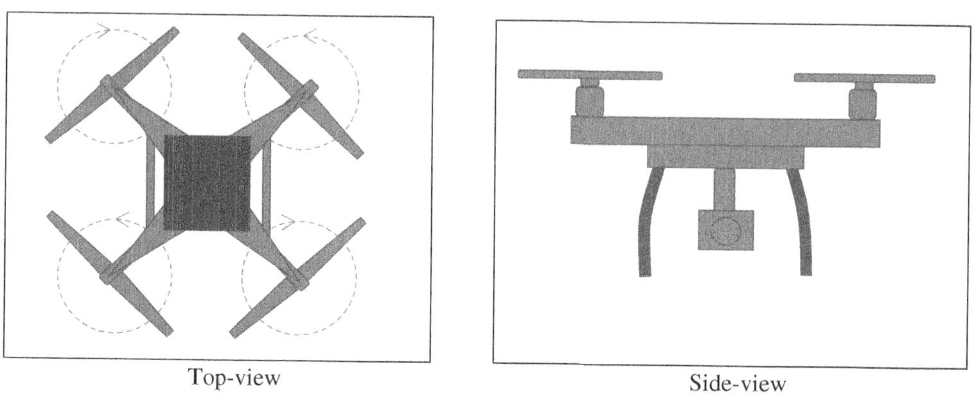

Figure 2.5 Quadcopter configuration.

The vehicle can carry its payload (e.g., camera) under the main body, for missions such as aerial photography and videography. Since four engines have no covers, the main body is not aerodynamically optimal, and landing gear is fixed, then a quadcopter has a higher drag. Hence, this configuration has a slow maximum airspeed compared with other configurations. Table 2.9 provides the technical features and configuration specifications of a quadcopter.

The world's leader [13] in commercial and civilian small UAV industry is "DJI," accounting for 72% of the quadcopter market share. Its quadcopter products (including flying platforms, flight controllers, helicopters accessories, aerial and handheld gimbals and ground stations) have been employed globally for the music, television, and movie industries. The DJI's quadcopters range from the hulking $5000 Hollywood-caliber Matrice 600 pro to the palm-size $400 Spark.

Figure 2.6 illustrates a DJI Phantom 4 Pro quadcopter. This quadcopter has a mass of 1380 kg, and has the capability of auto takeoff and auto return home. The maximum speed is 45 mph its flight time is 28 minutes. The Phantom 4 Pro has a titanium/magnesium alloy construction increasing the strength of its frame and reducing its weight. It is equipped with a camera with a 1-in. 20 megapixel complementary metal oxide semiconductor (CMOS) sensor which is capable of shooting 4000 at 60 frames per second video and Burst Mode stills at 14 frames per second. The Phantom 4 Pro features 5 directions of obstacle avoidance along with three intelligent autonomous flight modes (i.e., position, sport, and attitude). Three sets of dual vision sensors form a 6-camera navigation system that works constantly to calculate the respective speed and distance between the aircraft and an object.

A number of sections in other chapters of this book also provides various design aspects of quadcopters. For instance, Section 5.7 provides the quadcopter dynamic model including electric engine's thrust formulation. Moreover, Section 4.7 is dedicated to the aerodynamic design of quadcopters.

Example 2.4 Wing and Engine Sizing

Problem Statement: In the preliminary design phase of a very small fixed-wing UAV with a single electric engine, the maximum takeoff weight is determined to be 2 lb and the aircraft C_{Do} is determined to be 0.03. The airport is located at a city with the

Table 2.9 Technical features of a typical quadcopter.

No	Feature	Specification
1	Number of engines	Four
2	Engines type	Electric
3	Engines configuration	All upward in x-shape
4	Propeller rotation	Two clockwise, two counterclockwise
5	Takeoff and landing	Vertical
6	Hover	Yes
7	Landing gear	Fixed – skid
8	Speed	Slow
9	Drag	High

Figure 2.6 Quadcopter DJI Phantom 4 Pro.

elevation of 5000 ft. By using the matching plot technique, determine wing area (S) and engine power (P) of the aircraft that is required to have the following performance capabilities:

a) Maximum speed: 25 Knot True Air Speed (KTAS) at 5000 ft
b) Stall speed: less than or equal to 12 Knot Equivalent Air Speed (KEAS)
c) Rate of climb: more than or equal to 150 fpm at sea level
d) Takeoff run: less than or equal to 80 ft (on a dry concrete runway, μ = 0.04)
e) Absolute ceiling: more than or equal to 10 000 ft
f) Range: 2 km
g) Endurance: 30 minutes

The UAV has no flap, and the landing gear is fixed. Assume any other parameters that you may need for this aircraft

Solution

First, it must be noted that range and endurance requirements do not have any effect on the engine power or wing area, so we ignore them at this design phase. The air density at 5000 ft is 0.002 048 slug/ft^3 and at 10 000 ft is 0.001 755 slug/ft^3.

The matching plot is constructed by deriving five equations:

1) **Stall speed**

The stall speed is required to be greater than 120 KEAS. The wing sizing based on stall speed requirements is represented by Eq. (2.38). From Table 2.6, the aircraft maximum lift coefficient is selected to be 1.5.

$$\left(\frac{W}{S}\right)_{V_s} = \frac{1}{2}\rho V_s^2 C_{L_{max}} = \frac{1}{2} \times 0.002378 \times (12 \times 1.688)^2 \times 1.5 = 0.732 \frac{lb}{ft^2} \quad (2.38)$$

or (E-1)

where 1 knot is equivalent to 1.688 ft/s.

2) Maximum speed

The maximum speed is required to be greater than 25 KTAS at 5000 ft. The wing and engine sizing based on maximum speed requirements for a prop-driven aircraft are represented by Eq. (2.40).

$$\left(\frac{W}{P_{SL}}\right)_{V_{max}} = \frac{\eta_P}{\frac{1}{2}\rho_o V_{max}^3 C_{D_o}\frac{1}{\left(\frac{W}{S}\right)} + \frac{2K}{\rho\sigma V_{max}}\left(\frac{W}{S}\right)} \tag{2.40}$$

From Table 2.6, the wing aspect ratio (AR) is selected to be 7. From Table 2.6, the Oswald span efficiency factor is considered to be 0.85. Thus:

$$K = \frac{1}{\pi \cdot e \cdot AR} = \frac{1}{3.14 \times 0.85 \times 7} = 0.053 \tag{4.4}$$

The air relative density (σ) at 5000 ft is 0.002 048/0.002 378 or 0.861. The substitution yields:

$$\left(\frac{W}{P_{SL}}\right)_{V_{max}} = \frac{0.7 \times 550}{0.5 \times 0.002378 \times (25 \times 1.688)^3 \times 0.03\frac{1}{\left(\frac{W}{S}\right)} + \frac{2 \times 0.053}{0.002048 \times 0.861 \times (25 \times 1.688)}\left(\frac{W}{S}\right)} \tag{E-2}$$

where the whole term is multiplied by 550 to convert lb/(lb·ft/sec) to lb/hp. It is assumed that the propeller is variable-pitch, so, the prop efficiency is 0.7.

3) Takeoff run

The takeoff run is required to be greater than 80 ft at the elevation of 5000 ft. The wing and engine sizing based on takeoff run requirements for a prop-driven aircraft are represented by Eq. (2.42). Recall that the air density at 3000 ft is 0.002 175 slug/ft^3.

$$\left(\frac{W}{P}\right)_{S_{TO}} = \frac{1 - \exp\left(0.6\rho g C_{D_G} S_{TO}\frac{1}{W/S}\right)}{\mu - \left(\mu + \frac{C_{D_G}}{C_{L_R}}\right)\left[\exp\left(0.6\rho g C_{D_G} S_{TO}\frac{1}{W/S}\right)\right]}\frac{\eta_P}{V_{TO}} \tag{2.42}$$

The takeoff speed is assumed to be:

$$V_{TO} = 1.1V_s = 1.1 \times 12 = 13.2 KEAS$$

Takeoff lift and drag coefficients are:

$$C_{L_{TO}} = C_{L_C} + \Delta C_{L_{flapTO}} = 0.3 + 0 = 0.3$$

where the aircraft lift coefficient C_{L_C} is (from Table 2.6) assumed to be 0.3. Since the UAV has no flap, $\Delta C_{L_{flapTO}}$ is 0. The landing gear is fixed, from [8]:

$$C_{D_{oLG}} = 0, \quad C_{D_{oHLD\_TO}} = 0$$

$$C_{D_{oTO}} = C_{D_o} + C_{D_{oLG}} + C_{D_{oHLD\_TO}} = 0.03 + 0 + 0 = 0.03$$

$$C_{D_{TO}} = C_{D_{oTO}} + KC_{LTO}^2 = 0.03 + 0.053(0.3)^2 = 0.035$$

The variable C_{D_G} is:

$$C_{D_G} = C_{D_{TO}} - \mu C_{L_{TO}} = 0.035 - 0.04 \times 0.3 = 0.023$$

The takeoff rotation lift coefficient is:

$$C_{L_R} = \frac{C_{L_{max}}}{(1.1)^2} = \frac{C_{L_{max}}}{1.21} = \frac{1.5}{1.21} = 1.24$$

The substitution yields:

$$\left(\frac{W}{P}\right)_{S_{TO}} = \frac{\left[1 - \exp\left(0.6 \times 0.002048 \times 32.2 \times 0.023 \times 80 \frac{1}{(W/S)}\right)\right]}{\begin{array}{l} 0.04 - \left(0.04 + \dfrac{0.023}{1.24}\right)\left[\exp\left(0.6 \times 0.002048 \times 32.2 \times 0.023 \times 80 \dfrac{1}{(W/S)}\right)\right] \\ \left(\dfrac{0.7}{13.2 \times 1.688}\right) \times 550 \end{array}} \quad \text{(E-3)}$$

Again, the whole term is multiplied by 550 to convert lb/(lb·ft/sec) to lb/hp.

4) Rate of climb

The rate of climb run is required to be greater than 150 fpm (or 2.5 ft/s) at sea level. The maximum lift-to-drag ratio is

$$\left(\frac{C_L}{C_D}\right)_{max} = \frac{1}{2\sqrt{0.053 \times 0.03}} = 12.48$$

The wing and engine sizing based on rate of climb requirements for a prop-driven aircraft are represented by Eq. (2.44).

$$\left(\frac{W}{P}\right)_{ROC} = \frac{1}{\frac{ROC}{\eta_P} + \sqrt{\frac{2}{\rho\sqrt{\frac{3C_{D_o}}{K}}}\left(\frac{W}{S}\right)}\left(\frac{1.155}{(L/D)_{max}\eta_P}\right)} \quad (2.44)$$

The substitution yields:

$$\left(\frac{W}{P}\right)_{ROC} = \frac{1 \times 550}{\frac{150}{60 \times 0.7} + \sqrt{\frac{2}{0.002378\sqrt{\frac{3 \times 0.03}{0.053}}}\left(\frac{W}{S}\right)}\left(\frac{1.155}{12.48 \times 0.7}\right)} \quad \text{(E-4)}$$

And again, the whole term is multiplied by 550 to convert lb/(lb·ft/sec) to lb/hp.

5) **Absolute ceiling**

The absolute ceiling is required to be greater than 10 000 ft. The wing and engine sizing based on service ceiling requirements for a prop-driven aircraft are represented by Eq. (2.48). At absolute ceiling, the rate of climb is required to be 0. At 10000 ft altitude, the air density is 0.001755 slug/ft^3; so the relative air density is 0.738. The substitution yields:

$$\left(\frac{W}{P_{SL}}\right)_C = \frac{\sigma_C}{\sqrt{\frac{2}{\rho_C\sqrt{\frac{3C_{D_o}}{K}}}\left(\frac{W}{S}\right)}\left(\frac{1.155}{(L/D)_{max}\eta_P}\right)} \quad (2.48)$$

$$\left(\frac{W}{P}\right)_C = \frac{0.738 \times 550}{\sqrt{\frac{2}{0.001755\sqrt{\frac{3 \times 0.03}{0.053}}}\left(\frac{W}{S}\right)}\left(\frac{1.155}{12.48 \times 0.7}\right)} \quad \text{(E-5)}$$

And again, the whole term is multiplied by 550 to convert lb/(lb·ft/s) to lb/hp.

6) **Construction of matching plot**

Now, we have five equations of E-1, E-2, E-3, E-4, and E-5. In all of them, power loading is defined as functions of wing loading. When we plot all of them in one graph, the Figure 2.7 will be produced. Recall in this example, that the unit of W/S in lb/ft^2, and the unit of W/P is lb/hp.

Now, we need to recognize the acceptable regions. The region below each graph satisfies the performance requirements. In other words, the region above each graph does not satisfy the performance requirements. For the case of stall speed, the region in the left side of the graph satisfies stall speed requirements (see Figure 2.7). Hence,

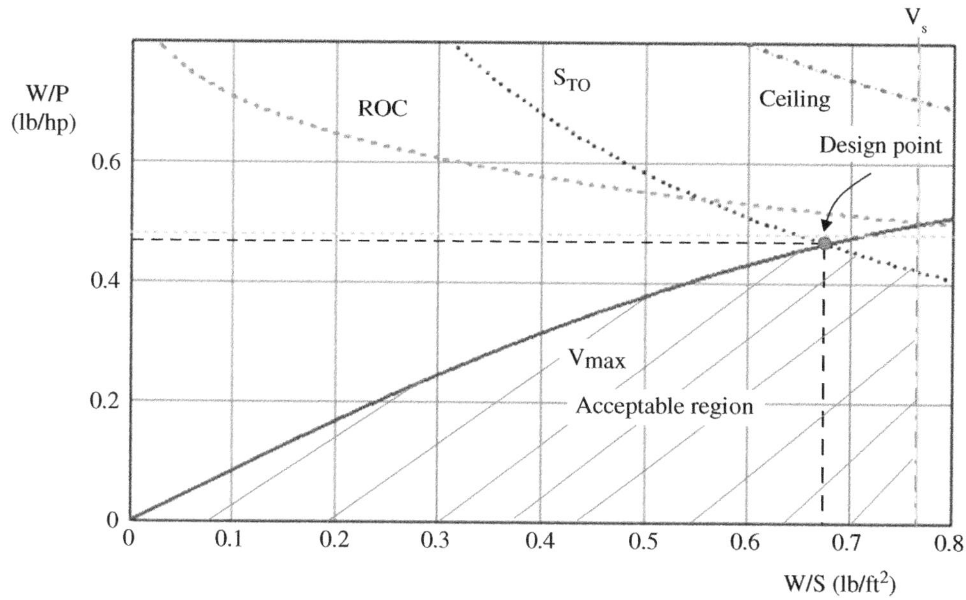

Figure 2.7 Matching plot for example problem 2.4.

the region between the graphs of maximum speed, takeoff run and stall speed is the target area.

In this region, we are looking for the smallest engine (lowest power) that has the lowest operating cost. Thus, the highest point of this region is the design point. Therefore, the wing loading and power loading will be extracted from Figure 2.7 as:

$$\left(\frac{W}{P}\right)_d = 0.465$$

$$\left(\frac{W}{S}\right)_d = 0.676$$

Then, the wing area and engine power will be calculated as follows:

$$S = W_{TO} \Big/ \left(\frac{W}{S}\right)_d = \frac{2lb}{0.676\frac{lb}{ft^2}} = 2.96\text{ft}^2 = 0.275\text{m}^2 \tag{2.49}$$

$$P = W_{TO} \Big/ \left(\frac{W}{P}\right)_d = \frac{2lb}{0.465\frac{lb}{hp}} = 4.3\text{hp} = 3.21\text{kW} \tag{2.50}$$

Therefore, the wing area and engine power will be:

$S = 2.96\,\text{ft}^2,\ P = 4.3\ \text{hp}$

Questions

1 What parameters are determined in the preliminary design phase for a fixed-wing UAV?

2 What parameters are determined in the preliminary design phase for a quadcopter UAV?

3 What is the typical engine propeller efficiency (η_P)?

4 Briefly describe the technical features of a quadcopter.

5 What are the weaknesses of a quadcopter?

6 What are four elements in building the maximum takeoff weight for a quadcopter UAV with electric engines?

7 What are four elements in building the maximum takeoff weight for a fixed-wing UAV with piston engine?

8 What are typical payloads for a quadcopter for a aerial filming mission?

9 What is the payload weight ratio of Predator B SkyGuardian UAV?

10 What is the payload weight ratio of Black Eagle 50 UAV?

11 What is the payload weight ratio of A-160 Hummingbird UAV?

12 What is the payload weight ratio of RQ-7B Shadow UAV?

13 What is the weight of TL300MN FPV optical camera?

14 What is the weight of DJI Zenmuse X5 optical camera?

15 What is the weight of Garmin GSX 70 radar used in a NASA Global Hawk?

16 What is the weight of ALE-50 decoy used in ECM?

17 What is the weight of AN/ALR-67 radar warning receiver?

18 What is the weight of BOD 63M-LA01-S115 laser range finder?

19 What is the weight APM 2.8 ArduPilot?

20 Draw a typical flight mission for a UAV in an aerial filming mission for a distance location.

21 What percent of fuel weight is consumed in the climb segment for a UAV with piston engine?

22 What percent of battery energy is consumed in the climb segment for a UAV with electric engine?

23 What is the relation between the minimum power speed for most prop-driven aircraft and the stall speed?

24 What is the typical relation between the cruising speed for the cruise segment and the maximum speed?

25 Define energy density for a battery.

26 What are typical values for AR for UAVs?

27 What are typical values for the zero-lift drag coefficient for UAVs with fixed landing gear?

28 What are typical values for the maximum lift-to-drag ratio for UAVs with retractable landing gear?

29 What percent of battery energy is consumed for a pro-driven engine, in non-cruising flight segments (e.g., takeoff and climb)?

30 How the empty weight fraction of a UAV is modeled?

31 Describe the matching plot technique?

32 What aircraft performance requirements are utilized to preliminary size the wing and engine of a UAV?

33 Draw a typical matching plot for a prop-driven aircraft.

34 Draw a typical matching plot for a jet aircraft.

35 List five main reasons for the popularity of quadcopters.

36 Draw top-view of a typical quadcopter.

37 Why a quadcopter has a higher drag compared with a fixed-wing UAV?

38 What is the gross mass of a DJI Phantom 4 Pro quadcopter?

Problems

1 A fixed-wing UAV with a prop-driven electric engine (with Lithium-ion batteries) is in the preliminary design phase. You are required to determine its battery weight ratio. The UAV is required to be able to fly to a distance of 8 km and return back. Assume a C_L/C_D of 9, and a prop efficiency of 0.7.

2 A fixed-wing RC plane with a prop-driven electric engine (with Lithium-ion batteries) is in the preliminary design phase. You are required to determine its battery weight ratio. The UAV is required to be able to fly to a distance of 4 km and return back. Assume a C_L/C_D of 7, and a prop efficiency of 0.65.

3 A small fixed-wing RC UAV with a mass of 2 kg is equipped with a pack of Li—Po (Lithium Polymer) batteries for its electric motor. If the energy of batteries is just consumed in a loiter mission (for electric motor), for a maximum endurance of 30 minutes, determine the mass of batteries pack. Assume that the electric motor is creating 5 N of thrust, with a prop efficiency of 0.73, and the UAV is flying at a speed of 35 m/s.

4 A small fixed-wing RC UAV with a mass of 3 kg is equipped with a pack of Li—Po (Lithium Polymer) batteries for its electric motor. If the energy of batteries is just consumed in a loiter mission (for electric motor), for a maximum endurance of 40 minutes, determine the mass of batteries pack. Assume that the electric motor is creating 6 N of thrust, with a prop efficiency of 0.77, and the UAV is flying at a speed of 40 m/s.

5 The design of a small RC plane for an aerial filming mission is in the preliminary design phase. A Sony CC-1XHRM camera has been selected. An APM 2.8 ArduPilot and a single electric engine with Lithium– polymer batteries have been chosen. The total desired range is 15 km. Estimate the UAV maximum takeoff weight. Assume a cruise C_L/C_D of 9, and a prop efficiency of 0.72.

6 The design of a small RC plane for an aerial filming mission is in the preliminary design phase. A Foxeer 16:9 RCX09-459-NTSC camera (See Table 12.2 for the weight) has been selected. An Omnibus F4+ (Plus) autopilot unit and a single electric engine with Lithium– polymer batteries have been chosen. The total desired range is 3 km. Estimate the UAV maximum takeoff weight. Assume a cruise C_L/C_D of 7, and a prop efficiency of 0.68.

7 The design of an MALE UAV for surveillance mission is in the preliminary design phase. A surveillance sensor system and a payload with a total weight of 1500 lb has been selected. An autopilot unit with a weight of 30 lb, and a single piston engine have been chosen. The total desired range is 1000 km. Assume, a typical mission profile with no loiter, a $(C_L/C_D)_{max}$ of 12, a prop efficiency of 0.81, and specific fuel consumption of 0.5 lb/hr/hp. Estimate the UAV maximum takeoff weight.

8 The design of a quadcopter for an aerial filming mission is in the preliminary design phase. A Foxeer 16:9 RCX09-459-NTSC camera (See Table 12.2 for the weight) has been selected. An APM 2.8 ArduPilot and Lithium– ion batteries have been chosen. The total desired range is 5 km. Estimate the UAV maximum takeoff weight. Assume a C_L/C_D of 2, and a prop efficiency of 0.6.

9 The design of a quadcopter for an aerial filming mission is in the preliminary design phase. A DJI Zenmuse X5 Camera (See Table 12.2 for the weight) with gimbal has been selected. An APM 2.8 ArduPilot and Lithium– ion batteries have been chosen. The total desired range is 4 km. Estimate the UAV maximum takeoff weight. Assume a C_L/C_D of 4, and a prop efficiency of 0.64.

10 In the preliminary design phase of a fixed-wing RC plane, the maximum takeoff weight is determined to be 3 lb. The target region for aerial photography is located at 300 ft away from the operator. By using the matching plot technique, determine wing area (S) and engine power (P) of the aircraft that is required to have the following performance capabilities:

a Maximum speed: 30 KTAS at 500 ft altitude
b Stall speed: less than 10 KEAS
c Rate of climb: more than 500 fpm at sea level
d Takeoff run: less than 50 ft (on a dry concrete runway)
e Absolute ceiling: more than 5,000 ft.
f Endurance: 20 minute

The UAV is equipped with a piston-prop engine and the vehicle C_{Do} is determined to be 0.03. Other features of the UAV are considered as follows:

$K = 0.03$, $\eta_P = 0.7$, $C_{DG} = 0.06$, $C_{LTO} = C_{LG} = 0.8$, $\mu = 0.04$

11 In the preliminary design phase of a MALE UAV, the maximum takeoff weight is determined to be 2000 lb. The target region for aerial photography is located at 400 km away from the operator. By using the matching plot technique, determine wing area (S) and engine power (P) of the aircraft that is required to have the following performance capabilities:

a Maximum speed: 120 KTAS at 12000 ft altitude
b Stall speed: less than 50 KEAS
c Rate of climb: more than 1200 fpm at sea level
d Takeoff run: less than 1000 ft (on a dry concrete runway)
e Service ceiling: more than 12 000 ft.
f Endurance: 20 hours

The UAV is equipped with a turbo-prop engine and the vehicle C_{Do} is determined to be 0.025. Other features of the UAV are considered as follows:

$K = 0.04$, $\eta_P = 0.75$, $C_{DG} = 0.05$, $C_{LTO} = C_{LG} = 0.7$, $\mu = 0.05$

12 In the preliminary design phase of a small fixed-wing UAV, the maximum takeoff weight is determined to be 20 lb. The target region for aerial photography is located at 1000 ft away from the operator. By using the matching plot technique, determine wing area (S) and engine power (P) of the aircraft that is required to have the following performance capabilities:

a Maximum speed: 50 KTAS at 1000 ft altitude
b Stall speed: less than 20 KEAS
c Rate of climb: more than 700 fpm at sea level
d Takeoff run: less than 100 ft (on a dry concrete runway)
e Service ceiling: more than 3000 ft.
f Endurance: 45 minute

The UAV is equipped with a piston-prop engine and the vehicle C_{Do} is determined to be 0.035. Other features of the UAV are considered as follows:

$K = 0.06$, $\eta_P = 0.6$, $C_{DG} = 0.04$, $C_{LTO} = C_{LG} = 0.9$, $\mu = 0.044$

3

Design Disciplines

CONTENTS

Educational Outcomes

After study of this chapter, the reader will be able to:

1) Discuss functions of major components of a Unmanned Aerial System (UAS)
2) Discuss functions of major components of a Unmanned Aerial Vehicle (UAV)
3) Evaluate structural design of a UAV

Design of Unmanned Aerial Systems, First Edition. Mohammad H. Sadraey.

Companion website: www.wiley.com/go/sadraey/unmanned-aerial-systems

4) Evaluate propulsion system design of a UAV
5) Evaluate propulsion system design of a UAV
6) Evaluate mechanical and power transmission design of a UAV
7) Develop electric system design of a UAV
8) Evaluate control surfaces design of a UAV
9) Assess likely failure modes of sub-systems/components
10) Provides components installation guidelines

3.1 Introduction

As presented in Chapter 1, there are three major design phases in a UAV design project: (a) conceptual design; (b) preliminary design; and (c) detail design. The conceptual design techniques were briefly reviewed in Chapter 1. The preliminary design technique is fully provided in Chapter 2. However, the detail design techniques of UAS components are described in Chapters 3 through 13.

Moreover, there are several design disciplines which work in parallel within a UAV design project. Some examples are:

1) Aerodynamic design.
2) Structural design.
3) Propulsion system design.
4) Power transmission system design.
5) Mechanical system design.
6) Control surfaces design.
7) Ground control station.
8) Launch and recovery system.

This chapter briefly covers the first six disciplines; the other two disciplines are presented in Chapters 11 and 10 respectively. Due to the limited volume of the book, only the basic fundamentals are presented. The interested reader should refer to other references such as [8] for more details.

Table 3.1 shows the UAV major components and their primary functions. The main function of the wing is to generate the aerodynamic force of lift to keep the aircraft airborne. The primary function of the fuselage is to accommodate the payload which includes camera and sensors and other useful loads communication system. The horizontal tail's primary function is to generate an aerodynamic force to provide longitudinal trim/stability of the aircraft. Likewise, the vertical tail's primary function is to generate an aerodynamic force to provide directional trim/stability of the aircraft. The engine is the main component in the aircraft propulsion system to generate the power thrust. The primary function of the landing gear is to facilitate take-off and landing operations. The primary function of each component is the driving force in its detail design.

In addition, due to popularity and uniqueness of quadcopters in aeronautics, these UAVs are specially treated in two other Chapters (2 and 5). The technical features, and advantages of quadcopters as well as their applications are presented, too, while the design of the communication system is discussed in Chapter 13, as a subsystem of ground control station.

Table 3.1 Unmanned aerial vehicle (UAV) major components and their functions.

No	Component	Primary function	Major areas of influence
1	Fuselage	Accommodations (Payload, systems)	UAV performance, longitudinal stability, lateral stability, cost, stability
2	Wing	Generation of lift	UAV performance, lateral stability
3	Horizontal/ Vertical tail	Longitudinal/ Directional stability	Longitudinal/Directional trim and control, stealth
4	Payload	Sense, measure, release/drop a store	UAV weight, drag, performance, power consumption
5	Engine	Generation of thrust/ power	UAV performance, stealth, cost, control
6	Landing gear	Facilitate take-off and landing	UAV performance, stealth, cost
7	Control surfaces	Create rolling, pitching, and yawing moment	Maneuverability, cost
8	Autopilot	Control, guidance, and navigation	Maneuverability, stability, cost, flight safety
9	Mechanical Systems	Power Transmission	UAV weight, power consumption
10	Recovery system	Recovery / Landing	UAV weight, longitudinal trim control
11	Launch/TO system	Accommodate takeoff	Longitudinal trim and control
12	Ground system	Parallel control, command, receive data	Cost, communication system
13	Communication system	Communicate with ground station	UAV weight, detectability, sense and avoid system

3.2 Aerodynamic Design

The primary aerodynamic function of the UAV components (e.g., wing and tail) is to generate sufficient lift force or simply lift (L). However, they have two other aerodynamic products, namely drag force or drag (D), and nose-down pitching moment (M).

The aerodynamic forces of lift and drag are functions of the following factors:

1) Aircraft configuration
2) Aircraft angle of attack
3) Aircraft geometry (e.g., wing planform area [S])
4) Airspeed (V)
5) Air density (ρ)
6) Reynolds number of the flow
7) Air viscosity.

$$L = \frac{1}{2}\rho V^2 S C_L \tag{3.1}$$

$$D = \frac{1}{2}\rho V^2 S C_D \tag{3.2}$$

where C_L and C_D are the lift and drag coefficients respectively.

While a UAV designer is looking to maximize the lift, the other two (drag and pitching moment) must be minimized. In fact, a wing is considered as a lifting surface that lift is produced due to the pressure difference between lower and upper surfaces. Aerodynamics textbooks are a good source to consult for information about mathematical techniques for calculating the pressure distribution over the wing and for determining the flow variables.

During the aerodynamic design process, several parameters must be determined. For instance, for a wing, they are:

1) Wing reference (or planform) area
2) Number of the wings
3) Vertical position relative to the fuselage (high, mid, or low wing)
4) Horizontal position relative to the fuselage
5) Cross section (or airfoil)
6) Aspect ratio (AR)
7) Taper ratio (λ)
8) Tip chord (C_t)
9) Root chord (C_r)
10) Mean Aerodynamic Chord (MAC or C)
11) Span (b)
12) Twist angle (or washout) (α_t)
13) Sweep angle (Λ)
14) Dihedral angle (Γ)
15) Incidence (i_w) (or setting angle, α_{set})
16) High lifting devices such as flap.

The wing aerodynamic design technique is presented in Chapter 4.

One of the necessary tools in the wing design process is an aerodynamic technique to calculate wing lift, wing drag, and wing pitching moment. With the progress of the science of aerodynamics, there are variety of techniques and tools to accomplish this time consuming job. A variety of tools and software based on aerodynamics and numerical methods have been developed in the past decades. The Computational Fluid Dynamics (CFD)[1] Software based on the solution of Navier—Stokes equations, vortex lattice method, thin airfoil theory, and circulation are available in the market. The application of such software packages; which is expensive and time-consuming; at this early stage of wing design seems un-necessary.

Wing is a three dimensional component, while the airfoil is two dimensional section. Because of the airfoil section, two other outputs of the airfoil, and consequently the wing, are drag and pitching moment. The wing may have a constant or a non-constant cross-section across the wing. There are two ways to determine the wing airfoil section:

(i) Airfoil design (ii) Airfoil selection. The design of the airfoil is a complex and time consuming process and needs expertise in fundamentals of aerodynamics at graduate level. Since the airfoil needs to be verified by testing it in a wind tunnel, it is expensive too.

Selecting an airfoil is a part of the overall wing design. Selection of an airfoil for a wing begins with the clear statement of the flight requirements. For instance, a subsonic flight's design requirements are very much different from a supersonic flight's design objectives. On the other hand, flight in the transonic region requires a special airfoil that meets Mach divergence requirements. The designer must also consider other requirements such as airworthiness, structural, manufacturability, and cost requirements. For fundamentals of aerodynamics, please refer to references such as [14, 15]. The UAV aerodynamic design is discussed in Chapter 4.

3.3 Structural Design

The structure of a conventional fixed-wing UAV consists of five principal units: fuselage, wings, horizontal tail, vertical tail, and control surfaces. The landing gear is also part of the structure, but will be covered in section 3.5. Engine pylon, engine inlet (for supersonic UAVs), fairings (and fillets), and landing gear bay doors are also assumed to be part of aircraft structure. The primary functions of the structure are:

1) To keep the aerodynamic shape of the UAV.
2) To carry the loads.

Airframe structural components are constructed from a wide variety of materials. The earliest aircraft were constructed primarily of wood. Steel tubing and the most common material — aerospace aluminum – have followed. Many newly designed UAVs are built from advanced light-weight composite materials, such as epoxy/glass, honeycomb, and carbon fiber.

Structural members of a fuselage mainly include stringers, longerons, bulkheads, and skin. The structural members in a wing/tail are spar, rib, stiffeners, and skin. The fuselage/wing/tails skin can be made from a variety of materials, ranging from impregnated fabric to plywood, aluminum, or composites. Under the skin and attached to the structural components are the many components that support airframe function. The entire airframe and its components are joined by rivets, bolts, screws, and other fasteners. Welding, adhesives, and special bonding techniques are also employed.

The most common form of UAV structure is semi-monocoque (single shell) which implies that the skin is stressed/reinforced. The structural members are designed to carry the flight loads or to handle stress without failure. In designing the structure, every square inch of wing and fuselage, must be considered in relation to the physical characteristics of the material of which it is made. Every part of the structure must be planned to carry the load which is applied on it.

The structural designer will determine flight loads, calculate stresses, and design structural elements such as to allow the UAV components to perform their aerodynamic functions efficiently. This goal will be considered simultaneously with the objective of the lowest structural weight. The most common tool in structural analysis is the use of finite element methods (FEM). One of the earliest and the most well-known computer software is NASTRAN, developed by National Administration for Aeronautics and Astronautics (NASA) in the mid-1960s. The stress analysis is the basic calculation to determine the safety factor. There are five major stresses [16] to which structural members are subjected: (a) tension; (b) compression; (c) torsion; (d) shear; and (e) bending. A single member of the structure is often subjected to a combination of stresses.

The fuselage usually consists of frame assemblies, bulkheads, and formers. The skin is reinforced by longitudinal members called longerons. Often wings/tails are of full cantilever design. In general, wing construction is based on one of three fundamental designs: (a) monospar; (b) multispar; and (c) box beam. Spars are the principal structural members of the wing. They correspond [17] to the longerons of the fuselage. Spars (Figure 3.1) run parallel to the lateral axis of the aircraft, from the fuselage toward the tip of the wing, and are usually attached to the fuselage by a beam, or a truss. Generally, a wing has two spars. One spar is usually located at the maximum thickness, and the other about two-thirds of the distance toward the wing trailing edge (in front of flap/control-surface). The main spar is responsible for the protection of the maximum thickness of the wing at various stations. Ribs are attached to the main spar (Figure 3.1), and are repeated at frequent intervals. The ribs geometries are exactly the same as the shape of airfoil sections at each station.

Honeycomb structured wing panels are often used in composite wings. Nacelles (i.e., pods) are streamlined enclosures used primarily to house the engine and its components. Engine mounts are also found in the nacelles. These are the assemblies to which the engine is fastened. They are usually constructed from chrome/molybdenum steel tubing in light UAV and forged chrome/nickel/ molybdenum assemblies in larger UAVs. Cowling are the detachable panels covering those areas into which access must be gained regularly, such as the engine and its accessories. In the design of airframe, several factors such as ultimate load, aerodynamic loads (pressure distribution), weight loads (e.g., fuel and engine), weight distribution, gust load, load factor, propulsion loads, landing loads (e.g., brake) and aero-elasticity effects must be considered.

One of the design requirements for some military UAVs is stealth. In the concept of stealth, the three basic methods of minimizing the reflection of pulses back to a receptor are:

1) To manufacture appropriate areas of the UAV from radar-translucent material such as Kevlar or glass composite as used in radomes, which house radar scanners.

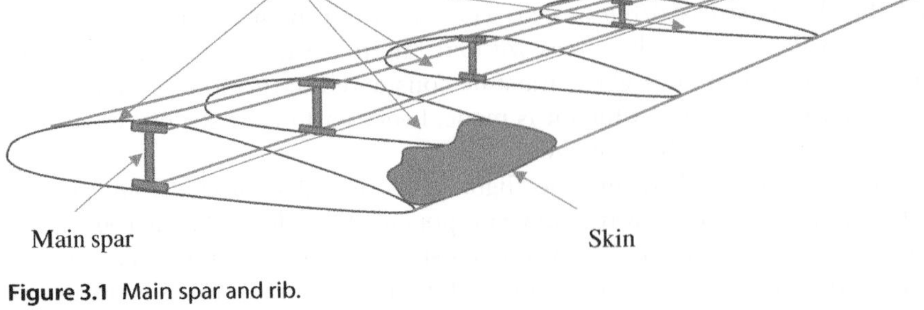

Figure 3.1 Main spar and rib.

2) To cover the external surfaces of the aircraft with RAM (radar absorptive material).
3) To shape the aircraft externally to reflect radar pulses in a direction away from the transmitter. The acoustic (i.e., noise) wavelength (signature) range for detecting an air vehicle is 16 m–2 cm.

It is recommended to manufacture appropriate areas of the UAV (e.g., radomes to house radar scanners and a SatCom radar dish antenna) from radar-translucent materials (See Figure 3.2), such as honeycomb, Kevlar or glass composite. In either a down-looking radar, or an up-looking satellite communications, or a forward-looking seeker, the radome should allow radio waves to pass through skin without any distortion.

The operating flight loads limits on a UAV are usually presented in the form of a V-n diagram. Structural designers will construct this diagram with the cooperation of the flight dynamics group. The diagram will determine the structural failure areas, and area of structural damage/failure. The UAV should not be flown out of the flight envelope, since it is not safe for the structures. The UAV structural design is beyond the scope of this text, the interested reader is recommended to consult with references such as [16].

3.4 Propulsion System Design

A heavier-than-air craft (UAV) requires a propulsion system in order to have a sustained flight. Without a proper aero-engine or powerplant, a heavier-than-air vehicle can only glide for a short time. The contribution of a powerplant to an aircraft is to generate the most influential force in the aircraft performance; that is, the propulsive force or thrust. The secondary function of the propulsion system is to provide power/energy to other subsystems such as hydraulic system, electric system, pressure system,

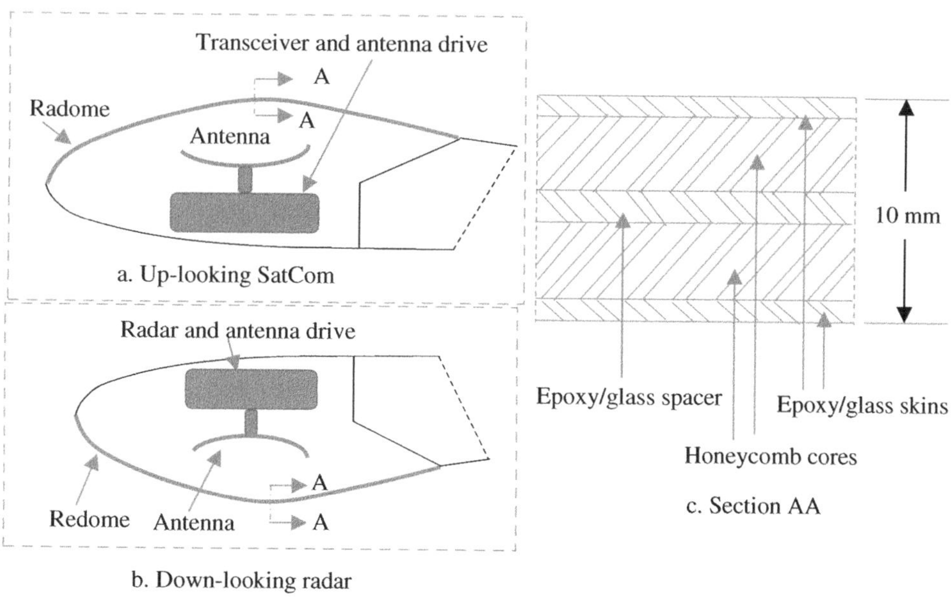

Figure 3.2 A typical cross-section of a low distortion radome.

air conditioning system, and avionics. These subsystems rely on the engine power to operate. Due to the popularity of electric engines in small UAVs (particularly in quadcopters), this type of engine is specifically treated in section 3.4.2.

3.4.1 General Design Guidelines

Soon after the design requirements and constraints are identified and prioritized, the propulsion system designer will begin to select the type of engine. There are a number of engine types available in the market for flight operations. They include: (a) electric (battery); (b) solar-powered; (c) piston-prop; (d) turbojet; (e) turbofan; (f) turboprop; (g) turboshaft; (h) ramjet; (i) rocket engines.

A jet engine (e.g., turbojet, turbofan) directly generates thrust, while prop-drivel engines (e.g., electric, solar-powered, piston-prop, and turboprop) employ a propeller to convert the engine power to thrust. In a cruising flight, the thrust (T) is related to engine power (P) as below:

$$T = \frac{P\eta_p}{V} \tag{3.3}$$

where η_P denotes propeller efficiency, and V is the level-flight airspeed.

Example 3.1

A small UAV with a piston engine is cruising at sea level with a speed of 50 knot. The propeller efficiency is 0.7. The wing area is $2\,m^2$, and the drag coefficient is 0.04. How much power the engine should generate for this flight?

Solution

$$D = \frac{1}{2}\rho V^2 S C_D = \frac{1}{2} \times 1.225 \times (50 \times 0.514)^2 \times 2 \times 0.04 = 32.4\,\mathrm{N} \tag{3.2}$$

$$T = D = 32.4\,\mathrm{N} \tag{2.35}$$

$$T = \frac{P\eta_p}{V} \Rightarrow P = \frac{TV}{\eta_p} = \frac{32.4 \times 50 \times 0.514}{0.7} = 1191\,W = 1.19\,\mathrm{kW} \tag{3.3}$$

Propeller is a means to convert the engine power to engine thrust. The aerodynamic equations and principles that govern the performance of a wing are generally applied to a propeller. Hence, the propeller may be called a rotating wing. It simply creates the lift (i.e., thrust) with the cost of drag. For this reason, the prop efficiency never could reach 100%. In a cruising flight and in the optimum angle of twist (the best propeller pitch), the propeller efficiency (η_P) may operate at around 75–85%. Propeller design to determine parameters such as the blade airfoil section and twist angle is out of scope of this text.

An important issue in designing UAV engine is the type of fuel consumed by the engine. There are various requirements for the fuel, such as its density, boiling and freezing temperatures. Two issues for fuel when an aircraft is flying at high altitudes are: (a) it evaporates; and (b) it freezes. Ignoring these two significant problems may result in the mission failure.

In general, the job of an aircraft designer is to determine/design/select the following items:

1. Select engine type.
2. Select number of engines.
3. Determine engine location.
4. Select engine from manufacturers' catalogs.
5a. Size propeller (if prop-driven engine).
5b. Design inlet (if jet engine).
6. Design engine installation.
7. Iterate and optimize.

Items 5a and 5b are design activities for an aircraft designer which must be carried along and parallel with the engine design team. This is to emphasize that the aircraft designer has the final say in these two propulsion system parameters. MQ-1A Predator (Figure 8.5) A was equipped with a piston engine – max speed of 80 knot, while the MQ-9B Predator B is equipped with a turboprop engine – max speed of 260 knot.

In general, piston engine propeller-driven UAVs are difficult to detect with an infrared (IR) sensor because their engines are running at much lower temperatures than turbine engines. Table 3.2 shows the propulsion systems of a number of current UAVs.

Global Hawk is equipped with an AE 3007H Rolls-Royce turbofan engine. The engine is mounted on the top surface of the rear fuselage section with the engine exhaust between the V-shaped tail section. The RQ-7A Shadow 200 has a fuel capacity of 40 l l. Its pusher rotary engine is driving a two-bladed propeller.

The MQ-5B Hunter is powered by two heavy fuel diesel engines developed by Northrop Grumman. One engine is installed at the fuselage nose (tractor function), and one at the rear of fuselage (pusher function). Due to two engines, the ceiling is higher than 20 000 ft, and endurance is 15 hours.

The author has presented the propulsion system design for an air vehicle in Chapter 8 of [8], so the materials are not repeated in this text. You are encouraged to refer to [8, 18, 19] to study the details of propulsion system design.

Table 3.2 Propulsion systems of a number of Unmanned aerial vehicles (UAVs).

No	UAV	Engine type	Engine	Thrust/power
1	Global Hawk	Turbofan	Rolls-Royce AE 3007	34 kN
2	General Atomics Avenger	Turbofan	Pratt & Whitney PW500	17.7 kN
3	MQ-1 Predator A	Piston	Rotax 914	115 hp
4	MQ-9 Reaper	Turboprop	Honeywell TPE333–10	900 hp
5	RQ-7 Shadow 200	Piston	UEL AR 741	28.3 kW
6	Northrop Grumman Firebird	Piston	Lycoming TO-540	260 kW
7	DJI Mavic 2 Pro quadcopter	Electric	4820 mAh at 15.2 V	4 × 73 W
8	AeroVironment RQ-11 Raven	Electric	Aveox brushless 27/26/7	250 W
9	Arcturus UAV T-20	Piston	Avgas 110 LL	7.5 kW
10	Aurora Perseus B	Piston	Rotax 94	73 kW

3.4.2 Electric Engines

All-Electric or Hybrid-Electric propulsion systems are becoming more and more efficient and attractive as battery technologies improve. There should be significant attention given to the pros and cons and "how-to" content on how to design UAVs with all-electric or hybrid-electric propulsion systems.

An electric propulsion system includes electric motor, battery, and propeller. So, in electric-powered UAVs, the powerplant is a battery-powered electric motor. An electric motor converts electrical energy to mechanical energy. Most model or radio-controlled small aircraft (wing span of less than two meters), and all quadcopters are utilizing electric propulsion system. Other features of a typical electric powered aircraft are low speed (less than 30 knot), low range (less than 50 km), low endurance (less than one hour), low cost (from a few hundred to a few thousand dollars), compact size, constant center of gravity, and quietness.

The highest feasible power that a battery or group of batteries may provide is typically about less than 100 horse power for less than an hour. The main advantages of an electric-powered engine are its independence from fuel, and from mechanical engine, no vibration, the near silent motor, as well as a low cost. However, the main disadvantages originate from a limit in the electric energy storage. However, if solar cells are utilized, the range and endurance is dramatically increased.

There are two main different motor types used in model aircraft: (a) brushed; (b) brushless. From another aspect, electric motors are divided into two groups: (a) Direct Current (DC) motors; and (b) Alternating Current (AC) motors. The electric motor of an aircraft with an electric propulsion system is usually of DC type.

The battery-powered electric motor is very popular for remotely (or radio) controlled (RC) airplane. The typical engine-power-to-airplane-mass ratio is 100—200 W/kg. The typical voltage of electric engine that are employed by RC aircraft is 10—12 V. The maximum engine power is determined via the power of the battery which provides the electric energy to the engine. The electric motor is frequently operated at a much higher voltage which a single battery is providing. Hence, a couple of batteries are usually connected in series to add up the voltages. For instance, if an electric motor requires 12 V to operate; 6 batteries of 2 V can be connected in series. Moreover, as the number of batteries is increased, the maximum power that can be provided to an electric motor will be increased.

The power of an electric engine is determined [20] by multiplying the electric current (I) it is consuming by the terminal voltage (V).

$$P = IV \tag{3.4}$$

When the unit of V is voltage (V), and the unit of current is Ampere (A), the unit of power will be Watt (W). Furthermore, the power is the rate of change (consumption) of energy with respect to time.

$$P = \frac{dE}{dt} \tag{3.5}$$

where the unit of time (t) is second (s). Thus, the energy/work supplied by a battery to an electric engine is:

$$E = \int_{0}^{E} IVdt \tag{3.6}$$

where E is the duration of flight operation (i.e., endurance). When the current and voltage are constant (for instance in a cruising flight), this equation can be reduced to:

$$E = IVt \tag{3.7}$$

where the unit of energy will be in Joule (J). For a given battery, the total energy is fixed. In addition, for a given electric engine, the voltage is usually fixed. Hence, the duration in which a battery can supply energy is based on the power which is consumed by the engine. This implies that the current; which is drawn by the electric engine; can be adjusted.

The required engine power varies based on the flight conditions and flight operation. For instance, the engine power required in a takeoff is much higher than that in a cruising flight. Since the electric engine voltage is usually constant, the electric current will vary. In addition, the mechanical power is equal to the torque times the rotational speed (P = T.ω). Therefore, it can be loosely concluded that a DC motor converts the electric current into torque and the voltage into revolutions per minute (RPM). Moreover, the engine torque is a function of the rotor mass moment of inertia.

Example 3.2

Determine the total energy supplied by a 4200 mAh 11.1 Volts Lithium Polymer battery, and the maximum power that can be provided to an electric motor.

Solution

Based on the description of the battery, a maximum current of 4.2 A (i.e., 4200 milli-Amps) at 11.1 Volt is provided for one hour (which is 3600 seconds). Thus:

- Total energy

$$E = IVt = 4.2 \times 11.1 \times 3600 = 167832\,\mathrm{J} = 167.8\,\mathrm{kJ} \tag{3.7}$$

- Maximum power

$$P_{max} = IV = 4.2 \times 11.1 = 46.6\,\mathrm{W} \tag{3.4}$$

3.5 Landing Gear Design

Another aircraft major component that is needed to be designed is the landing gear (also referred to as undercarriage). The landing gear is the component that supports the vehicle on the ground and facilitates the taxi, takeoff, and landing. In fact, landing gear design tends to have several interferences with other components (e.g., wing, tail, and fuselage) and the aircraft structural design. In general, the following are the landing gear parameters which are to be determined:

1) Type (e.g., nose gear [tricycle], tail gear, bicycle).
2) Fixed (faired, or un-faired), or retractable, or partially retractable

3) Height
4) Wheel base
5) Wheel track
6) The distance between main gear and aircraft cg
7) Strut diameter
8) Tire sizing (diameter, width)
9) Landing gear compartment if retracted
10) Load on each strut.

The landing gear usually includes wheels, but some aircraft are equipped with skis for snow or float for water. In the case of a vertical take-off and landing aircraft such as a helicopter, wheels may be replaced with skids. The landing gear is divided into two sections:

1) Main gear or main wheel,[2]
2) Secondary gear or secondary wheel.

The main gear is the gear which is the closest to the aircraft center of gravity (cg). The primary functions of a landing gear are as follows:

1) To keep the aircraft stable on the ground and during loading, unloading, and taxi.
2) To allow the aircraft to freely move and maneuver during taxing.
3) To provide a safe distance between other aircraft components such as wing and fuselage while the aircraft is on the ground position to prevent any damage by the ground contact.
4) To absorb the landing shocks during landing operation.
5) To facilitate take-off by allowing aircraft acceleration and rotation with the lowest friction.

In order to allow for a landing gear to function effectively, the following design requirements are established:

1) Ground clearance requirement
2) Steering requirement
3) Take-off rotation requirement
4) Tip back prevention requirement
5) Overturn prevention requirement
6) Touch-down requirement
7) Landing requirement
8) Static and dynamic load requirement
9) Aircraft structural integrity
10) Ground lateral stability
11) Low cost
12) Low weight
13) Maintainability
14) Manufacturability

In general, there are nine configurations for a landing gear as follows: (a) single main; (b) bicycle; (c) tail-gear; (d) tricycle or nose-gear; (e) quadricycle; (f) multi-bogey;

(g) releasable rail; (h) skid; and (i) seaplane landing device. In order to select the best landing gear configuration, the designer must perform a trade-off study using a comparison table.

Another design aspect of the landing gear is to decide what to do with it after take-off operation. In general, there are four alternatives:

1) Landing gear is released after take-off.
2) Landing gear hangs underneath the aircraft (i.e., fixed).
3) Landing gear is fully retracted inside aircraft (e.g., inside wing, or fuselage).
4) Landing gear is partially retracted inside aircraft.

In order to formulate the design requirements, the designer is encouraged to develop several equations and relations based on the numerical requirements and solve them simultaneously.

The landing gear design is affected by the location of the payloads such as cameras, radars, and antennas. For instance, when the tail (e.g., vertical tail) is underneath the fuselage (e.g., Predator MQ-9), the landing gear height will be longer. The vertical tail may be moved down to allow the satellite radar to easily see the satellite and communicate with it by sending signal and receiving signal. This selection requires a longer landing gear than normal.

For such cases, the takeoff rotation can be compromised, even when no rotation is required in takeoff operation. So, the liftoff happens without rotation (the UAV will liftoff when the lift is greater than UAV weight while the wing has the cruising angle of attack).

Table 3.3 shows the features for landing gear of a number of current UAVs. The landing gear of Global Hawk is supplied by Heroux Company of Canada. The nose gear which is a derivative of the F-5 design is height adjustable to suit the runway characteristics. The landing gear automatically retracts at an altitude of 4000 ft.

There are various similarities between landing gear design process of a manned aircraft and unmanned aircraft. The author has presented the landing gear design for an air vehicle in Chapter 9 of [8], so the materials are not repeated in this text. You are encouraged to study [8] to find the details of landing gear design.

Table 3.3 Landing gear of a number of Unmanned aerial vehicles (UAVs).

No	UAV	Type	Fixed/retractable
1	Global Hawk	Tricycle	Retractable
2	General Atomics Avenger	Tricycle	Retractable
3	MQ-1 Predator	Tricycle	Fixed
4	RQ-7 Shadow 200	Tricycle	Fixed
5	Northrop Grumman Firebird	Tricycle	Retractable
6	MQ-5B Hunter	Tricycle	Fixed with fairing
7	DJI Mavic Pro Quadcopter	Four-leg skid	Fixed
8	Yamaha RMAX	Skid	Fixed

3.6 Mechanical and Power Transmission Systems Design

One of the UAV subsystems that has a great influence on the control surfaces' design is the power system. The power required to operate a UAV includes that to drive the control surfaces' deflection, which is usually accomplished either hydraulically or electrically. There are several power systems employed in UAVs; each has advantages and disadvantages. They are: (a) electro-mechanical; (b) pneumatic; (c) hydraulic; and (d) electro-hydrostatic. In this section, advantages and disadvantages these systems, plus their influence on the design of the control surfaces, are briefly discussed. Commands can be sent to actuate the control surfaces using either a fly-by-wire (FBW) or a fly-by-optic system. Figure 3.3 demonstrates an Insitu ScanEagle UAV on a ground launcher; it is launched by the pneumatic power. This UAV has a maximum takeoff weight (MTOW) of 48.5 lb, a wing area of 14.7 ft^2, and a piston engine with maximum power of 1.5 hp.

Numerous changes have been made in aircraft flight control systems. Initially, flight control systems were purely mechanical, which was ideal for smaller, slow-speed, low-performance aircraft because they were easy to maintain. However, more control-surface force is required in modern high performance airplanes. Thus, during the twentieth century a hydraulic power boost system was added to the mechanical control. This modification maintained the direct mechanical linkage between the pilot and the control surface.

As aircraft became larger, faster, and heavier, and had increased performance, they became harder to control because the pilot could not provide the necessary power to directly operate the control surfaces. Thus, the entire effort of moving the control surface had to be provided by the actuator. A stability augmentation system (SAS) was added to the hydraulic boosted mechanical regulator system to make the aircraft

Figure 3.3 Insitu ScanEagle on a ground launcher.

flyable under all flight configurations. Motion sensors were used to detect aircraft perturbations and to provide electric signals to a SAS computer, which, in turn, calculated the proper amount of servo actuator force required.

The new step in the evolution of flight control systems is the use of a FBW control system. In this design, all autopilot commands are transmitted to the control-surface actuators through electric wires. Thus, all mechanical linkages from the control stick to the servo actuators are removed from the UAV. The FBW system provided the advantages of reduced weight, improved survivability, and decreased maintenance.

Originally the flight control computers were analog (such as the early F-16 aircraft computers), but these have been replaced by digital computers. In addition, the controller consists of a digital computer which accepts the pilot commands and signals from the sensors (position and rate gyros) and accelerometers, and sends commands to the actuators. This is now referred to as a digital flight control system (DFCS). For twenty-first century aerospace vehicles the use of hydraulics has essentially been eliminated in favor of an all-electric system incorporating the use of digital computers.

Mechanical flight control systems are the most basic design. They are used in a variety of aircraft. Currently electro-mechanical systems are used in small UAV where the aerodynamic forces are not excessive. Pure mechanical flight control systems use a collection of mechanical parts such as rods, cables, and pulleys to transmit commanded movements to the control surfaces from the actuators. Mechatronic (combination of electrical/electronic and mechanical systems) systems are becoming more and more popular, due to their attractive features such as lighter weight, and smaller size.

The complexity and weight of a mechanical flight control system increases considerably with the size and performance of the airplane. Hydraulic power overcomes many of these limitations. A hydraulic flight control system has two parts: (a) the mechanical circuit; and (b) the hydraulic circuit. A control command causes the mechanical circuit to open the matching servo valves in the hydraulic circuit. The hydraulic circuit powers the actuators which then move the control surfaces. However, hydraulic systems can require long hydraulic lines which may need to be redundant. These lines can develop leaks. Electro-hydrostatic actuators use local hydraulic reservoirs and pumps to avoid the need for these long hydraulic lines. Replacing the mechanical or hydraulic linkages with electrical power lines and linkages can save weight and improve reliability.

Electronic FBW systems can respond more flexibly to changing aerodynamic conditions. Fully electronic systems would require less maintenance, whereas mechanical and hydraulic systems require lubrication, tension adjustments, leak checks, fluid changes, etc. Digital flight-control systems were able to incorporate "multi-mode" flight control laws with different modes, each optimized to enhance maneuverability and controllability for a particular phase of flight. Furthermore, placing such control circuitry between the operator and the aircraft could provide various safety systems. For example, the control system could prevent a stall, or limit a maneuver which could overstress the structure.

Fly-by-optics systems are sometimes used instead of FBW systems because they can transfer data at higher speeds, and are nearly immune to electromagnetic interference (EMI). In a simplistic view, the electrical cables of the FBW system are replaced by fiber optic cables.

3.7 Electric Systems

3.7.1 Fundamentals

Large UAVs such as Global Hawk employ large amounts of electrical power for a variety of systems. A variety of electrical systems are required to fly the UAV, such as flight control system, engine control, communication, and navigation systems, and cameras, and sensors. In a large UAV, an electric power of up to 100 kW may be required. Moreover, in the ground control station for a large UAV, the electric power requirement is so high that, a separate large truck is utilized to carry a large generator to generate necessary electric energy.

Virtually, all UAVs (from micro to large) contain some forms of an electrical system. Large UAVs have a complex electrical system that provides electric energy to all electric devices. In general, large UAVs are equipped with an electric energy generator, while small UAVs are employing batteries. Even large UAVs need a battery to provide a secondary power supply, since in the event of an alternator (or generator) failure, battery circuit is used to supply power.

An airborne wide-area, nearly all-weather surveillance radar for an UAV — sponsored by the Army and Defense Advanced Research Projects Agency (DARPA) — was developed by Lincoln Laboratory. Table 3.4 indicates [12] the breakdown of the power requirements for various components in the radar and communications system. As observed this payload system consumes about 1400 W of prime power. The required power should be generated by the UAV electric system. Five primary sources of power in a UAV are: (a) main engine; (b) battery; (c) auxiliary power unit (APU); (d) solar cells; and (e) generator.

Table 3.4 Unmanned aerial vehicle (UAV) radar and support equipment power breakdown.

No	System	Component	Power (W)
1	**Radar**	Transmitter	355
2		Receiver and exciter	95
3		Processor	400
4		Antenna system	75
5		Cables and connectors	25
		Subtotal	**950**
6	**Support Equipment**	Inertial Navigation System (including heat sink)	50
7		Data link	185
8		Altimeter and Global positioning system (GPS) receiver	20
9		Support structure	N/A
10		Cooling fans	150
		Subtotal	**405**
Total			**1355**

The Kwearfott KN-4072 INS unit used in Global Hawk consumes 35 W power, while RCCT Micro INS requires 3.5 W. In contrast, Freewave F-Series communications system with a mass of 0.44 lb consumes 30 W power, while L-3 Mini TDCL Transceiver requires about 60 W. The UHF RX/TX transmitter/receiver of the Global Hawk communications system requires 150 W. The high voltage power supply (HVPS) of the Global Hawk communications system generates 1800 W.

In a UAV where the air breathing engine is used, there is only one electrical system powered by the engine-driven alternator or generator. The aircraft battery is used for emergency power and engine starting. Electrical power is typically distributed through one or more common points known as an electrical bus (or bus bar).

Military UAVs are loaded with sensors, antennas, processors, cameras, telemetry system, radios, radars, navigation system, jammers, cooling fans, and power supplies. This will require a staggering array of wires, cables, harnesses, and connectors, which accounts for a significant share of vehicle weight, and have a major impact on both operational cost and mission performance. To minimize the vehicle's weight, the electric system interconnections should be optimized (e.g., use smaller diameter wires). In wiring modules, it is recommended to place the wiring to one side of the module rack, so that, the other side will be free and available for temporary hookups.

3.7.2 Safety Recommendations

An electrical circuit must be protected from faults (open- and short-circuit) that can occur in the system. An open circuit is an electrical fault that occurs when a circuit becomes disconnected. A short circuit is an electrical fault that occurs when one or more circuits create an unwanted connection. The most dangerous short circuit occurs when a positive wire creates an unwanted connection to a negative connection or ground. This is typically called a short to ground.

There are two ways to protect electrical systems from faults: (a) mechanically; and (b) electrically. Wires and electric elements are mechanically protected from abrasion and excess wear through proper installation and by adding protective covers and shields. In contrast, wires are often electrically protected using circuit breakers and fuses. The circuit breakers and fuses protect electric elements in the event of a short circuit.

Large UAVs utilize an external power circuit that provides a means of connecting electrical power from a ground source to the UAV. External power is usually used for starting the engine or maintenance activities on the UAV. The external power system allows operation of various electrical systems without discharging the onboard battery.

Large UAVs often employ a separate power distribution bus (often referred to as an avionics bus) specifically for electronic equipment. Since modern avionic devices employ sensitive circuit boards, it is often advantageous to disconnect (using a separate power switch) all avionics from main electrical power to protect their circuits.

The two most generally used conductors (for wires) are copper and aluminum. Copper has a higher conductivity, is more ductile, has relatively high tensile strength, and can be easily soldered. Copper is more expensive and heavier than aluminum. Although aluminum has only about 60% of the conductivity of copper, it is used extensively.

At high altitude (e.g., above 30 000 ft) where the temperature is below −50 °C, temperature-sensitive electric elements such as crystal oscillators, batteries, capacitors, and some semiconductor devices like A/D and D/A converters may not perform normally. The solution to heat the electric unit to warm up to the normal operating temperature.

3.7.3 Wiring Diagrams

Electrical wiring schematic diagrams should be included in most UAV service manuals and specify information, such as the size of the wire and type of terminals to be used for a particular application. Furthermore, wiring diagrams typically identify each component within a system by its part number and its serial number, including any changes that were made during the production run. Wiring diagrams are much needed used for troubleshooting electrical malfunctions.

In a pictorial diagram, figures/pictures of components are used instead of the conventional electrical symbols. A pictorial diagram helps the maintenance technician visualize the operation of a system.

Payload integration of an instrumented Yamaha RMAX unmanned helicopter (Figure 1.1) unmanned helicopter is implemented by [21]. The power distribution system employs an onboard generator with 12 V DC outputs. It includes a hot-swappable connection to use external power. Each component has a dedicated circuit breaker for boosting safety. Figure 3.4 shows the wiring diagram which includes power wiring, a typical configuration of RS-232, Ethernet, multiple serial lines, power regulation, air cooling, and EMI shielding. External wiring of the modules consists of: (a) RS-232 Serial; (b) Ethernet; and (c) 12 V DC power.

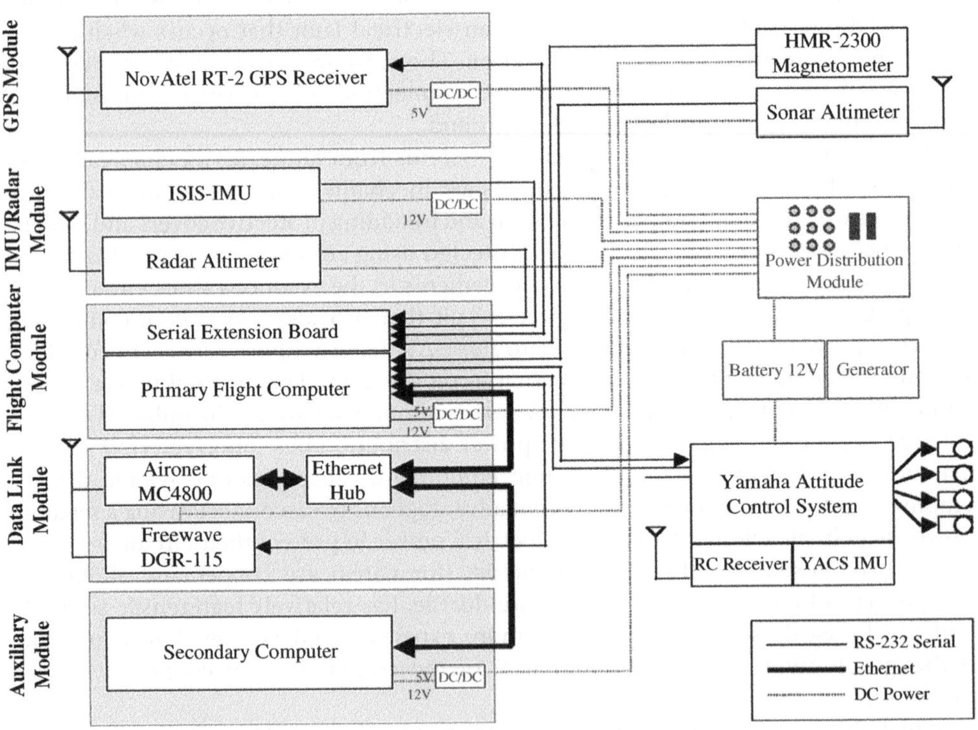

Figure 3.4 Wiring diagram of an instrumented Yamaha RMAX unmanned helicopter.

3.7.4 Wire Insulation and Shielding

Wiring and electric devices must be insulated and strongly recommended to be shielded for safety and effectiveness reasons. Two fundamental properties of insulation materials are insulation resistance and dielectric strength. Insulation resistance is the resistance to current leakage through and over the surface of insulation materials. Dielectric strength is the ability of the insulator to withstand potential difference and is usually expressed in terms of the voltage at which the insulation fails because of the electrostatic stress. Insulation materials are currently made of Tefzel, Teflon/Kapton/Teflon and PTFE/Polyimide/PTFE.

With the increase in number of sensitive electronic devices (such as transmitter and receiver) on large UAVs, it has become very important to ensure proper shielding for sensitive electric circuits. Shielding is the process of applying a metallic covering to wires to eliminate EMI. EMI is caused when electromagnetic waves induce high frequency voltages in a wire or component. The induced voltage can cause system inaccuracies or even failure. Use of shielding with 85% coverage or greater is recommended. Wherever appropriate, coaxial, twinaxial, triaxial, or even quadraxial cables should be utilized, with their shields connected to ground. Note that coaxial and triaxial cables are vulnerable to certain types of damage.

3.7.5 Batteries

There are basically two types of batteries: (a) primary; and (b) rechargeable (i.e., secondary). A primary battery is designed to be used once and discarded, and not recharged with electricity and reused like a secondary cell. A rechargeable battery is one which can be charged, discharged into a load, and recharged many times, as opposed to a disposable or regular battery which is supplied fully charged and discarded after use. Rechargeable batteries are produced in many different shapes and sizes.

Primary batteries are useful when long periods of storage are desired, and are also more cost-efficient. Rechargeable cells are generally more economical to use than primary cells. Their initially higher cost and the purchase cost of a charging system can be spread out over many use cycles.

The lead–acid battery is the oldest type of rechargeable battery. Despite having a very low energy-to-weight ratio and a low energy-to-volume ratio, its ability to supply high surge current implies that the battery has a relatively large power-to-weight ratio. Since they are inexpensive compared to newer technologies, lead–acid batteries are very popular and are widely used. Most of the world's lead-acid batteries are used in auto industries for automobile starting.

A single Li-ion cell is charged in two stages: (a) Constant current (CC); and (b) Constant voltage (CV). Because lithium-ion batteries can have a variety of positive and negative electrode materials, the energy density and voltage vary accordingly.

Large UAV batteries are usually identified by the material used for the plates. The most common types of battery used are: (a) lead-acid; (b) Ni-Cd; (c) lithium-ion; (d) lithium polymer; (e) nickel metal hydride; (f) alkaline. The lead-acid batteries, also known as flooded or wet batteries, are dry charged. A Ni-Cd battery consists of a metallic box (usually stainless steel, plastic-coated steel, painted steel, or titanium) that contains a number of individual cells. These cells are connected in series to obtain higher voltages.

Aircraft that are outfitted with Ni - Cd batteries often have a fault protection system that monitors the condition of the battery. The battery charger is equipped with a system that monitors the condition of the battery. The following conditions must be monitored:

1) Overheat condition
2) Low temperature condition (below –40 °F)
3) Cell imbalance
4) Open circuit
5) Shorted circuit.

If the battery charger finds a fault, it turns off and sends a fault signal to the Electrical Load Management System (ELMS). The quadcopter DJI Mavic 2 Pro with a weight of 2 pounds requires 3 Lithium Polymer batteries.

Table 3.5 provides [22] technical features – including energy density — of some rechargeable batteries. A high energy density lithium-ion Panasonic NCR battery with a mass of 46 g, a length of 65.3 mm, a diameter of 18.5 mm has a rated capacity of 2700 mAh at 3.6 V. This battery has a gravimetric energy density of 214 Watt-hour per kg (Wh/kg), and a volumetric energy density of 577 Watt-hour per liter (Wh/L). It is interesting to note that in an iPhone, the battery can typically provide 3.7 V at 3 Amps for one hour. The regular electric energy consumption (i.e., current) is much less than 3 Amps. So, the time of battery discharge will be much more than one hour. For instance, a Venom 7.2 V 5000 mAh 6-Cell Ni-MH battery will provide an electric energy of 5 A of current at 7.2 volts for one hour. Moreover, a 22.2 V 3200 mAh 6S 30C Smart Li-Po battery can generate an electric energy of 3.2 A of current at 22.2 volts for one hour.

In selection of battery for a UAV, a number of criteria are influential: (a) cost; (b) energy density; (c) life; (d) safety; (e) weight; (f) dimensions; (g) charge/discharge efficiency; (h) self-discharge rate; (i) cycle durability; (j) nominal cell voltage; and (k) specific energy.

3.7.6 Generator

Large UAVs employ a low-power AC electrical system to power certain instruments such as satellite radar, and transmitter. The AC generator/alternator (powered by APU) often produces three-phase 115 V AC at 400 Hz. A device called an inverter may also be used to supply AC power by inverting DC into AC. Many inverters provide more than one voltage output. Two common voltages found on large UAV inverters are 26 V AC and 115 V AC. Whenever AC is used, a distribution circuit separated from the DC

Table 3.5 Technical features of some rechargeable batteries.

No	Battery type	Energy density, E_D (Wh/kg)	Specific power	Short name	Nominal cell voltage
1	Lithium-ion battery	100—250	300—900 kJ/kg	Li-ion	3.6—3.8
2	Lithium polymer	100—265	360—950 kJ/kg	LiPo	
3	Nickel–cadmium	40—60	150 W/kg	NiCd	1.2
4	Lead–acid	30—42	180 W/kg		2.1
5	Nickel metal hydride	60—120	250—1000 W/kg	Ni-MH	1.2
6	Alkaline	160 Wh/L		L	1.5

system must be employed. If the AC generator is lost, the UAV battery will continue to supply DC power to operate vital systems for a safe landing. Onboard electrical systems of Shadow 200 are powered by a GEC/Plessey 28 V, DC, 2 kW generator.

3.8 Control Surfaces Design

Two primary prerequisites for a safe flight are stability and controllability. The controllability requirements will influence the design of control surfaces and create variety of design constraints. Flight stability is defined as the inherent tendency of an aircraft to oppose any input and return to original trim condition if disturbed. When the summation of all forces along each three axes, and summation of all moments about each three axes are zero, an aircraft is said to be in trim or equilibrium. In this case, aircraft will have a constant linear speed and/or a constant angular speed. Control is the process to change the aircraft flight condition from an initial trim point to a final or new trim point. This is performed mainly by autopilot through moving the control surfaces/engine throttle. The desired change is basically expressed with a reference to the time that takes to move from initial trim point to the final trim point (e.g., pitch rate and roll rate).

Maneuverability is profoundly significant for fighter aircraft and missiles and is a branch of controllability. Control systems should be designed with sufficient redundancy to achieve two orders of magnitude more reliability than some desired level. Aircraft controllability is a function of a number of factors including control surfaces.

In general, control surfaces may be broadly classified into two types: (a) conventional; and (b) non-conventional. Conventional control surfaces are divided into two main groups: (a) primary control surfaces; and (b) secondary control surfaces. The primary control surfaces (Figure 3.5) are in charge of control of flight route, and usually in a conventional aircraft are as follows: (a) ailerons; (b) elevator; and (c) rudder.

The primary control surfaces of aileron, elevator, and rudder are respectively utilized for lateral control, longitudinal control and directional control. However, they also largely contribute to lateral trim, longitudinal trim and directional trim of aircraft. In majority of aircraft configurations, lateral and directional motions are coupled; hence, the aileron is also affecting directional motion and rudder affects the lateral motion. Conventional primary control surfaces are like plain flap, but their applications are different. When control surfaces are deflected, cambers of their related lifting surfaces (wing, horizontal tail, or vertical tail) are changed. Thus, the deflection of a control varies the aerodynamic forces; and consequently a resultant moment will influence the aircraft motion.

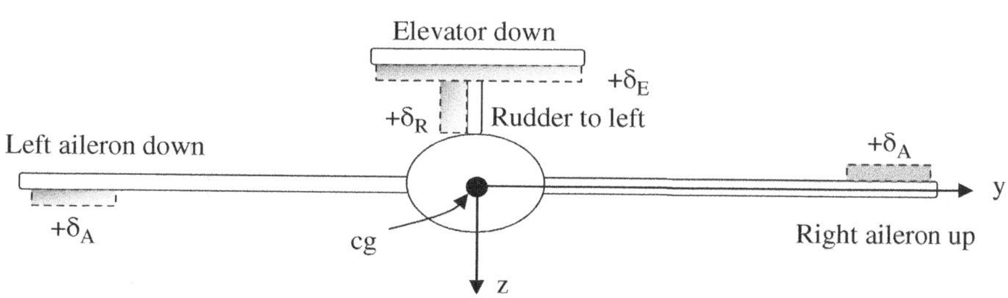

Figure 3.5 Convention for positive deflections of control surfaces (Back-view).

To analyze the aircraft control, a coordinate system must be defined. There are four coordinate systems: (a) Earth-fixed; (b) body-fixed; (c) wind axes system; and (d) stability axes system. Here, for the purpose of control, a body-fixed coordinate system is adopted; where there are three orthogonal axes which follow the right-hand rule. The x-axis is along the fuselage (body) centerline passing through the aircraft center of gravity; the y-axis is perpendicular to x-axis and to the right (from top-view); and z-axis is perpendicular to xy plane (i.e., downward). Figures 3.5 illustrate the convention for positive directions of axes of aircraft. Positive roll is defined as a clockwise rotation about x axis as seen from the pilot seat (when in cruise; right wing down, left wing up). Similarly, positive pitch is defined as a clockwise rotation about y axis as seen from the pilot seat (nose up). Finally, positive yaw is defined as a clockwise rotation about z axis as seen from the pilot seat (nose to right). Figure 3.6 illustrates the conventional control surfaces. These conventions are significant and are used to develop design techniques in this book. Figure 3.7 depicts the axes and positive rotations convention.

An aircraft is capable of performing various maneuvers and motions; they may be broadly classified into three main groups: (a) longitudinal motion; (b) lateral motion; and (c) directional motion.

Figure 3.8 illustrates a flowchart that represents the control surfaces design process. In general, the design process begins with a trade-off study to establish a clear line between stability and controllability requirements and ends with optimization. During the trade-off study, two extreme limits of flying qualities are examined and the border line between stability and controllability is drawn. For instance, a fighter can sacrifice the stability to achieve a higher controllability and maneuverability. Then, an automatic flight control system may be employed to augment the aircraft stability. In the case of a civil airliner, the safety is the utmost goal; so the stability is clearly favored over the controllability.

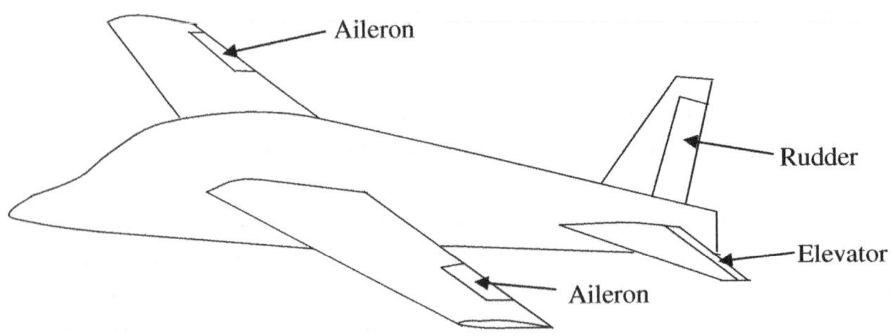

Figure 3.6 Primary control surfaces

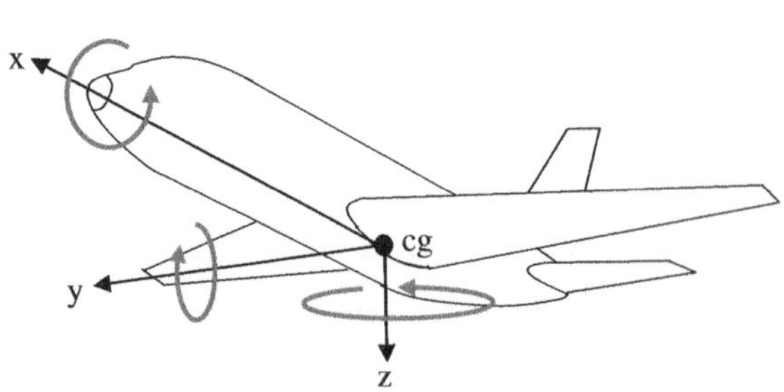

Figure 3.7 Axes and positive rotations convention.

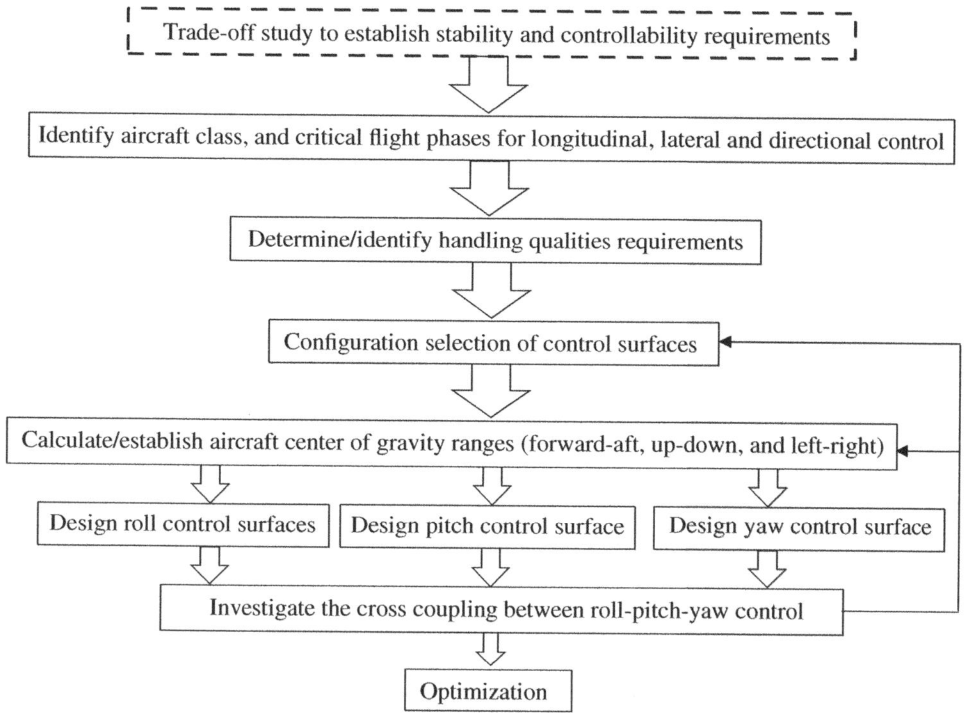

Figure 3.8 Control surfaces design process.

The first step in the design of control surfaces is to select the control surface configuration. The primary idea behind the design of flight control surfaces is to position them so that they function primarily as moment generators. They provide three types of rotational motions (roll, pitch, and yaw). A conventional configuration includes elevator, aileron and rudder. Variations to this classical configuration lead to some variations in the arrangements of these control surfaces. Table 3.6 represents several control surface configuration options. Some types of control surfaces are tied with particular aircraft configuration; they must be selected for specific aircraft configuration. Table 3.6 also illustrates a few aircraft examples.

The control surface configuration selection is a function of aircraft configuration (e.g., wing, tail, and engine), cost, performance, controllability power transmission, and operational requirements. The consequence of some aircraft configurations is to have a particular type of control surfaces. For instance, when a V-tail configuration is selected during the aircraft conceptual design phase, a ruddervator is the best candidate to control both yawing and pitching moments. Another example is when the designer decides to have a delta wing without aft tail. In such a case, an elevon is a great candidate to perform as a means of control power to control pitch rate and roll rate.

The final decision on the control surface configuration will be the output of a trade-off study to balance and satisfy all design requirements in an optimum way. In general, unconventional control surfaces are more challenging to design, more complex to manufacture, and also harder to analyze. However, unconventional control surfaces are more efficient when a higher control power is required in a challenging design environment.

Table 3.6 Control surface configuration optionssco.

No	Control Surface Configuration	Aircraft Configuration
1	Conventional (aileron, elevator, rudder)	Conventional (e.g., RQ-11 Raven)
2	All moving horizontal tail, rudder, aileron	Horizontal tail and elevator combined
3	All moving vertical tail, elevator, aileron	Vertical tail and rudder combined
4	Flaperon, Elevator, Rudder	Flap and aileron combined
5	Taileron, Rudder	All moving horizontal tail (elevator) and aileron combined
6	Elevon, Rudder (or equivalent)	Aileron and elevator combined (e.g., ScanEagle)
7	Ruddervator, Aileron	V-tail (e.g., Global Hawk, Predator, and Shadow)
8	Drag-Rudder, Elevator, Aileron	No vertical tail (X-45 UCAV, RQ-170 Sentinel)
9	Canardvator, Aileron	Elevator as part of canard, plus aileron
10	Four control surfaces	Cross (+ or ×) tail configuration (e.g., most missiles)
11	Aileron, Elevator (or equivalent), Split Rudder	No vertical tail. Aileron-like surfaces that is split into top and bottom sections
12	Spoileron, Elevator, Rudder	Spoiler and aileron combined
13	Thrust vector control	Augmented or no control surfaces (e.g., Quadcopters)

Based on performance requirements, a UAV will need multiple control surfaces such as elevators, ailerons, rudders, flaperons, ruddervators, and/or elevons. The number and the type of control surfaces depend on several factors such as the UAV mission, cost, and controllability requirements. In this paper, we concentrate on a conventional configuration that includes three control surfaces; namely the elevator, aileron, and rudder. The three coefficients of interest are: C_l (roll), C_m (pitch), and C_n (yaw). Data requirements may be included which can be used to minimize a particular objective as a function of control surface deflections. Other data requirements specific to constrained control allocation techniques include control minimum and maximum position limits and actuator rate limits.

This data may also be dependent on other variables. As an example, many control laws have constraints implemented in software, determined by variables such as the dynamic pressure, which are imposed on the available surface deflections. For example, the commanded control actuator rate may need to change to maintain a constant aircraft rotation rate across the entire flight envelope, since hinge moments and other aerodynamic factors normally change with flight condition.

One flight control effector innovation is the use of thrust vectoring. One such flight control design is the F-15 ACTIVE vehicle which used thrust vectoring to produce pitch, roll, and yaw controlling moments. Other examples of such vehicles exist. One of the first operational aircraft to use thrust vectoring was the Harrier, flown by the British Royal Air Force, British Royal Navy, and the U.S. Marines.

NASA research explored thrust vectoring at extremely high angle-of-attack on the High Alpha (angle-of-attack) Research Vehicle (HARV), a modified F-18. The F-16

Multi-Axis Thrust Vectoring (MATV) research program made significant contributions to understanding thrust vectoring design requirements and agility benefits. The X-31 research aircraft also demonstrated the benefits of thrust vectoring. The F-22 and the F-35 are more recent examples of aircraft using thrust vectoring for control. Figure 3.9 presents the flight control system with conventional control surfaces.

The primary idea behind the design of flight control surfaces is to position them so that they function primarily as moment generators. They provide three types of rotational motion (roll, pitch, and yaw). Variations to this classical configuration lead to some variations in the arrangements of these control surfaces.

With the control effectiveness known (or estimated) for the surfaces controlling each of the three axes mentioned above, then the classical three control/three degrees of freedom system can be defined by an algebraic problem with three equations (the commanded moments) and three unknowns (the required control deflections). Assuming that the mathematical system of equations is consistent, a unique control configuration exists for any desired vector of control generated moments. To apply this methodology, the following information regarding the candidate UAV is needed:

1) UAV configuration and the layout of the major components.
2) Mass properties: center of gravity travel, weight and inertia variations.
3) Extreme performance objectives: maximum Mach number versus altitude; maximum load factor and maximum and minimum thrust limits.
4) Operational and controllability requirements.
5) Type of power transmission.
6) Other systems engineering requirements (e.g., budget constraints, maintainability requirements, production requirements, and reliability requirements).

The design criteria of the control surfaces are determined through the definition of system operational requirements, which in turn evolve from the UAV mission. The control surfaces may impose limits on the automatic flight control system (e.g., saturation). Although the designed control surfaces may satisfy controllability requirements, several system engineering factors must also be taken into account. These factors include structural considerations (fatigue, flutter, and aero-elasticity), manufacturability, the power to deflect the control surfaces, total cost (manufacturing cost and operating cost), control input saturation, maintainability, reliability, and producibility.

Some parameters must be minimized, some must be maximized, while other ones must be evaluated to ensure that they are acceptable. In some cases, the design of the

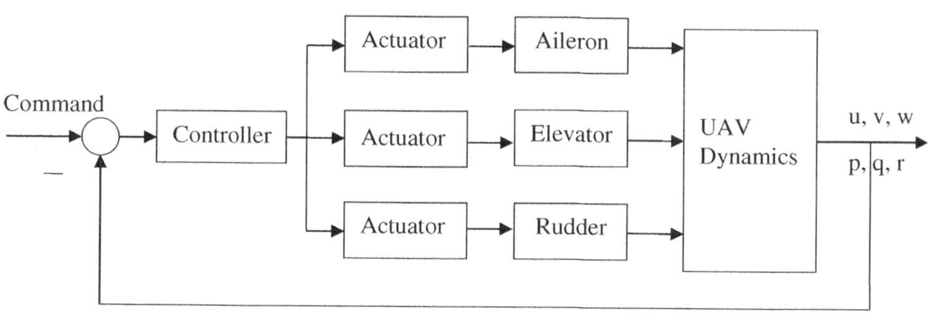

Figure 3.9 Flight control systems with conventional control surfaces.

control surfaces may impose slight to considerable changes to the UAV configuration during the conceptual design process.

Integrated flight control refers to the control and optimization techniques to achieve or improve overall system performance by taking into account subsystem interactions, uncertainties, and even failures for all control effectors. The flight control integration task specifically addresses the issues of interacting impacts between flight control systems and other aircraft functional or structural systems. As an example, conventional control surfaces lose control effectiveness at low speeds and also are ineffective in the post-stall flight regime. Thrust vectoring control has been developed to expand the flight envelope of modern military aircraft.

The integration of system engineering principles with the analysis-driven design process demonstrates that a higher level of integrated control can be attained, identifying the requirements/functional/physical interfaces and the complimentary technical interactions which are promoted by this design process. The objective is to assess the influences of the conceptual design choices and modeling uncertainties on the control system configuration and not so much to predict accurately the absolute value of the control torques.

One specific area in which the systems engineering approach has been applied to an integrated flight control system design is the design of integrated flight and propulsion control systems. For example, in hypersonic vehicle design, the coupling between the propulsion system's dynamics and the aerodynamics is significant enough that an approach designing the propulsion control system and the flight control system separately is not adequate. The strong relationship between the analysis and the influencing parameters allow definite, traceable relationships to be constructed. In the case of control surface design, the major parameters are drawn almost completely from the operational and controllability requirements.

The author has presented the conventional control surfaces design for an air vehicle in Chapter 12 of [8], so the materials are not repeated in this text. You are encouraged to study [8] to see the full details of conventional control surfaces design.

3.9 Safety Analysis

It is a fundamental duty of any UAV chief designer to ensure that the design is safe, and all major systems/components are away from failure. A process must be established to perform safety engineering analysis, either: (a) before that design process begins; (b) during the design process; and (c) following each major design milestone/phase. Each member of the design staff in all departments may be expected to conduct research and establish a list of safety features. However, a more efficient approach is to establish the "Safety Analysis Department". In such cases, the "Safety Engineer" has the authority to request a redesign, when its associated cost is justified.

The Safety Analysis Department (SAD) will primarily:

1) Identifyy and list UAV and system level safety level requirements.
2) Determine if the design can be expected to meet identified safety requirements and objectives.
3) Derive safety requirements for the design of subsystems (e.g., components such as wing). A hazard assessment is a collaborative and multidisciplinary effort which requires a great deal of domain knowledge and insight.

The Federal Aviation Administration (FAA) has published a handbook [23] to help launch and reentry vehicle operators conduct flight safety analyses. However, recommendations and procedures provided by this reference are also applicable to the UAV design process. In this section, two topics related to safety analysis are briefly presented:

1) Lessons learned from the past design projects (either from failures or successes).
2) Likely failure modes of subsystems and components.

Reference [24] provides guidelines and methods for conducting the safety assessment process on civil airborne systems and equipment. The interested reader is referred to [23, 24] for further introduction to the topic.

3.9.1 Design Lessons Learned

There are a lot of valuable lessons learned from industry on the design mis-steps and mistakes made on the legacy vehicles flying today. The failure—success experiences of UAVs and UASs are very valuable, and they play an important role in teaching young designers. An important feature of the design process is the lessons learned in the past. The lesson will be utilized in improving the next generation.

Based on Murphy's Law, whatever can go wrong, will go wrong, at the wrong time, and in the wrong place. A study on the historical experience [25, 26] of other vehicles confirms this law. This law will open the eyes of the UAV designer about design challenges, and the values of the historical experience of other vehicles. The failure—success experience of the successful UAVs (either crash or mishap) such as Global Hawk and Predator providee free lessons for UAV designers.

Here, a few lessons learned for a number of current/recent UAVs are presented. The interested reader is encouraged to study the literature, and consult with respective references such as [27–29].

The major setback during Phase II flight testing of Global Hawk (Figure 1.2) was the destruction of air vehicle 2 on March 29, 1999, during the program's 18th sortie. The loss of air vehicle 2 and its payload was estimated at $45 million. Of more importance, was the fact that the program lost its only integrated sensor suite.

The crash was due to a lack of proper frequency coordination between the Nellis Air Force Base and EAFB flight test ranges. Essentially, Nellis officials who were testing systems in preparation for Global Hawk's first planned D and E exercise were unaware that Global Hawk was flying over China Lake Naval Air Weapons Station, which is within EAFB's area of responsibility. Thus, many changes have been applied in the design of Northrop Grumman RQ-4B Global Hawk as compared with RQ-4A. For instance, the Northrop Grumman RQ-4B Global Hawk has a 50% payload increase, larger wingspan (130.9 ft) and longer fuselage (47.6 ft), and new generator to provide 150% more electrical output.

On 15 August 2011 a US Air Force C-130 cargo plane collided with a RQ-7 Shadow while on approach to FOB Sharana in Afghanistan [30]. The cargo plane made an emergency landing with damage to two engines and one wing, while the RQ-7 was destroyed. The collision caused the C-130 was grounded for several months for repair, while the RQ-7 wreckage was never recovered.

In July 2015, an unarmed Gray Eagle crashed in Iraq after a communications failure in a flight operation. In December 1999, a software problem caused a Global Hawk to veer off Edwards' main runway. The nose gear collapsed causing $5.3 million worth of damage to the sensor suite. An error in software code to coordinate between the mission planning system and the aircraft commanded the vehicle to taxi at 155 knot.

On 3 April 2014, a Shadow 200 participating in training exercises at Fort Indiantown Gap crashed near an elementary school in Pennsylvania, and was then hit by a civilian vehicle destroying the drone. On 2 Feb, 2017, A Shadow 200 launched from an airfield near Fort Huachuca, AZ disappeared. After a long (nine-day) search, the wreckage was found in the mountains west of Denver. There are a number of other lessons learned which are placed throughout the text.

The Predator's small size and lack of overall structural strength dictated that the RQ-1 could only carry a missile that weighed less than 175 pounds. This made the Hellfire missile almost the only choice. Originally, there were some concerns [27] about firing a helicopter-borne antitank missile from this lightweight UAV. The question was: Would firing the missile break up the Predator, either by wrenching a wing from its mounts or knocking off the vertical stabilizer? Operators witnessed a test which consisted of chaining a Predator to a concrete pad and shooting the missile. No parts fell off the Predator.

A number of lessons learned from the Perseus B (Figure 3.10), which is a remotely piloted aircraft developed as a design-performance testbed under NASA's Environmental Research Aircraft and Sensor Technology (ERAST) project. Perseus B is approximately 25 ft long, has a wingspan of 71.5 ft, and stands 12 ft high, and is powered by a Rotax 914 piston engine. Two Perseus A aircraft were built and made 21 flights in 1993—1994. Although one of the Perseus A aircraft was destroyed in a crash after a vertical gyroscope failed in flight, the other aircraft completed its test program. Perseus B first flew in 1994, and made two flights in 1996 before being damaged in a hard landing on the dry lakebed after a propeller shaft failure. A series of modifications were made including external fuel pods on the wing that more than doubled the fuel capacity to 100 gal, and extending the original 58.5-ft wingspan to 71.5 ft to enhance high-altitude performance.

Engine power was increased by more than 20% by boosting the turbocharger output. Fuel consumption was reduced with fuel control modifications and a leaner fuel-air mixture that did not compromise power. The aircraft again crashed on 1 October 1999, suffering moderate damage to the aircraft but no property damage, fire, or injuries in the area of the crash. A flight termination system, required for all remotely piloted aircraft being flown in military-restricted airspace, includes a parachute system deployed on command plus a C-Band radar beacon and a Mode-C transponder to aid in location.

Figure 3.10 National Administration for Aeronautics and Astronautics (NASA) Perseus B.

Table 3.7 illustrates reliability and mishap rates for several manned and unmanned aircraft. Current accident rate for a typical UAVs is about 50 times greater than that of a General Dynamics fighter F-16 Fighting Falcon. Autopilot is responsible for considerable amount of failures/mishaps. In the autonomy, operator tells the system what is desired from the mission (not how to do it) with the flexibility of dynamic changes to the mission goals being possible in flight with minimal operation re-planning.

There are a number of other lessons learned which are placed throughout the text.

3.9.2 Likely Failure Modes of Sub-Systems/Components

SAD is responsible for a systematic examination of the proposed design to examine how failures can lead to the functional hazards, and how safety requirements can be met. Each failure condition identified by the SAD should be addressed in qualitative or quantitative terms. Failure assessment is recommended to be performed at the beginning of UAV development. The main objective is to "identify and classify failure conditions associated with the system by their likelihood and severity." Calculated hazard risks can be compared to acceptable thresholds to determine when the design acceptable or a redesign is necessary.

A likelihood indicates that a failure, or accident is possible even though it may occur rarely over a period of time, or during a considerable number of operations. Reliability predictions are based upon subsystems/equipment being operated within prescribed parameters over a specific period of time. Hence, when the UAV's operational parameters or equipment's environment exceeds the design limits, the validity of the prediction is not valid. Human error (in a remotely piloted vehicle) can have damaging effects even when equipment reliability is high.

In identification and evaluation of failures, the following environmental constraints and operating conditions must be considered: drop, shock, vibration, extreme temperatures, low pressure, rain, ice formation, humidity, bird strike, noise, fire, electrostatic discharge, lightning, and electromagnetic environmental effects.

The identification of the likely failure modes/conditions is vital to establish the safety objectives. In this regards, one must consider how things may go wrong, and compromise

Table 3.7 Reliability and mishap rates for several manned/unmanned aircraft.

No	Aircraft	Mis-hap rate (per 100 000 hours)	Reliability (%)
1	General Aviation (e.g., Cessna 172)	1.22	N/A
2	AV-8B Harrier II	10.7	N/A
3	Lockheed U-2 Dragon Lady	3	96.1
4	General Dynamics F-16 Fighting Falcon	3.5	96.6
5	McDonnell Douglas F/A-18 Hornet	3.2	N/A
6	Boeing 747	0.013	98.7
7	Boeing 777	0.013	99.2
8	General Atomics RQ—1 Predator	32	89
9	Northrop Grumman RQ—4 Global Hawk	160	

the safety of operators or others who find themselves in the vicinity of the UAV crash site. There are a variety of likely failure modes; some major ones are:

1) Loss of UAV control
2) Navigation system failure
3) Propulsion system failure
4) Communication system failure
5) Launch failure
6) Recovery failure
7) Failure of structure
8) Target recognition failure
9) Loss of electric power
10) Stability failure
11) Payload (e.g., camera) operation failure
12) Software bug.

In each case, failure condition must be analyzed, the root cause needs to be identified, and then, develop success-failure predictions. This approach will render and improve the possibility of a zero failure design.

Most major failures will lead to a crash; the UAV may collide with another air vehicle, or may have an unplanned descend to a crowded area. For instance, if the longitudinal stability is insufficient, the UAV will likely stall and crash soon after takeoff. Moreover, loss of control may have a number of reasons including:

1) Engine failure
2) loss of communication with the operator in ground station
3) Hardware fault in avionic
4) Electrical system disconnection
5) Control surface not moving (movement blocked by external obstacles [jam], or servomotor failure, or broken mechanical linkage).

A very likely failure in the UAVs is the autopilot failure; which results in the loss of control, and therefore the loss of UAV itself. A safe flight requires the control system to maintain adequate authority over the UAV throughout the flight operation. Two external common reasons for the autopilot failure are: (a) loss of communication with the ground operator; and (b) loss of electric power. Internal causes for such failure may arise from elements within the autopilot including: (a) gyros; (b) accelerometers; and (c) computer hardware. These three elements may fail, if they are exposed to a significant level of vibration (mainly from internal combustion engines). One effective solution is to fit multiple isolation mounts to the engine, and also to suspend sensitive elements of the autopilot on dedicated and tuned anti-vibration systems.

In the cases, where the failure is due to only one element, a plan for a degree of redundancy is a simple (not optimal) solution. For instance, any electric wire may be supported by a parallel one. Moreover, a twin-engine UAV will have less probability of the engine's failure. To improve the communication robustness, it is recommended to fit dual receivers/transmitters, so if one receiver/transmitter fails, the second receiver/transmitter handles the received/transmitted signal. Predator B is equipped with dual-redundant flight control surfaces and triple-redundant avionics and flight controls.

To prevent structural failure, the UAV must be flown within the recommended flight envelope. Any maneuver, (e.g., varying the bank/pitch angle) should be such that the load factor is less than the maximum allowable value. For example, if the maximum allowable load factor is 2, the bank angle in a turn must not exceed 60°. To prevent the loss of visibility by operator, when the UAV is in the line of sight, the UAV must be painted with non-white (due to cloud), and non-blue (due to blue sky) colors.

Most likely but unexpected failures will be identified during the flight tests operations. For instance, an underpowered UAV will have a longer runway, a less maximum speed, a shorter endurance, and a lower climb rate than expected. Most potential structural issues will be experienced after the first few flight tests. If the UAV structure (e.g., wing) passes the flight tests, but failed in a flight afterward, the main reason for the failure is the operator's mishandling (i.e., a flight operation out of flight envelope).

One conclusion apparent from the available data is that, for most of the systems examined, electrical and mechanical reliability play as much or more of a role in the accidents as human error. Mishaps attributed at least partially to aircraft failures [31] range from 33% (Global Hawk) to 67% (Shadow) in the data reported here.

An improvement in electromechanical reliability of current active UAVs [32] will come only through an increase in the cost of the aircraft. However, a reduction of human errors leading to accidents might not necessarily entail increased costs if suggested changes can be incorporated early in the design process. In the systems analyzed, human factors issues were present in 21% (Shadow) to 67% (Predator) of the accidents.

3.10 Installation Guidelines

In this section, a number of installation guidelines for Global Positioning System (GPS)/Compass, Inertial measurement unit (IMU), and electric motors are provided. These points are not direct design lessons; however, they will impact UAV design techniques, and communications between design team members. Preparing installation guideline will prevent uncontrollable flight, injury to UAV operators, and damage to equipment.

3.10.1 GPS/Compass

Since, the GPS/Compass is sensitive to magnetic interference, they should be installed far away from any electronic devices and power supplies. Moreover, the GPS is sensitive to vibration interference, so position the GPS bracket at a good distance from any rotor. If you are uncertain whether materials near the GPS/Compass module are magnetic or not, you may use a compass or magnet to check it. Insert a special indication line (e.g., orientation arrow pointing forward) for mounting for the GPS/Compass in packaging.

3.10.2 IMU

In a quadcopter, the IMU is best positioned near the vehicle's center of gravity, where vibration is relatively low. Always use double-sided foam/bracket tape for secured installation. The sides of the IMU should be precisely parallel to the vehicle. Install the IMU such that the arrow marked on the IMU faces the sky, and points directly forward, backward, left or right.

3.10.3 Electric Motor

In a UAV, the order of motors isis crucially important, since the rotation direction of each motor should be such that their torques are initially nullified. Wrong connection between the main controller, servos, electronic speed controller, and motors will lead to a dangerous accident. Mounting location for main controller should be such that shorter extension wires are needed as possible. All ports should be accessible when installing the main controller, so as to facilitate wiring and software configuration. Propellers should have a positive angle of attack in order to generate thrust. Wrong propeller installation will lead to incorrect thrust direction, so the UAV will not follow the operator's commands.

Questions

1 What design disciplines will work in parallel within a UAV design project?

2 During the aerodynamic design process, what parameters must be determined?

3 What are the criteria for selection of an airfoil for a wing?

4 What is the primary function of a fuselage?

5 What are the main structural members of a fuselage?

6 What are the main structural members of a wing?

7 What are the three fundamental designs in the wing construction?

8 What is the primary function of an autopilot?

9 In designing the high lift device for a wing, what parameters must be determined?

10 What are the MTOW, and engine power, and wing area Boeing Insitu ScanEagle?

11 In the concept of stealth, what are the three basic methods of minimizing the reflection of pulses back to a receptor?

12 What is the acoustic (i.e., noise) wavelength (signature) range for detecting an air vehicle?

13 Provide at least three main differences between RQ-4A and RQ-4B.

14 What are the two issues for fuel when an aircraft is flying at high altitude?

15 What landing gear parameters need to be determined in the design process?

16 What are the typical engine types for a UAV propulsion system?

17 What are the typical five major stresses to which structural members are subjected?

18 What is the typical propeller efficiency (η_P)?

19 What are the nine configurations for a landing gear?

20 Name typical mechanical/power transmission systems options.

21 What are the conventional control surfaces?

22 Name four unconventional control surfaces.

23 What do FEM and CFD stand for respectively?

24 What is Nacelle?

25 What are the primary control surfaces of Global Hawk and Predator?

26 Briefly describe the technical features of a quadcopter.

27 What are the weaknesses of a quadcopter?

28 What is Murphy's Law?

29 What installation location is recommended for the motor?

30 What installation location is recommended for the GPS?

31 What installation location is recommended for the IMU?

32 Why radome should be designed and fabricated such that it has a low distortion?

33 What lessons were learned from NASA Perseus program?

34 Why did a Global Hawk crash on 29 March 1999, during the program's 18th sortie?

Design Questions

1 You are the lead engineer of the propulsion system design group for a UAV with the following desired features. The UAV is a quadcopter with a payload mass of 100 g, and is required to fly for 30 minutes for aerial filming below 500 ft. The desired cruise airspeed is 25 knots. Analyze the mission requirements, and conceptually design the propulsion system. Provide the following items: (a) engine type;

(b) estimated design power/thrust; (c) estimated fuel weight or battery features; and (d) engine locations. Provide the analysis for your design.

2 You are the lead engineer of the propulsion system design group for a UAV with the following desired features. The UAV is a small fixed-wing vehicle with a payload mass of 500 g, and is required to fly for 90 minutes for aerial filming below 5000 ft. The desired cruise airspeed is 50 knot. Analyze the mission requirements, and conceptually design the propulsion system. Provide the following items: (a) engine type; (b) estimated design power/thrust; (c) estimated fuel weight or battery features; (d) number of engines; and (e) engine(s) locations. Provide the analysis for your design.

3 You are the lead engineer of the propulsion system design group for a UAV with the following desired features. The UAV is a fixed-wing vehicle with a payload mass of 200 kg, and is required to fly for five hours for tracking ground moving targets at 20000 ft. The desired cruise airspeed is 160 knot. Analyze the mission requirements, and conceptually design the propulsion system. Provide the following items: (a) engine type; (b) estimated power/thrust; (c) estimated fuel weight or battery features; (d) number of engines; and (e) engine(s) locations. Provide the analysis for your design.

4 You are the lead engineer of the propulsion system design group for a UAV with the following desired features. The UAV is a fixed-wing vehicle with a payload mass of 1100 kg, and is required to fly for 30 hours for tracking ground moving targets at 60 000 ft. The desired cruise airspeed is 300 knot. Analyze the mission requirements, and conceptually design the propulsion system. Provide the following items: (a) engine type; (b) estimated engine power/thrust; (c) estimated fuel weight or battery features; (d) number of engines; and (e) engine(s) locations. Provide the analysis for your design.

5 You are the lead engineer of the landing gear design group for a UAV with the following desired features. The UAV is a quadcopter with a payload mass of 100 g, and is required to fly for 30 minutes for aerial filming below 500 ft. The desired cruise airspeed is 25 knots. Analyze the mission requirements, and conceptually design the propulsion system. Provide the following items: (a) landing gear type; (b) fixed or retractable; and (c) a sketch/drawing for the design. Provide the analysis for your design.

6 You are the lead engineer of the landing gear design group for a UAV with the following desired features. The UAV is a small fixed-wing vehicle with a payload mass of 500 g, and is required to fly for 90 minutes for aerial filming below 5000 ft. The desired cruise airspeed is 50 knot. Analyze the mission requirements, and conceptually design the propulsion system. Provide the following items: (a) landing gear type; (b) fixed or retractable; and (c) a sketch/drawing for the design. Provide the analysis for your design.

7 You are the lead engineer of the landing gear design group for a UAV with the following desired features. The UAV is a fixed-wing vehicle with a payload mass of 200 kg, and is required to fly for five hours for tracking ground moving targets at 20 000 ft. The desired cruise airspeed is 160 knot. Analyze the mission

requirements, and conceptually design the propulsion system. Provide the following items: (a) landing gear type; (b) fixed or retractable; and (c) a sketch/drawing for the design. Provide the analysis for your design.

8 You are the lead engineer of the landing gear design group for a UAV with the following desired features. The UAV is a fixed-wing vehicle with a payload mass of 1100 kg, and is required to fly for 30 hours for tracking ground moving targets at 60 000 ft. The desired cruise airspeed is 300 knot. Analyze the mission requirements, and conceptually design the propulsion system. Provide the following items: (a) landing gear type; (b) fixed or retractable; and (c) a sketch/drawing for the design. Provide the analysis for your design.

Problems

1 A large UAV with a piston engine is cruising at 10 000 ft altitude with a speed of 150 knot. The propeller efficiency is 0.73. The wing area is 25 m^2, and the drag coefficient is 0.036. How much power the engine should generate for this flight?

2 A small UAV with an electric engine is cruising at sea level with a speed of 50 knot. The propeller efficiency is 0.6. The wing area is 1.5 m^2, and the drag coefficient is 0.045. How much power the engine should generate for this flight?

3 Determine the total energy supplied by a 3000 mAh 10 V Li–Po Lithium Polymer battery, and the maximum power that can be provided to an electric motor.

4 Determine the total energy supplied by a 8000 mAh 6 V Li–Po Lithium-ion battery, and the maximum power that can be provided to an electric motor.

5 Determine the total energy supplied by a 10 000 mAh 6 V Li–Po Nickel–cadmium battery, and the maximum power that can be provided to an electric motor.

6 Determine the total energy supplied by a 2000 mAh 6 V Li–Po Alkaline battery, and the maximum power that can be provided to an electric motor.

7 A small UAV is using a NACA 63–215 airfoil for its wing section. The wing chord at root and tip are 0.8 m and 0.6 m respectively. Determine maximum thickness of the wing at the root and the tip.

8 A small UAV is using a NACA 23 012 airfoil for its wing section. The wing chord at root and tip are 1.2 m and 0.8 m respectively. Determine maximum thickness of the wing at the root and the tip.

9 A mini UAV with an electric engine is cruising with a velocity of 20 knot at sea level. The wing area is 0.22 m^2, the engine prop efficiency is 0.72, and the UAV drag coefficient is 0.042. The batteries of the UAV provide a 12-volt electric energy; determine the electric current consumed by the engine.

10 A small UAV with an electric engine is cruising with a velocity of 25 knot at 5,000 ft altitude. The wing area is 0.54 m^2, the engine prop efficiency is 0.74, and the UAV drag coefficient is 0.038. The batteries of the UAV provide a 24-volt electric energy; determine the electric current consumed by the engine.

11 A small UAV with an electric engine is cruising with a velocity of 22 knot at 5,000 ft altitude. The wing area is 0.62 m^2, the engine prop efficiency is 0.65, and the UAV drag coefficient is 0.035. The electric energy is provided by a Venom 7.2 V 5000 mAh 6-Cell Ni-MH battery. How long this flight will last?

12 A small UAV with an electric engine is cruising with a velocity of 25 knot at sea level altitude. The wing area is 1.2 m^2, the engine prop efficiency is 0.68, and the UAV drag coefficient is 0.041. The electric energy is provided by a 22.2 V 3200 mAh 6S 30C Smart Li-Po battery. How long this flight will last?

Notes

1 Computational Fluid Dynamics.
2 The term "wheel" is often used to mean the entire wheel/brake/tire assembly.

4

Aerodynamic Design

CONTENTS

Design of Unmanned Aerial Systems, First Edition. Mohammad H. Sadraey.

Companion website: www.wiley.com/go/sadraey/unmanned-aerial-systems

Educational Outcomes

After study of this chapter, the reader will be able to:

1) Conduct wing aerodynamic design
2) Conduct horizontal tail aerodynamic design
3) Conduct vertical tail aerodynamic design
4) Conduct fuselage aerodynamic design
5) Provide aerodynamic guideline to accommodate antennas

4.1 Introduction

The aerodynamics is the science that involves the study of the behavior (i.e., dynamics) of air when confronting a moving object (e.g., Unmanned Aerial Vehicle [UAV]). The UAV has a number of components which are characterized by their aerodynamic outputs (e.g., lift), two of which are wing and tail. A wing/tail is considered as a lifting surface in which the lift is produced due to the pressure difference between lower and upper surfaces. In contrast, surfaces such as aileron, rudder, and elevator are referred to as the control surfaces. Lifting surfaces are generally fixed while control surfaces are always moving up/down. Both lifting and control surfaces are functions of their aerodynamic features.

The fuselage is not fundamentally considered as an aerodynamic component based on its function. However, fuselage has a considerable role in creating drag, while it generates a little lift. The primary aerodynamic function of the wing and tail is to generate sufficient lift force or simply lift (L). However, they have two other undesirable products, namely drag (D) and nose-down pitching moment (M). In this chapter, the aerodynamic design of UAV components which have a direct contact with moving air is presented. The primary objective of aerodynamic design is to optimize the UAV outer shape such that the lift is maximized, while the drag and pitching moment are minimized.

Table 4.1 illustrates the UAV aerodynamic components and their major aerodynamic influences. For each aerodynamic component, the designer should aim to maximize their lift and minimize the drag. This chapter presents the aerodynamic design of four components of an UAV; namely wing, vertical tail, horizontal tail, and fuselage. Aerodynamics textbooks (e.g., [14, 33, 34]) are a good source to consult for information about mathematical techniques for calculating the pressure distribution over the wing and for determining the flow variables. During the aerodynamic design process, several parameters such as airfoil section and wing aspect ratio must be determined. The aerodynamic features of the UAV would affect the UAV performance, stability, control, and structural design.

Table 4.1 Unmanned Aerial Vehicle (UAV) aerodynamic components and their functions.

No	Component	Primary function(s)	Major aerodynamic influence
1	Wing	Generation of lift	Lift
2	Horizontal tail	Longitudinal stability, longitudinal trim	Horizontal tail lift
3	Vertical tail	Directional stability, directional trim	Vertical tail lift, side-force
4	Fuselage	Payload accommodations	Drag

4.2 Fundamentals of Aerodynamics

The aerodynamics is an applied science that involves the study of the behavior (i.e., dynamics) of flow of air; when it has a relative speed with an object (e.g., unmanned aerial vehicle). In order to aerodynamically design a UAV, one must be able to calculate and analyze the aerodynamic outputs of the vehicle. One of the necessary tools in the aerodynamic design process is the aerodynamic theories to calculate aerodynamic forces and moments (e.g., lift, drag, and pitching moment). The aerodynamic forces on an object in an airflow (e.g., wing) can be calculated from pressure distribution around the object.

The aerodynamic forces of lift and drag of a lifting surface (e.g., wing and tail) are functions of the following factors: (a) surface angle of attack; (b) geometry (e.g., planform area; S); (c) airfoil section; (d) airspeed (V); (e) air density (ρ); and (f) Reynolds number of the flow. The lift and drag are determined by:

$$L = \frac{1}{2}\rho V^2 S C_L \tag{4.1}$$

$$D = \frac{1}{2}\rho V^2 S C_D \tag{4.2}$$

where C_L and C_D are lift and drag coefficients respectively. The drag coefficient is a function of several parameters particularly UAV configuration. A mathematical expression for the variation of drag coefficient versus lift coefficient is:

$$C_D = C_{D_o} + K C_L^2 \tag{4.3}$$

This equation is sometimes referred to as aircraft "drag polar." The variable K is referred to as the induced drag correction factor. It is obtained from:

$$K = \frac{1}{\pi \cdot e \cdot AR} \tag{4.4}$$

where e is the Oswald span efficiency factor, and AR is the wing aspect ratio. The variable AR is introduced later in this chapter.

The zero-lift drag coefficient (C_{Do}) of a well-designed fixed-wing UAV; in clean configuration; is about 0.015–0.025. However, for a quadcopter (i.e., rotary-wing configuration), we regard the total drag coefficient (C_D) as a parameter, and Eq. (4.3) is not

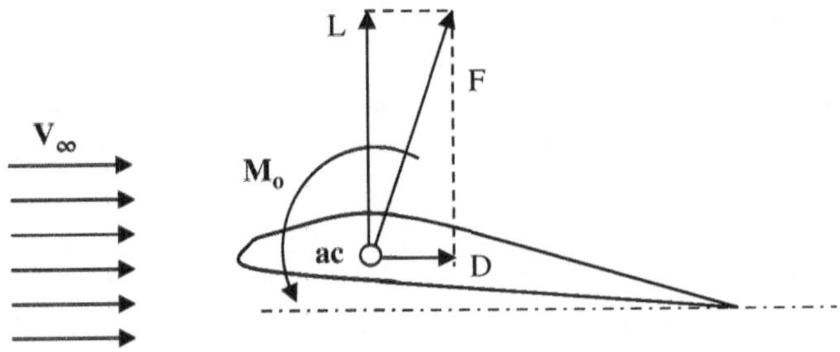

Figure 4.1 Aerodynamic lift, drag, and pitching moment.

employed. For a quadcopter, the typical value for C_D is about 0.2–0.4. The technique to calculate the C_{Do} of UAV components is beyond the scope of this book, the interested reader is referred to references such as [9].

The lift is always defined as the component of the aerodynamic force perpendicular to the relative wind (Figure 4.1). The drag is always defined as the component of the aerodynamic force parallel to the relative wind.

The aerodynamic center (ac) which is frequently selected to be the center of lift, is located nearly at the quarter chord. The wing three-dimensional (3d) lift coefficient (C_L) is directly a function of the airfoil two-dimensional (2d) lift coefficient, c_l. In turn, the variable c_l is a function of variations of pressure coefficients on the upper and lower surfaces. For the case of a small angle of attack (less than 5°), 2d lift coefficient is calculated from

$$c_l = \frac{1}{c}\int_0^c \left(C_{pl} - C_{pu}\right) dx \tag{4.5}$$

where c_{pl} and c_{pu} are pressure coefficients at the lower and upper surfaces respectively, and c is the airfoil chord. Thus, the lift is produced due to the pressure difference between lower and upper surfaces of the wing/tail. References such as [14, 15] provides techniques to determine the pressure distribution around any lifting surface such as the wing. The variations of lift coefficient versus angle of attack is often linear below stall angle. The UAVs are usually flying at an angle of attack below stall angle.

With the progress of the science of aerodynamics, there are variety of techniques and tools to accomplish this time consuming job. A variety of tools and software based on aerodynamics and numerical methods have been developed in the past decades. The CFD[1] Software based on the solution of Navier–Stokes equations, vortex lattice method, and thin airfoil theory are available in the market. The application of such software packages is expensive and time-consuming, and will accelerate the wing/tail design process. As Figure 1.5 (in Chapter 1) illustrates, a design group; formed by several aerodynamic engineers; should be devoted for conducting such design operation. For fundamentals of aerodynamics, please refer to references such as [14, 33].

4.3 Wing Design

This section is devoted to the aerodynamic design of the wing, and presents the design procedure, wing parameters, airfoil selection process, and design technique. During the wing aerodynamic design process, the following parameters must be determined:

(a) wing reference (or planform) area; (b) number of the wings; (c) vertical position relative to the fuselage (high; mid, or low wing); (d) horizontal position relative to the fuselage; (e) cross section (or airfoil); (f) aspect ratio (AR); (g) taper ratio (λ); (h) tip chord (C_t); (i) root chord (C_r); (j) Mean Aerodynamic Chord (MAC or C); (k) span (b); (l) twist angle (α_t); (m) sweep angle (Λ); (n) dihedral angle (Γ); (o) incidence (i_w) or setting angle (α_{set}); (p) high lifting devices (e.g., flap); and (q) aileron.

The author has presented the wing design for an air vehicle in Chapter 5 of [8], so the materials are not fully repeated in this text. You are encouraged to study this reference to find the details of wing design. The wing fundamental design requirements are lift, aerodynamic efficiency, aircraft performance, stability, producibility, operational requirements, cost, and flight safety.

4.3.1 Wing Design Procedure

Figure 4.2 provides the wing design procedure. It implies that the wing design is an iterative process which means that the selections/calculations are usually repeated several times. A wing designer should aim to maximize the lift, while minimize the drag

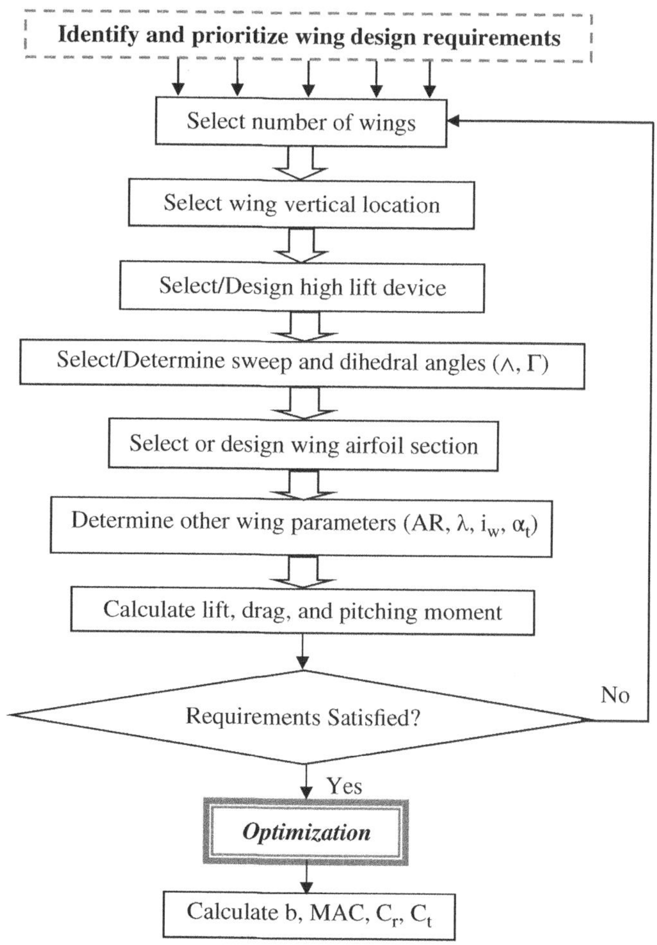

Figure 4.2 Wing design procedure.

and pitching moment. In practice, a number of wing parameters (e.g., wing location, aspect ratio, and airfoil section) are selected, while others (e.g., wing area, wing span, and mean aerodynamic chord) are calculated.

4.3.2 Airfoil Selection/Design

One of the critical tasks in the wing aerodynamic design is to determine its cross section (i.e., airfoil). Wing is a three dimensional component, while the airfoil is two dimensional section. Because of the airfoil section, two other outputs of the airfoil, and consequently the wing, are drag and pitching moment. The wing may have a constant or a non-constant cross-section across the wing. There are two ways to determine the wing airfoil section: (a) airfoil design; and (b) airfoil selection. The design of the airfoil is a complex and time consuming process and needs expertise in fundamentals of aerodynamics at graduate level. Since the airfoil needs to be verified by testing it in a wind tunnel, it is expensive too.

If you are not ready to design your own airfoil, you are recommended to select a proper airfoil from the previously designed and published airfoil sections. Three reliable airfoil resources are (a) NASA[2]; (b) NACA[3]; and (c) Eppler. The aerodynamic characteristics of Eppler airfoils; which names begin with the letter "E" followed by three numbers; have been published in [35]. The details of NACA airfoils [36] have been published in a book published by Abbott and Von Donehoff [37]. In addition, a summary of low-speed airfoil data are provided by Ref. [38]. In 1950s, airfoils have been classified by the NACA, the forerunner of present NASA, and have been cataloged using a four/five digits code. Figure 4.3 shows the 6-series NACA 63_3–218 airfoil section with a thickness-to-chord ratio of 18%.

Selecting an airfoil is a part of the overall wing design. Selection of an airfoil for a wing begins with the clear statement of the flight requirements. For instance, a subsonic flight design requirements are very much different from a supersonic flight design objectives. On the other hand, flight in the transonic region requires a special airfoil that meets Mach divergence requirements. The designer must also consider other requirements such as airworthiness, structural, manufacturability, and cost requirements.

Several graphs indicate the behavior of an airfoil in a flow field; they are employed in the selection of the best airfoil. Aerodynamic theories are applied or wind tunnel tests are conducted to plot such figures for each airfoil. Figure 4.4 shows the typical variations of drag coefficient as a function of lift coefficient for a laminar airfoil; such as in 6-series NACA airfoils. This is one of the significant graphs which are utilized to decide about the wing airfoil section. In general, the following are the criteria to select an airfoil for a wing with a collection of design requirements:

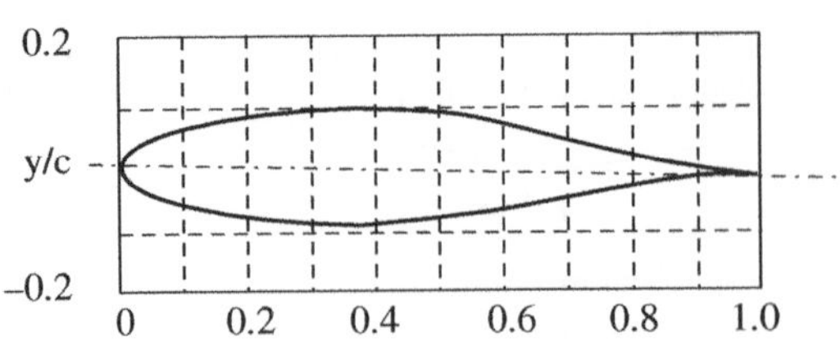

Figure 4.3 A 6-series National Advisory Committee for Aeronautics (NACA) 63_3–218 airfoil section.

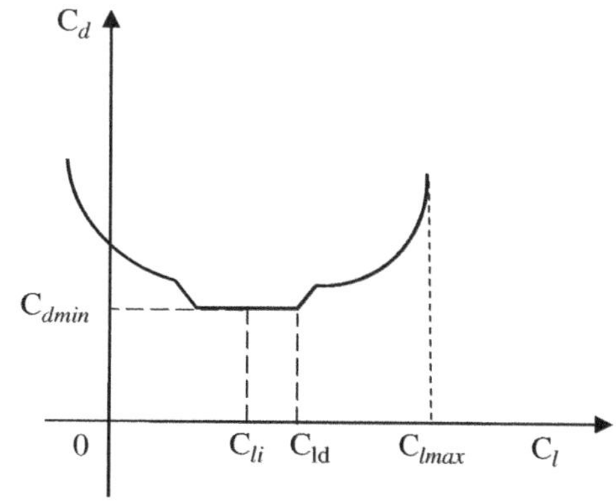

Figure 4.4 Typical variations of C_l versus C_d for a laminar airfoil.

1) The airfoil with the highest maximum lift coefficient (Cl_{max}).
2) The airfoil with the proper ideal or design lift coefficient (C_{l_d} or C_{l_i}).
3) The airfoil with the lowest minimum drag coefficient ($C_{d_{min}}$).
4) The airfoil with the highest lift-to-drag ratio ($(C_l/C_d)_{max}$).
5) The airfoil with the highest lift curve slope ($C_{l_{\alpha_{max}}}$).
6) The airfoil with the lowest (closest to zero; negative or positive) pitching moment coefficient (C_m).
7) The proper stall quality in the stall region (the variation must be gentle, not sharp).
8) The airfoil must be structurally reinforceable. The airfoil should not that much thin that spars cannot be placed inside.
9) The airfoil must be such that the cross section is manufacturable.
10) The cost requirements must be considered.
11) Other design requirements must be considered. For instance, if the fuel tank has been designated to be places inside the wing inboard section, the airfoil must allow the sufficient space for this purpose.
12) If more than one airfoil is considered for a wing, the integration of two airfoils in one wing must be observed.

The NACA 4- and 5-digit airfoil sections are not very efficient, but 6-series airfoils are the best (and more popular) among the NACA airfoils. The four- and five-digit airfoil sections (e.g., 2412, and 23015) where designed simply by using parabola and line. They were not supposed to satisfy major aerodynamic design requirements, such as laminar flow and no flow separation. However, the six series airfoils were designed to maintain laminar flow over a large part of the chord, thus they maintain lower C_{dmin} compared with four- and five-digit airfoils. The 6-series NACA airfoils are designated by five main digits and begin with number 6 (e.g., 63–209).

The two recommended NASA airfoil sections are the Medium-Speed "MS(1)" and the Low-Speed "LS(1)" series. Like the old NACA 4- and 5-digit and 6-series, there are any number of these to choose from. Two typical NASA airfoil designations are MS(1)-xxyy and LS(1)-xxyy, where the xx is the design lift coefficient, and the yy is the thickness-to-chord ratio. The NASA LS and MS airfoil sections have proven to be well-behaved in wind tunnel testing.

For instance, MS(1)-0313 was the original selection for the AR of 11 wing for AGM-154 Joint Standoff Weapon (JSOW). The JSOW with a 500 kg mass and a 4.1 m length is the product of a joint venture between the United States Navy and Air Force to deploy a standardized medium range precision guided weapon. However, the airfoil was replaced with the MS(1)-0315 to get the additional 2% of thickness. A higher thickness (13%) airfoil was selected to address the flutter concern of the wing. Furthermore, this airfoil has a higher stall angle.

Another popular NASA airfoil also used for several smaller vehicles is the LS(1)-0313. This is a little better than the MS series, except for MDD. For the tail and control surfaces (e.g., elevator), the symmetrical sections LS(1)-0010 and LS(1)-0013 are recommended. They are kind of a hybrid between the original NASA general aviation airfoils and the supercritical. They have a little "kick" at the back that gives them the high C_{lmax}, and trailing edge does not come to a point, it's actually flat, to get the pressure gradient. They demonstrate about 40% laminar flow with very smooth surface finishes, and no laminar separation.

It is reminded that the Reynolds (Re) number has a huge impact on airfoil and wing aerodynamic performance. This is becoming more and more important in UAV industries, which operate at much lower Re than three to nine million where most of the data are found. In general, the Eppler airfoils are generally for very low Reynolds number, Wortman for sailplane-ish Reynolds number, and the NASA Low-Speed airfoils (e.g., LS(1)-0413) and Mid-Speed airfoils (e.g., MS(1)-0313) are for "moderate" Reynolds numbers. Computationally designed airfoils should always be wind tunnel tested before used in flight.

It is assumed that an airfoil section data base (such as *NACA* or *Eppler*) is available and the wing designer is planning to select the best airfoil from a collection. The procedure to select an airfoil for the wing is recommended as follows:

1) Determine the average aircraft weight (W_{avg}) in cruising flight:
2) Calculate the aircraft ideal cruise lift coefficient (C_{L_C}).
3) Calculate the wing cruise lift coefficient ($C_{L_{C_w}}$).
4) Calculate the wing airfoil ideal lift coefficient (C_{l_i}).
5) Calculate the aircraft maximum lift coefficient ($C_{L_{\max}}$):
6) Calculate the wing maximum lift coefficient ($C_{L_{\max_w}}$).
7) Calculate the wing airfoil gross maximum lift coefficient ($C_{l_{\max_{gross}}}$).
8) Select/Design the high lift device (HLD) (type, geometry, and maximum deflection).
9) Determine the HLD contribution to the wing maximum lift coefficient ($\Delta C_{l_{HLD}}$).
10) Calculate the wing airfoil "*net*" maximum lift coefficient ($C_{l_{\max}}$)
11) Identify airfoil section alternatives that deliver the desired C_{li} (step 4) and C_{lmax} (step 10). This is a very essential step.
12) Among several acceptable alternatives, select the optimum airfoil section by using a comparison table.

4.3.3 Wing Design Technique

This section briefly presents the techniques/recommendations/guidelines to determine wing parameters.

1) **Wing vertical location**

   One of the wing parameters that could be determined at the early stages of wing design process is the wing vertical location relative to the fuselage centerline. This wing parameter will directly influence the design of other aircraft components including tail, and landing gear. In principle, there are four options for the vertical location of the wing. They are: (a) high wing (e.g., ScanEagle); (b) mid wing; (c) low wing (e.g., Global Hawk and Predator); and (d) parasol wing.

In term of advantage, the high wing makes the aircraft laterally more stable, and will produce more lift compared with mid and low wing. A high wing configuration is also recommended for a UAV which is taking off from a launcher. Moreover, the aerodynamic shape of the fuselage lower section can be smoother. In term of disadvantage for the high wing, the landing gear is longer if connected to the wing. This makes the landing gear heavier and requires more space inside the wing for retraction system. This will further make the wing structure heavier. For a complete comparison between the four alternatives, the reader is referred to other references such as [8].

2) **Wing Incidence**

The wing incidence (i_w) is the angle between fuselage center line and the wing chord line at root. It is sometimes referred to as the wing setting angle (α_{set}). The fuselage center line lies in the plane of symmetry and is usually defined parallel to the cabin floor. The wing must be able to generate the desired lift coefficient during cruising flight. This design requirement naturally matches with the wing airfoil angle of attack corresponding to the airfoil ideal lift coefficient. Therefore, as soon as the wing ideal lift coefficient (found from aircraft cruising speed) is determined, a reference to C_l-α graph demonstrates the wing setting angle. The typical number for wing incidence for majority of aircraft is between 1–4°.

3) **Aspect Ratio**

Aspect ratio (AR) is defined as the ratio between the wing span; b and the wing Mean Aerodynamic Chord (MAC or $\bar{C}$).

$$AR = \frac{b}{\bar{C}} \tag{4.6}$$

The Mean Aerodynamic Chord, root chord, tip chord, wingspan, and Aerodynamic Center in a straight wing is illustrated in Figure 4.5. The wing planform area for a rectangular or straight tapered shape is defined as the span times the mean aerodynamic chord:

$$S = b \cdot \bar{C} \tag{4.7}$$

The wing aspect ratio has several influences over the aircraft features. From the aerodynamic points of view, as the AR is increased, the aerodynamic features of a three-dimensional wing (such as $C_{L\alpha}$, α_o, α_s, C_{Lmax}, C_{Dmin}) are getting closer to its

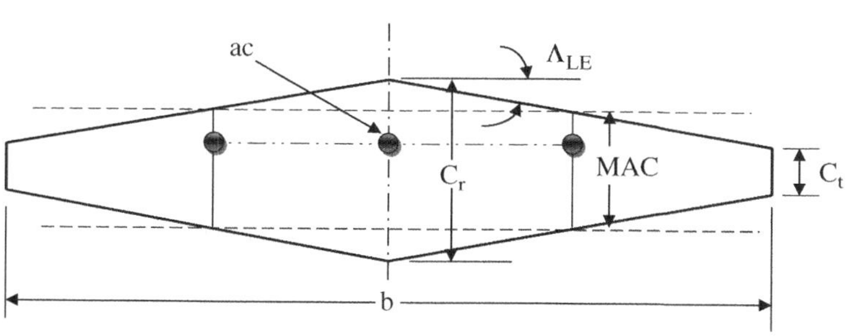

Figure 4.5 Mean Aerodynamic Chord and Aerodynamic Center in a straight wing.

two-dimensional airfoil section (such as $C_{l\alpha}$, α_o, α_s, C_{lmax}, C_{dmin}). As the AR is increased, the aircraft maximum lift-to-drag ratio is increased, while the wing will be structurally heavier. Moreover, as the AR is increased, the effect of wing tip vortex on the horizontal tail is decreased.

A shorter wing needs lower cost to build compared with a long wing. As the AR increases, the aircraft mass moment of inertia around x-axis will be increased. This means that it takes longer to roll. In another word, this will reduce the maneuverability of aircraft in roll. As the aspect ratio is increased, the wing stiffness around the y-axis is decreased.

For some design requirements, a low aspect ratio wing is favorable, while for other design requirements, a high aspect ratio wing is desirable. The exact value of the AR will be determined through a thorough investigation and lots of calculation over aircraft performance, stability, control, manufacturability, and cost. The typical aspect ratio for UAVs is 6–25. The aspect ratio of Global Hawk is 25.

4) Taper ratio

Taper ratio (λ) is defined as the ratio of the tip chord (C_t) to the root chord (C_r).

$$\lambda = \frac{C_t}{C_r} \tag{4.8}$$

The geometric result of taper is a smaller tip chord. In general, the taper ratio varies between zero and one. Three major planform geometries relating to taper ratio are rectangular, trapezoidal and delta shape.

In general, a rectangular wing planform is aerodynamically inefficient, while it has a few advantages, such as performance, cost and ease of manufacture. A wing with a rectangular planform has a larger downwash angle at the tip than at the root. Therefore, the effective angle of attack at the tip is reduced compared with that at the root. Thus, the wing tip will tend to stall later than the root. The spanwise lift distribution is far from elliptical; where it is highly desirable to minimize the induced drag. Hence, one of the reasons to taper the planform is to reduce the induced drag.

The wing taper will change the wing lift distribution; an elliptical one is desired. The wing taper will increase the cost of the wing manufacture, since the wing ribs will have different shapes. The taper will reduce the wing weight, since the center of gravity of each wing section (left and right) will move toward fuselage center line. The wing taper decrease the wing mass moment of inertia about x-axis (longitudinal axis). Consequently, this will improve the aircraft lateral control.

For a constant-taper and constant-sweep angle (trapezoidal) planform, Mean Aerodynamic Chord is determined [39] as follows:

$$\bar{C} = \frac{2}{3} C_r \left(\frac{1+\lambda+\lambda^2}{1+\lambda} \right) \tag{4.9}$$

Example 4.1

A straight tapered wing has the following features

$$AR = 6, \lambda = 0.5, S = 2\,m^2$$

Determine tip chord, root chord, mean aerodynamic chord, and wing span.

Solution

$$AR=\frac{b}{\bar{C}} \tag{4.6}$$

$$S=b\cdot\bar{C} \tag{4.7}$$

Multiplying numerator and denominator of Eq. (4.6) by a b and then, using Eq. (4.7), we obtain:

$$AR=\frac{b}{\bar{C}}\frac{b}{b}=\frac{b^2}{S}\Rightarrow b=\sqrt{S.AR}=\sqrt{2\times6}=3.46m$$

$$AR=\frac{b}{\bar{C}}\Rightarrow\bar{C}=\frac{b}{AR}=\frac{3.46}{6}=0.577m \tag{4.6}$$

$$\bar{C}=\frac{2}{3}C_r\left(\frac{1+\lambda+\lambda^2}{1+\lambda}\right)\Rightarrow C_r=\frac{\frac{3}{2}\bar{C}}{\frac{1+\lambda+\lambda^2}{1+\lambda}}=\frac{\frac{3}{2}\times0.577}{\frac{1+0.5+0.5^2}{1+0.5}}\Rightarrow C_r=0.742\text{m} \tag{4.9}$$

$$\lambda=\frac{C_t}{C_r}\Rightarrow C_t=C_r\lambda=0.742\times0.5=0.371\text{m} \tag{4.8}$$

5) Sweep Angle

The angle between the wing leading edge and the lateral axis perpendicular to the aircraft centerline (y-axis) of the aircraft is called leading edge sweep (Λ_{LE}). Basically, a wing is being swept for the following four design goals:

I) Improving the wing aerodynamic features (lift, drag, pitching moment) at transonic, supersonic and hypersonic speeds by delaying the compressibility effects.
II) Adjusting the aircraft center of gravity.
III) Improving static lateral stability.
IV) Impacting longitudinal and directional stability.

The sweep angle has several advantages and disadvantages such that can only be decided through a compromise. If the aircraft needs to have different sweep angles at various flight conditions, an ideal option is to select the variable sweep wing that may be swept back and then returned to its original position during flight. In some cases, one sweep angle cannot satisfy all design requirements. For instance, a very high sweep angle wing satisfies the high speed cruise requirements, however, at low subsonic speed, the aircraft is not satisfactorily controllable or laterally stable. One solution is to divide the wing sections into inboard plane, and outboard plane; each having different sweep angles.

6) Twist Angle

If the wing tip is at a lower incidence than the wing root, the wing is said to have negative twist or simply twist (α_t) or washout. This indicates that wing angle of attack is reduced along the span. The wings on a number of modern aircraft have different airfoil sections along the span, with different values of zero lift angle of attack; this is called aerodynamic twist. When the tip incidence and root incidence are not the same, the twist is referred to as *geometric twist*. However, if the tip airfoil section and root airfoil section are not the same, the twist is referred to as *aerodynamic twist*. Both types of twist have advantages and disadvantages by which the designer must establish a selection that satisfies the design requirement. There are two major goals for the employing the twist in wing design process:

1) Avoiding tip stall before root stall.
2) Modification of lift distribution to an elliptical one.

The typical value for the geometric twist is between – 1 to –4° (i.e., negative twist). The exact value of the twist angle must be determined such that the tip stalls after root as well as the lift distribution be elliptic.

7) **Dihedral Angle**
When you look at the front view of an aircraft, the angle between the chord-line plane of a wing with the "xy" plane is referred to as the wing dihedral (Γ). The chord line plane of the wing is an imaginary plane that is generated by connecting all chord lines across span. If the wing tip is higher than the xy plane, the angle is called positive dihedral or simply dihedral, but when the wing tip is lower than the xy plane, the angle is called negative dihedral or anhedral. There are several advantages and disadvantages for dihedral angle. The primary reason of applying the wing dihedral is to improve the lateral stability of the aircraft. The typical dihedral angle is a value between – 15° to + 10°. In the wing design, one must be careful to determine the wing parameters such that satisfy both stability and controllability requirements.

8) **High Lift Device**
Two primary objectives of HLDs are: (a) maximizing the payload weight; and (b) minimizing the stall speed (V_s). The primary applications of HLDs are during take-off and landing operations during which the aircraft speed is slightly greater than stall speed. Since the airspeed is very low compared with cruising speed, the wing must produce more lift coefficient. At the airfoil level, a HLD deflection tends to cause the following changes in the airfoil features: (a) lift coefficient (C_l) is increased; (b) maximum lift coefficient (C_{lmax}) is increased; (c) zero-lift angle of attack (α_o) in changed; (d) stall angle (α_s) is changed; (e) pitching moment coefficient is changed; (f) drag coefficient is increased; and (g) lift curve slope is increased.
In designing the HLD for a wing, the following parameters must be determined: (a) HLD location along the span; (b) the type of HLD; (c) HLD chord (C_f); (d) HLD span (b_f); and (e) HLD maximum deflection (down) (δ_{fmax}). Two main groups of HLDs are: (a) leading edge high lift device (LEHLD); and (b) trailing edge high lift devices (TEHLD or flap). Six basic types of leading edge HLD are: (a) plain flap; (b) split flap; (c) single slotted flap; (d) double slotted flap; (e), triple slotted flap; and (f) Fowler flap.
When the low cost is the number one priority, select the least expensive HLD that is the plain flap. If the performance is the number one priority, select the HLD that satisfies the performance requirements. If only one HLD such as a single slotted flap does not satisfy the performance requirements, add another HLD such as leading edge flap to meet the design requirements.

4.3.4 Wing Design Steps

The wing fundamental design requirements are lift, aerodynamic efficiency, aircraft performance, stability, producibility, operational requirements, cost, and flight safety. The practical steps in a wing design process are recommended as below:

1) Select number of wings.
2) Select wing vertical location.
3) Select wing configuration (e.g., straight, swept, tapered, delta).
4) Calculate average aircraft weight at cruise:
5) Calculate required aircraft cruise lift coefficient (with average weight):
6) Calculate the required aircraft take-off lift coefficient:
7) Select the HLD type and its location on the wing.
8) Determine HLD geometry (span, chord, and maximum deflection).
9) Select/Design airfoil.
10) Determine wing incidence or setting angle (i_w).
11) Select sweep angle (Λ) and dihedral angle (Γ).
12) Select other wing parameter such as aspect ratio (AR), taper ratio (λ) and wing twist angle (α_{twist}).
13) Calculate lift distribution at cruise (without flap, or flap up).
14) Check the lift distribution at cruise that must be elliptic. Otherwise, return to step 13 and change few parameters.
15) Calculate the wing lift at cruise (C_{Lw}).
16) The wing lift coefficient at cruise (C_{Lw}) must be equal to the required cruise lift coefficient (step 5). If not, return to step 10 and change wing setting angle.
17) Calculate wing lift coefficient at take-off.
18) The wing lift coefficient at take-off must be equal to take-off lift coefficient (step 6). If not, first, vary flap deflection (δ_f), and geometry (C_f, b_f) to correct wing lift coefficient; otherwise, return to step 7 and select another HLD.
19) Calculate wing drag (D_w).
20) Vary wing parameters to minimize wing drag and wing pitching moment
21) Draw the final design with dimensions.

4.4 Tail Design

4.4.1 Design Procedure

The tail in a conventional aircraft has often two components of horizontal tail and vertical tail and carries two primary functions: (a) trim (longitudinal and directional); and (b) stability (longitudinal and directional). Since two conventional control surfaces (i.e., elevator and rudder) are indeed parts of the tails to implement control, it is proper to add the following item as the third function of tails: (c) Control (longitudinal and directional).

The majority of parameters are finalized through technical calculations, while a few parameters are decided via an engineering selection approach. Figure 4.6 illustrates the block diagram of the tail design process. The aircraft design is an iterative process; therefore this procedure will be repeated several times until the optimum aircraft

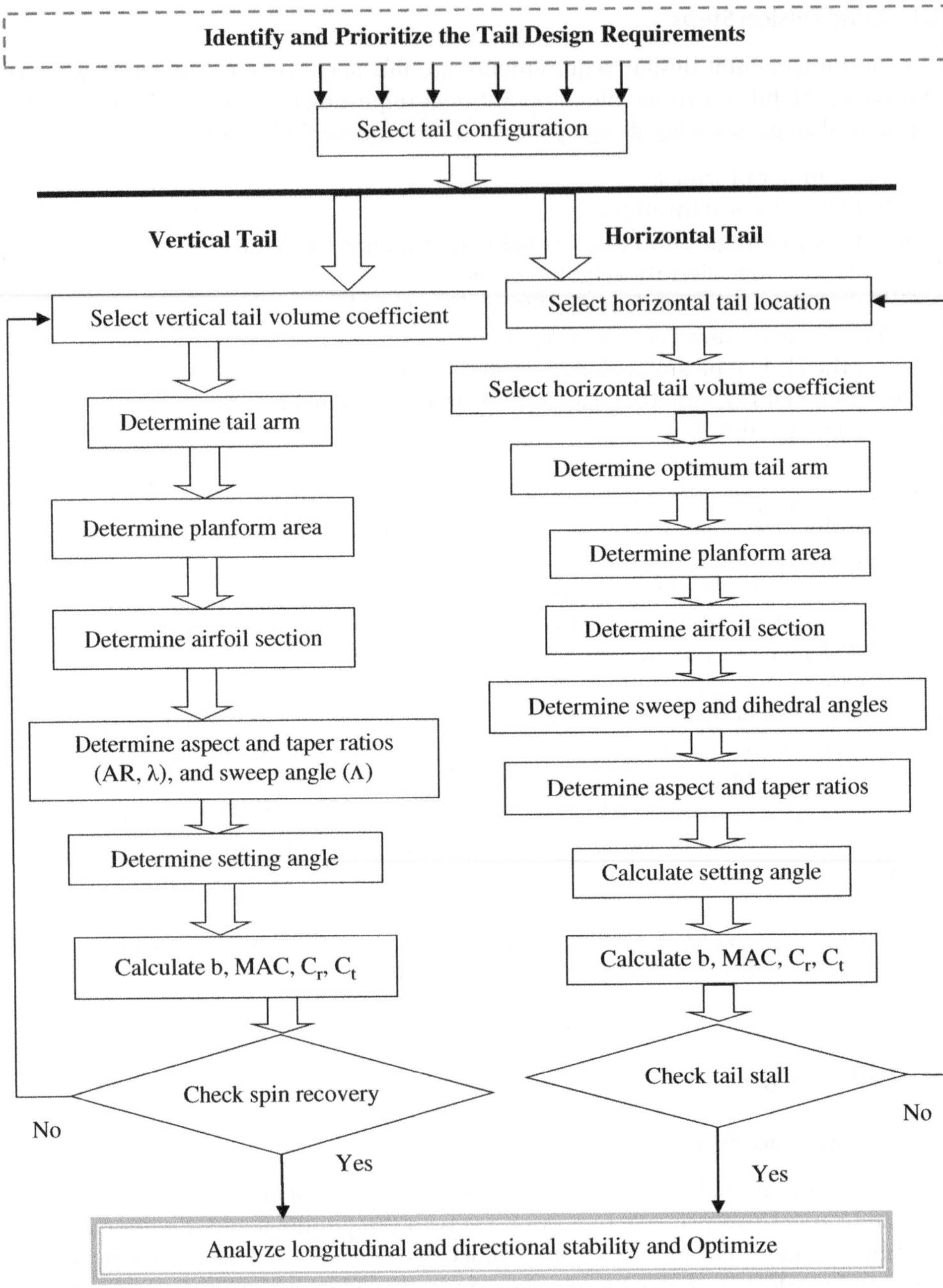

Figure 4.6 Tail design flowchart.

configuration has been achieved. The design of vertical and horizontal tails might be performed almost in parallel. However, there is one step in the vertical tail design (i.e., spin recovery) that the effect of horizontal tail into vertical tail is investigated. The details on each step will be introduced in the later sections.

4.4.2 Tail Configuration

The term "*tail*" in this section means the combination of horizontal and vertical tails. The first step in the tail design is the selection of the tail configuration. The choice of the tail configuration is the output of a selection process, not the result of a mathematical calculation. The decision for the selection of the tail configuration must be made based on the reasoning, logic and evaluation of various configurations against design requirements.

The general tail design requirements that must be considered and satisfied in the selection of tail configurations are as follows: (a) longitudinal trim; (b) directional trim; (c) lateral trim; (d) longitudinal stability; (e) directional stability; (f) lateral stability; (g) manufacturability and controllability; (h) handling qualities; (i) stealth (only in some specific military UAVs); (j) operational requirements; (k) airworthiness; (l) cost; (m) competitiveness; and (n) size limits.

In general, the following tail configurations are available that are capable of satisfying the design requirements in one way or another: (a) aft tail and one aft vertical tail; (b) aft tail and twin aft vertical tail; (c) canard and aft vertical tail; (d) canard and twin wing vertical tail; (e) triplane; (f) tailless (delta wing with one vertical tail); and (g) no formal tail (also known as "*flying wing*").

The first configuration has several sub-configurations that will be examined in more details. In the first three configurations, vertical tail is installed at the aft of fuselage, while in the fourth configuration; two vertical tails are installed at the wing tips. The aft tail configurations are as follows: (a) conventional; (b) T-shape; (c) cruciform; (d) H-shape; (e) triple-tail; (f) V-tail; (g) inverted V-tail; (h) improved V-tail; (i) Y-tail; (j) twin vertical tail; (k) boom-mounted; (l) inverted boom-mounted; (m) twin T; and (n) U-tail. For instance, the Global hawk (Figure 1.2) has a V-tail, MQ-9 Reaper (Figure 5.12) has a Y-tail, ScanEagle (Figure 3.3) has vertical tails at wingtips, and Arcturus UAV has T-tail (Figure 6.38).

A new tail configuration for UAVs is that the vertical tail is underneath the fuselage (e.g., Predator MQ-9). The main reason behind moving the vertical tail down is to allow the satellite radar (which is located inside fuselage) to directly see the satellite and communicate with it by sending and receiving signals. The satellite radar is more effective when there is a line-of-sight between radar dish and the satellite.

For such cases, the takeoff rotation can be compromised, even, no rotation is required in takeoff operation. So, the liftoff happens without rotation (the UAV will liftoff when the lift is greater than UAV weight while the wing has the cruising angle of attack).

One of the critical decisions in the design of the horizontal tail is the selection of the location of horizontal tail. The options are: (a) aft tail (or sometimes referred to as tail aft); and (b) foreplane or Canard (sometimes referred to as tail-first). Both aft tail and canard are capable of satisfactorily fulfilling tail design requirements. However, there are several aspects of flight features that are influenced differently by either of these two options.

The canard avoids deep stall 100%. A canard has a higher efficiency when compared with aft tail. An aircraft with a canard configuration tends to have a smaller static margin compared with an aircraft with a conventional aft tail configuration. Often times the canard generates a positive lift while a conventional tail often produces a negative lift. In general, the canard aerodynamic and stability analysis techniques are considerably more complicated than the technique to evaluate the aerodynamic feature and stability analysis of the conventional tail configuration aircraft. A canard is more efficient for fulfilling the longitudinal trim requirements, while a conventional tail tends to be more efficient for fulfilling the longitudinal control requirements.

In the Global Hawk, a V-tail was selected to mount the engine above the fuselage for better shielding the engine exhaust hot gas from below IR sensors, and also shield from the side without negatively affecting aerodynamic efficiency. A V-tail will result in a lower maneuverability, but this is not significant for this UAV, since the UAV is flying mostly straight and level in reconnaissance missions.

The pros and cons of the canard versus a conventional tail configuration are numerous and complex and it is hard to say which is superior without considering a specific design requirement. One must employ systems engineering technique to compromise and to decide the tail configuration. Ref. [8] has provided a full description for each tail configuration, and their pros and cons; the interested reader may refer to this reference for more details.

4.4.3 Horizontal Tail Design Technique

In general, tail is designed based on the trim requirements, but later revised based on stability and control requirements. In the early stage of horizontal tail, the design is performed without considering the elevator. The criterion is to design a horizontal tail to satisfy the cruising flight longitudinal trim requirements. The reason is that the aircraft spends the majority of the flight mission time in the cruising flight. There are various similarities between tail design process of a manned aircraft and unmanned aircraft. Chapter 6 of Ref. [8] presents the detail design of tail of an air vehicle. In this section, main differences are discussed.

After selecting tail configuration, the following tail parameters need to be determined during the design process: (a) horizontal tail horizontal location with respect to fuselage (aft tail or canard); (b) planform area (S_h); (c) tail arm (l_h); (d) airfoil section; (e) aspect ratio (AR_h); (f) taper ratio (λ_h); (g) tip chord (C_{h_tip}); (h) root chord (C_{h_root}); (i) Mean Aerodynamic Chord (MAC_h or C_h); (j) span (b_h); (k) sweep angle (Λ_h); (l) dihedral angle (Γ_h); (m) tail installation; and (n) incidence (i_h). The tail design procedure is recommended as provided below:

1) Select tail configuration
2) Select horizontal tail location (aft, or forward [canard])
3) Select the horizontal tail volume coefficient; $\overline{V}_H$
4) Calculate optimum tail moment arm (l_{opt}) to minimize the aircraft drag and weight
5) Calculate horizontal tail planform area; S_h.
6) Calculate wing-fuselage aerodynamic pitching moment coefficient
7) Calculate cruise lift coefficient (C_{Lc})
8) Calculate horizontal tail desired lift coefficient at cruise from trim equation
9) Select horizontal tail airfoil section
10) Select horizontal tail sweep angle and dihedral
11) Select horizontal tail aspect ratio and taper ratio
12) Determine horizontal tail lift curve slope; $C_{L_{\alpha_h}}$
13) Calculate horizontal tail angle of attack at cruise
14) Determine downwash angle at the tail
15) Calculate horizontal tail incidence angle; i_h
16) Calculate tail span, tail root chord, tail tip chord and tail mean aerodynamic chord

17) Calculate horizontal tail generated lift coefficient at cruise
18) If the horizontal tail generated lift coefficient (step 17) is not equal to the horizontal tail required lift coefficient (step 8), adjust tail incidence
19) Check horizontal tail stall
20) Calculate the horizontal tail contribution to the static longitudinal stability derivative ($C_{m\alpha}$)
21) Optimize the horizontal tail
22) Draw the final design with dimensions.

4.4.4 Tail Planform Area and Tail Arm

Two tail parameters of area (S_h) and arm (*l*) are interrelated; they are determined simultaneously. The UAV longitudinal trim requirement is satisfied when the following non-dimensional equation is set to zero:

$$C_{m_{owf}} + C_L\left(h - h_o\right) - \bar{V}_H C_{L_h} = 0 \tag{4.10}$$

This equation provides a critical tool in the design of the horizontal tail. An important non-dimensional parameter in the horizontal tail design is $\bar{V}_H$, and is referred to as the *Horizontal tail Volume Coefficient*. This parameter includes two important tail variables: (a) Tail planform area (S_h); and (b) tail arm (*l*).

$$\bar{V}_H = \frac{lS_h}{\bar{C}S} \tag{4.11}$$

A typical value for "$\bar{V}_H$" is about 0.4–1.2. The variable "h_o" denotes the non-dimensional wing-fuselage aerodynamic center $\left(\frac{X_{ac_{wf}}}{\bar{C}}\right)$ position. A typical value for "h_o" is about 0.25 for majority of aircraft configurations. The parameter "h" denotes the non-dimensional aircraft center of gravity (cg) position $\left(\frac{X_{cg}}{\bar{C}}\right)$. The value for "h" must be known prior to the horizontal tail design.

A typical value for "h" is about 0.1 to 0.3. The variable $C_{m_{owf}}$ in Eq. (4.10) is the wing-fuselage pitching moment coefficient. A typical value is about −0.06. The parameter C_L in Eq. (4.10) is the aircraft cruise lift coefficient that is determined by the lift equation (i.e., Eq. (4.1)).

The UAV static longitudinal stability is examined through the sign of the longitudinal stability derivative Cm_α or the location of the aircraft neutral point. For an UAV with a fixed aft tail, the aircraft static longitudinal stability derivative is determined [40] as:

$$C_{m_\alpha} = C_{L_{\alpha wf}}\left(h - h_o\right) - C_{L_{\alpha h}}\eta_h \frac{S_h}{S}\left(\frac{l}{\bar{C}} - h\right)\left(1 - \frac{d\varepsilon}{d\alpha}\right) \tag{4.12}$$

When the derivative Cm_α is negative (i.e., when the neutral point is behind the aircraft cg), the UAV is said to be statically longitudinally stable. In the design of horizontal tail, this equation should be examined too. If the UAV is found to be statically longitudinally unstable, the tail should be redesigned. A significant parameter in satisfying longitudinal trim (Eq. 4.10), and static longitudinal stability (Eq. 4.12) is the tail arm (*l*); which is the distance between tail aerodynamic center and the UAV cg.

The value of tail arm directly influences the tail area, the latter is responsible for generation of the tail lift. As the tail arm is increased, the tail area must be decreased, while as the tail arm is reduced, the tail area must be increased. Both short arm and long arm are capable of satisfying longitudinal trim requirements, given the appropriate necessary tail area. To find the optimum tail arm, one needs to derive a relation between tail geometry and other aircraft general design requirements such as the low weight and low drag.

4.4.5 Tail Airfoil Section

The fundamentals of airfoil selection process are provided in Section 4.3; when the wing airfoil selection was discussed. There are two distinct features for the tail that should be incorporated in the tail airfoil section selection. Basically, the tailplane airfoil lift curve slope ($C_{L_{\alpha_t}}$) must be as large as possible along with a considerably wide usable angles of attack. Since the aircraft center of gravity moves during the cruising flight, the airfoil section must be able to create sometimes a positive ($+L_h$) and sometimes a negative lift ($-L_h$). This requirement necessitates the tailplane to behave similarly in both positive and negative angles of attack. For this reason, a symmetric airfoil section is a suitable candidate for horizontal tail. Figure 4.7 shows the characteristics graphs of a symmetric

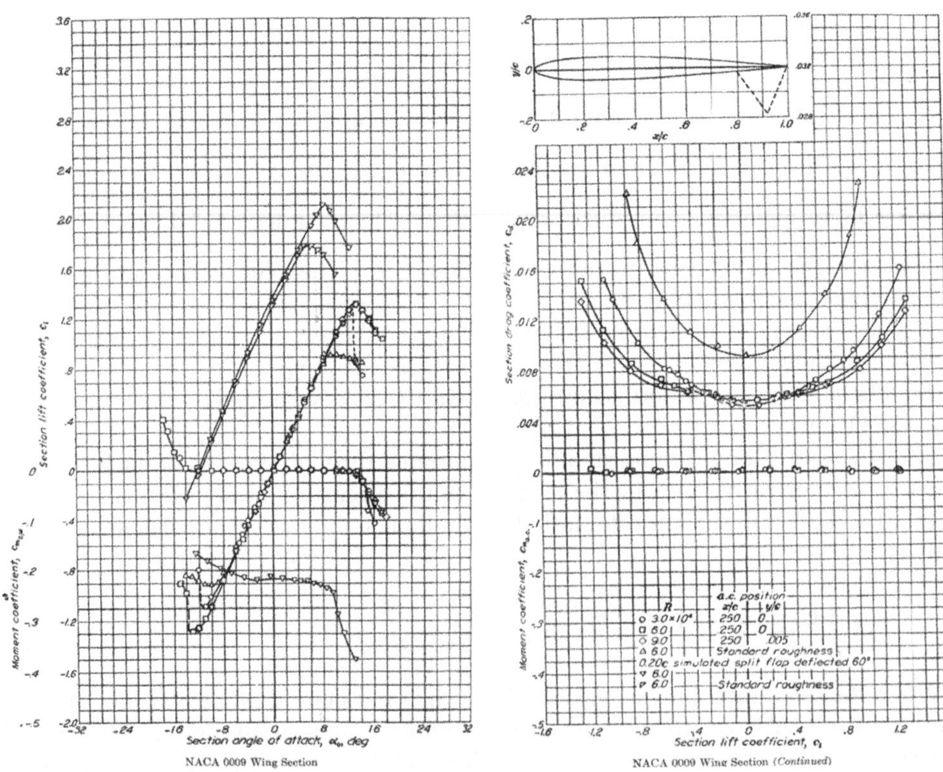

Figure 4.7 C_l-α, C_m-α, and C_d-C_l graphs of National Advisory Committee for Aeronautics NACA 0009 airfoil section [36].

airfoil; the NACA 0009 airfoil section. In a five-digit and six-series airfoils, a zero for the third digit indicates a symmetric airfoil.

In addition, another tail requirement is that the horizontal tail must be clean of compressibility effect. In order for the tail to be out of the compressibility effect, the tail lift coefficient is desired to be less than the wing lift coefficient. To insure this requirement, the flow Mach number at the tail must be less than the flow Mach number at the wing. This objective will be realized by selecting a horizontal tail airfoil section to be thinner (say about 2% of MAC) than the wing airfoil section.

4.4.6 Tail Incidence

When a fixed tail configuration is adopted, the horizontal tail setting angle (i.e., tail incidence); i_h; must be determined. The tail setting angle (i_h) primary requirement is to nullify the pitching moment about cg at the cruising flight. This is the longitudinal trim requirement through which the tail is generating a lift to counteract all other aircraft pitching moments. Tail incidence is determined to satisfy trim design requirement when no control surface (i.e., elevator) is deflected.

The desired tail lift coefficient is calculated through Eq. (4.8), and then the tail incidence will be determined by using the C_L-α graph of the tail airfoil section.

$$C_{L_{\alpha_h}} = \frac{C_{L_h}}{\alpha_h} \Rightarrow \alpha_h = \frac{C_{L_h}}{C_{L_{\alpha_h}}} \tag{4.13}$$

This is an initial value for the setting angle and will be revised in the later design phases. The typical value would be about −1°. Due to the presence of the downwash at the horizontal tail location, the tail effective angle of attack is defined as follows:

$$\alpha_h = \alpha_f + i_h - \varepsilon \tag{4.14}$$

where α_f is the fuselage angle of attack and ε is the downwash at the tail. The fuselage angle of attack is defined as the angle between the fuselage center line and the aircraft flight path (V_∞). The technique to determine downwash is presented by aerodynamics textbooks such as [14, 33].

4.4.7 Other Horizontal Tail Parameters

Other tail parameters such as aspect ratio, taper ratio, sweep angle, and dihedral angle are governed by similar fundamentals to that of the wing. The final values for these tail parameters are calculated when the design requirements such as longitudinal stability, cost, efficiency, and safety are examined.

4.5 Vertical Tail Design

4.5.1 Parameters

The third lifting surface in a conventional aircraft is the vertical tail; which is sometimes referred to as vertical stabilizer or fin. The vertical tail tends to have two primary functions: (a) directional stability; and (b) directional trim. Moreover, the vertical tail is a major contributor in maintaining directional control, which is the primary function of the rudder. In the design of the vertical tail, the following parameters must be determined: (a) planform area (S_v); (b) tail arm (l_v); (c) airfoil section; (d) aspect ratio (AR_v); (e) taper ratio (λ_v); (f) tip chord (C_{vt}); (g) root chord (C_{vr}); (h) Mean Aerodynamic Chord (MAC_v or C_v); (i) span (b_v); (j) sweep angle (Λ_v); (k) dihedral angle (Γ_v); (l) incidence (i_v); and (m) vertical tail location. Some of the vertical geometry parameters are shown in Figure 4.8.

4.5.2 Vertical Tail Location

In order to maintain the directional stability, the only location for the vertical tail is aft of the aircraft center of gravity. Three possible candidates are (a) aft of fuselage; (b) wingtips; and (c) boom(s). If a single aft horizontal tail has been selected, two places for the vertical tail is on the top/bottom of the rear fuselage. The vertical tail cannot be placed in front of the fuselage (i.e., forward of the aircraft cg), since it makes the aircraft directionally unstable. Other two options, namely wingtips and boom, are appropriate for some special purposes.

4.5.3 Vertical Tail Moment Arm (l_v)

In the early stage of the vertical tail design, where other aircraft components have not been designed, the vertical tail moment arm (l_v) is selected to be equal to the horizontal tail moment arm (l). This assumption means that the vertical tail is located at the same distance to the wing as the horizontal tail. The assumption will be modified in the later design stage when other aircraft components are designed and the aircraft directional and lateral stability, control and trim are analyzed. An experimental rule for the vertical tail effectiveness to achieve a recoverable spin is as follows: at least 50% of the vertical tail planform area must be out of the horizontal tail wake region to be effective in the case of a spin. Examining the directional stability, trim, and control will provide an accurate value.

4.5.4 Planform Area (S_v)

The vertical tail area must be large enough to satisfy directional stability, control, and trim requirements. Two most effective parameters in the directional stability are Cn_β

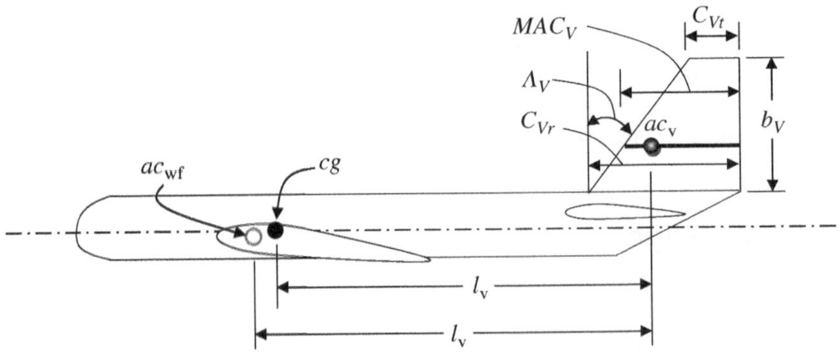

Figure 4.8 The vertical tail parameters.

and Cn_r. The major contributor to the static directional stability derivative (Cn_β) is the vertical tail [40]:

$$C_{n_\beta} \approx C_{n_{\beta_V}} = K_f C_{L_{\alpha V}} \left(1 - \frac{d\sigma}{d\beta}\right) \eta_V \frac{l_V S_V}{bS} \tag{4.15}$$

where $C_{L_{\alpha V}}$ denotes the vertical tail lift curve slope, $\frac{d\sigma}{d\beta}$ is the vertical tail sidewash gradient, and η_V is the dynamic pressure ratio at vertical tail. The parameter K_f represents the contribution of fuselage to aircraft Cn_β and depends strongly on the shape of the fuselage and its projected side area. The fuselage contribution to directional static stability tends to be strongly negative. The typical value of K_f for a conventional aircraft is about 0.65–0.85. The value of C_{n_β} for a statically directionally stable aircraft is positive. A higher value for Cn_β implies a more directionally statically stable aircraft. A typical value for Cn_β is about 0.2–0.4 1/rad. The parameter S_v can be calculated by employing Eq. (4.15).

An important parameter in Eq. (4.15) is a variable referred to as the vertical tail volume coefficient ($\bar{V}_V$):

$$\bar{V}_V = \frac{l_V S_V}{bS} \tag{4.16}$$

A typical value for $\bar{V}_V$ is 0.05–0.1. The parameter S_v can be calculated by employing Eq. (4.16) too. It is interesting to note that a typical value for the ratio between vertical tail area and the wing area for a conventional UAV is about 0.1–0.15. Another criterion for the vertical tail area is that it must be small enough such as to minimize the manufacturing cost and the aircraft weight. The design of the vertical tail is one of the difficult tasks for aircraft designers, since theoretical and experimental results may not match concerning the features of the vertical tail. It is often the case for several aircraft that the vertical tail area is found, in flight test, insufficient to satisfy lateral-directional stability requirements.

If the aircraft is in the manufacturing stage, and the initial vertical tail design may not be changed, one solution to increase the vertical tail area is to employ a dorsal fin. A dorsal fin is generally a flat plate (i.e., no airfoil section) installed in front of the original vertical tail with a greater sweep angle. The other benefit of a dorsal fin is to reduce the minimum control speed (V_{mc}) during take-off operation. In addition, it provides a hidden antenna feature that allows the com antennas to be located under the fin for further drag reduction. Another approach to solve the small vertical tail area problem is to employ the ventral fin.

4.5.5 Incidence (i_v)

The vertical tail incidence is defined as the angle between the vertical tail chord line and the aircraft x-z plane (when look at the aircraft from top view). The vertical tail is responsible for the generation of the vertical tail lift coefficient (C_{LV}). One of the basic aircraft design objectives is the symmetricity about the x-z plane. Hence, if the engines, wing, horizontal tail, and fuselage are designed to be symmetric about x-z plane, the vertical tail is not required to produce any lift to maintain directional trim in a normal flight condition. For this reason, the vertical tail incidence must be initially zero.

However, in a prop-driven aircraft with one single engine (or with odd number of prop-driven engines), the lateral trim is disturbed by the revolution of the propeller and engine shaft about x-axis. To nullify this rolling moment, the vertical tail is required to generate a lift and a moment. The vertical tail in most single engine prop-driven aircraft have about 1–2° of incidence to insure the prevention of aircraft roll in a reaction to propeller revolution.

4.5.6 Other Vertical Tail Parameters

Other vertical tail parameters such as airfoil section, aspect ratio, taper ratio, sweep angle, and dihedral angle are governed by similar fundamentals as those of the horizontal tail. The final values for these vertical tail parameters are determined when the design requirements such as directional stability, efficiency, cost, and safety are examined.

4.5.7 Vertical Tail Design Technique

The vertical tail design requirements are as follows: (a) directional trim; (b) directional control; (c) directional stability; (d) manufacturability; (e) handling qualities; (f) stealth (only in some specific military UAVs); (g) operational requirements; (h) airworthiness; (i) cost; (j) competitiveness; and (k) height limit. The vertical tail design procedure for a UAV with a conventional configuration is recommended as provided below:

1) Select vertical tail configuration (e.g., conventional, twin vertical tail, vertical tail at swept wing tip, V-tail).
2) Select the vertical tail volume coefficient; $\bar{V}_V$.
3) Determine/select the vertical tail moment arm (l_v).
4) Calculate vertical tail planform area; S_v.
5) Select vertical tail airfoil section.
6) Select vertical tail aspect ratio; AR_v.
7) Select vertical tail taper ratio; λ_V.
8) Determine the vertical tail incidence angle.
9) Determine the vertical tail sweep angle.
10) Determine the vertical tail dihedral angle.
11) Calculate vertical tail span (b_v), root chord (Cv_{root}), and tip chord (Cv_{tip}), and mean aerodynamic chord (MAC_v).
12) Check the spin recovery.
13) Adjust the location of the vertical tail relative to the horizontal tail by changing l_v, to satisfy the spin recovery requirements.
14) Analyze directional trim.
15) Analyze directional stability.
16) Modify to meet the design requirements.
17) Optimize the vertical tail.
18) Draw the final design with dimensions.

The vertical tail design is an iterative process. When the other aircraft components (such as fuselage and wing) are designed, the aircraft dynamic longitudinal-directional stability needs to be analyzed, and based on the outcome; the tail design may need some adjustments.

4.6 Fuselage Design

The third component that considerably impacts UAV aerodynamics is the fuselage. The primary function of the fuselage is to accommodate/house the payload, and autopilot. This section is devoted to the aerodynamic design of a fuselage.

4.6.1 Fuselage Design Fundamentals

Since an UAV has no human onboard, there will be no cockpit and cabin; hence, the human factor is not considered in the fuselage design. Moreover, a pressurized cabin/place is often not required. However, some avionic and payload may not operate efficiently at low pressure, thus, sometimes a pressurized compartment is required. For instance, the Global Hawk fuselage is made out of aluminum, and contains pressurized payload and avionics compartments to house the heat-generating avionics. Two cooling techniques of convection and conduction require a pressurized compartment. The radiation heat transfer is operational in vacuum (or low pressure), but will not be effective in cooling of a high temperature sensor.

By definition, payload is the useful load that the aircraft is intended to carry. For a non-military UAV, the payload basically includes such items as autopilot, measurement devices, sensors, and communication equipment. For the case of a military UAV, the payload will have such items as rocket, missile, and bomb. The fuel is not considered as a payload.

In addition, in order to reduce aircraft drag, a few other major components and systems such as landing gear, engine, fuel system, and power transmission system are highly likely to be enclosed by the fuselage.

There are a few expectations that are recommended to be considered during the fuselage design process. The expectations include low weight, low drag, positive contribution to the lift generation, external symmetricity, and safety against environmental hazards such as lightning. The fuselage drag usually contributes 30–50% of the aircraft zero-lift drag (C_{Do}). Furthermore, the fuselage may be aerodynamically designed such that it provides as much as 20% of the total lift. In is interesting to note that, in a basic configuration, only as high as about 5% of the aircraft lift is produced by the fuselage.

Two major fuselage parameters that must be determined during design process are: (a) fuselage length (L_f); and (b) maximum diameter (D_f). The fuselage configuration as well as these two parameters are functions of several design requirements.

4.6.2 Fuselage Aerodynamics

The aerodynamic design of a fuselage is a function of its configuration and internal arrangement. After identification of the payload and design requirements, the first fuselage design step is to decide about the fuselage configuration and determine the fuselage internal arrangement. This decision is very important and will influence all fuselage parameters. Fuselage configuration design is a conceptual design but at the fuselage level, and does not involve detail calculations. Indeed, the configuration design of the fuselage requires several skills and long experience. The fuselage configuration is a function of internal arrangement. In order to specify the location for each internal item, one must first identify and decide what item/component is supposed to be accommodated.

A conventional UAV fuselage may constitute by the following sections: autopilot station, load compartment, nose section, rear section, fuel tanks, communication system, radar, internal systems (i.e., electrical, mechanical, and hydraulic), and engine(s). Each section needs to be designed separately, since each has a unique design requirement. However, at this stage of design, the locations of these sections relative to each other need to be determined.

The volume and the external shape of the fuselage are functions of what is desired to be stored inside. In general, there are seven basic rules for internal arrangement and to locate the allocated items inside the fuselage:

1) Keep the fuselage as small and compact as possible.
2) Arrangement to be symmetric from top-view as much as possible.
3) There must be sufficient space to accommodate all equipment/devices.
4) The useable loads such as fuel must be close to the aircraft center of gravity.
5) The satellite radar dish must be allocated in an upper/lower place such that to have a line of sight with the communication satellite and ground station, and to enable the communication system to communicate without any disruption.
6) Arrangement must be such that the aircraft center of gravity is close to the wing-fuselage aerodynamic center.
7) If necessary, any piece of equipment could be placed/installed out of the fuselage. The best example is the camera (for photography and taking video) which is best to be underneath the fuselage nose to have 360° access to the target area.

Figure 4.9 illustrates the side-view of fuselages for two typical unmanned aircraft with their internal arrangements; a low-speed long endurance civil UAV, and a high-speed

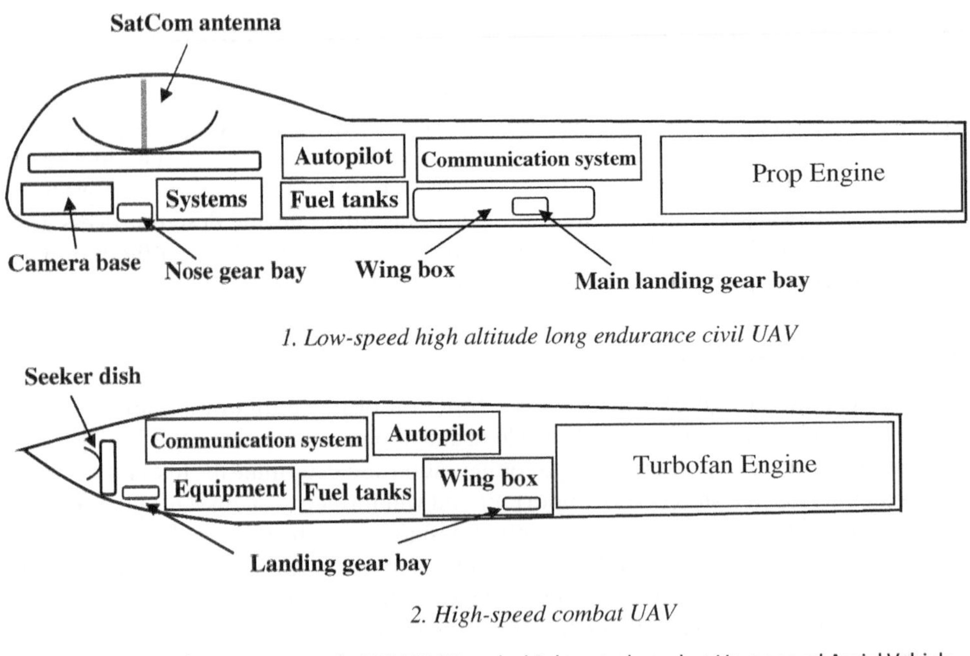

Figure 4.9 Internal arrangement of a HALE UAV, and a high-speed combat Unmanned Aerial Vehicle (UAV) (side-view).

combat UAV. The volume and the external shape of the fuselage are functions of what is desired to be stored inside. Three important components that significantly contribute to the shape of the fuselage are: (a) largest component; (b) engine (if is housed by fuselage); and (c) radar dish. The largest component (e.g., engine) naturally determines that maximum diameter of the fuselage. For instance, the satellite communications antenna of Global Hawk has a diameter of 48 in. The radar dish influences the curvature of the adjacent portion of the fuselage for perfect communication efficiency. In some cases, a configuration may look desirable but would not be feasible. Whenever all fuselage sections are designed, the geometry and dimensions of each section will be finalized.

The aerodynamic characteristics of the aircraft have a fundamental relationship with the cross-section of the fuselage and the shape of the nose. The final solution will be the output of a trade-off study that yields the optimum design.

- **Fuel Tanks**

An aircraft must carry its own fuel by internal fuel tanks. In the aircraft conceptual design phase, the best space for storing fuel is determined. Two most conventional locations for fuel tanks are wing, and fuselage. The total fuel may be divided into several smaller fuel tanks. In case the fuel tanks are considered to be accommodated by the fuselage, the required fuel volume should be calculated and then, sections of the fuselage volume must be allocated for fuel tanks. The required fuel volume of the fuel tanks is obtained by dividing fuel mass (m_f) by the fuel density. The fuel mass is a function of aircraft mission including range and endurance. The fuel mass to fly a specified range/endurance is given by Breguet range equations [9].

- **Wing Box**

UAV structural designers prefer to have the wing main spar to carry through the fuselage to maintain the aircraft structural integrity. This is due to the fact that the wing lift force generates a large bending moment where the wing attaches to the fuselage. The technique by which this moment is carried across the fuselage is a key structural consideration. The wing carry through structure must be designed to minimize the bending stress.

For a fuselage designer, this requirement is satisfied by allocating a section of the fuselage volume to the wing spar (i.e., wing box). The wing box is a structural component and is a section of the fuselage between the wing roots. This section is the strongest structural area of the aircraft, and carries the largest bending moment and stresses.

The center fuselage of the X-45A (Figure 4.10) is blended into a swept lambda wing, with a small exhaust outlet. The weapons bay door is under the fuselage, which also houses almost all communication equipment, autopilot, and a turbofan engine.

Figure 4.10 X-45A UCAV. By NASA/Dryden Flight Research Center/Jim Ross.

4.6.3 Autopilot Compartment

The detail design of the autopilot is presented in Chapters 5 through 9; this section is devoted to the design of the autopilot compartment. The major elements of an autopilot (e.g., navigation, control, and guidance), measurement devices, and avionics contribute to the geometry of an autopilot. In order to control the aircraft and flight, the autopilot must deflect control surfaces and move the engine throttle. In a conventional UAV, three basic control surfaces are elevator, aileron, and rudder. Moreover, an autopilot needs to measure and observe the flight parameters to ensure a successful flight. Flight variables measurements are carried out through various avionic instruments.

The following basic flight and navigation instruments must be installed and included in the autopilot compartment: free air temperature indicator, heading/direction indicator, airspeed indicator, altimeter, rate-of-climb indicator (vertical speed), gyroscopic rate-of-turn indicator, bank and pitch indicator, GPS, accelerometer, potentiometer, glide slope indicator, transponder, magnetometer, engine instruments: (rpm, fuel, exhaust gas temperature, and turbine inlet temperature), compass, weather radar, and radio. Some measurement devices such as pitot tube are located out of the compartment due to the need to be in contact with free air. However, their outputs will end up in the autopilot compartment via various kinds of connections such as wire/pipe/hose. There are several other devices such as interface, analog-to-digital convertor, digital-to-analog convertor, and cooling system by which the autopilot may operate normally. The complete list is provided in Chapters 5 through 9.

The detail design of the autopilot involves the integration of components, assemblies, and instruments that are under the control of various functional disciplines. For a small remotely controlled aircraft, the autopilot could be as small as a chip with the size of 30 mm × 50 mm × 30 mm. However, for a large UAV with a high performance, the automatic control function and processing operation increases. This requirement results in an increase in the size, type and number of sensors, and the integration of an onboard computer to process the increased demand. For such UAV with a full autonomy, the size of the autopilot could be as big as 20 cm × 40 cm × 30 cm.

4.6.4 Optimum Length-to-Diameter Ratio

Two of the main fuselage design outer parameters are the fuselage length (L_f) and the maximum diameter (D_f). These two fuselage parameters are producing the fuselage volume, wetted area, and weight. The fuselage optimum length-to-diameter ratio (or slenderness ratio) may be determined based on a number of design requirements. The design objectives for determining the fuselage length-to-diameter ratio are:

1) results in the lowest zero-lift drag
2) creates in the lowest wetted area
3) delivers the lightest fuselage
4) provides the maximum internal volume
5) generates the lowest mass moment of inertia
6) contributes the most in the aircraft stability
7) requires the lowest cost to fabricate.

The first and second requirements concern the aircraft performance requirements. The third objective aims for the weight requirements, and the fourth one satisfies the operational requirements. The goal of the fifth one is the controllability requirements; and the sixth one satisfies the stability requirements. Finally, the last objective targets the lowest cost of manufacturing the fuselage. Depending upon the aircraft mission and the design priorities, one of these objectives becomes the most significant. For instance, a short fuselage with a low weight and a high drag is more desired for minimizing weight, but a long fuselage with a high weight and a low drag is more desirable for performance optimization. This is a fundamental question of cost versus performance.

For the objectives two to six, the designer is expected to develop a formulation to mathematically express the requirement in terms of fuselage length and diameter. Then, differentiate the formula with respect to fuselage length or diameter. When the result of the differentiation is set equal to zero, the final solution yields the optimum fuselage length and diameter. In order to optimize all-at-once, the multidisciplinary design optimization (MDO) techniques [41] can be utilized to reach a global optimum.

Table 4.2 demonstrates the fuselage geometry of a number of UAVs including diameter and length. For a cylindrical fuselage, the optimum slenderness ratio (L/D) for the lightest fuselage [8] is one. The same is true for the lowest wetted area. In both cases, the fuselage will create a large frontal area and very high drag. Due to significance of aircraft performance, a fuselage length-to-diameter ratio of 1 is not recommended. For most low-speed UAVs, a fuselage slenderness ratio between 5 and 10 is recommended.

4.6.5 Fuselage Aerodynamics

In the design of fuselage external shape, there are several objectives that drive the design; including aerodynamic considerations, radar detectability (or stealth) and operational requirements. Aerodynamic considerations for fuselage external shape include goals such as low drag, low pitching moment, zero rolling moment, low yawing moment, and sometimes generating lift as much as possible. The zero rolling moment requirement necessitates a symmetric fuselage about xz plane. Hence, form top-view, the fuselage must be symmetric.

In order to create a low drag, the fuselage designer needs to select a section which resembles an airfoil shape. From side-view, the streamlining must be considered.

Table 4.2 Fuselage geometry of a number of Unmanned Aerial Vehicles (UAVs).

No	UAV	Type	Engine	TO mass (kg)	b (m)	L (m)	D_{max} (m)
1	Northrop Grumman RQ-4 Global Hawk	Surveillance	Turbofan	14628	39.9	14.5	2.32
2	General Atomics MQ-1 Predator	Surveillance	Piston	1020	14.8	8.22	1.13
3	Reaper	Surveillance	Turboprop	4760	20	11	1.5
4	Lockheed Martin RQ-170 Sentinel	Stealth reconnaissance	Turbofan	–	19.99	4.5	–
5	NASA Perseus B	Testbed	Piston	816	21.8	7.6	0.75
6	AeroVironment RQ-11 Raven	Aerial filming	Electric	1.9	1.37	0.915	0.11

For this end, one may add a semi-circle to the nose and apply a cone (or semi-cone) to the rear fuselage. In order to avoid any yawing moment by the fuselage, the top view of the fuselage is recommended to follow a symmetrical airfoil.

The ultimate solution for lift generation is to have a blended wing-fuselage (e.g., X-45 and X-47). The form is composed of distinct wing and body structures, though the wing is smoothly blended into the fuselage. This enables the entire craft to generate lift, and reduces the weight, wetted area, and drag of the UAV. Reference [42] discusses the design techniques for the blended wing body for subsonic transport aircraft. At supersonic speeds, the oblique shock waves at the fuselage edges are contributing in generating a great deal of lift.

From the structural standpoint, the most efficient (strongest) shape for a slender pressurized fuselage is a cylinder (circular cross section) with hemispherical end caps. Hence, the external shape of a fuselage – with symmetry with respect to XZ plane – can be represented by the following [43] equation:

$$Y = Z^{0.5}(1-Z)^{0.5} \quad Z\varepsilon[0,1] \tag{4.17}$$

Reference [43] provides a code for formulation of fuselage cross-section geometry optimization problem, in which the objective function is based on geometrical calculations.

When an existing UAV is required to accommodate a new larger payload, the fuselage must be redesigned and elongated. The Avenger ER utilizes a modified fuselage design (completed in 2012) extended by 4 ft (to 13.4 m [44 ft]) to accommodate larger payloads and more fuel.

4.6.6 Lofting

After all internal sections of the fuselage are designed (e.g., their geometry is determined) and allocated, it is then time to design the fuselage external shape, or lofting. Lofting is the process to determine the external geometry of a vehicle; here, the UAV fuselage. Lofting is performed on the fuselage to improve the overall aerodynamic performance of the UAV.

The objective is to minimize fuselage drag, while producing a fair amount of lift by fuselage. During lofting process, it is desirable to design a cross-sectional area along the length of the fuselage to minimize any sharp edges. The main location of a sharp change is at the wing-fuselage, tail-fuselage and pylon-fuselage junctions. Then, the fuselage is fabricated to be smoothly lofted for an optimal design.

Another goal in lofting is to provide the best environment for communication system (e.g., satellite communications antenna) for sending/receiving signals. For instance, the distinctive appearance (bulge) at the top front surface of the Global Hawk fuselage houses the 48 in. wideband satellite communications antenna.

The fuselage cross section also contributes to aircraft control/stability (e.g., lateral stability and spin recovery characteristics). Damping provided by various parts of aircraft such as fuselage can counter the yawing moment of the wings during a spin. So, provision of a large amount of damping in yaw for fuselage is an effective means for the prevention of a spin. The aerodynamic yawing moment due to rotation of the fuselage about spin axes is largely dependent on the fuselage shape and its cross section. Therefore, the aircraft designer can reduce the spin recovery load on the rudder by a careful design of the fuselage and a proper aircraft weight distribution. Moreover, the side-area of fuselage will considerably contribute to the UAV lateral stability.

To avoid large regions of boundary layer separation and the associated drag increase in the fuselage, the length of the rear fuselage is often two to three times the diameter of the cylindrical section. Where the wing is connected to the fuselage, a filleting is required to avoid flow separation and turbulence. The exact shape may be determined by wind tunnel experiments.

In order to have an airfoil shape, the rear fuselage should be tapered to a zero diameter. If an engine is podded by aft fuselage, the fuselage diameter should be reduced from the diameter at mid-section to the engine diameter. If the engine is not accommodated by the rear fuselage, the diameter of the fuselage must be reduced from maximum diameter to almost zero. Caution must be taken not to taper the rear fuselage at too large an angle, otherwise flow separation will occur. For a subsonic aircraft, the taper angle should be no greater than about 20°. For ease of fabrication, part of the rear fuselage may be conical. The transition from the cylinder to the cone should be smooth with sufficiently large radius of curvature. If the rear fuselage is spacious and the aircraft cg limits allow, portion of the fuel or systems may be stored in it.

Another requirement which affects the rear fuselage design is the takeoff and landing rotation clearance requirement. During the rotation of takeoff and landing, the rear fuselage must be such that the there is no tail strike with the ground. The primary solution to avoid the striking incident is often through an increase in the landing gear height. Another common solution to this problem is to cut the rear fuselage by an upsweep angle (or taper angle).

4.6.7 Fuselage Design Steps

In general, the following are the fuselage design requirements: (a) accommodation requirement; (b) operational and mission requirements; (c) airworthiness requirement; (d) crashworthiness requirements; (e) aerodynamic requirement; (f) aircraft stability requirement; (g) low weight; (h) low wetted area and low side area; (i) symmetricity; (j) structural integrity and strength; (k) maintainability; (l) manufacturability; (m) cost; (n) long life; (o) radar detectability; and (p) accessibility. Unlike a manned aircraft, the fuselage of an unmanned aircraft is free of seats, stick/yoke, pedal, windshield, lavatories, washroom, drink, meal, luggage, air conditioning, and pressurization. These differences make the design of fuselage for an unmanned aircraft considerably less challenging.

The fuselage design is initiated by identifying fuselage design requirements and the process is ended by the optimization. As in the case of other aircraft components, there is no unique design solution to satisfy the fuselage requirements. Each acceptable solution will deliver a pool of advantages and disadvantages which must be decided based on the systems engineering technique.

It must be emphasized that there is no unique solution to satisfy the customer and airworthiness requirements in designing a fuselage. Several fuselage designs may satisfy the requirements, but each will have unique advantages and disadvantages. The following are the recommended fuselage design steps for a UAV:

1) Identify and list the fuselage design requirements.
2) Identify the payloads which are desired to be accommodated by fuselage.
3) Identify the communications system elements, and their geometries and operation requirements.
4) Identify the largest payload and component housed by the fuselage, and obtain its dimensions.
5) Determine the dimension of the autopilot compartment.
6) Determine the size of pressurized compartment (if any).
7) Select the fuselage configuration and internal arrangement.
8) Design the fuselage lay out (internal); side-view, front-view, and top-view.
9) Determine required flight measurement instruments.
10) Establish fuselage optimum length-to-diameter ratio.
11) Design communications system compartment.
12) Design payload bay.
13) Determine the required volume for other components (e.g., fuel, batteries, landing gear).
14) Check if the available fuselage space for other components is sufficient.
15) Calculate the fuselage maximum diameter.
16) Design fuselage nose section.
17) Design fuselage rear section.
18) Determine upsweep angle.
19) Calculate fuselage overall length.
20) Apply lofting.
21) Check if the fuselage design satisfies the design requirements.
22) If any design requirement is not met, return to the relevant design step and recalculate the corresponding parameter.
23) Optimize the fuselage.
24) Draw the final design with dimensions.

The author has presented the fuselage aerodynamic design for an air vehicle in [8], so it is not repeated in this text. You are encouraged to study [8] to find the details of fuselage aerodynamic design.

4.7 Antenna

Aircraft antennas perform a broad range of functions on modern aircraft, from transmitting data to helping locate the aircraft via the GPS network. There are two categories of antennas: (a) fixed; (b) mechanically operated. In general, the fixed antenna has a smaller size compared with moving antennas. Technical features of various types of antennas are presented in Section 13.5.

4.7.1 Fixed Antenna

The fixed antenna such as VHF (Very High Frequency) radio antenna is a mono-pole type and recommended to be covered by a plate to look like a vertical tail. This initiative will improve the directional stability of the UAV. The aerodynamic coverage of fixed antennas will reduce the UAV drag, so the flight performance will be enhanced.

The Global Hawk has more than five antennas protruding from its fuselage. To keep the UAV symmetry (both aerodynamically and center of gravity point of view), try to install all antenna at the x-z plane.

4.7.2 Radar Dish Antenna

Another piece of equipment that a UAV is equipped with is a communication device such as radar. The communication radar is an object-detection system which uses electromagnetic waves to determine the range, altitude, direction, or speed of both moving and fixed objects. Some UAVs may employ a weather radar to locate precipitation, calculate its motion, estimate its type (rain, snow, hail, etc.), and forecast its future position and intensity. Modern weather radars are mostly doppler radars, capable of detecting the motion of rain droplets in addition to intensity of the precipitation.

The radar dish, or antenna, transmits pulses of radio waves or microwaves which bounce off any object in their path. The object returns a tiny part of the wave's energy to the dish. In modern aircraft, various types of radar are employed including weather radar. The fuselage's nose of most modern large UAV contains a radar dish that communicates weather systems to the autopilot and ground station.

There are several locations to place the radar antenna or dish. The location for antenna must have open view to transmit and receive radar signals. One of the best locations is the fuselage nose (e.g., Global Hawk and Predator). For the best line of sight, the vertical tail is recommended to be placed such that it will not obscure the sight of the radar.

In is recommended to design fuselage section of the UAV in which houses radar scanners and SatCom dish antenna, such that it allows the best sight for the radar/antenna. In either a down-looking radar, or an up-looking satellite communications, or a forward-looking seeker, the radome (Figure 4.11) should allow radio waves to pass through skin without any distortion and degradation.

A radome is a structural, weatherproof enclosure that protects a radar antenna and is transparent to radar waves. Radome protects the antenna surfaces from the environment (e.g., wind, rain, ice, and sand), and conceals antenna electronic equipment from public view. Thus, in the fuselage design, the nose may be considered for the radar dish.

4.7.3 Satellite Communication Antenna

The materials provided for the installation of radar dish are all true for the satellite communication antenna (Figure 4.11b). Thus, they are not repeated here. During entire flight operation, the satellite communication radar should have a line-of-sight with a satellite, so the radome must have a curvature to provide the maximum signal efficiency.

In is recommended to design the redome which covers SatCom dish antenna, such that it allows the best sight for the antenna. The radome should allow communications signal to pass through the skin without any distortion and degradation. The satellite communication dish of Northrop Grumman Firebird which is about 30 in. is located at the center of fuselage. Figure 13.6 exhibits the SatCom dish antenna of a NASA Altair UAV at the middle of the fuselage.

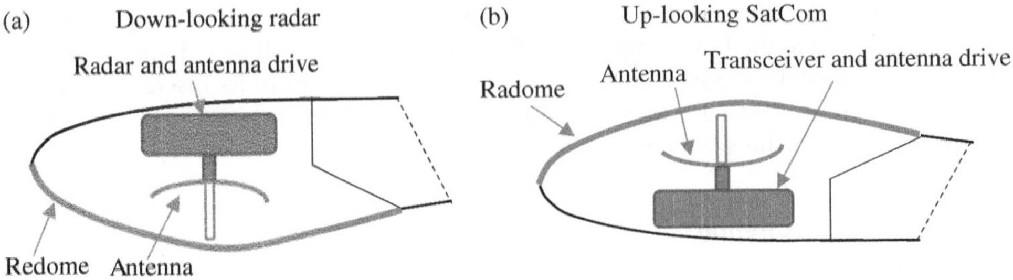

Figure 4.11 Typical low drag, low distortion radome.

4.7.4 Antenna Design/Installation

A few hints on antenna design/installation are worth mentioning. The antenna may be enclosed inside the receiver's case; attached to the outside of the receiver, or mounted separately and connected to the receiver by a cable, as with satellite dishes.

Unless an omnidirectional antenna is employed in the UAV, it is necessary to mount the antenna in a rotatable turret in order for LOS to be maintained between GCS/Satellite and UAV during various maneuvers. The mono-pole antenna should not be installed near an electronic circuit to avoid interference. If it is outside of the UAV, it will be more effective, but it will create more drag. A mono-pole antenna should be installed along the fuselage centerline (i.e., x-axis) to reduce the drag. The dish antenna on a UAV should be covered to reduce the drag. Figure 13.6 illustrates a NASA Altair UAV satellite antenna (inside its fuselage).

4.8 Aerodynamic Design of Quadcopters

In the past several years, the popularity of quadcopter UAVs has skyrocketed. Section 2.10 presents the quadcopter configuration including technical features, and advantages and disadvantages of the quadcopter configuration. Moreover, Section 5.7 provides the quadcopter dynamic model including electric engine's thrust formulation. This section is dedicated to the aerodynamic design of quadcopters.

A quadcopter does not have a fixed wing and a fixed tail, hence conventional techniques for the wing/tail design do not apply here. However, the general requirements of low vehicle drag, with a high aerodynamic efficiency are still applicable.

In the design of the body of a quadcopter, the cost requirements as well as the minimum drag must be simultaneously considered. A well-designed outer shape can reduce the drag of a body as much as 50%. Moreover, a well-designed body will reduce the flow interferences with propellers. If the battery pack has been designated to be placed inside the main body, the shape must allow a sufficient space for this purpose.

For a quadcopter (i.e., rotary-wing configuration), Eq. (4.3) is not employed; since parameters K and C_L are primarily are defined for a wing. Thus, we regard the total drag coefficient (C_D) as a parameter. For this coefficient, the reference area (S in Eq. (4.2)) is often the projected frontal (i.e., cross-sectional) area of the entire UAV. For a quadcopter, the typical value for C_D (Eq. 4.2) is about 0.2–0.4. A number of symmetrical shapes having drag values at low speed are illustrated in Table 12.9. The drag coefficient values in this table are based on the frontal area.

Keep the aerodynamic drag of the UAV as low as possible commensurate with the practical installation and operation of the payload, and radio antennae. Where the payload is connected to the fuselage, some forms of filleting (e.g., fillet) are required to avoid airflow separation and turbulence. Design techniques (e.g., lofting) should be employed to reduce body drag. The exact shape may be determined by wind tunnel experiments.

The study of Ref. [44] is recommended as a good reference for aerodynamic design of quadcopters.

It provides a modern treatment of the aerodynamic principles of helicopters and other rotating-wing vertical lift aircraft, such as tilt rotors and autogiros, as well as advanced methods of helicopter aerodynamic analysis.

4.9 Aerodynamic Design Guidelines

The primary objective of aerodynamic design is to optimize the UAV outer shape such that the lift is maximized, while the drag and pitching moment are minimized. Figure 4.12 provides the UAV aerodynamic design flowchart. Keep the aerodynamic drag of the UAV as low as possible commensurate with the practical installation and operation of the payload, and radio antennae.

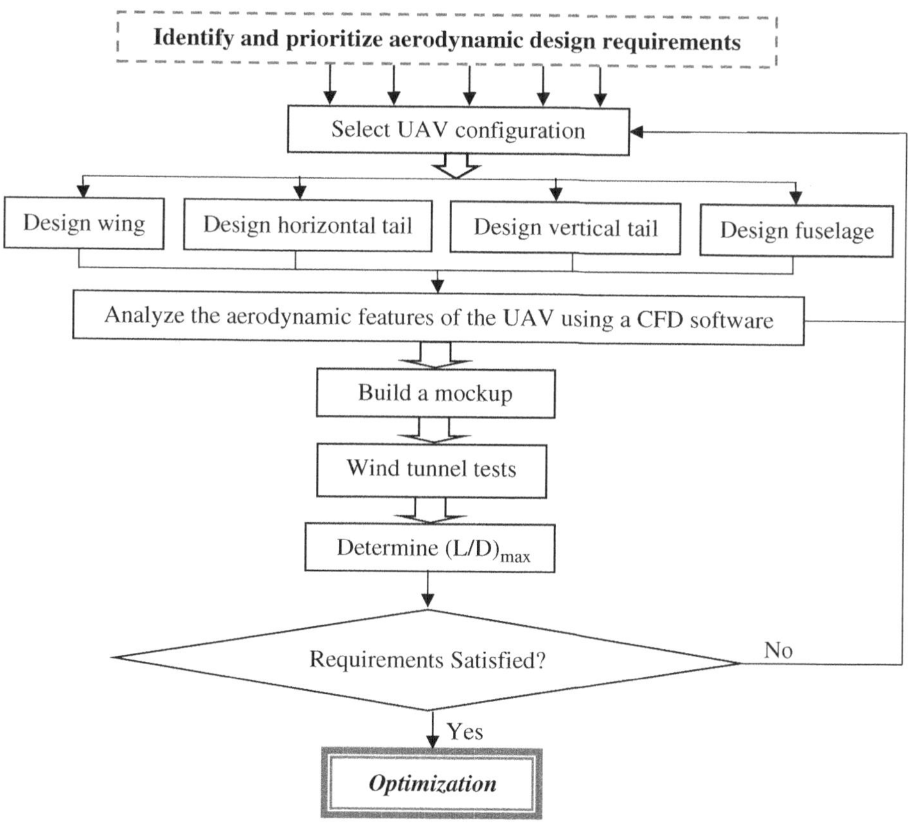

Figure 4.12 Unmanned Aerial Vehicle (UAV) aerodynamic design flowchart.

In the aerodynamic design of a fixed-wing UAV and quadcopter, the cost requirements as well as the minimum drag must be simultaneously considered. A well-designed outer shape can reduce the drag of a UAV as much as 50%.

The first step is to identify and list the aerodynamic design requirements. Next, the designer has to design each component (wing, fuselage, tail, body) individually. When the specifications of each component are determined, a systems engineering approach is needed to evaluate the aerodynamic efficiency of the entire vehicle as one body. Using CFD software packages such as Fluent and SimScale, one can fully analyze the aerodynamic efficiency of the vehicle.

When theory confirm the general aerodynamic features of the vehicle, the mockup and model of the UAV can be built for wind tunnel tests. During various wind tunnel tests, the real values for UAV lift coefficient, UAV drag coefficient, $(L/D)_{max}$, and aerodynamic forces will be calculated (indeed measured). The last step is to check if the design requirements are met. If not, the design corrections will be applied. This process is continued until the aerodynamic design requirements are satisfied.

Questions

1 What is the primary function of a wing?

2 What is the primary function of a horizontal tail?

3 What is the primary function of a vertical tail?

4 What is the primary function of a fuselage?

5 What are two primary aerodynamic forces?

6 Where is the typical location of the aerodynamic center (ac) with respect to its chord?

7 Name three reliable airfoil resources.

8 What parameters must be determined, in the detail design of a wing? Provide name of at least 10 parameters.

9 What parameters must be determined, in the detail design of a horizontal tail? Provide name of at least 10 parameters.

10 What parameters must be determined, in the detail design of a vertical tail? Provide name of at least 10 parameters.

11 What parameters must be determined, in the detail design of a fuselage? Provide name of at least 5 parameters.

12 What airfoil section originally selected for the wing of AGM-154 JSOW?

13 What is the horizontal tail volume coefficient, and its typical values?

14 List criteria to select an airfoil for a wing. Provide at least five items.

15 List criteria to select an airfoil for a horizontal tail. Provide at least two items.

16 What is the typical value for C_D of a quadcopter?

17 What is redome?

18 Name at least two CFD software package.

19 Describe the aerodynamic design process of a wing.

20 Describe the aerodynamic design process of a UAV.

21 Describe the aerodynamic design process of a horizontal tail.

22 Describe the aerodynamic design process of a vertical tail.

23 Describe the aerodynamic design process of a vertical fuselage.

24 What is drag polar?

25 What is the typical value for C_D for a quadcopter?

26 What theories are often used in establishing CFD Software packages?

27 What are two ways to determine the wing airfoil section?

28 What does NACA stand for?

29 What does NASA stand for?

30 What is thickness-to-chord ratio of the NACA 63_3–218 airfoil section?

31 Describe features of 6-series NACA airfoils.

32 Name a UAV with a low wing.

33 Define wing incidence.

34 Define wing aspect ratio.

35 Define tail taper ratio.

36 Define wing sweep angle.

37 Define wing twist angle.

38 Define wing geometric twist.

39 Define wing aerodynamic twist.

40 Define wing dihedral angle.

41 Draw top view of a wing, and show the wing aerodynamic center.

42 What are the effects of deflecting a HLD in a wing?

43 Name at least five HLDs.

44 Name at least five aft tail configurations.

45 What are the advantage of canard as compared with an aft tail?

46 What are two important requirements is selecting tail airfoil?

47 Discuss the basic rules for internal arrangement and to locate the allocated items inside the fuselage.

48 Draw side-view of a fuselage for a HALE UAV. Show its internal arrangement.

49 Draw side-view of a fuselage for a high-speed combat UAV. Show its internal arrangement.

50 What are the design objectives for determining the fuselage length-to-diameter ratio?

51 What is lofting?

Problems

1 A straight tapered wing of a large UAV has the following features

$AR = 25, \lambda = 0.3, S = 69\,m^2$

Determine tip chord, root chord, mean aerodynamic chord, and wing span.

2 A straight tapered wing of a large UAV has the following features

$AR = 19, \lambda = 0.3, S = 11.5\,m^2$

Determine tip chord, root chord, mean aerodynamic chord, and wing span.

3 A straight tapered horizontal tail of a small UAV has the following features

$AR = 5, \lambda = 0.9, S = 0.8\,m^2$

Determine tip chord, root chord, mean aerodynamic chord, and tail span.

4 A straight tapered vertical tail of a UAV has the following features

$AR = 1.4, \lambda = 0.4, S = 0.2\,m^2$

Determine tip chord, root chord, mean aerodynamic chord, and vertical tail span.

5 Identify C_{li}, C_{dmin}, C_m, $(C_l/C_d)_{max}$, α_o (deg), α_s (deg), C_{lmax}, a_o (1/rad), $(t/c)_{max}$ of the NACA 0009 airfoil section (flap-up). You need to indicate the locations of all parameters on the airfoil graphs as shown in Figure 4.7.

6 Select a NACA airfoil section for the wing for a prop-driven small UAV with the following characteristics:

$m_{TO} = 0.6\,kg$, $S = 0.22\,m^2$, $V_c = 41\,km/h$ (500 ft), $V_s = 20\,km/h$ (@sea level)

7 Select a NACA airfoil section for the wing for a prop-driven MALE UAV with the following characteristics:

$m_{TO} = 1000\,kg$, $S = 12\,m^2$, $V_c = 325\,km/h$ (at 15 000 ft), $V_s = 112$ km/hr (@sea level)

8 Select a NACA airfoil section for the wing for a jet HALE UAV-not equipped with a high lift device-with the following characteristics:

$m_{TO} = 15\,000\,kg$, $S = 130\,m^2$, $V_c = 600\,km/h$ (at 50 000 ft), $V_s = 70$ knot (@sea level)

9 A designer has selected a NACA 63_2–215 for an UAV wing during a design process. Determine the recommended wing setting angle.

10 Determine and plot the lift distribution for a small UAV with a wing with the following characteristics. Divide the half wing into 12 sections.

$S = 1.3\,m^2$, $AR = 6$, $\lambda = 0.7$, $i_w = 3\,°$, $\alpha_t = -1\,°$, airfoil section: NACA 63-209

If the aircraft is flying at the altitude of 1000 ft with a speed of 30 knot, how much lift is produced?

11 Design a wing for a small UAV with the following features:

$S = 2\,m^2$, $m = 5\,kg$, $V_C = 40$ knot (@ 5000 ft), $V_S = 22$ knot (@ sea level)

The UAV has a monoplane low wing and employs no flap. Determine airfoil section, aspect ratio, taper ratio, tip chord, root, chord, MAC, span, twist angle, sweep angle, dihedral angle, incidence, and wing angle of attack at take-off. Plot lift distribution at cruise and sketch the wing including dimensions.

12 The angle of attack of a horizontal tail for a MALE UAV is $-1.5\,°$. Other tail features are as follows:

$S_h = 12\,m^2$, $AR_h = 4$, $\lambda_h = 0.7$, tail airfoil section: NACA 64–208, $\eta_h = 0.96$

If the aircraft is flying at an altitude of 20 000 ft with a speed of 120 knot, determine how much lift is generated by the tail. Assume that the tail has no twist.

13 An aft horizontal tail is supposed to be designed for a single piston engine UAV. The aircraft with a mass of 1000 kg is cruising with a speed of 140 knot an altitude of 22 000 ft. The aircraft center of gravity is at 19% MAC and the wing-fuselage aerodynamic center is located at 24% MAC.

$$S = 15\,m^2,\ AR = 8,\ S_h = 3\,m^2,\ l = 3.5\,m,\ C_{m_{owf}} = -0.06,\ \eta_h = 1$$

Determine the horizontal tail lift coefficient that must be produced in order to maintain the longitudinal trim.

14 A reconnaissance UAV with a wing area 30 m² and wing aspect ratio of 15 has the following parameters:

$$C_{L_{\alpha V}} = 4.5\ 1/rad;\ \frac{d\sigma}{d\beta} = 0.3;\ \eta_V = 0.96;\ K_f = 0.7;\ l_V = 15\,m$$

The UAV is desired to be statically directionally stable with a Cn_β of 0.3 1/rad. Determine the vertical tail area to satisfy this requirement.

15 A reconnaissance UAV with a wing area 10 m² and wing aspect ratio of 7 has the following parameters:

$$C_{L_{\alpha V}} = 5.2\ 1/rad;\ \frac{d\sigma}{d\beta} = 0.4;\ \eta_V = 0.94;\ K_f = 0.82;\ l_V = 6\,m$$

The UAV is desired to be statically directionally stable with a Cn_β of 0.3 1/rad. Determine the vertical tail area to satisfy this requirement.

16 A commercial fixed-wing small UAV with a TO mass of 5 kg, wing area 2 m² and a wing aspect ratio of 12 has the following parameters:

$$C_{L_{\alpha h}} = C_{L_{\alpha wf}} = 4.9\ 1/rad;\ C_{m_{owf}} = -0.07;$$
$$h = 0.22;\ h_o = 0.25;\ \frac{d\varepsilon}{d\alpha} = 0.4;\ \eta_h = 0.96;\ l = 1.5\,m$$

The UAV is desired to be statically longitudinally stable with a Cm_α of −1.4 1/rad. The aircraft is cruising with a speed of 30 knot at 2000 m altitude. Determine the horizontal tail area to satisfy this requirement. Then, calculate the horizontal tail lift coefficient for this cruising flight.

17 A small fixed-wing small UAV with a TO mass of 1200 kg, wing area 14 m² and a wing aspect ratio of 9 has the following parameters:

$$C_{L_{\alpha h}} = C_{L_{\alpha wf}} = 4.6\ 1/rad;\ C_{m_{owf}} = -0.05;\ h = 0.18;$$
$$h_o = 0.24;\ \frac{d\varepsilon}{d\alpha} = 0.3;\ \eta_h = 0.94;\ l = 5\,m$$

The UAV is desired to be statically longitudinally stable with a Cm_α of −2.6 1/rad. The aircraft is cruising with a speed of 120 knot at 7000 m altitude. Determine the horizontal tail area to satisfy this requirement. Then, calculate the horizontal tail lift coefficient for this cruising flight.

18 Design a wing for a fixed-wing surveillance UAV with the following features:

$S = 20\,m^2$, $m = 2200\,kg$, $V_C = 150$ knot (@ 20 000 ft), $V_S = 65$ knot (@ sea level)

The aircraft has a monoplane low wing and employs the plain flap. Determine airfoil section, aspect ratio, taper ratio, tip chord, root, chord, MAC, span, twist angle, sweep angle, dihedral angle, incidence, high lifting device type, flap span, flap chord, flap deflection and wing angle of attack at take-off. Plot lift distribution at cruise and sketch the wing including dimensions.

19 Design a wing for very small fixed-wing UAV with the following features:

$S = 1.2\,m^2$, $m = 2\,kg$, $V_C = 35$ knot (@ 500ft), $V_S = 15$ knot (@ sea level)

The aircraft has a monoplane high wing and employs a split flap. Determine airfoil section, aspect ratio, taper ratio, tip chord, root, chord, MAC, span, twist angle, sweep angle, dihedral angle, incidence, high lifting device type, flap span, flap chord, flap deflection and wing angle of attack at take-off. Plot lift distribution at cruise and sketch the wing including dimensions.

20 Design a horizontal tail for a small fixed-wing UAV with the following characteristics:

$m_{TO} = 1200\,kg$, $D_{fmax} = 0.8\,m$, $V_c = 130$ knot (at 20 000 ft), $\alpha_f = 1.5\,°$ (at cruise)

The wing has a reference area $12\,m^2$ and the following features:

$AR = 8$, $\lambda = 0.6$, $i_w = 2.4\,°$, $\alpha_{twist} = -1.3\,°$, $\Lambda_{LE} = 37\,°$, $\Gamma = 3\,°$, NACA 65_2–415

The aircraft has a low wing and an aft conventional tail configuration, and the aerodynamic center of the wing-fuselage combination is located at 22% of MAC. In cruising flight condition, the aircraft center of gravity is located at 42% of the fuselage length. Assume that the aircraft cg is 12 cm ahead of the wing-fuselage aerodynamic center.

The following tail parameters must be determined: airfoil section, S_h, C_{h_tip}, C_{h_root}, b_h, i_h, AR_h, λ_h, Λ_h, Γ_h. At the end, draw a top-view of the aircraft that shows fuselage, wing and horizontal tail (with dimensions).

21 Design a horizontal tail for a very small fixed-wing UAV with the following characteristics:

$m_{TO} = 1.4\,kg$, $D_{fmax} = 0.15\,m$, $V_c = 15$ knot (at 2000 ft), $\alpha_f = 1.5\,°$ (at cruise)

The wing has a reference area of $0.12\,m^2$ and the following features:

$AR = 5$, $\lambda = 01$, $i_w = 2\,°$, $\alpha_{twist} = -1\,°$, $\Lambda_{LE} = 10\,°$, $\Gamma = 5\,°$, NACA 63_2–215

The aircraft has a low wing and an aft conventional tail configuration, and the aerodynamic center of the wing-fuselage combination is located at 24% of MAC. In cruising flight condition, the aircraft center of gravity is located at 40% of the fuselage length. Assume that the aircraft cg is 4 cm ahead of the wing-fuselage aerodynamic center.

The following tail parameters must be determined: airfoil section, S_h, C_{h_tip}, C_{h_root}, b_h, i_h, AR_h, λ_h, Λ_h, Γ_h. At the end, draw a top-view of the aircraft that shows fuselage, wing and horizontal tail (with dimensions).

22 Design a vertical tail for the UAV in problem 20 to satisfy the static directional stability requirements.

23 Design a vertical tail for the UAV in problem 21 to satisfy the static directional stability requirements.

Notes

1 Computational Fluid Dynamics.
2 National Administration for Aeronautics and Astronautics.
3 National Advisory Committee for Aeronautics.

5

Fundamentals of Autopilot Design

CONTENTS

Design of Unmanned Aerial Systems, First Edition. Mohammad H. Sadraey.

Companion website: www.wiley.com/go/sadraey/unmanned-aerial-systems

Educational Outcomes

After study of this chapter, the reader will be able to:

1) Develop nonlinear fixed-wing Unmanned Aerial Vehicle (UAV) dynamic model
2) Develop linear fixed-wing UAV dynamic model
3) Develop rotary-wing and quadcopter UAV dynamic model
4) Discuss autopilot categories
5) Implement numerical methods in solving nonlinear differential equations
6) Assess flying qualities of UAVs
7) Be familiar with the current autopilot design techniques
8) Evaluate primary subsystems of an autopilot

5.1 Introduction

The most important subsystem in a UAV compared with a manned aircraft is the autopilot, since there is no human in a Remotely piloted vehicle (RPV)/UAV. The autopilot is a vital and required subsystem within unmanned aerial vehicles. Since there is no human pilot involved in a UAV, an autopilot is an electro-mechanical device that must be capable of accomplishing all types of controlling functions including automatic take-off, flying toward the target destination, perform mission operations (e.g., surveillance) and automatic landing. The autopilot has the responsibility to: (a) stabilize the UAV; (b) track commands; (c) guide the UAV; and (d) navigate.

The autopilot is responsible for considerable amount of failures/mishaps in the UAV flight operations [45]. The functions of the control and stability of a UAV will depend in nature on the different UAV configurations and the characteristics required of them.

A typical UAV autopilot design problem statement is summarized as "Design and evaluate, an autopilot that permits the UAV to *autonomously/remotely-controlled* fly a complete mission." This control system must account for uncertainties in the UAV model, disturbances in the atmosphere, and noise in the measurement equipment. A UAV designer must be familiar with basic fundamentals such as trim, control, and stability.

Some manufacturer (e.g., ArduPilot) offer a complete autopilot "ready-made" or at least custom-built for individual applications. In either case, the UAV designer must evaluate the product's qualities such as accuracy, reliability, life, power consumption, environmental features and weight, before making decision. In a typical autopilot, three laws are governing simultaneously in three subsystems:

1) Control system → Control law
2) Guidance system → Guidance law
3) Navigation system → Navigation law

Autopilot is the integrated software and hardware that serve the three functions as defined. In the design of an autopilot, all three laws need to be selected/designed. The autopilot has the responsibility to: (a) stabilize under-damped or unstable modes; and (b) accurately track commands generated by the guidance system.

The autopilot design begins with the analysis of performance and stability of the vehicle. Any lack in stability or weakness in the performance should be considered in the synthesizing autopilot. This analysis requires a mathematical description of the dynamic of the UAV, and numerical analysis. Autonomous vehicle technologies for small fixed-wing UAVs were presented by Ref. [46]. This chapter covers fundamentals of design for autopilot, including dynamic modeling, autopilot categories and design simulation.

5.1.1 Autopilot and Human Operator

A manned-unmanned aircraft team have at least two distinct piloting members, namely a human pilot, and an autopilot. There are various technical differences between capabilities of human pilot and autopilot where a teaming protocol must consider. Human has various limitations (power, size, tolerance). For instance, human pilot needs rest, is not very accurate, and may get sick. Moreover, human is [47] incapable of securely storing high-quality cryptographic keys, and they have unacceptable speed and accuracy when performing cryptographic operations. They are also large, expensive to maintain, difficult to manage, and they pollute the environment. It is true that these human elements continue to be created and deployed. But they are sufficiently pervasive that we must design our protocols around their limitations.

Table 5.1 provides a technical comparison between features of a human pilot, and of an autopilot and avionics. There are some weaknesses for a human pilot (e.g., short endurance) and a number of strengths (e.g., survivability) lacking by an autopilot. However, there are some weaknesses for an autopilot (e.g., no verbal communication) and a number of strengths (e.g., IR/radar sensors) lacking by a human pilot. A teaming framework must be designed such that these two players can overlap their roles. An efficient teaming will create an environment such that both parties operate within their limits, while generating an unachievable goal by one party. The weaknesses/strengths of each team member must be included in the determination of teaming category, and also in the evaluation of the mission efficiency.

This chapter will introduce primary sections of an autopilot, and will review some fundamental topics such as dynamic modeling, UAV dynamics, forces and moments,

Table 5.1 Comparison technical between features of a human pilot and of an autopilot and avionics.

No	Attribute	Human pilot	Autopilot and avionics
1	Mathematical operation	Pilot has a limited ability to perform calculations in flight	Autopilot's computer can perform millions of calculations per second.
2	Pressure limit	0.75–1 atm	Mechanical-electric-avionic equipment can perform under any pressure; even in vacuum.
3	Temperature limit	50–120 °C	−20–180 °C
4	Acceleration	1–9 g	0–50 g
5	Orientation	Upright (on the seat)	No limit (upright, up down, or sideward).
6	Upgradability	Pilots cannot be upgraded.	Major upgrades in autopilot and avionics occur every few years.
7	Survivability	Pilots are able to react to any unexpected situations, and to survive.	Autopilot can only perform as programmed, and will crash in undefined flight conditions.
8	Visual performance	The eye can see only in visual bands (400–700 nm)	Can operate in day/night using optical/IR sensors and radars.
9	Recognition/ detection	Able to recognize objects only in short distance.	Barely able to recognize objects. Can detect some objects in long distance.
10	Endurance	8–12 hours, rest is required.	Can work for months/years.
11	Communication	Talk, hear, see, and type commands. Incapable of detecting radio frequencies and radar signals.	No verbal communication (sound sensors). Radio frequencies are employed.
12	Facing risk	Avoids dangerous arenas.	Does not care about dangerous situations.

transfer function, state space, and linearization. Moreover, autopilot categories are presented.

5.1.2 Primary Subsystems of an Autopilot

The primary functions of an autopilot are to: (a) accurately track commands generated by the command system in parallel with the guidance system; (b) guide the UAV to follow the trajectory; (c) stabilize under-damped or unstable modes; and (d) determine UAV coordinates (i.e., navigation). Therefore, an autopilot consists of: (a) command subsystem; (b) control subsystem; (c) guidance subsystem; and (d) navigation subsystem. These four subsystems must be designed simultaneously to satisfy the UAV design requirements. These subsystems include particular hardware (i.e., physical equipment), as well as the separate software (code) section. All required equipment must be designed and built, while all codes need to be written, compiled, and uploaded into the microcontroller.

In a conventional autopilot, three laws are simultaneously governing the three subsystems: (a) control system through a control law; (b) guidance system via a guidance law;

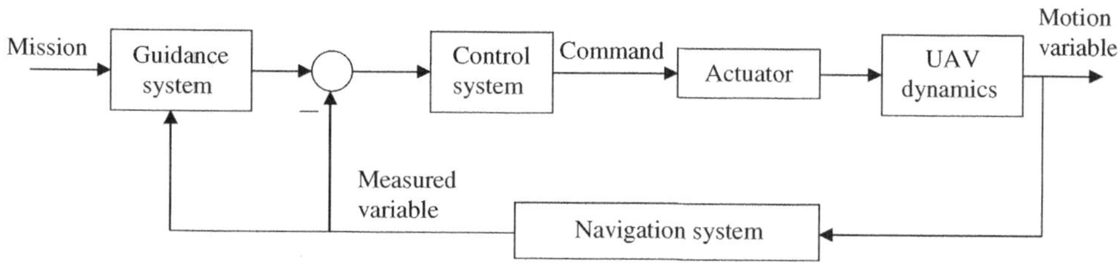

Figure 5.1 Control, guidance and navigation systems in an autopilot.

and (c) navigation system through a navigation law. In the design of an autopilot, all three laws need to be developed/designed. The design of the control law is at the heart of autopilot design process. The relation between the control system, the guidance system, and the navigation system is shown in Figure 5.1. This figure is not illustrating the command system, it will be covered later.

The UAV dynamics is presented in Section 5.3. The design of control subsystem, guidance subsystem, and navigation subsystem are discussed in Chapters 6, 7, and 8 respectively. The design and development of microcontroller will be described in Chapter 9.

5.1.3 Autopilot Design or Selection

There are basically two methods for making the autopilot of a UAV:

1) Autopilot design
2) Autopilot selection

The design of an autopilot is a complex and time-consuming process and needs expertise in fundamentals of flight dynamics and control at graduate level. Since the autopilot needs to be verified by testing it in a real situation, it is expensive, time-consuming and labor-intensive. Hardware-in-the-loop (HIL) simulations is an effective technique to measure performance of the control/guidance algorithm of a designed autopilot. Large UAV production companies such as Northrop Grumman (designer and manufacturer of RQ-4 Global Hawk) and General Atomics (designer and manufacturer of MQ-1 Predator) have sufficient human expert and budget to design their own autopilots. With the advent of high speed and powerful microprocessors, microcontrollers, Micro-Electro-Mechanical Systems (MEMS); the design of autopilot is not as hard as 20 years ago.

However, RC modelers, UAV amateurs, experimental homebuilt UAV builders, small UAV companies, and university students in their capstone design projects do not afford to design their autopilots. Instead they select the best or a suitable commercial autopilot among the current available autopilots that are found in the market. Usually commercial autopilots are not reconfigurable, which means that variations of onboard software (including guidance algorithm) is not feasible.

ArduPilot is an open-source, UAV autopilot software suite. ArduPilot was originally developed by hobbyists to control model aircraft, but later has evolved into a full-featured and reliable autopilot. The ArduPilot software suite consists of navigation software, along with ground station controlling software including mission planner.

If you have enough time, budget and manpower; you can decide to design an autopilot for your UAV. If not, you need to focus on the aerodynamic, and structural design of your UAV, and purchase a commercial autopilot. But remember the autopilot design is a design project for itself, and this needs to be integrated into the UAV design process properly. If you are not ready to design your own autopilot, you are recommended to select a proper autopilot from the previously designed and available autopilots. Two popular autopilot resources are DJI and Arduino.

A middle approach between autopilot design and autopilot selection is the custom-made technique. In order to reduce development costs and time, you may design and build a custom-made autopilot, which utilizes available elements such as Global Positioning System (GPS) for navigation. For such purpose you may employ an open-source software such as Arduino. A significant characteristic of an open architecture is the possibility or reprogramming in flight and real time telemetry. GPS systems are increasingly used to create and use waypoints in navigating UAVs. GPS receivers used in aviation have databases which contain named waypoints, radio navigation aids, and airports.

5.2 Dynamic Modeling

5.2.1 Modeling Technique

The design of autopilot requires a variety of mathematical equations and technical information including the UAV dynamics. The first step to analysis and to design an autopilot is to describe the dynamic behavior of the vehicle in a mathematical language. The quantitative mathematical description of a physical systems is known as dynamic modeling. There are several ways for mathematical descriptions. The most widely used method is the differential equation.

Description of the behavior and components of a dynamic system with a mathematical language is referred to as the dynamic modeling. The dynamic behavior (i.e., UAV dynamics) of a flight vehicle is based on the Newton's second law: When a force is applied on an object, the acceleration of the object is directly proportional to the magnitude of the force, in the same direction as the force, and inversely proportional to the mass of the object.

Once a physical system (e.g., UAV) has been described by a set of mathematical equations, they are manipulated to achieve an appropriate mathematical format. There are mainly two techniques to model a dynamic system: (a) transfer function; and (b) state space representation. The first one is described in s-domain (frequency); the second one is presented in time domain. A dynamic system may also by represented pictorially; this method is using a block diagram. An UAV is a dynamic system and its dynamic behavior will be modeled by either way. Using a UAV dynamic model, one is able to design automatic flight control system (AFCS) to satisfy the design requirements.

The essential feature of an AFCS is the existence of a feedback loop to give a good performance. This is a closed loop system; if the measured output is not compared with the input the loop is open. Usually it is required to apply a specific input to a system and for some other part of the system to respond in the desired way. The error between the actual response and the ideal response is detected and fed back to the input to modify

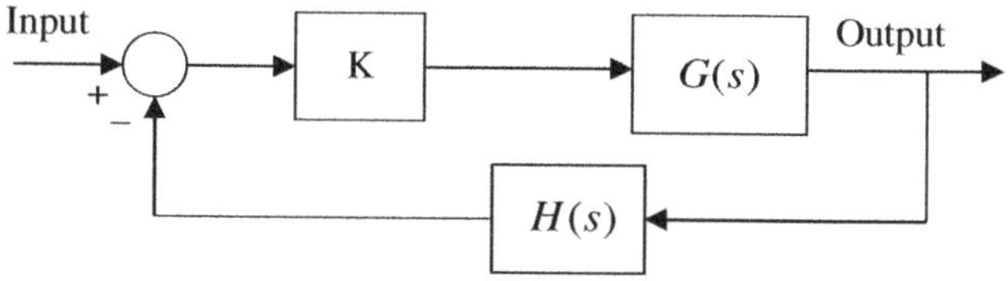

Figure 5.2 A basic closed-loop system.

it so that the error is reduced. The simplest linear closed loop system incorporates a negative feedback, and will have one input and one output variable. Figure 5.2 illustrates a basic (SISO[1]) closed-loop system, where K, G(s), and H(s) represent controller, Plant, and measurement device respectively. In general, both input and output vary with time, and the control system can be mechanical, pneumatic, hydraulic and electrical in operation, or any combination of these or other power sources.

Various effective AFCSs have been designed for UAVs using traditional design methods. The synthesis of any AFCS is often costly and requires time-consuming endeavor. This chapter presents the dynamic modeling of the UAV, and enables a UAV designer to use the mathematical tools in the design process.

The control system design process requires the UAV mathematical model as the basis for the design. The UAV modeling process basically consists of: (a) dynamic modeling; (b) aerodynamic modeling; (c) engine modeling; and (d) structural modeling. An UAV is basically a nonlinear system. Furthermore, its dynamics and equations of motion are also nonlinear. The UAV dynamic model is could be linearized or decoupled. The nonlinear coupled equations of motion are the most complete dynamic model. The flat-Earth vector equations of motion are often used, and when they are expanded the standard 6-DOF (6 Degrees of Freedom) equations used for UAV control design and flight simulation are obtained. Reference [48] presents the set of nonlinear coupled equations of motion for a UAV. Other UAV dynamic models are: (a) linear decoupled equations of motion; (b) hybrid coordinates; (c) linear coupled equations of motion; and (d) nonlinear decoupled equations of motion.

The aerodynamic forces and moments (aerodynamic model) of the complete UAV are defined in terms of dynamic pressure, aircraft geometry and dimensionless aerodynamic coefficients. The aerodynamic coefficients are assumed to be the linear functions of state variables and control inputs. These forces and moments are used in equations of motion (dynamic model) as part of control system design, as well as the flight simulation.

The propulsion model is based on the propulsion system powering the UAV (e.g., prop-driven or jet engine). In case of a jet engine, the engine thrust (T) is modeled in terms of throttle setting (δ_T). In case of a prop-driven engine, the engine power is a function of required engine thrust, aircraft speed, and propeller efficiency. Several control system design techniques may be found in the literature. The choice of the type of control system depends on a variety of factors including the system's characteristics and the design requirements. The heart of the control system is the controller. A summary of the different systems and the different control systems design techniques is shown in Figure 6.3 (in Chapter 6). There are several textbooks and papers regarding controller design in the literature; Refs. [49, 50] present several controller design techniques.

The real UAV behavior is nonlinear and has uncertainties. It is also important to note that all measurement devices (including gyros and accelerometers) have some kind of noise that must be filtered. It is well known that the atmosphere is a dynamic system that produces lots of disturbances throughout the aircraft's flight. Finally, since fuel is expensive and limited, and actuators have dynamic limitations, optimization is necessary in control system design. Therefore, it turns out that only a few design techniques, such as robust nonlinear control, are able to satisfy all safety, cost, and performance requirements. In order to select the best controller technique, one must utilize a trade-off study and compare the advantages and disadvantage of all candidate controllers.

Virtually all dynamic systems are nonlinear; yet an overwhelming majority of operational control laws have been designed as if their dynamic systems were linear and time-invariant. As long as the quantitative differences in response are minimal (or at least acceptable in some practical sense), the linear time invariant model facilitates the control system design process. This is because of the direct manner in which response attributes can be associated with model parameters.

A small error in modeling, a small error in the control system design, or a small error in the simulation may each result in problems in flight, in the worst case might even result in the loss of an unmanned aerial vehicle. As long as the dynamic effects of parameter variations are slow in comparison to state variation, control design can be based on an ensemble of time-invariant dynamic models. Fast parameters may be indistinguishable from state components, in which case the parameters should be included in an augmented state vector for estimation. The UAV field is one where extensive use is made of modeling and simulation technologies.

The complete aircraft systems and dynamics model incorporates different subsystem models (e.g., aerodynamics, structures, propulsion, and control subsystems) that have interdependent responses to any input. These subsystems also interact with the other subsystems. The dynamic modeling of an aircraft is at the heart of its simulation. The response of an aerial vehicle system to any input, including commands or disturbances (e.g., wind gusts), can be modeled by a system of ordinary differential equations (i.e., the equations of motion). Dealing with the nonlinear, fully coupled differential equations of motion is not an easy task.

The dynamics of an aircraft can be modeled in different ways. The equations of motion take several forms (Figure 5.3) including: (a) nonlinear fully coupled; (b) nonlinear semi-coupled; (c) nonlinear decoupled; (d) nonlinear reformulated; (e) linear coupled; (f) linear decoupled; and (g) linear time-variant.

To develop a computer simulation to evaluate the performance of an aerial vehicle (manned or unmanned) including its control system, we must invariably use a nonlinear fully coupled model. In order to design a control system, one of the above models are utilized. Each of these models has advantages and disadvantages. These include precision, accuracy, complexity, and credibility. The use of flight simulation tools to reduce risk and flight testing for an aerial vehicle system reduces the overall program schedule.

5.2.2 Fundamental Model

The Newton's Second Law states that the time derivatives of linear momentum (m.V) and angular momentum (I.α) are equal to the externally applied forces (F) and moments (M) respectively.

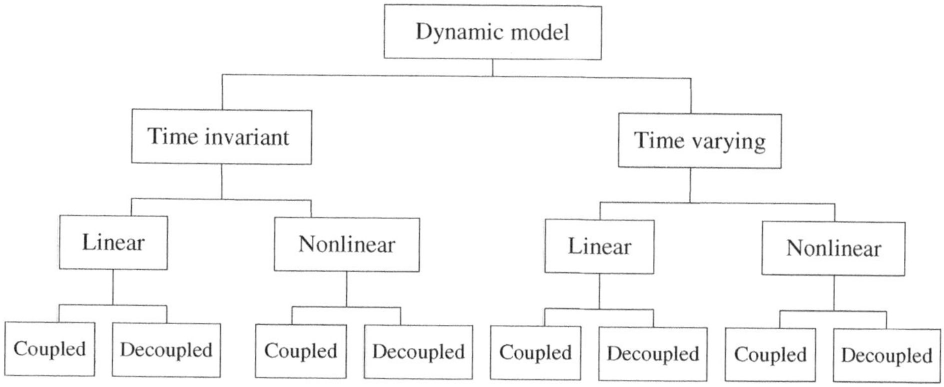

Figure 5.3 Classes of dynamic models.

$$\sum F = \frac{d}{dt}(mV_T) \tag{5.1}$$

$$\sum M = \frac{d}{dt}(I\omega) \tag{5.2}$$

where m, V_T, I, and ω are UAV mass, total airspeed, UAV mass moment of inertia, and angular speed respectively. As Figure 5.4 illustrates, an aircraft has three axes (x, y, and z), hence the vehicle has six degree-of-freedom (DOF), three linear displacement along three axes, and three angular displacement about the three axes. In contrast, there are three linear velocities (U, V, and W), and three angular velocities (P, Q, and R). Three angular velocities represent roll rate about x-axis (P), pitch rate about y-axis (Q), and yaw rate about z-axis (R).

Therefore, there are three groups of forces, including aerodynamic forces (F_A) and thrust forces (F_T), and three aerodynamic moments. The aerodynamic forces D, Y, and L, are the drag, side-force, and lift, and L_A, M_A, N_A, are the aerodynamic moments (See Figure 5.4). The aerodynamic forces and moments are reviewed in the next section. Applying forces, moments, and velocities into Eqs. (5.1) and (5.2), the following standard body-axis six first-order nonlinear differential equations are obtained [40].

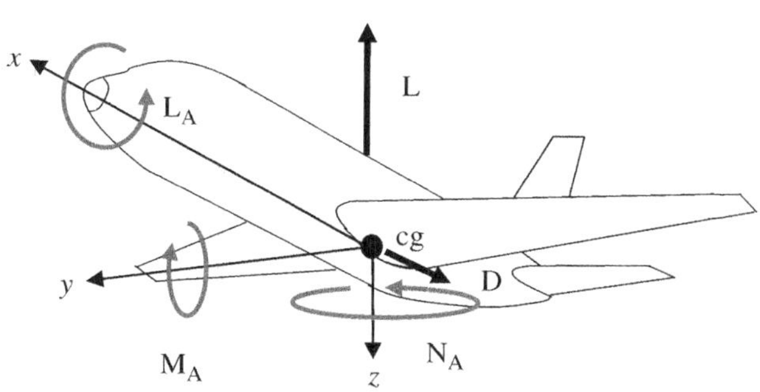

Figure 5.4 Aerodynamic axes, forces and moments.

Force equations:

$$m\left(\dot{U}-VR+WQ\right)=mg_x+F_{A_x}+F_{T_x} \tag{5.3}$$

$$m\left(\dot{V}+UR-WP\right)=mg_y+F_{A_y}+F_{T_y} \tag{5.4}$$

$$m\left(\dot{W}-UQ+VP\right)=mg_z+F_{A_z}+F_{T_z} \tag{5.5}$$

Moment equations:

$$\dot{P}I_{xx}-\dot{R}I_{xz}-PQI_{xz}+\left(I_{zz}-I_{yy}\right)RQ=L_A+L_T \tag{5.6}$$

$$\dot{Q}I_{yy}+\left(I_{xx}-I_{zz}\right)PR+I_{xz}\left(P^2-R^2\right)=M_A+M_T \tag{5.7}$$

$$\dot{R}I_{zz}-\dot{P}I_{xz}+\left(I_{yy}-I_{xx}\right)PQ+I_{xz}QR=N_A+N_T \tag{5.8}$$

Equations (5.3) through (5.5) govern linear motion along x, y and z axes, and constitutes force equations of the equations of motion. In contrast, Eqs. (5.6) through (5.8) govern angular motion about x, y and z axes, and constitutes the moment equations of the equations of motion.

Many UAVs are symmetric about xz plane, so, the cross-product of inertia (I_{xz}) may be assumed zero. Furthermore, the thrust moments (L_T, M_T, N_T) are also negligible. Applying these two assumptions provides a simpler version of the moment equations as:

$$\dot{P}=\frac{I_{yy}-I_{zz}}{I_{xx}}QR+\frac{L_A}{I_{xx}} \tag{5.6a}$$

$$\dot{Q}=\frac{I_{zz}-I_{xx}}{I_{yy}}RP+\frac{M_A}{I_{yy}} \tag{5.7a}$$

$$\dot{R}=\frac{I_{xx}-I_{yy}}{I_{zz}}PQ+\frac{N_A}{I_{zz}} \tag{5.8a}$$

These moment equations are sometimes referred to as the Euler's equations. In Section 5.5, six force and moment equations are reformatted and simplified for various applications.

5.2.3 Transfer Function

One format of the mathematical model of a control system is the transfer function. The control systems are dynamic in nature, and the mathematical models are usually differential equations. If these equations are linearized, then the Laplace transform can be utilized to develop transfer function and control law. The Laplace transform method is originally used to facilitate and systematize the solution of ordinary constant-coefficient differential equations.

Laplace transform is a tool to convert a time-domain function into a frequency-domain one in which information about frequencies of the function can be captured.

It is often much easier to solve problems in frequency-domain with the help of Laplace transform. The Laplace Transform of the function f(t), denoted by F(s) is a function of the complex variable s. The Laplace transform is defined by:

$$\mathcal{L}\{f(t)\} = F(s) = \int_0^{\infty} f(t)e^{-st}dt \tag{5.9}$$

If the system differential equation is linear, the ratio of the Laplace transform of output variable to the Laplace transform of input variable, is called the transfer function. In general, the format of a transfer function is the ratio of two polynomials in s:

$$F(s) = \frac{b_1 s^m + b_2 s^{m-1} + \ldots + b_{m+1}}{s^n + a_1 s^{n-1} + \ldots + a_n} \tag{5.10}$$

Moreover, when the roots of numerator (i.e., zeros; z), and the roots of denominators (i.e., poles; p) are determined, the transfer function F(s) could be characterized by its poles and zeros:

$$F(s) = K\frac{(s-z_1)\ldots(s-z_i)\ldots(s-z_m)}{(s-p_1)\ldots(s-p_i)\ldots(s-p_n)} \tag{5.11}$$

Poles and zeros may be real or complex conjugate. For more details, the reader may refer to calculus textbooks such as [51]. Any high-order transfer function may be decomposed into a number of first-order and second-order transfer functions. A first-order TF is modeled as:

$$F(s) = \frac{k}{as+b} \tag{5.12}$$

A proper second-order system is modeled as

$$F(s) = \frac{N(s)}{as^2 + bs + c} \tag{5.13}$$

where N(s) represents the numerator; which in this case, is a first-order polynomial. The standard form for a pure second-order dynamic system is a function of two parameters of damping ratio (ξ) and natural frequency (ω_n) as:

$$F(s) = \frac{k}{s^2 + 2\xi\omega_n s + \omega_n^2} \tag{5.14}$$

The desired damping ratio and natural frequency are employed in the design process to satisfy the control system design requirements.

The denominator of a transfer function is referred to as the characteristic polynomial/equation, since it reveals the stability characteristics of the dynamic system.

A UAV is dynamically stable, when all poles of the characteristic equation have negative real parts. If there is at least one zero for the real part of poles of the characteristic equation, the UAV is dynamically neutrally stable. When there is at least one positive real part in poles of the characteristic equation, the UAV is dynamically unstable.

For instance, the longitudinal transfer functions (response to an elevator input) are $\frac{u(s)}{\delta_E(s)}, \frac{\alpha(s)}{\delta_E(s)}, \frac{\theta(s)}{\delta_E(s)}, \frac{q(s)}{\delta_E(s)}$. Similarly, the lateral transfer functions (response to an aileron input) are $\frac{\psi(s)}{\delta_A(s)}, \frac{\beta(s)}{\delta_A(s)}, \frac{p(s)}{\delta_A(s)}, \frac{r(s)}{\delta_A(s)}, \frac{\phi(s)}{\delta_A(s)}$. In contrast, the directional transfer functions (response to a rudder input) are $\frac{\psi(s)}{\delta_R(s)}, \frac{\beta(s)}{\delta_R(s)}, \frac{p(s)}{\delta_R(s)}, \frac{r(s)}{\delta_R(s)}, \frac{\phi(s)}{\delta_R(s)}$. All of these transfer functions are of fourth order where each represents two to three oscillatory modes.

5.2.4 State-Space Representation

In general, the state space is defined as the n-dimensional space in which the components of the state vector represent its coordinate axes. A mathematical model in time-domain based on matrix operation is referred to as state space model. The state space is a mathematical tool in modern control which is able to cover any multi-input-multi-output (MIMO) system. In contrast, the transfer function belongs to era of classical control and only is able to cover a SISO system.

Basic matrix properties are used to introduce the concept of state and the method of writing and solving the state equations. The state of a system is a mathematical structure containing a set of n variables $x_1(t)$, $x_2(t)$, …, $x_n(t)$, called the state variables, m inputs, $u_1(t)$, $u_2(t)$, …,$u_m(t)$, and p outputs, $y_1(t)$, $y_2(t)$, …, $y_p(t)$. In general, a state space model includes a set of n linear first-order differential equation, plus p linear algebraic equation:

$$\begin{aligned} \dot{x} &= Ax + Bu \\ y &= Cx + Du \end{aligned} \tag{5.15}$$

where A_{nxn}, B_{nxm}, C_{pxn}, and D_{pxm} are matrices. The state equations of a system are a set of n first-order differential equations, where n is the number of independent states. In building matrices, the algebra rules and matrix operations requirements need to be considered.

A UAV is dynamically stable, when all eigenvalues of matrix A have negative real parts. If there is at least one zero for the real part of eigenvalues of matrix A, the UAV is dynamically neutrally stable. When there is at least one positive real part in eigenvalues of matrix A, the UAV is dynamically unstable.

A dynamic system with two inputs (u_1, u_2), three states (x_1, x_2, x_3), and one output (y) will have the following state-space form:

$$\begin{bmatrix} \dot{x}_1 \\ \dot{x}_2 \\ \dot{x}_3 \end{bmatrix} = \begin{bmatrix} a_{11} & a_{12} & a_{13} \\ a_{21} & a_{22} & a_{23} \\ a_{31} & a_{32} & a_{33} \end{bmatrix} \begin{bmatrix} x_1 \\ x_2 \\ x_3 \end{bmatrix} + \begin{bmatrix} b_{11} & b_{12} \\ b_{21} & b_{22} \\ b_{31} & b_{32} \end{bmatrix} \begin{bmatrix} u_1 \\ u_2 \end{bmatrix} \tag{5.16a}$$

$$y = \begin{bmatrix} c_1 & c_2 & c_3 \end{bmatrix} \begin{bmatrix} x_1 \\ x_2 \\ x_3 \end{bmatrix} + \begin{bmatrix} d_{11} & d_{12} \\ d_{21} & d_{22} \end{bmatrix} \begin{bmatrix} u_1 \\ u_2 \end{bmatrix} \tag{5.16b}$$

The UAV dynamics may be represented by state space model. In this case, the input variables (u) are such variables as the elevator (δ_E), aileron (δ_A), and rudder (δ_R) deflections, as well as the engine throttle (δ_T). In addition, the output variables are usually motion variables such as linear speeds (u, v, w), angular velocities (p, q, r), and angles such as angle of attack (α), and pitch angle (θ). Thus, typical output variables (y) to be controlled are: α, β, θ, γ, ϕ, ψ, P, Q, R, U, V, W, x, y, h, M, n_x, n_y, n_z. Due to the orders of vehicle's dynamic models, the A matrix of the state-space representation of the longitudinal dynamics, and the lateral-directional dynamics is 4×4. For instance, a typical state space model for a linearized longitudinal motion for a UAV is presented in Section 5.5.5.

5.3 Aerodynamic Forces and Moments

5.3.1 Forces and Moments Equations

When the air is passing around an object (e.g., an UAV), the aerodynamic forces and moments are generated. The aerodynamic forces (D, Y, and L) and moments (L_A, M_A, and N_A) are functions of airspeed, air density, wing area, and UAV configuration:

$$D = \bar{q} S C_D \tag{5.17}$$

$$Y = \bar{q} S C_y \tag{5.18}$$

$$L = \bar{q} S C_L \tag{5.19}$$

$$L_A = \bar{q} S C_l b \tag{5.20}$$

$$M_A = \bar{q} S C_m C \tag{5.21}$$

$$N_A = \bar{q} S C_n b \tag{5.22}$$

where S is the wing reference area, C is the wing mean aerodynamic chord, b is the wing span, and $\bar{q}$ is the dynamic pressure.

$$\bar{q} = \frac{1}{2} \rho V^2 \tag{5.23}$$

With the force and moment expressed as linearized functions of the state and control, the nonlinear inverse can be constructed based on these linearized functions and the nonlinear kinematics. The parameters, C_D, C_y, C_L, C_l, C_m, and C_n are drag coefficient, side-force coefficient, lift coefficient, rolling moment coefficient, pitching moment coefficient, and yawing moment coefficient respectively.

For the case of a quadcopter, where there is no fixed-wing, the lift is the vertical component of the engine thrusts.

$$L = 4T \cos(\alpha) \tag{5.24}$$

The quadcopter drag is calculated in a similar fashion as with the fixed-wing aircraft. However, the S will be the frontal area of the vehicle, and the C_D will be computed differently. The method for the calculation of the quadcopter C_D is presented in the Section 12.9.3.

5.3.2 Stability and Control Derivatives

The drag, side-force, lift, rolling moment, pitching moment, and yawing moment coefficients are all functions of air density, airspeed, geometry, and configuration (coefficients). The aerodynamic forces and moments coefficients are modeled with respect to a new group of variables referred to as the stability and control derivatives:

$$C_D = C_{D_o} + C_{D_\alpha}\alpha + C_{D_q} Q\frac{C}{2U_1} + C_{D_{\dot\alpha}}\dot\alpha\frac{C}{2U_1} + C_{D_u}\frac{u}{U_1} + C_{D_{\delta_e}}\delta_e \quad (5.25)$$

$$C_y = C_{y_\beta}\beta + C_{y_p} P\frac{b}{2U_1} + C_{y_r} R\frac{b}{2U_1} + C_{y_{\delta_a}}\delta_a + C_{y_{\delta_r}}\delta_r \quad (5.26)$$

$$C_L = C_{L_o} + C_{L_\alpha}\alpha + C_{L_q} Q\frac{C}{2U_1} + C_{L_{\dot\alpha}}\dot\alpha\frac{C}{2U_1} + C_{L_u}\frac{u}{U_1} + C_{L_{\delta_e}}\delta_e \quad (5.27)$$

$$C_l = C_{l_\beta}\beta + C_{l_p} P\frac{b}{2U_1} + C_{l_r} R\frac{b}{2U_1} + C_{l_{\delta_a}}\delta_a + C_{l_{\delta_r}}\delta_r \quad (5.28)$$

$$C_m = C_{m_o} + C_{m_\alpha}\alpha + C_{m_q} Q\frac{C}{2U_1} + C_{m_{\dot\alpha}}\dot\alpha\frac{C}{2U_1} + C_{m_u}\frac{u}{U_1} + C_{m_{\delta_e}}\delta_e \quad (5.29)$$

$$C_n = C_{n_\beta}\beta + C_{n_p} P\frac{b}{2U_1} + C_{n_r} R\frac{b}{2U_1} + C_{n_{\delta_a}}\delta_a + C_{n_{\delta_r}}\delta_r \quad (5.30)$$

All of these equations are employed in the design and simulation process to determine the aerodynamic forces and moments coefficients. Thus, all stability and control derivatives must be determined in advance; a reliable reference for this purpose is DATCOM [52].

5.3.3 Non-dimensional Stability and Control Derivatives

In the Eqs. (5.25) through (5.30), parameters such as $C_{L\alpha}$, C_{Du}, $C_{n\beta}$, C_{lp}, are referred to as non-dimensional stability derivatives. In addition, parameters such as $C_{m_{\delta_e}}$, $C_{n_{\delta_a}}$ and $C_{l_{\delta_r}}$ are referred to as non-dimensional control derivatives. Each derivative is the partial derivative of an aerodynamic force or moment coefficient with respect to a motion variable or control variable. All non-dimensional stability and control derivatives have the same unit of 1/rad or 1/deg.

For instance,

$$C_{L_q} = \frac{\partial C_L}{\partial q} \quad (5.31)$$

$$C_{y_\beta} = \frac{\partial C_y}{\partial \beta} \tag{5.32}$$

A group of the variables which are widely used in the design of control surfaces are control derivatives. The control derivatives are simply the rate of change of aerodynamic forces and moments (or their coefficients) with respect to a control surface deflection (e.g., elevator). Control derivatives represent how much change in an aerodynamic force or moment acting on an aircraft when there is a small change in the deflection of a control surface. The greater a control derivative, the more powerful is the corresponding control surface. Three most important non-dimensional control derivatives are $C_{l_{\delta A}}$, $Cm_{\delta E}$, $Cn_{\delta R}$. The derivative $C_{l_{\delta A}}$ is the rate of change of rolling moment coefficient with respect to a unit change in the aileron deflection (Eq. 5.33). The derivative $Cm_{\delta E}$ is the rate of change of pitching moment coefficient with respect to a unit change in the elevator deflection (Eq. 5.34). The derivative $Cn_{\delta R}$ is the rate of change of yawing moment coefficient with respect to a unit change in the rudder deflection (Eq. 5.35).

$$C_{l_{\delta A}} = \frac{\partial C_l}{\partial \delta_A} \tag{5.33}$$

$$C_{m_{\delta E}} = \frac{\partial C_m}{\partial \delta_E} \tag{5.34}$$

$$C_{n_{\delta R}} = \frac{\partial C_n}{\partial \delta_R} \tag{5.35}$$

After the aircraft main components (e.g., wing, tail, and landing gear) are designed, the control power requirements may be expressed and interpreted in terms the control derivatives. For instance, a rudder is designed to satisfy the requirement of $Cn_{\delta R} < -0.4$ 1/rad for a fighter. In contrast, an elevator is designed to satisfy the requirement of $Cm_{\delta E} < -2$ 1/rad for a transport aircraft. It is a challenging task for a flight dynamic engineer to determine the UAV derivatives accurately. Wind tunnel tests are a beneficial technique to calculate the stability and control derivatives. The list of all non-dimensional stability and control derivatives (about 33 longitudinal, lateral-directional derivatives) may be found in [52].

5.3.4 Dimensional Stability and Control Derivatives

Another method to expand the aerodynamic forces and moments is by using dimensional stability and control derivatives. For instance, the aerodynamic forces and moment in x and z directions (i.e., drag, lift and pitching moment) are expanded as:

$$X = X_\alpha \alpha + X_u u + X_q q + X_{\dot\alpha} \dot\alpha + X_{\delta_e} \delta_e \tag{5.36}$$

$$Z = Z_\alpha \alpha + Z_{\dot\alpha} \dot\alpha + Z_u u + Z_q q + Z_{\delta_e} \delta_e \tag{5.37}$$

$$M = M_\alpha \alpha + M_{\dot\alpha} \dot\alpha + M_u u + M_q q + M_{\delta_e} \delta_e \tag{5.38}$$

Thus, the important dimensional longitudinal stability and control derivatives are X_u, Z_u, M_u, X_α, Z_α, M_α, X_{δ_e}, Z_{δ_e}, M_{δ_e}, $Z_{\dot{\alpha}}$, $M_{\dot{\alpha}}$, Z_q, and M_q. For instance, the definition of a few dimensional derivatives are provided below:

$$Z_q = \frac{\partial Z / \partial q}{m} = \frac{-\bar{q}S\bar{C}C_{L_q}}{2mU_1} \tag{5.39}$$

$$M_\alpha = \frac{\partial M / \partial \alpha}{I_{yy}} \tag{5.40}$$

$$N_r = \frac{\partial N / \partial r}{I_{zz}} = \frac{-\bar{q}Sb^2 C_{n_r}}{2I_{xx}U_1} \tag{5.41}$$

$$L_\beta = \frac{\partial L / \partial \beta}{I_{xx}} \tag{5.42}$$

For calculations of these dimensional derivatives, refer to references such as [40, 53].

5.3.5 Coupling Stability Derivatives

In the nonlinear differential equations of motion, the coupling of states is observed in the left hand side of the Eqs. (5.3) through (5.8), but the coupling of forces is in the right hand side. The coupling of the state variables is easily observable by looking at the equations of motion. Coupling forces are represented using stability derivatives. There are more several coupling derivatives that represent the interactions between the lateral-directional and longitudinal motions. Decoupling is performed by using the assumption that any rolling or yawing motion does not produce any motion in pitch, and vice versa.

One of the neglected modeling areas in the flight dynamics literature is the coupling of forces in dynamical modeling (Refs. [40, 54]). Among the more than 20 coupling derivatives, two of them are the most important in representing the effects of lateral-directional motion on longitudinal motion. These are the sideslip coupling derivative (C_{D_β}) and the bank angle coupling derivative (C_{L_φ}). Other derivatives are negligible compared to these two derivatives. The main interactions between longitudinal, lateral, and directional motions (besides interactions between the lateral and directional motions) are of three types:

1) A pure longitudinal motion has no direct effect on the lateral-directional motions.
2) The main effect of directional motion on the longitudinal motion is to increase the axial force (mainly drag), and hence to reduce the longitudinal speed of the airplane. Therefore, the main derivative is C_{D_β}.
3) The main effect of the lateral motion on the longitudinal motion is to reduce the normal force (in other words, the lift force) on the aircraft. Therefore, the main derivative is C_{L_φ}. Based on these coupling derivatives, the normal and axial forces affected by the lateral–directional components are as follows:

$$X = X_\alpha \alpha + X_u u + X_q q + X_{\dot{\alpha}} \dot{\alpha} + X_{\delta_e} \delta_e + X_\beta \beta \tag{5.43}$$

$$Z = Z_{\alpha}\alpha + Z_{\dot{\alpha}}\dot{\alpha} + Z_u u + Z_q q + Z_{\delta_e}\delta_e + Z_{\varphi}\phi \tag{5.44}$$

Reference [55] has presented the full derivation of these two coupling derivatives. Two major dimensional coupling derivatives are:

$$X_{\beta} = \left[sign(\beta)\right] q_1 S_{vt} \frac{2}{\pi} \tag{5.45}$$

$$Z_{\varphi} = 0.1mg\left[sign(\phi)\right] \tag{5.46}$$

where "sign (x)" is just a "+" or a "−." In the semi-coupled equations of motion, the last terms in Eqs. 5.58 and 5.59 are neglected.

If the influence of the lateral-directional motion in the longitudinal variables is neglected, two sets of independent equations are obtained (Section 5.5.5).

5.4 Simplification Techniques of Dynamic Models

The UAV dynamics model is the equations of motion based on Newton's second law. Working with the nonlinear, fully coupled equations of motion can be extremely time consuming and complicated. This is why many designers resort to techniques to reduce the complexity of the equations and to make the design and simulation process easier.

Two basic simplification techniques are linearization and decoupling. They may be applied individually or simultaneously. Linearization is performed using: (a) Taylor series approach; (b) direct method; and (c) feedback linearization. Decoupling is performed by assuming that any rolling or yawing motion does not produce any motion in pitch, and vice versa. In other words, the longitudinal motion is independent of the lateral-directional motion.

In each case, the designer faces different limits of application. As an example, the linearized equations of motion are valid only in the vicinity of the trim or a singular point. In other words, this is when the perturbed motion is very close to the steady state motion (trim condition). Decoupling is also valid when the interaction between the longitudinal and lateral-directional motion can be ignored.

5.4.1 Linearization

In reality, the UAV dynamics is nonlinear. However, dealing with a nonlinear system is not as easy task. Nonlinear systems are difficult to deal with. One way to remove or handle the nonlinearity is using a linearization technique to linearize it. The linearization converts a nonlinear model to a linear one by considering small deviation from equilibrium. Thus, the linearization is canceling or ignoring nonlinearity.

Linearization is the cancelation of the nonlinearities within a system model. There are basically two types of linearization: (a) Jacobian linearization (Taylor series); and (b) feedback linearization (also called Dynamic Inversion). Jacobian linearization literally neglects the model's nonlinearities, but feedback linearization actively suppresses the nonlinearity. However, to derive the linear equations using a Taylor series, the nonlinear aerodynamic coefficients are replaced by terms involving the stability and control derivatives.

The two methods of feedback linearization are: (a) input–output linearization: This technique is restricted to systems in which the so-called zero dynamics are stable. The input-output map is linearized, while the state equation may be only partially linearized; (b) state-space, input-state, or full-state linearization: In this technique the state equation is completely linearized. This Section presents the mathematical technique to linearize a nonlinear equation.

Any nonlinear equation is linearized at only an equilibrium (i.e., trim) point. The linearization is valid only at the vicinity of the equilibrium point. The trim point is a point or a motion condition that there is no acceleration (i.e., where the velocity is constant). Linearization/approximation will improve the quality of control system.

A few flight operations which have trim points are: Cruise with constant speed, climb with constant speed, and turn with constant speed. Take-off and landing do not have the trim conditions, since the airspeed is not constant. A linear system satisfies the properties of superposition and homogeneity. The general form of a linear equation is $y = \mathrm{m}\, x$, where m is referred to as the slope and is a constant number.

A linear term is one which is first degree in the dependent variables and their derivatives. A linear equation is an equation consisting of a sum of linear terms. The linearization about the equilibrium Point can be accomplished two ways:

1) Using Taylor series
2) Directly substitute

5.4.1.1 Taylor Series

Using Taylor series expansion, for the case of a function with a single independent variable, F(X) can be represented by the following infinite series:

$$F(X) = F(X_0) + \left.\frac{dF}{dX}\right|_{X=X_0} (X - X_0) + \left.\frac{d^2F}{dX^2}\right|_{X=X_0} \frac{(X - X_0)^2}{2!} + \cdots \tag{5.47}$$

Linearization using Taylor series expansion, ignores the higher order terms. Thus, one will obtain the following linear function:

$$F(X) = \left.\frac{dF}{dX}\right|_{X=X_o} X \tag{5.48}$$

Basically, the linear version of a nonlinear equation (algebraic or differential) with one independent variable is equal to the slope (m) at the trim point, multiplied by the variable (F(x) = m x). With the same token, an equation with two independent variables (say x and y) is modeled with:

$$F(X,Y) = \left.\frac{dF}{dX}\right|_{\substack{X=X_o \\ Y=Y_o}} X + \left.\frac{dF}{dY}\right|_{\substack{X=X_o \\ Y=Y_o}} Y \tag{5.49a}$$

This technique can be extended to the nonlinear dynamic model (equation) of a UAV, since there are six independent variables (u, v, w, p, q, r).

5.4.1.2 Direct Technique

The direct technique is based on the small perturbation theory [14]. In the direct technique, first substitute all independent variables (e.g., X) with two terms (the initial trim value; X_o, plus a small perturbation, x).

$$X = X_o + x \tag{5.49b}$$

Then, expand the non-linear equations using mathematical operations. The last step is to assume x is small, which implies retaining only linear terms (e.g., mx). Other terms such as x^n, log(x), are removed. However, there are two exceptions. In using the small perturbation theory, for angular variables such as x, the sin(x) is replaced with x, and any cos(x) is replaced with 1.

$$\cos(x) = 1 \tag{5.49c}$$

$$\sin(x) = x \tag{5.49d}$$

For the proof, see mathematical texts such as [56].

Example 5.1

Linearize the following first-order nonlinear differential equation (i.e., force Equation in x-direction; Eq. (5.3)), using the direct substitution.

$$m(\dot{U} - VR + WQ) = -mg\sin(\text{theta}) + F_{A_x} + F_{T_x}$$

Solution

The solution is found in three steps:

1) Substitution:

$$m\left[(\dot{U}_1 + \dot{u}) - (V_1 + v)(R_1 + r) + (W_1 + w)(Q_1 + q)\right] =$$
$$-mg\sin(\Theta_1 + \theta) + (F_{A_{X1}} + f_{A_X}) + (F_{T_{X1}} + f_{T_X})$$

2) Expanding:

$$LHS = m\left[\dot{U}_1 - V_1R_1 + W_1Q_1 + \dot{u} - V_1r - R_1v - vr + Q_1w + W_1q + wq\right]$$

$$RHS = -mg\left[\sin\Theta_1\cos\theta + \sin\theta\cos\Theta_1\right] + F_{A_{X1}} + f_{A_X} + F_{T_{X1}} + f_{T_X}$$

3) Nonlinear terms (e.g., vr) are removed. Moreover, the sin(θ) is replaced with θ, and cos(θ) is replaced with 1. Thus, the linearized equation will be:

$$m(\dot{u} - V_1r - R_1v + Q_1w + W_1q) = -mg\cos\Theta_1\theta + f_{A_X} + f_{T_X}$$

5.4.2 Decoupling

Another technique to simplify a nonlinear equation is applied by ignoring the coupling between various parameters. It is widely accepted that there is subtle coupling between lateral-directional and longitudinal motions. In many cases, these two motions are decoupled to simplify the analysis task. In this section, the root causes of coupling are introduced. Then, the decoupling techniques and benefits are presented.

There is a strong coupling between lateral and directional state variables. Any lateral moment (L) may produce directional motion (y). Any directional moment (N) may produce lateral motion (ϕ). The reason lies behind the existence of lateral moment arm and directional moment arm at the same time. In other word, any deflection in aileron (δ_A) or rudder (δ_R) will generates state variables of β, ϕ, ψ, P, R.

1) **Effect of directional motion on lateral motion:** In a conventional airplane, the vertical tail is located at the top the fuselage. Thus, the aerodynamic center of the vertical tail (ac_V) is above (Figure 5.5) the center of gravity (cg) of the airplane. Therefore, a side force that is intended to produce a yawing motion (directional motion), has a secondary effect which is producing a rolling (lateral) motion.
2) **Effect of lateral motion on directional motion:** Assume that an aircraft is supposed to roll (Figure 5.6). The autopilot is moving the stick/actuator to deflect ailerons in order to generate a rolling motion (lateral motion). Any rolling motion produces a sideslip angle (β). A side slip angle generates a side force (Y). By the same token, this side force is producing a yawing (directional) motion. Furthermore, when both ailerons deflect with the same amount, the drag of left- and right-wing sections will be different. This also creates a yawing moment.
3) **Effect of lateral motion on directional motion:** When a UAV has any lateral (roll), or directional (yaw) motion, it will create a number of effects on the longitudinal motion. For instance, when a UAV rolls (changes the bank angle; ϕ), the aircraft drag will be increased; which results in a reduction in the airspeed; V (i.e., longitudinal variable). Moreover, if the engine thrust is not used to compensate such airspeed reduction, the aircraft begins descending; or changes the altitude; h (i.e., a longitudinal variable). Another case is the influence of a change in heading angle (ψ) to an increase in the drag, which results in a reduction in the airspeed; V. Any roll or yaw requires an increase in engine thrust to keep the airspeed (compared with the cruising flight).

To consider the influence of lateral-directional motion on the longitudinal motion; there are basically two types of aircraft dynamic coupling: (a) coupling of the state

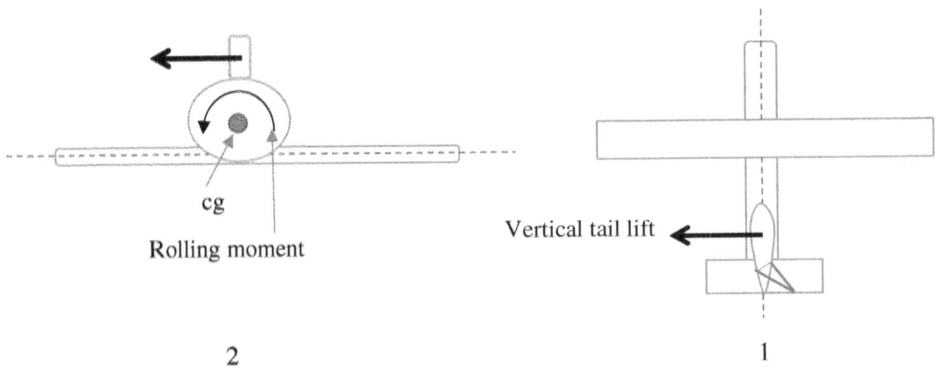

Figure 5.5 Effect of directional motion on lateral motion.

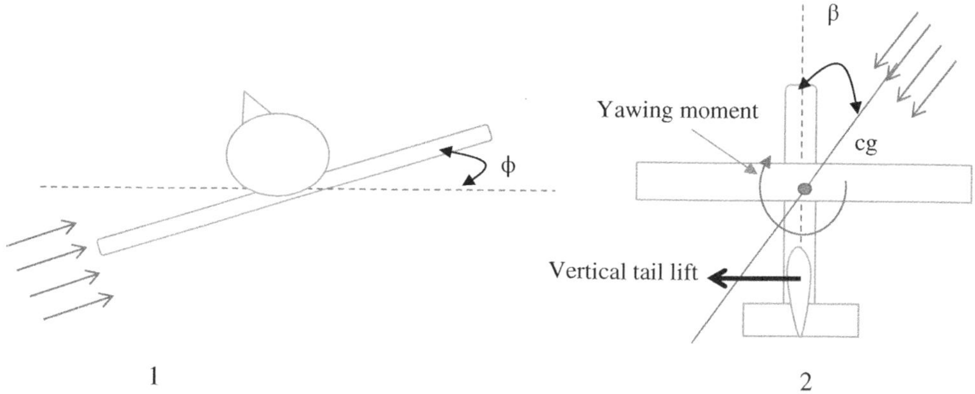

Figure 5.6 Effect of lateral motion on directional motion.

variables; and (b) coupling of the forces. These two cases are discussed in detail in Section 5.5.

When longitudinal motion is decoupled from lateral-directional motions, the following independent longitudinal state Eq. (5.50) and lateral-directional state Eq. (5.51) are derived.

$$\begin{bmatrix} \dot{u} \\ \dot{\alpha} \\ \dot{\theta} \end{bmatrix} = A_{lon} \begin{bmatrix} u \\ \alpha \\ \theta \end{bmatrix} + B_{lon} \begin{bmatrix} \delta_E \\ \delta_T \end{bmatrix} \tag{5.50}$$

$$\begin{bmatrix} \dot{\beta} \\ \dot{\phi} \\ \dot{\psi} \end{bmatrix} = A_{lat-Dir} \begin{bmatrix} \beta \\ \phi \\ \psi \end{bmatrix} + A_{lat-Dir} \begin{bmatrix} \delta_A \\ \delta_R \end{bmatrix} \tag{5.51}$$

For the longitudinal and lateral-directional A and B matrices, references such as [40, 53, 54] are suggested.

5.5 Fixed-Wing UAV Dynamic Models

The original aircraft model consists of a set of coupled first-order nonlinear ordinary differential equations that provide an accurate representation of the dynamic behavior of the aircraft. Depending on the numerical solution and design criteria including cost, simplicity, efficiency, et cetera, the equations of motion for the aircraft may have five different forms: (a) nonlinear fully coupled; (b) nonlinear semi-coupled; (c) nonlinear decoupled; (d) linear coupled; and (e) linear decoupled equations. The first form represents the fundamental equations of motion. The other forms are derived from these equations using two simplification techniques (linearization and decoupling). In this section, these forms are presented.

In a literature survey, designers were seen to have employed a variety of different dynamic models. A majority of simulation software products, including MathWork's [57] Matlab/Simulink simulation utilize semi-coupled nonlinear equations of motion. Most of these software packages do not include coupling derivatives. Ref. [58] utilizes the nonlinear decoupled equations of motion to design a full envelope missile autopilot. A different version of the nonlinear decoupled equations of motion was employed in Ref. [59] to synthesize a full envelope nonlinear autopilot for an air-to-air missile. For the longitudinal motion of a hypersonic aircraft, the nonlinear decoupled longitudinal (three DOF) equations of motion are used in Ref. [60] to synthesize a robust flight control system. Ref. [61] has employed the nonlinear semi-coupled equations of motion to design a nonlinear missile autopilot using time-scale separation. In a similar way, Ref. [62] used the nonlinear, semi-coupled equations of motion to design the control law for a supermaneuverable aircraft.

5.5.1 Nonlinear Fully Coupled Equations of Motion

The standard body-axis nonlinear fully coupled equations of motion include three force- and three moment- first-order differential equations (in state-space model) as follows (Ref. [52]):

$$\dot{U} = RV - WQ - g\sin\theta + \frac{1}{m}\left[-D + T\cos\alpha\right] \tag{5.52}$$

$$\dot{V} = -UR + WP + g\sin\varphi\cos\theta + \frac{1}{m}\left[Y + T\cos\alpha\sin\beta\right] \tag{5.53}$$

$$\dot{W} = UQ - VP + g\cos\varphi\cos\theta + \frac{1}{m}\left[-L - T\sin\alpha\right] \tag{5.54}$$

$$\dot{P} = (c_1R + c_2P)Q + c_3(L_A + L_T) + c_4(N_A + N_T) \tag{5.55}$$

$$\dot{Q} = c_5PR + c_6\left(P^2 - R^2\right) + c_7M \tag{5.56}$$

$$\dot{R} = (c_8P - c_2R)Q + c_4(L_A + L_T) + c_9(N_A + N_T) \tag{5.57}$$

In the above equations, the parameters c_i are functions of the UAV moments of inertia, and these can be calculated from the formula introduced in Ref. [54]. The parameters U, V, W are the linear velocity components, and P, Q, R are the corresponding angular rates. The aerodynamic forces D, Y, and L are the drag, side-force, and lift, and L_A, M_A, N_A, are the aerodynamic moments. The variables α, β, ϕ, and θ represent angle of attack, sideslip angle, bank angle, and pitch angle respectively. The motions of a UAV are: (a) longitudinal; (b) lateral; and (c) directional.

5.5.2 Nonlinear Semi-Coupled Equations of Motion

The nonlinear semi-coupled equations of motion are similar to the nonlinear fully coupled equations with an exception (5.48 through 5.53). The normal (Z or F_z) and axial

(X or F_x) aerodynamic forces from Eqs. (5.41) and (5.42) do not contain the very last terms (the coupling derivatives). In another word:

$$X = X_\alpha \alpha + X_u u + X_q q + X_{\dot{\alpha}} \dot{\alpha} + X_{\delta_e} \delta_e \tag{5.58}$$

$$Z = Z_\alpha \alpha + Z_{\dot{\alpha}} \dot{\alpha} + Z_u u + Z_q q + Z_{\delta_e} \delta_e \tag{5.59}$$

Thus, in this model, the motions are coupled, but the forces are not coupled.

5.5.3 Nonlinear Decoupled Equations of Motion

When only decoupling technique (not linearization) is applied to Eqs. (5.3) through (5.8), the equations of motion are still nonlinear, but decoupled as shown in the following two sections.

1) Longitudinal
   Assuming a constant mass, no roll rate, a zero roll angle, no sideslip, and no yaw rate, the nonlinear longitudinal equations of motion reduce to two forces equations and one moment equation.

$$\dot{U} = -WQ - g\sin\theta + \frac{1}{m}\left[-D + T\cos\alpha\right] \tag{5.60}$$

$$\dot{W} = UQ + g\cos\theta + \frac{1}{m}\left[-L - T\sin\alpha\right] \tag{5.61}$$

$$\dot{Q} = \frac{M}{I_Y} \tag{5.62}$$

2) Lateral-Directional
   A ssuming zero pitch rate, constant pitch attitude, constant air speed, and constant altitude, the nonlinear lateral-directional equations of motion reduce to one force equation and two moment equations.

$$\dot{V} = g\sin\varphi\cos\theta_1 + W_1 P - U_1 R + \frac{1}{m}\left[Y + T\cos\alpha_1 \sin\beta\right] \tag{5.63}$$

$$\dot{P} = c_3\left(L_A + L_T\right) + c_4\left(N_A + N_T\right) \tag{5.64}$$

$$\dot{R} = c_4\left(L_A + L_T\right) + c_9\left(N_A + N_T\right) \tag{5.65}$$

where c_3, c_4, and c_9 are introduced in Ref. [54].

5.5.4 Linear Coupled Equations of Motion

By applying only linearization (i.e., not the decoupling) to the nonlinear equations of motion (Eqs. 5.3 through 5.8), the linearized coupled equations are produced. The linear coupled equations of motion may be shown into two forms: (a) differential equation; and (b) state-space representation.

a) **Basic differential equations**
The linear coupled equations of motion in the form of differential equation may be divided into two groups: (a) three force equations; and (b) three moment equations. They are provided below:
Force equations:

$$m(\dot{u}-V_1r-R_1v+W_1q+Q_1w)=-mg\cos\Theta_1\theta+f_{A_X}+f_{T_X} \tag{5.66}$$

$$m(\dot{v}+U_1r+R_1u-W_1p-P_1w)=-mg(\sin f_1\sin\Theta_1\theta+\cos f_1\cos\Theta_1\varphi)+f_{A_Y}+f_{T_Y} \tag{5.67}$$

$$m(\dot{w}-U_1q-Q_1u+V_1p+P_1v)=-mg(\cos f_1\sin\Theta_1\theta-\sin f_1\cos\Theta_1\varphi)+f_{A_Z}+f_{T_Z} \tag{5.68}$$

Moment equations:

$$I_{xx}\dot{p}-I_{xz}\dot{r}-I_{xz}(P_1q+Q_1p)+(I_{zz}-I_{yy})(R_1q-Q_1r)=l_A+l_T \tag{5.69}$$

$$I_{yy}\dot{q}+(I_{xx}-I_{zz})(P_1r+R_1p)+I_{xz}(2P_1p-2R_1r)=m_A+m_T \tag{5.70}$$

$$I_{zz}\dot{r}-I_{xz}\dot{p}+(I_{yy}-I_{xx})(P_1q+Q_1p)+I_{xz}(Q_1r+R_1q)=n_A+n_T \tag{5.71}$$

where the subscript 1 refers to the initial trim point. For the derivation details, the interested reader may refer to Ref. [40].

b) **State–space representation**
By incorporating stability and control derivatives, and converting the linear coupled equations of motion (Eqs. 5.66 through 5.71) into state-space model, the following linearized equations are produced.

$$\begin{aligned} E\dot{x} &= A^{*}x+B^{*}u \\ y &= Cx+Du \end{aligned} \tag{5.72}$$

where

$$A^{*}=\begin{bmatrix} X_u & X_\alpha & -g\cos\gamma & 0 & X_\beta & 0 & 0 & 0 \\ Z_u & Z_\alpha & -g\sin\gamma & Z_q+V_T & 0 & Z_\varphi & 0 & 0 \\ 0 & 0 & 0 & 1 & 0 & 0 & 0 & 0 \\ M_u & M_\alpha & 0 & M_q & 0 & 0 & 0 & 0 \\ 0 & 0 & 0 & 0 & Y_\beta & g\cos\theta_o & Y_p & Y_r-V_T \\ 0 & 0 & 0 & 0 & 0 & 0 & \dfrac{\cos\gamma_o}{\cos\theta_o} & \dfrac{\sin\gamma_o}{\cos\theta_o} \\ 0 & 0 & 0 & 0 & \mu L_\beta+\sigma N_\beta & 0 & \mu L_p+\sigma N_p & \mu L_r+\sigma N_r \\ 0 & 0 & 0 & 0 & \mu N_\beta+\sigma L_\beta & 0 & \mu N_p+\sigma L_p & \mu N_r+\sigma L_r \end{bmatrix} \tag{5.73}$$

$$B^{*} = \begin{bmatrix} X_{\delta_t}\cos\alpha & X_{\delta_e} & 0 & 0 \\ X_{\delta_t}\sin\alpha & Z_{\delta_e} & 0 & 0 \\ 0 & 0 & 0 & 0 \\ M_{\delta_t} & M_{\delta_e} & 0 & 0 \\ 0 & 0 & Y_{\delta_a} & Y_{\delta_r} \\ 0 & 0 & 0 & 0 \\ 0 & 0 & \mu L_{\delta_a} + \sigma N_{\delta_a} & \mu L_{\delta_r} + \sigma N_{\delta_r} \\ 0 & 0 & \mu N_{\delta_a} + \sigma L_{\delta_a} & \mu N_{\delta_r} + \sigma L_{\delta_r} \end{bmatrix},$$

$$E = \begin{bmatrix} 1 & 0 & 0 & 0 & 0 & 0 & 0 & 0 \\ 0 & V_T - Z_{\dot{\alpha}} & 0 & 0 & 0 & 0 & 0 & 0 \\ 0 & 0 & 1 & 0 & 0 & 0 & 0 & 0 \\ 0 & -M_{\dot{\alpha}} & 0 & 1 & 0 & 0 & 0 & 0 \\ 0 & 0 & 0 & 0 & V_T & 0 & 0 & 0 \\ 0 & 0 & 0 & 0 & 0 & 1 & 0 & 0 \\ 0 & 0 & 0 & 0 & 0 & 0 & 1 & 0 \\ 0 & 0 & 0 & 0 & 0 & 0 & 0 & 1 \end{bmatrix}$$

The eight states and four control variables are $x = [V_t, \alpha, \theta, Q, \beta, \phi, P, R]^T$, and $u = [\delta_T, \delta_E, \delta_A, \delta_R]^T$. For the derivation details, the interested reader may refer to Ref. [55].

5.5.5 Linear Decoupled Equations of Motion

Decoupling is based on the assumption that the longitudinal motion is independent of the lateral-directional motion. When decoupling (no coupling effects) and linearization techniques are applied simultaneously to Eqs. (5.3) through (5.8), the state-space equations [53] are split into two groups, each having four states, two inputs, and four outputs.

A fundamental point when applying this form is that they are reliable only in the vicinity of the initial *trim point*. The validity of these equations is conversely related to the distance from the trim point. As the flight condition get further from trim point, the validity of the result is reduced. The linear decoupled equations of motion are obtained as:

$$\begin{aligned} \dot{x} &= Ax + Bu \\ y &= Cx + Du \end{aligned} \tag{5.74}$$

where the A, B, C, and D matrices are presented [53] below.

a) **Longitudinal state-space model**

$$\begin{bmatrix} \dot{u} \\ \dot{w} \\ \dot{q} \\ \dot{\theta} \end{bmatrix} = \begin{bmatrix} X_u & X_w & 0 & -g \\ Z_u & Z_w & u_o & 0 \\ M_u^{'} & M_w^{'} & M_q^{'} & 0 \\ 0 & 0 & 1 & 0 \end{bmatrix} \begin{bmatrix} u \\ w \\ q \\ \theta \end{bmatrix} + \begin{bmatrix} X_{\delta_E} \\ Z_{\delta_E} \\ M_{\delta_E}^{'} \\ 0 \end{bmatrix} \delta_E \tag{5.75}$$

b) **Lateral-directional state-space model**

$$\begin{bmatrix} \dot{\beta} \\ \dot{p} \\ \dot{r} \\ \dot{\varphi} \end{bmatrix} = \begin{bmatrix} \frac{Y_\beta}{u_o} & \frac{Y_p}{u_o} & -1+\frac{Y_r}{u_o} & \frac{g\cos\theta}{u_o} \\ L_\beta & L_p & L_r & 0 \\ N_\beta & N_p & N_r & 0 \\ 0 & 1 & 0 & 0 \end{bmatrix} \begin{bmatrix} \beta \\ p \\ r \\ \varphi \end{bmatrix} + \begin{bmatrix} 0 & \frac{Y_{\delta_A}}{u_o} \\ L_{\delta_A} & L_{\delta_R} \\ N_{\delta_A} & N_{\delta_R} \\ 0 & 0 \end{bmatrix} \begin{bmatrix} \delta_A \\ \delta_R \end{bmatrix} \tag{5.76a}$$

In other words, the longitudinal motion is independent of the lateral-directional motion. In both longitudinal (Eq. 5.75) and lateral-directional (Eq. 5.76a) models, the C is a 4×4 identity matrix, and D is a 4×2 matrix where all elements are zero. However, these two models may be reformatted and combined into one model.

$$\begin{aligned} E\dot{x} &= A^* x + B^* u \\ y &= Cx + Du \end{aligned} \tag{5.76b}$$

where

$$A^* = \begin{bmatrix} X_u & X_\alpha & -g\cos\gamma & 0 & 0 & 0 & 0 & 0 \\ Z_u & Z_\alpha & -g\sin\gamma & Z_q + V_T & 0 & 0 & 0 & 0 \\ 0 & 0 & 0 & 1 & 0 & 0 & 0 & 0 \\ M_u & M_\alpha & 0 & M_q & 0 & 0 & 0 & 0 \\ 0 & 0 & 0 & 0 & Y_\beta & g\cos\theta_o & Y_p & Y_r - V_T \\ 0 & 0 & 0 & 0 & 0 & 0 & \frac{\cos\gamma_o}{\cos\theta_o} & \frac{\sin\gamma_o}{\cos\theta_o} \\ 0 & 0 & 0 & 0 & \mu L_\beta + \sigma N_\beta & 0 & \mu L_p + \sigma N_p & \mu L_r + \sigma N_r \\ 0 & 0 & 0 & 0 & \mu N_\beta + \sigma L_\beta & 0 & \mu N_p + \sigma L_p & \mu N_r + \sigma L_r \end{bmatrix} \tag{5.77}$$

$$B^* = \begin{bmatrix} X_{\delta_t}\cos\alpha & X_{\delta_e} & 0 & 0 \\ X_{\delta_t}\sin\alpha & Z_{\delta_e} & 0 & 0 \\ 0 & 0 & 0 & 0 \\ M_{\delta_t} & M_{\delta_e} & 0 & 0 \\ 0 & 0 & Y_{\delta_a} & Y_{\delta_r} \\ 0 & 0 & 0 & 0 \\ 0 & 0 & \mu L_{\delta_a} + \sigma N_{\delta_a} & \mu L_{\delta_r} + \sigma N_{\delta_r} \\ 0 & 0 & \mu N_{\delta_a} + \sigma L_{\delta_a} & \mu N_{\delta_r} + \sigma L_{\delta_r} \end{bmatrix},$$

$$E = \begin{bmatrix} 1 & 0 & 0 & 0 & 0 & 0 & 0 & 0 \\ 0 & V_T - Z_{\dot{\alpha}} & 0 & 0 & 0 & 0 & 0 & 0 \\ 0 & 0 & 1 & 0 & 0 & 0 & 0 & 0 \\ 0 & -M_{\dot{\alpha}} & 0 & 1 & 0 & 0 & 0 & 0 \\ 0 & 0 & 0 & 0 & V_T & 0 & 0 & 0 \\ 0 & 0 & 0 & 0 & 0 & 1 & 0 & 0 \\ 0 & 0 & 0 & 0 & 0 & 0 & 1 & 0 \\ 0 & 0 & 0 & 0 & 0 & 0 & 0 & 1 \end{bmatrix}$$

Thus, the state and control variables are $x = [V_t, \alpha, \theta, Q, \beta, \phi, P, R]^T$, and $u = [\delta_T, \delta_E, \delta_A, \delta_R]^T$. For calculations of dimensional stability derivatives such as M_q and $M_{\dot{\alpha}}$, and dimensional control derivatives such as $M_{\delta E}$, you may consult with references such as [40, 53, 54].

By solving state space equations, one can derive various longitudinal, lateral, and directional transfer functions. For instance, using Cramer's rule, the pitch rate (q) to elevator deflection (δ_e) transfer function will be:

$$\frac{q(s)}{\delta_e(s)} = \frac{\begin{vmatrix} (s - X_u - X_{T_u}) & -X_\alpha & X_{\delta_e} \\ -Z_u & \{s(U_1 - Z_{\dot{\alpha}}) - Z_\alpha\} & Z_{\delta_e} \\ -(M_u + M_{T_u}) & -(M_{\dot{\alpha}} s + M_\alpha + M_{T_\alpha}) & M_{\delta_e} \end{vmatrix}}{\begin{vmatrix} (s - X_u - X_{T_u}) & -X_\alpha & g\cos\Theta_1 \\ -Z_u & \{s(U_1 - Z_{\dot{\alpha}}) - Z_\alpha\} & \{-(Z_q + U_1)s + g\sin\Theta_1\} \\ -(M_u + M_{T_u}) & -(M_{\dot{\alpha}} s + M_\alpha + M_{T_\alpha}) & (s^2 - M_q s) \end{vmatrix}}$$

These transfer functions are used to simulate the flight, analyze the UAV stability, and to characterize the flight modes (e.g., phugoid).

5.5.6 Reformulated (Nonlinear Semi-Coupled) Equations of Motion

These equations are basically the same as the nonlinear ones, except that they are reformulated to be appropriate for some special applications (e.g., two-DOF controller). There is a relationship between speed in z-direction (i.e., w) and angle of attack (α), and a relationship between speed in y-direction (v) and sideslip angle (β) as:

$$\sin(\alpha) = \frac{w}{U_o} \tag{5.78}$$

$$\tan(\beta) = \frac{v}{U_o} \tag{5.79}$$

where U_o is the initial trim airspeed. By assuming small angles, and applying the linearization technique (small perturbation theory), these two equations may be simplified to:

$$\alpha = \frac{w}{U_o} \tag{5.80}$$

$$\beta = \frac{v}{U_o} \tag{5.81}$$

Differentiating both sides yields

$$\dot{w} = \dot{\alpha} U_o \tag{5.82}$$

$$\dot{v} = \dot{\beta} U_o \tag{5.83}$$

Using these two differential equations, the $\dot{w}$ is replaced with $\dot{\alpha} U_o$, and $\dot{v}$ is replaced with $\dot{\beta} U_o$. There are various forms of implementing for these two replacements. Here, one example is introduced. The original dynamics model of the UAV is reformulated as two lower-order systems by invoking time-scale separation between the vehicle's translational and rotational dynamics. The slow or primary controlled variables are the forward speed, climb angle, bank angle, and sideslip angle (U, γ, ϕ, β). The fast or secondary controlled variables are the three angular rates plus the angle of attack (α, Q, P, R). The four controls are the throttle, elevator, aileron and rudder (δ_T, δ_E, δ_A, δ_R).

Slow dynamics:

$$\dot{U} = \frac{1}{m}\left[-D + Y\sin\beta - mg\sin\gamma + T\cos\beta\cos\alpha\right] \quad (5.84)$$

$$\dot{\gamma} = \frac{1}{mV}\left[L\cos\varphi - mg\cos\gamma - Y\sin\varphi\cos\beta\right] + \frac{T}{mV}\left[\sin\varphi\sin\beta\cos\alpha + \cos\varphi\sin\alpha\right] \quad (5.85)$$

$$\dot{\Phi} = P + \tan\theta\left(Q\sin\Phi + R\cos\Phi\right) \quad (5.86)$$

$$\dot{\beta} = P\sin\alpha - R\cos\alpha + \frac{1}{mV}\left[mg\cos\gamma\sin\varphi + Y\cos\beta - T\sin\beta\cos\alpha\right] \quad (5.87)$$

Fast dynamics:

$$\dot{P} = \left(c_1 R + c_2 P\right)Q + c_3\bar{L} + c_4 N \quad (5.88)$$

$$\dot{Q} = c_5 PR - c_6\left(P^2 - R^2\right) + c_7 M \quad (5.89)$$

$$\dot{R} = \left(c_8 P - c_2 R\right)Q + c_4\bar{L} + c_9 N \quad (5.90)$$

$$\dot{\alpha} = Q - \tan(\beta)\left[P\cos\alpha + R\sin\alpha\right] + \frac{1}{mV\cos\beta}\left[-L + mg\cos\gamma\cos\varphi - T\sin\alpha\right] \quad (5.91)$$

In Chapter 6, Section 6.2.4, the concept and application of slow and fast dynamics are presented.

5.5.7 Un-powered Gliding Equations of Motion

If an unpowered UAV is considered a point mass, its gliding motion in a vertical plane is governed by:

$$\dot{V} = -\frac{D}{m} - g\sin(\gamma) \quad (5.92)$$

$$\dot{\gamma} = \frac{L}{mV} - \frac{g}{V}\cos(\gamma) \quad (5.93)$$

$$\dot{h} = V\sin(\gamma) \quad (5.94)$$

$$\dot{x} = V\cos(\gamma) \quad (5.95)$$

These are governing equations of motion in xz-plane.

5.6 Dynamic Model Approximation

In a six DOF motion, there are various lateral-directional-longitudinal motions in three planes which are usually coupled. If we decouple these motions, we have to assume zero influence from each motion to another one. In this section, the dynamic model approximation for a number of isolated motions are presented: (a) pure pitching motion; (b) pure rolling motion; (c) pure yawing motion; and (d) longitudinal modes.

5.6.1 Pure Pitching Motion Approximation

In a pure pitching motion, the sum of all pitching moments (M) will be applied about aircraft cg to change the pitch angle (θ). The variables are (a) angle of attack (α); (b) pitch angle; (c) rate of change of angle of attack; and (d) rate of change of pitch angle. By assuming "the change in α and θ are identical," one can derive the following approximate governing differential equation for longitudinal motion:

$$\ddot{\alpha} - \left(M_q + M_{\dot{\alpha}}\right)\dot{\alpha} - M_{\alpha}\alpha = M_{\delta E}\delta_E \tag{5.96}$$

where M_q and $M_{\dot{\alpha}}$ are two-dimensional longitudinal stability derivatives, and $M_{\delta E}$ is a dimensional longitudinal control derivative.

5.6.2 Pure Rolling Motion Approximation

The approximate for the bank-angle-to-aileron-deflection transfer function is modeled by a second order system as:

$$\frac{\phi(s)}{\delta_A(s)} = \frac{L_{d\delta A}}{s^2 - sL_p} \tag{5.97}$$

where L_p and $L_{\delta A}$ are dimensional roll damping (stability), and roll control derivatives respectively. Since the L_p is often negative, and there is a free "s" in the denominator, this motion is neutrally stable.

5.6.3 Pure Yawing Motion Approximation

The approximate yawing moment equation can be derived from the approximation of the Dutch roll motion [53]:

$$\ddot{\psi} - N_r\dot{\Psi} + N_{\beta}\Psi = N_{\delta R}\delta_R \tag{5.98}$$

where N_r and N_{β} are two-dimensional yaw stability derivatives, and $N_{\delta R}$ is the dimensional yaw control derivative.

Applying the Laplace transform, this differential equation is readily converted to a transfer function:

$$\frac{\psi(s)}{\delta_R(s)} = \frac{N_{\delta R}}{s^2 - sN_r + N_{\beta}} \tag{5.99}$$

5.6.4 Longitudinal Oscillatory Modes Approximation

In the longitudinal motion, there are mainly two oscillatory motion modes: (a) long-period or phugoid mode; and (b) short-period mode.

a) **Phugoid mode**
The phugoid mode is considered as a gradual interchange of potential and kinetic energy about the equilibrium altitude and airspeed. Its characteristic equation [53] in s-domain is:

$$s^2 - sX_u - \frac{gX_u}{U_o} = 0 \tag{5.100}$$

where X_u is a dimensional longitudinal stability derivative.

b) **Short-period mode**
An approximation to the short-period mode of longitudinal motion can be obtained by assuming no change in the airspeed. The approximate characteristic equation for this mode is:

$$s^2 - s\left(M_q + M_{\dot{\alpha}} + \frac{Z_\alpha}{U_o}\right) + M_q \frac{Z_\alpha}{U_o} - M_\alpha = 0 \tag{5.101}$$

Using these two approximate characteristic equations, one can design the feedback system to control the longitudinal motion.

5.7 Quadcopter (Rotary-Wing) Dynamic Model

Quadrotor UAVs are becoming increasingly popular. They offer flight characteristics comparable to traditional helicopters, namely, stationary, vertical, and lateral flights in a wide range of speeds, with a much simpler mechanical structure. With their small-diameter propellers driven by electric motors, these multirotor platforms are safer to operate than helicopters in indoor environments. Also, they have sufficient payload transporting capability and flight endurance for various missions.

This section presents technical features, and advantages and disadvantages of the quadcopter configuration. A number of sections in other chapters of this book also provides various design aspects of quadcopters. For instance, Section 2.10 presents the quadcopter configuration including technical features, and advantages and disadvantages of the quadcopter configuration. Moreover, Section 4.8 provides the aerodynamic design of quadcopters.

5.7.1 Overall Thrust of Four Motors

The brushless electric DC motors are standard motors utilized in quadcopters. The required electric power is provided by batteries.

The magnitude of the thrust force (F) generated by each propeller is proportional to the prop's rotational speed (ω) squared:

$$F = k_1 \omega^2 \tag{5.102}$$

where k_1 is a proportionality constant. When ω is in rad/s, and F is in N, a typical value for k_1 for a small quadcopter is 10^{-6} N/(rad/s)^2. For four similar motors, this force mathematical model will be employed.

$$F_1 = k_1\omega_1^2 \tag{5.103}$$

$$F_2 = k_1\omega_2^2 \tag{5.104}$$

$$F_3 = k_1\omega_3^2 \tag{5.105}$$

$$F_4 = k_1\omega_4^2 \tag{5.106}$$

For any non-hovering flight, the rotational velocities of motors will be different. This is to create a desired torque for an intended maneuvering flight operation. It needs to be reminded, that the quadcopters, unlike fixed-wing aircraft, do not have any conventional control surfaces (e.g., elevator, rudder, and ailerons). Thus, for any rotational motion, an external torque is created instead of aerodynamic moments (e.g., rolling, pitching, and yawing moments). To include the direction of rotation in the thrust force calculation, the following force model is suggested:

$$F_i = k_1\left|\omega_i\right|\omega_i \tag{5.107}$$

This format guarantees that for both counterclockwise (say, positive) rotation, and clockwise (say negative) rotation, the motor torque direction is correctly incorporated. The subscript i refers (Figure 5.7) to the motor number (i.e., i = 1, 2, 3 or 4). To ensure that the overall force (all thrust forces in the level configuration) are upward, in summing all force thrusts, another negative sign are added to two negatives values.

$$F = F_1 + F_2 + F_3 + F_4 = k_1\left[\left|\omega_1\right|\omega_1 - \left|\omega_2\right|\omega_2 + \left|\omega_3\right|\omega_3 - \left|\omega_4\right|\omega_4\right] \tag{5.108}$$

The Eq. (5.108) is valid for level flight; no pitch angle (θ), and no bank angle. In such configuration, the z and x components (Figure 5.8) of this overall force are;

$$F_z = F.\cos(\Theta) \tag{5.109}$$

$$F_x = F.\sin(\Theta) \tag{5.110}$$

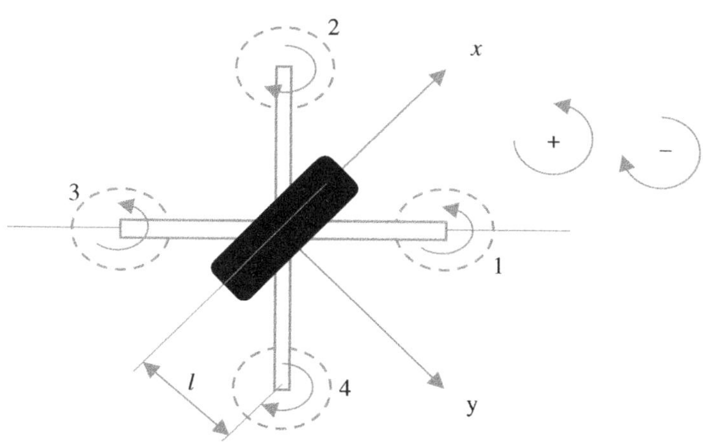

Figure 5.7 Direction of propeller rotation for each motor (Top-view).

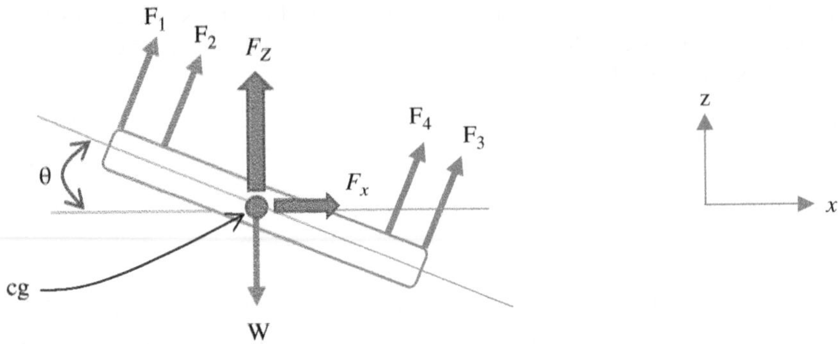

Figure 5.8 Lift and thrust forces (side-view with a pitch angle); wind-axis coordinate system.

Example 5.2

A small quadcopter with a mass of 0.5 kg has four electric motors, has a coefficient, k_1 of $10^{-6}\,N/(rad/s)^2$.

a) If all motors are turning with a rotational speed of 8000 rpm, determine the overall upward force generated by all four engines.
b) The quadcopter has a pitch angle of 10°. Determine the force components in z and x directions.

Solution

a) Overall upward force

$$F = k_1\left[|\omega_1|\omega_1 - |\omega_2|\omega_2 + |\omega_3|\omega_3 - |\omega_4|\omega_4\right] \tag{5.108}$$

$$F = 10^{-6}\left[8000\times 8000 - |-8000|\times(-8000) + 8000\times 8000 - |-8000|\times(-8000)\right]\times 0.105^2$$

$$F = 10^{-6}\times 4\times\left[8000^2\right]\times 0.105^2 = 2.807\,N$$

Please note that the 0.105 is used to convert rpm to rad/sec.

b) Force components

$$F_z = F\cos(\Theta) = 2.807\times\cos(10) = 2.76\,N \tag{5.109}$$

$$F_x = F\sin(\Theta) = 2.807\times\sin(10) = 0.487\,N \tag{5.110}$$

Figure 5.9 presents the variations of thrust of an electric motor versus its rotational speed as measured in the lab. This motor is employed in mini UAVs such as AscTec Hummingbird. This test thrust (N) curve can be fitted with an expression as a square function of motor angular speed (rpm).

Figure 5.10 provides the variations of torque of an electric motor versus its rotational speed as measured in the lab. This test torque (Nm) curve can also be fitted with an expression as a square function of motor angular speed (rpm).

Rotation of the motor will induce a torque on quadcopter frame. The direction of the torque is always opposite to that of the spinning prop. If the prop is rotating

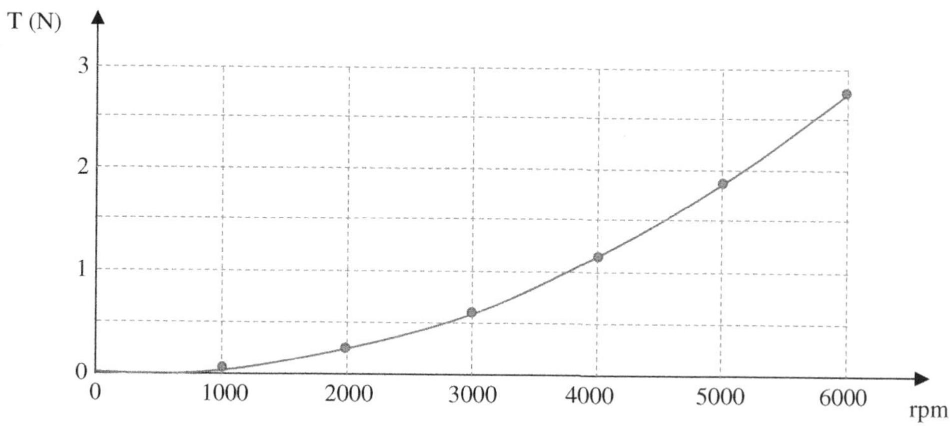

Figure 5.9 Variations of thrust of a small electric motor versus its rotational speed.

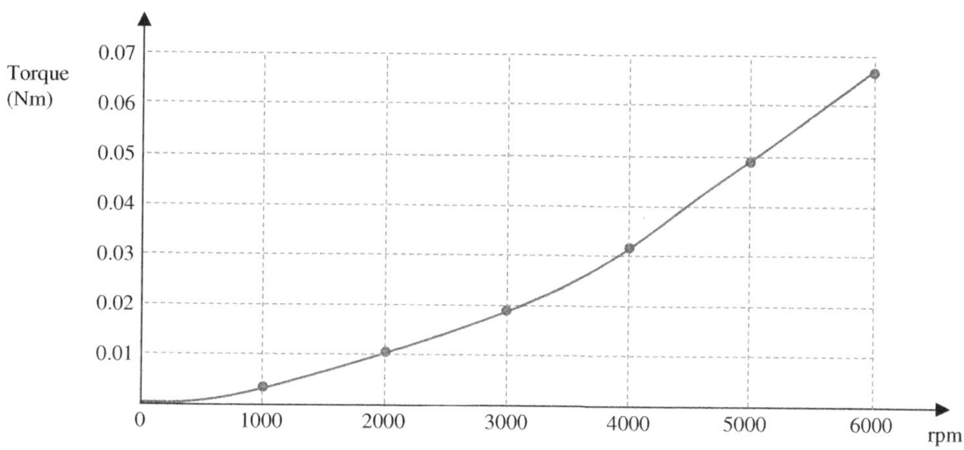

Figure 5.10 Variations of torque of a small electric motor versus its rotational speed.

counterclockwise, then the induced torque is clockwise, acting on the quadcopter main frame. To cancel out such torques, two propellers should rotate clockwise, while the other two are rotating clockwise. The prop angle of attack should be such that all propellers are generating upward force. In this model, it is assumed that the magnitude of the generated internal torque (T) is proportional to the rotational speed squared with a proportionality constant k_2.

$$T_i = -k_2 \omega_i^2 \tag{5.111}$$

The negative sign indicates that the internal torque is opposite to the direction of each prop. In order to incorporate the direction of motor torque, the ω^2 is written as $|\omega|\ \omega$. Each engine thrust force will create an external torque about the quadcopter center of gravity. It is determined by multiplying the thrust forces (F) and the length (l) between the engine center and the quadcopter center of gravity.

$$T_{x,y} = F.l \tag{5.112}$$

By employing the above mentioned trick to the expression for the external torque, we will have the following equation for the torque generated by the first motor:

$$T_{x,y} = -k_1 |\omega_1| \omega_1 l \qquad (5.113)$$

Other torque equations are written in the same format. All four engines are often located at the same distance from the quadcopter center, so the arm (l) of all four motors have the same value.

5.7.2 Dynamic Model

The mathematical model to describe the dynamics of the quadcopter is derived by employing the conservation of linear and angular momentum, i.e., Newton's second law for rigid bodies. There are five rigid bodies in the system, namely quadcopter frame and four identical motors. The internal torque generated by each rotating motor plays a significant role in the quadcopter governing equation. Thus, for a quadcopter, we will have a system of six seconds order non-linear differential equations.

The Lagrangian framework [63] is utilized to represent the position and orientation of all four motors. Each motor has only one DOF with respect to the copter frame, while the quadcopter has six DOF. The control forces to the system are the thrust forces and motor torques. Hence, the dynamic system is simplified down to only the quadcopter frame, with four external forces and 4τ. The combinations of these forces and torques will determine the status of the quadcopter.

Indeed, there are three mutually orthogonal unit vectors [64] representing the forward/backward direction, the left/right direction, and the up/down direction for the rigid body; that is a frame for the body.

$$\ddot{x} = \frac{1}{m} k_1 \left[|\omega_1| \omega_1 - |\omega_2| \omega_2 + |\omega_3| \omega_3 - |\omega_4| \omega_4 \right]_x \qquad (5.114)$$

$$\ddot{y} = \frac{1}{m} k_1 \left[|\omega_1| \omega_1 - |\omega_2| \omega_2 + |\omega_3| \omega_3 - |\omega_4| \omega_4 \right]_y \qquad (5.115)$$

$$\ddot{z} = \frac{1}{m} k_1 \left[|\omega_1| \omega_1 - |\omega_2| \omega_2 + |\omega_3| \omega_3 - |\omega_4| \omega_4 \right]_z \qquad (5.116)$$

$$\ddot{\phi} = \frac{1}{I_{xx}} k_1 \left[|\omega_4| \omega_4 - |\omega_2| \omega_2 \right] l \qquad (5.117)$$

$$\ddot{\Theta} = \frac{1}{I_{yy}} k_1 \left[|\omega_3| \omega_3 - |\omega_1| \omega_1 \right] l \qquad (5.118)$$

$$\ddot{\psi} = -\frac{1}{I_{zz}} k_2 \left[|\omega_1| \omega_1 + |\omega_2| \omega_2 + |\omega_3| \omega_3 + |\omega_4| \omega_4 \right] l \qquad (5.119)$$

where ϕ, θ, ψ are the orientations of the frame about x, y, and z axes respectively. Equations (5.114) through (5.119) may be solved by numerical integration. However, it is required to know the orientation of the quadcopter frame to determine the force components. To get the orientation of the frame as a function of time, we need to first solve Eqs. (5.117) through (5.119).

The electric motor behavior exhibits a first order dynamic system, it has an exponential variation with a time constant of τ. A typical electric motor has a time constant of about one to three seconds. Therefore, the dynamics of an electric motor is modeled as:

$$Thrust = \frac{F_{rated}}{\tau s + 1} \tag{5.120}$$

The F_{rated} is the rated force that the prop is producing when the angular rotation reaches its desired value.

5.7.3 Simplified Dynamic Model

The six DOF simplified dynamics of a quadcopter UAV [65] under the body-fixed frame is given by:

$$\ddot{x} = \frac{F_T}{m}\left(\sin\phi\sin\phi + \cos\phi\sin\theta\cos\phi\right) \tag{5.121}$$

$$\ddot{y} = \frac{F_T}{m}\left(-\cos\phi\sin\phi + \sin\phi\sin\theta\cos\phi\right) \tag{5.122}$$

$$\ddot{z} = -g + \frac{F_T}{m}\left(\cos\theta\cos\phi\right) \tag{5.123}$$

$$\dot{p} = \frac{I_{yy}-I_{zz}}{I_{xx}}qr - \frac{J_{TP}}{I_{xx}}q\Omega + \frac{\tau_\phi}{I_{xx}} \tag{5.124}$$

$$\dot{q} = \frac{I_{zz}-I_{xx}}{I_{yy}}pr + \frac{J_{TP}}{I_{yy}}p\Omega + \frac{\tau_\theta}{I_{yy}} \tag{5.125}$$

$$\dot{r} = \frac{I_{xx}-I_{yy}}{I_{zz}}pq + \frac{\tau_\varphi}{I_{zz}} \tag{5.126}$$

where F_T is the total thrust (of four motors), m is the quadcopter mass, τ_θ is the pitching torque, τ_ϕ is the rolling torque and τ_ψ is the yawing torque, and J_{TP} is the rotor inertia. Moreover, Ω denotes the overall propellers speed, and is obtained from:

$$\Omega = \omega_2 + \omega_4 - \omega_1 - \omega_3 \tag{5.127}$$

where ω_i (i = 1, 2, 3, 4) is the rotational speed of each rotor. The attitude angular speed relation between the body frame and the wind axis can be written as:

$$\begin{bmatrix} \dot{\phi} \\ \dot{\theta} \\ \dot{\varphi} \end{bmatrix} = \begin{bmatrix} 1 & \sin\phi\tan\theta & \cos\phi\tan\theta \\ 0 & \cos\phi & -\sin\phi \\ 0 & \sin\phi/\cos\theta & \cos\phi/\cos\theta \end{bmatrix} \begin{bmatrix} p \\ q \\ r \end{bmatrix} \tag{5.128}$$

This equation allows the engineer to move the angular speeds between body axis and the wind axis.

5.8 Autopilot Categories

The fundamental function of an AFCS is to control the flight variables to ensure the UAV is flying in the predetermined trajectory. In a UAV, since there is no human pilot involved, an autopilot is a device that must be capable of accomplishing all types of controlling functions including automatic take-off, flying toward the target destination, and automatic landing.

In general, there are five categories/functions for autopilots: (a) stability augmentation; (b) hold functions; (c) navigation functions; (d) command augmentation; and (e) combined category. Due to three aircraft axes (Figure 5.4), there are three stability augmentation systems (SAS), namely: (a) roll damper; (b) yaw damper; and (c) pitch damper.

Often, the aircraft lateral and directional motions are coupled. Hence, there are two basic groups of hold functions: (a) longitudinal hold functions; and (b) lateral-directional hold functions. In longitudinal plane, there are primarily three functions: (a) pitch attitude hold; (b) altitude hold; and (c) control wheel steering mode. However, in Lateral-directional mode, three hold functions are dominant: (a) bank angle hold, or wing leveler; (b) heading angle hold; and (c) turn rate mode at constant speed and altitude.

Similarly, for the navigation functions, there are two groups of hold functions: (a) longitudinal navigation functions; and (b) lateral-directional navigation functions. In longitudinal plane, there are primarily four modes: (a) automatic flare mode; (b) glide slope hold; (c) terrain following; and (d) automatic landing. However, in Lateral-directional mode, three hold functions are dominant: (a) localizer; (b) Very High Frequency Omni-Directional Range (VOR) hold; and (c) tracking a series of waypoints.

In the category of command augmentation, three basic systems are available: (a) command tracking; command generator tracker; and (c) normal acceleration CAS. Table 5.2 summarizes typical common categories of autopilots.

The features of the first four categories are explained in the following sections. In the fifth category (i.e., combined), various functions are combined for a complex mission. For instance, the autopilot of General Atomics Reaper MQ-9B (a Certifiable version of its Predator B; Figure 5.12) had been configured in an Intelligence, Surveillance, and Reconnaissance (ISR) for the entire mission. This UAV, on 25 May 2017, had an endurance of 48.2 hours. Section 9.8 introduces a number of open source commercial autopilots.

5.8.1 Stability Augmentation

For a lightly stable UAV, the autopilot must provide the augmented stability. As the name implies, an SAS is to augment the stability of an open-loop plant. It improves aircraft stability, and even stabilizes an unstable aircraft. The SAS can also be simultaneously employed with manual control (Ground-system/operator command). There are primarily three SAS, namely: (a) roll damper; (b) yaw damper; and (c) pitch damper.

The widening flight envelope (e.g., Figure 6.10 in Chapter 6) creates a need to augment the stability of the UAV's dynamics over some parts of the flight envelope. Because

Table 5.2 Autopilot categories.

Category:	1	2		3		4
		Hold functions		**Navigation functions**		
No	**Stability Augmentation Systems (SAS)**	**Longitudinal**	**Lateral-Directional**	**Longitudinal**	**Lateral-Directional**	**Command Augmentation Systems (CAS)**
1	Roll damper	Pitch Attitude hold (θ)	Bank angle hold (ϕ); wing leveler	Automatic flare mode	Localizer	Command tracking 1) Pitch rate CAS 2) Roll rate CAS
2	Yaw damper	Altitude hold (h)	Heading angle hold (ψ)	Glide slope hold	VOR hold	Command generator tracker[a] (Model following)
3	Pitch damper	Control wheel steering mode	Turn rate mode at constant speed and altitude	Terrain Following	Tracking a series of waypoints	Normal acceleration CAS (n_z)
4	–	Speed/Mach hold	–	Automatic landing	–	–

[a] Time varying trajectories.

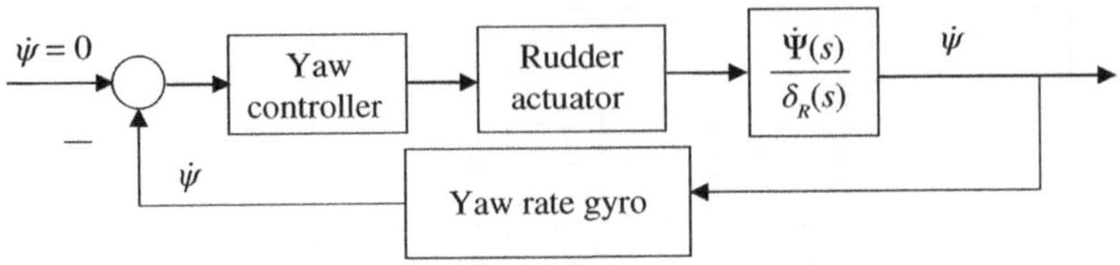

Figure 5.11 Block diagram of a yaw damper.

of the large changes in the UAV's dynamics, a dynamic model that is stable and adequately damped in one flight condition may become unstable, or at least inadequately damped in another flight condition. For such cases, an autopilot is as an automatic control system that also provides stability augmentation.

In this section, one stability augmentation application of an autopilot; yaw damper, is presented. The yaw damper is to augment the directional stability of an UAV by damping the yawing oscillation, known as Dutch-roll. The purpose of the yaw-damper feedback is to use the rudder to generate a yawing moment that opposes any yaw rate that builds up from the Dutch roll mode. Most large UAVs with high-altitude cruising flight are equipped with such mode. The reason is that at high altitude, the turbulence is strong, and gust hits the fuselage nose, and pushes the nose to the left and right almost continuously. The autopilot mode will keep the heading in the desired direction.

Figure 5.11 depicts the block diagram of a yaw damper. The reference yaw rate is desired to be zero ($\dot{\psi} = 0$), so the yaw damper will drive any undesired yaw rate to zero. A rate gyro will measure the yaw rate, and so provide a feedback for the system. The controller will produce a controlling signal which is implemented through an actuator (i.e., rudder servo). The yaw damper controller can be as simple as a gain, K. When the UAV tends to turn, and has a bank angle (ϕ_1), the yaw damper will try to fight with the constant bank angle turn. This is due to the fact that the computed yaw rate is a function of bank angle:

$$R_1 = \dot{\psi}_1 \cos(\theta_1)\cos(\phi_1) \tag{5.129}$$

where subscript 1 stands for steady-state value. One solution for such case is to use a washout filter with a time constant of about four seconds. This will create a lag in the response of the rate gyro. More details on this SAS is presented in Chapter 6 (Section 6.5).

5.8.2 Hold Functions

Often, aircraft lateral and directional motions are coupled. Thus, there are two basic groups of hold functions: (a) longitudinal hold functions; and (b) lateral-directional hold functions. In longitudinal plane, there are primarily three functions: (a) pitch attitude hold; (b) altitude hold; and (c) control wheel steering mode. However, in Lateral-directional mode, three hold functions are dominant: (a) bank angle hold, or wing leveler; (b) heading angle hold; and (c) turn rate mode at constant speed and altitude.

The longitudinal hold function is an efficient control method of cruising flight for a long period. Due to the burning fuel, the weight of aircraft at the beginning of the flight and at the end of flight is significantly different (usually about 20%). However, for a long endurance UAV such as Global Hawk, the change in the UAV weight is as much as 50%.

Figure 5.12 General Atomics MQ-9 Reaper. *Source:* Leslie Pratt.

At any weight and altitude, the lift force must be equal to aircraft weight in a straight-level flight:

$$W = L = \frac{1}{2}\rho V^2 S C_L \tag{5.130}$$

The Eq. (5.130) has four independent parameters: (a) UAV weight (W); (b) airspeed (V); (c) altitude or its corresponding air density (ρ); and (d) angle of attack (α), or its associated lift coefficient (C_L).

Since the fuel is consumed during flight, the aircraft weight is constantly decreased during the flight. In order to maintain a level flight, we must decrease the lift as well. Of the many possible solutions, only three alternatives are more practical and usually examined. In each case, two flight parameters will be held constant throughout cruise. The three options of interest for continuous decrease of the lift during cruise are (Figure 5.13):

1) Decreasing flight speed (Constant-altitude, constant-lift coefficient flight)
2) Increasing altitude (Constant-airspeed, constant-lift coefficient flight)
3) Decreasing angle of attack (Constant-altitude, constant-airspeed flight)

In the first option, the velocity must be reduced with the same rate as the aircraft weight is decreased. In the second solution, the air density must be decreased; in another word, the flight altitude must be increased. The third option offers the reduction of aircraft angle of attack; i.e., the reduction of lift coefficient. In terms of autopilot operation, the first option is applied through the engine throttle; and the third option is implemented through control stick. In the second option, no action is needed by the autopilot; the aircraft will gradually gain height (climbs).

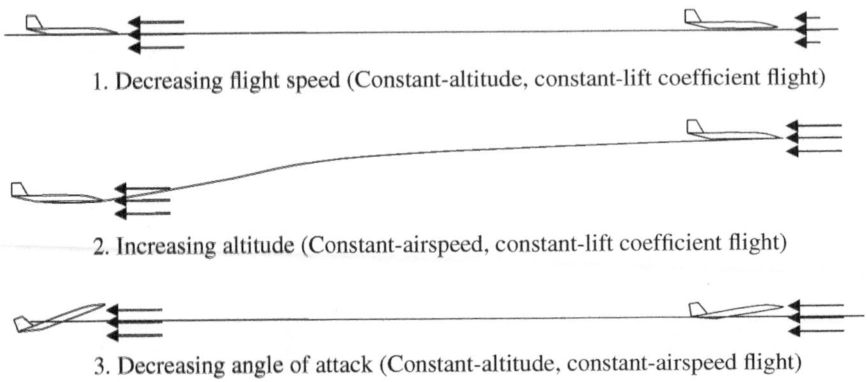

Figure 5.13 Three options of interest for a continuous decrease of the lift during cruise.

Based on the safety regulations and practical considerations, the second option is of interest for the majority of aircraft. In general, when flight is conducted under the jurisdiction of Federal Aviation Regulations, the accepted flight program is mainly the constant altitude–constant airspeed flight program.

In terms of autopilot mode, the first and second flight operations are controlled through the pitch attitude hold mode. However, the second and third flight operations are controlled through the altitude hold mode.

5.8.3 Navigation Functions

For the navigation functions, there are two groups of hold functions: (a) longitudinal navigation functions; and (b) lateral-directional navigation functions. In longitudinal plane, there are primarily four modes: (a) automatic flare mode; (b) glide slope hold; (c) Terrain Following; and (d) automatic landing. However, in Lateral-directional mode, three hold functions are dominant: (a) localizer; (b) VOR hold; and (c) Tracking a series of waypoints.

BTW, the term VOR, which stands for Very-high-frequency Omni-directional Range, is a type of radio navigation system for aircraft. The VOR station broadcasts a VHF radio signal, and data that allows the airborne receiver to derive the magnetic bearing from the station to the aircraft. This line of position is referred to as the "radial." Pilots then utilize this information to determine their position and navigate the aircraft to their destination. Chapter 8 which presents the navigation system design, also covers the navigation functions in more details.

5.8.4 Command Augmentation Systems

Some autopilots improve the stability of the aircraft (SAS), while some augment the response to a control input; Command Augmentation System (CAS). The slow modes (e.g., phugoid in longitudinal motion, and spiral in lateral-directional motion) are controllable by a human pilot in GCS. But since it is undesirable for a pilot to have to pay continuous attention to controlling these modes, an automatic control system is needed to provide "pilot relief." Reference [54] presents AFCS design and provides several detailed examples.

Figure 5.14 NASA ALTUS II.

In the category of command augmentation, three basic systems are available: (a) command tracking; (b) command generator tracker; and (c) normal acceleration CAS. The command tracking system are mainly divided into two modes: (a) pitch rate tracking; and (b) roll rate tracking. Command generator tracker is known as the model following system, since it creates time varying trajectories.

Figure 5.14 illustrates the General Atomic ALTUS II, remotely piloted aerial vehicle, developed under NASA ERAST program, for upper atmospheric science missions. The ALTUS II is a civilian variant of the MQ-1 Predator UAV designed for scientific research missions. The UAV with a wingspan of 16.5 m, aspect ratio of 24, and a wing area of 12.2 m^2 is equipped with a Rotax-912 piston engine. It has a maximum speed of 100 knot, and endurance of 24 hours and a service ceiling of 65 000 ft.

5.9 Flight Simulation – Numerical Methods

In the design, development, and verification of the autopilot (including the flight control law), the flight simulation of the UAV is an important tool. In general, there are two methods for flight simulation: (a) ground-based equipment simulator (Figure 5.15); and (b) just computer software. A Sub-method of the first method is known as the hardware-in-the-loop (HIL) simulation. HIL simulation is a type of real-time simulation.

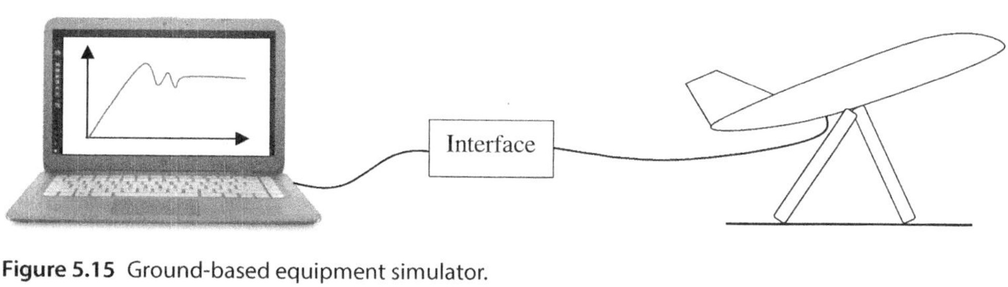

Figure 5.15 Ground-based equipment simulator.

The applications and variations for these simulations are vast. In this section, both techniques are briefly described.

If just software is utilized to simulate the flight, dynamic models of the vehicle needs to be developed. If the nonlinear differential equations are employed as the basis for the dynamics, the numerical method is the most viable and effective approach. In simulating an aerial vehicle system, all significant aspects of the dynamics must be modeled. An aircraft model includes: (a) aerodynamics; (b) propulsion system; (c) structures; (d) dynamics; (e) other mechanical, electrical, and hydraulic subsystems; and (f) the atmospheric flight condition. Depending on the application, one may choose to which level each subsystem is to be modeled.

5.9.1 Numerical Integration

The key component in a simulation software package is the UAV dynamic model represented as a set of nonlinear ordinary differential equations (ODEs). The nonlinear ordinary differential equations (such as equations of motion (5.3) through (5.8)) do not have a closed-form solution. However, there are various techniques to integrate them through numerical analysis. One popular technique (i.e., approximate solutions) is the *Runge–Kutta* method [56] that is a family of implicit and explicit iterative methods, which include the well-known routine called the Euler method. The methods; also knows as numerical integration; were developed around 1900 by the German mathematicians C. Runge and M. Kutta (RK).

Consider the following ODE initial-value problem:

$$\frac{dx}{dt} = f(x,t),\ x(t_o) = x_o \tag{5.131}$$

The simplest RK method to find the discrete solution is the Euler integration:

$$x_E(t_o + T) \approx x(t_o) + Tf(x(t_o), t_o) \tag{5.132}$$

where T is the time step, and subscript E indicates Euler. By far the most often used is the classical fourth-order Runge–Kutta formula [56] which requires four evaluations of the right-hand side per step T. The most popular RK technique is the Fourth-order Runge–Kutta method. In each step the derivative is evaluated four times: once at the initial point, twice at trial midpoints, and once at a trial endpoint. From these derivatives the final function value is calculated.

5.9.2 Matlab/Simulink

MATLAB is a multi-purpose numerical computing environment. The programming language developed by MathWorks, allows matrix operation, plotting the results of calculations, implementation of algorithm, and interfacing with programs written in other languages such as C and Python. The MATLAB software package may be employed to numerically analyze (i.e., simulate) the nonlinear ODEs. In order to solve the nonlinear coupled first-order differential equations (DEs) system numerically, two separate files must be produced in MATLAB: one to introduce the DEs (as a function), and one to solve them (via ODE45). These two are explained in the following. An Example (5.3) illustrates the application of the technique. Applied numerical methods including Runge–Kutta formula with MATLAB is presented by [66].

1) ode45

   This command solves non-stiff differential equations, medium order method. The format is:

```
[Tout, Yout] = ode45(ODE_FUN, Tspan, Yo) with TSPAN
= [To T_final]
```

   It integrates the system of differential equations y' = f(t,y) from time To to T_final with initial conditions Yo. ODE_FUN is a function handle. For a scalar T and a vector Y, ODE_FUN (T,Y) must return a column vector corresponding to f(t,y). Each row in the solution array Yout corresponds to a time returned in the column vector Tout. To obtain solutions at specific times To, T1,...,T_final (all increasing or all decreasing), use TSPAN = [To T1 ... T_final].

2) function

   This command adds a new function. New functions may be added to MATLAB's vocabulary if they are expressed in terms of other existing functions. The commands and functions that comprise the new function must be put in a file whose name defines the name of the new function, with a filename extension of ".m." At the top of the file, there must be a line that contains the syntax definition for the new function.

   For example, the existence of a file called stat.m with:

```
function [mean, stdev] = stat(x)
n = length(x);
mean = sum(x) / n;
stdev = sqrt(sum((x - mean).^2)/n);
```

   defines a new function called stat that calculates the mean and standard deviation of a vector. The variables within the body of the function are all local variables.

Example 5.3

Solve the following three nonlinear coupled first-order differential equation system numerically on the time interval [0 15] seconds with initial conditions of $y_1(0) = 1$, $y_2(0) = 4$, and $y_3(0) = -5$.

$$\dot{y}_1 = 3\sin(y_2)$$
$$\dot{y}_2 = 0.2 - [\cos(y_1)]^2$$
$$\dot{y}_3 = y_1 y_2 - y_3$$

Then, plot the results.

Solution

To solve the three nonlinear coupled differential equations system, we need one m file and one function. The function file defines the differential equations, but the m file solves them via ODE45 command. This m files calls the function and reads the DEs inside that function.

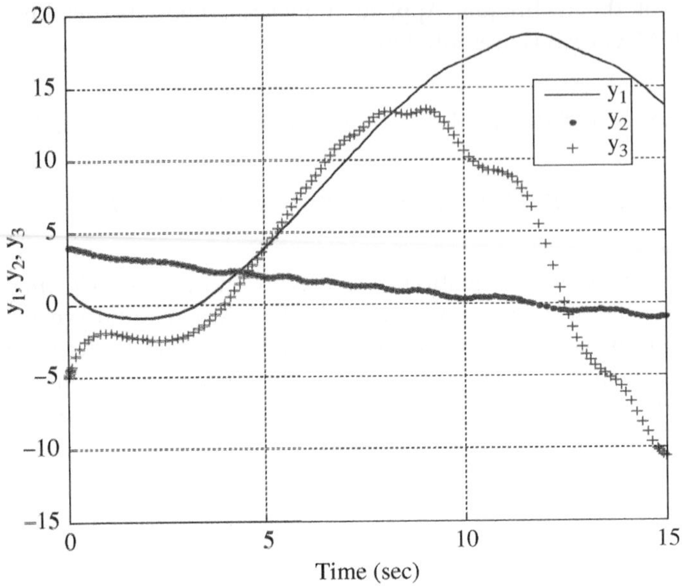

Figure 5.16 Results of simulation.

The m-file (with any name):

```
clear all
clc
[t,y] = ode45('f1',[0 15],[1; 4 ; -5]);
plot(t,y(:,1),'-',t,y(:,2),'.', t,y(:,3),'+');
xlabel ('Time (sec)');
ylabel ('y_1, y_2, y_3')
grid
legend ('y_1', 'y_2', 'y_3')
```

The function (must be saved with the name of f1):

```
function func = f1(t,y)
y1 = 3*sin(y(2));
y2 = 0.2-cos(y(1))^2;
y3 = y(1)*y(2)-y(3)
func = [y1 ; y2 ; y3];
```

Note that the name of function file (here: f1) must be used two more times. Figure 5.16 demonstrates the simulation results.

5.9.3 Hardware-In-the-Loop Simulation

In HIL simulation, a real-time computer as a virtual representation of the UAV model, and a real version of the controller are employed. The development hardware also contains an interface with which to control the virtual input to the UAV. When a ground-based test equipment is employed, one can evaluate the effectiveness of the designed control system and its influence on the vehicle's stability characteristics. The controller hardware contains the controller software that is generated from the controller model. The real-time processor (i.e., hardware) contains a code for the physical system that is generated from the UAV model. In this case, the hydraulic servo actuators are attached to the bottom of the simulator vehicle and are driven by computers to produce the desired motion.

A reliable, high fidelity pilot-in-the-loop or *hardware-in-the-loop (HIL)* simulation requires a more complex dynamic model. The goal of minimizing execution time must be traded against the precision and accuracy of the simulation. The availability of special-purpose simulation languages, massive computing capabilities at decreased cost, and advances in simulation methodologies have made simulation one of the most widely used and accepted tools in flight operations research and UAV system analysis.

5.10 Flying Qualities for UAVs

5.10.1 Fundamentals

In the design of a UAV autopilot, the designer must consider the flying qualities specifications and criteria. In the past 20 years, typical guidance used in developing UAV platform system specifications for flying qualities has been MIL-F-8785, MIL-HDBK-1797, MILHDBK-516, and ADS-33-PRF. Current flying qualities standards and specifications are deficient for UAVs and rely heavily on manned air vehicle standards and criteria; since they do not take into account sensor and payload mission requirements.

Many of the flying or handling qualities specification for manned aircraft (e.g., stick force) do not directly apply to UAVs and autopilot design. However, for the case of a UAV which is remotely controlled by an operator in the ground station, there are various requirements which influences the stability (e.g., phugoid, short period, dutch roll, roll convergence, and spiral modes) and controllability of the UAV. Moreover, to allow for tradeoffs between the air vehicle and sensor/payload subsystems, flying qualities specifications must provide flexibility and guidance in tailoring air vehicle requirements.

Moreover, special consideration must be given to the way in which the autopilot is engaged and disengaged, so that dangerous transient motions are not generated. Another area of interest for handling qualities in the design of autopilot is the process through which the autopilot mode is switched from one to another. The differences between manned aircraft and UAV flying qualities mainly relate to primary and secondary flight control systems, data link time delays, system failure states, and flight display requirements. The flying qualities of an airplane are strongly connected to its control surfaces, mainly in size of area, arm, and defection angle.

Unlike manned civil aircraft, remotely piloted UAVs are subject to time delays produced by the communication system, which can have an adverse effect on system handling qualities. Compliance with the requirements can be demonstrated through analysis, simulation and flight tests. This section discusses a proposed standard specifying the requirements for UAV handling qualities.

5.10.2 Classes, Categories, and Acceptability Levels

Flying qualities standards and criteria are currently deficient for UAVs and rely heavily on manned air vehicle standards and guidance. UAV specific missions, and payload requirements require development of new standards and criteria formulation. Naval Air Systems Command [67] has initiated a project to develop flying qualities standards and criteria for UAVs. Modifications to the existing manned flying qualities specifications should remove requirements for stick force gradients, and pilot comfort. However, most basic stability and control requirements of the vehicle will remain intact.

The top requirements in MIL-F-8785C and MIL-STD-1797B are for aircraft to have Level 1 flying qualities in the Operational Envelope, and Level 2 flying qualities in the Service Envelope. The detailed criteria in the lower level requirements define Level 1, 2, or 3 flying qualities. Detailed performance to meet Level 1, 2, or 3 flying qualities generally varies as a function of aircraft Class and flight phase.

For manned aircraft, there are four classes which are based on size, maneuverability, and missions. The flight envelope usually covers altitude, airspeed, and normal acceleration. Class I is for small and light aircraft, Class II is for medium-weight, low to medium maneuverability aircraft, Class III are large and heavy airplanes such as large transport aircraft, and Class IV is for high maneuverability aircraft such as fighters.

However, for UAVs, these should be replaced with I. micro UAVs, II. mini UAVs, III. small UAVs, IV. medium weight UAVs, V. heavy weight UAVs, and VI. UCAVs. These are for fixed-wing UAVs; a new category must also be defined for quadcopters (i.e., class VII).

Flight phases are divided into three Categories: A, B, and C. Category A includes flight phases that require precision/rapid maneuvering such as combat, or terrain following. Category B includes flight phases that are not terminal and do not require precision maneuvering such as cruise and loiter. Category C contains the terminal flight phases, such as takeoff and landing. These categories could be kept the same for UAVs.

A flying qualities specification for UAVs have to cover a wider range of aircraft that do not fit into the existing Classes. Aircraft Classes for UAVs must account for requirements that will vary depending on expendability of the airframe and payload, mission performance objectives, potential for damage to people or property, and airspace integration. Without humans onboard, higher loss rates than piloted aircraft is acceptable.

A new set of specifications should be prepared for the flight modes: a. remotely controlled through line-of-sight, b. remotely controlled through display, c. single mode autopilot (e.g., altitude-hold mode), d. fully autonomous. Each of these flight modes should have a unique flying qualities specification, due to the level of autonomy.

5.10.3 Force Restrictions

The requirements; involving forces to carry out maneuvers; restricted by mechanical/electric servos limitations should be considered. The control surfaces are deflected through electric/mechanical servo/actuators, so the stick force gradients requirements are removed. References [47, 68] are helpful resources in introduction, principles, fundamentals, and useful data for various aspects of human factors including human manual force limits. Moreover, MIL-STD-1472G [69] establishes general human engineering criteria for design and development of military systems, equipment and facilities.

5.11 Autopilot Design Process

In the preceding sections, primary subsystems of an autopilot are introduced. Moreover, dynamic modeling using two techniques (transfer function, and state-space model) including aerodynamic forces and moments have been discussed. In this section, the design process of an autopilot is briefly presented.

The primary requirements for the design of an autopilot are as follows: (a) manufacturing technology; (b) required accuracy; (c) stability requirements; (d) structural stiffness; (e) load factor; (f) flying quality requirements; (g) maneuverability; (h) reliability; (i) cost; (j) UAV configuration; (k) maintainability; (l) weight; (m) communication system; (n) aerodynamic considerations; (o) processor; and (p) complexity of trajectory.

Based on existing handling qualities for general aviation aircraft, and also on past experiences with UAV flight dynamics, the following requirements are suggested in the design process:

1) **General requirements:**
   1) Stability of the overall system (minimum requirement)
   2) Output (or state tracking) performance
   3) Accuracy from command to response
   4) Minimization of a certain performance index (to follow)
2) **Specific design requirements (considering uncertainty):**
   a) **Step response specifications:**
   Typical step response specifications are as follows:
      1) Overshoot <5%
      2) Steady state error <1%
      3) Rise time <1 second
      4) Minimal cross channel coupling

   b) **Robustness specifications:**
      5) The autopilot must be robust with respect to weight, center of gravity, and dynamic model uncertainties.
      6) The autopilot should reject disturbances arising from measurement and sensor error models, along with environmental variations.

Figure 5.17 presents the general design process of an autopilot. The design has an iterative nature and begins with the design requirements. Three major sections of control system, guidance system, and navigation system are designed in parallel. In designing these systems, three laws (control law, guidance law, and navigation law) must be selected. Moreover, each of these three systems require design or selection of respective equipment.

Chapters 6, 7, and 8 will present detailed designs for control, guidance, and navigation subsystems respectively. At any stage of the design procedure, the output is checked with the design requirements, and a feedback is taken. The process is repeated until the design requirements are met. A significant step in designing an autopilot, is the integration of control system, navigation system, and guidance system in one unit. This phase will generate various design challenges which must be carefully treated.

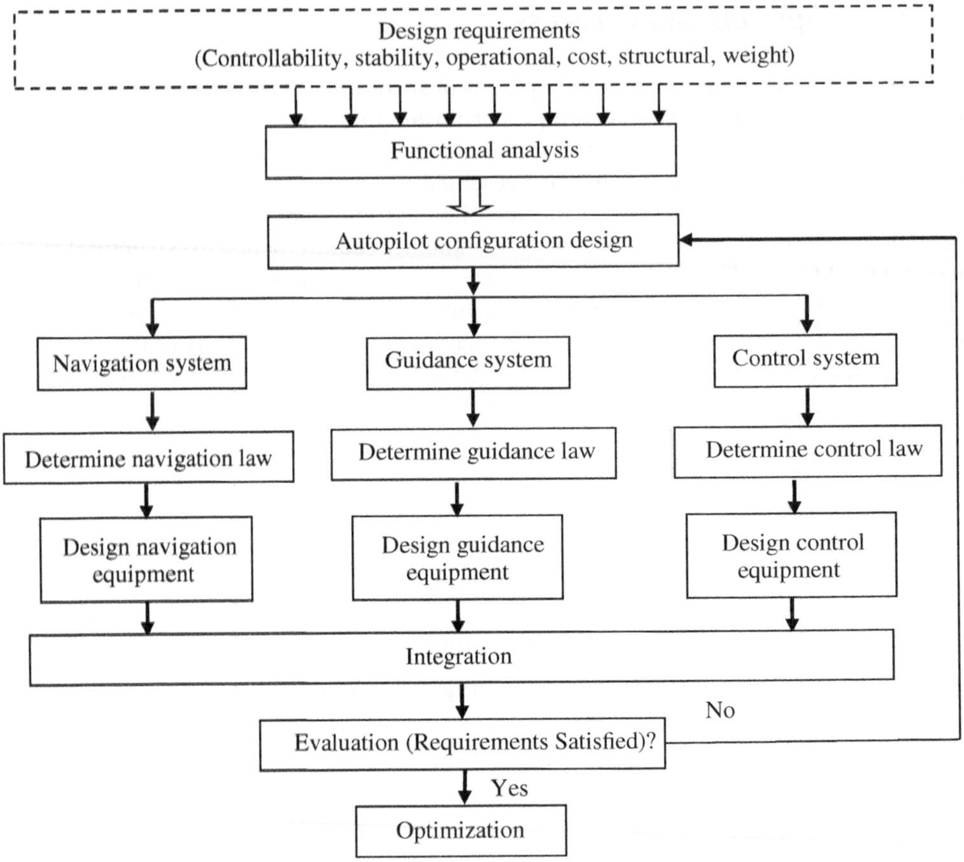

Figure 5.17 Autopilot design process.

Questions

1 Name primary subsystems of an autopilot.

2 Compare an autopilot and a human operator in at least three aspects.

3 What is the function of the control system?

4 What is the primary function of the guidance system?

5 What is the primary function of the navigation system?

6 Define Laplace transform.

7 Define transfer function.

8 What are the poles and zeros of a transfer function?

9 What is dynamic modeling?

10 What are two methods for simplifying a nonlinear coupled dynamic model?

11 What is name of the technique for linearization?

12 Describe the state space model.

13 What is the control derivative?

14 What is the stability derivative?

15 What are the main aerodynamic forces?

16 What are the main aerodynamic moments?

17 List various types of dynamic model of a UAV.

18 Write the linear dynamic model (state-space) for longitudinal mode for a fixed-wing UAV.

19 Write the linear dynamic model (state-space) for lateral-directional mode for a fixed-wing UAV.

20 Write the dynamic model (differential equations) for a quadcopter UAV.

21 Describe autopilot design process.

22 What are the primary requirements for the design of an autopilot?

23 What does the UAV dynamic model consist of?

24 Describe the Runge–Kutta method for numerical integration.

25 Describe the human-in-the-loop simulation.

26 By referring to Eq. (5.42), identify the unit of dimensional derivative L_β.

27 What is the unit of non-dimensional derivative $C_{n\beta}$?

28 What is the unit of the stability derivative C_{mq}?

29 Draw a UAV, and show all aerodynamic forces and moments in this figure.

30 What are the outputs for decoupling longitudinal motion from lateral-direction motions?

31 Based on Figure 5.9, what is the thrust produced by the electric motor employed in the AscTec Hummingbird UAV at 5000 rpm?

32 Based on Figure 5.10, what is the torque produced by the electric motor employed in the AscTec Hummingbird UAV at 6000 rpm?

33 List three SAS in autopilots.

34 List four longitudinal hold functions in autopilots.

35 List three lateral-directional hold functions in autopilots.

36 List four longitudinal navigation functions in autopilots.

37 List three CAS in autopilots.

38 What are the three options of interest for continuous decrease of the lift during a cruising flight?

39 When the UAV measurement devices/processor have measured/calculated three airspeed components (u, v, and w), how is the UAV angle of attack (α) determined?

40 When the UAV measurement devices/processor have measured/calculated three airspeed components (u, v, and w), how the UAV sideslip angle (β) is determined?

Problems

1 Using the Taylor series, linearize the following nonlinear equation

$$F(x) = 10x^2 + 3x - 12$$

for the trim point $x = 3$.

2 Using the Taylor series, linearize the following nonlinear equation

$$F(x, y) = 10yx^2 + 3xy^4 + 60x - 30y + 20$$

for the trim point $x = 5$, $y = 2$.

3 The longitudinal dynamic model of a UAV has the following matrix A in the state-space model.

$$\begin{bmatrix} 3 & -4 & 9.8 \\ 2 & 24 & -3.1 \\ 9 & 1.8 & 0 \end{bmatrix}$$

Is this UAV, longitudinally dynamically stable?

4 The airspeed-to-elevator-deflection transfer function of a UAV is

$$\frac{u(s)}{\delta_E(s)} = \frac{s^2+3s+4}{125s^4+308s^3+354s^2+30s+29}$$

Is this UAV, longitudinally dynamically stable?

5 The bank-angle-to-aileron-deflection transfer function of a UAV is

$$\frac{\phi(s)}{\delta_A(s)} = \frac{150s^2+35s+130}{870s^4+630s^3+790s^2+440s+6}$$

Is this UAV, laterally-directionally dynamically stable?

6 A dynamic system has the following transfer function.

$$F(s) = \frac{3}{s^2+2s+10}$$

Calculate the damping ratio and natural frequency of this system.

7 A dynamic system has the following transfer function.

$$F(s) = \frac{12}{s^2+7s+4}$$

Calculate the damping ratio and natural frequency of this system.

8 A UAV has the following dimensional derivatives

$N_{\delta R} = -2.5\ 1/\text{sec}^2$, $N_r = -0.4\ 1/\text{sec}$, $N_\beta = 3.5\ 1/\text{sec}^2$

Approximate the heading-angle-to-rudder-deflection transfer function.

9 A UAV has the following dimensional derivatives

$M_q = -6\ 1/\text{s}$, $M_{\dot{\alpha}} = -3.2\ 1/\text{s}$, $M_\alpha = -6\ 1/\text{s}^2$, $M_{\delta E} = -42\ 1/\text{s}^2$

a Determine the approximate angle of attack governing differential equation.
b Derive the angle-of-attack-to-elevator-deflection transfer function

10 A UAV is cruising with an airspeed of 150 knots. If the vertical component of the airspeed if 8 knots, determine the UAV angle of attack.

11 A UAV is touching down with an airspeed of 60 knots, while crosswind of 5 knots is blowing from the right. Determine the sideslip angle.

12 Solve the following three nonlinear coupled first-order differential equation system numerically on the time interval [0 2] seconds with initial conditions of $y_1(0) = 2$, $y_2(0) = 6$, and $y_3(0) = -3$. Then, plot the results.

$$\dot{y}_1 = 2\cos(y_2)$$
$$\dot{y}_2 = 1.4y_3 - [\sin(y_2)]^2$$
$$\dot{y}_3 = y_1 y_2 - 2y_3$$

13 Solve the following three nonlinear coupled first-order differential equation system numerically on the time interval [0 10] seconds with initial conditions of $y_1(0) = -3$, $y_2(0) = 1$, and $y_3(0) = 4$. Then, plot the results.

$$\dot{y}_1 = -3\cos(y_2)$$
$$\dot{y}_2 = 2.8 + [\sin(y_1)]^3$$
$$\dot{y}_3 = y_1 y_3 + 4y_2$$

14 Linearize the following first-order nonlinear differential equation (i.e., force Equation in y-direction; Eq. (5.4)), using the direct substitution.

$$m(\dot{V} + UR - WP) = mg_y + F_{A_y} + F_{T_y}$$

15 Linearize the following first-order nonlinear differential equation (i.e., force Equation in z-direction; Eq. (5.5)), using the direct substitution.

$$m(\dot{W} - UQ + VP) = mg_z + F_{A_z} + F_{T_z}$$

16 A small quadcopter with a mass of 2 kg has four electric motors, has a coefficient, k_1 of 2×10^{-6} N/(rad/s)^2.

- **a** If all motors are turning with a rotational speed of 10 000 rpm, determine the overall upward force generated by all four engines.
- **b** The quadcopter has a pitch angle of 15°. Determine the force components in z and x directions.

17 A small quadcopter with a mass of 5 kg has four electric motors, has a coefficient, k_1 of 1.5×10^{-6} N/(rad/s)^2.

- **a** If all motors are turning with a rotational speed of 12 000 rpm, determine the overall upward force generated by all four engines.
- **b** The quadcopter has a pitch angle of 20°. Determine the force components in z and x directions.

18 A small UAV with a takeoff mass of 1200 kg, a wing area of 15 m^2, and a wing mean aerodynamic chord of 1.6 m is cruising with a speed of 130 knot at 5000 ft altitude. The non-dimensional stability derivative C_{L_q} is 4.2 1/rad. Calculate dimensional stability derivative Z_q.

19 A mini UAV with a takeoff mass of 2 kg, a wing area of 0.8 m^2, and a wing mean aerodynamic chord of 0.3 m is cruising with a speed of 30 knot at 2,000 ft altitude.

The non-dimensional stability derivative C_{nr} is -0.12 1/rad. Calculate dimensional stability derivative N_r.

20 A large UAV has the following dimensional roll stability and control derivatives:

$L_{\delta_A} = 1.4 \ 1/sec^2 \ L_p = -0.43 \ 1/sec$

Approximate the bank-angle-to-aileron transfer function. Then, plot the bank angle response of the UAV to a 20 deg aileron input for 3 seconds.

21 A small UAV has the following dimensional roll stability and control derivatives:

$L_{\delta_A} = 63 \ 1/sec^2 \ L_p = -14 \ 1/sec$

Approximate the bank-angle-to-aileron transfer function. Then, plot the bank angle response of the UAV to a -5 deg aileron input for 1.5 seconds.

22 A UAV electric engine with a propeller is able to generate a rated thrust of 12 N. The engine is modeled as a first order system with a time constant of 0.3 seconds. Plot the variations of engine thrust as a function of time for 3 seconds.

23 Consider the UAV in problem 18. Suppose an upward gust hits under the nose of the UAV and causes a pitch with a rate of 20 deg/sec. What vertical force is generated by the UAV as a response/reaction to this perturbation? What is its directions?

24 Consider the UAV in problem 19. Suppose a gust hits on the side of the vertical tail and causes a yaw with a rate of 30 deg/sec. What yawing moment is generated by the UAV as a response/reaction to this perturbation? What is its directions? Does this behavior indicate a directionally stable UAV?

Note

1 Single-Input-Single-Output.

6

Control System Design

CONTENTS

Design of Unmanned Aerial Systems, First Edition. Mohammad H. Sadraey.

Companion website: www.wiley.com/go/sadraey/unmanned-aerial-systems

Educational Outcomes

After study of this chapter, the reader will be able to:

1) Mathematically model a control system using differential equations.
2) Linearize a nonlinear system using Taylor series technique.
3) Explain elements of control system.
4) Derive transfer function of a linear system using Laplace transform technique.
5) Use block diagrams to derive the transfer function of a system.
6) Derive state space representations of a control system.
7) Convert state space representation of a system to transfer function(s) and vice versa.
8) Evaluate the performance characteristics of a control system.
9) Employ root locus technique to design the compensation for a control system.
10) Employ frequency response technique to determine the steady-state response of a system to a sinusoidal input.
11) Apply various design techniques to design a controller for a control system.
12) Conduct Unmanned Aerial Vehicle (UAV) control system design.
13) Discuss various categories of autonomy.
14) Develop flight control requirements.
15) Design various autopilot modes for a given set of control requirements.
16) Select and implement an appropriate servo/actuator for a given control requirement.

6.1 Introduction

One of the main subsystems within an auto-pilot (see Figure 5.1 in Chapter 5) is the control system. The control system controls the direction of the motion of the vehicle or simply the orientation of the velocity vector. The control system is used to keep a UAV on a predetermined course or heading, necessary for the mission. The control system uses the vehicle state information provided by the on-board sensors to drive the control surface actuators (i.e., servos).

Despite poor and gusty weather conditions, the UAV must maintain a specified heading and altitude in order to reach its destination safely. In addition, in spite of rough air, the trip must be made as smooth as possible for a low load factor. The problem is considerably complicated by the fact that the UAV has six degrees of freedom. This fact makes control more difficult than the control of a ship, whose motion is limited to the surface of the water.

A successful control designer needs not only a good understanding of aerodynamics and flight dynamics, but also a good understanding of the systems engineering approach. All UAVs must meet controllability requirements to be certified for commercial use or adopted by the military. Many military UAVs such as Unmanned combat air vehicle (UCAVs) also have additional maneuverability requirements. A UAV's ability to meet these requirements is often limited by the amount of control authority available. Thus, it is essential for designers to evaluate the control authority of candidate concepts early in the conceptual design phase.

An essential activity in the detail design is the development of a functional description of the control system components to serve as a basis for identification of the resource necessary for the system to accomplish its mission. Such a function may ultimately be accomplished through the use of equipment, software, facilities, data, or various combinations thereof. The control system design must meet the controllability requirements prescribed for maneuvers as desired by the customer or standards. The UAV standards have not been yet finalized, but military specifications (Military Standards (MIL-STD)-1797 or UAV certification guidelines), or FAR Part 23 (Ref. [70]) or Federal Aviation Regulations (FAR) Part 25 (Ref. [71]) may be referenced. With advances in computer technology, and the introduction of new mathematics theories in nonlinear systems, more applications of advanced control system design techniques such as robust and nonlinear control are seen in the literature.

6.2 Fundamentals of Control Systems

The UAV control system must fundamentally be of a closed-loop form, which employs a negative feedback. In this section, fundamentals of control systems, control concepts, basic definitions, major elements, and general design techniques are presented.

6.2.1 Elements, Concepts and Definitions

A closed-loop system is generally made up of four (Figure 6.1) basic elements: 1. Plant, 2. Controller, 3. Actuator (or servo), and 4. Measurement device. The block diagram of a control system with negative feedback can often be simplified to the form shown in Figures 6.1 and 6.2 where the standard (see Refs. [49, 50]) symbols and definitions used in feedback systems are indicated. In this feedback control system, the output, Y is the controlled variable. One of the reasons for using negative feedback is to regulate the output of the plant, that is, to hold the output constant at a "set point," as in an aircraft "altitude hold" autopilot mode.

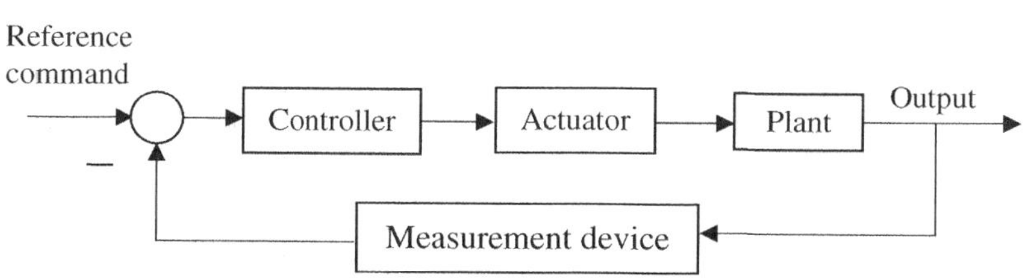

Figure 6.1 Block diagram of a closed-loop control system.

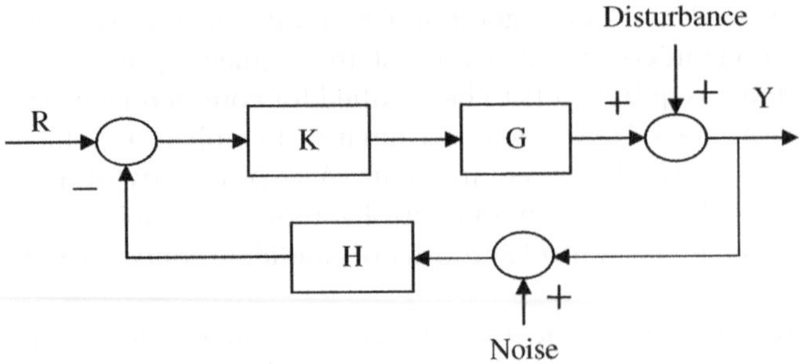

Figure 6.2 Block diagram of a control system including disturbance and noise.

This output is measured by a feedback element (Measurement device, H) to produce the primary feedback signal, which is then compared with the reference input. The difference between the reference input (R) and the feedback signal (i.e., error), is the input to the controller (K). Then, the controller is generating a signal to the actuator based on a control law.

The controller signal is applied to the plant (G) via an actuator (e.g., a mechanical jack), actuating the signal. The error signal is defined as the ideal or desired system value minus the actual system output. The ideal value establishes the desired performance of the plant. Disturbance is the unwanted signal (e.g., gust) that tends to affect the controlled variable. The disturbance may be introduced into the system at many places. Noise is the unwanted signal (e.g., engine vibration) that tends to affect the measured variable. A filter may be utilized to get rid of noise in the measurement device. For the block diagram given in Figure 6.2, the input-to-output transfer function is obtained [50] as:

$$\frac{Y(s)}{R(s)}=\frac{KG}{1+KGH} \tag{6.1}$$

In terms of the UAV application, a flight controller in a feedback control system is used to control the aircraft motion. Two typical signals to the system are the reference flight path, which is set by the autopilot, and the level position (attitude) of the airplane. The ultimately controlled variable is the actual course and position of the UAV. The output of the control system, the controlled variable, is the aircraft heading.

The controller is designed based on a control law. Some typical control laws are: (a) linear; (b) nonlinear; (c) optimal; (d) adaptive; and (e) robust. A summary of the dynamic systems; and the different control system design techniques is shown in Figure 6.3. In implementing control law, there are two main approaches: (a) analog control; (b) digital control.

The real aircraft behavior is nonlinear and has uncertainties. A UAV is basically a nonlinear system. Furthermore, its dynamics and equations of motion are also nonlinear. In this section, some common nonlinear systems phenomena are presented. Later, they are applied to the design of the UAV autopilot. Systems containing at least one nonlinear component are called nonlinear systems. Basically, there are two types of nonlinearities: (a) continuous; and (b) discontinuous (hard). Hard nonlinearities include: (a) coulomb friction; (b) saturation; (c) dead-zone; (d) backlash; and

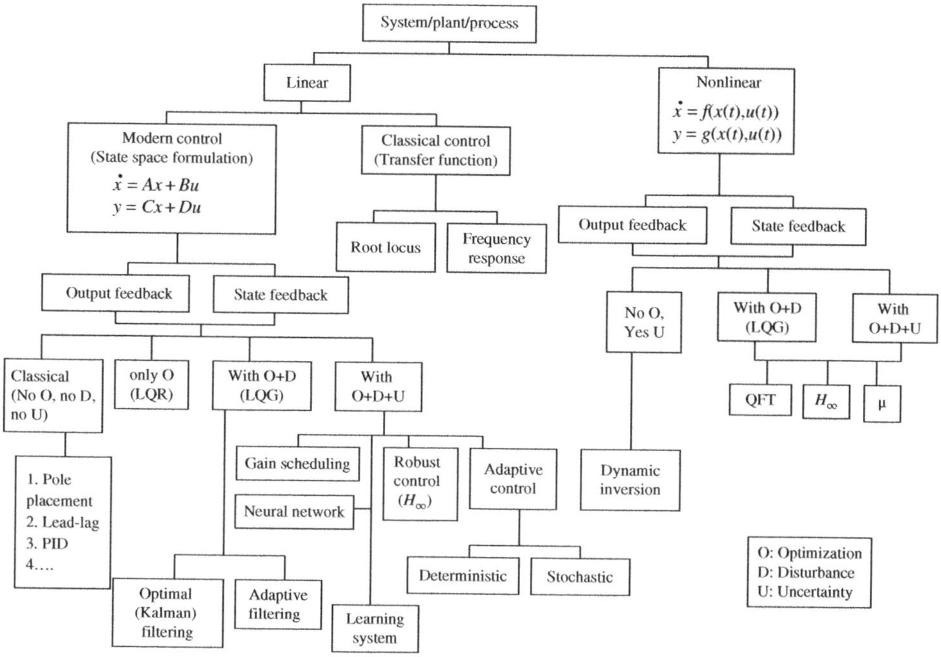

Figure 6.3 Control system design techniques.

(e) hysteresis. In another classification, nonlinearities could be inherent (natural) or intentional (artificial).

It is also important to note that all measurement devices (including gyros and accelerometers) have some kinds of noise that must be filtered. It is well known that the atmosphere is a dynamic system that produces lots of disturbances throughout the aircraft's flight. Finally, since fuel is expensive and limited, and actuators have dynamic limitations, optimization is necessary in control system design. Therefore, it turns out that only a few design techniques, such as robust nonlinear control, are able to satisfy all safety, cost, and performance requirements. However, due to cost, and complexity, many UAV designers adopt a more conventional control architecture. Two conventional controller design tools/techniques are: (a) root locus technique; and (b) frequency domain techniques. The interested reader may refer to Refs. [49, 50] for more details.

6.2.2 Root Locus Design Technique

A very effective and graphical tool in the design of control systems is the root locus. In a transfer function, closed loop poles are the roots of the characteristic equation of a system. In a state-space representation, this is equivalent to the eigenvalues of the state matrix (i.e., A). The root locus is the plot of variations of the closed loop poles as a function of feedback gain (K in Figure 6.2), or a compensator parameter.

By using the root locus method, the designer can predict the effects on the closed-loop poles locations of varying the gain value, or on adding a pole or a zero. Recall that any controller/compensator is made up of some poles and zeros. One clear application of the root locus is to visualize the acceptable values of a gain; which implies the stable system (i.e., when all poles are in the left-half plane).

The design of control system by the root locus method is determining the controller gain value such that all poles are within the target area, and all design requirements are met. The target area is constructed by applying all design requirements such as rise time, settling time, maximum overshoot, natural frequency, and damping ratio. The addition of a pole to the open loop Transfer Function (TF) has the effect of pulling the root locus to the right, tending to lower the system's relative stability and to slow down the settling of the response. The zero has the opposite effect.

In practice, this technique may indicate that the desired performance cannot be achieved just by a simple gain controller. Moreover, in some cases, we may find out that the system will not be stable for all values of the gain. This implies that a simple gain is not a solution, and a more advanced controller should be designed. A series of calculations are involved in determining poles and zeros of the compensator (e.g., lead compensation). One of the primary goals in the root locus technique is to have the dominant closed loop poles at the desired location in the s plane, so that the performance specifications are met. The MATLAB has the command "rlocus" to construct the root locus. Moreover, the MATLAB command "sisotool" is creating the root locus as well as the time response of the system.

6.2.3 Frequency Domain Design Technique

The frequency response approach is very useful in designing the control systems. The designer will determine the best compensator that most closely produces the desired closed-loop response. Particularly, in dealing with high-frequency noises, the frequency-response approach in more convenient than other approaches. There are basically two approaches in the frequency-domain design. One is the polar approach, and the other one is the Bode diagram approach. In general, the second approach is simpler, and in this section, this approach is briefly introduced.

A complex number has two parts, real and imaginary ($s = \sigma \pm j\omega$). When this number is illustrated in the s–plane, it can also be represented by its magnitude (M), and its phase angle (ϕ). A useful graphical representation of a transfer function is obtained by replacing s with $j\omega$, and (after converting it into two-part number) plotting the magnitude and phase of the resulting complex number versus frequency, ω. For instance, for the transfer of

$$G(s)=\frac{Num(s)}{Den(s)}=\frac{k(s+z_1)(s+z_2)\ldots(s+z_m)}{(s+p_1)(s+p_2)\ldots(s+p_n)} \tag{6.2}$$

We can obtain

$$G(j\omega)=\frac{Num(j\omega)}{Den(j\omega)}=M(\omega)\angle\phi(\omega) \tag{6.3}$$

where M is the amplitude ratio of the output and input sinusoids and ϕ is the phase shift between the input sinusoid and output sinusoid. In Eq. (6.2), the parameter k is referred to as the "dc gain"; where dc equals zero frequency (the transfer function gain when $s = 0$).

Such a graphical representation is called a *Bode plot.* Thus, a Bode plot has a two-part plot of M and ϕ versus ω. In the Bode plot, the input variable is a function of frequency; not function of time (often sinusoidal). It is customary to convert the present the magnitude in dB unit, and the frequency in the log format.

Although the Bode plot is originally developed for electric systems, but it has a lot of useful applications in the aeronautical systems. There is a wealth of information which can be deducted from a Bode plot, some of which are used in the design of a control system. Some important parameters in the Bode plots are: (a) phase margin; (b) gain margin; (c) phase-crossover frequency; (d) bandwidth; (e) cutoff rate; (f) corner frequency; (g) resonant peak; and (h) resonant frequency. In MATLAB, the command "bode" is used to plot the frequency response of a system. Moreover, the MATLAB command "sisotool" is creating the Bode plot as well as the root locus and time response of the system.

Example 6.1

Consider the following system:

$$G(s)=\frac{4s+7}{s^2+2s+10}$$

Plot the Bode diagram of this transfer function.

Solution

The following MATLAB code is used to draw the bode diagram:

```
n=[4 7]
d=[1 2 10]
bode(n,d)
grid
```

Figure 6.4 demonstrates the Bode diagram.

A general approach to the design, based on the Bode plot, is that, first the open-loop gain is adjusted so that the requirement on the steady-state accuracy is met. Then, the magnitude and phase curves of the uncompensated open-loop (with the open loop gain just adjusted) are plotted. If the design specifications (e.g., gain margin and phase margin) are not satisfied, then a suitable compensator is added. The open loop gain and the compensator are determined through a series of calculations.

6.2.4 Controller Configurations and Control Architectures

In general, there are six basic controller configurations in control system compensations: (a) series or cascade compensation (Figure 6.1); (b) feedback compensation; (c) state-feedback control (Figure 6.5); (d) series-feedback (Figure 6.6) compensation (two-degree of freedom [2DOF]); (e) forward compensation with series compensation (2DOF); and (f) forward compensation (2DOF). Each configuration is best for a particular circumstance, and has unique features including cost, complexity, performance, and effectiveness.

For instance, one application for the 2-DOF controller is the *time-scale separation*; where the flight (state) variables are divided into two groups: (a) slow-states; and (b) fast

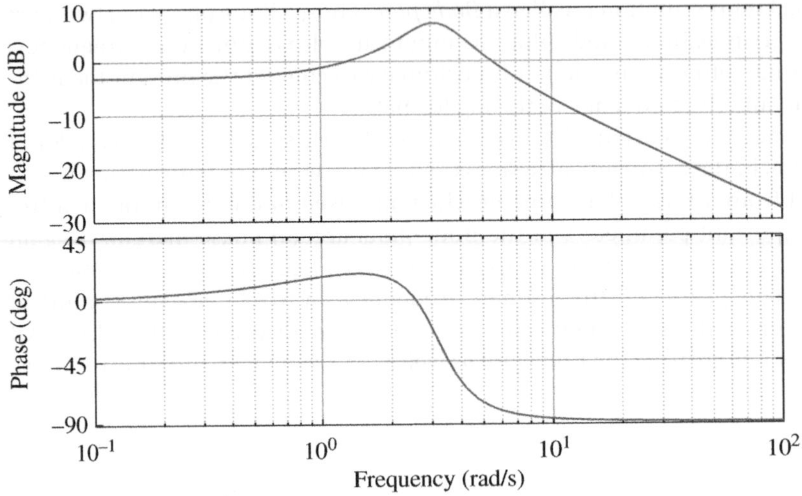

Figure 6.4 Bode diagram of system given in example 6.1.

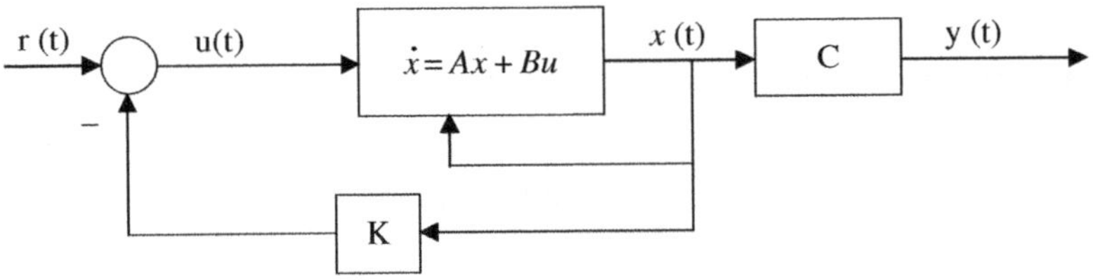

Figure 6.5 State-feedback control configuration.

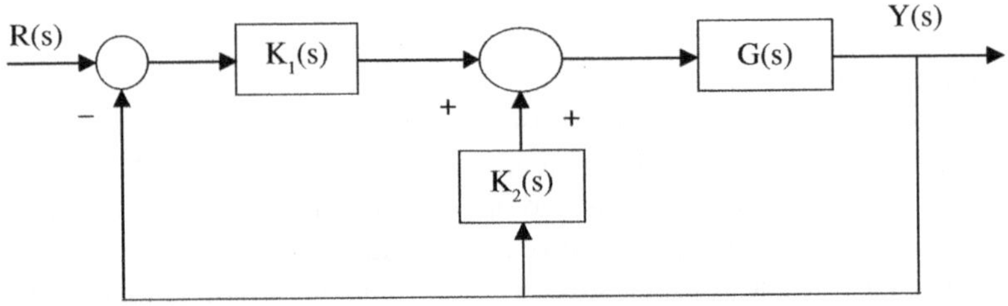

Figure 6.6 Series-feedback compensation (two-degree of freedom (2DOF)).

states. In this technique, the dynamics to be controlled are separated into fast and slow variables. In application to flight vehicle dynamics, the control technique (e.g., Linear Quadratic Regulator (LQR)) modifies the problem into a two time-scale problem. The two time-scales are estimated by assuming that there is a significant frequency separation between the fast and slow states. The number of slow and fast states depends on the number of controls. If there are three control surfaces, roll, yaw, and pitch rates (p, q, r) are defined as fast dynamic variables, which are controlled by three inputs: aileron, elevator, and rudder.

Angles of attack, sideslip angle, and bank angle (α, β, ϕ) are defined as slow dynamic variables. The speed, pitch angle, and heading angle are defined as very slow dynamic

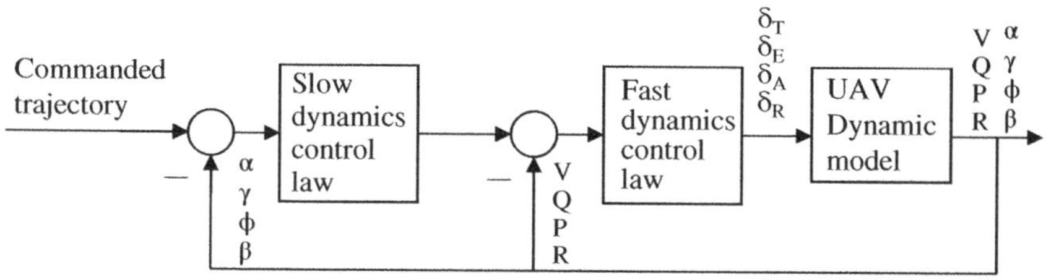

Figure 6.7 Time-scaled flight control system architecture.

variables. The assumption is that the very slow states are not time-varying when compared with the fast and slow dynamic variables.

In the two time-scale approach, the slow states are constant compared to the fast states. At the same time, the fast states are assumed to track rapidly, so that the effect of their rates of change on the slow states is zero. Therefore, the design of the controller for the fast dynamics is independent of that for the slow dynamics. It can be designed using control technique (e.g., LQR) directly. After designing a fast state controller, a separate control technique procedure is executed to design the slow state controller for α, β, and ϕ using the commands for p, q and r as inputs.

If the UAV has four control inputs, four variables should be chosen in each time scale. The slow or primary controlled variables are the angle of attack, climb angle, bank angle, and sideslip angle (α, γ, ϕ, β). The fast or secondary controlled variables are the three angular rates plus the forward speed (V, Q, P, R). Since one of the commanded variables in the pre-determined trajectory is the forward speed (V), the corresponding α equation is in fact replaced with the V equation.

The outer control loop involves the translational dynamics. In response to position and/or velocity commands, it produces the δ command, together with an attitude reference signal, for the inner-loop to track. The inner control loop involves the rotational dynamics, and tracks the attitude reference by determining the δ_T, δ_E, δ_A, and δ_R commands. A time-scaled flight control system architecture with four control inputs is shown in Figure 6.7.

6.3 Servo/Actuator

6.3.1 Terminology

Another important element in the UAV flight control is the actuator which deflects (pushes and pulls) the control surfaces (e.g., elevator). There are a few related terms used by various groups/disciplines which is briefly reviewed here. A servo is a rotary/linear actuator that allows for precise control of angular/linear position, velocity and acceleration. The term servo is mainly found in the electrical engineering terminology (also servomotor), as well as the radio/remotely controlled airplane modelers' community. Optical encoders are usually used in servo systems to detect the position and control the movement of power drives and sensors such as radar.

The term servomotor is used when an electric motor is coupled with a sensor for position feedback.

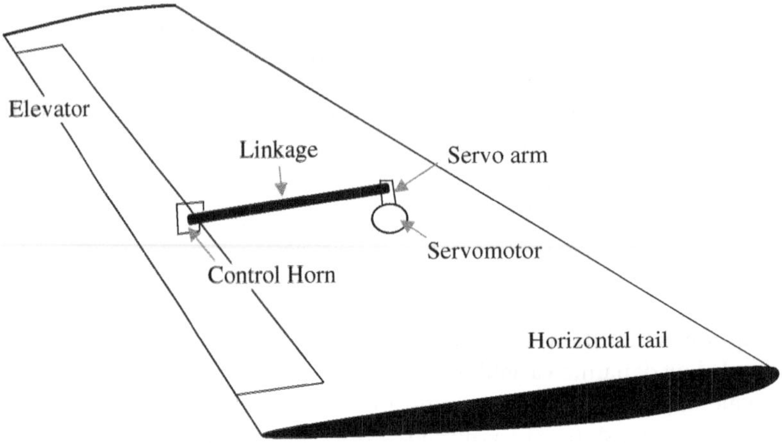

Figure 6.8 Sketch of a servomechanism for an elevator.

A servomechanism is a feedback control system in which the controlled variable is a (mechanical) position, velocity, torque, frequency, etc. It consists of an electric motor, mechanical linkage (Figure 6.8), and a microcontroller. An actuator is a mechanical/hydraulic device used as a mechanism to translate mechanical motion (often rotary) into linear motion. The term "actuator" is mainly found in the aerospace engineering terminology.

When a signal is received from the microcontroller/autopilot to deflect a control surface, the actuator will implement this command. The source of actuator power is either an electric or a mechanical or a hydraulic or a pneumatics device. For mini to medium small UAVs (and also quadcopters), an electric motor is frequently used. However, for a large and heavy UAV (e.g., Global Hawk), a hydraulic actuator is more effective. For the case of hydraulic/pneumatic power, the servo is a hydraulic/pneumatic cylinder (or valve); not an electric motor. A comparison between a number of features for hydraulic actuators and servomotors are provided in Table 6.1.

Small servos are of two types: (a) analog (frequency of 50 Hz); and (b) digital (frequency of 100 Hz). Digital servo looks the same as analog servo (motor, gear, and potentiometer), but has a microprocessor. Digital servo is much faster than the analog servo, but analog servo has overshoot. Digital servo is more expensive, more accurate, and are faster, thus, it consumes more electric energy. Small servos are available for various output torques (so, various sizes and weights).

For instance, a HS-322HD Delux servo (from HiTEC Company) generates a 30 N.cm torque and have the following characteristics: voltage: 4.8 V, mass: 43 g, size: 40 mm × 20 mm × 37 mm, operating speed: 60° in 0.19 seconds (i.e., 52.6 rpm). Moreover, the actuator DA 30-HT from Volz Servo – features a brushless motor and a contactless wear free position sensor – has a rated torque of 18 Nm/160 lbf-in and uses 24–32 V. The DA 30-HT – with a mass of 670 g, and a dimension of 132 mm × 67 mm × 30 mm – can be equipped with a standard PWM input (Pulse Width Modulation).

6.3.2 Electric Motors

Other electromechanical devices that are used to deflect UAV control surfaces (as actuators) are electric motors. An electric motor is an electromechanical device [72]

Table 6.1 Features of two types of servos.

No	Feature	Electric engine	Hydraulic actuator
1	Power	Electric	Hydraulic
2	Motion	Rotation	Linear
3	Typical physical form		
4	Unmanned aerial vehicle (UAV) types	Quadcopter, micro UAV, mini UAV, small UAV	Large UAV, Heavy UAV, Unmanned combat air vehicle (UCAV)
5	Output	Torque	Force
6	Capability	Torque range: 0.1–100 N.m	Force range: 10–1000 N
7	Overall weight	Lighter	Heavier
8	Medium	Wire	Tube/pipe
9	Signal	Electricity	Oil
10	Maintenance	More labor-intensive	Less labor-intensive

that uses electric energy to produce mechanical rotational motion. It may be powered by direct current, DC (e.g., a battery), or by alternating current (AC). Electric motors may be classified by the source of electric power, by their internal construction, and by their application. Two main parts of an electric motor is rotor coil and magnet. A well-designed motor can convert over 90% of its input energy into useful power.

A *servomotor* consists of an electric motor coupled with a sensor to measure the position/speed to create a feedback, (and may have a microprocessor/microcontroller). It also requires a controller and measurement device (e.g., encoder or potentiometer) to create a closed-loop control system. The input to its control is a signal (either analog or digital) representing the position commanded for the output shaft. The very simplest servomotors use position-only sensing via a potentiometer and bang-bang control of the motor. More accurate servomotors use optical rotary encoders to measure the angular speed of the output shaft to control the motor speed.

In general, there are four types of electric motors: (a) brushless DC motors; (b) brushed DC motors; (c) brushless AC motors; and (d) induction AC motors. The simplest one is brushed DC motor which has a permanent magnet and is of low cost. The brushless DC motor is synchronous and is a good candidate for small industrial application and is typically electronically commutated. It uses electronic commutation system rather than a mechanical commutator and brushes. In brushless DC motors, the current to torque, and voltage to rpm have linear relationships. The AC motors are used in large heavy UAVs and UCAVs, while DC motors are used in micro to small UAVs, as well as quadcopters.

A servo-motor, also referred to as smart motor, is a combination of an electric motor, an electromechanical sensor, an encoder, and a controller as one package. Servomotors are generally used as a high-performance alternative to the stepper motors (or step motors) which have some inherent ability to control position. A main difference between servomotor and stepper motor is the feedback, so servomotors have a closed-loop control system (due to encoder and controller), while step motors have an open-loop system.

6.3.3 Hydraulic Actuator

Another type of actuator which is employed for the deflection of control surfaces are hydraulic actuators (i.e., jack). The hydraulic actuator is a cylinder that converts hydraulic power into useful mechanical (often linear) motion. Such an actuator uses hydraulic pressure as a mechanical force to move objects like controls. It consists of two basic mechanisms: a control device (variable throttles, gates or valves) and an actuating device (e.g., the piston of an actuating mechanism). In a hydraulic actuator the working fluid (i.e., oil) from a pressure line enters the control system through the constant throttles and is carried to the variable throttles and to the actuator chambers.

The electrical input signal when passed through an electromechanical converter, controls the position of the gate slide. The displacement of this slide changes the cross-sectional ratio of the actuating openings through which the pressure fluid passes. Simultaneously, the pressures in the actuator chambers are changed, thus causing a displacement of the slide valves. A comparison between a number of features for hydraulic actuators and servomotors are provided in Table 6.1.

The hydraulic actuators compared with other types of power amplifiers (e.g., electromechanical or pneumatic amplifiers), offer the advantage of a low metal weight per unit of power, often not exceeding 50 g/kW of output power. The power amplification of a hydraulic actuator is very high (about 100 000). Hydraulic actuators with a load feedback substantially improve the dynamic features, and the efficiency of hydraulic control systems.

Other candidates for linear actuators are electromechanical and pneumatic ones. Both of these actuators have advantages and disadvantages compared with hydraulic actuators. The design of linear actuators is beyond the scope of this text, the interested reader may consult with references such as [73, 74]. For instance, the reliability of hydraulic actuators is higher than that of pneumatic ones, which have a higher cost.

6.3.4 Delay

To represent the actuator's delay (i.e., lag) due to a physical limitation, we may use a transfer function or state-space model. A regular aircraft actuator for control surfaces (e.g., aileron servo) is modeled with a first-order transfer function (lag), G_A, which is of the form:

$$G_A(s) = \frac{K}{s+K} \tag{6.4}$$

where K represents the inverse of the time constant (τ) of the actuator.

$$\tau = \frac{1}{K} \tag{6.5}$$

A typical value for the time constant of an UAV control surface actuator is about 0.02–0.1 seconds, so the K frequently varies between 10 and 50 ($10 < K < 50$). Recall that

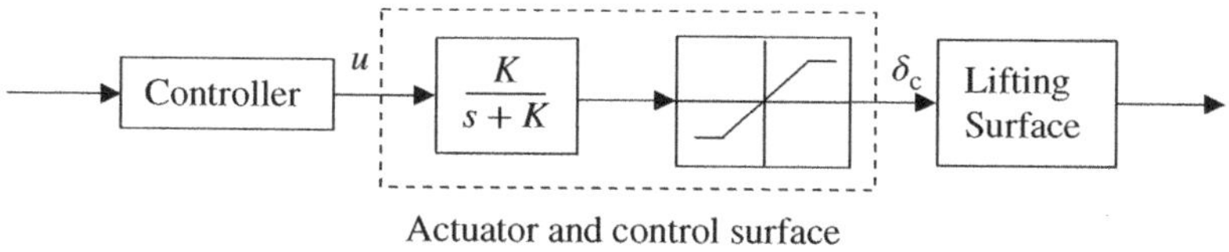

Figure 6.9 Scheme of the control surface actuator.

the time constant is defined as the time that takes (i.e., delays) the response of an element to reach 63% of the steady-state value. The smaller the time constant, the faster (more desired) is the actuator. The dynamic behavior of an electric motor (used in most quadcopters) may also be represented by a first order model.

In terms of state-space representation, the transfer function (Eq. 6.4) which represents the delay, is converted to the following form:

$$\dot{x} = -Kx + Ku \tag{6.6}$$

$$y = x$$

Either of the models may be utilized in the control system design and simulation.

6.3.5 Saturation

All control surfaces have a limit for maximum deflection (about ±30°). To prevent a control surface (e.g., aileron) not to exceed the desired limit, a physical stop (*limiter*) is arranged. For instance, the stops for the ailerons are just a piece of aluminum bar riveted to the aileron hinge bracket. This hard stop is modeled in MATLAB/Simulink with a saturation block. Hence, a control surface and its actuator are modeled (Figure 6.9) with a limiter plus a first order system (Eq. 6.4). Thus, we need to bound the signal entering the actuator.

6.4 Flight Control Requirements

There are primarily three flight control requirements: (a) longitudinal control requirements; (b) lateral/Roll control requirements; and (c) directional control requirements. In this section, these requirements are presented. These requirements must be met by the flight control system. The description and analysis of aircraft modes show that Automatic Flight Control Systems (AFCSs) can be divided into different categories. One category includes modes that involve mainly the rotational degrees of freedom, whereas some categories involve the translational degrees of freedom.

6.4.1 Longitudinal Control Requirements

An aircraft must be longitudinally controllable, as well as maneuverable within the flight envelope. In a conventional aircraft, the longitudinal control is primarily applied through the deflection of elevator (δ_E), and engine throttle setting (δ_T). There are two groups of requirements in the aircraft longitudinal controllability: (a) required actuator force (in linear actuator) and torque (in servomotor); and (b) aircraft response to the control surface deflection input. In order to deflect the elevator, the actuator must apply a force and hold it (in the case of an aircraft with a stick-fixed control system). In an aircraft with a stick-free control system, the actuator force is amplified through such devices as tab and spring.

The aircraft response in the longitudinal control is frequently expressed in terms of pitch rate (q). However, the forward speed and angle of attack would be varied as well. The most critical flight condition for pitch control is when the aircraft is flying at a low speed. Two flight operations which feature a very low speed are take-off and landing. Take-off control is much harder than the landing control due to the safety considerations. A take-off operation is usually divided into three sections: (a) ground section; (b) rotation or transition; and (c) climb. The longitudinal control in a take-off is mainly applied during the rotation section, in which the nose is pitched up by rotating the aircraft about main gear.

The control surfaces must be designed such that the aircraft possesses acceptable flying qualities anywhere inside the operational flight envelope; as well as allowable cg range, and allowable aircraft weight. The operational flight envelopes define the boundaries in terms of speed, altitude, and load factor within which the aircraft must be capable of operating in order to accomplish the desired mission. A typical operational flight envelope for a large aircraft is shown in Figure 6.10.

6.4.2 Roll Control Requirements

Roll or lateral control requirements govern the aircraft response to the aileron deflection; thus, the requirements are employed in the design of aileron. It is customary to specify roll power in terms of the change of bank angle achieved in a given time in

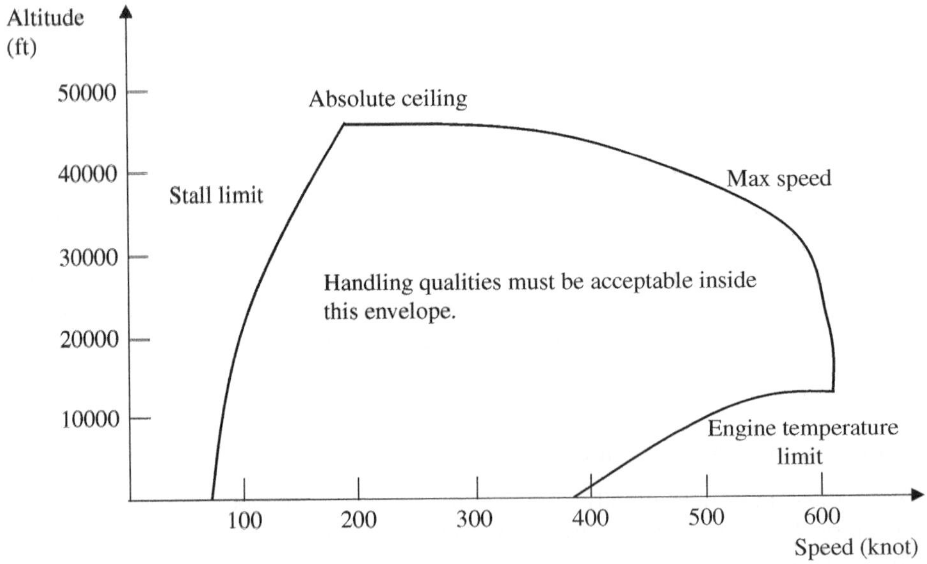

Figure 6.10 A typical operational flight envelope.

response to a step function in roll command. Thus, the aircraft must exhibit a minimum bank angle within a certain specified time in response to aileron deflection. The required bank angles and time are specified in Tables for various aircraft classes and different flight phase (see Refs. [75, 76]).

Roll performance in terms of a bank angle change ($\Delta\phi$) in a given time (t) is specified in Tables (Ref. [76]) for Class I through IV aircraft. The notation "60 degrees in 1.3 seconds" indicates the maximum time it should take from an initial bank angle (say 0°) to reach a bank angle which is 60° different than the initial one, following the full deflection of aileron. It may also be interpreted as the maximum time it should take from a bank angle of −30° to +30°. For class IV aircraft, for level 1, the yaw control should be free. For other aircraft and levels, it is permissible to use the yaw control to reduce any sideslip which tends to retard roll rate. Such yaw control is not permitted to induce sideslip which enhances the roll rate.

6.4.3 Directional Control Requirements

In a conventional aircraft, directional control is usually maintained by the use of aerodynamic controls (e.g., rudder) alone at all airspeeds. There are a number of cases that directional control must be achievable within a specified limits and constraints. In this section, most important ones are presented. Directional control characteristics shall enable the pilot to balance yawing moments and control yaw and sideslip. Sensitivity to yaw control pedal forces shall be sufficiently high that directional control and force requirements can be met, and that satisfactory coordination can be achieved without unduly high pedal forces, yet sufficiently low that occasional improperly coordinated control inputs will not seriously degrade the flying qualities.

In a multi-engine aircraft, at all speeds above 1.4 V_s with asymmetric loss of thrust from the most critical factor while the other engine(s) develop normal rated thrust, the airplane with yaw control pedals free may be balanced directionally in steady straight flight. The trim settings shall be those required for wings-level straight flight prior to the failure. When an aircraft is in directional trim with symmetric power/thrust, the trim change of propeller-driven airplanes with speed shall be such that wings-level straight flight can be maintained over a speed range of ±30% of the trim speed or ±100 knots equivalent airspeed, whichever is less (except where limited by boundaries of the Service Flight Envelope) with yaw-control-device (i.e., rudder). In the case of one-engine-inoperative (asymmetric thrust), it shall be possible to maintain a straight flight path throughout the Operational Flight Envelope with yaw-control-device (e.g., rudder) not greater than actuator's maximum force, without re-trimming.

6.5 Control Modes

Two primary prerequisites for a safe flight are stability and controllability. The control system is not only able to control the UAV, but also is sometimes expected to provide/augment stability. Flight stability is defined as the inherent tendency of an aircraft to oppose any input and return to original trim condition if disturbed. When the summation of all forces along each three axes, and summation of all moments about each three axes are zero, an aircraft is said to be in trim or equilibrium. In this case, aircraft will have a constant linear speed and/or a constant angular speed.

Control is the process to change the aircraft flight condition from an initial trim point to a final or new trim point. This is performed mainly by autopilot by moving the control surfaces/engine throttle. The desired change is basically expressed with a reference to the time that it takes to move from initial trim point to the final trim point (e.g., pitch rate, q, and roll rate, p).

6.5.1 Coupled Control Modes

An aircraft is capable of performing various maneuvers and motions; they may be broadly classified into three main groups: (a) longitudinal control; (b) lateral control; and (c) directional control. In the majority of aircraft, longitudinal control does not influence the lateral and directional control. However, lateral and directional control are often coupled; any lateral motion will often induce a directional motion; and; any directional motion will often induce a lateral motion. The definition of these motions is as follows:

1) *Longitudinal control*: Any rotational motion control in the x-z plane is called longitudinal control (e.g., pitch about y-axis, plunging, climbing, cruising, pulling up, and descending). Any change in lift, drag and pitching moment have the major influence on this motion. The pitch control is assumed as a longitudinal control. Two major longitudinal control inputs are the elevator (δ_E), and engine throttle (δ_T).
2) *Lateral* control: The rotational motion control about x-axis is called lateral control (e.g., roll about x-axis). Any change in wing lift distribution and rolling moment have the major influence on this motion. The rolling control is assumed as a lateral control. The primary lateral control input is the aileron deflection (δ_A). However, the rudder (δ_R) deflection has an indirect influence on this motion too.
3) *Directional* control: The rotational motion control about z-axis and any motion along y-axis is called directional control (e.g., yaw about z-axis, side-slipping, and skidding). Any change in side-force and yawing moment have the major influence on this control. The yaw control is assumed as a directional control. A level turn is a combination of lateral and directional motions. The primary directional control input is the rudder deflection (δ_R). However, the aileron deflection (δ_A) has an indirect influence on this motion too.

In a conventional UAV, three primary control surfaces are used to control the physical three-dimensional attitude of the UAV, the elevators, rudder, and ailerons. In a conventional UAV, the longitudinal control (in the x-z plane) is performed through a longitudinal control surface or elevator. The directional control (in the x-y plane) is performed through a directional control surface or rudder. The lateral motion (rolling motion) is executed using aileron. Hence, there is a direct relationship between the control system and control surfaces. A graphical illustration of these three control motions is sketched in Figure 6.11. An UAV has six degrees of freedom (i.e., three linear motions along x, y, and z; and three angular motions about x, y, and z). Thus, there are normally six outputs in an actual flight; six examples are three linear velocities (u, v, w), and three angular rates (p, q, r).

The primary idea behind the design of flight control surfaces is to position the surfaces; so that they function primarily as moment generators. They provide three types

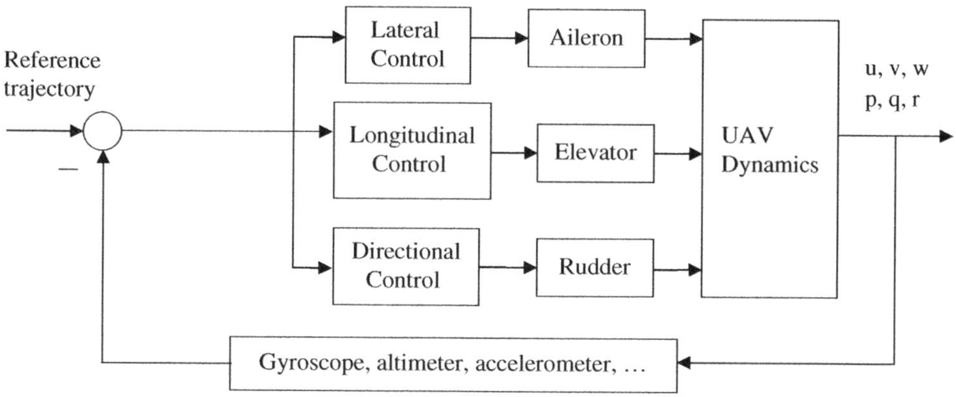

Figure 6.11 Flight control system with conventional control surfaces.

Table 6.2 Autopilot inner loops.

No	Loop	Command variable	Control surface	Constraints
1	Normal acceleration command	n_z	Elevator	$-30° \leq \delta_E \leq +30°$ $-2 \leq n_z \leq +5$
2	Bank angle command	ϕ	Aileron	$-30° \leq \delta_A \leq +30°$ $-2 \leq n_z \leq +5$
3	Turn coordinator	Lateral acceleration (n_y)	Rudder	$-30° \leq \delta_R \leq +30°$ $n_y = 0$

of rotational motion (roll, pitch, and yaw). Variations to this classical configuration lead to some variations in the arrangements of these control surfaces. Table 3.6 (in Chapter 3) represents several control surfaces configurations.

There are various measurement devices to measure the flight variables such as airspeed, pitch angle, heading angle, bank angle, linear accelerations (Normal, Lateral, and Longitudinal), angular rates (Pitch, Roll, and Yaw rates), altitude, and position. The measured flight data are recorded, and may be stored in a data storage element; which may be conveniently read by the user (in real-time, or off-line). If an interface with telemetry data systems is available, a user in the ground station may access the data in real time. Typical measurement devices (sensors) are: (a) gyroscope; (b) rate gyroscope; (c) pitot-tube; (d) altimeter; (e) magnetometer; (f) compass; (g) accelerometer; and (h) Global Positioning System (GPS). In Chapter 12, the features of these instruments will be presented.

The directional gyroscope is used as the error-measuring device. Two gyros must be used to provide control of both heading and attitude (level position) of the airplane. The error that appears in the gyro as an angular displacement between the rotor and case is translated into a voltage by various methods, including the use of transducers such as potentiometers. Augmented stability for the UAV may be desired in the control system by rate feedback. In other words, in addition to the primary feedback, which is the position of the airplane, another signal proportional to the angular rate of rotation of the airplane around the vertical axis is fed back in order to achieve a stable response. A rate gyro is used to supply this signal.

6.5.2 Cruise Control

The UAV cruise control is assumed to be the easiest control, while this has several alternatives. In the lift Eq. (3.1), the independent parameters involved are: UAV weight (W), airspeed (V), altitude or its corresponding air density (ρ), and angle of attack, or its associated lift coefficient (C_L). Since the fuel is consumed during flight, the weight of the aircraft constantly decreases during the flight. In order to maintain a level flight, we must decrease the lift as well. Of the many possible solutions only three are more practical and will be examined. In each case, two flight parameters will be held constant throughout cruise. The three options of interest for continuous decrease of the lift during cruise are:

1) Decreasing flight speed (Constant-altitude, constant-lift coefficient flight)
2) Increasing altitude (Constant-airspeed, constant-lift coefficient flight)
3) Decreasing angle of attack (Constant-altitude, constant-airspeed flight)

For each flight program, a separate controller is designed. In the first option, the velocity must be reduced with the same rate as the aircraft weight is decreased. In the second solution, the air density must be decreased; in other words, the flight altitude must be increased. The third option offers the reduction of aircraft angle of attack, i.e., the reduction of lift coefficient. In terms of autopilot operation, the first option is applied through the throttle, and the third option is implemented through stick/yoke/wheel. In the second option, no action is needed by the pilot; the aircraft will gradually gain height (climbs).

Based on the safety regulations and practical considerations, the second option is the option of interest for the majority of aircraft. In general, when a flight is conducted under the jurisdiction of FAR, the accepted flight program is the constant altitude–constant airspeed flight program.

In the first option, for the view-point of autopilot control, there are three drawbacks to this flight program:

1) It has the need to continuously compute the airspeed along the flight path and to reduce the throttle setting accordingly.
2) By reducing the airspeed, the flight times increases.
3) This is a fact that air traffic control rules require "constant" true airspeed for cruise flight, currently constant means ±10 knots.

The good news is that, the autopilot has solved part of this problem and performs continuous calculations.

The second flight program is commonly referred to as cruise–climb flight. In this option, the air density will be automatically reduced as the aircraft weight is decreased. No autopilot intervention is necessary. Therefore, cruise–climb flight requires no computations or efforts by the pilot. In an aircraft equipped with an autopilot, the aircraft cruise control will be implemented by the autopilot. After establishing the desired cruise airspeed, the pilot simply engages the Mach-hold mode (or constant-airspeed

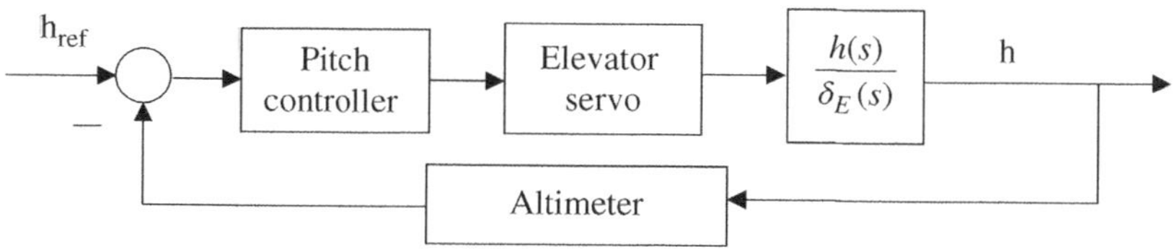

Figure 6.12 Block diagram of an altitude control system.

mode) on the autopilot; and the aircraft will slowly climb at the desired flight-path angle as the fuel is burned.

Figure 6.12 illustrates the block diagram of altitude control system. The purpose of the altitude hold autopilot is to hold altitude at a desired height during specified flight phases (e.g., cruise). For maintaining the constant cruise altitude, either use of elevator or throttle may suffice. One option is to keep the throttle contestant (engine thrust). Thus, the elevator is employed to change the angle of attack for varying UAV weight. The pitch controller could be as simple as a PID (Proportional-Integral-Derivative) or to use a more complicated one. The measurement device is an altimeter (either thorough GPS, or pitot-tube, or radar altimeter). In either case, the UAV will maintain constant altitude through a longitudinal control system.

6.5.3 Pitch-Attitude Hold

Another method for cruise control is the pitch-attitude hold (Figure 6.13), which is frequently used when the vehicle is in wing-level flight. This autopilot hold function is also suitable in a climbing flight for keeping the climb angle constant. The controlled variable is pitch angle (θ), which is the sum of climb angle (γ) and angle of attack (α). Pitch attitude is one of the state variables involved in the phugoid (long-period) mode. The measurement device for the control system is an attitude gyroscope. Note that, the controller does not hold the climb angle constant, because the angle of attack changes with time due to fuel burn and decreasing vehicle weight.

If the engine thrust is increased, the angle of attack tends to decrease, and the vehicle will climb. As the aircraft weight decreases, the angle of attack will decrease too, which causes a gradual climb.

If the preset is the climb angle, it will level out as air density is gradually decreased. This consequently will increase the angle of attack. The pitch-attitude hold is normally employed as the inner loops for other autopilot mode such as altitude-hold and automatic landing. The pitch controller will guarantee a zero steady-state error, with a desired transient response.

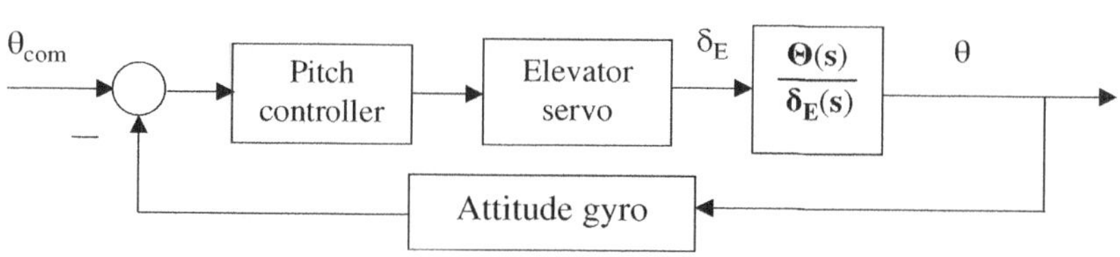

Figure 6.13 Block diagram of a pitch-attitude control system.

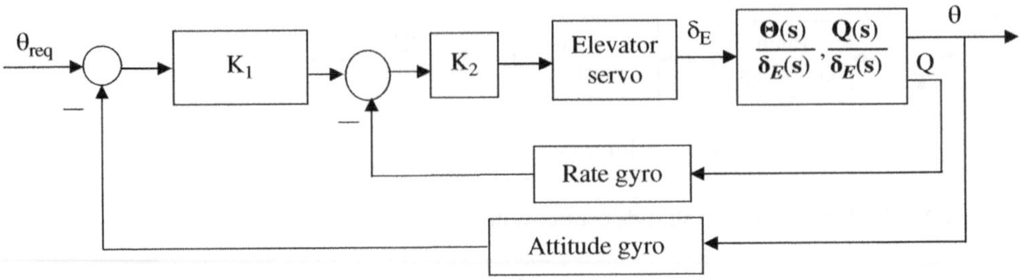

Figure 6.14 Block diagram of a pitch angle control system with two feedbacks.

An *attitude gyro* provides an error signal proportional to the deviation from a preset orientation in inertial space. An inner- loop pitch rate (q) feedback can be added as the second feedback to provide a good short-period damping. The pitch rate is measured by *rate gyro*. The features of attitude gyro and rate gyro are discussed in Chapter 11. Hence, another more effective method to control the pitch angle is to have two feedbacks (Figure 6.14) which requires two sensors. In this control system, the pitch angle is measured with a regular attitude (i.e., vertical) gyro, while the pitch rate (Q) is measured by a rate gyro. This technique is augmenting the longitudinal stability, and provides a steadier flight. Two independent controllers (K_1, K_2) are changing the natural frequency and damping ratio of the pitching motion independently.

6.5.4 Wing Leveler

Wings-level is, by definition, a zero bank angle ($\phi = 0$). Many forces will tend to disrupt a state of wings-level, such as structure's asymmetry, engine torque, atmospheric turbulence, and fuel slosh. In flying along in gusty conditions, one side of the wing (left or right) will always drop. Moreover, even in the smoothest air, eventually a wing will dip. So, without the autopilot intervention, the aircraft always tend to roll. This implies that no air vehicle naturally has roll stability, while they may have lateral stability. The earliest autopilot mode (in 1930s) was the wing-leveler used to reduce the pilot's need to control the bank angle in a long cruising flight.

A means is always necessary (Figure 6.15) to keep your wings level through roll control. The primary control surface for such a goal is the ailerons. Without a feedback loop, the wing level will not be activated; so, a measurement device (e.g., attitude gyro) is needed to sense the bank angle. To get to a zero bank angle (assuming ϕ is initially non-zero) we will need to attain a non-zero roll rate. To get the roll rate, we will need to attain a roll acceleration.

In addition to the autopilot's command of the servo, there is an additional command to limit aileron hunting and over-controlling. Modern autopilot systems incorporate

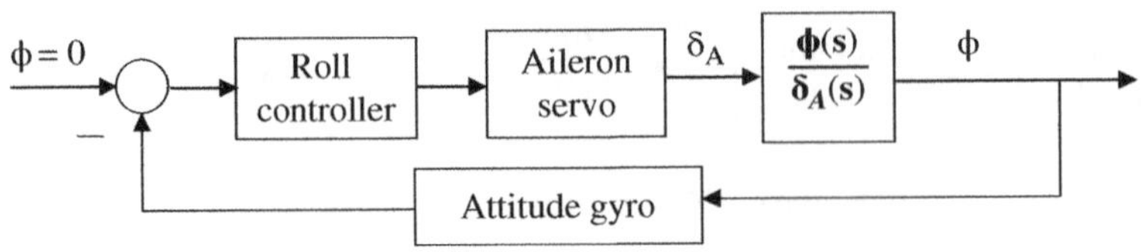

Figure 6.15 Block diagram of bank angle control system.

this servo-position feedback loop to limit control deflection under certain conditions. Example 6.2 will demonstrate the application of optimal control to wing-leveler.

The approximate for the bank-angle-to-aileron-deflection transfer function is modeled by a second order system as:

$$\frac{\phi(s)}{\delta_A(s)} = \frac{L_{\delta A}}{s^2 - sL_p} \tag{6.7}$$

where L_p and $L_{\delta A}$ are dimensional roll damping (stability), and roll control derivatives respectively.

Another more effective method to control the bank angle is to have two feedbacks (Figure 6.16) which requires two sensors. In this control system, the bank angle is measured with a regular attitude gyro, while the roll rate (P) is measured by a rate gyro. This technique augments the directional stability, and provides a steadier flight. Two independent controllers (K_1, K_2) are changing the natural frequency and damping ratio of the rolling motion independently.

The differential governing equation of motion for a pure rolling motion [53] is

$$\frac{-1}{L_P}\frac{dP}{dt} + P = \frac{L_{\delta A}}{L_p}\delta_A \tag{6.8}$$

Applying Laplace transform, and deriving the transfer function will yield:

$$\frac{-1}{L_P}sP + P = \frac{L_{\delta A}}{L_p}\delta_A \tag{6.9}$$

$$\frac{P(s)}{\delta_A(s)} = \frac{\dfrac{L_{\delta A}}{L_p}}{\dfrac{-1}{L_P}s+1} = \frac{-L_{\delta A}}{s - L_P} \tag{6.10}$$

Recall that for a fixed-wing UAV, the damping derivative L_P is negative, so the roll-rate-to-rudder-deflection is a stable system. This transfer function, plus the bank angle transfer function will be employed in the development of the bank angle control system.

6.5.5 Yaw Damper

The yaw damper is a directional control system that will use the rudder to further damp any disturbance in yaw beyond the output of natural directional stability. This yaw control system can be employed to augment the directional stability characteristics of UAVs. Thus, it benefits the vertical tail design, when the yaw is lightly damped. In practice, a UAV with a yaw damper will need a smaller vertical tail; compared with a UAV without a yaw damper.

The system ensures a good flying quality over the entire flight regime. Hence, the UAV designer will achieve an improvement in directional stability without compromising the level of directional flying qualities. The actuator for this damping is the rudder

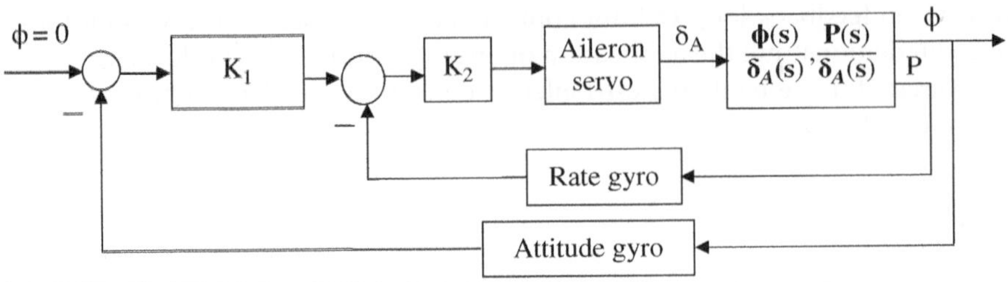

Figure 6.16 Block diagram of a bank angle control system with two feedbacks.

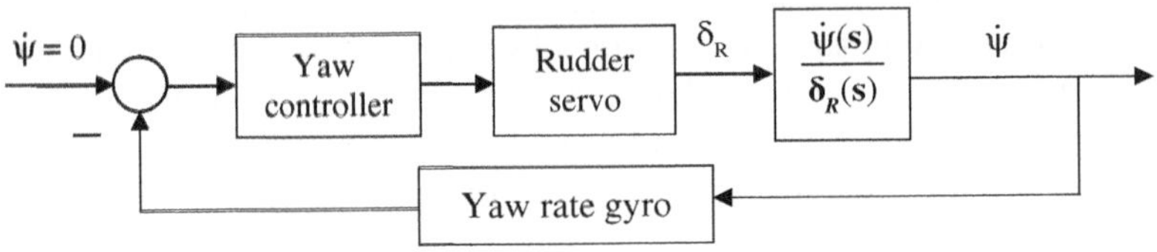

Figure 6.17 Block diagram of a yaw damper (control system).

(Figure 6.17), and the measurement device is the yaw rate gyro. Any disturbance in yaw angle (ψ) will be damped, so the yaw angle and yaw rate will go to zero.

Many air vehicles are provided with a yaw rate damper to artificially provide adequate damping in Dutch roll. If the UAV is required to perform a precise mission (e.g., taking picture), the yaw damper will facilitate the mission.

The approximate yawing moment equation [53] is:

$$\ddot{\psi} - N_r \dot{\Psi} + N_\beta \Psi = N_{\delta R} \delta_R \tag{6.11}$$

where N_r and N_β are dimensional yaw stability derivative, and $N_{\delta R}$ is the dimensional yaw control derivative. Applying the Laplace transform, this differential equation is readily converted to a transfer function:

$$\frac{\psi(s)}{\delta_R(s)} = \frac{N_{\delta R}}{s^2 - sN_r + N_\beta} \tag{6.12}$$

By comparing this transfer function with the standard form for a second-order system (i.e., Eq. (5.14)), one can conclude:

$$s^2 - sN_r + N_\beta = s^2 + 2\xi\omega_n s + \omega_n^2 \tag{6.13}$$

All corresponding terms in both sides should be equivalent, which results in the following:

$$-N_r = 2\xi\omega_n \tag{6.14}$$

$$N_\beta = \omega_n^2 \tag{6.15}$$

Or

$$\omega_n = \sqrt{N_\beta} \tag{6.16}$$

$$\xi = \frac{-N_r}{2\omega_n} = \frac{-N_r}{2\sqrt{N_\beta}} \tag{6.17}$$

A low damping ratio would result in a poor damping, and a large overshoot; which is very difficult for the autopilot to control the flight direction in xy plane. A reasonable design objective is to provide a damping ratio $0.35 < \xi < 1$, with a natural frequency $0.1 < \omega_n < 1$ rad/s. These two objectives may be applied to Eqs. (6.16) and (6.17) to design a feedback control system such that the rudder deflection is proportional to the yaw rate:

$$\delta_R = -k\dot{\Psi} \tag{6.18}$$

Substituting the control deflection expression into Eq. (6.11) yields:

$$\ddot{\psi} - N_r\dot{\Psi} + N_\beta\Psi = N_{\delta R}\left(-k\dot{\Psi}\right) \tag{6.19}$$

By rearranging, one can obtain:

$$\ddot{\psi} - \left(N_{\delta R} - kN_r\right)\dot{\Psi} + N_\beta\Psi = 0 \tag{6.20}$$

By proper selection of factor k, one can satisfy the yaw damping requirement.

6.5.6 Auto-Landing

There are a number of methods for automatic landing. The auto-landing works using two components, a localizer and a glideslope transmitter. An important part of this operation is governed by the localizer hold mode of the autopilot, which employs the heading command system and the horizontal guidance system.

The localizer is a navigation instrument which is used to guide the vehicle along the axis of the runway. It is a transmitter with a horizontal antenna array located at the opposite end of the runway, and the UAV receives the broadcasted signal through its receiver antenna. The localizer hold mode incorporates the current UAV heading angle (ψ), and localizer error angle (λ) to create a reference heading angle (ψ_{ref}). The reference command is governed by the intercept beam and localizer dynamics. The hold mode will create a command until the localizer error angle is zero. To have the desired path and slope, all three control surfaces (aileron, elevator, and rudder) are utilized. The block diagram of the localizer hold mode has a number of subsystems; Figure 6.18 illustrates the main block diagram for automatic landing system.

The localizer error angle, λ, is determined from UAV distance (d) from the intended path (top-view) and slant range (R) as:

$$\lambda = 57.3\frac{d}{R} \tag{6.21}$$

The distance, d, in s-domain is determined from the integration of its rate of change

$$d(s) = \frac{1}{s}\dot{d}(s) \tag{6.22}$$

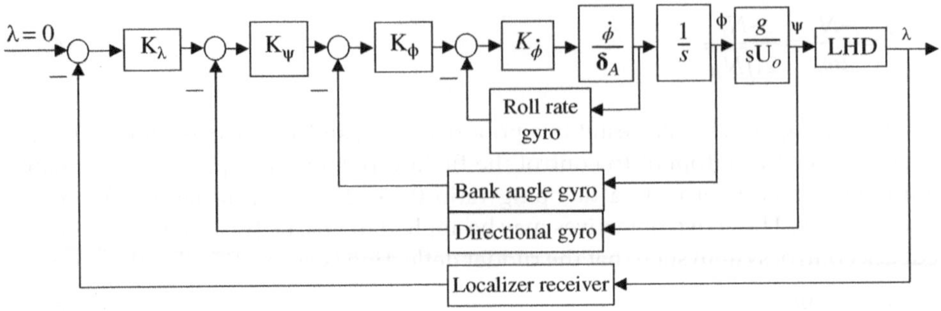

Figure 6.18 Block diagram of localizer hold mode.

where

$$\dot{d} = U_o \sin(\psi - \psi_{ref}) \approx U_o(\psi - \psi_{ref}) \tag{6.23}$$

Therefore

$$\dot{d}(s) \approx U_o(\psi(s) - \psi_{ref}(s)) \tag{6.24}$$

This equation is employed to derive the localizer hold dynamics; LHD (i.e., transfer function). The task of the control system designer is to determine the values for gains K_λ, K_ψ, K_ϕ, and $K_{\dot{\phi}}$ to satisfy the auto-land requirements.

6.5.7 Turn Coordinator

Turn is a flight maneuver which leads to a change in the heading angle (ψ); it is usually performed by simultaneous rolling and yawing motions (so, bank-to-turn). In rare cases (e.g., some missiles), a turn [77] only involves a yawing motion (skid-to-turn). The most desired type of a level turn is the one which is coordinated. It has a few advantages including a constant radius, no lateral load factor, and better environment for mechanical instruments. In the case of a UAV which is monitoring a ground target, a coordinated turn allows it to perform the mission successfully.

Every modern aircraft has a turn coordinator which has an instrument consisting of a rate gyro to indicate the rate of yaw; and an accelerometer to determine the centripetal acceleration. In a coordinated turn, the turn radius (R) is kept constant. This type of turn is accomplished when the aerodynamic side force is zero. Thus, a coordinated turn features no slip and no skid. Hence, the aircraft is traveling in a circular path with a constant radius and a constant airspeed (V). Hence, the horizontal component of the lift (L) is equal to the centrifugal force, while its vertical component equals the vehicle's weight.

$$L \sin\phi = m\frac{V^2}{R} \tag{6.25}$$

$$L \cos\phi = W \tag{6.26}$$

where ϕ is the bank angle. The turn coordinator is a lateral autopilot which provides a body lateral load factor, n_y, to a commanded load factor, n_{yc} (often zero). There are various configurations for a turn coordinator; most include an inner loop (yaw rate damper) and an outer loop. The outer loop uses an accelerometer feedback and has the capability of moment arm feedback, if the accelerometer is located at a different position than the UAV center of gravity.

The ratio between the lift and the aircraft weight (W) is called the load factor (n):

$$n = \frac{L}{W} \tag{6.27}$$

The relationship [9] between turn radius, bank angle, and airspeed in a coordinated turn is:

$$R = \frac{V^2}{g \tan\phi} \tag{6.28}$$

When the sensors in the turn coordinator remind the flight computer that the airplane is turning, the computer sends a signal to the roll servo (an electric motor, or a hydraulic jack). Then, through a bridle cable (if a mechanical system), grips one of the aileron cables. As the roll servo gently applies the aileron against the turn, the flight computer monitors the progress, eventually removing the command when the turn coordinator signals that the wings are once again level. This loop works continuously, many times a second.

Table 6.2 presents the inner loops of a turn coordination mode of an autopilot. In a coordinated turn, all three control surfaces (in a conventional fixed-wing UAV) are simultaneously employed. The elevator is deflected to maintain the desired angle of attack to compensate the lift. The aileron is deflected to develop the bank angle. The rudder has the primary role in transferring a regular turn to a coordinated one. In a bank-to-turn ($\phi > 0$) maneuver, due to an increase in the angle of attack, the normal load factor (n_z) will be more than one. The relation [9] between normal load factor and the vehicle bank angle is:

$$n_z = \frac{1}{\cos(\phi)} \tag{6.29}$$

Hence, as the bank angle is increased, the load factor is increased too.

Inner loop augmentation and control modes should be designed to improve UAV damping in all axes. The turn coordinator frequently satisfies a commanded load factor and a commanded bank angle (The block diagram is similar to that of a wing leveler). The outer loops should be designed to provide altitude hold, and vertical flight path angle hold. Moreover, if the guidance system is involved, the guidance loops should be designed to provide proportional navigation in elevation and azimuth. For a non-UCAV, the lateral acceleration (a_y) is desired to be zero; which is sensed by a lateral accelerometer. To develop an easier design environment for the turn coordinator, it is recommended to install the lateral accelerometer at the vehicle center of gravity.

The block diagram representing the turn coordinator mode of an autopilot is shown in Figure 6.19. There are three feedbacks (lateral acceleration, yaw rate, and rate of yaw

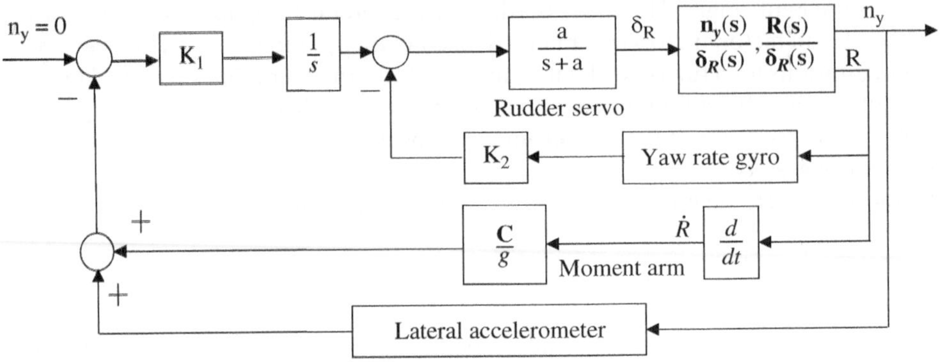

Figure 6.19 Block diagram of a turn coordinator.

rate), and two sensors (Lateral accelerometer and yaw rate gyro). The parameter "C" in the second loop represents the moment arm for the accelerometer. If the accelerometer is located at the vehicle's center of gravity (c.g.), C will be assumed zero. Various controllers may be employed to satisfy the control requirements; the simplest ones are two gains. When all transfer functions are derived, the design problem statement is to determine two gains K_1 and K_2 to satisfy the turn coordination requirements.

The components of load factor in y and z directions may be expressed [40] as the function of flight variables. The normal load factor as a function of rate of angle of attack ($\dot{\alpha}$) is:

$$n_z = \frac{V_T}{g}[Q - \dot{\alpha}] \tag{6.30}$$

where V_T is the total airspeed. The lateral load factor (n_y) as a function of yaw rate (R) is:

$$n_y = \frac{Y}{mg} = \frac{V_T}{g}\left[\dot{\beta} + R\right] \tag{6.31}$$

The yaw-rate-to-rudder-deflection transfer function [40] is given by:

$$\frac{R(s)}{\delta_R(s)} = \frac{V_T N_{\delta_R} s + \left(N_\beta Y_{\delta_R} - Y_\beta N_\beta\right)}{V_T s^2 + s\left(V_T N_r - Y_\beta\right) + \left(Y_\beta N_r + V_T N_\beta - Y_r N_\beta\right)} \tag{6.32}$$

Using the Eq. (6.26), the transfer function for lateral-acceleration-to-rudder-deflection is obtained:

$$\frac{n_y(s)}{\delta_R(s)} = \frac{V_T}{g} \frac{Y_{\delta_R} s^2 + s\left(Y_r N_{\delta_R} - Y_{\delta_R} N_r\right) + \left(Y_{\delta_R} N_\beta - N_{\delta_R} Y_\beta\right)}{V_T N_{\delta_R} s + \left(N_\beta Y_{\delta_R} - N_{\delta_R} Y_\beta\right)} \tag{6.33}$$

Note that an n_y of zero results in a coordinated turn.

Example 6.2
The yaw-rate-to-rudder-deflection and lateral-acceleration-to-rudder-deflection transfer functions for an UCAV with a weight of 2200 lb, and a wing area of 120 ft^2 at an airspeed of 839 ft/s are given below:

$$\frac{R(s)}{\delta_R(s)} = \frac{-14.5(s+0.11)}{s^2+0.245s+14.5}$$

$$\frac{\mathbf{n}_y(s)}{\delta_R(s)} = \frac{-0.14(s^2-0.01s-20.9)}{s+0.11}$$

Assume that the accelerometer is installed at the UAV's center of gravity (cg), and the rudder actuator has a time constant of 0.025 seconds. Design a turn coordinator, such that the damping ratio and natural frequency satisfy the following requirements:

$$0.5 \le \zeta \le 1$$

$$5 \le \omega_n \le 15 \ (rad/\sec)$$

Solution
Since the accelerometer is installed at the UAV's center of gravity, the moment arm (C) is zero. Due to a time constant of 0.025 second, the rudder actuator transfer function (using Eq. (6.4)) becomes:

$$G_R(s) = \frac{40}{s+40}$$

Using the block diagram given in Figure 6.19 for a turn coordinator, the resulting one is obtained as shown in Figure 6.20.

Now, the control system design is performed by determining two controller's gains; K_1 and K_2. From experience, a K_2 of 0.4 is selected. By combining the transfer functions, and simplifying the block diagram, the block diagram is further simplified to the one illustrated in Figure 6.21.

The root locus for this block diagram is shown in Figure 6.22. Using the sisotool in MATLAB, one can vary the gain K_1 to see the damping ratio and natural frequency of the dominant poles. Selecting a gain K_1 of 0.25 yields characteristics as follows:

$$\xi = 0.532$$

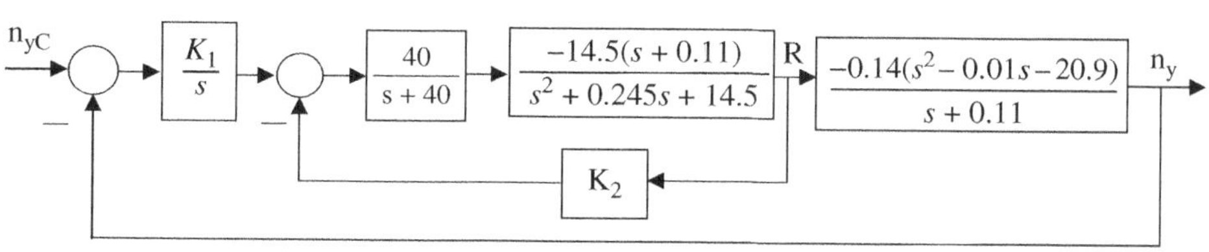

Figure 6.20 Resulting block diagram for the turn coordinator.

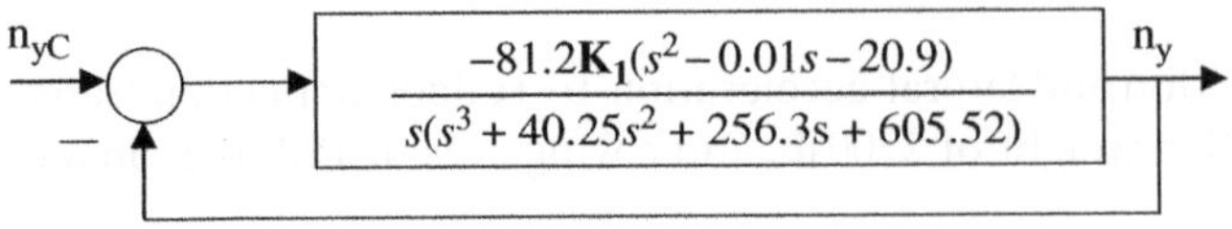

Figure 6.21 Resulting block diagram for the turn coordinator.

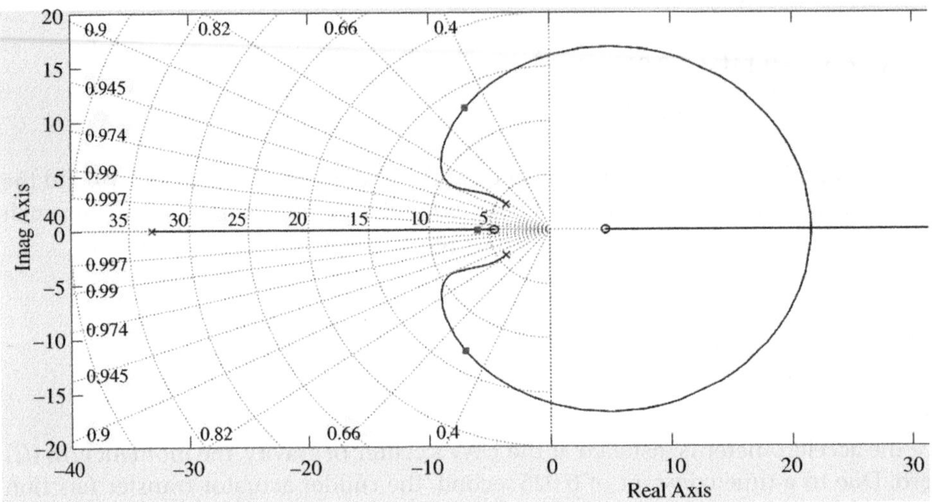

Figure 6.22 Root locus for the turn coordinator.

$$\omega_n = 13.2\, rad/\sec$$

This design may be simulated by MATLAB to observe the performance of the control system. Figure 6.23 shows the response of the turn coordinator for a step aileron input. Note that the lateral acceleration returns to zero.

Although, the design requirements are met, but there is a steady state error. If another controller such as PID has been employed, we could make the steady state error to be zero.

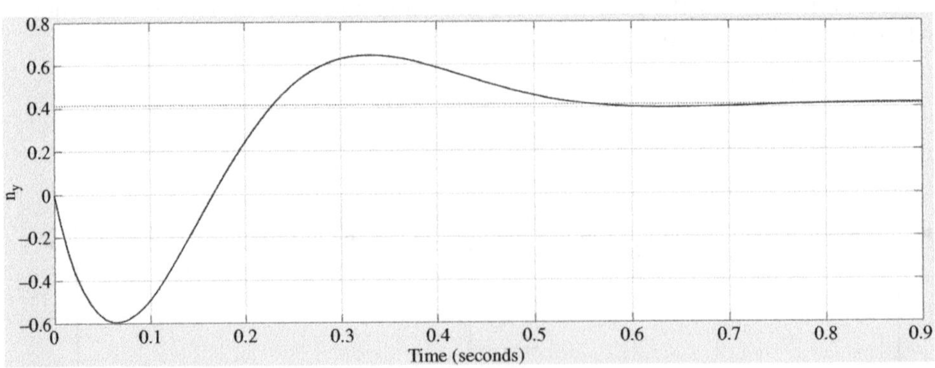

Figure 6.23 Response of the turn coordinator for a step aileron input.

6.6 Controller Design

6.6.1 PID Controller

One form of controller widely used in industrial process control called a three-term, or PID (Proportional-Integral-Derivative) controller. In this controller, three operations are applied on the error signal: (a) Proportionally (P) amplified; (b) Integrated (I); and (c) Differentiated (D). Thus, the control signal, u(t), in time domain is:

$$u(t) = K_P\left(e(t)\right) + K_I \int e(t)dt + K_D \frac{de(t)}{dt} \tag{6.34}$$

Hence, the controller has three terms: 1. Proportional, 2. Integral, 3. Derivative. In s-domain, this controller has a transfer function:

$$G_c(s) = K_P + \frac{K_I}{s} + K_D s \tag{6.35}$$

Various performance deficiencies may be corrected by employing the right values of PID gains. This type of controller is effective, low cost, and easy to apply. Thus, it is even used in aircraft autopilot. References [49, 50] provide a technique to determine PID gains.

Example 6.3

A control system (Figure 6.24) is modeled by the following block diagram:

where $G(s) = \frac{s+10}{2s^2+2s+9}; H(s) = \frac{10}{s+11}$

Design a PID controller (C(s)) such that the response to the unit-step reference input satisfies the following performance specifications:

1) Steady state error less than 10%
2) Maximum overshoot less than 20%
3) Settling time less than three seconds
4) Rise time less than one second.

Solution

A Simulink model (by MATLAB) is created (see Figure 6.25). The PID gains are determined to be: 1. $K_p = 5$, $K_i = 0.1$, $K_d = 1$.

The simulation results (Figure 6.26) illustrates that all four design requirements are met.

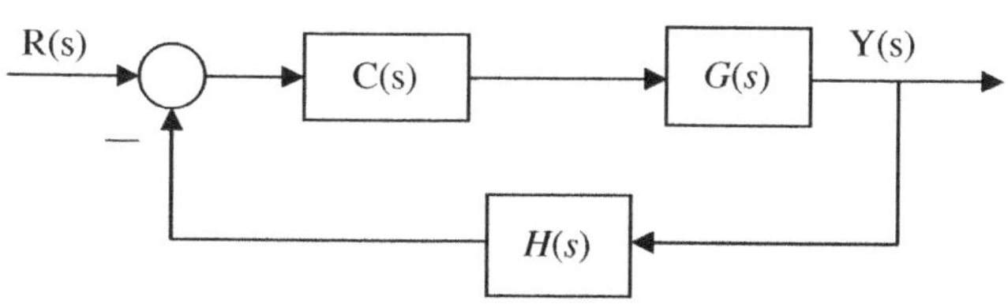

Figure 6.24 The control system of example 6.3.

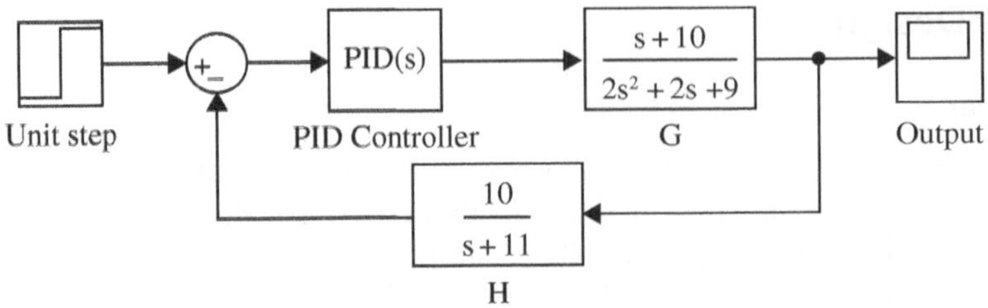

Figure 6.25 Simulink model of the closed-loop system.

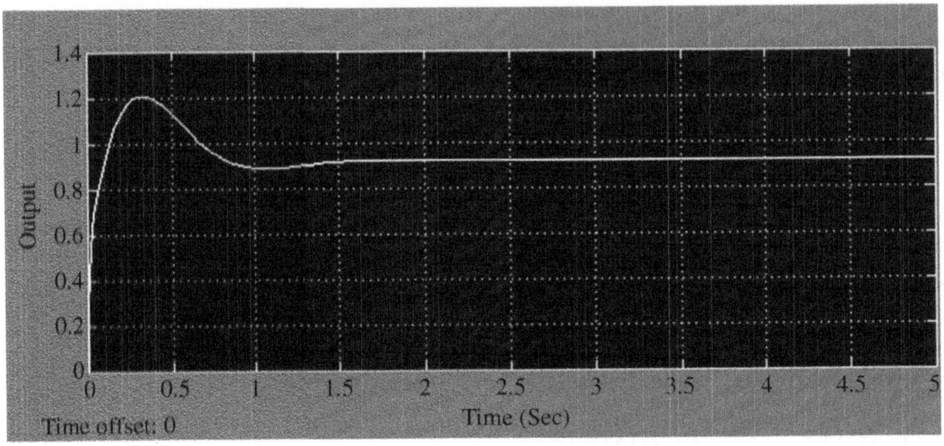

Figure 6.26 The response of the system to a unit step input.

6.6.2 Optimal Control – LQR

The optimal control [78] is based on the optimization of some specific performance criterion; or Performance Index; J. In this technique, no disturbance, noise, or uncertainty is considered. The performance of a control system, written in terms of the state variables, can be expressed as:

$$J = \int_0^{t_f} g(x,u,t)\,dt \tag{6.36}$$

We are interested in minimizing the error of the system; any deviation from equilibrium point is considered an error. To this end, an error-squared performance index is defined. For a system with one state variable, x_1, we have

$$J = \int_0^{t_f} [x_1(t)]^2\,dt \tag{6.37}$$

An optimization technique for a dynamic system in state-space format is defined. The LQR is an optimal controller. The LQR problem is simply defined as follows. The system of interest is of the form:

$$\dot{x} = Ax + Bu$$
$$y = Cx + Du, x(0) = x_{o.} \tag{6.38}$$

Given the *weighting matrices* Q and R, the design task is to find the optimal control signal u(t) such that the quadratic cost function:

$$J = \frac{1}{2}\int_{0}^{\infty}\left(x^T Qx + u^T Ru\right)dt \tag{6.39}$$

is minimized. The solution to this problem is:

$$u = -Kx \tag{6.40}$$

where

$$K = R^{-1}B^T P \tag{6.41}$$

and P is the unique, positive semi-definite solution to the *Algebraic Riccati Equation* (ARE):

$$PA + A^T P + Q - PBR^{-1}B^T P = 0. \tag{6.42}$$

Based on this technique, the LQR gains are calculated using a MATLAB code, and then a control system is designed. The engineering judgment skill must be utilized in the selection of Q and R. Tuning techniques are recommended in the determination of design parameters. For instance, Q and R must be such that the detectability (i.e., $(\sqrt{Q}, A)$ must be detectable) and observability requirements are met.

In MATLAB, the command and its format is:

$$\mathrm{K} = LQR(\mathrm{A,B,Q,R})$$

The LQR command only requires the state matrix A, control matrix B, and weighting matrices Q and R to determine the optimal feedback gain matrix K. Note that if the system is not "state controllable," the closed-loop characteristic equation (includes K) will have a positive pole; or a pair of complex conjugates with a positive real part. This implies that no gain (K) can stabilize the system (i.e., no stabilizing solution). In such cases, the MATLAB solution will be "NaN" which means "Not a number."

The flight simulation may be executed by a SIMULINK model to analyze the response. The matrices Q and R are the weight for state and input variables respectively. They are determined based on the cost function.

The LQR is a popular optimal control technique that has been successfully applied to control several UAV configurations. The LQR was successfully used [79] in stabilizing a RMAX helicopter (Figure 1.1) as well as for accurate orientation and position control of MIT's Raven quadrotors.

Example 6.4

As an application, consider a dynamic system with two state variables, one input, and one output. The state space model is:

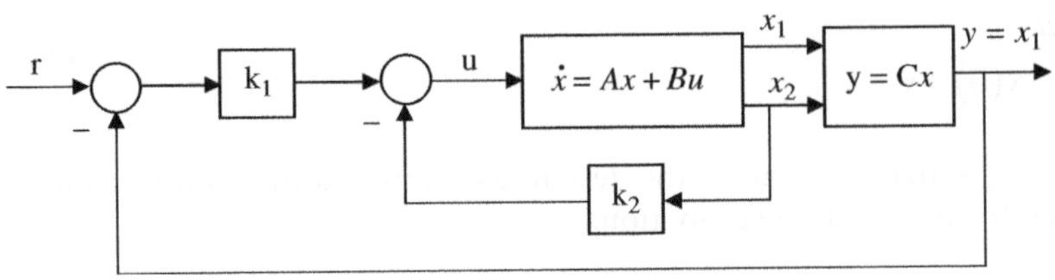

Figure 6.27 Quadratic optimal regulator system for a two-state model.

$$A = \begin{bmatrix} a_{11} & a_{12} \\ a_{21} & a_{22} \end{bmatrix},\ B = \begin{bmatrix} b_1 \\ b_2 \end{bmatrix}, C = \begin{bmatrix} 1 & 0 \end{bmatrix}, D = 0$$

We know that state and output variables are:

$$x = \begin{bmatrix} x_1 \\ x_2 \end{bmatrix} \text{and } y = x_1$$

The control signal u is given (Figure 6.27) by

$$u = k_1 (r - x_1) - k_2 x_2 = k_1 r - (k_1 x_{1+} k_2 x_2)$$

For a zero reference input (i.e., r = 0):

$$u = -k_1 x_{1-} k_2 x_2 = -Kx$$

To determine the state-feedback gain matrix K, where

$$K = \begin{bmatrix} k_1 & k_2 \end{bmatrix}$$

The modified state-space representation is developed by incorporating two feedbacks into the original plant:

$$\dot{x} = Ax + Bu = Ax + B(-Kx + k_1 r) = (A - BK)x + Bk_1 r$$

Therefore, the revised state-space model matrices are:

A1 = A – B*K
B1 = B*K(1)
C1= C
D1= D

The new state-space model can be used in the simulation (e.g., step [A1,B1,C1,D1]) to determine the behavior of the system. Example 6.5 demonstrates the application of LQR for a wing leveler autopilot mode.

Example 6.5
A fixed-wing UAV with a mass of 1300 kg, a wing area of 17 m^2 has a piston engine which delivers a power of 250 hp. At the cruising speed of 120 knot, the vehicle has the following dimensional stability and control roll derivatives:

$L_p = -13\ 1/s, L_{\delta A} = 75\ 1/s^2.$

Design a control system for keeping the wing level at cruise.

Solution

$$\frac{\phi(s)}{\delta_A(s)} = \frac{L_{\delta A}}{s^2 - sL_p} = \frac{75}{s^2 - s(-13)} = \frac{75}{s^2 + 13s}$$

Using a MATLAB code, this transfer function is converted to state space model:

```
N=75
D=[1 13 0]
A,B,C,D]=tf2ss(N,D)
```

The outcome is:

$$A = \begin{bmatrix} -13 & 0 \\ 1 & 0 \end{bmatrix},\ B = \begin{bmatrix} 1 \\ 0 \end{bmatrix},\ C = \begin{bmatrix} 0 & 75 \end{bmatrix},\ D = 0$$

To get a fast response, q_{11} must be sufficiently large compared with q_{22}, and R. thus, we tentatively select:

$$q_{11} = 10, q_{22} = 10, R = 1.$$

The following program lines represent out selection in MATLAB code:

```
Q = [10 0
     0 10]
R = 1
K = LQR(A,B,Q,R)
```

which yields the following solution (i.e., gains) to this problem:

$$K = \begin{bmatrix} 0.6134 & 3.1623 \end{bmatrix}$$

Now, we will investigate the step-response characteristics of the designed system using the optimal gains which determined via a MATLAB code. There are two ways for simulation: (a) a MATLAB code using the modified state-space representation; and (b) Simulink model. The first method is presented here.

1) MATLAB code
   The modified state-space representation is developed by incorporating two feedbacks into the original plant:

$$\dot{x} = Ax + Bu = Ax + B\left(-Kx + k_2 r/75\right) = \left(A - BK\right)x + Bk_2 r/75$$

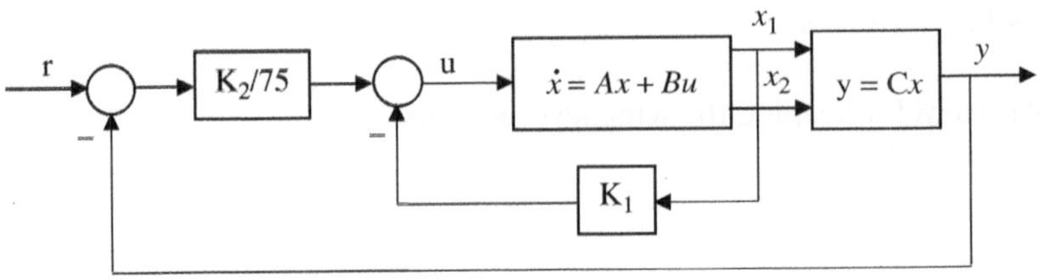

Figure 6.28 Wing leveler quadratic optimal regulator system.

Figure 6.28 illustrates the quadratic optimal regulator system for this two-state model. The revised state-space model matrices are:

```
A1 = A-B * K
B1 = B*K(2)/C(2)
C1 = C
D1 = D
```

The MATLAB code for simulating the closed-loop system using the revised state-space model is:

```
[y,x,t] = impulse(A1,B1,C1,D1)
Plot(t,y)
```

which produces the response of the output (i.e., y) as depicted in Figure 6.29. Since, this is a wing leveler, an impulse input is applied to see if the wing is going back to its original level condition.

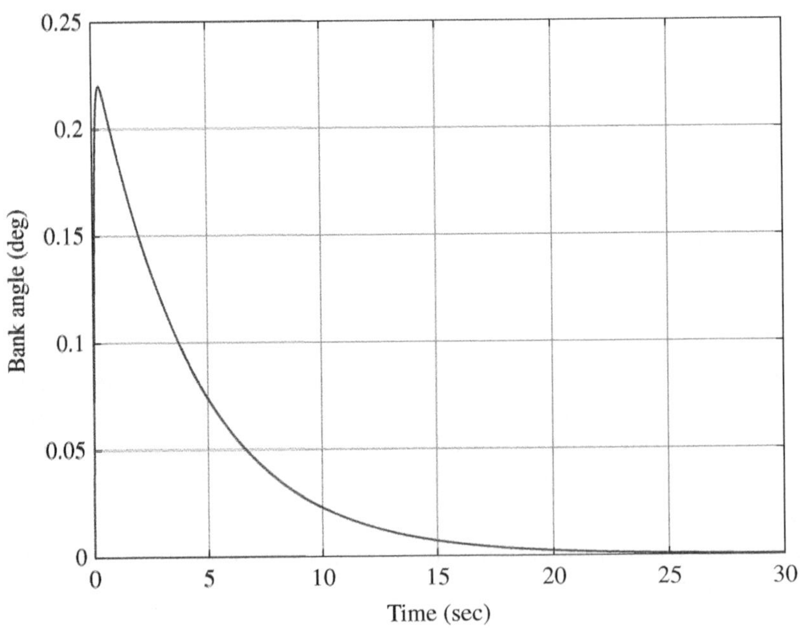

Figure 6.29 The wing bank angle response to an impulse input.

The response confirms that the variations of the output is minimized, and it is stabilizing. Note that by selecting a new set of weighting functions, a slightly different output is generated. The following the complete MATLAB code which has more details.

```
clc
clear all
Num=75
Den=[1 13 0]
[A,B,C,D]=tf2ss(Num,Den)
Q = [10 0
     0 10]
R = 1
K = lqr(A,B,Q,R)
A1 = A - B*K
B1 = B*K(2)/C(2)
C1 = C
D1 = D
t = 0:0.01:30;
[y,x,t] = impulse(A1,B1,C1,D1,1,t);
plot(t,y)
grid
ylabel('Bank angle (deg)')
xlabel('Time (sec)')
```

6.6.3 Gain Scheduling

An effective control law for a time-varying dynamic model is gain scheduling, at which the gains should be scheduled as a function of time. A few for an UAV examples are:

1) when the fuel is rapidly burned, so both vehicle weight and its center of gravity are varying with time.
2) when the flight Mach number is considerably changing, so the aerodynamic model is varying with time.
3) when the flight altitude is changing, so the engine power/thrust are significantly varying with time.
4) when the stability and control derivatives are consequently varying with time as a function of airspeed, altitude, vehicles center of gravity.

For each of such cases, one must determine a unique gain, which could employ PID, LQR, or any suitable techniques. The, the gains are tabulated (Table 6.3) based on parameters such as time, altitude, or Mach number. To accomplish this goal, the nonlinear vehicle differential equations are linearized at several equilibrium flight conditions over the desired flight envelope to obtain various linear models (one for each trim point).

If the state–space representation is utilized, unique A, B, C, and D matrices are obtained. For instance, if the LQR is used for controller design, the software will be run

Table 6.3 Schedules of gains for a long flight.

Trim points	Cruising flight conditions	PID gains	UAV mass (kg) and X_{cg}/C)
1	V = 120 knot	0.1, 1.2, 0.01	1200 kg, 0.26
2	V = 120 knot	0.1, 1.4, 0.02	1100 kg, 0.23
3	V = 120 knot	0.15, 1.7, 0.015	1000 kg, 0.18
4	V = 150 knot	0.2, 0.8, 0.015	1200 kg, 0.26
5	V = 150 knot	0.2, 1.1, 0.01	1100 kg, 0.23
6	V = 150 knot	0.25, 1.4, 0.03	1000 kg, 0.18
…			

to determine the optimal gains for each set of A, B, C, and D matrices. The optimal gains for one point in the gain schedule may be used as initial stabilizing gains in the algorithm for the next point. The gain scheduling strategy has been implemented on a Yamaha RMAX helicopter, as well as a STARMAC quadrotor [79] and successfully demonstrated for a backflip maneuver.

Example 6.6

Consider the UAV in Example 6.2 in a cruising flight. The aircraft begins the flight (constant airspeed, constant altitude) with a mass of 1300 kg, but ends it with a mass of 1100 kg. Due to the reduction (variations) in mass, the center of gravity will move from 25% to the 20% of MAC. The variations of mass and center of gravity will change the dynamic model; an example is shown in Table 6.4 for the variations of two roll (one stability and one control) derivatives. Use the gain scheduling technique based on the optimal control method, determine and tabulate the gains for each flight condition.

Solution

The solution to this problem is generally similar to the solution to problem in example 6.2. The only difference is the variations in two roll derivatives, which varies the UAV

Table 6.4 Values of two roll derivatives at each flight condition.

Flight condition	Mass (kg)	Altitude (ft)	Airspeed (knot)	L_p (1/s)	$L_{\delta A}$ (1/s^2)
1	1300	10 000	120	−13	75
2	1280	10 000	120	−12.7	73
3	1260	10 000	120	−12.3	71
4	1240	10 000	120	−12	69
5	1220	10 000	120	−11.7	67
6	1200	10 000	120	−11.3	65
7	1180	10 000	120	−11	63
8	1160	10 000	120	−10.7	61
9	1140	10 000	120	−10.3	59
10	1120	10 000	120	−10	57

Table 6.5 Values two optimal gains at each flight condition.

Flight condition	UAV mass (kg)	k_1	k_2
1	1300	0.6134	3.1623
2	1280	0.6272	3.1623
3	1260	0.6466	3.1623
4	1240	0.6619	3.1623
5	1220	0.6780	3.1623
6	1200	0.7006	3.1623
7	1180	0.7186	3.1623
8	1160	0.7374	3.1623
9	1140	0.7641	3.1623
10	1120	0.7854	3.1623

transfer function and state space model in each flight condition. Therefore, the MATLAB program which is utilized in example 6.2 is executed here 10 times. Each time, we pick a new set of values for two dimensional roll derivatives (L_p and $L_{\delta A}$) to determine two optimal gains (k_1 and k_2) for a flight condition given in Table 6.3. The results are presented in Table 6.5.

It is noted that the values of k_2 is not changing, while the values of k_1 is not changing considerably (about 28%). In practice, the gains are scheduled to be picked from this table for each flight condition. Recall, the stability and control roll derivatives L_p and $L_{\delta A}$ are [40] functions of mass moment of inertia about x-axis (I_{xx}) which in turn is a function of vehicle's mass.

$$L_p = \frac{QSb^2\, C_{l_p}}{2U_o I_{xx}} \tag{6.43}$$

$$L_{\delta_A} = \frac{QSb\, C_{l_{\delta A}}}{I_{xx}} \tag{6.44}$$

where C_{l_p} is the non-dimensional roll stability derivative, and $C_{l_{\delta A}}$ is the non-dimensional roll control derivative. If the cruising flight is taken 5 hours (300 minutes), each set of gains are applied for 30 minutes.

6.6.4 Robust Control

The robust control approach belongs to the family of model-based design methods that can handle the parametric uncertainty and unmodeled dynamics. The linear robust control technique (known as H_∞) can be applied to any linear system, either Jacobian or feedback linearized. In this approach, disturbances, noise, and uncertainty; ΔG (see Figure 6.30) are considered. Furthermore, an optimization technique is employed to minimize the infinity norm of the error transfer function. Consider a system described by the state-space equations:

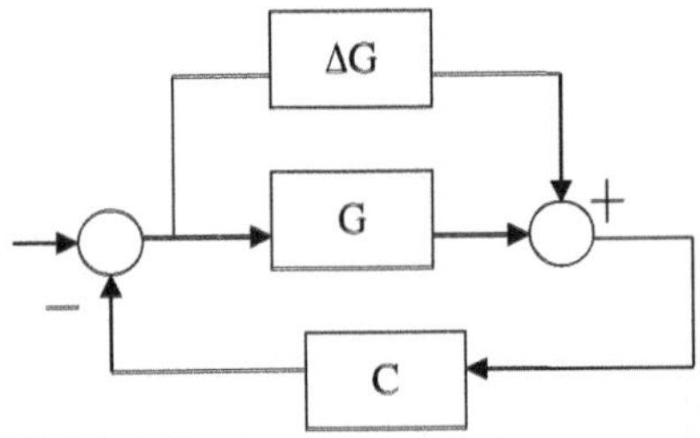

Figure 6.30 Closed-loop system with additive and multiplicative perturbations.

$$\begin{aligned} \dot{x} &= Ax + B_1 w + B_2 u \\ z &= C_1 x + D_{12} u \\ y &= C_2 x + D_{21} w \end{aligned} \quad (6.45)$$

The desire is to design the feedback control u = K(s) y, such that $\|T_{zw}(s)\|_\infty < \gamma$ for a given positive number γ. Note that γ is a function of the maximum singular value of the unstructured uncertainty (in fact $\bar{\sigma}[\Delta G_\infty] = \frac{1}{\gamma}$). The controller (solution) is given (Ref. [80]) by the transfer function:

$$K(s) = -F\left(sI - \hat{A}\right)^{-1} ZL \quad (6.46)$$

where

$$\hat{A} = A + \frac{1}{\gamma^2} B_1 \hat{B_1^T} X + B_2 F + ZLC_2 \quad (6.47)$$

and

$$F = -B_2^T X,\ L = -YC_2^T,\ Z = \left(I - \frac{1}{\gamma^2} YX\right)^{-1} \quad (6.48)$$

where X and Y are solutions of pairs of AREs. The closed loop transfer function matrix $T_{zw}(s)$ from the disturbance w to the output z is given by:

$$T_{zw}(s) = G_{11} + G_{12} K \left(I - G_{22} K\right)^{-1} G_{21}, \quad (6.49)$$

where

$$G(s) = \begin{bmatrix} 0 & D_{12} \\ D_{21} & 0 \end{bmatrix} + \begin{bmatrix} C_1 \\ C_2 \end{bmatrix} (sI - A)^{-1} (B_1, B_2) = \begin{bmatrix} G_{11} & G_{12} \\ G_{21} & G_{22} \end{bmatrix} \quad (6.50)$$

A robust controller can handle uncertainty, disturbance, and noise. The implemented control structure of the Yamaha R-50 helicopter [79] is based on a cascaded architecture using H∞ loop-shaping technique. This controller demonstrated good tracking performance during a set of maneuvers. Its inner Multiple-input multiple-output (MIMO) loop stabilizes the helicopter's attitude, and four independent Single-Input Single-Output (SISO) controllers are responsible for trajectory tracking.

6.6.5 Digital Control

In the early history of AFCSs, all aspects of the flight control were analog, including the controller. With microprocessors so fast, flexible, light, and inexpensive, the control laws could be implemented in digital form. With the introduction of computers and microprocessors in the 1970s, the modern aircraft take advantage of digital control. Digital control is a branch of control theory that employs computers/microcontrollers to act as controllers. In digital control, a computer is responsible for the analysis and implementation of the control algorithm. A digital control system may also use a microcontroller to an application-specific integrated circuit. Since a digital device only accepts a digital signal, a sampler (a kind of switch) is required to take the samples of the continuous signal. The samples are in the forms of zero (0), and one (1).

Usually, a digital control system consists of three main elements: (a) an analog-to-digital (A/D) conversion for converting analog input to digital format for the machine; (b) a digital-to-analog (D/A) conversion for converting digital output to a form that can be the input for a plant; and (c) a digital controller in the form of a computer, microcontroller or a programmable logic controller. The digital controller is implemented using a software code in a computer.

The schematic of a digital control system is shown in Figure 6.31, where z is the Z-transform variable.

The hold device is a D/A converter the discrete control samples (K(z)) into the continuous-time control. The sampler with sample period T is an A/D converter that takes the samples $y_k = y\ (kT)$ of the output of G(s). In the digital control, the transfer functions are in z-domain (i.e., discrete). In a discrete (digital) system, the Laplace transform is replaced with the z-transform. The relationship between variable s and variable z is:

$$z = e^{sT} \tag{6.51}$$

where T is the sample rate (e.g., 0.01 second). The approximation of the exponential function is:

$$e^{sT} \approx \frac{1 + sT/2}{1 - sT/2} \tag{6.52}$$

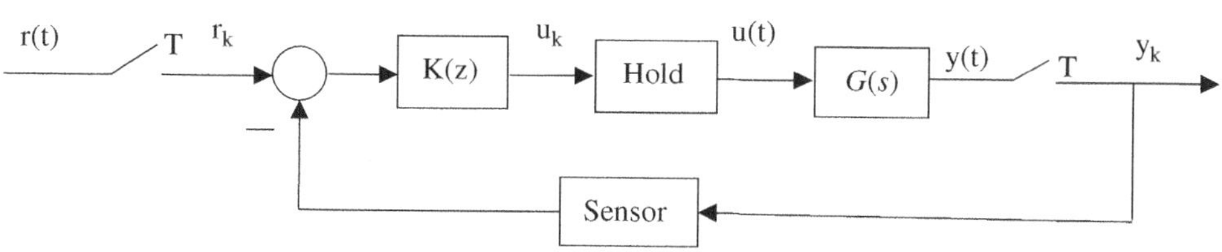

Figure 6.31 Digital control system.

This is referred to as bilinear transformation or Tustin's approximation. Inverting this transformation yields:

$$s = \frac{2}{T}\frac{z-1}{z+1} \tag{6.53}$$

In the digital control, an approximate discrete equivalent of every transfer function (plant, sensor, and controller) are obtained by this transformation technique. As the sampling rate (the average number of samples, T) is higher, the approximation gets more accurate. Reference [81] presents the analysis and design of digital control systems. Software packages such as MATLAB[1] is recommended for the simulation of digital control system. The features and applications of the microcontroller are presented in Chapter 9.

6.7 Autonomy

A typical full flight consists of the following phases: (a) ground taxi, including ground collision avoidance; (b) take-off; (c) climb; (d) en-route cruise; (e) turn and maneuver; (f) descent and (g) landing; (h) ground operation at the destination; and (i) handling of emergencies in any of these sectors. Autonomy is defined with respect to every flight phase, and with various levels. Moreover, some high-level autonomy such as detect-and-avoid, fault monitoring, and automated recovery will be addressed in this section.

6.7.1 Classification

In general, there are four ways for piloting a UAV: (a) remote control; (b) autopilot-assisted control (i.e., Automated); (c) semi-full autonomy; and (d) full autonomy. Under full autonomous control, the reality is that the on-board computer is in control not a human operator. A minimal autopilot system includes attitude sensors and onboard processor. Reference [79] divides autonomy into 11 (from 0 to 10) levels as a gradual increase of Guidance-Navigation-Control (GNC) capabilities: 0. Remote control, 1. Automatic Flight Control, 2. external system independence Navigation, 3. Fault/Event Adaptive, 4. Real time Obstacle/Event Detection and Path Planning, 5. Real time Cooperative Navigation and Path Planning, 6. Dynamic Mission Planning, 7. Real time Collaborative Mission Planning, 8. Situational Awareness and Cognizance, 9. Swarm Cognizance and Group Decision Making, 10. Fully Autonomous.

Due to the nonlinearities and uncertainties of aerial vehicle dynamics, a lot of advanced control techniques, such as neural network, fuzzy logic, sliding mode control, robust control, and learning systems have been used in autopilot systems to guarantee a smooth desirable flight mission. Nowadays, technological advances in wireless communication and micro electromechanical systems, make it possible to use inexpensive small autopilots. We need to first distinguish between these four modes. Here a definition for each mode is provided:

1) In the remote control mode, a human operator is controlling the UAV from a ground station. He/she is making the decision, applying input, and providing commands.

2) In the automated or automatic system, in response to feedbacks from one or more sensors, the UAV is programmed to logically follow a pre-defined set of rules in order to provide an output. Knowing the set of rules under which it is operating means that all possible outputs are predictable. A modern autopilot allows the vehicle to fly on a programmed flight paths without human interference for almost all the mission, without an operator doing anything other than monitoring its operation.
3) Currently, there is no consensus for the definition of autonomy in the UAV community. In this book, "autonomy" is the ability of an agent to carry out a mission in an independent fashion without requiring human intervention. Autonomy should include a kind of artificial intelligence, since the decision-making is performed by the autopilot. An autonomous vehicle is capable of understanding higher level intents and directions. Such a vehicle is able to take appropriate action to bring about a desired state/trajectory, based on this understanding and its perception of the environment, It is capable of deciding a course of action, from a number of alternatives, without depending on human oversight and control, although these may still be present (for monitoring). Although the overall activity of an autonomous UAV is predictable, individual actions may not be. An autonomous UAV is able to monitor and assess its status (e.g., altitude, airspeed) and configuration (e.g., flap deflection); and command and control assets onboard the vehicle.

The core components of autonomy are command, control, navigation, and guidance. Higher levels of autonomy, which reduce operator workload, include (in increasing order) sense-and-avoid, fault monitoring, intelligent flight planning and reconfiguration. An autonomous behavior [79] includes observe, orient, decide, and act. The aviation industry objective is that eventually autonomous UAV will be able to operate without human intervention across all flight sectors. Such an objective requires advances in various technologies including guidance system, navigation system, control system, sensors, avionics, communication systems, infrastructures, and software, microprocessors.

6.7.2 Detect (i.e., Sense)-and-Avoid

Collision avoidance is a primary concern to the Federal Aviation Administration (FAA) regarding aircraft safety. UAVs are seen as potential key airspace users in the future of air transportation, which necessitates additional research and study of safety measures. One of the major limitations to the widespread use of unmanned vehicles in civilian airspace has been the detect (or see or sense)-and-avoid problem. In manned civilian aviation, "see-and-avoid"[2] is the primary mechanism by which piloted aircraft avoid collisions with each other. Obviously, this is impractical for the widespread use of unmanned vehicles, so they must achieve an equivalent level of safety/reliability/assurance to that of manned aircraft operations. There is currently a large amount of research projects being conducted in the area of detect-and-avoid. Based on FAA, a human equivalent sense and avoid system must be able to reliably detect and avoid intruding aircraft within a 3-mile radius with a field of regard of 270 × 30°.

Active solutions include the use of machine vision, and GPS/radar to detect collision threats, and precision control to avoid collision. However, the current adapting technologies are not at such reliable level [83]. In addition, high computational requirement is another obstacle. A major design issue of UAVs is that any black box is additional weight, and weight restrictions for some smaller UAVs may restrict UAV functionality or the inclusion of cooperative systems.

6.7.3 Automated Recovery

Another challenge in the execution of automated control is automated recovery. Here, the recovery could be either regular automated landing or to recover by some means such as a net. Since physical pilot control is not present, there is a high potential for unsuccessful recovery. The UAV must have a number of fail-safes in place in case of any elemental failure. Another source of failure is communication failure or link loss. In the event that command and control links have been completely severed between an UAV and the ground station, the UAV should be switched to pre-programmed mode to attempt for some fixed period of time to re-establish communications, or to independently complete the mission. This is another big area of research for several research institutes and UAV industries.

6.7.4 Fault Monitoring

Unscheduled UAV maintenance creates a lot of issues and cost for large UAV operator units, because spare parts are not always available at any place and sometimes have to be shipped across the world. Moreover, if a flight mission has to be canceled or even delayed, the UAV causes significant costs. Hence, reducing the amount of unscheduled maintenance is a great cost factor for UAV operators. For this objective, and to ensure the integrity of the UAV systems, fault monitoring must be continually conducted on flight. Fault monitoring ensures that undetected system faults will not lead to a catastrophic failure of the aircraft's systems which may eventually lead to human casualties on the ground.

Failure prediction is the combination of condition (i.e., health) monitoring and condition prediction to forecast when a failure will happen. The objectives of Fault Monitoring are: 1. Reduction of unscheduled maintenance, 2. Advanced failure prediction, 3. Condition monitoring, 4. Ability to better plan maintenance, and finally 5. Prevent failures. In the event of system faults, the UAV must have the capability to reconfigure itself and re-plan its flight path in a fail-safe manner. The predictive UAV health monitoring is able to predict failures so that maintenance can be planned ahead.

6.7.5 Intelligent Flight Planning

An intelligent UAV system must have the ability to plan and re-plan its own flight path, to cope with undesired situations. This results in the requirement for advance sensors, and a high-level computing environment where flight planning algorithms can be executed. The intelligent flight planning requires a significant improvement in software and hardware performance. The flight planning process requires knowledge of the UAV's surroundings; including airspace, terrain, other traffic, weather, restricted areas obstacles and closest airfield. The UAV must plan the optimal route for its mission, considering the local environment, to minimize the flight time and fuel usage. Intelligent planning will detect any incoming aircraft for collision avoidance.

To achieve the full potential of unmanned systems at an affordable cost, efforts must be conducted to implement technologies and evolve tactics, techniques and procedures that improve the teaming of unmanned systems with the manned aircraft. The functions of a UAV in a team with manned aircraft depend in nature on the different UAV configurations and their characteristics. To this end, the critical challenges must be identified for further growth to fulfill expanding UAV roles in supporting the aviation safety goals. Moreover, new technologies need to be developed, and new regulations must be prepared.

6.8 Manned–Unmanned Aircraft Teaming

6.8.1 Need for Teaming

A very challenging topic in the UAV control system design is the manned–unmanned aircraft teaming. Today's aircraft inventory includes a diverse mix of manned and unmanned systems. UAVs, are a prime candidate for the teaming with manned aircraft in performing complex/dangerous missions. The statistics are growing exponentially. Unmanned aircraft systems are subject to regulation by the FAA to ensure the safety of flight, and the safety of people and property on the ground. Incidents involving unauthorized and unsafe use of small, remote-controlled aircraft have risen dramatically. Pilot reports of interactions with suspected unmanned aircraft have increased from 238 sightings in all of 2014 to 780 through August of 2015. One of the main goals for the manned–unmanned teaming is to provide flexible flight operations. Teaming a UAV system with manned systems will offer advantage to both.

To achieve the full potential of unmanned systems at an affordable cost, efforts must be conducted to implement technologies and evolve tactics, techniques and procedures that improve the teaming of unmanned systems with the manned aircraft. The functions of a UAV in a team with manned aircraft depend in nature on the different UAV configurations and their characteristics. To this end, the critical challenges must be identified for further growth to fulfill expanding UAV roles in supporting the aviation safety goals. Moreover, new technologies need to be developed, and new regulations must be prepared.

In flight teaming cases, overall system designers should engineer a platform to support the human in the loop and maximize their chances of performing their critical functions successfully. A framework needs to be developed to analyze the root cause of failures that have been attributed to "human error," or system error. An efficient teaming will create an environment such that both parties operate within their limits, while generating an unachievable goal by one party.

6.8.2 Teaming Problem Formulation

Formulation of a manned–unmanned teaming problem basically requires mathematical modeling of UAV flight dynamics, a human decision-making process, and communication between human and autopilot. Figure 6.32 demonstrates the functional block diagram of a teaming flight operation. In principle, there are two independent decision makers: 1. Autopilot for UAV, and 2. Human pilot for the manned aircraft. Moreover,

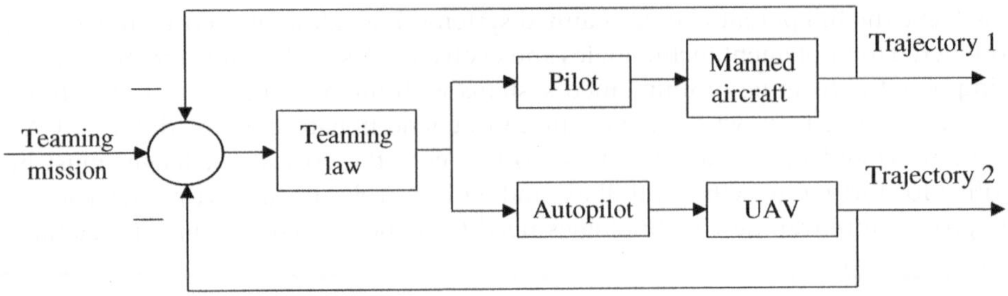

Figure 6.32 Functional block diagram of a teaming flight operation.

there are two separate trajectories, and two feedbacks. The teaming law creates a command for both manned and unmanned aircraft. There is one group of input (mission parameters) and two outputs (i.e., trajectories). Both trajectories are fed back to the same point for comparison with the mission input. Any difference will create an error signal for the teaming law block. The teaming law will generate two signals; one for the pilot of the manned aircraft, and one for the autopilot of the UAV.

Figure 6.32 contains information concerning dynamic behavior, but it does not include any information on the physical construction of the team. The mathematical model of the aircraft/UAV (dynamics model), and autopilot have been provided by Ref. [54]. The mathematical model for the decision-making process of the pilot is discussed in Section 6.8.3. In addition, a teaming law will be presented in Section 6.8.5. In general, there are three categories of teaming, each governed by a distinct law: 1. UAV-leader, manned-aircraft-follower; 2. manned-aircraft-leader, UAV-follower; and 3. mixed leader-follower.

Each teaming case has various advantages and disadvantages, and is suited for specific applications and flight missions. For instance, the teaming category 1 (i.e., UAV-leader, manned-aircraft-follower), is appropriate for a flight mission where the operation involves some hazards to human. Two examples for teaming category 1 are: 1. Observing a volcano, 2. Monitoring a target in the enemy zone for a military mission. In such a mission, the UAV takes the lead and the manned aircraft will follow suit. If any hazard arises, the UAV will be the first to face and handle it. This category will guarantee the safety of human pilot in the manned aircraft. A pictorial representation of the functions performed by each team member in category 1 is illustrated in Figure 6.33.

The UAV flight parameters are measured by both UAV avionics and manned aircraft measurement devices. Thus, the manned aircraft has two feedbacks; one from the UAV, and one from its own flight. The UAV will fly to accomplish the trajectory as the leader, while the manned aircraft will be guided and controlled based on the teaming law. The pilot decision making process has an overlap with the teaming law, as he/she uses the eyes as a navigation instrument.

However, the teaming category 2 is appropriate for a flight mission where the UAV acts as a reserve and no hazard is involved to human pilot. The teaming law for this category may be based on many already developed techniques and guidance laws. One simple and effective guidance law is the line of sight [84].

In the second category (Figure 6.34), the manned aircraft flight parameters are measured by both UAV avionics and manned aircraft measurement devices, as well as the pilot's eyes. Thus, the UAV has two feedbacks; one from the manned-aircraft-leader

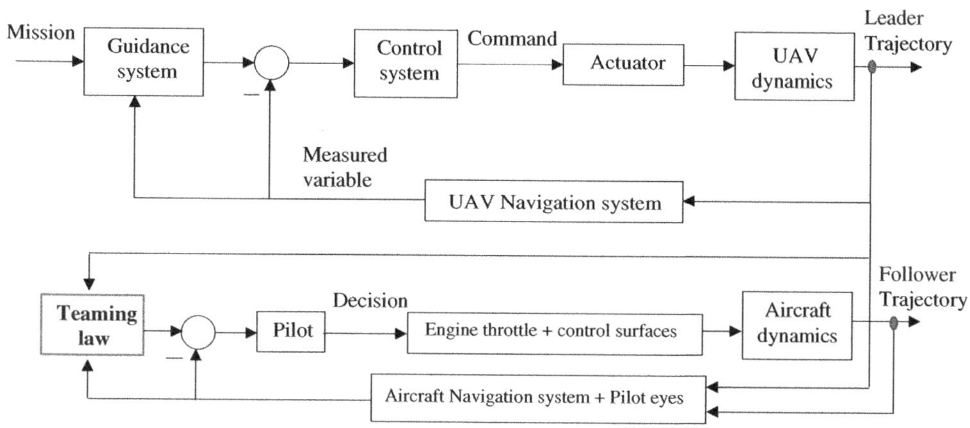

Figure 6.33 Unmanned aerial vehicle (UAV)-Leader, manned-aircraft-follower teaming block diagram.

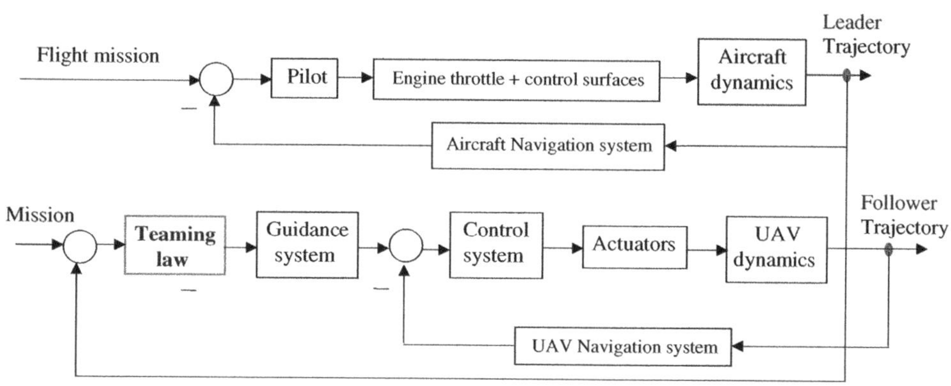

Figure 6.34 Manned-aircraft-leader, Unmanned aerial vehicle (UAV)-follower teaming block diagram.

and one from its own flight. The manned aircraft (human pilot) will fly to accomplish the mission trajectory as the leader, while the UAV will be guided and controlled based on the teaming law. The pilot decision-making process could be independent from the teaming law, as he/she plays the role of the leader.

The third teaming category is the most challenging one and requires the development of a new communication system between human pilot and autopilot. Theory, concepts, challenges of human communication is presented by Ref. [85]. The mathematical formulations of control systems, guidance systems, and navigation systems are presented by many books and papers including Ref. [54]. This book mainly concentrates on the formulation of communication system between manned and unmanned aircraft as well as the teaming law.

6.8.3 Decision Making Process

In the manned–unmanned teaming, the UAV autopilot will perform based on the control law and guidance law which have been programmed for. However, the pilot of the

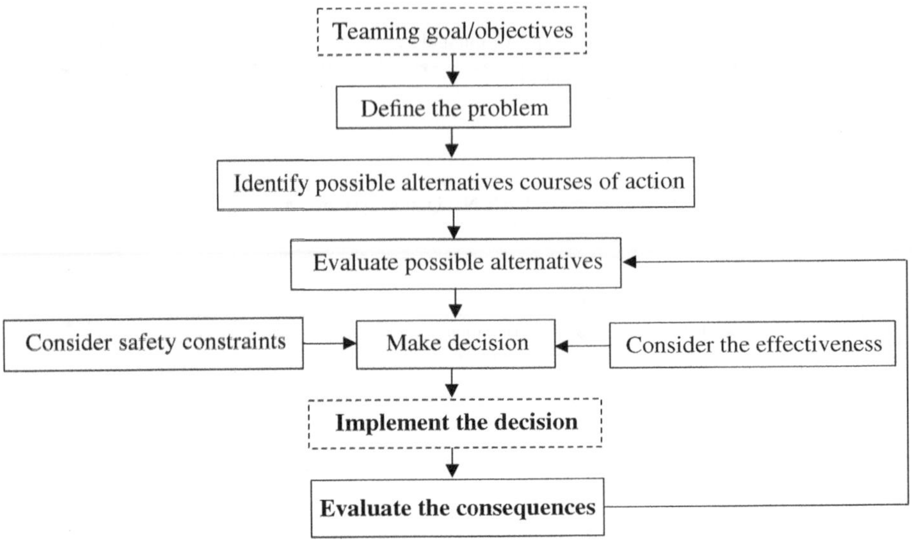

Figure 6.35 Decision making process for pilot as the follower of an Unmanned aerial vehicle (UAV).

manned aircraft should make a decision at every instance before any Implementation. Decision-making is a vital skill in the pilot side of the teaming, particularly when the manned aircraft is the follower. Following a logical procedure along with being aware of common challenges, can help ensure both best decision-making and a safe flight. The following are the basic steps (Figure 6.35) of the decision-making process.

1) **Define the problem.** The first step in making the right decision is recognizing the teaming problem and deciding to address it. Determine why this decision will make the flight safe and successful.
2) **Collect data.** Second step is to gather flight data information so that pilot can make an aircraft control decision based on real flight data. This requires making a value judgment, determining what information is relevant to the decision, along with where to find it.
3) **Identify alternatives.** Once the pilot has a clear understanding of the issue, it is time to identify the various solutions at your disposal. It's likely that you have many different options when it comes to making your decision, so it is important to come up with a range of options. This helps you determine which course of action is the best way to achieve your objective.
4) **Weigh the evidence.** Now, the pilot will need to evaluate the feasibility and acceptability of various alternatives. The pilot should weigh up the pros and cons, then select the option that has the highest chances of success.
5) **Choose among alternatives.** When it is time to make the decision, make sure that you understand the risks involved with your chosen course of action.
6) **Implement the decision.** You should create a plan for implementation. This involves identifying what components and equipment are utilized.
7) **Evaluate the consequences.** You need to observe the consequences; and to evaluate the decision for effectiveness.

The decision-making process which is illustrated in Figure 6.35 should be formulated in the control system design phase. Common challenges of decision making are: 1. having not enough information, 2. misidentifying the problem, 3. Overconfidence in the outcome.

6.8.4 Teaming Communication Process

In Section 6.8.1, three possible configurations were considered in the maned–unmanned aircraft teaming. In all three cases, there must be a communication system available and working in entire flight mission. An important aspect of a manned-unmanned aircraft teaming is the communication between UAV autopilot and the pilot of manned aircraft. In this section, the communication process for the third teaming configuration (i.e., mixed leader-follower) is explored. As a basic case, a team of only two members; one UAV, and one manned aircraft is considered.

Figure 6.36 demonstrates the block diagram including the communication process for a team of two members (one manned, and one unmanned aircraft) in the mixed leader–follower configuration. Both team members have the same mission (e.g., engaging with one target), but each member has a separate but coordinated trajectory. The coordination is created through an effective communication system based on the teaming law. The process involves four feedbacks; two for manned aircraft, and two for UAV. Moreover, there are four error signals; two signals in the UAV control/guidance system, and two signals in the manned aircraft control/guidance system.

As expected, the human pilot talks (creates a voice signal) to communicate with the UAV, but the UAV creates a written text to communicate with the manned aircraft. A Voice-to-Text (V/T) converter in the manned aircraft will convert the voice of human pilot to a text. Then, a transmitter sends this text out to UAV as a radio wave. The UAV receiver will then receive this signal, and the autopilot will process and employ this signal in controlling the UAV. The processor of such signal will be in the microcontroller.

In the meantime, when the UAV creates a signal for delivering to the manned aircraft, the signal will be transmitted through the UAV transmitter to the manned aircraft. When the manned aircraft receiver receives such signal, it will process the signal and

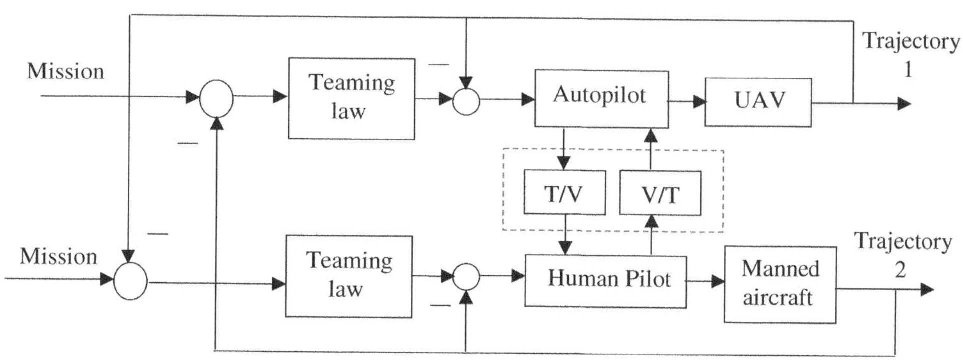

Figure 6.36 Communication between manned and unmanned aircraft.

creates a text. The last step is for the manned aircraft Text-to-Voice (T/V) converter to convert the UAV signal to a voice. When the pilot hears the voice, he/she will process/make a decision/react to the signal according to the teaming law and the required mission. The necessary technology for V/T and T/V has long been developed and practiced. The challenges and opportunities of speech-to-text conversion in real-time have been discussed in [86].

The UAV must have a number of fail-safes in place in case of any elemental failure. A source of failure is communication failure or link loss. In the event that links have been completely severed between an UAV and the manned aircraft, the UAV should be switched to pre-programmed mode to attempt for some fixed period of time to re-establish communications, or to independently complete the mission.

6.8.5 Teaming Laws

In order to begin the synthesis of the teaming law, the design requirements relative to both parties must be technically established. Based on handling qualities [76], and also airworthiness standards [70, 71], the followings are typical design requirements to be used in the design process: stability of the overall teaming system); output (or state tracking) performance; accuracy from command to response; overshoot; steady state error; rise time; settling time. In addition, the law must be robust with respect to aircraft types, communication elements, and missions.

Even with the current advanced and sophisticated computers and software packages, some tasks rely on human knowledge that is currently difficult for a computer to reason about or process. For example, humans are much better than computers at recognizing faces in crowds or noticing a target in an environment with similar objects. Therefore, the team law must consider the human factors as well as the weaknesses and strengths of all team members.

There are fundamental principles which govern an efficient teaming law; some of which are presented in this section. As the most important principle, the safety of the manned aircraft (in fact, the human pilot) is of much higher priority compared with the UAV airworthiness. Thus, the collision avoidance and see-and-detect are two primary concerns to teaming success. Moreover, when the leader aircraft is out of sight of the follower, the follower aircraft must circle around to detect the leader. In the UAV–Leader, manned–aircraft–follower teaming, the teaming law may dictate the human pilot not to follow the UAV.

The state–space approach may be utilized to model the team of (n) UAVs and (m) manned aircraft. If a linear model is considered, the state–space representation of the team will be:

$$\begin{aligned} \dot{x} &= Ax + Bu \\ y &= Cx + Du \end{aligned} \tag{6.54}$$

where x, y, and u denote the state, output, and input variables. In addition, A, B, C, and D denote state, input, output, and transmission matrices. When the team is comprised of one UAV and one manned aircraft, there are two inputs and two outputs. Hence, all four matrices are two-by-two.

$$\begin{bmatrix} \dot{x}_1 \\ \dot{x}_2 \end{bmatrix} = \begin{bmatrix} a_{11} & a_{12} \\ a_{21} & a_{22} \end{bmatrix} \begin{bmatrix} x_1 \\ x_2 \end{bmatrix} + \begin{bmatrix} b_{11} & b_{12} \\ b_{21} & b_{22} \end{bmatrix} \begin{bmatrix} u_1 \\ u_2 \end{bmatrix}$$

$$\begin{bmatrix} y_1 \\ y_2 \end{bmatrix} = \begin{bmatrix} c_{11} & c_{12} \\ c_{21} & c_{22} \end{bmatrix} \begin{bmatrix} x_1 \\ x_2 \end{bmatrix} + \begin{bmatrix} d_{11} & d_{12} \\ d_{21} & d_{22} \end{bmatrix} \begin{bmatrix} u_1 \\ u_2 \end{bmatrix} \tag{6.55}$$

The output y_1 represents the flight trajectory of the UAV, and the output y_2 represents the flight trajectory of the manned aircraft. Using state and output feedbacks, a nominal linear control signal will be:

$$u = -K(x, y) => \begin{bmatrix} u_1 \\ u_2 \end{bmatrix} = \begin{bmatrix} k_{11} & k_{12} \\ k_{21} & k_{22} \end{bmatrix} \begin{bmatrix} x_1 & y_1 \\ x_2 & y_2 \end{bmatrix} \tag{6.56}$$

In order to optimize the performance of the team, an optimal control law may be used to minimize the performance index, J:

$$J = \int_0^{\infty} \left(x^* Qx + u^* Ru + y^* Sy \right) dt \tag{6.57}$$

where Q, R, and S are the symmetric weighting functions of state, input, and output variables respectively. In determining the elements of the weighting functions, the priority will go to the performance and safety of the manned aircraft. Substituting Eq. (6.55) into Eq. (6.54) will deliver an optimization problem. Employing an optimization technique such as solving the Riccati equation will yield the control signals. Since two aerial vehicles are flying simultaneously, the teaming laws are solved within each vehicle separately; and two set of results will be created; one for the UAV and one for the manned aircraft. The autopilot of the UAV will follow one solution, and human pilot will follow the second solution. The teaming law will guarantee the success of the formation flight.

6.9 Control System Design Process

In this Section, the overall control system design process is described. In general, the primary criteria for the design of a control system are as follows: 1. manufacturing technology, 2. required accuracy, 3. stability requirements, 4. structural stiffness, 5. load factor, 6. flying quality requirements, 7. maneuverability, 8. reliability, 9. life-cycle cost, 10. UAV configuration, 11. stealth requirements, 12. Maintainability, 13. communication system, 14. aerodynamic considerations, 15. Processor, 16. complexity of trajectory, 17. compatibility with guidance system, 18. compatibility with navigation system, and 19. weight.

Among the four subsystems of an autopilot, the design of the control subsystem is the most challenging one. The guidance and navigation subsystems feed the control system to provide a successful UAV flight mission. Figure 6.37 illustrates a flowchart that represents the control system design process. In general, the design process begins with a

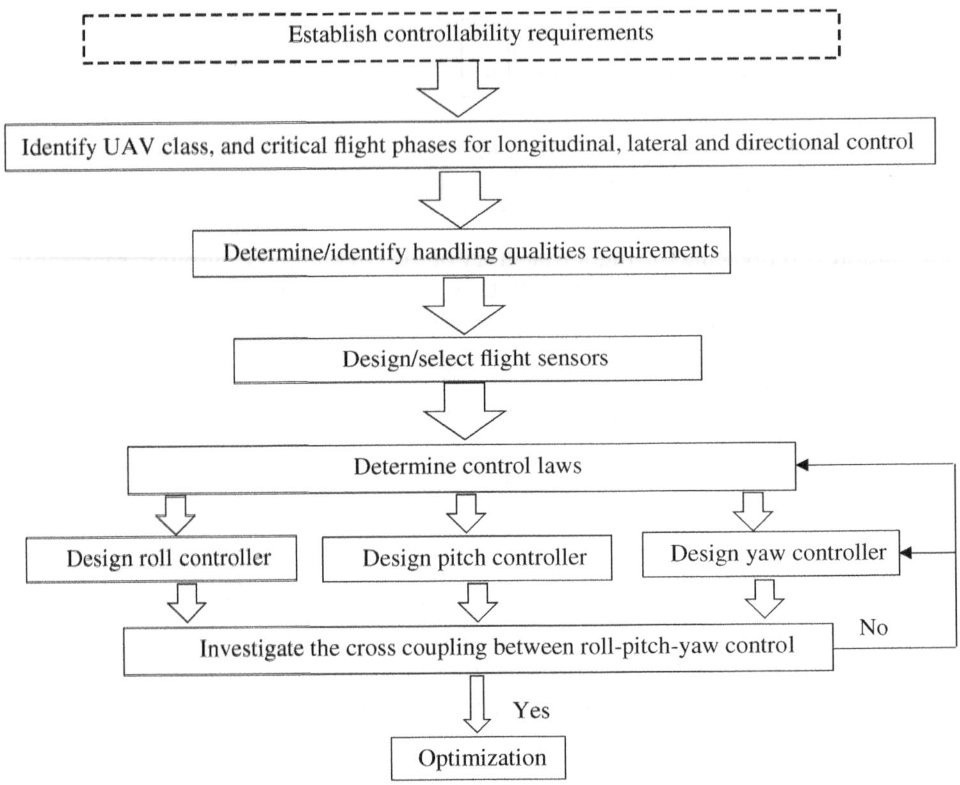

Figure 6.37 Control system design process.

trade-off study to establish a clear line between stability and controllability requirements and ends with optimization.

During the trade-off study, two extreme limits of flying qualities are examined and the border line between stability and controllability is drawn. For instance, a fighter UAV can sacrifice the stability to achieve a higher controllability and maneuverability. Then, an AFCSs may be employed to augment the aircraft stability. In the case of a civil UAV, the safety is the utmost goal; so, the stability is clearly favored over the controllability.

The results of this trade-off study will be primarily applied to establishing the most aft and the most forward allowable location of aircraft center of gravity. Three roll control, pitch control, and yaw control are usually designed in parallel. Then the probable cross coupling between three controls is studied to ensure that each control is not negating controllability features of the aircraft in other areas. If the cross-coupling analysis reveals an unsatisfactory effect on any control plane, one or more control systems must be redesigned to resolve the issue. Flight control systems should be designed with sufficient redundancy to achieve two orders of magnitude more reliability than some desired levels. In general, the control system performance requirements are: 1. Fast response, 2. small overshoot, 3. Zero steady-state error, 4. Low damping ratio, 5. Short rise time, 6. Short settling time. If the overshoot is large, a large load factor will be applied on structure (due to an increase in acceleration).

In a lightly stable UAV, or a completely unstable UAV, the stabilization of the UAV is another control system requirement. This additional and necessary stabilization may make the control system design a more challenging problem.

Consider an UAV which its heading is controlled by a regular rudder position in the presence of a crosswind. In this system, the heading that is controlled is the direction the airplane would travel in still air. The autopilot will normally correct this heading, depending on the crosswinds, so that the actual course of the airplane coincides with the desired path. Another control included in the complete UAV control system controls the ailerons and elevators to keep the airplane in level flight. Hence, the design of a heading control system involves the simultaneous application of all three control surfaces.

In FAR 23 [70], one level of redundancy for control system is required (i.e., power transmission line). The lines of power transmission (i.e., wire and pipe) should not be close to each other, should not be close to fuel tanks, and should not be close to hydraulic lines. In most Boeing aircraft, there are three separate hydraulic lines. If there is a leak in the hydraulic lines or if the engines get inoperative, there is an extra hydraulic system which is run independently. For example, the transport aircraft Boeing 747 has four hydraulic systems. These design considerations provide a highly safe and reliable aircraft.

The main element in a control system is the controller. The major requirements to be accomplished in the design of the controller are: (a) system stability; (b) reference tracking; (c) disturbance rejection; (d) noise attenuation; (e) control energy reduction; (f) robust stability; and (g) robust performance. Not all dynamic models have the ability to fully satisfy all of the requirements.

Figure 6.38 exhibits an Arcturus T-20 UAV with a mass of 84 kg, and a wingspan of 5.3 m, which is a fully composite aircraft capable of internal and external payloads. It is launched from a portable catapult, and can be recovered with a shipboard landing system, or belly land on unimproved surfaces. The T-20 carries a retractable gimbal mounted digitally stabilized, Electro-Optic/Infra-Red (EO/IR) camera that relays video in real time via a C-band Line-of-sight (LOS) data link to the Ground control station (GCS). This UAV has a Maximum speed of 139 km/h, a range of 89 km, an endurance of 24+ hours, and a service ceiling of 25 000 ft.

Figure 6.38 Arcturus T-20 Unmanned aerial vehicle (UAV).

Table 6.6 Typical deficiencies of a system and the suitable compensator to compensate.

No	Deficiency of a dynamic system	Necessary Compensator
1	Steady state error not zero	Proportional-Integral (PI)
2	Large overshoot; long rise time, long settling time; low bandwidth	Proportional-Derivative (PD)
3	Steady state error not zero; large overshoot; long rise time	PID Controller
4	Slow response	Lead–lag (Phase-lead)
5	Fast response	Lag–lead (Phase-lag)
6	Low damping ratio (too much oscillation)	Rate feedback
7	Sensitive to noise & disturbance	Wash-out filter
8	Unwanted mode	Zero-pole cancelation
9	A pole is not on the desired location	Pole placement
10	System response is not optimum	Quadratic Optimal Regulator
11	System's dynamic model includes uncertainty	Robust control

Table 6.6 demonstrates typical deficiencies of a system and the suitable controller/compensator to compensate. The deficiency is observed, when a unit step input is applied to the system. The selection of type of a controller is a function of many parameters including design requirements, UAV mission, cost, and effectiveness.

Questions

1 What are the primary criteria for the design of control system?

2 Define transfer function.

3 What is the typical format of state-space representation of a dynamic system?

4 Define time constant.

5 An actuator with a first order system model has a time constant of 0.1 seconds. What is its transfer function?

6 Name four control laws.

7 Name four basic elements of a closed-loop system.

8 Name three nonlinearities of a control system.

9 What are the two conventional controller design tools/techniques?

10 Write the typical mathematical model for a regular actuator for control surfaces.

11 What is a typical value for the time constant of an UAV actuator?

12 Define longitudinal control.

13 Define lateral control.

14 Define directional control.

15 Describe the flight envelope.

16 Name typical UAV measurement devices (sensors).

17 Name three conventional control surfaces.

18 Describe PID Controller.

19 Describe optimal control

20 Describe robust control.

21 Describe digital control.

22 What do A/D and D/A stand for?

23 List autopilot categories/modes.

24 What is the function of a yaw damper?

25 Describe the role of a wing leveler.

26 What are the main issues concerning integration of UAVs in civil airspace?

27 What type of compensator is recommended for a control system, where the system is sensitive to noise & disturbance?

28 What type of compensator is recommended for a control system, where the steady state error of the response to a unit step input is not zero?

29 What type of compensator is recommended for a control system, where the steady state error of the response to a unit step input is not zero, it has a large overshoot; and a large rise time?

30 What type of compensator is recommended for a control system, where the system's dynamic model includes uncertainty?

31 What type of compensator is recommended for a control system, where the system's response is not optimum?

32 What is the torque range of an electric motor as servo?

33 What is the force range of a hydraulic actuator as servo?

34 What are four types of electric motors used as servos?

35 Discuss features of a servo-motor.

36 Why a saturation block is needed in modeling control surfaces in Simulink?

37 Why a fixed body may have up to six degrees of freedom? What are they?

38 Draw the block diagram for the UAV flight control system with conventional control surfaces.

39 Draw the block diagram for the pitch-attitude control system of a UAV.

40 Draw the block diagram for the yaw damper of a UAV.

41 What is a localizer?

42 Draw the block diagram for the bank angle control system of a UAV.

43 In a cruising flight, what element is often used to change the airspeed?

44 In a cruising flight, what element is often used to change the angle of attack/pitch angle?

45 In a cruising flight, what element is often used to change the altitude?

46 Explain how a turn is coordinated.

47 What is the range of deflections for an elevator?

48 What is the range of deflections for a rudder?

49 What is the range of deflections for an aileron?

50 Describe gain scheduling control technique.

Problems

Perform problems 1 to 14 for a reconnaissance UAV with a maximum takeoff mass of 1100 kg, a wing area of 15 m^2, and a wing aspect ratio of 7. The non-dimensional longitudinal and lateral-directional stability and control derivatives are given in Table 6.3 for an airspeed of 100 knot ($\alpha = 2$ deg) at sea level. You may assume the derivatives remain constant throughout the flight. Other UAV parameters are:

$$C_{Do} = 0.024,\ K = 0.05,\ C_{Lo} = C_{mo} = 0,\ I_{xx} = 1220\ \text{kg m}^2,\ I_{yy} = 1760\ \text{kg m}^2,\ I_{zz} = 2500\ \text{kg m}^2$$

In addition, the engine thrust (T) is modelled as $T = 22\ \delta_T$ where T is in N and δ_T (throttle setting) is in degrees. The maximum deflections of control surfaces and throttle setting are ±30 degrees and 45 degrees respectively.

Table 6.3 Stability and control derivatives for a reconnaissance UAV.

Longitudinal derivatives	1/rad	Longitudinal derivatives	1/rad	Lat-dir derivatives	1/rad	Lat-dir derivatives	1/rad
C_{L_α}	5.2	C_{D_u}	0.1	C_{l_β}	−0.08	C_{n_r}	−0.05
C_{m_α}	−1.9	C_{L_q}	8.4	C_{y_β}	−0.31	$C_{l_{\delta A}}$	0.23
C_{D_α}	0.06	C_{m_q}	−14	C_{n_β}	0.04	$C_{y_{\delta A}}$	0.0
$C_{L_{\dot\alpha}}$	1.6	C_{D_q}	0.0	C_{l_p}	−0.53	$C_{n_{\delta A}}$	−0.03
$C_{m_{\dot\alpha}}$	−5.1	$C_{L_{\delta E}}$	0.6	C_{y_p}	−0.08	$C_{l_{\delta R}}$	0.009
$C_{D_{\dot\alpha}}$	0.0	$C_{m_{\delta E}}$	−1.6	C_{n_p}	−0.06	$C_{y_{\delta R}}$	0.18
C_{L_u}	0.012	$C_{D_{\delta E}}$	0.01	C_{l_r}	0.16	$C_{n_{\delta R}}$	−0.07
C_{m_u}	0.002			C_{y_r}	0.12		

1 Calculate the following dimensional longitudinal stability and control derivatives:

$$X_u, Z_u, M_u, X_\alpha, Z_\alpha, M_\alpha, X_{\delta_e}, Z_{\delta_e}, M_{\delta_e}, Z_{\dot\alpha}, M_{\dot\alpha}, Z_q, M_q.$$

2 Determine the aircraft longitudinal transfer functions: $\frac{u(s)}{\delta_e(s)}, \frac{\alpha(s)}{\delta_e(s)}, \frac{\theta(s)}{\delta_e(s)}$.

3 Is the aircraft longitudinally dynamically stable? Why?

4 Apply −2° of elevator deflection,

- **a** Plot the response (forward speed, angle of attack, and pitch angle) of the open-loop aircraft system to this input (for 200 seconds). You may use MATLAB for simulation.
- **b** What are the new steady state values for angle of attack, pitch angle, and aircraft speed?
- **c** Is this aircraft still cruising or climbing?
- **d** Determine the rise time; settling time; and maximum overshoot of each response (α, θ, and u).

5 Calculate the state-space longitudinal dynamic model of the aircraft.

6 Determine eigen values of matrix A. Is the aircraft longitudinally dynamically stable? Why?

7 Using state-space longitudinal model, apply −1° of elevator deflection,
- **a** Plot the response (forward speed, angle of attack, and pitch angle) of the open-loop aircraft system to this input (for 300 seconds). You may use MATLAB for simulation.
- **b** What are the new steady state values for angle of attack, pitch angle, and aircraft speed?
- **c** Is this aircraft still cruising or climbing?

8 Design a closed-loop control system including controllers to follow the following trajectory (Figure 6.39). Provide your Simulink model; and the simulation results including a comparison between the required trajectory and the flown trajectory. Hint: Employ both elevator and throttle setting.

9 Calculate the lateral-directional state-space dynamic model of the aircraft (use Table 6.3).

10 Determine the lateral-directional eigenvalues of matrix A. Is the aircraft laterally-directionally dynamically stable? Why?

11 Apply −1° of aileron and +2° of rudder deflections simultaneously,
- **a** Plot the response (bank angle, sideslip angle, heading angle, roll rate, and yaw rate) of the open-loop aircraft system to this input (for 10 seconds). You may use MATLAB for simulation.
- **b** What are the values for bank angle, sideslip angle, heading angle after 3 seconds?

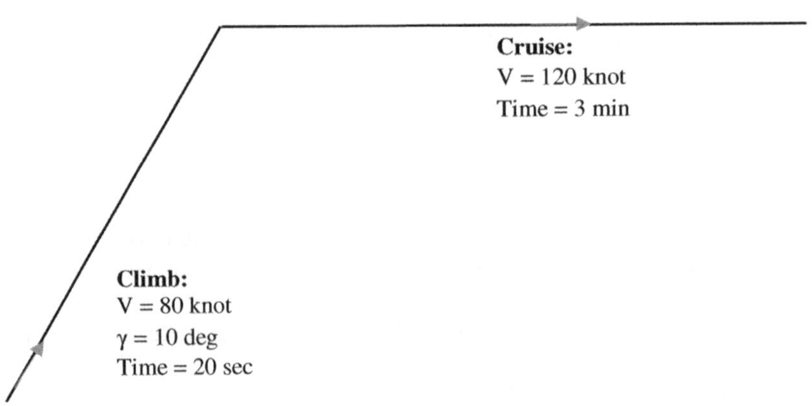

Figure 6.39 Required trajectory.

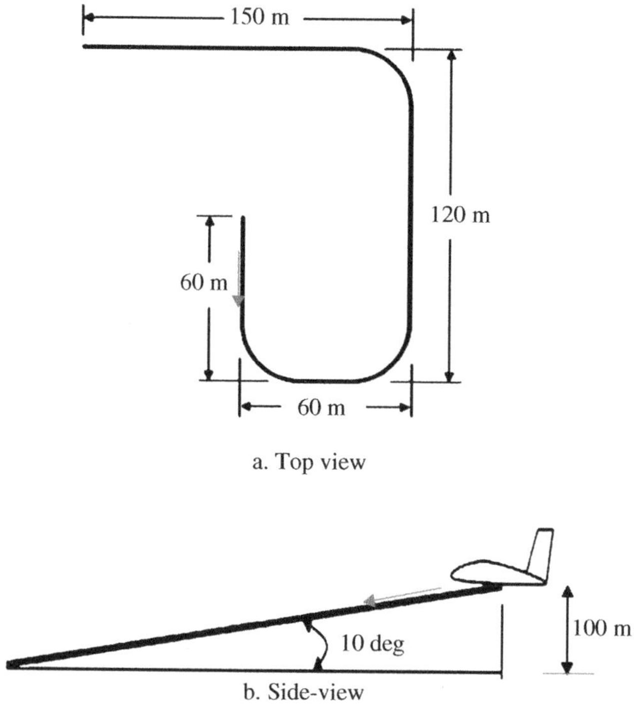

Figure 6.40 The descent and landing trajectory.

12 For the given UAV, design a closed-loop control system including controllers to follow the trajectory shown in Figure 6.40. Provide your Simulink model; and the simulation results including a comparison between the required trajectory and the flown trajectory.

13 Determine the aircraft full longitudinal-lateral-directional state space model for four inputs, eight state variables, and five outputs (use Table 6.3). The input variables are $u = [\delta_T, \delta_E, \delta_A, \delta_R]^T$ and the state variables are $x = [V_T, \alpha, \theta, q, \beta, \phi, p, r]^T$. The output variables are forward speed, angle of attack, pitch angle, pitch rate, sideslip angle, bank angle, roll rate and yaw rate.

14 Apply simultaneous deflections of −2° of elevator, 3° of aileron, and + 2° of rudder (only all for 1 second). Plot the response (forward speed, angle of attack, pitch angle, pitch rate, sideslip angle, bank angle, roll rate, yaw rate and heading angle) of the open-loop aircraft system to these inputs (for 10 seconds). You may use MATLAB for simulation.

15 The yaw-rate-to-rudder-deflection and lateral-acceleration-to-rudder-deflection transfer functions for a small UAV with a weight of 2,000 lb, and a wing area of 120 ft^2 at an airspeed of 200 knot are given below:

$$\frac{R(s)}{\delta_R(s)} = \frac{-10(s+0.15)}{s^2+0.3s+17}$$

$$\frac{n_y(s)}{\delta_R(s)} = \frac{-0.18(s^2-0.02s-23)}{s+0.07}$$

Assume that the accelerometer is installed at the UAV's center of gravity (cg), and the rudder actuator has a time constant of 0.04 seconds. Design a turn coordinator, such that the damping ratio and natural frequency satisfy the following requirements:

$$0.5 \le \zeta \le 0.8$$

$$0.4 \le \omega_n \le 0.9 \ (rad/\text{sec})$$

16 The following figure shows part of the block diagram for a roll control system:

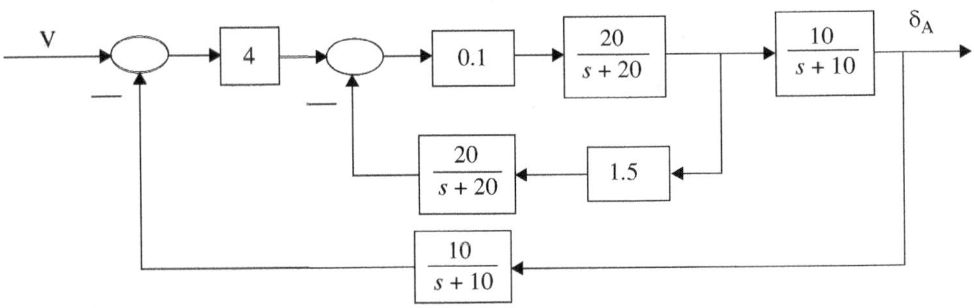

The input is a voltage (12 volts for 2 seconds). Simulate this system, and plot the variations of the aileron deflection with time.

17 The following figure shows the block diagram for a bank angle control system:

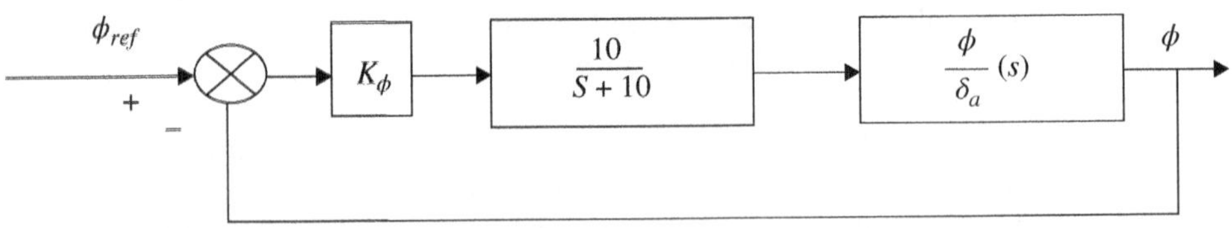

The bank-angle-to-aileron-deflection transfer function is:

$$\frac{\phi}{\delta_a}(s) = \frac{1.437\left((s-0.073)^2+(0.7214)^2\right)}{(s+0.7333)(s-0.0293)\left((s-0.0470)^2+(1.0402)^2\right)}$$

Plot the variations of the bank angle with time to a 5° bank angle reference command. What is the overshoot, and steady state error?

18 The following figure shows the block diagram for a dynamic system:

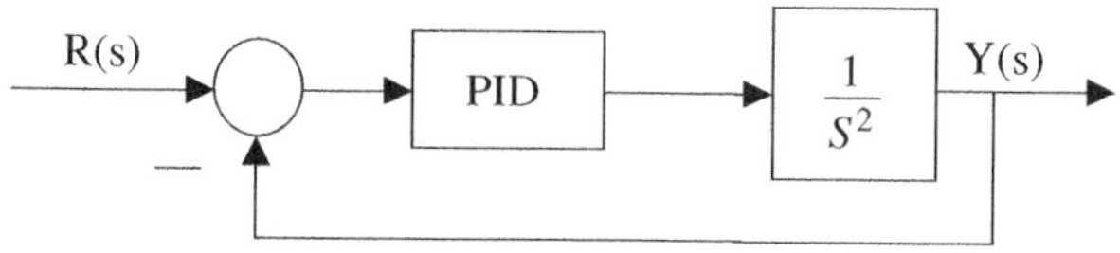

Design a controller (PID) such that the response to the unit-step reference input satisfies the following performance specifications:

a Steady state error: 0
b Maximum overshoot less than 20%
c Settling time less than two seconds
d Rise time less than 0.5 seconds.

19 A control system is modeled by the following block diagram:

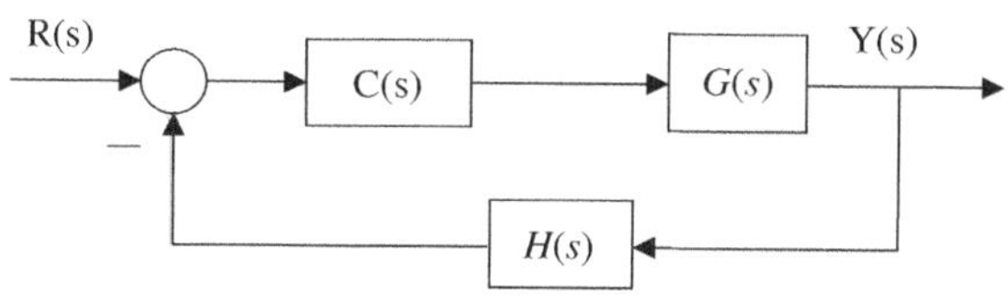

where

$$G(s)=\frac{s+10}{2s^2+2s+9};H(s)=\frac{10}{s+11}$$

Design a controller (C(s)) such that the response to the unit-step reference input satisfies the following performance specifications:

a Steady state error less than 10%
b Maximum overshoot less than 20%
c Settling time less than three seconds
d Rise time less than one second.

20 Repeat problem 19 with $G(s)=\frac{12}{2s^2+3s+10};H(s)=\frac{100}{s+100}$.

Notes

1 http://www.mathworks.com.
2 Currently, the traffic alert and collision avoidance system (TCAS) is the primary cooperative collision avoidance system and is in use by a variety of airspace users [82].

7

Guidance System Design

CONTENTS

Design of Unmanned Aerial Systems, First Edition. Mohammad H. Sadraey.

Companion website: www.wiley.com/go/sadraey/unmanned-aerial-systems

Educational Outcomes

After study of this chapter, the reader will be able to:

1) Discuss Unmanned Aerial Vehicle (UAV) guidance process
2) Conduct UAV guidance system design
3) Explain elements of guidance system
4) Develop a guidance system for a set of given mission requirements
5) Select and implement a guidance law to satisfy design requirements
6) Provide guideline for a sense and avoid system
7) Provide guideline for a formation flight
8) Implement motion planning and trajectory design
9) Evaluate seeker features for a guidance mission
10) Discuss the current guidance techniques and technologies
11) Evaluate UAV maneuverability features

7.1 Introduction

Guidance is fundamentally defined as the process for guiding the path of a vehicle toward a given point. In another word, guidance is the means by which a UAV steers, or is steered, to a target. One of the primary functions of an autopilot is to guide the UAV to follow the trajectory. The guidance system determines a path to follow based on commanded signals, such as waypoints, altitude, and airspeed. A guidance system can be defined as the "driver" of a UAV that exercises planning and decision-making functions to achieve an assigned flight mission.

Please note that, in the past, guidance and navigation were used interchangeably. Back to the old days of marine exploration in the eighteenth and nineteenth century, navigation was defined as the technique of monitoring and controlling the movement of a vehicle from one place to another. However, with the advance of the science of guidance, there is a clear difference between navigation, control, and guidance. Thus, navigation is a pre-requisite for the guidance, and guidance is a pre-requisite for the control.

Most UAVs are planned to follow a predetermined and fixed trajectory. However, there are some classes of UAVs, mainly military ones; which their mission is to follow a moving target. In either case, the generation of a trajectory is the primary function of the guidance system. For the case of a fixed trajectory, the guidance system will be simple and straight-forward. However, for the case of following a moving target, the autopilot requires a complex guidance system to generate the trajectory and to create the command for the control system. A UAV – through guidance and control systems – must generate the steering commands and subsequent control surface deflections to adequately adjust its way along the chosen flight path. A guided UAV is guided according to a certain guidance law.

Guidance is mainly responsible for measuring and controlling the flight variables including the aircraft's angles, the rate of change of the angles, and the body axis accelerations. The navigation system then calculates the location of the aircraft, compares it with the pre-determined reference trajectory, and modifies the autopilot commands to drive the error to zero. The guidance subsystem often produces an acceleration command. Thus, the guidance subsystem makes the necessary correction to keep the vehicle on course by sending the proper command to the control system of autopilot. Another important UAV operation where the guidance system is necessary is formation flight of multiple UAVs and teaming with manned aircraft.

An important application of guidance system is the detect/sense-and-avoid a collision. This is and will be one of the biggest limitations to the widespread use of unmanned vehicles in civilian airspace. In manned civilian aviation, see-and-avoid is the primary mechanism by which piloted aircraft avoid collisions with each other. This is not practical for widespread use of unmanned vehicles, so they must achieve an equivalent level of safety to that of manned aircraft operations. There is currently a large amount of research being conducted on the UAV detect-and-avoid problem. Active solutions include the use of a sensor (e.g., radar) to detect collision threats, however this requires high amounts of electrical power, and is a bit heavy. Passive solutions include the use of machine vision (e.g, camera), which reduces the power requirement, however, has a high computational requirement. This chapter will describe various aspects of guidance system, elements of from guidance, guidance laws. In addition, three popular guidance laws are presented in detail.

7.2 Fundamentals

7.2.1 Guidance Process

Guidance is the process of producing a trajectory based on what is received from the command subsystem and the feedback from the navigation system. The guidance system is originally developed by the missile industries. Reference [87] reviews the guidance laws for short-range tactical missiles. In a guided missile, the guidance system is only used in the homing phase of flight (last phase) to impact the target. For a full autonomous operation, the vehicle needs to have: (a) navigation; (b) guidance; and (c) control system of which the guidance system is the key element. The relation between the control system, the guidance system, and the navigation system is shown in Figure 7.1. Based on the guidance law, the guidance subsystem then produces the desired states which go to the control subsystem. The output of the guidance subsystem is sent to the control system, and the control system implements this command through actuators driving control surfaces such as the elevator, aileron, and rudder.

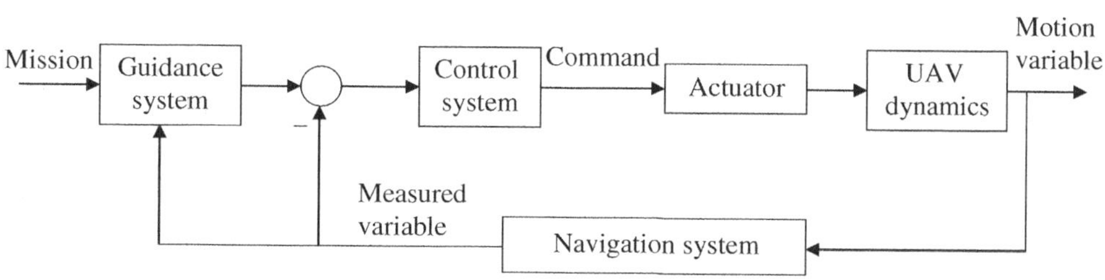

Figure 7.1 Control, guidance, and navigation systems in an autopilot.

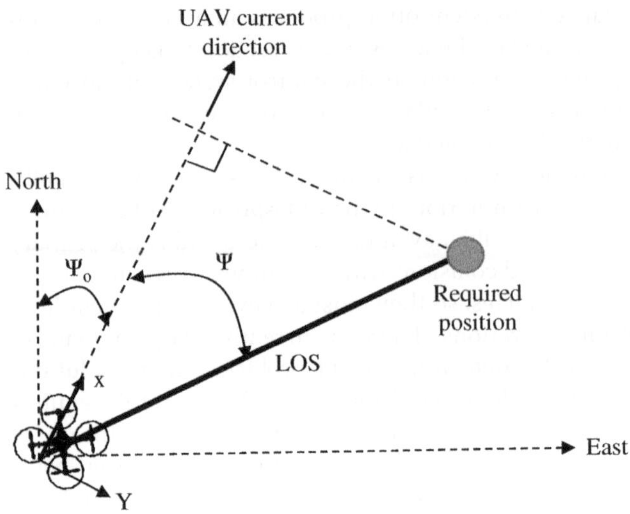

Figure 7.2 A case where a guidance system is required.

The Predator A is originally designed to be flown in fully manual mode; so, it did not require a guidance system. However, in the later version, it had a guidance system such that it can be programmed to automatically return to base if the datalink is lost. Moreover, the Global Hawk (Figure 1.2) can be operated in a fully autonomous mode, thus it requires a guidance system.

In UAVs, there are many cases where the guidance system is needed. A few examples are:

1. The current heading (Ψ_0) of the UAV does not lead to the desired position (Figure 7.2).
2. An UAV is seeking a fixed target; in order to make some measurements (e.g., take pictures).
3. An UAV is following a moving ground target.
4. A follower UAV is following a leader UAV; for a formation flight.

One of the sub-items of case 1 is the "sense/detect and avoid" capability which means the capability of an unmanned aircraft to remain a safe distance from and to avoid collisions with other airborne aircraft. General Atomics Aeronautical Systems had the successful flight test of a prototype of its Due Regard Radar, an air-to-air radar system that supports GA-ASI's overall airborne sense and avoid architecture for Predator B (Figure 5.12). The main sensors of the Warrior (a variant of the MQ-1 Predator), are an SAR/GMTI (Synthetic Aperture Radar/Ground Moving Target Indicator) in the large fairing.

7.2.2 Elements of Guidance System

The most important subsystem in a UAV compared with a manned aircraft is the autopilot. The autopilot is a vital and required subsystem within unmanned aerial vehicles. The autopilot has the responsibility to: (a) stabilize under-damped or unstable modes;

(b) accurately track commands generated by the guidance system; (c) guide the UAV to follow the trajectory; and (d) determine UAV coordinates (i.e., navigation). Therefore, an autopilot consists of: (a) command subsystem; (b) control subsystem; (c) guidance subsystem; and (d) navigation subsystem. These four subsystems must be designed simultaneously to satisfy the UAV design requirements.

7.2.3 Guidance Components

The guidance subsystem generates command for control system (e.g., generates suitable trajectories to be followed). It has mainly four elements: (a) the guidance measurement device (e.g., radar/seeker); (b) guidance law; (c) hardware (e.g., processor); (d) software/programming. Chapter 8 covers microprocessors which is part of the hardware for the guidance system.

The type of guidance system used depends upon the type and mission of the UAV being controlled, and they can vary in complexity from an inertial guidance system for long range UAV to a simple system where an operator visually observes the UAV (e.g., remote controlled aircraft) and sends guidance commands via a radio link. In any case, the guidance command serves as the input to the UAV control system. The command may be in the form of a heading or attitude command, a pitching or turning rate command, or a pitch or yaw acceleration command, depending upon the type of guidance scheme used.

Hardware equipment in the guidance system typically includes a homing head (or seeker), measurement device, transmitter, and processor. The homing head might be active, passive, or semi-active. There are several guidance laws and several guidance system categories. The important point is to make sure that the guidance law matches the guidance system category. The guidance system categories include: (a) Line Of Sight (LOS); (b) Navigation Guidance (e.g., Inertial Navigation, Global Positioning System (GPS)); and (c) homing (e.g., radar, infra-red, and television). The typical guidance law includes: (a) collision; (b) proportional navigation (PN) guidance; (c) constant beam course; (d) pursuit guidance; (e) three point guidance; (f) optimal control guidance; (g) lead guidance; (h) lead angle; (i) preset.

The principal elements that comprises a UAV guidance system to follow a moving target are illustrated in Figure 7.3. Two inputs to the guidance system are: (a) LOS and/or range to the target (provided by the seeker); and (b) the position of the UAV (coming

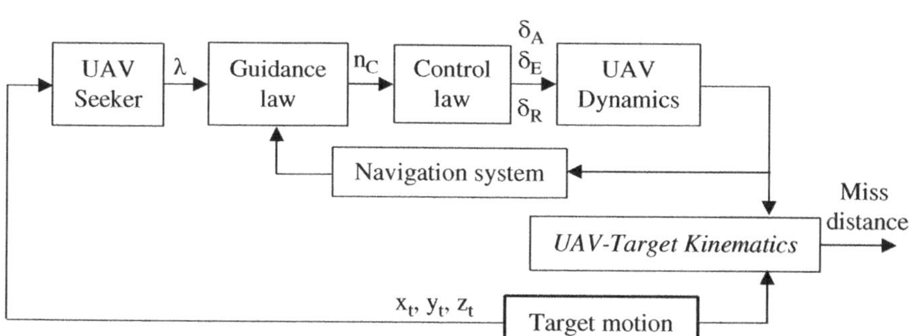

Figure 7.3 Major subsystems of a unmanned aerial vehicle (UAV) guidance system with a moving target.

Table 7.1 Flight and control variables in four subsystems.

No	Subsystem	Input	Output
1	**Command subsystem**	X, Y, Z	V_t, γ, ψ, R
2	**Navigation subsystem**	U, W, W, P, Q, R, θ, ϕ, ψ	X, Y, H, V_t
3	**Guidance subsystem**	**UAV:** x, y, z **Target:** x, y, z	n_{Cx}, n_{Cy}, n_{Cz}, or θ_C, ϕ_C, Ψ_C
4	**Control subsystem**	n_{Cx}, n_{Cy}, n_{Cz}, (or θ_C, ϕ_C, Ψ_C)	δ_T, δ_E, δ_A, δ_R
5	**UAV dynamic model**	δ_T, δ_E, δ_A, δ_R	α, γ, ϕ, β, V, Q, P, R

from navigation system). The output of the guidance system is a rotational speed/acceleration command; either a turn (to the left or right), or a pitch (up or down).

The guidance command may be in the form of: (a) centripetal acceleration (a_C); (b) non-dimensional acceleration (n_C); or (c) angular velocity (ω_C). The guidance command will be sent to the control system for implementation. The point of closest approach of the UAV and target is known as the *miss distance*. This distance is required for a successful UAV mission as in a sense and avoid function for a formation flight.

An important UAV operation where the guidance system is necessary is formation flight. A simple but effective design method is preferable for such mission. One benefit of this approach is that the guidance and flight control design process is integrated. There is a possibility that the guidance law becomes complex as the number of UAVs is high. Table 7.1 represents the state and control variables that each subsystem receives and produces.

7.2.4 Target Detection

A pre-requisite of a guidance operation is the perception. Perception is the ability to use inputs from sensors to build a model of the environment within which the vehicle is flying, and to assign entities, events, and situations to classes. Perception may be divided into various functions such as: (a) mapping; (b) object recognition; and (c) obstacle and target detection. One of the applications of the target detection is the collision avoidance.

The target detection system is composed of three main elements: (a) detection sensor; (b) processor; and (c) algorithm. The data gathered by the sensor will be processed based an algorithm. The targets have variety of classifications; two large groups are: (a) fixed target; and (b) moving target. Furthermore, sensors are categorized into two groups: (a) active; and (b) passive.

A detecting system can have three operating modes: (a) alert; (b) standby; and (c) active. In the alert mode, the sensor is enabled, while computation is disabled. In the Standby mode, the transmission/receiving/sensing capabilities, and also computation are all disabled, so that no energy is consumed by the sensing process or by outgoing communications. In the active mode, the sensor can send and receive signals, the processor can perform computation. Given a location in the monitored area, the energy of the signal expected by a sensor at a position from a target is, a direct function of the energy emitted by the target, and inversely function of the geometric distance between target and the sensor.

There are various theories to identify a target among many objects. For instance, *Game theory* [88] with a sophisticated decision procedure has the capability to observe a large range of probable scenarios before performing the action, and detecting a target. An application is a localized minimal coverage path algorithm such that adjacent nodes must be updated, where path information can be considered online in an easier and more competent method. Another application is to autonomously sense and avoid, where current regular civil UAVs are unable to perform. Small UAVs represent a significant mid-air collision hazard to other aircraft operating in the same airspace. They frequently operate in close proximity to the ground, e.g., buildings, towers, and bridges.

There are two phases involved in detecting a target by a target detection system: (a) inspection/surveillance/ observation; and (b) target recognition. These two phases are briefly described in this section.

1) The first phase is surveillance, the purpose of which is to search a predetermined area for a target and identify its presence. In the phase one (surveillance), the behavior, and performance of the target is observed. This may be accomplished actively, by transmitting energy out into the medium, and waiting for the reflected energy to return, as in radar. The observation may be performed passively by receiving energy being emitted by the target. For such a goal, a large area must be swept and searched. The success of this phase depends upon camera's accuracy or radar range.

   Cameras are in creasingly gaining popularity for mapping areas, corridors, construction sites and cities. Cameras with their wide field of regard, low power and weight provide a feasible alternative to active sensors (e.g., radars) for detecting other aircraft within a certain radius. A UAV may also use lidar for obstacle detection and avoidance to navigate safely through environments. The capability of a target detection sensor is measured by the maximum range at which it can reliably detect a target and its ability to distinguish individual target in a multi-target group.
2) In the **second phase**, the target must be identified, recognized, and then classified. In addition, the target's behavior should be interpreted in order to determine its category, type, size, and most importantly identity. The sensor must scan the detection range until the target detected. In order to increase the sensor life and reduce cost, it is required to select a detection algorithm to exploit energy resources in the most efficient way. The communication is the most expensive operation for a target detection in terms of power consumption.

The computations for a detection algorithm can be simplified by some assumptions [89] such as the following:

1) Until detected, the object can be at any location in the monitored area with equal probability.
2) The sensor may be randomly deployed on the monitored area with uniform probability distribution.
3) The detection period is infinitesimal.

The recognition process involves comparing what is observed with the UAVs a priori knowledge. The target nature may be classified in a number of groups, such as another aircraft, moving car, human, fixed object, and building. In order to recognize a target, a mathematical (e.g., digital) model must be defined for the processor. If the target is moving, a dynamical model of the target is needed. In the literature, various target recognition techniques have been developed and examined. For instance, Ref. [90] proposes a multi-image matching method to estimate semantic correspondences across multiple images. The interested reader may consult with [91] for studying principles, algorithms, and applications in *computer vision*. Sensors such as radar, camera, and lidar are commonly used onboard UAV to provide raw measurements for state estimation and perception algorithms. The estimation of the target motion – using measured data – is frequently performed through a Kalman filter.

7.2.5 Moving Target Tracking

There are basically two tracking functions for a UAV: (a) tracking a reference command/trajectory; and (b) tracking a moving target. Tracking a given trajectory could be a pure control problem (Chapter 6), while tracking a moving target will involve both control and guidance system. In the first case, the autopilot must make the output track (i.e., follow) a changing command, as in making an aircraft track the autopilot's pitch-rate command or a polynomial input of degree n. The command may be either constant or time varying. Practical command inputs may change with time, so that the tracking error may grow and then decrease again. In Section 7.7.4.2, fundamentals and techniques of trajectory tracking is presented.

In a tracker problem, it is desired for an aircraft to follow or track a command signal. For this purpose, we are often interested not in regulating the state near zero, but in following a nonzero reference command signal to guarantee perfect tracking. In a task requiring precision, tracking the autopilot's performance in the tracking task will be influenced by the function of a sighting device (e.g., camera/radar) and the response of the UAV to turbulence.

A fully autonomous UAV should be capable of tracking a time-parameterized reference trajectory. However, for trajectory tracking there exist fundamental performance constraints that cannot be overcome by any control system. Moreover, to meet temporal specifications, the airspeed profile often needs to be controlled independently. To overcome this challenge, temporal constraints are not frequently imposed in path-following problems, and the vehicle is allowed to converge to and follow a path without imposing any temporal specifications. This will result in a smoother convergence to the path, and the control signals are less likely to be saturated. This approach must also avoid collision in multi-vehicle cooperative missions. In parallel, the control law should generate control surfaces deflections, and engine throttle commands that are sent to the motors.

In the flying qualities, there is a Category B where the nonterminal flight phases are normally accomplished using gradual maneuvers without precision tracking, although accurate flight-path control may be required. However, Category A includes nonterminal flight phases that require maneuvering, precision tracking, or precise flight-path control. Category C involves terminal flight phases are normally accomplished using gradual maneuvers and usually require accurate flight-path control. In two cases of precise flight-path control and precision tracking, the guidance system is actively involved. For instance, in high-performance UAVs for precision tracking of moving targets, the aircraft may have to maneuver to its performance limits. Another mission example is in an automatic terrain-following mode, where an autopilot must track a randomly changing input of quite wide bandwidth, without significant overshoots in its response.

To successfully follow a target, updates as to the target's position and its velocity relative to the UAV must be known or estimated continuously. In order to obtain target trajectory information, methods must be devised to enable the sensor to monitor the performance of the target. This information is used to predict the target's future position, so the target can be accurately followed.

Sensing the presence of a target (by a sighting means) is an essential first step to the moving target tracking problem. In order to measure or accurately localize the target's position, a series of measurements should be made to estimate its behavior or motion relative to the UAV. This is accomplished by repeatedly determining the target's range, heading, and altitude. The features and performance of a seeker is discussed in Section 7.11.

The concept of noise filtering is important to the guidance system designer. For instance, in a UAV that follows a moving target, the seeker measurement of the LOS rate signal, required for the implementation of PN guidance, may not be perfect, and could be corrupted by noise. Moreover, it is not always possible to accurately measure every variable of a target that we want to track. In writing navigation code for uploading to the microcontroller, a filtering method must be employed. A very popular method; the Kalman filtering is a mathematical technique that uses a series of measurements observed over time, subject to errors, and produces estimates of unknown variables. The basic fundamentals of Kalman filtering is briefly discussed in Chapter 8.

7.3 Guidance Laws

In a conventional autopilot, three laws are governing simultaneously in three subsystems: (a) control system through a control law; (b) guidance system via a guidance law; and (c) navigation system through a navigation law. In the design of an autopilot, all three laws need to be determined/designed. The determination of these laws is at the heart of the autopilot design process. A guided UAV is guided according to a certain guidance law. The guidance subsystem generates command signal for the control system. The guidance command is usually generated in a very short period of time. In practice, it can be assumed to be continuously generated. The guidance law is the basis for generation of guidance signal.

Guidance is the means by which an UAV steers, or is steered, to a target. A fully autonomous UAV is guided according to a certain guidance law. We may classify guidance laws as classical and modern guidance laws. The modern guidance laws are derived from optimal control theory, differential games, and singular perturbation theory.

There are various types of guidance laws. In this section, three more effective and popular ones are presented. Four classical guidance laws for UAVs are: (a) PN guidance; (b) pursuit guidance; (c) waypoint navigation guidance; and (d) LOS guidance (i.e., command guidance). A survey of advances in guidance systems of unmanned rotorcraft is presented by [79]. Figure 7.4 depicts a graphical representation of three basic guidance laws by comparing the flight paths of a UAV that targets another vehicle.

The classical guidance laws are originally developed and employed in missiles for decades and are designed using rather simple ideas. Of these, the PN guidance laws form the boundary between the classical and the modern approach. These laws each have various variants, which are not reviewed here. In summary [92], "pursuit

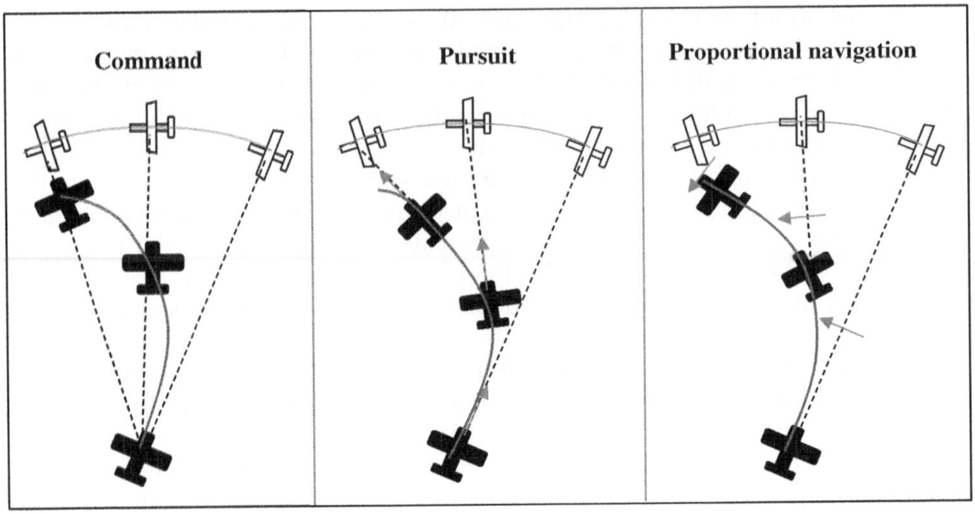

Figure 7.4 Graphical representation of three basic guidance laws.

guidance" basically points the UAV at the target at all times, while the concept of "proportional navigation guidance" is to point the UAV at a point ahead of the target so that the UAV will lead the target. LOS guidance by definition requires an external source (such as ground control station (GCS)) to establish the LOS between the target and the source along which the UAV is then guided to travel.

The selection of guidance law dictates the selection of hardware, and it influences the overall cost. Moreover, the type of guidance law will affect the control system, and navigation system. Thus, a systems engineering vision is required to select the best guidance law for a specific mission. It is the guidance law which in principle, distinguishes an unguided projectile from a guided UAV. The majority of available guided missiles use these guidance laws or their variants. They have the advantage of standard equipment, easy mechanization, and minimal information requirement. Their disadvantage lies in the fact that their accuracy suffers against maneuvering and intelligent targets. The design and analysis of guidance laws has been an active area of research for the past seven decades.

Here, a short description of these three laws is provided; in future sections, the details are presented:

1) The basic principle in LOS guidance law is to guide the UAV on a LOS course in an attempt to keep it on an imaginary line connecting the target and the GCS (tracking line).
2) However, for PN guidance law, the UAV is commanded to turn at a rate proportional to the angular velocity of the LOS (the imaginary line connecting the UAV and the target).
3) The basic idea for the pursuit guidance law is to keep the UAV pointed toward the target.
4) In the waypoint guidance, the UAV is guided along an assigned path defined by a series of waypoints.

In the following waypoints, the fight path is reconfigured in the presence of disturbances, with a computational effort. Different guidance algorithms can be implemented depending on the mission the UAV has been designed for. Most algorithms must have two functions: a. path planning, b. trajectory smoothing and tracking.

Reference [84] compares classical guidance laws, and it provides details for modern guidance laws. Among these four laws, the waypoint navigation guidance is the most popular in UAVs. The advantage of waypoint guidance law lies in the fact that it is easy to mechanize, and it requires easily obtainable information. For instance, thousands of waypoints can easily be programmed in an UAV software.

7.4 Command Guidance Law

The command guidance law (also referred to as LOS law) is originally defined for missile interception; but here, it is slightly revised and applied to a UAV to follow a moving target or a leader aircraft. In addition, the revised law may be employed as a tool for sense and avoid. The basic principle in guidance law is to guide the UAV on a LOS course in an attempt to keep it on a line joining the target and the ground station (tracking line). The command guidance law is sometimes referred to as the line-of-sight guidance law.

An imaginary line between a UAV (see Figure 7.5) to a target is referred to as line-of-sight. Thus, the revised law is to correct any deviation of the UAV from LOS, while staying away from the target by a desired distance. In the original command guidance law for missiles, the LOS is measured remotely by a ground station, while in a UAV application, the LOS is measured by an onboard sensor.

A formation flight; where, there is a leader UAV and a number of (say four) follower UAVs requires a guidance system. For such a case, the LOS is defined as the line joining the follower UAV and the leader UAV. In addition, the leader UAV may follow a moving ground target. For this law, the target-tracking radar acquires the target shortly after take-off and then guides the UAV into the beam of the target-tracking radar.

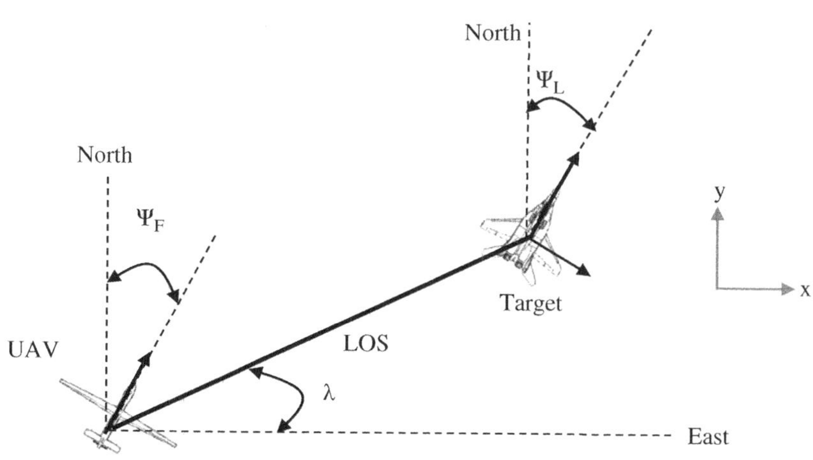

Figure 7.5 Line-of-sight (LOS)(Top view).

For the guidance command the actual distance from the tracking line to the UAV is required. If the line-of-sight guidance law is continuously employed, it results in an interception. This law is basically used by homing missiles for to hit a target. However, the law may be used to move closer to a target in order to no to lose it. Hence, when a UAV is close enough to a target to follow it, another law must be adopted. An appropriate application of this law is in a formation flight for the purpose of the sense and avoid approach. This may be implemented by assuming the goal of the follower UAV to be at a desirable distance behind or at the side of the leader UAV.

The LOS angle (λ) can be readily determined by forming a right triangle, when putting follower UAV and target, each at a corner. Then, the hypotenuse is along the LOS. The LOS angle is calculated by trigonometry from Figure 7.5 as:

$$\lambda = \tan^{-1}\left(\frac{y_T - y_U}{x_T - x_U}\right) \tag{7.1}$$

where x_T and x_U represent the distance between target and UAV to a reference line along x-axis, and y_T and y_U represent the distance between target and UAV to a reference line along y-axis. Recall that, the LOS angle is not the difference between heading angles of the UAV and the other aircraft. If the reference is selected to be at the UAV location, both y_U and x_U will be zero. The instantaneous distance between UAV and the target will be:

$$D_{TU} = \sqrt{(y_T - y_U)^2 + (x_T - x_U)^2} \tag{7.2}$$

The closing velocity (V_c) – the negative rate of change of separation between UAV and target is obtained [93] by:

$$V_c = \frac{-\left(V_{TUx}(x_T - x_U) + V_{TUy}(y_T - y_U)\right)}{D_{TU}} \tag{7.3}$$

where V_{TUx} and V_{TUy} are components of the relative velocity and are given by

$$V_{TUx} = \dot{x}_T - \dot{x}_U \tag{7.4}$$

$$V_{TUy} = \dot{y}_T - \dot{y}_U \tag{7.5}$$

The instantaneous line-of-sight rate can be computed by taking the derivative of the Eq. (7.1):

$$\dot{\lambda} = \frac{V_{TUy}(x_T - x_U) - V_{TUx}(y_T - y_U)}{D_{TU}^2} \tag{7.6}$$

In the line-of-sight guidance law, the velocity of the follower UAV (V_n) perpendicular to the LOS should be equal to the LOS rate at that point. It is assumed that the LOS value is available from the use of onboard vision sensors (e.g., radar).

$$V_n = D_{TU}\dot{\lambda} \tag{7.7}$$

where $\dot{\lambda}$ is the rate of change of the LOS angle, and D_{LF} denotes the distance between the follower UAV and the target or leader UAV. Moreover, V_{Fn} is velocity of the follower UAV perpendicular to the LOS. Hence, the guidance command is perpendicular to the LOS.

The LOS guidance scheme [94] can be mechanized in two ways: command line-of-sight (CLOS), and beam rider (BR) guidance scheme. In CLOS guidance scheme there is an uplink to transmit guidance signal from a ground station to the flower UAV. Hence, it is required by the ground station to track the UAV as well as the Leader UAV and target. So, the follower UAV is guided so as to remain on the commanded LOS.

In the BR guidance scheme, an electro-optical beam is directed at the leader UAV from follower UAV. There are sensors inside the follower UAV which senses the deviation of the follower UAV from the centerline of the beam. The autopilot of the follower UAV generates appropriate guidance commands to annul this deviation. Thus, the follower UAV is required to constantly track the leader UAV. Performance of UAVs using LOS guidance is good in a low maneuver formation flight at short ranges and moderate speed. However, these UAVs suffer from certain disadvantages. A major disadvantage is that the commanded speed required for approaching leader UAV becomes very high toward the end, since UAV has limit for max speed. In addition, the performance degrades at high speeds and with maneuvering targets.

The command guidance law must be employed along a sense and avoid function, for a successful formation flight. Thus, the UAV will always be on the LOS, while staying at a distance from the leader UAV or a target. The output of the command guidance law is a roll, yaw, or pith; while the command the sense and avoid law is on the engine throttle to adjust the distance.

Figure 7.6 shows general three main blocks of a guidance system. LOS is used for calculating desired heading angle (Ψ), the kinematic controller has been used for desired pitch angle (θ) and a first-order low-pass filter for airspeed. The specifications of the filter and smoother are a function of the UAV dynamics and its maneuverability. For instance, the smoother determines a climb angle (λ) based of the maximum rate of climb and the maximum climb angle. Since the UAV angle of attack is already a known value, the pitch angle is calculated:

$$\theta = \alpha + \gamma \tag{7.8}$$

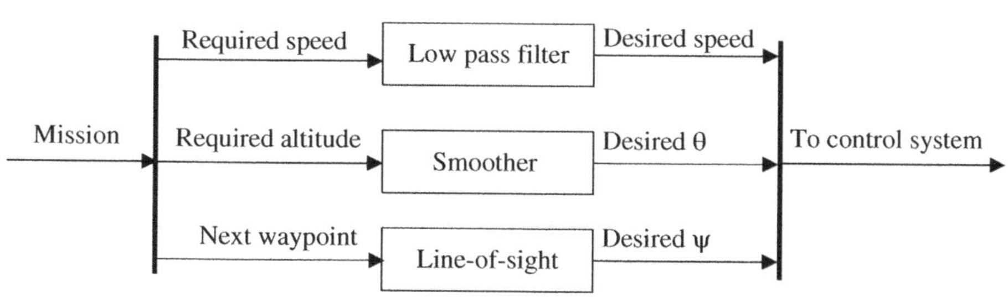

Figure 7.6 Three major blocks of a guidance system.

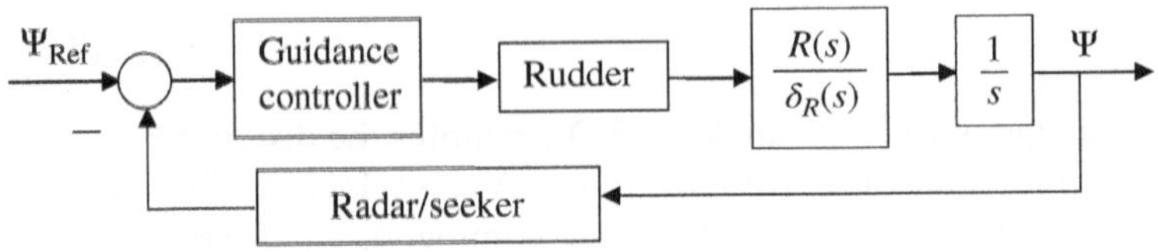

Figure 7.7 Block diagram of the command line of sight (LOS) guidance system.

In the control system, the airspeed is controlled via the engine throttle, the pitch angle via elevator, and heading angle via rudder and aileron. A smoother is usually utilized when a command (e.g., airspeed, pitch angle) cannot be a step input. A general block diagram of the command (LOS) guidance plus control systems is depicted in Figure 7.7.

Example 7.1

A UAV is flying toward east at a speed of 60 m/s, while a target aircraft is flying northward at a speed of 40 m/s. They are at a distance with the geometry shown below (Figure 7.8):

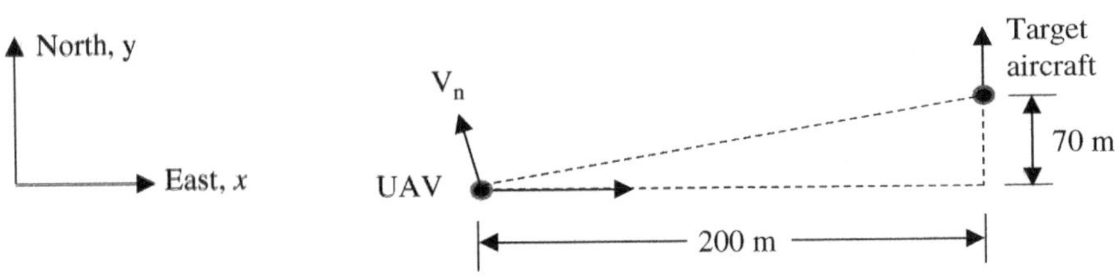

Figure 7.8 Geometry of the unmanned aerial vehicle (UAV) and target aircraft.

The UAV guidance system is governed by the command guidance law. It is required to move into and fly along the LOS. What command will be generated by the guidance system at this moment?

Solution

For simplicity, the reference point is selected to be the location of the UAV. Hence; $x_U = 0$; $y_U = 0$. In addition, based on the given geometry, the component velocities are:

$$\dot{x}_T = 0; \dot{y}_U = 0$$

Thus,

$$V_{TUx} = \dot{x}_T - \dot{x}_U = 0 - 60 = -60\, m/s \tag{7.4}$$

$$V_{TUy} = \dot{y}_T - \dot{y}_U = 40 - 0 = 40\, m/s \tag{7.5}$$

The instantaneous distance:

$$D_{TU} = \sqrt{(y_T - y_U)^2 + (x_T - x_U)^2} = \sqrt{(70-0)^2 + (200-0)^2} = 211.9\, m \tag{7.2}$$

The instantaneous line-of-sight rate:

$$\dot{\lambda} = \frac{V_{TUy}(x_T - x_U) - V_{TUx}(y_T - y_U)}{D_{TU}^2} = \frac{40(200-0)-(-60)(70-0)}{211.9^2}$$

$$= 0.272\frac{rad}{sec} = 15.51\frac{\deg}{sec} \quad (7.6)$$

$$V_n = D_{TU}\dot{\lambda} = 211.9 \times 0.272 = 57.6\frac{m}{s} \quad (7.7)$$

Figure 7.7 shows the direction of this commanded normal velocity. This command will be implemented via aileron and rudder deflations (i.e., a turn).

7.5 PN Guidance Law

Proportional navigation (PN) is known to be the most widely used guidance law in (analogous to proportional control) UAVs and modern missiles. The PN guidance law is originally defined for homing missile interception; but here, it is slightly revised and applied to a UAV to follow a moving target or a leader aircraft. In addition, the revised law may be employed as a tool for sense and avoid function in a swarming mission. Please note that, PN has nothing to do with navigation, it is just a guidance law used to guide air vehicles. This misnomer dates back to the early days of navigation by ships.

In PN guidance law, the UAV is commanded to turn at a rate proportional to the angular velocity of the LOS, and in the same direction.

$$\omega_C = N'\dot{\lambda} \quad (7.9)$$

The commanded normal acceleration (a_C) is given [93] as:

$$a_C = N'V_C\dot{\lambda} \quad (7.10)$$

where V_C is the closing velocity (Eq. 7.3), N' a unit-less designer-chosen gain (usually in the range of 2–5) known as the effective navigation ratio, and the acceleration command (in m/s^2) is perpendicular to the instantaneous UAV-target line-of-sight (λ). Recall that, the load factor is defined as the ratio between UAV lift and its instantaneous weight.

$$n = \frac{L}{W} \quad (7.11)$$

The two-dimensional commanded acceleration (n_C) is frequently non-dimensionalized and is expressed in terms of g's as:

$$n_C = \frac{a_C}{g} \quad (7.12)$$

Please note that, this n is not the load factor. This is just a unit-less vale of a commanded centripetal (i.e., normal) acceleration. It is customary to non-dimensionalize this acceleration, by dividing it to the gravitational constant (i.e., g-load).

There are two classes of PN, pure and true. Pure PN applies the commanded acceleration with reference to the velocity vector of the UAV; whereas True PN applies the commanded acceleration with reference to the light of sight. In practical applications, implementing an acceleration vector perpendicular to the LOS as required by true PN is a challenge.

In practice, the commanded acceleration is converted to the UAV body axis (e.g., y-axis) as follows:

$$n_C = \frac{N' V_C \dot{\lambda}}{g\cos(\lambda_L)} \tag{7.13}$$

where λ_L is referred to as the lead angle. As Figure 7.9 demonstrates, the UAV, with a velocity magnitude V_U, is heading at an angle of λ_L with respect to the LOS. The lead angle is the theoretically correct angle for the UAV to be on the right path toward the target. In other words, if the UAV is on this path, no further acceleration commands are required for the UAV to hit the target/waypoint. The closing velocity V_C is defined as the negative rate of change of the distance from the UAV to the target ($V_C = -\dot{R}_U$).

If the LOS is in the xy plane, the a_C will be the acceleration along body y –axis (i.e., a_y) to yaw. However, if the LOS is in the xz plane, the a_C will be the acceleration along z – axis (i.e., a_z) to pitch. The guidance system output in xy plane (n_C) may be converted to a sideslip angle (β) command to control system by using equations 5.81 and 5.83 (in Chapter 5).

Recall that, for any turn, the centripetal acceleration a (in any plane), is related to the linear velocity (i.e., UAV airspeed) by:

$$a = \frac{V^2}{R} \tag{7.14}$$

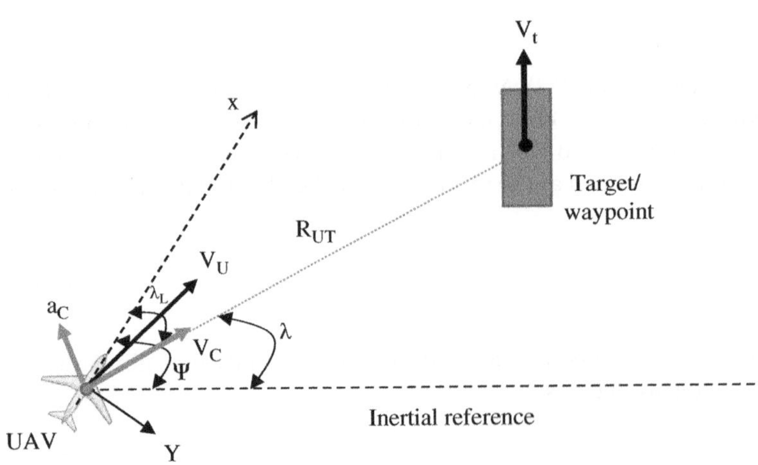

Figure 7.9 Unmanned aerial vehicle (UAV)-target engagement geometry.

In Figure 7.9, V_U is the UAV airspeed, R_{UT} is the distance between UAV and the target. The PN constant, N is the ratio of the commanded turn rate (ω_C) to the angular velocity of the LOS ($\dot{\Psi}_{LOS}$). The constant N is always greater than unity, and usually ranges from 1 to 5. If N = 1, then the UA is turning at the same rate as the LOS, or simply following on the target. When N > 1, the UAV will be turning faster than the LOS, and thus the UAV will build up a lead angle with respect to the LOS. The goal is to reach the desired difference between the heading angle of UAV and the target. Thus, N is a direct function of the target maneuverability (maximum yaw rate and pitch rate).

The direction of the LOS is stablished by the UAV seeker, in tracking the target. The component of UAV velocity perpendicular to the LOS (V_{Fn}) is:

$$V_{Fn} = V_F \sin\left[\Psi_{ref} - \left(\Psi_F - \Psi_L\right)\right] \tag{7.15}$$

which generates a positive LOS rotation. If the perpendicular components of follower UA and target velocity are equal and unchanging, there will be no rotation of the LOS. When the distance between the UAV and target is kept constant, the objective of guidance is met. Thus, in the steady state, the turn rate of the follower UAV is:

$$\omega_F = N\dot{\Psi}_{LOS} \tag{7.16}$$

This command will be fed into the control system. The block diagram for this guidance system (plus control system) is identical to Figure 7.7, the difference is the guidance controller; which is working on PN law. Recall that the objective is not to hit the target; it is to follow the target/ leader UAV. The LOS rate is obtained by measuring the rate of rotation of the UAV seeker tracking the target.

In practice, the UAV has to bank (using ailerons and rudder) to implement the commanded acceleration. When the UAV is banking, the load factor is increased, with the limit of ultimate allowable value. Thus, the commanded acceleration is applied through a bank angle (as well as a sideslip angle to make the turn coordinated). Thus, limited roll performance will influence the application of the guidance law. The turn governing equations and maneuverability of a UAV are presented in Section 7.7.5.

Example 7.2

A UAV is flying toward east at a speed of 40 m/s, while a target aircraft is flying northward at a speed of 30 m/s. They are at a distance with the geometry shown below (Figure 7.10):

The UAV guidance system is governed by the PN guidance law with an N' = 2. It is required to move into and fly along the LOS. What command will be generated by the guidance system at this moment?

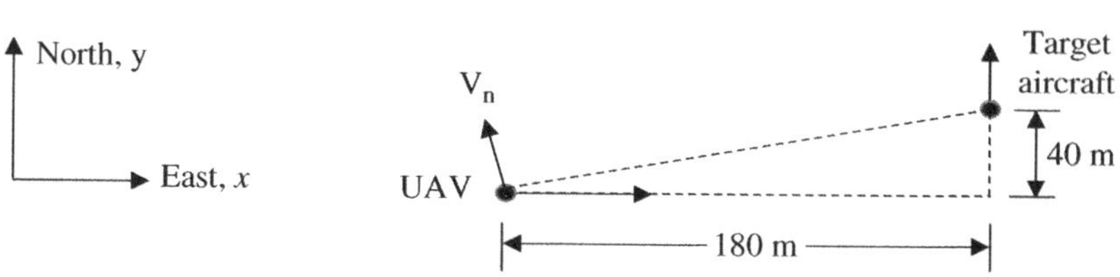

Figure 7.10 Geometry of the unmanned aerial vehicle (UAV) and target aircraft.

Solution

For simplicity, the reference point is selected to be the location of the UAV. Hence; $x_U = 0$; $y_U = 0$. In addition, based on the given geometry, the component velocities are:

$$\dot{x}_T = 0; \dot{y}_U = 0$$

Thus,

$$V_{TUx} = \dot{x}_T - \dot{x}_U = 0 - 40 = -40\,m/s \tag{7.4}$$

$$V_{TUy} = \dot{y}_T - \dot{y}_U = 30 - 0 = 30\,m/s \tag{7.5}$$

The instantaneous distance:

$$D_{TU} = \sqrt{(y_T - y_U)^2 + (x_T - x_U)^2} = \sqrt{(40-0)^2 + (180-0)^2} = 184.4\,m \tag{7.2}$$

The instantaneous line-of-sight rate:

$$\dot{\lambda} = \frac{V_{TUy}(x_T - x_U) - V_{TUx}(y_T - y_U)}{D_{TU}^2} = \frac{30(180-0)-(-40)(40-0)}{184.4^2}$$

$$= 0.206\frac{rad}{\sec} = 11.8\frac{\deg}{\sec} \tag{7.6}$$

The closing velocity:

$$V_c = \frac{-(V_{TUx}(x_T - x_U) + V_{TUy}(y_T - y_U))}{D_{TU}} = \frac{-[40(200-0)-(-60)(70-0)]}{184.4}$$

$$= 47.7\frac{m}{s} \tag{7.3}$$

The commanded normal acceleration:

$$a_C = N'V_C\dot{\lambda} = 2 \times 47.7 \times 0.206 = 19.7\frac{m}{s^2} \tag{7.10}$$

In terms of g's, the command is:

$$n_C = \frac{a_C}{g} = \frac{19.7}{9.81} = 2.004 \tag{7.12}$$

The two-dimensional commanded acceleration is 2 g.

7.6 Pursuit Guidance Law

The basic idea in pursuit guidance is to keep the UAV (i.e., pursuer) pointed toward the target (either a ground target or a leader UAV). The method operates by forcing the angular displacement error between a pursuer vehicle and its moving target to zero. Whenever there is an undesired angular deviation, a command is generated to annul the deviation. A direct heading toward a target is maintained [93] by driving the angle between the forward axis of the pursuer and the LOS to the target to zero. This is achieved by commanding the pursuing vehicle to maneuver at a rate equal to the rate of LOS.

As the target moves, the pursuer also moves to maintain a direct line to the target. The pursuit guidance algorithm is expressed mathematically as

$$x_C = N\dot{\lambda} \tag{7.17}$$

where x_C is the output command, N is a proportional gain, λ is heading (i.e., LOS) angle error, and is $\dot{\lambda}$ rate of LOS error angle. The guidance output command is converted by control system either into a rudder deflection for heading control, or to an elevator deflection for pitch control. The objective of the pursuer is to maintain a direct heading toward the target by driving the LOS error angle (λ) to zero. The pursuit guidance is not effective against fast moving targets, as it usually ends up in an energy-consuming tail chase scenario.

The greatest advantage of pursuit guidance law is in its simplicity. Only LOS error angle, and its rate needs to be measured. Then, an appropriate control surface command is generated for an interception course. In the pursuit guidance, the UAV longitudinal axis; attitude (or the UAV's velocity vector) is kept pointed at the target. In interesting to note that the yaw rate is also a function of angle of attack, since, the angle of attack is usually non-zero. A variation of the pursuit guidance law is the lead pursuit. As the name implies, lead pursuit means that the UAV is flying a course whereby it is in pursuit of a leading point just slightly ahead of the target.

The implementation is simple, and the basic idea is to strive to maintain a constant LOS angle between the UAV and the target. Pursuit guidance states that as long as the UAV is pointed at the target at all times, given enough kinetic energy, the UAV will meet the target.

Control commands are scaled by a proportional factor (N) of the current error, and are applied to direct the pursuer vehicle along the LOS. In designing the guidance system; the gain N is the only guidance parameter that needs to be pre-defined, which determines the input magnitude of magnification. The gain can be adjusted to allow for a safe turn performance, while enabling the vehicle to quickly converge upon the commanded path. This is achieved by gain scheduling to select gain values based on current UAV flight parameters. The selection of N is heavily dependent on the aircraft maneuverability, which may result in variable gains which provides optimal turn performance inside the flight envelope. For instance, if maximum allowable ultimate load factor is 2, the bank angle of the vehicle should be limited to ±60°. As a result, a schedule of gains (N) needs to be specified to allow for a safe turn within the flight envelope.

Figure 7.11 illustrates the trajectory of a UAV with a pursuit guidance against a moving target. It is observed that the UAV (i.e., pursuer) aims to maintain a direct heading

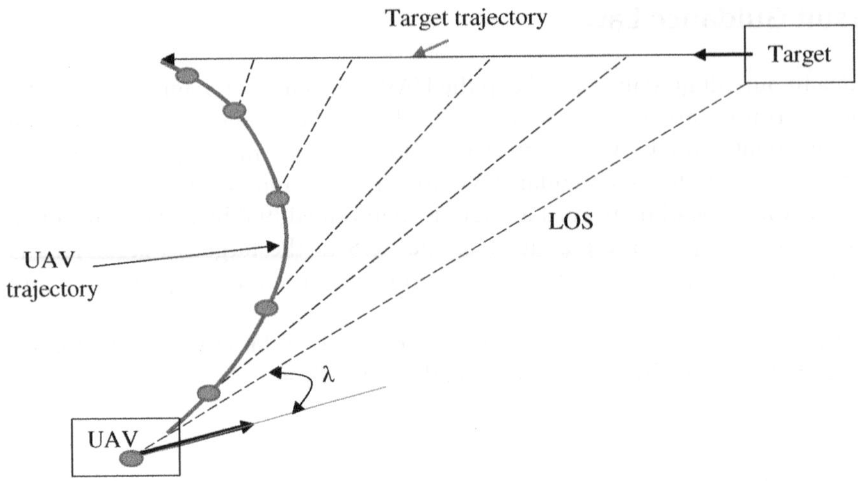

Figure 7.11 Pursuit guidance against a moving target.

toward the target by continuous alignment of its direction of flight with the LOS to the target. Hence, at the time of engagement, the LOS error angle (λ) is driven to zero. As the target moves, the LOS changes, which implies that the pursuer should adjust its heading accordingly. As the distance between the pursuer (UAV) and the target reduces, the pursuer converges on the targets position.

There is a further classification of this guidance law as: (a) pure pursuit; and (b) deviated pursuit. Pure pursuit makes the UAV point at the target while deviated pursuit makes the UAV point at a spot ahead of the target by a fixed angle. The idea behind deviated pursuit is to take advantage of the information of the target's direction.

7.7 Waypoint Guidance Law

7.7.1 Waypoints

In general, a waypoint is a term used to refer to an intermediate point or place on a route or line of travel, a stopping point or point at which course is changed. Waypoints are sets of coordinates that identify a point in physical space. The waypoints could be of two types: (a) pre-defined; and (b) predicted (i.e., synthetic). The waypoint is considered synthetic, when the UAV's position on the flight path is a projection of the position at which the UAV is expected to be within a specified time horizon. The waypoint synthetic is primarily adopted when the target is moving. When the vehicle reaches the designated waypoint, a set of new commands are applied to the vehicle to proceed to the next waypoint.

As explained in the introduction Eq. (7.2), about difference between navigation and guidance, some references/texts use the term "waypoint navigation," but indeed, they mean "waypoint guidance." To allow the control system to control an aerial vehicle, the guidance system must create the path in three dimensions; North-East-Down (NED) coordinates (e.g., x as North, y as East, and z as down or altitude). The path can be divided into two parts, one for horizontal motion, North-East, and one for vertical motion, altitude.

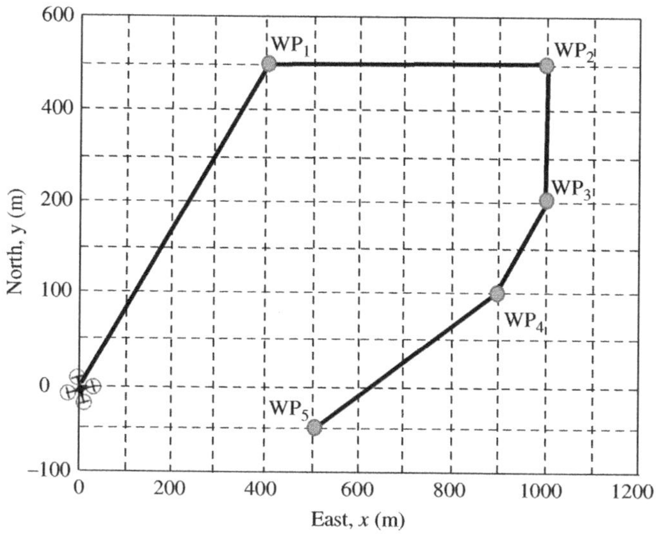

Figure 7.12 Predefined waypoints and desired trajectory.

Figure 7.12 illustrates a sample of five assigned waypoints and a 2d desired trajectory. This trajectory starts from the initial position for the UAV (0, 0). The vehicle is required to follow a path that passes through the waypoints WP_1 (400, 500), WP_2 (1000, 500), WP_3 (1000, 200), WP_4 (900, 100), and ends at WP_5 (500, −50).

It is a route and destination planner for the UAV. This approach instructs the UAV where to fly; at what height; the speed to fly at and it can also be configured to hover at each waypoint. Waypoint Navigation guidance allows an UAV to fly on its own with its flying destination or points pre-planned and configured into the remote control navigational software. When waypoints are programmed into the application, the UAV can go directly to the first point and proceed to each point in turn.

Thousands of waypoints can easily be programmed in an UAV software. In aviation, waypoints consist of a series of abstract GPS points (x, y, and z) that create artificial airways (i.e., highways in the sky). The coordinates of the way-points can be provided either before or after take-off. In this approach, the instruction is continually commanded to the control system to fly on a selected heading at a selected airspeed and altitude.

Even though the flight path itself may undergo rapid changes in heading angle, the trajectory taken by the vehicle should be smooth. In another word, the UAV must rapidly reacquire the new flight leg once the maneuver has been completed.

7.7.2 Types of Waypoint Guidance

In the waypoint guidance, the UAV is guided along an assigned path defined by a series of waypoints. A guidance technique for an UAV is to follow waypoints based on either a pre-planned trajectory, or a feasible reconfigured trajectory. If the atmosphere disturbs the trajectory, the fight path is reconfigured, based on an algorithm. Different guidance algorithms can be developed based on the mission the UAV has been designed for. A desired algorithm must have two functions: (a) path planning; (b) trajectory smoothing and tracking. In this guidance technique, the goal is to minimize the lateral offset between the desired flight path and the vehicle's heading, by issuing a heading command. When the vehicle reaches a waypoint, the guidance system issues a new command to guide the aircraft to the next waypoint. In the early 1970s, the CH-53A helicopter achieved autonomous waypoint navigation [79] using classical linear control techniques.

In relation to following a path between designated waypoints, there are two types of waypoint guidance laws: (a) direct-to-waypoint (DTW); and (b) track-to-waypoint (TTW) methods. The DTW approach creates heading commands to the vehicle based on the angular difference between the waypoint and UAV.

However, in the TTW method, the objective is to follow the track between waypoints. In this guidance method, the control system aims to minimize the lateral offset between the desired flight path and the aircraft's direction, issuing heading commands. Indeed, the TTW approach adds an additional constraint on a flight path that the vehicle must follow in order to reach the waypoint, rather than simply reaching the waypoint.

Both TTW and DTW approaches are far from optimal, due to a large deviation in transition between flight paths after reaching a waypoint. During any transition, the aircraft will generally have a large overshoot to correct its track. The deviation particularly increases when the flight-path transition angle is acute.

Various control techniques have been developed in the literature to minimize flight-path deviations. Such techniques include applying control methods such as model predictive control and robust control to predict flight-path changes, and to take control action before reaching a waypoint.

7.7.3 Segments of a Horizontal (Level) Trajectory

In this section, a guidance algorithm for a two-dimensional horizontal flight is presented. Various reconfigurable guidance algorithms are available and are proposed in the literature. For instance, the application of methods based on the direction field theory is developed by [95]. In general, a given set of waypoints should be considered, with assigned North, East, and altitude coordinates. In a two-dimensional horizontal flight mission, the guidance system will be generating the path to follow in North-East coordinates. This set of waypoints begins with a starting point, and ends with the final point.

There are three main segments in the waypoint guidance sequence: (a) approach to a waypoint; (b) turn around that waypoint; and (c) straight flight from the end of the current turn, to the beginning of the next turn. These three main segments of the guidance sequence (as depicted in Figure 7.13) are explained here. As a basic rule, it is assumed that the vehicle has reached a waypoint, when the vehicle enters the circular region (*circle of acceptance* or *proximity circle*) – centered in the waypoint WP_n, with radius equal to a defined *proximity distance* (b). The proximity distance (e.g., 5 m) is determined based on the constraints of vehicle dynamics and structure.

The proximity (or look-ahead) distance is defined as the minimum distance of the vehicle from the current waypoint (1) when it begins turning toward the next waypoint (2). When the distance between the vehicle and the current waypoint (1) is less than this distance, the waypoint (1) is reached and the vehicle can move to the next waypoint (2).

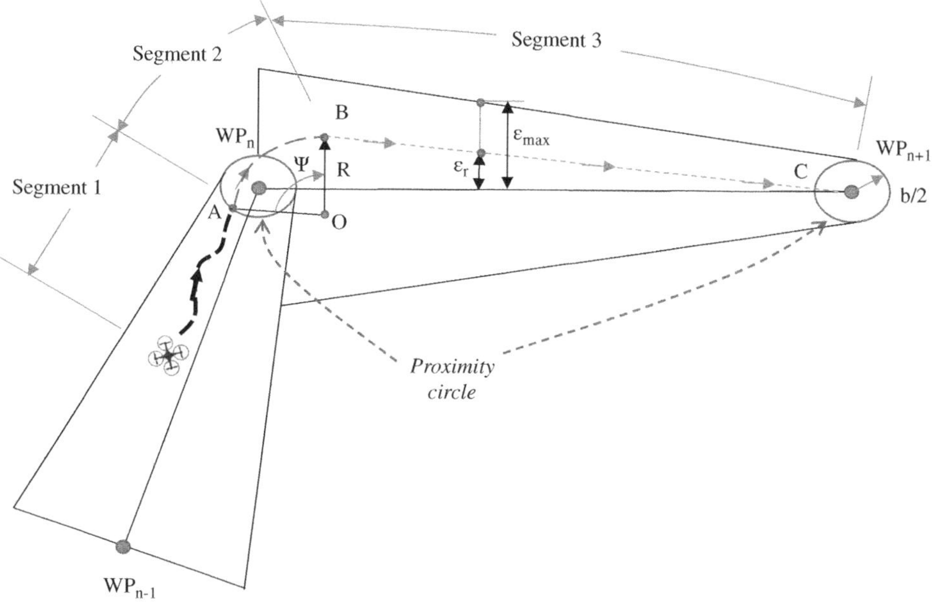

Figure 7.13 Guidance algorithm scheme (Top-view).

- **Segment 1. Approach to a waypoint**
  In the first phase, the UAV flies toward a starting waypoint with a theoretical straight flight and gets ready to begin turning around that waypoint. In Figure 7.13, this phase is identified by the dotted line before the point A. The first phase is from the waypoint WP_{n-1} to the waypoint WP_n with an assigned airspeed. This segment is ended as soon as the vehicle enters the *proximity circle* of that waypoint. From this moment, the UAV can begin turning around this waypoint toward the next waypoint WP_{n+1}. At the end of this segment, the vehicle will have a heading angle (Ψ_1).
- **Segment 2. Turn around the beginning waypoint**
  In this phase, the UAV first turns around the beginning waypoint in the direction of the next waypoint WP_{n+1}. The turn is performed according to UAV maneuverability constraints (e.g., turn rate, load factor, bank angle, airspeed, and turn radius). In Figure 7.13, this segment is identified by the curve between the points A and B.

  The turn begins when the distance of the vehicle from the waypoint WP_n (point A) is equal to the *proximity distance*, and it ends when the longitudinal axis of the vehicle is aligned with the line toward the next (final) waypoint, (i.e., WP_{n+1}).

  This partial turn is characterized by four parameters: (a) airspeed (V); (b) turn radius (R); (c) angular change in the heading angle ($\Delta\Psi$); and (d) bank angle (ϕ). At the end of this turn (i.e., segment), the vehicle will have a new heading angle (Ψ_2), which is the heading angle of the segment that connects the end point of the second segment (i.e., point B) to the next (final) waypoint (WP_{n+1}). This heading angle is the reference heading angle (Ψ_{ref}) for the new waypoint.
- **Segment 3. Straight-line flight toward the final waypoint**
  The third segment is a straight-line flight that starts at the end of the turn in segment 2, and ideally ends when the vehicle reaches the final waypoint. In reality, this segment is assumed done, when the distance of the vehicle from the waypoint WP_{n+1}

(point C) is equal to the *proximity distance* (b/2). In Figure 7.13, this segment is depicted by the line between the points B and C.

This segment may have a long distance, hence, atmospheric phenomena such as wind, gust, and disturbance may cause distraction to the flight. There should be an algorithm, in the guidance system, to correct the distraction, and to enable the vehicle to stay within the acceptable region.

7.7.4 Waypoint Guidance Algorithm

This section presents the waypoint guidance algorithm for a two-dimensional horizontal mission. For a three-dimensional trajectory, you may use the fundamentals presented in this section, and apply them to a 3D mission. The guidance system determines a trajectory to follow in the path from waypoint 1 (or WP_n) to waypoint 2 (or WP_{n+1}). When the UAV is not on a waypoint, it determines the desired path from the current location of the vehicle to the desired waypoint.

There are various approaches for following a path to prepare an algorithm. For instance, [96] presents an approach for a Multirotor UAV based on the Special Orthogonal group SO(3) that allows for independent adjustment of the vehicle's speed profile in order to satisfy desired temporal specifications.

The algorithm should guide the vehicle along the assigned trajectory defined by a series of waypoints. The waypoint guidance algorithm is divided into three distinct hierarchical layouts: (a) path planning; (b) trajectory smoothing; and (c) waypoints tracking. Thus, the guidance system works along with navigation system and control system for a successful flight operation. In a path following problem, the designer is required to develop an algorithm for a given path satisfying the given constraints such that the normalized error converges to a neighborhood of the zero.

The flight path may have a long distance, hence, atmospheric phenomena such as wind, gust, and disturbance may cause distraction to the flight. There should be an algorithm, in the guidance system, to correct the distraction and to enable the vehicle to stay within the acceptable region.

A reconfigurable guidance algorithm is proposed by the application of direction field theory by Ref. [97]. The vector field approach depends on the vehicle's perpendicular distance from the connection line of two waypoints and lookup tables from flight data. Reference [98] has implemented a waypoint-based guidance algorithm for a mini UAV via PID controllers for the outer loop navigation, and an adaptive controller for the inner loop variables.

7.7.4.1 Trajectory Smoother

When the waypoints are connected, a raw trajectory is generated. However, in many cases, such a trajectory may not be feasible, considering the UAV constraints and limits. Airspeed, altitude, turn rate, and structural considerations limit maximum achievable acceleration level. Various trajectory smoother techniques have been introduced in the literature in the past. Most of them aim to minimize the flight time with the optimum utility of the vehicle performance limits (e.g., maximum airspeed, and maximum load factor).

The trajectory must be smooth for two reasons: (a) UAV airworthiness; and (b) successful payload application (e.g., aerial photography). A well-designed smooth trajectory has ideally no abrupt and significant changes on the movement of the UAV. For example, a two-dimensional polynomial with m = 2 given as follows applies to most smooth trajectories in the planar space:

$$y = a + bx + cx^2 \tag{7.18}$$

Or, in terms of time:

$$x = a_1 + a_2 t + a_3 t^2 \tag{7.19}$$

$$y = b_1 + b_2 t + b_3 t^2 \tag{7.20}$$

where coefficients of a, b, c, a_1, a_2, a_3, b_1, b_2, and b_3 are all constants.

The trajectory smoother should apply changes to make the assigned trajectory kinematically feasible in terms of constraints. The trajectory smoother often transforms the waypoint path into a time-stamped feasible trajectory. A limitation of this algorithm is that the trajectory is composed of a number of time-stamped curves which specify the desired location of the UAV at a specified time.

Tracking the movement state estimation of an UAV basically concerns inferring the latent state of interest based on discrete time series noisy observations. The time of interest may be the past (namely, smoothing), the present (tracking), or the future (forecasting).

7.7.4.2 Trajectory Tracking

When the waypoints are connected, a raw trajectory is generated with the contribution of trajectory smoother. In the meantime, the error control is cross-tracked, to monitor the UAV position with respect to the reference path. The real position of the vehicle is stated in terms of the vehicle's East coordinate (E_V) and North coordinate (N_V). Moreover, the segment that connects two way-points are stated in terms of starting point; P_1 (E_1, N_1) and end point; P_2 (E_2, N_2). The *trajectory tracker* (part of navigation system) in conjunction with the control system, develops feasible inputs to control surfaces to stabilize the system parameters. When a waypoint is passed, the cross-track error control is applied in the straight-line flight toward the next waypoint.

The goal is to make the aircraft follow this straight line, by making the cross-track error (ε) as small as possible ($\lim_{t\to\infty} \varepsilon(t) = 0$). The cross-track error (between points 1 and 2) is given by:

$$\varepsilon(t) = -\left[x(t) - x_1\right]\sin(\varphi_c) + \left[y(t) - y_1\right]\cos(\varphi_c) \tag{7.21}$$

The angle φ_c is used to rotate the north-axis in the path-fixed reference frame with origin in P_1.

$$\varphi_c = a\tan\left[\frac{y_2 - y_1}{x_2 - x_1}\right] \tag{7.22}$$

To ensure that the cross-track error goes to 0 for both cases, various guidance principles (e.g., enclosure-based or look-ahead-based steering) can be used.

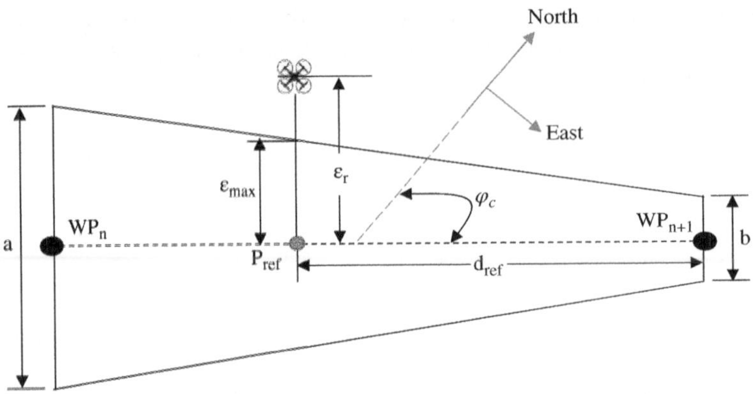

Figure 7.14 No corrections region, and cross-track error and reference distance.

However, to avoid continuous corrections of the trajectory (and avoiding continuous deflections of control surfaces), a *no corrections region* is defined between each two waypoints (Figure 7.14). This means that a correction to the heading angle is applied, only when the *cross-track error* is larger than an assigned value. Applying a look-ahead-based steering technique, the cross-track error is obtained from:

$$\varepsilon_r = \frac{|E_V - mN_V - (E_1 - mN_1)|}{\sqrt{m^2 + 1}} \tag{7.23}$$

where

$$m = \frac{E_2 - E_1}{N_2 - N_1} \tag{7.24}$$

where E_V, N_V, denote real east and north coordinates of UAV, and E_1, E_2, N_1, and N_2 stand for east and north coordinates of the first and second waypoints respectively. The cross-track error is controlled (nullified) by control subsystem.

The *no corrections region* is defined as an isosceles trapezoid (as shown in Figure 7.14), whose symmetry axis is defined by the line connecting two neighboring waypoints. The minor base (b) of the trapezoid is set equal to the diameter of the *proximity* circle around every (including the second) waypoint. The major base (a) is determined by considering the limits of turn radius, maximum speed, bank angle, and load factor.

The maximum acceptable cross-track error (ε_{max}) is variable, and this decreases along the path from semi-segment a/2 to semi-segment b/2.

$$\varepsilon_{max} = \frac{d_{ref}\left(\frac{a}{2} - \frac{b}{2}\right)}{d_s} + \frac{b}{2} \tag{7.25}$$

where d_s is the distance between two neighboring waypoints, and d_{ref} is the reference distance calculated as:

$$d_{ref} = \sqrt{\left(N_{WP\_n+1} - N_{ref}\right)^2 + \left(E_{WP\_n+1} - E_{ref}\right)^2} \tag{7.26}$$

where, N_{WP_n+1} and E_{WP_n+1} are the North and East coordinates of the waypoint n + 1 respectively. Moreover, N_{ref} and E_{ref} are the North and East coordinates of the reference point respectively, obtained by:

$$N_{ref} = N_V - \varepsilon_r \psi_{ref} \tag{7.27}$$

$$E_{ref} = E_V + \varepsilon_r \psi_{ref} \tag{7.28}$$

As already defined, ψ_{ref} (referred to as reference heading angle) is the heading angle of a segment that connects two consecutive waypoints.

When the UAV cross-track error (ε_r) is smaller than ε_{max} during the straight-line flight, no corrections of the heading angle of the vehicle are implemented. However, ε_r is greater than ε_{max}, the new reference heading angle is the heading angle of the direct line between the UAV position and the next waypoint. In such a case, the heading error is fed back to the control system, to deflect the rudder to correct the heading angle.

7.7.5 UAV Maneuverability Evaluation

One of the basic maneuvers to make a flight smooth, is to turn around a waypoint in the direction of the next waypoint. A turning flight has a couple of constraints, including: (a) maximum turn rate (ω_{max}); (b) minimum turn radius (R_{min}); (c) maximum load factor (n_{max}); (d) minimum and maximum airspeed (V_{min}, V_{max}); and (e) maximum bank angle (ϕ_{max}). These constraints define the UAV maneuverability, and indicate the maximum ability of a UAV to turn and maneuver.

The following set of equations [9] governs the relation between the performance parameters of a turning flight. In a level coordinated turn, the relationship between turn radius (R), load factor (n), and the airspeed (V) is governed by:

$$R = \frac{V^2}{g\sqrt{n^2 - 1}} \tag{7.29}$$

Furthermore, in a level coordinated turn, the relationship between turn rate (ω), load factor (n), and the airspeed (V) is expressed by:

$$\omega = \frac{g\sqrt{n^2 - 1}}{V} \tag{7.30}$$

The time to complete a circle (i.e., 360° turn) is obtained by:

$$t_{circle} = \frac{2\pi R}{V} \tag{7.31}$$

The allowable slowest airspeed in a level turn (V_{st}, stall speed in a turn) is a function of the maximum lift coefficient (C_{Lmax}) and bank angle (ϕ):

$$V_{S_t} = \sqrt{\frac{2mg}{\rho S C_{L_{max}} \cos(\varphi)}} \tag{7.32}$$

The maximum allowable bank angle in a turning maneuver is a function of ultimate load factor:

$$\phi_{max} = \cos^{-1}\left(\frac{1}{n_{max}}\right) \tag{7.33}$$

The corner speed (V^*) – usually the lowest airspeed for a turn to minimize the turn radius, and to maximize the turn rate – is:

$$V^* = \left[\frac{2n_{max} W}{\rho S C_{L_{max}}}\right]^{\frac{1}{2}} \tag{7.34}$$

The minimum allowable speed is about 10–20% greater than the stall speed. The maximum allowable bank angle should be such that: (a) it does not produce a load factor higher than the allowable one (Eq. 7.33); and (b) the payload (e.g., camera) is able to perform satisfactorily; for instance, has an acceptable view-angle.

Moreover, a turn must be coordinated in order to keep the radius of turn constant. For the requirements of a coordinated turn, you may refer to references such as [54]. The trajectory smoother must take into account all of these performance constraints to convert a series of waypoints into a smooth trajectory.

7.8 Sense and Avoid

7.8.1 Fundamentals

Collision avoidance is a primary concern [99] to the FAA regarding aircraft safety. A critical challenge in full integration of unmanned aircraft systems into the National Airspace System is the requirement to comply with CFR title 14 Part 91.113 (Right-of-way rules) to see and avoid other aircraft. Based on this regulation, when a rule of this section gives another aircraft the right-of-way, the pilot shall give way to that aircraft and may not pass over, under, or ahead of it unless well clear.

UAVs are seen as potential key airspace users in the future of air transportation, which necessitates additional research and study of safety measures. One of the major limitations to the widespread use of unmanned vehicles in civilian airspace has been the detect-and-avoid problem. There are many reports of civilian drones crashing into buildings, having hazardously close encounters with helicopters, peeping into residential windows, and being intentionally shot down. The likelihood that a RC plane will collide with an airline aircraft [100] is increasing. The guidance system must have a central role in preventing UAV-related accidents and collision avoidance.

The use of UAVs near airports and within controlled airspace poses a serious risk to aviation safety. In the United States, drones are typically restricted to flights within 500 ft of the ground and within sight of the operator. They also are supposed to stay clear of manned aircraft and airports. But in the thousands of FAA reports of possible drone safety incidents, many involved apparent illegal flights.

The purpose of a sense and avoid system is to detect and resolve certain hazards to a UAV flight. The UAV sense and avoid system must provide two services: (a) self-separation service; and (b) collision avoidance service. In general, there are six functions required in a sense and avoid system: (a) detect the intruder/obstacle; (b) track; (c) evaluate; (d) calculation; (e) command; and (f) execute.

The tracking function will employ the navigation system output to determine the trajectory of the UAV relative to the hazard/obstacle. Moreover, the microcontroller will be active in this period to timely process the data, calculate the positions, and generate the appropriate command. Thus, in a sense and avoid system, all three navigation, guidance, and control systems are working simultaneously. The avoid function will be covered in control system, while the sensing technique is described in the next section. Moreover, the trajectory prediction of the hazard has been discussed in Section 7.2.3.

There are currently several research projects [101] being conducted in the area of detect (or sense) and avoid. Active solutions include the use of machine vision, and GPS/radar to detect collision threats, and precision control to avoid collision. However, the current adapting technologies are not at such a reliable level [83]. In addition, the high computational requirement is another obstacle. A major design issue of UAVs is that any black box is an additional weight, and weight restrictions for some smaller UAVs may restrict UAV functionality or the inclusion of cooperative systems. The success of a full integration of unmanned aircraft systems into the National Airspace System [102] is a function of effective detect and avoid system.

The vision-based sense and avoid approaches/techniques can be categorized into four groups: (a) mapless-based approach; (b) simultaneous localization and mapping; (c) simultaneous mapping and planning (SMAP); and (d) safe landing area detection (SLAD).

The first approach can handle uncertainties and is implemented quickly that provides a fast reactive system which can prevent last-minute a collision. The second system is generally suitable for indoor operation of small quadrotors. The third technique builds local obstacle maps, while plans for a conflict-free trajectory using a control law. The SLAD approach takes a map from a range sensor, and this combines it with a set of landing point constraints. The algorithm should calculate a safe landing map, and then chooses the optimum landing point.

In the past 20 years, various algorithms were developed and tested extensively in different environments and scenarios, with a large number of successful obstacle avoidance runs at various commanded speeds. For instance, in 2006, a Yamaha RMAX helicopter (Figure 1.1), equipped with a Lidar, successfully demonstrated the SMAP algorithm. The vehicle flew from point A to point B at 2 m/s while avoiding obstacles. The trajectory was generated in real time at the GCS processor.

7.8.2 Sensing Techniques

The guidance system should determine the position of other aircraft or an obstacle with respect to the air vehicle. The temporary loss or lack of situational awareness is a causal factor in a collision. The sensors and processors of a sense and avoid system may be located onboard the aircraft, off-board, or distributed between both. The hazard (other

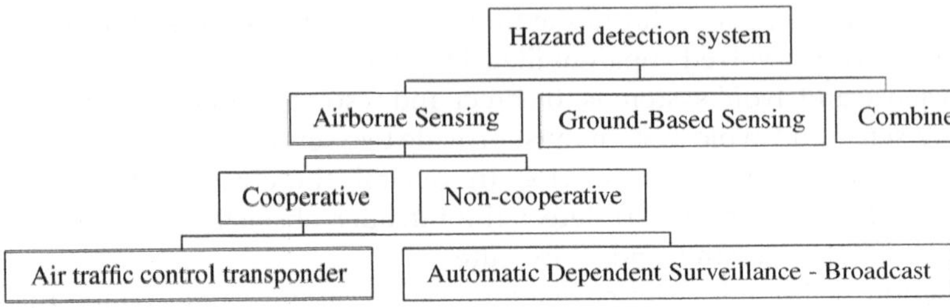

Figure 7.15 Hazard detection systems classification.

aircraft or obstacle) detection system – also referred to as the surveillance system – can be primarily implemented (Figure 7.15) in three forms. The first form is to carry the sensor onboard the UAV, while the second form is to employ sensing on the ground. The third form is a combined technique which will have a sensor (e.g., camera) at the UAV, and a human operator at the GCS. Current available technologies include regular cameras, electro-optic (EO) or infrared cameras, radar, lidar, and acoustic processing. Table 7.2 illustrates typical onboard obstacle detection sensors and their features.

A sensing system involves one or a group of devices that respond to a specific physical phenomenon or stimulus and generate signals that reflect some features of or information about an object. The ground sensing may be implemented via two methods: (a) radar; and (b) human operator in the ground station. Airborne detection sensing of other aircraft/object employ two groups of technologies: (a) cooperative or passive; and (b) non-cooperative or active. Cooperative technologies are those that wait to receive signals from another aircraft's onboard equipment. In a non-cooperative sensing, the UAV is equipped with a sensor (e.g., radar) to independently search for and sense/see the hazard. In the case of an onboard radar, the UAV requires both transmitter and receiver equipment. The prerequisite for effectiveness of a cooperative system is that the other aircraft should choose to cooperate with the UAV. The incoming signal, usually originating from a point on the earth's surface or from

Table 7.2 Typical onboard obstacle detection sensors and their features.

No	Obstacle detection sensor	Media	Output variables
1	Doppler radar	Radio wave	Relative velocity, distance, bearing (azimuth and elevation)
2	Laser radar	Infrared laser light	Relative velocity, distance, bearing (azimuth and elevation), geometry
3	Camera	Light	Identity
4	Acoustic wave sensor	Acoustic waves	Bearing (azimuth and elevation)
5	Microelectromechanical sensor (MEMS)	Acoustic waves	Bearing (azimuth and elevation)
6	Lidar	Pulsed laser light	Relative velocity, distance, bearing (azimuth and elevation), geometry
7	Receiver	Radio wave	Relative range, altitude, bearing (azimuth and elevation)

another aircraft, is called the uplink. The outgoing signal, usually sent to a region, is called the downlink.

There are mainly two types of cooperative sensing techniques: (a) using a transponder; and (b) using GPS. In the first type, the UAV carries a device called a transponder to see other aircraft. A *transponder* is a communication equipment that receives and automatically responds to an incoming signal. The term transponder is an abbreviation for the words, transmitter and responder. It receives an incoming signal over a range of frequencies, and then transmits the signal on a different frequency. For this system, other aircraft must transmit a signal to declare its presence. Currently, all large transport aircraft carry a transponder, to respond to ground-based secondary radar interrogations for air traffic control usage, and Traffic Alert and Collision Avoidance. The transponders usually receive interrogation from the ground surveillance radar on 1030 MHz and replies on 1090 MHz.

The second type of cooperative sensing technique is referred to as the Automatic Dependent Surveillance – Broadcast (ADS-B) system. It employs the GPS or an alternate navigation source, and automatically broadcasts its own aircraft position, velocity, and other data without needing to be interrogated. In the USA, those aircraft currently required to carry a transponder will be equipped with [103] ADS-B by 2020. Under European regulations, all aircraft operating in the region from June 2020 are mandated to be equipped with ADS-B out technology.

Table 7.2 illustrates typical onboard obstacle detection sensors and their features. The table also shows the media that a sensor is utilizing to sense the hazard/obstacle. The first six sensors are active which indicate they start searching for a hazard. However, a receiver only receives the incoming signal from transmitter (e.g., transponder) of another aircraft, if they cooperate. This sensor is not effective for detecting any object if it is without a transponder (e.g., a wall or tree).

Sensor technologies vary in their cost, range, field of view, and surveillance coverage. In selecting a surveillance system, a number of factors should be evaluated: (a) range; (b) timeliness (Sample/Update rate); (c) Field of view; (d) simplicity; (e) cost; (f) design challenge; (g) reliability; (h) accuracy; (i) size; (j) weight; (k) technology level; (l) flexibility; and (m) integration. These parameters are used in characterizing the performance of a sensing technique and evaluating the suitability of a candidate sensor.

A few technical parameters of a hazard detection sensor – which represent the performance characteristics – indicate its suitability to a specific sense and avoid system. Field of view is the angular sector within which the sensor makes search and measurements. When an object is outside the field of view, this sensor cannot detect it. Range is the distance originated from a sensor, within which an acceptable probability of object detection is expected. Accuracy is an indication of the difference between a measured value and a true value. In a sensor, it describes the uncertainty of the measurement data. The sensor should be capable of maintaining the track for a certain time and predicting the future motion (i.e., trajectory) of the object.

For a precision flight mission – to prevent a catastrophe in collisions – a high accuracy is required; otherwise, a collision is expected. For instance, the small bird strike to a large UAV may be considered as not crucial, while the strike of a larger bird will end up to a catastrophe. Sample or update rate is the interval at which a sensor conducts successive measurements and provides the output data. If it does not detect the target at an expected interval, its sample rate will not be high enough in collision avoidance.

The successive measurements are employed to form a track of another aircraft or a moving object.

The vision-based obstacle detection in unknown environments and automatic landing problems have been investigated [104] in the US Army's Precision Autonomous Landing Adaptive Control Experiment project. The detection algorithm applies a set of landing point constraints (slope, roughness, distance to obstacles) to the range map to find all safe landing regions, and then to choose the optimum landing point. A tradeoff study is needed for electing surveillance system to finalize the optimum sensor for a see and avoid system.

7.8.3 Collision Avoidance

A collision avoidance subsystem should be based on a set of prescribed rules. There are various choices for a UAV to avoid a collision [105] with an object or another aircraft. The fundamental principles of avoidance techniques would follow the customary right-of-way rules [103]. In encountering a UAV and another aircraft, the collision avoidance process must use a priority protocol to resolve the conflict (e.g., when both aircraft are selecting a climbing maneuver). The UAV maneuver to avoid a collision should be compatible with the maneuverability limits such as the maximum allowable load factor, and the minimum turn airspeed.

The need for resolution must begin with a defined volume to be avoided, predicting the time to collision, and to determine that a hazard poses a threat such that an avoidance maneuver is required. A "self-separation" initiative would act before a collision avoidance maneuver is needed, and this could result in an earlier and gentler maneuver. If the object is within the collision zone, a late, last-moment, and aggressive maneuver is necessary. It is important that the sense and avoid system distinguish a threatening hazard from a nonthreatening traffic.

If an intruder aircraft is within the collision avoidance zone of a UAV, a challenge is the ability of UAV to deal with maneuvers made by the intruder aircraft. Reference [106] suggests the "collision avoidance zone" to be a truncated cylinder ±100 ft in height and 500 ft in radius. The horizontal and vertical separation distance of this cylindrical airspace around the UAV is also called the safety zone.

The function of the collision detection system is to detect a collision and provide the resolution in the form of an evasion maneuver. A collision is detected based on the flight path obtained from the trajectory prediction. The sense and avoid system makes checks on the flight paths of both UAV and the other aircraft to evaluate whether the safety zone of the UAV has intruded. If so, the location and times of the probable collision are calculated.

Typical maneuvers to avoid a collision include: (a) left turn; (b) right turn (Figure 7.16); (c) climb; (d) descent; (e) slow down; and (f) speed up. These can be classified into three categories: (a) turn; (b) altitude change; and (c) speed change. The collision avoidance resolution is derived from a suitable guidance law (e.g., PN), and then executed by the control system of UAV's autopilot. For instance, [107] presents a PN-based optimal collision avoidance, while [108] proposes a method which produces an avoidance trajectory in both the horizontal and vertical planes of motion. The optimum scenario will be the one which delivers the highest chance of success or lowest probability of collision.

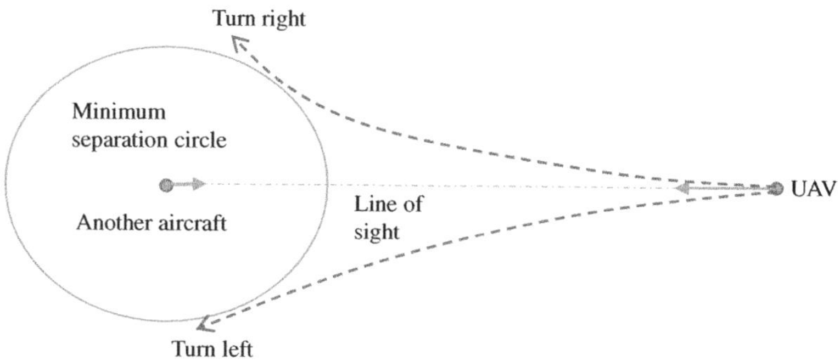

Figure 7.16 Two collision avoiding maneuvers (top-view).

Figure 7.17 demonstrates a collision geometry; a UAV with a ground speed of V_u, and an aircraft with a ground speed of V_a. If the UAV does not change its flight trajectory and ground speed, it will collide at some point along its path. For a successful avoidance, the UAV must from the aircraft with a distance greater than the prescribed miss distance.

If both UAV and the aircraft are assumed a point mass and are flying with different constant airspeeds, the coordinates of the next position are:

$$x_{t+1} = x_t + V(t)\cos\Psi(t)\cos\gamma(t).\Delta t \quad (7.35)$$

$$y_{t+1} = y_t + V(t)\sin\Psi(t)os\gamma(t).\Delta t \quad (7.36)$$

$$h_{t+1} = h_t + V(t)\sin\gamma(t).\Delta t \quad (7.37)$$

where Ψ and θ are heading and climb angles respectively. Heading angles of UAV and aircraft must have the same reference line, such as east. In Eqs. (7.35) through (7.37), the effect of the wind is neglected, so the wind speed is assumed to be zero.

The collision condition (for a two-dimensional condition) is that the time for both UAV (subscript u) and aircraft (subscript a) to reach to the collision point (t_C) are equal for both x and y components:

$$t_{xa} = t_{xu} = t_C \quad (7.38)$$

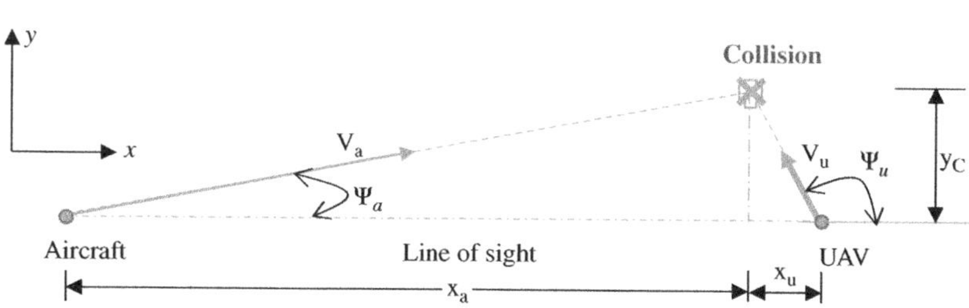

Figure 7.17 Collision geometry (Top-view).

$$t_{ya} = t_{yu} = t_C \tag{7.39}$$

Using speed and distance, the following equations are obtained:

$$\frac{x_a}{V_a \cos(\Psi_a)} = \frac{-x_u}{V_u \cos(\Psi_u)} \tag{7.40}$$

$$\frac{y_a}{V_a \sin(\Psi_a)} = \frac{y_u}{V_u \sin(\Psi_u)} \tag{7.41}$$

where Ψ_a, and Ψ_u are heading angles of UAV and aircraft respectively, and V_a and V_u are ground speeds of UAV and aircraft respectively. The sum of the distances traveled along the x-axis by the UAV and the aircraft is equal to the initial distance between the UAV and the aircraft (D).

$$D = x_a + x_u \tag{7.42}$$

However, in the y-axis, both aircraft and UAV have the same traveled distances.

$$y_C = y_a = y_u \tag{7.43}$$

where y_C is the distance traveled along y-axis until the collision. The distances traveled along x and y axes for both aircraft and UAV are obtained as:

$$x_a = V_a \cos(\Psi_a) t_C \tag{7.44}$$

$$x_u = V_u \cos(\Psi_u) t_C \tag{7.45}$$

$$y_a = V_a \sin(\Psi_a) t_C \tag{7.46}$$

$$y_u = V_u \sin(\Psi_u) t_C \tag{7.47}$$

where t_C is the time until the collision. Plugging corresponding values of distances along x axis from Eqs. (7.44) and (7.45) into Eq. (7.42), we obtain:

$$D = V_a \cos(\Psi_a) t_C + V_u \cos(\Psi_u) t_C \tag{7.48}$$

If the collision happens, the time until collision (based on x-component) will be obtained as:

$$t_C = \frac{D}{V_a \cos(\Psi_a) + V_u \cos(\Psi_u)} \tag{7.49}$$

We will have a similar result, when the y-components are considered:

$$y_a = y_u \Rightarrow V_a \sin(\Psi_a) t_C = V_u \sin(\Psi_u) t_C \tag{7.50}$$

By canceling the time (t_C), we obtain the new y-condition for a collision.

$$V_a = \frac{V_u \sin(\Psi_u)}{\sin(\Psi_a)} \tag{7.51}$$

If both conditions (7.42, 7.51) hold, a collision will happen. The UAV must employ its maneuverability power to avoid this collision with a minimum cost.

Example 7.3

An aircraft is flying toward northeast with a speed of 150 knot, while a UAV is flying at the same altitude toward northwest with a speed of 32 knot. The heading angles of the aircraft and UAV with respect with East are 10 and 120°. The initial distance between the two vehicles is 10 000 m.

a) Will there be a collision between UAV and the aircraft?
b) If not, determine the miss distance.

Solution

We first, assume there is a collision, thus, the time until the collision will be:

$$t_C = \frac{D}{V_a \cos(\Psi_a) + V_u \cos(\Psi_u)} = \frac{10000}{150 \times 0.514 \cos(10) + 32 \times 0.514 \cos(120)}$$
$$= 147.6 \sec \tag{7.49}$$

Now, we check both conditions in Eqs. (7.50).

$$y_a = V_a \sin(\Psi_a) t_C = 150 \times 0.514 \sin(10) \times 147.6 = 1977.5 m \tag{7.46}$$

$$y_u = V_u \sin(\Psi_u) t_C = 32 \times 0.514 \sin(120) \times 147.6 = 2103.9 m \tag{7.47}$$

After 147.6 seconds, each vehicle will have a different distance traveled in the y direction. Thus, these two vehicles will not collide. Since the assumption of collision was not true, these numbers are not accurate too. An accurate trajectory for both aircraft and UAV may be determined by a numerical simulation. The following is a MATLAB code to simulate the flight of both vehicles.

```
clc
clear all
Va=150*0.514; % m/s
Vu=32*0.514;
psi_a=10/57.3; %
psi_u=120/57.3; %
D=10000; % m
dt = 0.01; % time sec
```

```
xa(1)=0;%
ya(1)=0;%
xu(1)=0;%
yu(1)=0;%

for i = 1:1:11900
    xa(i+1)=Va*cos(psi_a)*dt+xa(i);
    xu(i+1)=Vu*cos(psi_u)*dt+xu(i);
    ya(i+1)=Va*sin(psi_a)*dt+ya(i);
    yu(i+1)=Vu*sin(psi_u)*dt+yu(i);
end
plot(xa,ya,'rO-',D+xu,yu,'g*-'); grid
xlabel ('X (m)')
ylabel ('Y (m)')
legend('Aircraft','UAV')
```

Figure 7.18 shows the simulation results for UAV and aircraft in example 7.3. As observed, the UAV is passing ahead of aircraft without a collision. From the simulation, the miss distance is determined to be 119 m.

When a conflict resolution algorithm is feasible, various guidance laws may be employed for a collision avoidance. For instance, the PN guidance with a PN constant less than one (i.e., $N < 1$). In such cases, the UAV will be turning slower than the LOS, thus continuously falling behind the target (i.e., another aircraft).

Another appropriate guidance law for a collision avoidance (as in a formation flight) is the LOS guidance law. This law may be implemented by assuming the goal (i.e., target) of the follower UAV to be constantly at a desirable distance behind or at the side of the leader UAV. In a GPS-denied environment, the visual odometry system can be used for vision-based flight control and obstacle avoidance.

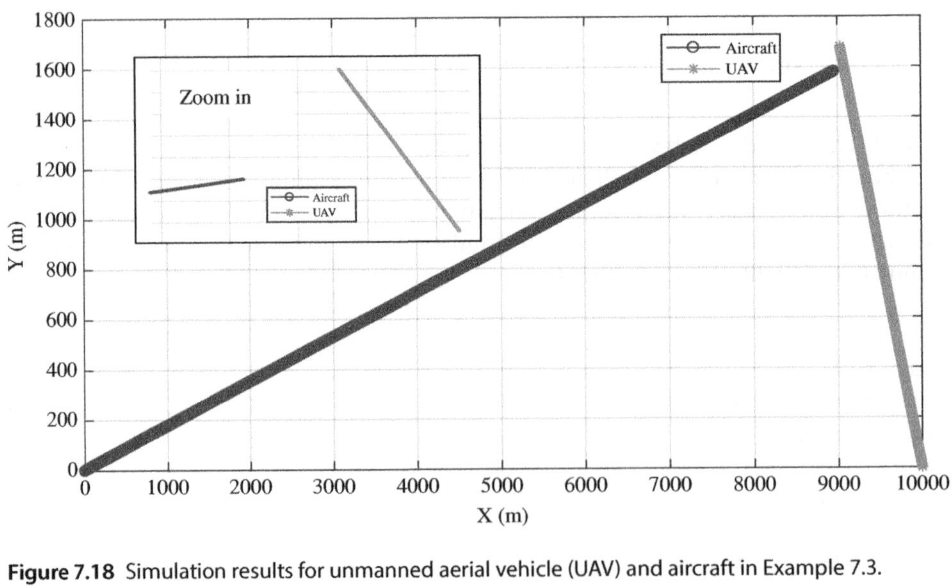

Figure 7.18 Simulation results for unmanned aerial vehicle (UAV) and aircraft in Example 7.3.

7.9 Formation Flight

Formation flight is the orchestrated flight of more than one aircraft. Some references (e.g., [109]) use the term "swarming" or "swarm formation flying" with the same meaning. One of the most state-of-the-art applications using multi-UAV platform is visual performance show. For example, in 2018, Intel succeeded to perform a record-breaking performance with 1200 Shooting Star quadrotors collaborated by orchestra (first flight of its kind) in Olympic Opening Ceremony[1] in PyeongChang, South Korea. The formation flight can be leaded by a leader UAV, or even without a leader (i.e., leaderless).

This type of flight originates from manned aircraft, and FAA regulates its minimum legal requirements. There are various arrangements; a few examples are: (a) parade formation; (b) standard formation turn; and (c) echelon turn. In order to prevent any collision in formation flight, the pilot should always keep all airplanes in front in sight, and be able to pass behind and under any airplane in front on your way to the outside. One of the most challenging problems in formation flight is guaranteeing collision avoidance.

One beautiful example of formation flight in nature is the flock of starlings which is a phenomenon of nature that amazes. Is this event, hundreds, even thousands of starlings flying together in a whirling, ever-changing pattern. As they fly coordinately, the starlings seem to be connected together. They climb, descent, and turn and change direction at a moment's notice.

With advances in science and technology, UAVs are able to conduct formation flight. One of difficult parts of formation flight is for a UAV to join a group of UAVs. The appropriate guidance law plus a precision control may guarantee the safety of formation flight. To maintain the format and a safe distance, there are a number of leader UAVs, and a number of follower UAVs. There is no unique standard formation flight; however, in each case, the follower UAV must maintain a lateral, a longitudinal, and a vertical distance from the leader UAV.

Now, as an example, consider four unmanned aerial vehicles are required to circle over a moving target (Figure 7.19) with extreme accuracy, in order to gather intelligence, transmit information instantly back to ground station. In the formation flight

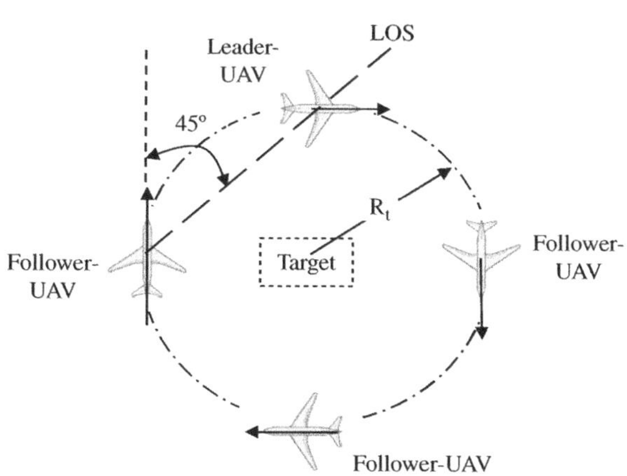

Figure 7.19 Leader-follower unmanned aerial vehicle (UAV) geometry.

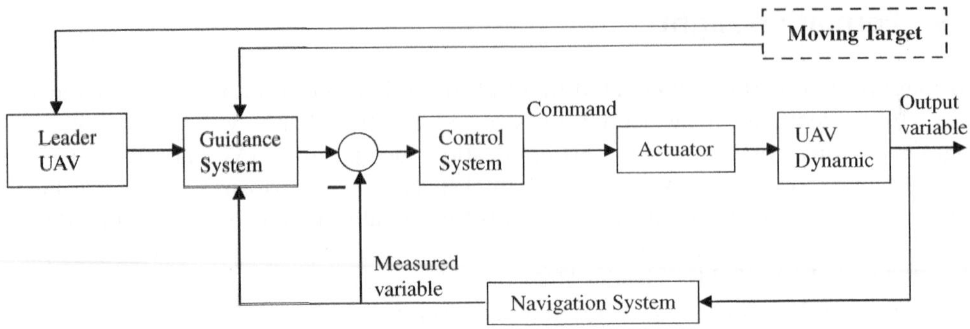

Figure 7.20 Control, guidance, and navigation systems in the follower unmanned aerial vehicle (UAV) autopilot.

system considered here, each *i*-th aircraft uses information from itself and the preceding (*i*-1)th aircraft to track a commanded relative position. When all four UAVs entered the target circle, each UAV is a leader as well as a follower UAV. Because each UAV is required to maintain a coordinated turn around the target, it will be assumed to be a leader UAV.

In addition, since each UAV needs to maintain a fixed spatial distance from another leading UAV, each UAV is assumed to be a follower as well. Thus, the integrated guidance and control system of all four UAVs are similar. The relation between the control system, the guidance system, and the navigation system for each UAV is shown in Figure 7.20.

Since the flight path is a circle, and there are four UAVs, the LOS or tracking line must always make 45° with the follower heading angle. When the follower UAV is turning around the target, the spatial distance with the leader UAV is achieved using the engine throttle. When the LOS angle (Ψ_{LOS}); the angle between the follower UAV heading angle (Ψ_F) and the leader UAV heading angle (Ψ_L) is more than 45°, the throttle is deflected such that to increase the engine thrust and accelerate the UAV. Based on a desired circular path, the following relationships may be derived by observation:

$$\Psi_F - \Psi_L = 90\text{deg} \tag{7.52}$$

And, the light of sight angle is:

$$\Psi_{LOS} = 45\text{deg} \tag{7.53}$$

In case the angle between LOS and the follower heading angle is less than 45°, the throttle is deflected such that to decrease the engine thrust and decelerate the UAV. As soon as the follower UAV reaches the commanded circle around the target and is stabilized, the guidance system will be activated to guide the aircraft such that to keep a constant 45° of line-of-sight angle. When the follower UAV's line-of-sight is different than the commanded LOS, the guidance system generates a yaw rate (R) and a change in the aircraft speed for control system:

$$R = k\left(\Psi_{LOS} - \Psi_{LOS_C}\right) \tag{7.54}$$

The constant k is determined in the design process of the guidance law. For this objective, the knowledge of transfer function between rudder deflection (δ_R) to the yaw rate (R); or ($R(s)/\delta_R(s)$) is required. A general block diagram of the LOS guidance plus control systems is already depicted in Figure 7.7.

7.10 Motion Planning and Trajectory Design

The guidance system functions heavily in the UAV mission, UAV performance features, flight plan, and trajectory. Some performance weaknesses of a UAV can be partially compensated by a proper flight plan. Motion planning refers to the derivation of a flight plan to guide the UAV from takeoff to target destination, and to return back to the air base. Constraint motion planning and trajectory optimization for UAVs are presented by Chapter 17 of Ref. [110]. If the path does not consist of straight lines between commanded waypoints, the path has to be parametrized.

A significant requirement for the flight planning is the accurate terrain maps. The terrain should be known, and digital elevation maps must be available for terrain modeling. If the trajectory is lively designed, the burden of massive data handling (e.g., interpolations through database to obtain the elevation at unknown geographic locations) for digital maps is a serious challenge. Some mathematical schemes such as spline provide a fast and smooth interpolations. The performance of the guidance system is frequently influenced by type of interpolation schemes, so the selection of the appropriate scheme must be accurately decided.

An optimal plan must be generated – using the available resources – while a certain cost function is minimized. In designing the guidance system, and planning the trajectory, one must consider the UAV performance capabilities, as well as the physical and environmental constraints. A typical UAV trajectory is shown in Figure 7.21.

If the guidance strategy does not rely on precomputed, stored reference profiles, a live trajectory planning system must be in place. Such system will perform onboard computation of a new reference trajectory based on available vehicle, environmental, and trajectory data. The trajectory-planning scheme may compute a reference flight profile by piecing together several flight segments that are defined by a small set of geometric parameters.

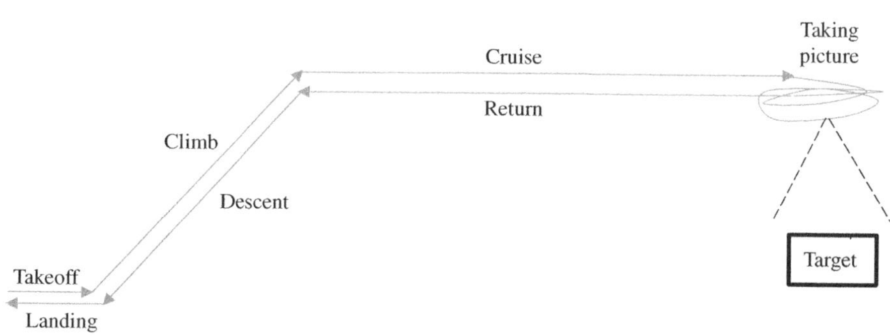

Figure 7.21 A typical unmanned aerial vehicle (UAV) trajectory.

The trajectory planning algorithm should be able to quickly generate new reference profiles for flight cases, if the flight environment experiences varying winds. Once a reference flight profile is determined, the guidance algorithm generates the closed-loop commands for tracking the reference trajectory. At this point, it is appropriate to evaluate the features of the proposed trajectory geometry before finalizing how the flight plan is generated. After the reference trajectory is determined, the guidance commands need to be generated.

For instance, for an unpowered approach and landing, a cubic polynomial altitude profile [111] may define the glideslope:

$$h_{ref} = a_o + a_1 x + a_2 x^2 + a_3 x^3 \tag{7.55}$$

where x is the ground track distance, that is, $x = 0$ at the initiation of approach. For this profile, the derivative of altitude with respect to ground track is the tangent of the glide angle:

$$\frac{dh_{ref}}{dx} = \tan\left(\gamma_{ref}\right) = a_1 + 2a_2 x + 3a_3 x^2 \tag{7.56}$$

Cubic polynomial coefficients a_o, a_1, a_2, and a_3 are computed from Eqs. (7.55) and (7.56) using initial conditions and ground track distance.

7.11 Guidance Sensor – Seeker

An important element in a UAV is the sensor. Some sensors are part of the payloads; while others are necessary for control/navigation/guidance systems. For instance, the altimeter radar is a sensor and is employed for navigation purpose; while a camera is a sensor which is configured as a payload. Ina addition, some radars are utilized as communication device.

Some radars are employed for visibility of ground activity, while some are utilized for seeking the target. Large UAVs usually carry optical sensors (e.g., TV cameras and forward looking infrared) which are less susceptible to detection than active devices such as radars. Under favorable conditions, optical sensors can supply high-quality images of the ground targets. However, optical sensors suffer from a limited field of view and from severely reduced performance in adverse weather and battlefield smoke and dust conditions.

A seeker is a significant part of the guidance system where the UAV is required to detect and follow a moving target. The seeker, by tracking the target, establishes the direction of the LOS, and the output of the seeker is the angular velocity of the LOS with respect to inertial space as measured by rate gyros mounted on the seeker. The LOS rate is obtained by measuring the rate of rotation of the UAV seeker tracking the target. The seeker, by tracking the target, establishes the direction of the LOS. The output of the seeker is the angular velocity of the LOS – measured by rate gyros mounted on the seeker.

Basically, the main function of the UAV seeker (in the missile terminology, it is known as *homing eye*) subsystem is to:

1) Provide the continuous measurements of target position and velocity.
2) Track the target with the antenna.
3) Measure the line-of-sight angular rate.

The antenna of the seeker is of various types; the most popular ones are: (a) radar; (b) infrared (IR); (c) laser; and (d) EO. The UAV radar system can detect and track any moving targets/vehicles such as cars, tanks, trucks, and low-flying helicopters. There are three kinds of seeker radar: (a) active (which has both transmitter and receiver); (b) passive (which only has a receiver); and (c) semi-active (which is using the signal from ground station). The way in which closing velocity and LOS rate data is obtained to mechanize guidance laws is a function of the type of seeker that is used; and how it is mounted to the UAV body.

An active seeker radar detects a target by sensing electromagnetic energy reflected from the target's surface. The source of the reflected radiation is a radar transmitter. The transmitter must beam electromagnetic radiation at the target, this radiation must travel to the target, reflect, travel back to the receiving antenna of the UAV. It is then amplified, demodulated, and analyzed to determine the direction of the target. This information enables the guidance computer to steer the UAV toward the target.

An effective seeker must have the ability to discriminate between the target's return and reflections from its background, i.e., the surface of the ground or sea. It should also be capable of resisting jamming or deception and be able to penetrate through adverse weather conditions. Figure 7.22 depicts the configuration of radar, antenna, and guidance system in a UAV. Figure 7.23 illustrates an MQ-4C UAV configuration with two seekers. Radar theory is a subject requiring a knowledge of electromagnetism and wave theory. Modern control theories used in radar tracking are discussed in Chapter 6.

Another sensor used for measuring the distances of nearby objects (in a sense and avoid system) is the laser range finder. Based on known calibration parameters and on the attitude estimated by the IMU, the endpoints of the laser in the global frame is projected. Then, from the projected laser beams, the x-y position and the heading of the vehicle in a 2D map are estimated. This sensor is briefly introduced in Chapter 12. The topic of guidance sensor is beyond the scope of this text, the interested reader is referred to references such as [93, 112, 113].

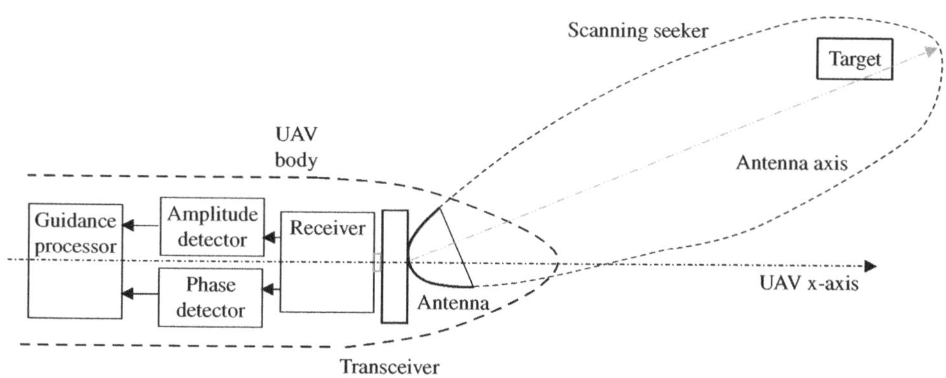

Figure 7.22 Configuration of radar, antenna, and guidance system.

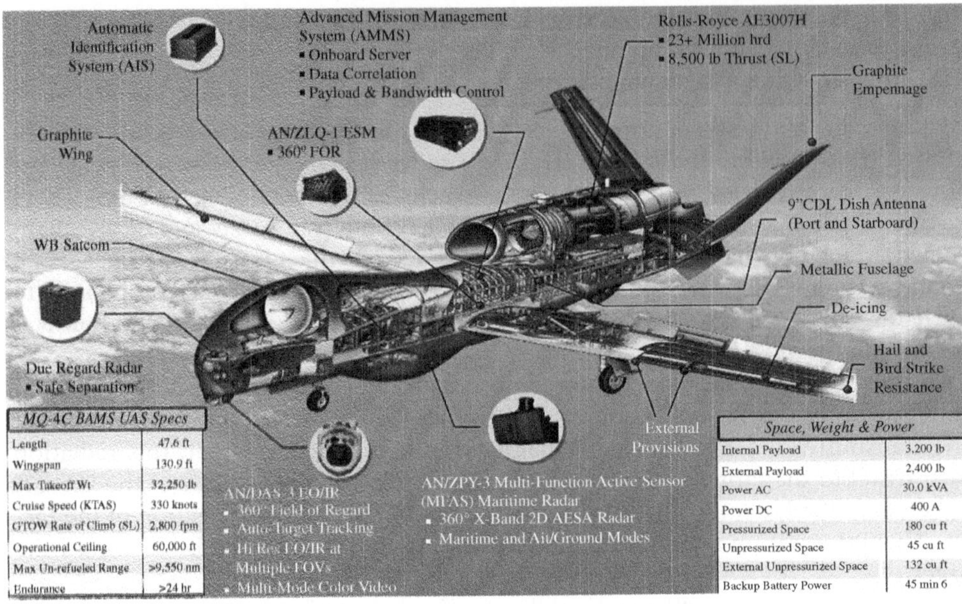

Figure 7.23 MQ-4C unmanned aerial vehicle (UAV) Global Hawk configuration (with two seekers).

The MQ-1C Gray Eagle STARLite Radar active electronically scanned radar is mounted on a rotating mechanical gimbal with a 360° field of regard, although the antenna itself has a 110° field of view.

7.12 Guidance System Design

In the preceding sections, various guidance systems and laws have been described. In designing a guidance system for a UAV, one must select the type of guidance system, guidance law, guidance devices, and sensors, and then conduct calculations and analysis. The design of the guidance system is conducted in parallel to the design of navigation and control systems.

In general, the primary criteria for the design of guidance system are as follows: (a) manufacturing technology; (b) required accuracy; (c) range; (d) weather; (e) reliability; (f) life-cycle cost; (g) field of view; (h) stealth requirements; (i) maintainability; (j) endurance; (k) communication system; (l) aerodynamic considerations; (m) processor; (n) complexity of trajectory; (o) compatibility with navigation system; (p) weight (q) target size; (r) target velocity; (s) compatibility with control system; and (t) target physical features. These requirements must be defined by some numbers and technical features. For instance, the required values for range (e.g., 1000 m) and field of view (e.g., 60°) must be known. The requirements must also be prioritized.

Figure 7.24 shows a flowchart that represents the guidance system design process as a subsystem of the autopilot. In general, the design process begins with a trade-off study to establish a clear line between cost of guidance equipment and performance (i.e., accuracy) requirements, and ends with optimization.

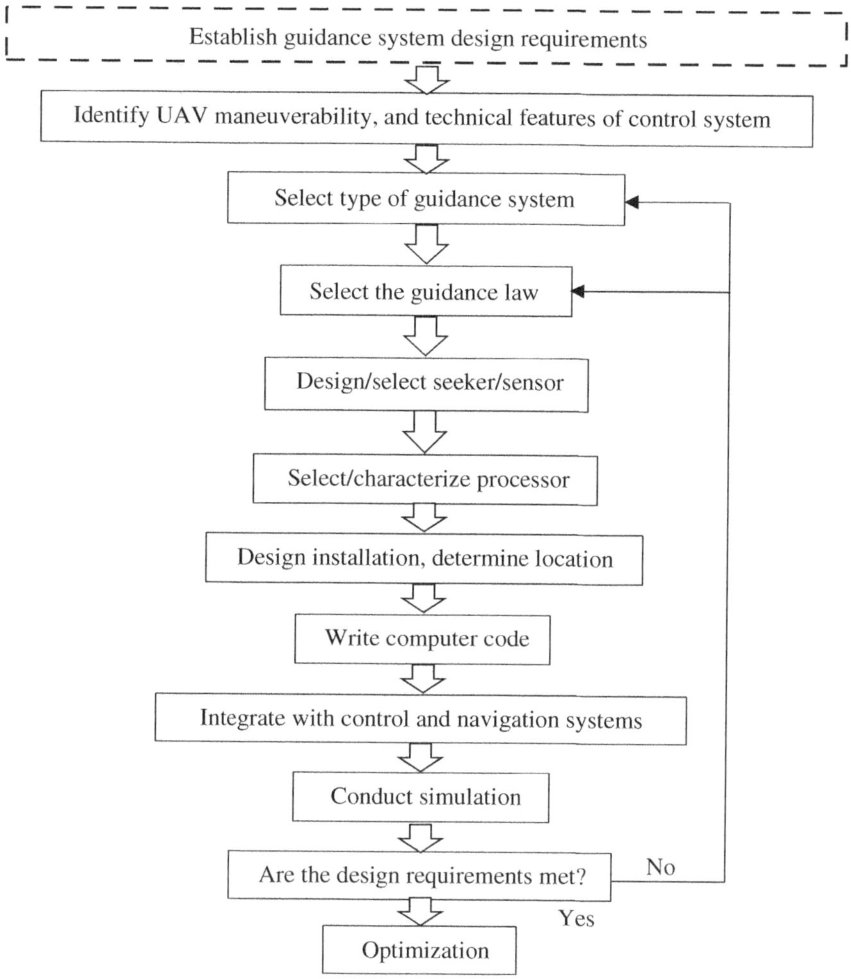

Figure 7.24 Guidance system design process.

The guidance system, guidance law and the guidance equipment (e.g., seeker) are selected. However, they all must be compatible, otherwise the integration and installation will be highly difficult. An effective guidance system should satisfy the design requirements with a minimum cost. Moreover, the features of guidance system should be such that they can be effectively integrated with the control system and navigation system.

The design of guidance system should be conducted in sync with the design of control and navigation system. For instance, in an approaching flight, the output of navigation (position in x and y) are used as references for guidance system. In addition, the altitude h is used as a feedback value for the control system. Hence, all three systems must perform efficiently while exchanging command/signal to each other.

Questions

1 Define guidance system.

2 Define guidance law.

3 Describe PN guidance law.

4 Describe pursuit guidance law.

5 Describe line-of-sight guidance law.

6 Name the main elements of a guidance system.

7 What is the main function of a seeker?

8 What are the three classical guidance laws?

9 What is the most widely used guidance law?

10 What requirements will influence the design of guidance system?

11 Define LOS by using a figure.

12 What are the functions of a seeker?

13 List four types of seekers.

14 Briefly describe waypoint guidance.

15 What is no corrections region?

16 What is cross-track error?

17 What is the function of a trajectory smoother?

18 What is the function of a trajectory tracker?

19 Describe the function of a transponder.

20 List a few typical onboard obstacle detection sensors.

21 What are the input and outputs of a transponder?

22 Describe the process and application of a correction to the UAV heading angle?

23 Draw a block diagram to indicate the relation between control, guidance, and navigation systems in an autopilot.

24 Draw a block diagram to indicate major subsystems of a UAV guidance system with a moving target.

25 Draw a block diagram to indicate three major blocks of a guidance system.

26 Draw block diagram of the command (LOS) guidance system.

27 Describe hazard detection systems classification.

28 What are the outputs of a Doppler radar as an obstacle detection sensor?

29 What are the outputs of a lidar as an obstacle detection sensor?

30 What are the outputs of a Doppler radar as an obstacle detection sensor?

31 In selecting a surveillance system, what factors should be evaluated?

32 Briefly describe the vision-based obstacle detection system.

33 Briefly describe how a collision avoidance system works.

34 Briefly describe how a collision detection system works.

35 What are typical maneuvers to avoid a collision?

36 Draw figure to demonstrate a collision geometry between to aerial vehicles.

37 What is the collision condition between two aerial vehicles? Under what conditions will a collision happen?

38 What is the formation flight?

39 What was the record-breaking swarm formation flight in 2018 by Intel?

40 Draw a block diagram to indicate control, guidance, and navigation systems in the follower UAV autopilot.

41 What does motion planning refer to?

42 Describe the trajectory planning algorithm.

43 List the most popular antennas of a seeker.

44 What are three kinds of seeker radar?

45 How does an active radar function?

46 Briefly describe the features Radar of a MQ-1C Gray Eagle.

47 List the primary criteria for the design of guidance system.

48 What constraints define the UAV maneuverability, and indicate the maximum ability of a UAV to turn?

49 What are five functions required in a sense and avoid system?

50 The vision-based sense and avoid approaches/techniques can be categorized into four groups. What are they?

Problems

1 An aircraft is flying toward northeast with a speed of 120 knot, while a UAV is flying at the same altitude toward northwest with a speed of 25 knot. The heading angles of the aircraft and UAV with respect with East are 12 and 135°. The initial distance between the two vehicles is 8000 m.
   - a Will there be a collision between the UAV and the aircraft?
   - b If not, determine the miss distance.

2 An aircraft is flying toward northwest with a speed of 200 knot, while a UAV is flying at the same altitude toward northeast with a speed of 32 knot. The heading angles of the aircraft and UAV with respect with East are 12 and 120°. The initial distance between the two vehicles is 7000 m.
   - a Will there be a collision between UAV and the aircraft?
   - b If not, determine the miss distance.

3 An aircraft is flying toward northeast with a speed of 150 knot, while a UAV is flying at the same altitude toward northwest. The heading angles of the aircraft and UAV with respect with East are 10 and 120°. The UAV has collided with the aircraft.
   - a Determine the initial distance between UAV and the aircraft, and the UAV speed.
   - b How long each aircraft and UAV have traveled?
   - c Determine the time of flight for both vehicles.

4 An aircraft is flying toward northwest with a speed of 145 knot, while a UAV is flying at the same altitude toward northeast. The heading angles of the aircraft and UAV with respect with East are 12 and 150°. The UAV has collided with the aircraft.
   - a Determine the initial distance between UAV and the aircraft, and the UAV speed.
   - b How long each aircraft and UAV have traveled?
   - c Determine the time of flight for both vehicles.

5 A UAV is flying toward east at a speed of 30 m/s, while a target aircraft is flying northward at a speed of 20 m/s. They are at a distance with the geometry shown below (Figure 7.25):

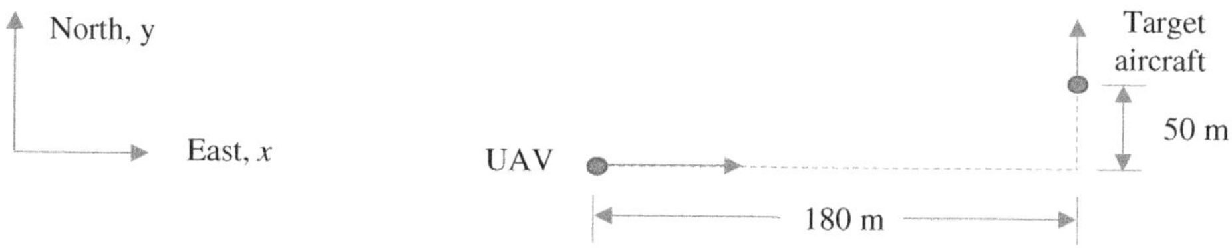

Figure 7.25 Geometry of the unmanned aerial vehicle (UAV) and target aircraft for problem 5.

The UAV guidance system is governed by the command guidance law. It is required to move into and fly along the line of sight. What commanded acceleration will be generated by the guidance system at this moment?

6 A UAV is flying toward west at a speed of 40 m/s, while a target aircraft is flying northward at a speed of 30 m/s. They are at a distance with the geometry shown below (Figure 7.26):

The UAV guidance system is governed by the command guidance law. It is required to move into and fly along the line of sight. What commanded acceleration will be generated by the guidance system at this moment?

7 The guidance system of the UAV in problem 5 is governed by the PN guidance law with an N' = 2.4. It is required to move into and fly along the LOS with the other aircraft. What commanded acceleration will be generated by the guidance system at this moment?

8 The guidance system of the UAV in problem 6 is governed by the PN guidance law with an N' = 3.1. It is required to move into and fly along the LOS with the other aircraft. What commanded acceleration will be generated by the guidance system at this moment?

9 Consider the team of UAV and target aircraft in problem 5. Determine the turn radius and turn rate of the UAV, when the guidance command is executed.

10 Consider the team of UAV and target aircraft in problem 6. Determine the turn radius and turn rate of the UAV, when the guidance command is executed.

11 Consider the team of UAV and target aircraft in problem 7. Determine the turn radius and turn rate of the UAV, when the guidance command is executed.

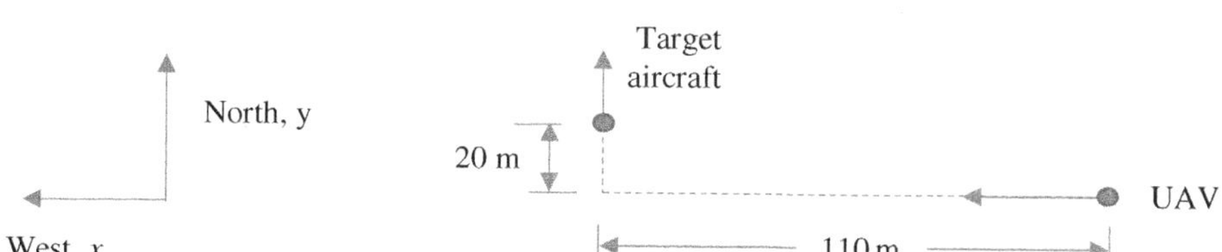

Figure 7.26 Geometry of the unmanned aerial vehicle (UAV) and target aircraft for problem 6.

12 Consider the team of UAV and target aircraft in problem 8. Determine the turn radius and turn rate of the UAV, when the guidance command is executed.

13 A UAV with a maximum takeoff mass of 1000 kg, a wing area of 15 m^2, has a maximum lift coefficient of 1.6. The UAV structure is designed to safely carry a maximum load factor of 1.8. Determine the UAV corner speed and maximum allowable bank angle.

14 A UAV with a maximum takeoff mass of 3 kg, a wing area of 1.6 m^2, has a maximum lift coefficient of 1.4. The UAV structure is designed to safely carry a maximum load factor of 1.4. Determine the UAV corner speed and maximum allowable bank angle.

15 The guidance system is commanded a UAV to turn at a radius of 50 m around a ground moving target. The maximum allowable bank angle is 45 degrees.
a determine the desired airspeed of the UAV during the turning flight.
b determine the turn rate in deg/sec.
c how long does it take to turn for a full circle?

16 The guidance system is commanded a UAV to turn at a radius of 100 m around a ground moving target. The maximum allowable bank angle is 60 degrees.
a determine the desired airspeed of the UAV during the turning flight.
b determine the turn rate in deg/sec.
c how long does it take to turn for a full circle?

17 For the purpose of an aerial filming, a UAV is following a moving ground target. At a particular moment, the lead angle is 30 deg, the closing velocity is 25 m/s. The rate of change of the LOS angle is required to be 15 deg/second. The UAV guidance system is using the PN guidance law with the N of 2.2. What acceleration will be commanded by the guidance system?

18 For the purpose of an aerial filming, a UAV is following a moving ground target. At a particular moment, the lead angle is 40 deg, the closing velocity is 30 m/s. The rate of change of the LOS angle is required to be 8 deg/second. The UAV guidance system is using the PN guidance law with the N of 1.7. What acceleration will be commanded by the guidance system?

19 In a teaming flight, a follower UAV (as introduced for problems 1-14 in Chapter 6) is missioned to follow a manned aircraft (leader). The leader aircraft is maneuvering only in the xz plane (i.e., cruising, climbing, cruising again, and then descending) with an airspeed of 140 knot. The UAV is initially flying 20 knot slower than the leader aircraft, and is at 200 m behind it. Develop a matlab code or a Simulink model as a simulation model for the guidance subsystem of the UAV. The objective for the UAV is to implement a guidance law such that the follower UAV to be constantly at a desirable distance (100 m) behind the leader UAV, and flying with the same speed as the leader aircraft.

20 In a teaming flight, a follower UAV (as introduced for problems 1-14 in Chapter 6) is missioned to follow a manned aircraft (leader). The leader aircraft is maneuvering only in the xy plane (i.e., cruising, level turning (180 degrees), and cruising back) with an airspeed of 140 knot. The UAV is initially flying 20 knot slower than the leader aircraft, and is at 200 m behind it. Develop a matlab code or a Simulink model as a simulation model for the guidance subsystem of the UAV. The objective for the UAV is to implement a guidance law such that the follower UAV to be constantly at a desirable distance (100 m) behind the leader UAV, and flying with the same speed as the leader aircraft.

Note

1 https://www.youtube.com/watch?v=fCd6P7Ya160

8

Navigation System Design

CONTENTS

Design of Unmanned Aerial Systems, First Edition. Mohammad H. Sadraey.

Companion website: www.wiley.com/go/sadraey/unmanned-aerial-systems

Educational Outcomes

After study of this chapter, the reader will be able to:

1) Manage a Unmanned Aerial Vehicle UAV navigation system design project
2) Conduct UAV inertial navigation system design
3) Discuss UAV navigation classifications
4) Provide pros and cons of various UAV navigation systems
5) Discuss current navigation techniques and technologies
6) Explain elements of a navigation system
7) Develop various systems for a navigation mission
8) Select appropriate navigation sensors for given mission requirements
9) Evaluate navigation sensors
10) Discuss navigation disturbances and their solutions
11) Implement navigation calculations
12) Explain Kalman filtering technique

8.1 Introduction

Three major sub-systems within an autopilot are: (a) control system; (b) guidance system; and (c) navigation system. The design of control system was presented in Chapter 6. The design of navigation system is provided in this chapter.

In general, navigation is the skill that involves the determination of position, orientation, and velocity of a moving object. More specifically, navigation is a field of study that focuses on the process of monitoring the movement of a vehicle from one place to another (Figure 8.1). A UAV navigation system is one that determines the position of the air vehicle with respect to some reference frame (i.e., situational awareness), for example, the Earth sea level. The temporary loss or lack of situational awareness is a causal factor in many flight collisions/accidents.

As a side note, in some references, you may see the term "Waypoint Navigation." Please note that, waypoint navigation has nothing to do with navigation, it is just a guidance law used to guide and track air vehicles. This misnomer roots back to non-aeronautical engineers who utilize this term to mean waypoint guidance (i.e., wayfinding). The waypoint navigation guidance is presented in Section 7.7.

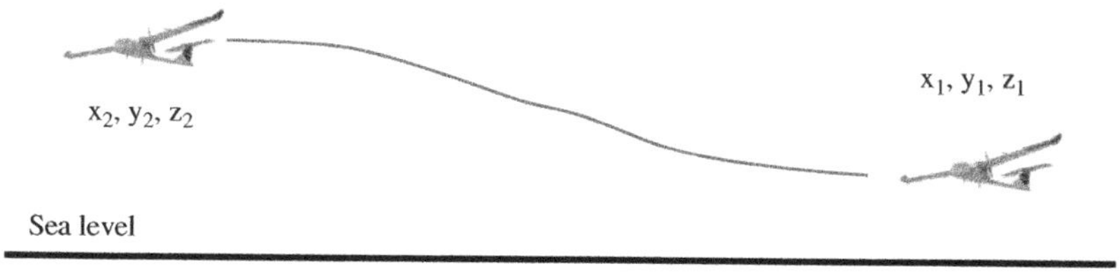

Figure 8.1 Coordinates of an unmanned aerial vehicle (UAV) are determined by navigation system at any time.

This chapter presents the basic fundamentals, classifications, and features of navigation systems. In addition, the two most popular navigation systems: (a) inertial navigation system; and (b) GPS are described with more detail. Some related topics such as coordinate systems, world geodetic system (WGS), avionics, gyroscope, and filtering reviewed in brief. Finally, design considerations for the navigation system is presented.

8.2 Classifications

Navigation systems may be entirely onboard a UAV, or they may be located elsewhere and communicate via radio or other signals with the UAV, or they may use a combination of these methods. Navigation is usually done using sensors such as gyros, accelerometers, altimeter, and the Global Positioning System (GPS).

Position has mainly three coordinates: (a) latitude; (b) longitude; and (c) altitude above sea level. The latitude of a place on Earth is its angular distance north or south of the equator. Latitude is usually expressed in degrees ranging from 0° at the equator to 90° at the North and South poles. Similarly, the longitude of a place on Earth is the angular distance east or west of the Greenwich meridian. Longitude is ranging from 0° at the Greenwich meridian to 180° east and west. Direction involves two main data: (a) Heading; and (b) pitch.

Calculation in a navigation system is performed based on a navigation law. Three major classes of navigation laws/systems are: (a) electronic navigation law/system; (b) inertial navigation law/system; and (c) celestial navigation law/system. Electronic navigation system may be performed by three electronic devices: (a) radio system; (b) radar; and (c) satellite. Hence, the GPS is a type of electronic navigation system.

The inertial navigation is classified as a type of dead reckoning (i.e., integration), which consists of extrapolation of a known position to some future time. Celestial navigation system is based on observation of the positions of the Sun, Moon, and other stars/plants. There is still a considerable amount of research [114, 115] on navigation signal conditioning, filtering, and proceeding in order to improve position setting and increase fault toleration.

A standard UAV navigation system often relies on GPS and inertial sensors (i.e., INS). If the GPS signal for some reason becomes unavailable or corrupted, the state estimation solution provided by the INS alone drifts in time and will be unusable after a few seconds (especially for small-size UAVs which use low-cost INS). The GPS signal also becomes unreliable when operating close to obstacles due to multi-path reflections. In addition, it is quite vulnerable to jamming (especially for a GPS operating on civilian

Table 8.1 Classification of navigation systems developed for Unmanned aerial vehicles (UAVs).

No	Level 1	Level 2	Level 3
1	**Sensing**	—	—
2	**State estimation**	**2.1. Conventional IMU/GPS systems**	
		2.2. Vision-based state estimation	a. On-Ground vision
			b. Visual Odometry
			c. Target Relative Navigation
			d. Terrain Relative Navigation
			e. Concurrent Estimation of Motion and Structure
			f. Bio-Inspired Optic Flow Navigation
		2.3. State estimation using range sensors	a. State estimation by ultrasonic and IR sensors
			b. Lidar-based SLAM
3	**Perception**	**3.1. Vision-based perception**	a. Target detection and tracking
			b. Mapless-based approaches
			c. Mapping-based methods *(SLAM, SMAP, SLAD)*
			d. Mission-oriented perception
		3.2. Lidar-based perception	a. Mapless-based obstacle detection
			b. Simultaneous Localization and Mapping *(SLAM)*
			c. Simultaneous Mapping and Planning *(SMAP)*
			d. Safe Landing Area Detection *(SLAD)*
4	**Situation awareness**	—	—

frequencies). By choosing the right of coordinates system, the navigation equations of motion (Eqs. 8.1 through 8.3) are integrated.

Table 8.1 provides the classification of existing navigation systems [79] developed for UAVs and three levels of navigation functions. The first level of classification (categories) is based on the autonomy level they provide, the second level (groups) is based on the sensing technology used, and the third level (classes) is based on the algorithmic approach or method used. The level of complexity in the design of the navigation system is directly a function of category/group/class of the desired navigation function. Hybrid approach is a solution to reduce the challenges of navigation in measuring orientation and translation (i.e., state estimation) in space. Vision sensors are suboptimal at determining orientation, but quite good at measuring location. On the other hand, accelerometers and conventional IMUs yield an increased error due to integration. Pairing IMU and vision sensors or GPS will provide a more precise measurement for a full six degree of freedom motion in space.

8.3 Coordinate Systems

Navigation of large UAVs over large areas of the Earth's surface, using nonlinear equations of motion to calculate the precise position, requires an accurate model of the Earth's shape, rotation, and gravity. A reliable coordinate system is an essential tool in a successful navigation system.

Prior to the description of the navigation laws and systems, we need to provide the definitions of a few related terms. The motion of a UAV can be described by means of vectors in three dimensions. Coordinate System is a measurement system for locating points in space, set up within a frame of reference. The frame of reference: a rigid body or set of rigidly related points that can be used to establish distances and directions. Inertial frame is defined as a frame of reference in which Newton's laws apply.

8.3.1 Fixed and Moving Frames

In general, there are two groups of frames and coordinate systems (or frames): (a) inertial frame (Earth-fixed) has its origin at the Earth's center of mass (nonrotating but translating with Earth); and (b) body frame. Three axes of the inertial frame are north, east, and down (NED). The body frame is fixed at the UAV, and is moving and rotating with the UAV. Three coordinate systems are classified as body frame: (a) body-fixed system; (b) wind-axes system; and (c) stability-axes system. The orthogonal axes are named x, y, and z. Figure 3.7 (Chapter 3) shows the axes and positive convention for body frame, and the body-fixed coordinate system. The inertial navigation system employs the earth-fixed coordinate system, so the position is always defined with respect a fixed point in Earth (e.g., ground system). In the short-range UAVs, we may loosely assume the Earth is flat (and neglect the effects of Earth's rotational motion) for easy calculations.

Figure 8.2 illustrates a NASA Altair UAV which employs both GPS and INS as the navigation system for improved accuracy. Altair has a 26 m long wingspan (6.1 m greater than current MQ-9s). The Altair has enhanced avionics to better enable flights in FAA-controlled airspace and demonstrate "over-the-horizon" command and control (C2) capability from the ground control station.

A local navigational system uses a geographic coordinate system which has its axes aligned east, north, and up (ENU), or NED, where "up" or "down" means along the spheroid normal at the system location.

Figure 8.2 NASA Altair UAV.

The Earth-centered, Earth-fixed (ECEF) Cartesian coordinate system has its origin at the center of the Earth, and rotates with the Earth. The X and Y axes lie in the plane of the Equator. The Z axis is perpendicular to X-Y plane, with positive values increasing northward. The X axis is passing through the prime meridian (0° longitude), and at the equator, the latitude is zero.

Latitude specifies the north–south position of a point on the Earth's surface, and is an angle which ranges from 0° at the equator to 90° at the poles. It is positive in the Northern Hemisphere. Longitude specifies the east–west position of a point on the Earth's surface, and is an angle which ranges from 0° at the equator to 180° at the prime meridian. Longitude is measured from the x-axis (in the equatorial plane) which points to the zero-longitude meridian. A prime meridian is arbitrary; the most widely used one is the Greenwich meridian in England. Geodetic height is the height above the spheroid, along the normal, while the elevation is above mean sea level (MSL). The elevation is usually obtained from a barometric altimeter, or from the land elevation plus the altitude above land (e.g., altimeter).

Two important parameters of the spheroid (radii of curvature) are required for navigation calculations: (a) meridian radius; and (b) prime vertical radius. These parameters will be presented in section 8.4.3.

8.3.2 World Geodetic System

Some knowledge of the Earth's gravitation and geometry are required to navigate and to simulate the motion of an UAV around the Earth. Two available and well-established resources are: (a) Geographic Information System (GIS); and (b) World Geodetic System (WGS). It this section, both systems, particularly the WGS is described.

The GIS is a framework for gathering, storing, managing, analyzing, and displaying data related to positions on Earth's surface. The GIS is rooted in the science of geography, many different types of information can be compared and contrasted using GIS. GIS systems can be used to produce three-dimensional images of natural structures such as mountains. It analyzes spatial location and organizes layers of information into visualizations using maps and 3D scenes. The location can be expressed in many different ways, such as latitude and longitude. With this unique capability, GIS reveals deeper insights into data, such as patterns, relationships, and situations. However, the GIS deals with flat map projections and does not accommodate the Earth's curvature, but for a UAV application, we need to accommodate both the curved and flat views of the world. Thus, GIS is not a suitable platform for UAV's navigation system and the flight simulation.

The WGS was originally developed in 1984 (thus referred to as WGS 84) by National Geospatial-Intelligence Agency. The GPS also uses the WGS as its reference coordinate system for the ephemerides of its satellites.

The WGS is rooted in the science of Geodesy which is a branch of mathematics that deals with the shape and area of the Earth. Unlike GIS, the WSG is incorporating such Earth's features as gravitation and curvature. Thus, using WGS, we can define UAV accurate positions and flight parameters on Earth.

The WGS 84 is an ECEF reference coordinate system and geodetic datum. WGS 84 is the standard U.S. system, and is based on a consistent set of constants and model parameters that describes the Earth's geometry, gravity, and geomagnetic fields. It is comprised of a reference ellipsoid, an orthogonal coordinate system, altitude data and a geoid.

WGS 84 geodetic coordinates are generated by using its reference ellipsoid. In 3-Dimensional WGS, the shape of the Earth is mathematically modeled by a spheroid

Table 8.2 WGS 84 defining parameters.

No	Parameter	Notation	Value
1	Semi-major axis (equatorial radius)	a	6378.137 km
2	Semi-minor axis (polar radius)	b	6356.752 km
3	Flattening factor	f	1/298.257 223 563
4	Mean angular velocity of the Earth	ω	7.292115×10^{-5} rad/s
5	Geocentric gravitational constant	GM	$3.986004418 \times 10^{14} m^3/s^2$
6	Eccentricity	e	0.081 819 19

(i.e., an ellipsoid of revolution. Since, the polar radius of the Earth is approximately 21 km less than the equatorial radius, the generating ellipse must be rotated about its minor axis, to generate an oblate spheroidal model.

The equipotential surface of the Earth's gravity field that coincides with the undisturbed MSL, extended continuously underneath the earth surface, is referred to as the *geoid*. Based on a 1° by 1° (latitude, longitude) worldwide grid, the root mean squares deviation of the geoid from the spheroid is about 30 m. To express the difference between the real earth and modeled earth, three types of heights are employed: (a) elevation (altitude from ground); (b) geoid height (altitude from sea level); and (c) geodetic height (the difference between 1 and 2).

The ellipse mathematical model [116] utilizes four defining parameters (Table 8.2) and constants such as the semi-major and semi-minor axes of the generating ellipse, flattening, and eccentricity. Two additional parameters to define the complete WGS 84; are a fixed Earth rotation rate, and the Earth's gravitational constant. In converting the distance traveled from British to SI systems, 1 nautical mile (nm) is equal to 1852 m. BTW, knot is a unit of speed and is equal to 1 nautical mile per hour.

8.4 Inertial Navigation System

8.4.1 Fundamentals

Inertial navigation is the oldest navigation technique which is originally developed by naval navigators. This technique is still applicable for the cases where the GPS signals are not available; or in GPS challenging Environments. This system may also be utilized as the parallel (redundant) system to GPS for increasing reliability. Inertial navigation is a type of dead reckoning that calculates the position of a moving object (e.g., UAV) based on onboard motion (i.e., inertial) sensors. Dead reckoning consists of extrapolation of a known position to a future period. It involves measurement of time, direction of motion, airspeed, and ground speed relative to inertial space (i.e., body coordinates).

The calculation is performed by taking the last known position and time, considering average speed and heading since then and the present time. The distance traveled is equal to the velocity times the duration of motion. So, the speed is multiplied by the time elapsed since the last position to get distance traveled. This can be added to the initial position to get the present position. The longitude and latitude are determined by using the North and East components of the ground speed. Once the initial latitude and longitude are established, the system receives signals from sensors that measure the acceleration along three axes. Hence, the current latitude and longitude are continually calculated.

Inertial navigation was used in a wide range of applications including the navigation of aircraft, missiles, spacecraft, submarines, and ships. Inertial navigation systems were in wide use until GPS became available. The inertial navigation system is still in common use on submarines, since GPS reception or other fix sources are not possible while submerged. Its main advantage over other navigation systems is that, once the starting position is set, it is independent, and does not require outside information. In addition, it is not affected by adverse weather conditions and it cannot be detected or jammed. Its main disadvantage is that since the current position is calculated solely from previous positions, its errors or deviations are increasing at a rate proportional to the time elapsed. Inertial navigation systems must therefore be frequently corrected with other devices/techniques.

Inertial navigation is a self-contained navigation technique in which measurements are used to track the position and orientation of an object relative to a known starting point, orientation, and velocity. Two main measurement devices for inertial navigation are: (a) accelerometers; and (b) gyroscopes. By processing signals from these devices, it is possible to calculate the position and orientation of a device. Recent advances in the construction of micro-machined electro-mechanical system devices have made it possible to manufacture small and light inertial navigation systems. For instance, *inertial measurement unit* (IMU) contains three orthogonal rate-gyroscopes and three orthogonal accelerometers, measuring angular velocity and linear acceleration respectively. To perform these functions, the system requires the following instruments: (a) three speed measuring devices (accelerometers, after integration); (b) three angular sensors (gyroscopes); (c) a clock; and (d) a processor.

IMU fundamentally fall into two categories: (a) stable platform systems; and (b) strap-down systems. In stable platform systems, the INS are mounted on a platform which is isolated from any external rotational motion, and therefore output quantities measured in the global frame. In strap-down systems the INS are mounted rigidly onto the UAV, and therefore output quantities measured in the body frame. Stable platform and strap-down systems are both based on the same underlying principles. Strap-down systems have reduced mechanical complexity and tend to be physically smaller than stable platform systems. These benefits are achieved at the cost of increased computational complexity. As the cost of computation has decreased strap-down systems have become the dominant type of IMU. The acceleration signals in both systems are similar, so when integrated, the same velocity.

To keep track of velocities, the signals from the accelerometers are integrated. The orientations are determined by the integration of the rate gyro signals. The platform mounted gyroscopes detect any platform rotations. These signals are fed back to torque motors which rotate the gimbals in order to cancel out such rotations, hence keeping the platform aligned with the initial frame. In order to improve overall noise and drift performance, gyros and accelerometers have rather high gains. The mechanism and governing laws of gyroscopes and accelerometers are presented in Section 9. For more information on technical details of inertial navigation, study of Ref. [117] is recommended.

8.4.2 Navigation Equations

A UAV may roll, pitch, and yaw while moving forward. However, the velocity of an UAV has three components of U, V, and W. Moreover, there are three attitude angles in the flight (i.e., φ, θ, Ψ). The navigation equations [54] which will be integrated to calculate the UAV position are:

$$\dot{x} = U\cos\theta\cos\Psi + V\left(-\cos\phi\sin\Psi + \sin\phi\sin\theta\cos\Psi\right) + W\left(\sin\phi\sin\Psi + \cos\phi\sin\theta\cos\Psi\right) \tag{8.1}$$

$$\dot{y} = U\cos\theta\sin\Psi + V\left(\cos\phi\cos\Psi + \sin\phi\sin\theta\sin\Psi\right) + W\left(-\sin\phi\cos\Psi + \cos\phi\sin\theta\sin\Psi\right) \tag{8.2}$$

$$\dot{h} = U\sin\theta - V\sin\phi\cos\theta - \cos\phi\cos\theta \tag{8.3}$$

where U, V, and W stand for velocity in x, y, and z direction; and θ, Ψ, and φ denote pitch, heading, and bank angles respectively. The integration delivers x, y, and h which are the UAV position in the global frame. The orientation parameters θ, Ψ, and φ are directly measured by gyros. In addition, the velocity parameters U, V, and W are calculated by integration of the signals measured by accelerometers. Figure 8.3 illustrates the navigation process block diagram, and the functions of its two example measurement devices.

The Navigation computer is solving Eqs. (8.1) through (8.3) to determine the UAV coordinates and position. The characteristics of accelerometers and gyroscopes are provided in Section 8.7. Eqs. (8.1) through (8.3) are nonlinear differential ones which model the navigation of a real (3d) air vehicle and delivers accurate results, as long as the measured data are reliable. Three velocity components of U, V, and W are separately calculated by solving three nonlinear differential equations of (5.3–5.5).

The solution of these equations often employs numerical techniques such as the Runge Kutta method (See Section 9 of chapter 5). In the next section, simple navigation equations for the case where the UAV is assumed as a point mass are presented.

8.4.3 Navigation Basic Calculations

The process of calculating a vehicle's current position by using a previously determined position is often referred to as the dead reckoning. This process requires known or estimated speed, and elapsed time.

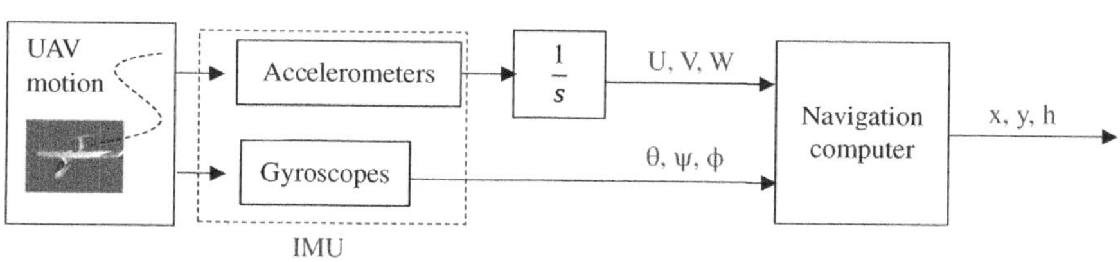

Figure 8.3 Navigation process.

In previous section, the UAV was considered as a rotating real body with three angles. In this section, a UAV is assumed as a point mass (without any rotation). This assumption causes errors in calculation, but it can be used as a side calculation to check the accuracy of real calculations.

In principle, the calculations of the UAV position are based on the ground speed (V), and the elapsed time (t). The velocity is defined as the time derivatives of the displacement. If the UAV is assumed as a *point mass*, and if the ground speed is kept constant, the new location (x) – in a linear motion – is simply determined by:

$$x = x_o + Vt \tag{8.4}$$

However, for an accelerated linear motion, the new location (x) is obtained by:

$$x = x_o + V_o t + \frac{1}{2}at^2 \tag{8.5}$$

where t represents time, V denotes ground speed, V_o denote initial ground speed, and a denotes linear acceleration. By solving these equations, these parameters are determined: (a) vehicle's heading; and (b) vehicle's ground speed, and 3. Time spent on each heading and at each speed. With this information, the operator would calculate the route and distance the vehicle had covered. The time is provided by a clock.

Example 8.1
A UAV is cruising eastward at sea level altitude with an airspeed of 120 knot. The UAV is located 10 km East a flight base, and it is assumed there is no wind. Determine the distance of the UAV traveled from this base after 20 minutes.

Solution

$$x = x_o + Vt = 10{,}000 + (120 \times 0.5144) \times (20 \times 60) = 84{,}080\, m = 84.08\, km \tag{8.4}$$

8.4.4 Geodetic Coordinates Calculations

The WGS as a fixed coordinate system for navigation calculation of an air vehicle on the Earth was introduced in Section 8.3.2. In using geodetic coordinates, two important parameters of the spheroid (radii of curvature) are required for navigation calculations: (a) meridian radius; and (b) prime vertical radius. They are primarily employed in the calculations of components of UAV velocity.

The meridian radius (R_M), is the radius of curvature in a meridian plane. From calculus texts, one can easily find the following equation to determine the radius of curvature for an ellipse. In terms of geodetic latitude, the meridian radius of curvature is given by:

$$R_M = \frac{a(1 - e^2)}{(1 - e^2 \sin^2 \phi)^{3/2}} \tag{8.6}$$

where "e" denotes eccentricity, a semi-major (equatorial radius), and ϕ denotes the geodetic latitude. This equation, integrated with respect to angle, gives the arc length. Both variables "a" and "e" are provided in Table 8.1.

The prime vertical radius (R_N), is the radius of curvature in a plane perpendicular to the meridian plane. It is equal to the distance along the normal, from the spheroid surface to the semi-minor axis, and is given by:

$$R_N = \frac{a}{\sqrt{1 - e^2 \sin^2 \phi}} \tag{8.7}$$

At geodetic height, h, the geographic-system north component of velocity over the Earth is determined by:

$$V_N = (R_M + h)\dot{\phi} \tag{8.8}$$

where $\dot{\phi}$ denotes the latitude rate.

When the component of prime vertical radius is taken parallel to the equatorial plane, the radius of a constant-latitude circle is obtained. Hence, the geographic-system east component of velocity over the Earth is obtained by

$$V_E = (R_N + h)\cos(\phi)\dot{l} \tag{8.9}$$

where $\dot{l}$ denotes the terrestrial longitude rate.

8.5 Kalman Filtering

It is not always possible to accurately measure every variable that we want to control (e.g., vehicle's attitude, altitude, velocity, and position). The noise is always present in the environment due to reflections from rain or the Earth's surface or as a result of radio interference. The noise is also produced within the electronic circuitry of the sensors. The guidance system sensor must be able to detect targets in a medium cluttered with noise, which is any energy sensed other than that attributed to a target. In order to improve the accuracy of the measured data, and to increase the efficiency of the autopilot, the noise must be removed.

State estimation is defined as the process of tracking the current vehicle 3D motion variables. State estimation concerns mainly the processing of raw sensor measurements to estimate variables that are related to the UAV's state, particularly those related to its motion, such as attitude, position, and velocity. The design of motion estimation algorithms is necessary for flight control and a crucial step in the development of autonomous flying machines. Based on the sensing technology, state estimation algorithms can be classified into three main categories: (a) conventional systems; (b) vision-based systems; and (c) systems that rely on active ranging sensors.

The Kalman filtering – also known as linear quadratic estimation – is a mathematical technique [118] that uses a series of measurements observed over time, subject to Gaussian perturbations or errors, and produces estimates of unknown variables. Localization is a particular case of state estimation that is limited to position estimation relative to some map or other reference locations. The Kalman filter provides a means for inferring the missing information from indirect and noisy measurements. The Kalman filter is has long been regarded as the optimal solution to many tracking and data prediction tasks such as tracking moving targets using radar (e.g., another aircraft, ground moving target).

The technique is initially developed for the separation of signal from noise, but later was expanded [118] for estimating the state of dynamic systems. Hence, the Kalman filter is also used for predicting the likely future courses of dynamic systems that is out of control, such as a moving target. Hence, Kalman filtering has applications in many areas including control, tracking, guidance, and navigation.

In tracking, the Kalman filter is the optimal estimator in case of linear dynamics and observations subject to Gaussian perturbations or errors. The main purpose of the filter is to retrieve the Gaussian laws of the state vector according to the observation series. The Kalman filter is an estimator for what is called the linear-quadratic problem, which is the problem of estimating the instantaneous "state" of a linear dynamic system perturbed by white noise – by using measurements linearly related to the state but corrupted by white noise.

The Kalman filter mathematical model is based on the assumption that the true state at time k is estimated from the state at time k-1 according to

$$x_k = F_k\, x_{k-1} + G_k\, u_k + v_k \qquad k = 0, 1, 2, \ldots \tag{8.10}$$

where $x(k)$ is the state vector, F_k is the state transition matrix; x_{k-1} is the known previous state, G_k is the known control input; u_k is the control vector, and v_k is zero mean white process noise. The calculations of the Kalman estimator may be seen from many different angles. For instance, it calculates the minimum of variance and the maximum of likelihood estimator, it is the least square solution of a linear quadratic problem, and also the solution of a linear variational problem.

The Kalman filtering has been developed and expanded over several decades. The first method for forming an optimal estimate from noisy data is the method of least squares. It minimizes the estimated measurement error in terms of its Euclidean vector norm. The detail description of the Kalman filtering is beyond the scope of this text, Ref. [119] provides various MATLAB codes for a number of tracking/navigation/control applications.

The nonlinear version of Kalman filter is referred to as the extended Kalman filter (EKF), which linearizes about an estimate of the current value. An EKF provides the most common approach to converting sensor data to estimate the UAV motion variables. The estimation can be done in one step by using one filter with all the state variables. Another approach is to employ two successive EKFs: (a) one filter for attitude and heading estimation (attitude and heading reference system) using IMU raw data; and (b) one filter for position and velocity estimation using GPS raw measurements. A third filter may also be used to enhance height estimation by incorporating altimeter measurements with vertical accelerations.

The SBG IG-500 N is one of the world smallest GPS enhanced Attitude and Heading Reference System. With its embedded EKF, the IG-500 N delivers precision for attitude and position measurements.

System integration and operation of an instrumented Yamaha RMAX (Figure 1.1) unmanned helicopter is implemented by [21]. Four major elements of the system are: (a) the UAV; (b) a modular avionics system; (c) baseline software including middleware, guidance, navigation, control, communications, and operator interface components; and (d) a set of simulation tools. The baseline navigation system running on the primary flight computer is a 17-state EKF. The states include vehicle position, velocity, attitude, accelerometer biases, gyro biases, and terrain height error. The system is all-attitude capable, and updates at 100 Hz.

8.6 Global Positioning System

8.6.1 Fundamentals

The GPS is a space-based navigation system [120] owned by the United States government and funded, controlled, and operated by the US Department of Defense. The GPS concept is based on time and the known position of originally 24 satellites. There are millions of civil users of GPS world-wide. (including UAVs and RC airplanes). By the introduction of GPS, the altitude measurement has become an easy task and rather more accurate. The GPS has almost no direct effect from atmospheric condition. No matter, if an aircraft is flying in an ISA or non-ISA condition, its altitude is measured precisely via GPS. Without radar altimeter or GPS, the measured altitude is subject to an error. This must be taken into consideration when we are calculating aircraft performance. Many new aircraft are currently equipped with a GPS receiver. So, their altitude measurement is accurate; and we do not need to consider the effect of air density variations.

The space segment of the system consists of the GPS satellites. These space vehicles send radio signals from space. The nominal GPS Operational Constellation consists of 24 satellites (Figure 8.4) that orbit the earth in 12 hours Ref. [117]. There are often more than 24 operational satellites as new ones are launched to replace older satellites. The satellite orbits repeat almost the same ground track once each day. The orbit altitude (20 200 km, 55° inclination) is such that the satellites repeat the same track and configuration over any point approximately each 24 hours.

There are six orbital planes (with nominally four space vehicles in each), equally spaced (60° apart), and inclined at about 55° degrees with respect to the equatorial plane. This constellation provides the user with between five and eight satellites visible from any point on the earth. The predictable accuracy is within a few meters horizontally and vertically and has a 200 ns time accuracy.

GPS provides specially coded satellite signals that can be processed in a GPS receiver, enabling the receiver to compute position (e.g., altitude), velocity, and time. Four GPS satellite signals are used to compute positions in three dimensions and the time offset in the receiver clock. Position in XYZ is converted within the receiver to geodetic latitude, longitude, and height above the ellipsoid. Velocity is computed from change in position over time, or the Doppler frequencies. The GPS satellites broadcast at the same two frequencies, 1.575 42 GHz (L1 signal) and 1.2276 GHz (L2 signal). Because the satellite signals are modulated onto the same L1 carrier frequency, the signals must be separated after demodulation.

The UAV position is calculated based on the time-difference observed in the received GPS satellites' signals. At least four satellites are required for UAV positioning. GPS has very high accuracy outdoors (5–50 m), but it is complicated in certain urban areas or indoor environment as it requires line of sight with a number of GPS satellites.

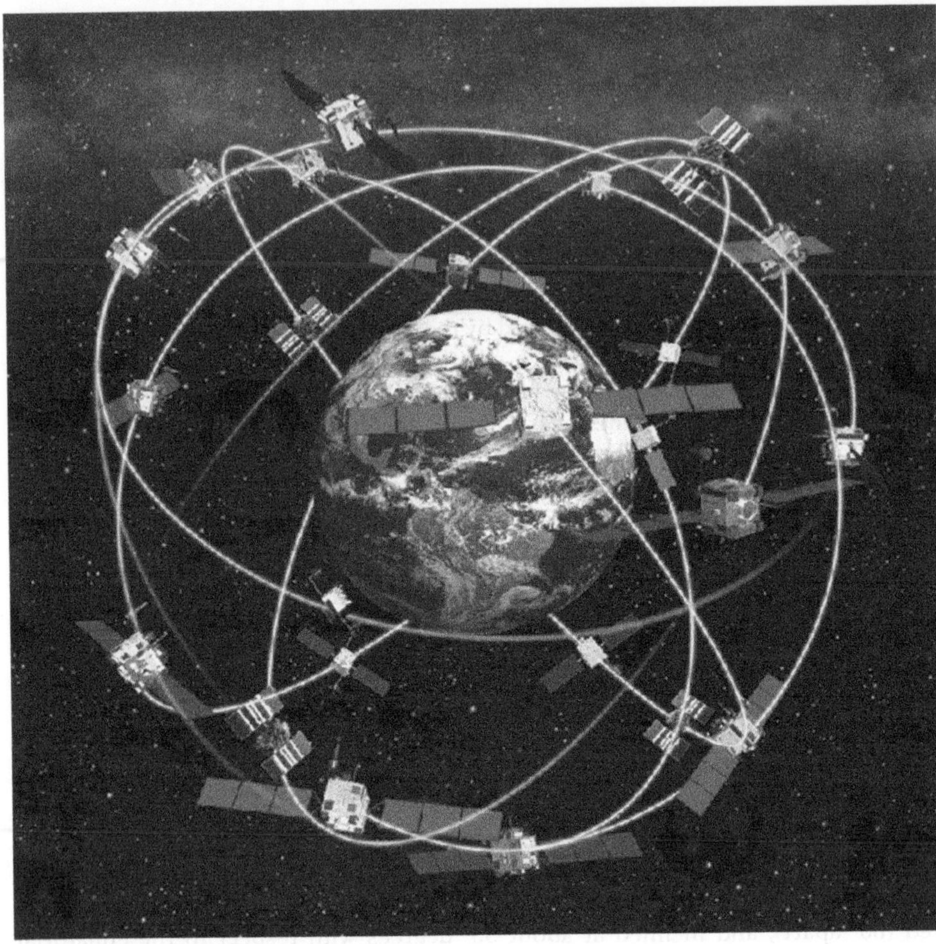

Figure 8.4 GPS satellites.

The availability of GPS has permitted UAV operation to be extended in range, enabling the operation of very long-range UAVs. However, the jamming of GPS signals in the event of hostilities is a major concern. In addition, there are times and areas of the earth, that the GPS signal is not available or weak. One solution is to integrate the GPS with an inertial navigation system. For instance, General Atomics RQ/MQ-1 Predator A (Figure 8.5) is equipped with a GPS-aided inertial navigation system.

The prime navigation system of the Global Hawk consists of: (a) an inertial navigation system; (b) a GPS. It is equipped with a monolithic ring laser gyro which operates in conjunction with an embedded differential ready GPS receiver for enhanced navigation performance and faster satellite acquisition. The aircraft is flown by entering specific way points into the mission plan. Way points can be changed in flight, as necessary. The Northrop Grumman Litton LN-100G Embedded GPS/INS – installed on the EO/IR payload bay - provides the aircraft attitude, position, velocity, and acceleration data.

Figure 8.5 General Atomics RQ/MQ-1 Predator A.

8.6.2 Earth Longitude and Latitude

The Earth rotates about its geographic axis, but maps are drawn based on meridians of longitude that pass through the geographic poles. UAVs which are equipped with GPS will receive their position data in terms of longitude and latitude. In this section, the calculation of the UAV position based on longitude and latitude are described.

Earth is in reality an ellipsoid, but at sea level, it may be assumed to be a sphere. Thus, the cross section of the earth at sea level is a circle. A circle is conveniently divided into 360°, and each degree is divided to 60 minutes. Each minute (1/60th of a degree) is approximately one nautical mile, and each nautical mile is 1,852 meter, 6,076.1 ft, and 1.1508 (statute) mile. Hence, one degree is 1/360 of a circle, and one minute of arc is 1/21 600 (since $60 \times 360 = 21\,600$) of a circle.

Latitude is measured (in degrees) north/south of the equator (Figure 8.6); and longitude is measured (in degrees) east or west of the prime meridian (a line running from north to south through Greenwich, England). The latitude of the equator is 0°, the latitude of the North Pole is −90° (90° N), and that of the South Pole is +90° (90° S). One half

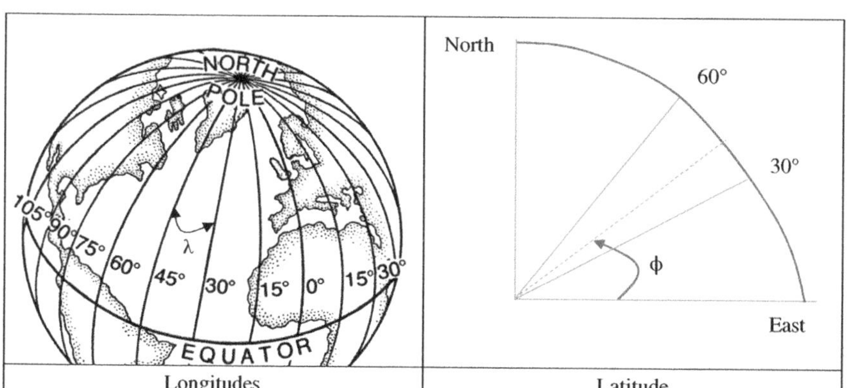

Figure 8.6 Longitude and latitude.

of the world is measured east longitude from 0 to + 180°, and the other half of west longitude from 0 to −180°.

Degrees of latitude are parallel, hence the distance between each degree remains almost constant. ***Latitude*** specifies the north–south position of a point on the Earth's surface, while ***longitude*** specifies the east–west position of the point. However, since degrees of longitude are farthest apart at the equator and converge at the poles, their distance reduces when moving toward poles. For instance, the latitude and longitude of city of Boston is 42.3601°, and − 71.0589°.

Each degree of latitude (ϕ) is 111.3 km (69.17 miles) apart. The range varies (due to the Earth's slightly ellipsoid shape) from 68.703 miles (110.567 km) at the equator to 69.407 miles (111.699 km) at the poles.

A chord length of one degree of longitude (λ) is widest at the equator at 69.172 miles (111.321 km) and gradually shrinks to zero at the poles. For instance, at 40° North (or South), the distance between a degree of longitude is 53 miles (85 km). The length of a change in longitude (ΔX_{lon}) and latitude (ΔY_{lat}) are:

$$\Delta Y_{lat} = 2\pi R.\left(\frac{\Delta\phi}{360}\right) \tag{8.11}$$

$$\Delta X_{lon} = \Delta X_{Equ}.\cos(\phi).(\Delta\lambda) \tag{8.12}$$

where ΔX_{Equ} denotes the length of one degree of longitude at equator.

$$\Delta X_{Equ} = 2\pi R.\left(\frac{1}{360}\right) \tag{8.13}$$

Hence, the length of one degree of longitude at equator is:

$$\Delta X_{lon} = 2\pi R\left(\frac{1}{360}\right) = \frac{2\times 3.14\times 6378.137\times 1}{360} = 111.319\,km = 69.171\,mile$$

where R denotes the radius of the Earth at equator (6371 km or 3959, miles from Table 8.1), and ϕ is the latitude. For instance, the length of one degree of longitude at the latitude of 30° is 96.3 km (i.e., 111.32 cos [30°] = 96.3). Eqs. (8.9) and (8.10) indicate that the length of one degree of longitude (ΔX_{lon}) is a function of latitude (ϕ), but the length of one degree of latitude (ΔY_{lon}) is independent of longitude (λ). In Eqs. (8.9) and (8.10), Δλ, ϕ, and Δϕ are in degrees.

Example 8.2

A UAV has flown from City of Boston with the GPS latitude and longitude of 42.361 145, and − 71.057 083 (42° 21' 40.1220″ N and 71° 3′ 25.4988″ W) to the city of New York with the latitude and longitude of 40.730 610, −73.935242 (40° 43' 50.1960″ N and 73° 56′ 6.8712″ W). Assume sea level altitude.

a) Determine the distance the UAV has flown in this flight.
b) Determine the flight direction of the UAV with respect to North.

Solution

a) Distance traveled.

We first calculate the distance along longitude and latitude.

$$\begin{aligned}\Delta Y_{lat} &= 2\pi R.(\Delta\phi)\\ &= 2\times 3.14\times 6378.137\times\frac{42.361145-40.730610}{360}\\ &= 181.51\,km\end{aligned} \tag{8.11}$$

$$\Delta X_{lon} = \Delta Y_{lat}.\cos(\phi)(\lambda_2-\lambda_1) = 111.32\text{ km} \tag{8.12}$$

$$\begin{aligned}\Delta X_{lon2} &= 111.319\times\cos(42.361145)\times(71.057083-73.935242)\\ &= -236.7\,km\end{aligned}$$

$$\begin{aligned}\Delta X_{lon1} &= 111.319\times\cos(40.730610)\times(71.057083-73.935242)\\ &= -242.8\,km\end{aligned}$$

The minus sign indicates that, the flight direction is eastward, so the sign can be ignored.

The chord length (Figure 8.7) along the longitude of New York City (between two latitudes) is 242.8 km, but along the longitude of Boston is 236.7 km. To simplify the problem, the average value is used in the triangle model.

$$\Delta X = \frac{242.8-236.7}{2} = 239.6\,km$$

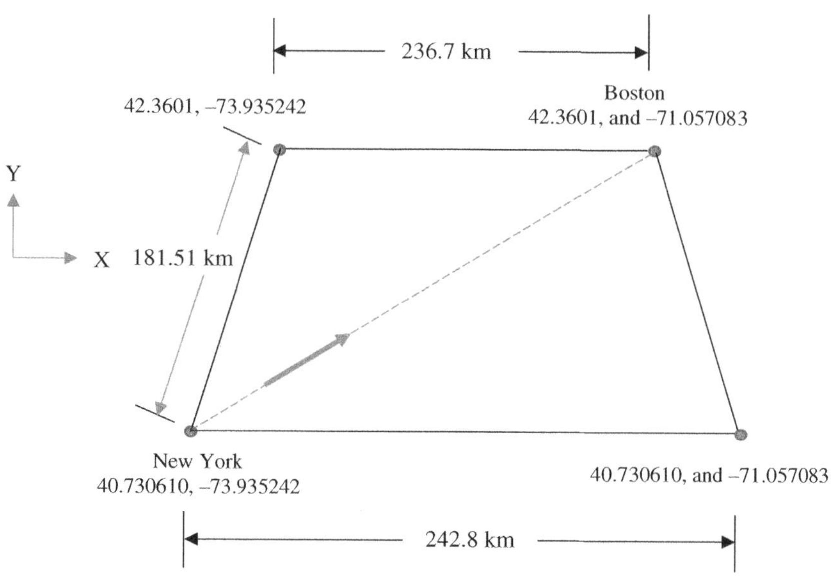

Figure 8.7 Longitudes and latitudes of the unmanned aerial vehicle (UAV) flight in Example 8.2.

Using Pythagorean equation, we obtain:

$$D = \sqrt{\Delta Y^2 + \Delta X_{avg}^2} = \sqrt{181.5^2 + 239.6^2} = 300.7\,km$$

Thus, the UAV has flown 300.7 km, in flight from New York City to Boston.

b) **Flight direction**
The UAV has flown 181.51 km toward North, while 239.6 km toward East. Employing trig equation:

$$\psi = \tan^{-1}\left(\frac{\Delta X}{\Delta Y}\right) = \tan^{-1}\left(\frac{239.6}{181.51}\right) = 52.8\,\text{deg}$$

The heading angle of the UAV with respect to North is 52.8°.

8.6.3 Ground Speed Versus Airspeed

The velocity that GPS is measuring is only the ground speed (not the airspeed). The airspeed is only measured via pitot-tube. So, to prevent an aircraft from stall, do not rely on GPS data, it is highly recommended to employ the pitot-tube. The application of GPS does not remove the necessity of pitot-tube in an air vehicle. The UAV ground track is determined by ground speed, not airspeed.

8.7 Position Fixing Navigation

The position fixing navigation is the determination of the position of the UAV (a fix) without reference to any former position. There are three basic methods of fixing position: (a) map reading; (b) celestial navigation; and (c) measuring range and/or bearing to identifiable points.

8.7.1 Map Reading

Map reading involves matching what can be seen of the outside world with a map. It is the traditional method of position fixing on land, and also by general aviation in clear weather. A camera is necessary to capture a picture of the outside world. Modern UAVs adopting this technique use a radar to obtain the electronic picture of the ground from the air and a computer matches it with a map stored in the form of a digital database. These systems are called terrain referenced navigation aids.

8.7.2 Celestial Navigation

Celestial navigation has been used by mariners for centuries. The basic principle of celestial navigation is that if the altitude of a celestial object (measured in terms of the angle between the line-of-sight and the horizontal) of a celestial object is measured; then the observer's position must lie on a specific circle (called the circle of position) on the surface of the earth centered on the point on the earth which is directly below the object. If the time of observation is noted and the celestial object is a star, then this circle can easily be found using astronomical charts and tables. Sightings on two or more such celestial objects will give two or more such circles of position, and their intersection will give the position of the UAV. This technique has been abandoned nowadays in favor of modern navigational aids. This technique may be employed in the areas that GPS signals are not available.

8.8 Navigation in Reduced Visibility Conditions

There are some applications in which the UAV needs to conduct a mission within an environment with reduced visibility; such as an environment with smoke or dust particles. In such conditions, being highly disturbed by noise induced in the measurement process, the regular sensors appropriate for normal condition, like laser range finders and cameras, cannot be used. On-board sensing is desirable for operation in such unknown environments. One approach is to use laser scanners to estimate the UAV position and motion, but it may require significant off-board processing. In the literature, a number of algorithms and techniques are introduced (e.g., Refs. [121, 122]) to deal with navigating an air vehicle within this condition.

8.9 Inertial Navigation Sensors

8.9.1 Primary Functions

The navigation system requires a number of measurement devices (sensors) to measure various flight parameters including velocity, positions, and attitude. When all sensors are located in one set, it is referred to as IMU. Some important sensors are (a) accelerometer; (b) gyroscope (or gyro); (c) rate gyro; and (d) magnetometer. The early models of IMUs were all mechanical, and had a very heavy weight (tens of kg). The new generations are all electronic, and weigh as low as a few grams. A typical current and modern IMU contain one accelerometer, rate gyro, and magnetometer per axis for each of the three vehicle axes. The IMUs are part of avionics that help a UAV to conduct a successful flight. Table 8.3 demonstrates the primary functions of a few navigation sensors. The first two sensors (i.e., accelerometers and gyroscopes) are described with more details in this section.

Table 8.3 Primary functions of a few navigation sensors.

No	Name	Measures	Remarks
1	Accelerometer	Linear acceleration	The linear acceleration is converted to linear velocity.
2	Rate gyro	Angular velocity	The angular velocity can be converted to angular positions.
3	Magnetometer	Attitude	e.g., heading
4	Magnetometer	Magnetic field	Magnetic north
5	Pitot tube	Altitude, airspeed	Using air pressure
6	Compass	Magnetic north	Heading angle is measure w.r.t. north
7	Altimeter	Altitude	Height

In general, there are two groups of sensors: (a) sensor as a payload; and (b) sensors necessary for navigation/guidance systems. In this chapter, only the sensors which are necessary for a regular flight (such as the ones used in navigation system) have been introduced. Payload sensors are addressed in Chapter 12.

In general, gyros are accurate in short term, but not reliable in long term due to drift. However, accelerometers are accurate in long term because of no drift; hence they can be used as complementary to gyro measurements. By fusing gyro and accelerometer data, a six Degree of freedom (DOF) sensor system is developed.

Basic (minimal) navigation sensors for a civil UAV include state estimation sensors, such as an IMU (three gyroscopes, three accelerometers, and three magnetometers) for attitude estimation, a global navigation system for position and velocity estimation, and an altimeter (barometric, Lidar, radar) for enhancing the altitude measurement. Most current commercial UAV autopilots contain these conventional navigation sensors.

8.9.2 Accelerometer

An accelerometer is a measurement device to measure linear acceleration of a motion. The final flight variable which is derived from the output signal of an accelerometer is the linear velocity. This convention is performed by the process of integration (either by an analog device or by a digital circuit). A secondary integration of such signal produces the position of the flight vehicle (i.e., x, y, and z). The process of integration will gradually result in an increased error. Thus, the accelerometer alone is not an accurate source of precise location information. A solution is to employ a hybrid approach which combines the accelerometer with a vision sensor or a magnetometer. Table 8.4 lists various accelerometers and their outputs.

The early version of mechanical accelerometer (Figure 8.8) was purely mechanical and worked based on the Newton's second law. It consists of a mass (referred to as proof mass), a suspension spring, a cylinder, a viscous damper for the motion of the mass, and a means to measure its displacement. The mass is constrained to move in one dimension only (i.e., along the cylinder).

Table 8.4 Accelerometers and their outputs.

No	Accelerometer	Axis	Acceleration
1	Along x-axis	Longitudinal axis	n_x
2	Along y-axis	Lateral axis	n_y
3	Along z-axis	Normal/vertical axis	n_z

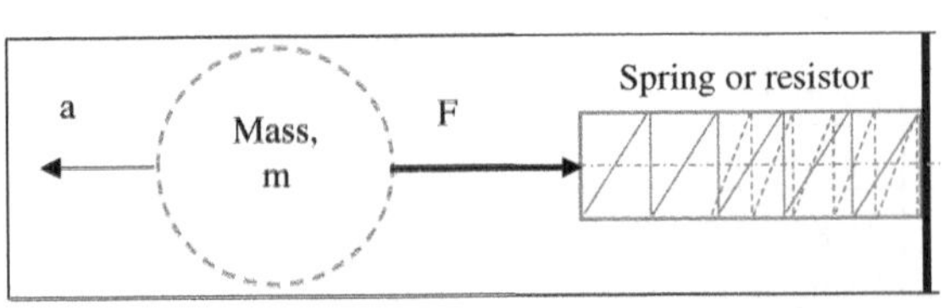

Figure 8.8 Elements of an early basic mechanical accelerometer.

Currently, there are variety of accelerometers employing laser, magnetic field, beam deflection, heat, and temperature, optics, frequency, micro-machined piezoelectric, and electric signal. In the purely mechanical accelerometer, the acceleration of the UAV creates an inertial force (F) on the opposite direction to the mass within the acceleration (a). Newton's second law is expressed in the expression below:

$$F = ma \tag{8.14}$$

When the moving mass (m) inside the accelerometer is known, and the force (F) is measured; the acceleration is simply the force divided by the mass. The force applied to the mass can also be measured by various elements such as a mechanical spring. In this case, the spring deflection is an indication of the applied force. The spring deflection can also be measured by a change in the resistance of a resistor, which can be represented by a voltage. The presentation of the theory behind non-mechanical accelerometers is out of scope of this text.

The value of the acceleration is frequently divided by g, to have a non-dimensional value. Hence, it measures accelerations exerted on airplane in gravity units (i.e., "g"s). The largest acceleration is often in the vertical direction (i.e., z-axis); so normal acceleration. The acceleration loads have to be carried by the aircraft structure (e.g., spar, rib, skin, and frames).

Today, various analog and digital accelerometers are available in the market that can measure the acceleration of up to 250 g. The maximum acceleration of an UAV is typically less than 20 g. The size of a modern accelerometer is about a few millimeters, and its price tag is a few dollars.

In an UAV, the accelerometer should be placed close to the UAV center of gravity, (i.e., aligned along the body axes). The output will be used as the feedback sensor for control of the elevator; hence, the z-axis g-load during a high-g maneuver may be controlled. An accelerometer is an internal sensor, which makes it less noisy and more reliable than an angle-of-attack sensor (external sensor). Therefore, an accelerometer can inherently stabilize an unstable short-period longitudinal mode, since its output contains a component proportional to alpha.

Usually 1 g is subtracted from the accelerometer output, in order to facilitate the cruise control. This initiative will create a zero-accelerometer output in a cruising flight. This implies that, the autopilot is keeping the vehicle in a level flight without applying any control input. If in a tight maneuver, (high load factor), autopilot fails to a force on the controls, the UAV will return to 1 g flight.

The variable position of the proof mass will cause a measurement error. This error is eliminated in high-sensitivity accelerometers by measuring the force required to maintain d (i.e., reading of the accelerometer) and its derivatives very close to zero. If the UAV (i.e., accelerometer) is in free fall above the Earth, the accelerometer reading will be zero. Accelerometers are commonly calibrated to read 1 g, when they are stationary on the Earth's surface; and having their axis parallel to the plumb-bob vertical. The accelerometer is recommended to be positioned near the UAV's center of gravity, where vibration is relatively low.

The normal acceleration (n_z) at a point is the component of acceleration in the negative-z, direction of the body axes. The UAV angular motion (e.g., pitch rate, q) has a contribution to the linear acceleration at a point away from the cg of a rigid body. For instance, in the body-z direction, the steady-state accelerometer reading – if it is located at a distance; x from the UAV center of gravity – is [54] a function of $(a_z' - g_z)$, where

$$a_z' = \left(a_z - \alpha_y x_R + x_R \omega_x \omega_z\right) \tag{8.15}$$

where ω is the angular rate, and α is the angular acceleration, and a is the translational acceleration.

The accelerometer output (f_n) in terms of pitch and roll angles with respect to the NED system is obtained as:

$$f_n = a_n + g\cos\theta\cos\phi \tag{8.16}$$

In level flight, at small angles of attack, the feedback signal for the control system is

$$f_n - 1 \approx a_n\left(\text{in g unit}\right) \tag{8.17}$$

This normal acceleration is approximately zero in steady level flight. Therefore, the accelerometer measurement is an approximate measurement of load factor; if α, β, θ, and ϕ are all small.

8.9.3 Gyroscope

Another important measurement device for a navigation system is gyroscope (or simply gyro). The gyroscope is composed of sets of nested wheels, each spinning at different axis. A gyroscope is a mechanical sensor which maintains the angular momentum, and it is utilized to measure the change in attitude (heading angle, bank angle, and pitch angle) and/or its rate of the UAV. A spinning wheel will resist any change in its angular momentum vector relative to inertial space. A gyro operates based on the *gyroscopic effect*, which states, a body rotating about one axis will resist about other axes (according to the law of conservation of angular momentum). Thus, the reaction to any incoming change to its axis will be to create a force on another axis. A navigation gyro uses this effect to track the orientation of a body with the aid of a gimbal-mounted flywheel that maintains constant orientation in space. There are two groups (See Table 8.5) of gyros:

Table 8.5 Gyros and their outputs.

No	Gyroscope	Axis	Measure	Reference plane	Symbol
1	Attitude gyro	About x-axis	Bank angle	y-z plane	ϕ
2		About y-axis	Pitch angle	x-z plane	θ
3		About z-axis	Heading angle	x-y plane	ψ
4	Rate gyro	About x-axis	Roll rate	y-z plane	p
5		About y-axis	Pitch rate	x-z plane	q
6		About z-axis	Yaw rate	x-y plane	r

1) Attitude gyro,
2) Rate gyro.

As the name implies, an attitude gyro is incorporating an attitude reference (e.g., bank angle and pitch angle), so it is used to sense deviations from a reference attitude in the aircraft y-z, x-y, and y-z planes. However, a rate gyro will measure the rate of change of an attitude (e.g. roll rate, pitch rate, and yaw rate) with respect to a reference plane. The pitch-rate gyroscope measures the (inertial) angular rate around the y axis (q). The roll-rate gyroscope measures the angular rate around the x axis (p). The yaw-rate gyroscope measures the angular rate around the z axis (r).

The attitude/rate is often measured by a change in the resistance of a resistor, which can be represented by a voltage. The output of the rate gyro is a signal proportional to the angular velocity of the case about its input axis with respect to inertial space. The sensitivity of the rate gyro is such that it cannot detect the component of earth's rate parallel to the input axis of the gyro. Thus, the gyro effectively measures the angular velocity of the case, and thus the aircraft, with respect to the earth. The gyro filter is usually necessary to remove noise and/or cancel structural mode vibrations.

Gyroscope is a spatial mechanism (as shown in Figure 8.9) and is composed of a spinning rotor, a fixed frame, and two gimbals (inner and outer). The rotor in spinning inside the inner gimbal, which in turn is free to rotate inside the outer gimbal. In Figure 8.9, the rotor is spinning about z-axis. When the plane of the rotor is rotated about one of its axes (say x), the gyroscope is creating a force along the other axis (y). The resultant force will be measures; and is an indication of the change in attitude.

The gyro is working based on the law of the gyro which can be derived from Newton's law in rotational form. The law states that, the time rate of change with respect to

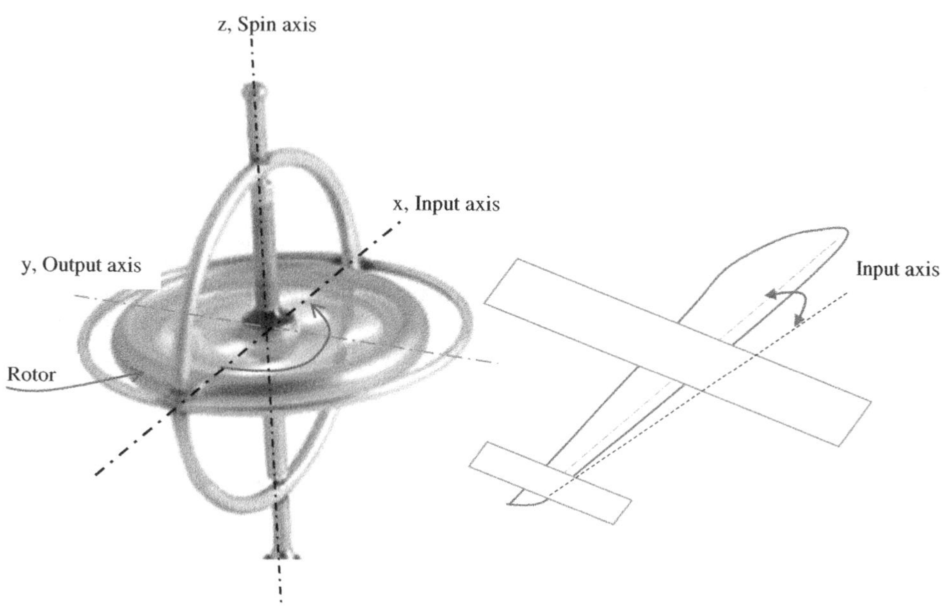

Figure 8.9 Gyroscope.

inertial space of the angular momentum (H) of an object about its center of gravity is equal to the applied torque (i.e., the angular momentum is conserved). This can be written in equation form as

$$M_{app} = \frac{d}{dt}H_I = I\frac{d}{dt}\omega \tag{8.18}$$

where I is the mass moment of inertia of the rotor, and ω is its angular speed. As Eq. (8.18) involves the time derivative of a vector with respect to inertial space, the equation of Coriolis must be applied.

$$\frac{d}{dt}H_I = \frac{d}{dt}H_E = \omega_{IE} \times H \tag{8.19}$$

where ω_{IE}, is the angular velocity of the earth with respect to inertial space ($0.072\,921\,15 \times 10^{-3}$ rad/s). The gyroscope itself can be mounted on a base that is moving with respect to the earth. In addition, the case of the gyroscope can be mounted on a platform so that it can rotate relative to the base. Finally, the inner gimbal can rotate relative to the case. Applying governing equations of these rotations, and then inserting the relevant terms into Eq. (8.18) and (8.19) yields the "law of the gyro."

In modern UAVs, a rate gyro is employed in conjunction with an attitude gyro for each axis, mainly for damping. For instance, in many cases, a roll-rate feedback is used, in conjunction with the roll-angle feedback to control the bank angle. Figure 8.10 demonstrates the block diagram of a roll-angle-hold autopilot which illustrates the control system along with the navigation system. In the navigation system, a rate gyro is utilized in the inner loop, while an attitude gyro is used in the outer loop. Having two feedbacks (roll rate, p; and bank angle, ϕ) will provide a precision control.

Because of the earth rotation, and of small accumulated errors caused by friction and imperfect balancing of the gyro, the heading indicator will drift over time. This drift will impact the accuracy of the navigation system, so the heading indicator is recommended to be set every 10–15 minutes. Accuracy and hardware fault are important factors in gyros. For example, a rate gyro with a full-scale reading of 200° per second, results in an attitude error of 20° in just 100 ms. This miscalculation could be devastating, especially in formation flight.

The necessity for the rate gyro depends upon the dynamics of the UAV, and its control system. If this system is planned to be used for controlling the pitch angle of the aircraft, an altitude or Mach hold control system can be used. The gyroscope is recommended to be positioned near the UAV's center of gravity, where vibration is relatively low.

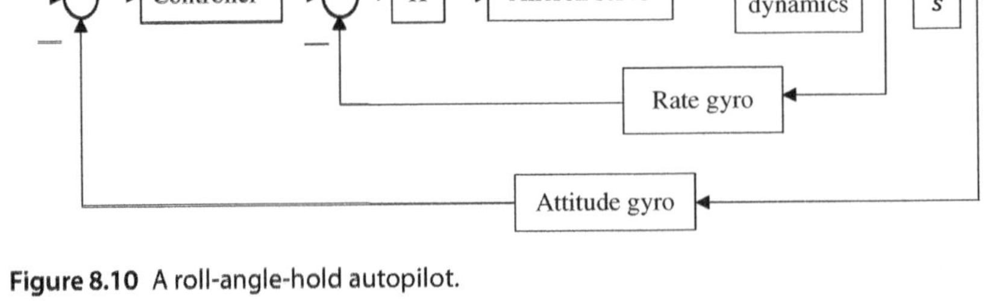

Figure 8.10 A roll-angle-hold autopilot.

A three Degree-Of-Freedom gyroscope is a 3-axis gyro that measures 3 orthogonal axes, assuming no crosstalk.

In the design of a rate gyro, one of the goals is the value of the gain (or sensitivity) of the rate gyro (e.g., pitch rate) to obtain the best overall system response after closing the outer loop. There is no simple rule to aid the engineer in selecting the final gain. The relation between inner and outer loops is such that, the higher the gain of the inner loop (rate feedback), the higher the allowable gain for stability for the outer loop (attitude feedback). Moreover, too high an inner loop gain results in higher natural frequencies and lower damping. Furthermore, it is desirable to obtain a Type 1 system to eliminate the steady-state error for a step input. The final values of gains usually result in a compromise between the desirability of fast response (short rise time) and the desire to have a low overshoot.

The traditional gyros are made of spinning wheels, but progress in microelectromechanical systems (MEMS) resulted in new micro-gyros which do not have spinning wheels. The MEMS gyroscopes measure changes in the forces acting on two identical masses that are oscillating and moving in opposite directions. Another advanced type of gyroscope is optical gyroscope-on-a-chip [123] with no moving parts and a greater degree of accuracy using a phenomenon called the Sagnac effect. This gyroscope (developed by Caltech) is smaller than a grain of rice – about 500 times smaller than the current device.

8.9.4 Airspeed Sensor

The GPS is a powerful tool in providing UAV ground speed, but it cannot measure the airspeed. The reason is that the atmosphere is a dynamic system, and often gust and wind are present. For safety reasons, the knowledge of airspeed is always crucial and necessary. A simple tool to measure airspeed is a standard pitot-static tube provided that it is suitably positioned to read accurate static pressure either as part of a combined unit ahead of any aerodynamic interference or as a separate static vent elsewhere on the aircraft. The errors involve in pressure reading include position error, installation error, and calibration issues.

There is an inaccuracy in the classic pitot-static tube in measuring airspeed, as well as the inability to record speeds below about 15 m/s. In addition, the fluctuating reading values from pitot tube can cause instability in the control system. This, it is recommended to rely on data from a system integrated with GPS that does not require knowledge of ambient static pressure.

Due to safety reason and FAA regulations, all aircraft must use a device called a pitot tube to measure the aircraft speed. Airspeed is measured by comparing the difference between the pitot and static pressure (Figure 8.11) and, through mechanical linkages, displaying the result on the airspeed indicator. A static port (tube) only measures the static pressure, since the hole is perpendicular to the air flow, so the flow must turn 90° to enter into the tube. In contrast, a pitot tube measures the dynamic pressure, since the hole is facing the airflow. When a pitot tube, has a static port, it is often referred to as the pitot-static tube.

By employment of the pitot tube and the static port, we are dealing with three types of aircraft speeds (i.e., airspeed): (a) indicated airspeed; (b) true airspeed; and (c) equivalent airspeed. Before explaining the difference between these three terms, it is beneficial to introduce how a pitot tube works.

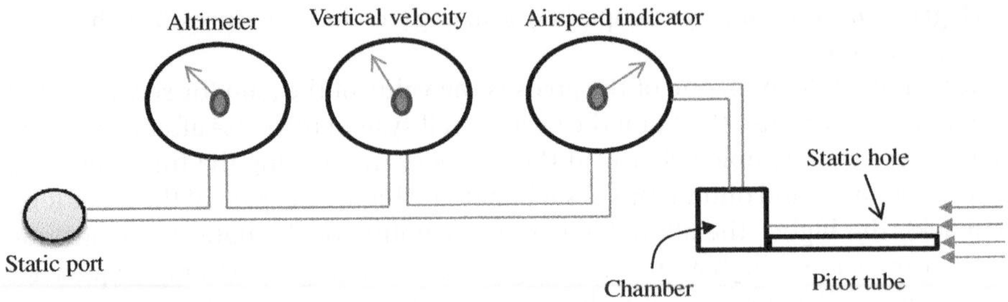

Figure 8.11 Pitot-static measurement device.

8.9.5 Altitude Sensor

An altitude sensor or altimeter is an active instrument used to measure the altitude of an object above a fixed level. Two onboard altimeters are: (a) the pressure altimeter, or aneroid barometer, which approximates altitude above sea level by measuring atmospheric pressure; and (b) the radio or radar altimeter, which measures absolute altitude (distance above land or water) based on the time required for a signal.

8.9.5.1 Radar Altimeter

Electronic sensors for measuring altitude and height, that is height above ground/sea-level, include those measuring distance by timing pulses of radio, laser, or acoustic energy from transmission to return. These vary in their accuracy, depending upon their frequency and power, Radio altimeters vary in their accuracy and range depending upon their antennae configuration. Laser systems may have problems in causing eye damage and precautions must be taken in their selection and use.

A radar altimeter is using electromagnetic waves to measure the distance of an aircraft (or other aerospace vehicles) above the ground. Radar altimeters are often used in aircraft during bad-weather landings. Rader altimeter is much more accurate, and more expensive than the pressure altimeters. They are an essential part of many blind-landing and navigation systems and are used over mountains to indicate terrain clearance. Special types are used in surveying for quick determination of profiles.

Radar altimeters have been in use on various spacecraft, starting with Skylab in 1973, to measure the shape of the geoid and heights of waves and tides over the oceans. The altimeter measures height by determining the time required for a radio wave to travel to and from a target. If the Earth were a perfectly flat horizontal plane, the signal would come only from the closest point, and would be a true measure of altitude. However, the Earth is not smooth, and energy is scattered back to the radar from all parts of the surface illuminated by the transmitter.

8.9.5.2 Mechanical Altimeter

Another altitude sensor is a mechanical device, which is based on pressure. The instrument employs a pitot tube and is less accurate than an electronic sensor. The pitot tube and static pressure holes are located at a suitable convenient position on the aircraft. Some convenient locations include (a) under the wing; (b) at the middle of fuselage nose; and (c) beside fuselage front or middle section. The location of the static tapping is very important because it is essential to select a position where the local static pressure is the same as that in the free stream. The location of the pitot tube is also very important because it is essential to select a position where the local airspeed is the same as that in the free stream, and also is not too sensitive to change in the aircraft angle of attack and sideslip angle. The pitot and static holes are normally heated to avoid icing at low temperatures and high altitudes. The location of the holes will usually induce 2–5% error in reading, so the pressure difference measuring device must be calibrated.

As an alternative to using separate pitot tube and static tapping, it is more convenient to use a combined device called the pitot-static tube (See Figure 8.12). The pitot-static tube consists of two concentric tubes. The inner one is simply a pitot tube, but the outer one is sealed at the front and has a few small holes in the side. By mounting it on under the wing or at the fuselage side it can be arranged so that it is well clear of interference from the flow around the aircraft. Both pitot-static tube and separate pitot and static tapings do the same job for accurate speed measurement.

For an incompressible flow (i.e., when $M < 0.3$), the summation of static pressure (P_s) and dynamic pressure (P_d) remains constant. Thus:

$$P_o = P + P_d = P + \frac{1}{2}\rho V^2 \tag{8.20}$$

Since, the total pressure is measured by Pitot-tube, we could determine the airspeed from Eq. (8.20) (assuming a known air density). However, for a compressible flow (i.e., when $M > 0.3$) including supersonic speeds, the Eq. (8.20) cannot be used, since the Bernoulli equation is not valid. In such flight conditions, the total pressure [14] is a function of Mach number and static pressure as:

$$P_o = P\left[1 + \frac{\gamma - 1}{2}M^2\right]^{\gamma/(\gamma-1)} \tag{8.21}$$

where γ is the ratio of specific heat, and is a constant (e.g., 1.4 for the air at sea level). Hence, the flight Mach number may be calculated by using a Pitot-tube which measures the total speed. In supersonic flight, there is a normal shock wave in front of a pitot-tube; hence, more calculations are required in determining airspeed. The total pressure drops across a normal shock wave.

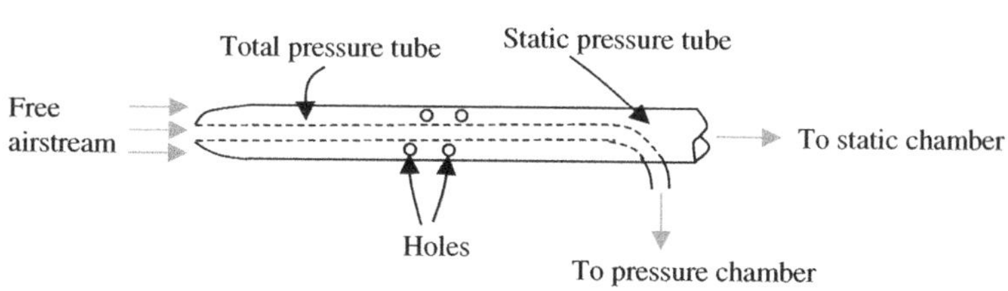

Figure 8.12 A pitot-static tube.

8.9.6 Pressure Sensor

One of the navigation sensors is the pressure sensor which measures atmospheric pressure. The pressure is measured from a number of objectives such as: (a) to determine altitude; (b) to provide a pressure feedback for pressurized cabin; and (c) to provide pressure feedback for fuel tank. Although there is no human in a UAV to breathe and in need of sea-level pressure at high altitude, but there are sensors and payloads which require sea-level pressure to operate regularly.

There are various types of pressure measurement devices from analog to digital versions. In Section 8.9.4, pitot-static as a pressure sensor was introduced; in this section, a few other types are presented. An electronic pressure sensor (i.e., barometer) a high precision digital pressure sensor, based on piezo-resistive technology.

Some new types combine advanced micromachining techniques, thin-film metallization, and bipolar semiconductor processing to provide an accurate, high-level analog/digital output signal that is proportional to applied pressure. Their interface allows for easy system integration with a microcontroller. A newer sensor module generation is based on leading MEMS technology and latest benefits, which have been widely used for over a decade. The sensing principle employed leads to very low hysteresis and high stability of both pressure and temperature signal.

8.9.7 Clock/Timer

A clock – one of the oldest human inventions – is an instrument used to measure, record, and indicate time. In a UAV, there is usually no need to indicate the time; however, the time must be measured and recoded for various navigation/guidance/control applications. The value of the variable time (t) in processing Eqs. (8.4) and (8.5) are provided by a clock/timer.

There are various types of clocks available in the market from digital to analog; and from electric to quartz to atomic. In the modern era which an electronic processor is employed in the microcontroller, a digital clock is appropriate. Most digital clocks use electronic mechanisms. Since the advent of digital clocks in the 1960s, the use of analog clocks has declined significantly. However, analog clocks are not prone to any hacking and jamming signals.

A commonly used format for recording time in digital clocks is the 24-hour notion with hours ranging from 00 to 23. After a reset, battery change, or power failure, a digital clock without a backup battery either start counting from 00, or to stay at 00. Hence, a backup battery is a must for a clock in a UAV.

A vehicle equipped with a GPS, no longer requires a timekeeping equipment, since the time is supplied by equipment on the satellites. However, for a UAV with an inertial navigation system, the time must be known for *dead reckoning*. Furthermore, for a UAV with celestial navigation (i.e., astronavigation), at a given time, any celestial body is located directly over one point on the Earth's surface. Thus, the location of the vehicle can be determined from tables in the nautical or air almanac for that year; if the time is known. The measurement of longitude (not latitude) requires accurate knowledge of time.

8.9.8 Compass

The knowledge of both true north and magnetic north is a must in an autonomous UAV to navigate its path toward the destination. The deviation between true north and magnetic north is discussed in section 8.10.6. A compass is a mechanical device to measure the direction of magnetic north. However, in trajectory design, the true north is the reference direction. True north is the direction along Earth's surface toward the geographic North Pole. Compass deviation can also be caused by local magnetic effects.

It is interesting to note that the airport runways are numbered based on magnetic north, instead of true north. The magnetic heading is used for designating runways. Runways are named by a number between 01 and 36, which is the magnetic azimuth of the runway's heading in deca-degrees. If an aircraft is directed to land in the runway 14, the aircraft magnetic heading will be 140° while landing. If a UAV is programmed to land without the remote pilot control, it must be equipped with a compass or magnetometer to determine the magnetic north in order to adjust its heading for landing. The runways in airports are numbered based on magnetic north.

It is important to note that GPS is unable to measure heading. Furthermore, there is no sensible way to show True North. GPS-enabled devices that do show an actual heading actually employ an internal magnetic compass. However, a GPS is equipped with a differential compass which is an alternative way to measure a bearing. When a UAV is flying, GPS periodically records its position. By comparing where it was to where it is now, the GPS-enabled processor can calculate which direction the UAV is heading, and it uses this to indicate the current bearing.

8.9.9 Magnetometer

Magnetometers are devices/sensors that originally measure the strength of magnetic fields and magnetic flux density. Magnetometers are digital compasses, and are classified into two basic types: (a) scalar; and (b) vector. A scalars magnetometer measures the scalar value of the magnetic flux, while a vector magnetometer measures the magnitude and direction of the magnetic field. A compass is a basic type of magnetometer that measures the direction of Earth's magnetic field. Any object or anomaly which distorts this magnetic field is also detected by magnetometer. Thus, a magnetometer can also be an alternative element to attitude gyro.

For instance, the 3-Axis Honeywell HMR-3000 magnetometer provides heading (0-360°), pitch (±40°), and roll (±40°) output for attitude reference systems. It has the following technical features: Accuracy: < 0.5 deg, Resolution: < 0.1 deg, Update Rate: 20 Hz, Power: 5 V DC, Weight: 0.75 oz, Current: 35 mA, size: 1.2 × 2.95 inches PCB.

3-axis gyro, 3-axis accelerometer, and 3-axis magnetometer can all be incorporated in one chip as an IMU (e.g., InvenSense MPU-9250). Magnetometers have been recently miniaturized to the extent that they can be incorporated integrated circuits and used as miniaturized compasses. A MEMS magnetic field sensor is a small-scale MEMS sensor for detecting and measuring magnetic fields. Many of these operate by detecting a change in voltage (i.e., electronically), or a mechanical displacement may be measured optically.

8.9.10 MEMS Inertial Module

An inertial navigation system basically requires three groups of sensors: (a) gyroscopes; (b) accelerometers; and (c) magnetometers. A sensors package composed of these three groups of sensors are usually referred to as the "inertial unit" or IMU. An IMU usually contains three orthogonal rate-gyroscopes and three orthogonal accelerometers, measuring angular velocity and linear acceleration respectively (and sometimes magnetometers). In the past, this unit mas made of all mechanical elements and was relatively heavy and large.

However, progress in miniaturization and microcontrollers has resulted in the integration of several sensors into one compact navigation unit (single chip), with an acceptable accuracy and affordable prices. MEMS are a new group of low-cost sensors for variety of applications from measurement of pressure/temperature to acceleration/attitude. These sensors combine the benefits of high-precision in electronic circuitry and high-load capability of mechanical systems in micro-size level. Furthermore, their I^2C interfaces allow for easy system integration with a microcontroller.

A few lightweight examples are: (a) Xsens MTi-G (with a mass of 58 g); (b) SBG IG-500N (45 g); and (c) Microstrain 3DMGX3-35 (23 g). For instance, the MTi-G[1] is (dimension: 58 mm ×58 mm ×22 mm) a GPS aided MEMS-based attitude and heading reference system and static pressure sensor.

The MPU-60X0 is the world's first integrated 6-axis MotionTracking device that combines a 3-axis gyroscope, 3-axis accelerometer, and a Digital Motion Processor (DMP), all in a small 4 mm × 4 mm × 0.9 mm package. Moreover, the new generation of InvenSense OIS specific 2-axis gyroscopes were specifically designed to meet the requirements of cameras in the smartphone market. At 3 mm × 3 mm, they are the smallest 2-axis gyroscopes in the market today and are available in heights as low as 0.75 mm allowing them to be integrated in the thinnest and most compact modules.

InvenSense[2] was the first company to deliver Motion Interface solutions with fully integrated sensors and robust MotionFusion firmware algorithms. They began with the world's first dual-axis MEMS gyroscopes in 2006 for the digital still camera market, the world's first integrated 3-axis motion processing solution for smart phones in 2009, the world's first single-chip integrated 6-axis MotionTracking device in 2010, the world's first integrated 9-axis MotionTracking device in 2012; and in 2014, the ICM-20728, the world's first integrated 7-axis (3-axis gyroscope + 3-axis accelerometer + pressure sensor) single-chip platform solution with onboard DMP.

As another example, the ISM330DLC (www.st.com) is a six-axis IMU (Inertial Measurement Unit) and contains a 3-axis accelerometer and a 3-axis gyroscope with user-selectable full-scale up to ±16 g and ±2000 dps respectively. It is an ultra-low power MEMS sensor for industrial applications, which combines high accuracy and robustness with long-term availability.

Figure 8.13 illustrates an Adafruit 9-DOF board which contains and IMU and an embedded temperature sensor. The dimensions are 33 mm × 20 mm × 2 mm, with a mass of only 2.3 g. The tag price is about $40. This board needs an analog supply voltage

Figure 8.13 Adafruit 9-DOF IMU (LSM9DS0) compared with the size of a coin.

of 2.4–3.6 V, and it is equipped with both SPI and I2C serial interfaces. The inertial module – available in a plastic land grid array package (LGA) – that contains a 3D digital linear accelerometer, digital 3D gyroscope, and a 3D digital magnetometer with a dimension of 4 mm × 4 mm × 1 mm is shown on the center.

8.9.11 Transponder

One of the navigation aids is the transponder that automatically emits a coded identifying signal in response to an interrogating received signal. A transponder mainly consists of a transmitter, a receiver, and a signal generator. This is an automated transceiver in a UAV to generate a radio frequency for identifying its vehicle to another aircraft or to a GCS. Chapter 7 provides more information about the characteristics and operations of transponders, as they are used in Guidance systems too.

8.10 Navigation Disturbances

The atmosphere is rarely calm, but is usually characterized by wind, gust, storm, and turbulence. There are a number of natural phenomena that tend to cause an air vehicle in flight to change direction or velocity or both. These phenomena are turbulence, gust, wind, drift (when considering spin-stabilized weapons), and the Coriolis Effect. The calculations involved in a practical trajectory procedure must take into account these natural factors. In general, drift can be removed from gyro via high-pass filter. Moreover, noise can be removed from the accelerometer via low-pass filter.

In this section, a number of navigation disturbances such as wind, gust, turbulence, drift, noise, and the Coriolis Effect are briefly presented. These factors must be considered in the design of the navigation system. Navigation disturbances have negative impacts on the UAV situational awareness and may cause an unsuccessful flight mission. The accuracy of UAV localization will be a direct factor of how navigation factors are dealt with, and they are resolved.

8.10.1 Wind

Wind is the air flowing horizontally with respect to Earth's surface. It is one of the almost permanent phenomena of the atmosphere. Wind is very significant in aircraft navigation and influences most aspects of performance (e.g., range, take-off run). Wind is controlled by a combination of (a) the pressure-gradient force; (b) the Coriolis force; and (c) friction. The prevailing wind is the wind that blows most frequently across a particular region. Different regions on Earth have different prevailing wind directions which are dependent upon the nature of the general circulation of the atmosphere and the latitudinal zones. Wind speed is lower at low altitude (about 50 km/h) and higher at high altitude (up to 200 km/h).

The effects of wind can be divided into two components: (a) range wind; and (b) cross wind. Range wind is that component of wind that acts in the xz plane (mainly along x-axis). It serves to either retard (if headwind) or aid (if tailwind) the ground speed, thus either increasing or decreasing range. If an aircraft encounters a headwind, its ground speed is decreased, however its airspeed does not change. On the other hand, if an aircraft encounters a tailwind, its ground speed is increased, however its airspeed does not change again.

Cross wind acts in the xy plane perpendicular to the airspeed (i.e., along y-axis); and serves to deflect the vehicle either to the right or left. The cross-wind changes the flight heading, so the pilot must correct its heading by using rudder. Every aircraft can tolerate a maximum amount of wind speed. If the wind is faster than that, the aircraft is not allowed to fly.

Although the wind will affect the speed of the airplane relative to the earth – i.e., its ground speed – it will not affect its speed relative to the air (i.e., airspeed). Wind acts on the vehicle throughout the time of flight; therefore, the total deviation in range and azimuth is a function of the time-of-flight. In terms of aircraft performance, wind will not influence the flight endurance, but changes the flight range. Thus, the wind speed will not influence aerodynamic forces and moment; it will change ground speed, aircraft heading, and flight range.

When a UAV employs a GPS for navigation, the coordinates will be determined correctly, so the wind effect is automatically corrected by the control system. However, for a UAV with inertial navigation system, the coordinates must be corrected via on-board devices. For a UAV with a GPS, the wind speed can be determined by comparing expected coordinates due to airspeed, with real coordinates due to wind. The compensation will be applied accordingly.

Thus, the wind speed will not influence aerodynamic forces and moments; it will change ground speed, aircraft heading, and flight range. The wind will affect the speed of the airplane relative to the earth – i.e., its ground speed (V_G) – it will not affect its speed relative to the air.

$$\overrightarrow{V_G} = \overrightarrow{V_A} \pm \overrightarrow{V_W} \tag{8.22}$$

where V_A is the airspeed, and V_W denotes the wind speed. The deviation due to wind is determined using a triangle (Figure 8.14) relationship.

$$\Psi = \tan^{-1}\left(\frac{V_W}{V_A}\right) \tag{8.23}$$

where Ψ is the heading deviation. Please note that Eq. (8.22) is in vector form, so each parameter has a value and a direction. For instance, if a UAV is experiencing a headwind, its ground speed will be decreased (i.e., $V_A - V_W$). However, if a UAV is experiencing a tailwind, its ground speed will be increased (i.e., $V_A + V_W$).

Example 8.3

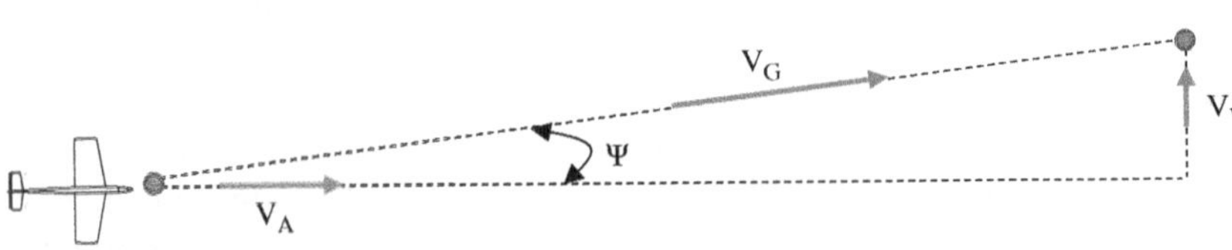

Figure 8.14 Wind influence on ground speed.

A UAV is planned to fly eastward at an airspeed of 100 knot for 5 minutes to reach a target. A cross wind is blowing from right (South) with a speed of 10 knot.

a) If the UAV does not apply any correction to its heading, what would be the miss distance?
b) Determine (a) the heading angle that the UAV needs to take; (b) new airspeed, in order to nullify the drift created by the wind.

Solution

a)

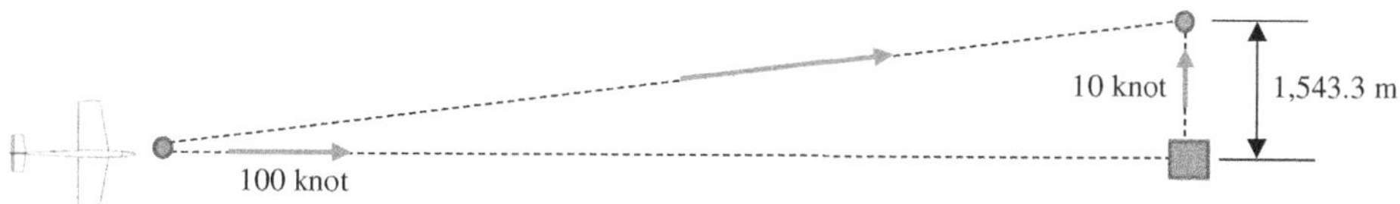

$$\Delta y = V_w T = 10 \times 0.514 \times 5 \times 60 = 1{,}543.3m$$

After five minutes, the UAV will be 1543.3 m north of the target. The miss distance is 1543.3 m.

b)

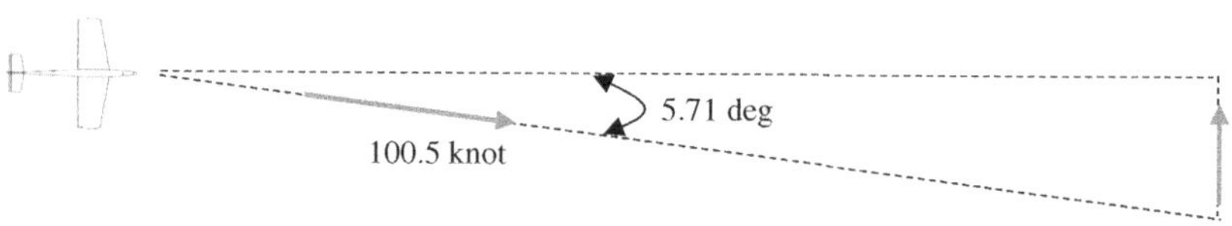

$$\varphi = \tan^{-1}\left(\frac{10}{100}\right) = 5.71\text{deg}$$

$$V_{new} = \frac{V}{\cos(\varphi)} = \frac{100}{\cos(5.71)} = 100.5knot$$

In order to reach the target in five minutes, the UAV must have a heading of −5.71° with a new (slightly faster) airspeed of 100.5 knot.

8.10.2 Gust and Disturbance

Another effective atmospheric phenomenon is the disturbance. The disturbances in the atmosphere can be described by the spatial and temporal variations in the gust components. One of the most important and well-known types of disturbances that every aircraft is experiencing is gust. In meteorology, gust is a sudden wind-speed increase of 4.6 m/s or more, to a peak speed of 8 m/s (about 18 miles per hour) or more. It is briefer than a squall and usually lasts 20 seconds or less. The air masses in the troposphere are in constant motion and the region is characterized by unsteady or gusting winds and turbulence. The troposphere extends from the Earth's surface to an altitude of approximately 11 km. A mathematical representation of gust can be found in the Military Specification MIL-F-8785C and Military Handbook MIL-HDBK-1797.

Air turbulence around an obstacle causes gusts; they occur frequently over buildings, irregular ground, and mountains, and are generally absent over water. The gust study is

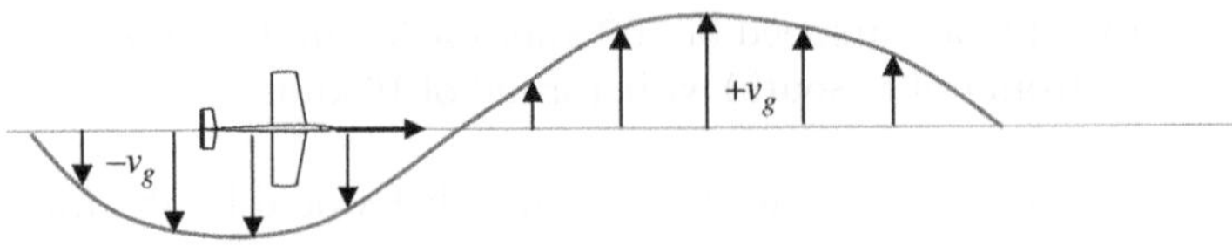

Figure 8.15 Sinusoidal horizontal gust.

mostly useful in aircraft stability and control, rather aircraft performance. In severe weather conditions, where the gust speed is high, aircraft flight is not assumed as safe, thus the flight must be postponed, or the route must be changed.

Figure 8.15 demonstrates a UAV experiencing a horizontal sinusoidal gust. The variation of the vertical gust velocity along the y axis of the airplane is similar to the velocity distribution created on a yawing airplane.

If the gust field wavelength is large in comparison with the aircraft, the gust produces a spanwise variation of velocity along the vehicle. The linear variation of velocity across the fuselage is the same as that produced on a yawing wing. The sinusoidal (rotary) gust velocity in terms of the gradient in the horizontal gust field can be express as a yawing moment:

$$r_g = \frac{dv_g}{dx} \tag{8.24}$$

In similar manner, the gust pitching moment (q_g) and gust rolling moment (p_g) can be developed.

A fixed-wing UAV usually exhibits directional stability, but not heading stability. So, part of the effect of the horizontal gust is nullified by the directional stability features. However, the change in heading angle is not inherently saved, unless a control system applies a correction through rudder deflection.

The turbulence models frequently used to evaluate aircraft performance are the Von Karman and Dryden models. The power spectra of these models are defined analytically in Ref. [124]. Each model consists of three velocity spectra, corresponding to the three axes associated with a body-fixed coordinate system. The frequency variable is spatially referenced to as a "frozen" turbulence field, a concept that has been shown to produce realistic results for normal aircraft speeds. The advantage of the spatially referenced spectra is that the turbulence characteristics are independent of the aircraft speed. However, for a typical dynamic simulation, it is usually desirable to generate the turbulence velocities as a function of time.

The usual individual gust or discrete gust idealization (Figure 8.16) consists of a one-minus-cosine pulse:

$$V_g = \frac{1}{2}U_{de}\left(1 - \cos\frac{2\pi x}{2H}\right) \quad \left(\text{in y direction}\right) \tag{8.25}$$

$$W_g = \frac{1}{2}U_{de}\left(1 - \cos\frac{2\pi x}{2H}\right) \quad \left(\text{in z direction}\right) \tag{8.26}$$

in which:

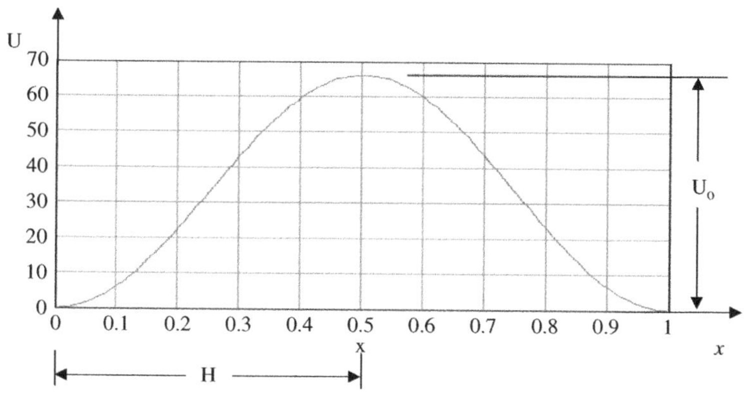

Figure 8.16 Discrete gust.

H: gradient distance (12.5 × $\bar{C}$ [times wing chord])
U_{de}: derived equivalent gust velocity, fps

According to FAR 23, the values of the derived equivalent gust velocity (U_{de}) are specified as functions of aircraft speed and altitude. At altitudes from sea level to 20 000 ft, the values are:

at V_B (design rough-air speed): U_{de}= 66 fps
at V_C (design cruise speed): U_{de}= 50 fps
at V_D (design dive speed): U_{de}= 25 fps

At altitudes above 20 000 ft, U_{de} decreases linearly to the following values at 50000 ft: at V_B, 38 fps; at V_C, 25 fps; and at V_D, 12.5 fps.

The discrete gust is usually considered either as an up-gust, or as a side-gust. In other words, the direction of the gust is perpendicular to the direction of flight. This gust changes the airspeed (in the y or z direction) of the aircraft suddenly.

8.10.3 Measurement Noise

Noise, by definition, is any undesired signal within the system, whether natural or man-made. The principal components of system noise are wide-band noise, 1/f noise, and interference. Wide-band noise consists of thermal noise, shot noise, and partition noise. Interference noise is man-made noise that can be reduced by proper circuit layout, shielding, and grounding techniques. The 1/f noise component, which exists in all natural processes, shows the accumulated effects of small changes. Different measurement devices have different noises from a variety of sources. In general, noise can be removed from an accelerometer via a low-pass filter.

The noise spectrum is white noise, meaning that all frequencies are present equally, at least throughout the radio spectrum. To estimate a signal of interest, or the state of a system in the presence of additive noise, a filter is used. Based on Ref. [57], a reasonable measurement noise covariance is (for normal acceleration n_z and pitch rate q):

$$R = \begin{bmatrix} \frac{1}{20} \\ \frac{1}{60} \end{bmatrix} \tag{8.27}$$

It is not always possible to accurately measure every variable that we want to control. Figure 8.17 illustrates the height measurement in a small UAV flying at 50 m altitude. As seen, the altitude is about 3–4 mm above the real flight altitude. There are many reasons for having non-accurate values in measuring altitude; one could be the engine propeller rotation.

In writing navigation code for uploading to the microcontroller, a filtering method must be employed. A very popular method, the Kalman filtering (briefly introduced in Section 5) is a mathematical technique that uses a series of measurements observed over time, subject to errors, and produces estimates of unknown variables. For example, the altitude (in Figure 8.17) is correctly estimated using the Kalman filter based on the state obtained from the altimeter measurement.

In an approaching flight, the output of navigation (position in x and y) are used as references for the guidance system. In addition, the altitude h used as a feedback value for control system.

8.10.4 Drift

Drift of a vehicle is defined as the lateral displacement (i.e., heading) from the intended original trajectory. Two primary sources of drift are: (a) wind; and (b) effect of the rotation of the rotor/propeller.

8.10.4.1 Drift Due to Rotation of Rotor/Propeller

Air vehicles are frequently equipped with propellers (for electric or piston engines) or turbine engines which act like spinning rotors. Such spinning rotors apply gyroscopic moments on the body to which they are attached. In many vehicles, these gyroscopic moments are canceled by counter-rotating propellers, and twin spool turbines which rotate in opposite directions. The principal cause of the drift lies in the gyroscopic properties of the spinning bodies. This phenomenon is also referred to as the *gyroscopic precession*.

The center of gravity of the vehicle, is usually following the curved path of the trajectory. A rotor seeks, first of all, to maintain the original direction of its axis in the direction of its line of departure. According to the law of the gyroscope; a rotating body will generate a force, if the direction of its axis of rotation is changed. This force will create a drift, if its center is not at the UAV's center of gravity.

This force creates a moment about UAV cg, so it tends to push the nose of the vehicle away from its intended trajectory. Therefore, the rotor's tendency to maintain its axis results in leaving this axis pointed slightly away from the tangent to the trajectory. However, a UAV is usually stable, so it resists against any disturbance, including this drift. Hence, this effect actually causes the axis of the vehicle to make oscillatory nutations about its flight path. The overall effect is to cause the vehicle to change its flight

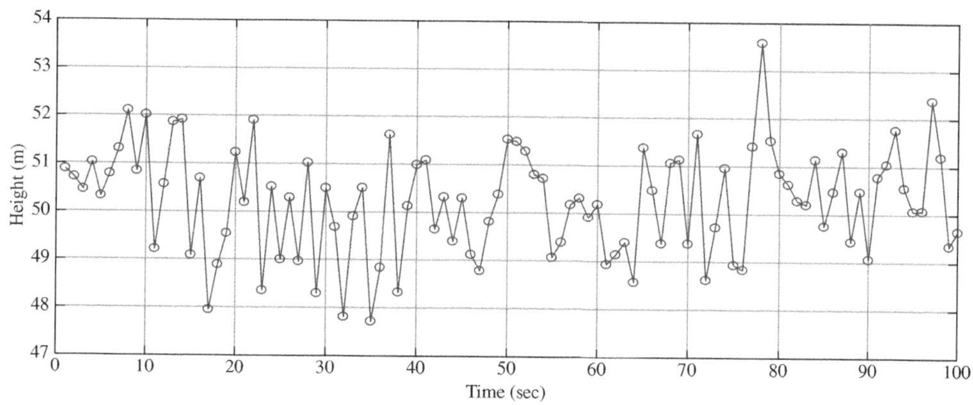

Figure 8.17 Height measurement in a small unmanned aerial vehicle (UAV) at 50 m altitude.

path, and the control system must correct this unintended change. The calculation of this type of drift is beyond the scope of this text, the interested reader is recommended to consult with references such as [54].

The navigation calculations in using inertial navigation system, i.e., dead reckoning, is subject to cumulative errors. It is subject to significant errors, since ground speed, and heading angle are usually not in all instances. The accuracy of dead reckoning is gradually decreased significantly, as the time passes, since each estimate of position is relative to the previous one. The primary source of error in navigation calculations is the error in three angular measurements by three gyros.

Using the format of the Euler's equations presented in (Section 5.2.2), the drift rate due to gyroscopic precession is obtained [54] by:

$$\dot{R} = \frac{I_{xx} - I_{yy}}{I_{zz}} PQ + Q\frac{H_r}{I_{zz}} \tag{8.28}$$

where N_g is the moment generated due to gyroscopic precession (i.e., created by rotor/propeller rotation), and H_r is the rotor angular momentum. This angular momentum is obtained by:

$$H_r = I_r \omega_r \tag{8.29}$$

where I_r is the rotor mass moment of inertia about its axis of rotation, and ω_r is the its rotational speed. This moment will cause a yaw rate (nose to the left or right). In deriving Eq. (8.28), it is assumed that all cross products; I_{xz}, I_{yz}, and I_{xy} are zero, and any changes in propulsion system and aerodynamic moments are neglected.

The drift direction follows the right-hand rule. When the rotor is operating about the x-axis (this is usually the case), a pitch up/down, will create a drift to the left/right. Similarly, a yaw to the left/right, will create a pitch up/down. Hence, in a UAV with a positive prop rotation, a positive pitch will generate a positive yaw. In contrast, in a UAV with a negative prop rotation, a positive pitch will generate a negative yaw.

In some references, together, pitch, and roll angles are known as *tilt*. When a UAV with a single engine is tilting, the prop/rotor will create a larger drift which must be nullified. One solution to nullify the drift due to rotation of rotor/propeller is to employ twin engines while their rotors/propellers rotate opposite each other (i.e., if the left engine is rotating clockwise, the right engine should rotate counterclockwise). To eliminate drift from a planned trajectory, the Synthetic Aperture Radar (SAR) has a potential to aid UAV navigation when INS measurements are not accurate enough.

Example 8.4
The propeller and rotor of a single electric engine of a fixed-wing UAV (with a weight of 100 lb) have a combined moment of inertia of 1 slug.ft^2 about the axis of rotation, and are rotating at 1200 rpm clockwise when viewed from in front. The roll, pitch, and yaw moments of inertia of the UAV are 130 slug.ft^2, 60 slug.ft^2, and 200 slug.ft^2 respectively. If the UAV rolls at 50°/s, while pitching at 5 deg/s, determine the angular acceleration in yaw due to gyroscopic precession. All inertias and angular rates are body-axes components.

Solution
The angular momentum of propeller and rotor of the engine is:

$$H_r = I_r\omega_r = 1\times 1{,}200\times\frac{2\pi}{60} = -125.6\,\text{slug}.ft^2/\text{sec} \tag{8.29}$$

The yaw angular acceleration is:

$$\dot{R} = \frac{I_{xx}-I_{yy}}{I_{zz}}PQ + Q\frac{H_r}{I_{zz}} = \frac{130-60}{200}\times 50\times 5 + 5\times\frac{(-125.6)}{200} = -1.614\frac{\text{deg}}{s^2} \tag{8.28}$$

Due to a positive roll and pitch, but negative prop angular rotation, the results would cause a negative yaw rate, which means the nose is yawed to the left.

8.10.4.2 Drift Due to Wind

Another source of heading deviation of an air vehicle from an intended original trajectory (i.e., drift) is a cross wind. A cross wind may be at right angle (i.e., 90°) to the flight direction, or an acute angle (i.e., less than 90°). In both cases, the wind will not change the airspeed, while it will vary the ground speed. The airspeed is employed in the calculations of aerodynamic forces (e.g., lift and drag) and moments (e.g., yawing moment). However, the ground speed is used in the navigation calculations (e.g., in Eqs. (8.1) through (8.5)).

The effect of a cross wind in navigation is presented in Section 8.10.4.2. The drift angle (i.e., deviation in heading angle) is readily calculated by treating both airspeed and wind speed as two vectors (Figure 8.18). Recall that, *course* is the direction an aircraft is moving, while heading is the direction the aircraft is pointing. The aircraft may be drifting a little due to a crosswind. Bearing is the angle between North and the direction to the destination.

One of the solutions to nullify the drift of an inertial navigation system is to enhance it with GPS. For instance, the SBG IG-500 N – a miniature inertial navigation system – provides precise drift-free attitude and position, even over aa long time, since it has been enhanced by GPS.

8.10.5 Coriolis Effect

Earth's rotation (as a moving frame) has a kinematic effect of any flying object (with a velocity of V_A); from aircraft to rocket. The Earth's angular velocity vector points

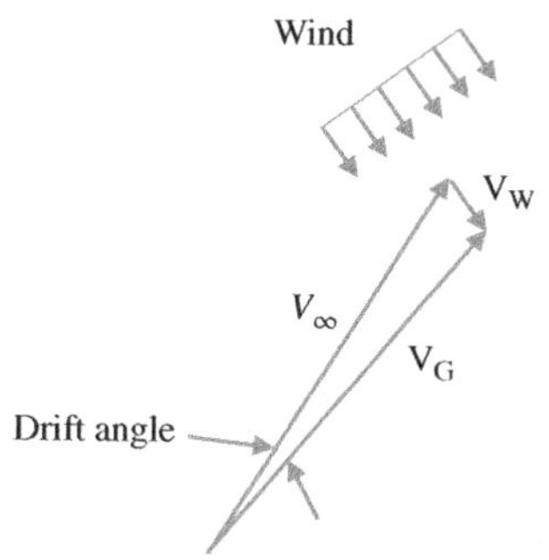

Figure 8.18 Drift due to the wind.

outward from the Earth's surface and, whichever way the velocity vector V_A is directed, the Coriolis force is directed to the right of V_A. In a frame of reference with clockwise rotation, the Coriolis force acts to the left of the motion of the object. This force tends to curve an object to the right in the Northern Hemisphere and to the left in the Southern Hemisphere.

The Coriolis force is an inertial force (object mass times object acceleration) that acts on objects which are in motion within a reference frame that rotates with respect to an inertial frame. Deflection of an object due to the Coriolis force is called the Coriolis Effect. The atmosphere rotates with the same rate as with the surface of the earth due to gravity. The Coriolis force is a direct function of (a) angular velocity of the Earth; (b) latitude; (c) velocity of the aircraft; and (d) UAV mass. The Coriolis acceleration of a flying object on the earth varies with its distance from the rotation axis (i.e., altitude). It is [125] obtained as:

$$F_C = ma_C \tag{8.30}$$

where Ω is the angular velocity of the Earth; which is 360° per about 24 hours or 7.292×10^{-5} rad/s, and V is the UAV ground speed. The $\times$ symbol represents the cross operator of two vectors. This indicates that the earth axis of rotation, the UAV velocity direction and the axis of the Coriolis force will create an orthogonal axis. As an example, if the axis of rotation is assumed as along the positive x, and the flight path as the positive y, the force is applied along the z-axis. The Coriolis force on a flying UAV is:

$$F_C = ma_C \tag{8.31}$$

where m is the mass of the UAV. The Coriolis force is applied on the cg of a UAV, and will change the aircraft ground speed, not airspeed. This effect is negligible for a small UAV with a short flight time. However, for a large UAV (such as Global Hawk) with a long flight duration (e.g., 40 hours), this effect is significant. For instance, a large UAV flying northward in the northern hemisphere, must tip its nose to the left to compensate, otherwise it will be deflected to the right. The Coriolis effect behaves the opposite way in the southern hemisphere.

Example 8.5

A large UAV with a mass of 15 000 kg is flying with a velocity of 300 knot northward in the northern hemisphere. The jet engine of this UAV is generating 30 kN of thrust in this flight. The UAV is planned to cruise for five hours at low altitude.

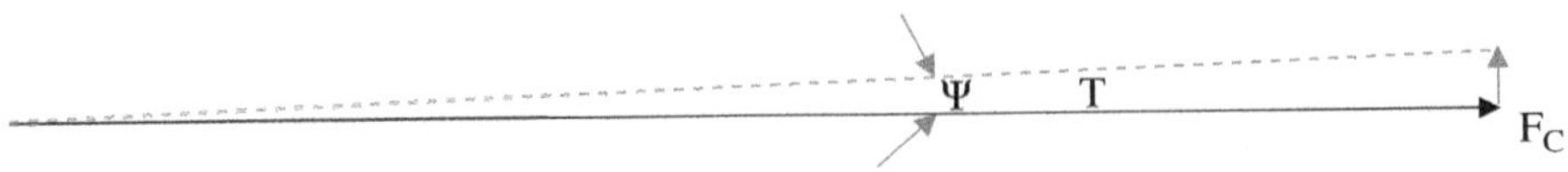

Figure 8.19 The drift due to Coriolis effect.

a) Determine the Coriolis force applied to the UAV.
b) Calculate the change in heading angle due to the Coriolis effect.
c) Assuming no wind, and if the flight direction is not corrected by the rudder, what distance will the UAV be away from its intended destination?

Assume the Coriolis force has a similar effect to the UAV motion as that of the engine thrust.

Solution

a) The Coriolis acceleration:

$$a_C = 2\Omega \times V = 2 \times 7.292 \times 10^{-5} \times 300 \times 0.5144 = 0.0225 \frac{m}{s^2} \tag{8.30}$$

The 0.5144 is used to convert the knot to m/s. The Coriolis force is:

$$F_C = ma_C = 15000 \times 0.0225 = 337.6\,N \tag{8.31}$$

b) Change in heading angle
There are two forces on the UAV: (a) engine thrust; and (b) Coriolis force. These two forces will push the UAV along a direction (Figure 8.19); which is readily calculated from trigonometric equations:

$$\Psi = \tan^{-1}\left(\frac{T}{F_C}\right) = \tan^{-1}\left(\frac{30{,}000}{337.6}\right) = 0.645\,\text{deg}$$

c) Distance away

$$X = Vt = 300 \times 0.5144 \times 5 \times 60 \times 60 = 2778\,km$$

$$Y = X\tan(\Psi) = 2778 \times \tan(0.645) = 31.26\,km$$

The UAV was planned to travel 2778 km northward. However, due to the Coriolis force, it will be 31.26 km to the left of its destination. This deviation should be corrected by applying a rudder.

8.10.6 Magnetic Deviation

The compass is a mechanical device to measure the direction of magnetic north. However, in trajectory design, the true north is the reference direction. True north (also called geodetic north) is the direction along Earth's surface toward the geographic North Pole. Compass deviation can also be caused by local magnetic effects.

Magnetic deviation[3] (expressed in degrees), sometimes called magnetic variation, is the angle between magnetic north and true north. The declination is positive when magnetic north is east of true north, and negative when it is to the west. Magnetic declination changes over time and with location. Maps of magnetic deviation are available for download from the National Oceanic and Atmospheric Administration (NOAA)[4] website.

As the compass points with local magnetic fields, declination value is needed to obtain true north. Hence, the actual north direction depends on latitude and longitude. The magnetic deviation must be nullified to have a correct flight trajectory. The total declination across the USA (See Figure 8.20) currently varies plus or minus 16 °.

8.11 Navigation System Design

In sections 8.4 through 8.10, various navigation techniques and equipment have been described. In designing navigation system for a UAV, one must select the type of navigation system, select the navigation devices, sensors, and then conduct calculations, simulation, and analysis. The design of the navigation system is conducted in parallel to the design of guidance and control systems.

8.11.1 Design Requirements

In general, the primary criteria for the design of navigation system are as follows: (a) manufacturing technology; (b) required accuracy; (c) range; (d) weather; (e) reliability; (f) life-cycle cost; (g) UAV configuration; (h) stealth requirements; (i) maintainability; (j) endurance; (k) communication system; (l) aerodynamic considerations; (m) processor; (n) complexity of trajectory; (o) compatibility with guidance system; (p) size; and (q) weight. Most of these requirements have conflicting effects; so, a kind of trade-off or multidisciplinary technique need to be employed in the final design.

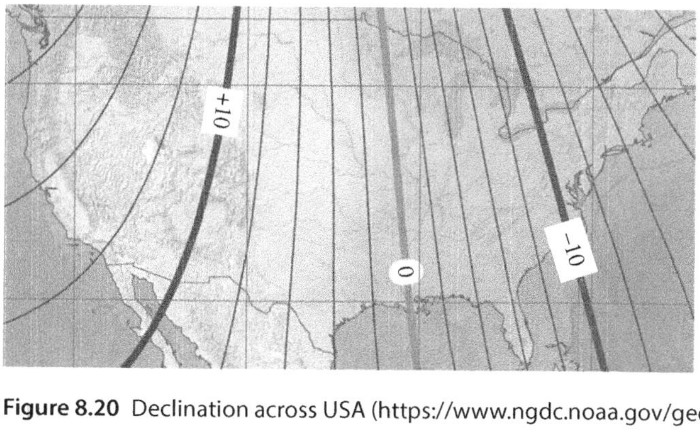

Figure 8.20 Declination across USA (https://www.ngdc.noaa.gov/geomag/WMM).

The level of complexity in the design of the navigation system is directly a function of the desired category/group/class/accuracy (Table 8.1) of navigation function.

8.11.2 Design Flowchart

In general, the design process begins with a trade-off study to establish a clear line between navigation equipment cost and performance (i.e., accuracy) requirements; and ends with optimization.

The designer must decide about two items: (a) select type of navigation system; (b) select the navigation sensors. After conducting the calculation process, it must be checked to make sure that the design requirements are met. A very crucial part of the design process is to integrate the navigation system with control and guidance systems (i.e., a consistent autopilot). If a complete navigation system (e.g., GPS) is selected/purchased, the

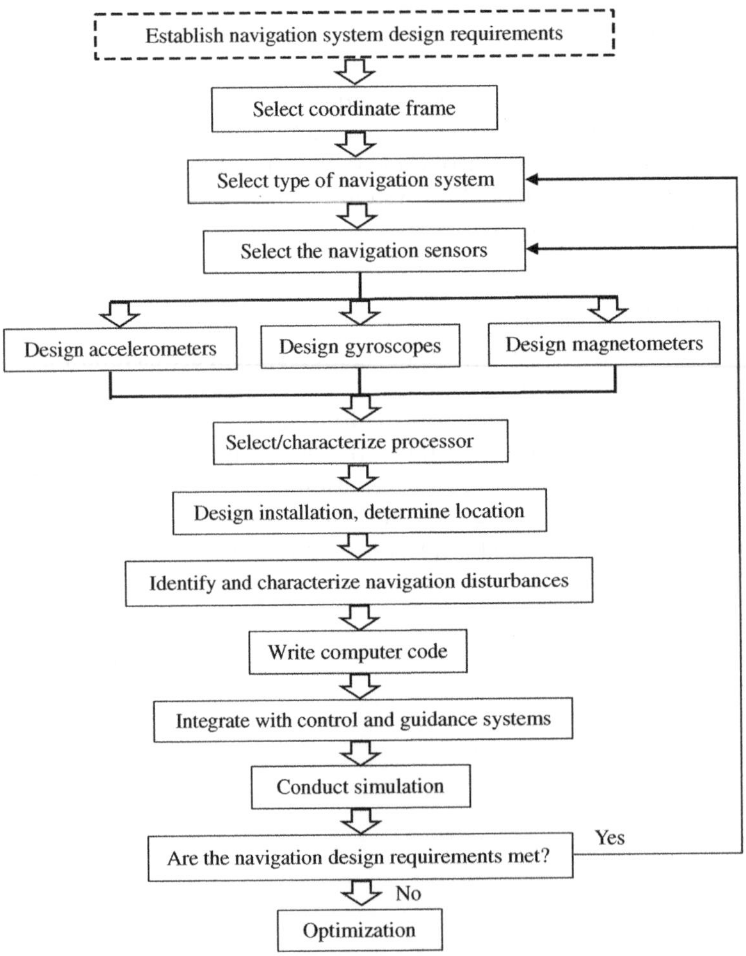

Figure 8.21 Inertial navigation system design process.

integration process must be still conducted. This includes activities such as matching frequencies, interfaces, and electric power requirements. More details are outside the scope of this work.

Figure 8.21 shows the inertial navigation system design process as a subsystem of the autopilot. The designer initially has to make three decisions: (a) select coordinate frame; (b) select type of navigation system; and (c) select the navigation sensors. Then, three basic navigation sensors should be designed/selected: (a) accelerometers; (b) gyroscopes; and (c) magnetometers. However, these three can be miniaturized in one small unit/chip (as in a MEMS device).

Another major step in the selection or characterization of the navigation processor, which can be the same processor of the autopilot. After, identifying and characterizing the probable navigation disturbances, a computer code should be written for compiling to the processor.

One of the challenging steps is the integration of navigation system with control and guidance systems. Compatibility of systems (inputs and outputs) and effectiveness of the overall system should be evaluated. By conduct a simulation (either a full numerical, or a human-in-the-loop version), one can figure out whether or not the navigation design requirements met. Some other guidelines for the navigation sensors installation, and more details of how to determine the best location, are provided in Chapters 1 and 12.

8.11.3 Design Guidelines

In designing navigation system, all design requirements must be examined. In general, the optimum navigation system is characterized by a tradeoff between the offered accuracy and the deployment costs. In principle, the higher the accuracy of a navigation system, the higher the deployment cost and the higher the value of the services that can be offered.

The design of navigation system should be conducted in sync with the design of control and guidance system. For instance, in an approaching flight, the output of navigation (position in x and y) are used as references for guidance system. In addition, the altitude h used as a feedback value for control system. Hence, all three systems must perform efficiently while exchanging command/signal to each other.

It is highly recommended to examine the performance of the navigation system by implementing in a hardware (for instance, on a single board computer) before testing in a real flight test. The inertial navigation is a low cost system, but provides less accurate data; hence it requires frequent checking of position from other sources such as GPS. Military inertial navigation systems cannot totally rely on GPS satellite availability for course correction due to probable blocking or jamming of signal, or flying in the areas with unavailable GPS signals. In order to lower costs and improve reliability, there are ways to integrate input from GPS into INS-based navigation with various levels of integration. In an integrated GPS/INS system, the GPS will pass on position data to update the inertial navigation system. If the GPS is jammed, the navigation computer will rely solely on the INS.

In integrated GPS/INS applications, Kalman filters of 15–24 state variables are optimal. The tightly coupled GPS/INS systems are commonly used in such applications.

In a typical GPS/INS application, two groups of state variables are: 1. INS error model (3 position error states, 3 velocity error states, 3 platform tilts, 3 gyroscope drift rate errors, and 3 accelerometer biases); 2. GPS error model (3 user position components, 3 user velocity components, 1 user clock bias, and 1 user clock bias rate). The final selection of the appropriate state variables will depend on parameters such as the mission requirements, computational load, accuracy, and cost.

In many applications, GPS/INS integration is necessary. This integration has proved to be a very efficient means of navigation, primarily because of the short term accuracy achieved by the INS and the long-term accuracy of the GPS fixes. Two versions of the GPS/INS integration are available. These are: 1. tightly coupled GPS/INS, and 2. loosely couple or modular GPS/INS. The tightly coupled version has the ability to perform optimal signal processing and allows the various errors and noise sources acting on both the GPS and INS units to be taken into account in a global way.

In designing a Kalman filter for GPS application, there is a trade-off between optimal implementation and computational cost. Reducing number of states lowers the computational load. For instance, an 11-state (3 acceleration, 3 velocity, 3 position, and 2 clock states) filter is the optimal Kalman implementation for a high speed flight. Moreover, an 8-state Kalman filter design would be sufficient for a low speed flight. However, the INS for high speed UAVs can be modeled with 10 states.

Questions

1 What is primary function of navigation system?

2 List level 3 classification of the vision-based perception.

3 List level 3 classification of the Lidar-based perception.

4 List level 3 classification of the vision-based state estimation.

5 List level 2 classification of the perception process.

6 List level 2 classification of the state estimation.

7 Draw a UAV and in that figure, show three body-fixed axes and their positive directions.

8 By using a figure, explain the ECEF Cartesian coordinate system.

9 What are two important parameters of the spheroid that are required for navigation calculations?

10 Briefly describe the GIS.

11 Briefly describe the World Geodetic System (WGS84).

12 To express the difference between the real Earth and modeled Earth, three types of heights are employed. What are they?

13 What is the value for the earth equatorial radius in semi-major axis?

14 What is the Earth Eccentricity?

15 What basic sensors are needed in an INS?

16 What is the inertial measurement stable platform system?

17 Draw a block diagram to indicate the flow of signals and input/output in a navigation process.

18 Describe the WGS.

19 What is EKF?

20 What information is needed in a Kalman filter?

21 What is SBG IG-500 N?

22 Briefly describe the system integration and operation of an instrumented Yamaha RMAX unmanned helicopter.

23 Draw a figure to show Earth Longitude and Latitude.

24 Define Longitude and Latitude.

25 Discuss the difference between the ground speed and airspeed.

26 Briefly describe the Celestial Navigation.

27 Briefly describe how navigation can be done in reduced visibility conditions.

28 Draw a block diagram to show a roll-angle-hold autopilot.

29 Briefly discuss the differences between the traditional gyros and microelectromechanical gyro systems.

30 Discuss how a radar altimeter functions.

31 What are the convenient locations to install pitot-static tube in a UAV?

32 What is dead reckoning?

33 Briefly describe the MEMS Inertial Module.

34 Discuss technical features of MPU-60X0 motion tracking device.

35 Discuss technical features of Adafruit 9-DOF IMU.

36 What are the two major sensors in inertial navigation?

37 What does IMU stand for?

38 What does GPS stand for?

39 Name three basic coordinate systems.

40 What is the function of an accelerometer?

41 What is the function of a rate gyroscope?

42 What is the function of a magnetometer?

43 Describe number of and the configuration of GPS satellites.

44 What is the process to convert the output signal of an accelerometer to a linear velocity?

45 What are the input and output parameters of navigation equations?

46 Describe the elements and the operation of an early mechanical accelerometer.

47 Describe the basic principle of working gyro.

48 What is the main difference between an attitude gyro and a rate gyro?

49 Describe the function of a roll-angle-hold autopilot.

50 What is the law of the gyro?

51 What are the name of three axes of a gyro?

52 Describe the inertial navigation system design process.

53 Name three activities included in the integration process of a navigation system.

54 What are the primary criteria for the design of navigation system?

55 What are the major navigation disturbances?

56 Briefly compare headwind, tailwind, and cross wind.

57 What parameters are measured in a rate gyroscope?

58 What is gust?

59 What is the Coriolis Effect?

60 What is the gyroscopic effect?

61 What is the direction of Coriolis force on the northern hemisphere?

62 Define gyro drift.

63 What are the reasons for drift in a flight?

64 How noise can be removed from a measured signal?

65 Define noise as a navigation disturbance.

66 What issue should be considered in installing gyroscopes?

67 What issue should be considered in installing accelerometers?

68 Describe basics of Kalman filtering.

69 Briefly introduce the discrete gust idealization which consists of a one-minus-cosine pulse.

70 Discuss how a drift due to the rotation of rotor/propeller is created.

71 What is the gyroscopic precession?

72 What is magnetic deviation?

73 How much is approximate value for the magnetic deviation in New York City?

74 Describe basics the inertial navigation system design process.

Problems

1 A UAV is planned to fly westward at an airspeed of 50 knot for two minutes to reach a target. A cross wind is blowing from right (South) with a speed of 7 knot.
- If the UAV does not apply any correction to its heading, what would be the miss distance?
- Determine 1. Heading angle that UAV needs to take, 2. New airspeed, in order to nullify the drift created by the wind.

2 A UAV is planned to fly northward at an airspeed of 150 knot for 30 minutes to reach a target. A cross wind is blowing from right (East) with a speed of 7 knot.
- If the UAV does not apply any correction to its heading, what would be the miss distance?
- Determine 1. Heading angle that UAV needs to take, 2. New airspeed, in order to nullify the drift created by the wind.

3 A UAV has flown from City of Philadelphia with the GPS latitude and longitude of 39.9526° and − 75.1652° to the city of New York with the latitude and longitude of 40.730610° and − 73.935242°. Assume sea level altitude.
- Determine the distance the UAV has flown in this flight.
- Determine the flight direction of the UAV with respect to North.

4 A UAV has flown from City of Los Angeles with the GPS latitude and longitude of 34.0522° N, 118.2437° W to the city of San Francisco with the latitude and longitude of 37.7749° N, 122.4194° W. Assume sea level altitude.
- Determine the distance the UAV has flown in this flight.
- Determine the flight direction of the UAV with respect to North.

5 The propeller and rotor of a single electric engine of a fixed-wing UAV (with a mass of 2 kg) have a combined moment of inertia of 0.06 kg.m^2 about the axis of rotation, and are rotating at 500 rpm counterclockwise when viewed from in front. The roll, pitch, and yaw moments of inertia of the UAV are 7 kg.m^2, 3 kg.m^2, and 12 kg.m^2 respectively. If the UAV rolls at 10 deg/s, while pitching at 3 deg/s, determine the angular acceleration in yaw due to gyroscopic precession. All inertias and angular rates are body-axes components.

6 The propeller and rotor of a single electric engine of a fixed-wing UAV (with a mass of 15 kg) have a combined moment of inertia of 0.6 kg.m^2 about the axis of rotation, and are rotating at 500 rpm counterclockwise when viewed from in front. The roll, pitch, and yaw moments of inertia of the UAV are 30 kg.m^2, 10 kg.m^2, and 50 kg.m^2 respectively. If the UAV rolls at 15 deg/s, while pitching at 5 deg/s, determine the angular acceleration in yaw due to gyroscopic precession. All inertias and angular rates are body-axes components.

7 The propeller and rotor of a single electric engine of a fixed-wing UAV (with a weight of 40 lb) have a combined moment of inertia of 0.4 slug.ft^2 about the axis of rotation, and are rotating at 1100 rpm clockwise when viewed from in front. The roll, pitch, and yaw moments of inertia of the UAV are 62, 28, and 110 slug.ft^2 respectively. If the UAV pitches at 4 deg/s, determine the angular acceleration in yaw due to gyroscopic precession. All inertias and angular rates are body-axes components.

8 The propeller and rotor of a single electric engine of a fixed-wing UAV (with a weight of 10 000 lb) have a combined moment of inertia of 120 slug.ft^2 about the axis of rotation, and are rotating at 2000, rpm counterclockwise when viewed from in front. The roll, pitch, and yaw moments of inertia of the UAV are 13 000, 8200, and 23 000 slug.ft^2 respectively. If the UAV pitches at 3 deg/s, determine the angular

acceleration in yaw due to gyroscopic precession. All inertias and angular rates are body-axes components.

9 A UAV is cruising eastward at sea level altitude with an airspeed of 150 knot. The UAV is located 12 km East of a flight base, and it is assumed there is no wind. Determine the distance of the UAV from this base after 30 minutes.

10 A small UAV is cruising westward at sea level altitude with an airspeed of 30 knot. The UAV is located 100 m West of a flight base, and it is assumed there is no wind. Determine the distance of the UAV from this base after 12 minutes.

11 A Global Hawk has flown from City of Atlanta with the GPS latitude and longitude of 33.7490° and 84.3880° to the City of Melbourne, Australia with the latitude and longitude of 37.8136° S, 144.9631° E. Assume 50 000 ft altitude.
   a Determine the distance the UAV has flown in this flight.
   b Determine the flight direction of the UAV with respect to North.
   c If the airspeed was 320 knots, while there was a wind of 60 knot (at 90° from right side of the UAV) along the flight path, determine the drift due to this wind.
   d How long was this UAV in the air? The drift has been canceled out with the new heading angle.

12 A UAV has flown from City of Boston with the GPS latitude and longitude of 42.361145, and – 71.057083 to the city of New York with the latitude and longitude of 40.730610,–73.935242. Assume 20 000 ft altitude.
   a If the airspeed was 180 knots, while there was a wind of 35 knot (at 90° from left side of the UAV) along the flight path, determine the drift due to this wind.
   b How long this UAV was in the air? The drift has been canceled out with the new heading angle.

13 A large UAV with a mass of 20 000 kg is flying with a velocity of 320 knot northward in the northern hemisphere. The jet engine of this UAV is generating 40 kN of thrust in this flight. The UAV is planned to cruise for seven hours at low altitude.
   a Determine the Coriolis force applied to the UAV.
   b Calculate the change in heading angle due to the Coriolis effect.
   c Assuming no wind, and if the flight direction is not corrected by the rudder, what distance will the UAV be away from its intended destination?

14 A large UAV with a mass of 10 000 kg is flying with a velocity of 220 knot northward in the northern hemisphere. The turbofan engine of this UAV is generating 18 kN of thrust in this flight. The UAV is planned to cruise for two hours at low altitude.
   a Determine the Coriolis force applied to the UAV.
   b Calculate the change in heading angle due to the Coriolis effect.
   c Assuming no wind, and if the flight direction is not corrected by the rudder, what distance will the UAV be away from its intended destination?

15 Write a MATLAB code for simulation of navigation system based on Eqs. (8.1) through (8.3). Assume all three components of velocities (U, V, and W) and all

three flight angles (ϕ, θ, ψ) are given. The goal is to find the instantaneous UAV location (i.e., x, y, and z).

16 Develop a MATLAB Simulink model for simulation of navigation system based on Eqs. (8.1) through (8.3). Assume all three components of velocities (U, V, and W) and all three flight angles (ϕ, θ, ψ) are given. The goal is to find the instantaneous UAV location (i.e., x, y, and z).

17 A UAV is flying at an altitude where its pitotp tube has measured the total pressure as 90 kPa, and static pressure to be 85 kPa. If the flow is assumed as incompressible, and air density is known to be 1.1 kg/m^3, what is the UAV flight velocity?

18 A UAV is flying at an altitude where its pitot tube has measured the total pressure as 98 kPa, and static pressure to be 96 kPa. If the flow is assumed as incompressible, and air density is known to be 1.15 kg/m^3, what is the UAV flight velocity?

19 The following is a supersonic flow that hits the leading edge of a wing of a UAV which has 15° leading edge angle. The static pressure on the surface of wing is measured to be P_2 =1.8 MPa. A pitot tube is placed near the surface, and it measures the total pressure to be P_o =10.152 MPa. Calculate the flight Mach number M_1. Note that, there is a normal shock wave in front of the pitot-tube.

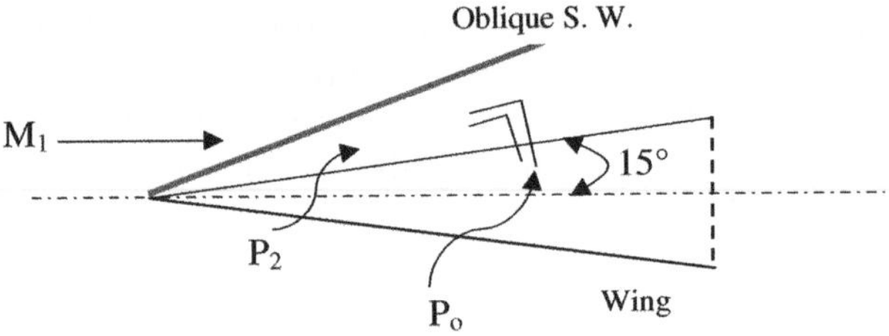

20 Develop a MATLAB code for implementation of the navigation Eqs. (8.1) through (8.3). The code should be written such that it can take outputs of accelerometers and gyroscopes to deliver UAV position, velocity and attitude angles.

Notes

1 www.xsens.com.
2 www.invensense.com.
3 Magnetic deviation is used by some references to mean the same as magnetic declination.
4 https://www.ngdc.noaa.gov/geomag/WMM/image.shtml.

9

Microcontroller

CONTENTS

Design of Unmanned Aerial Systems, First Edition. Mohammad H. Sadraey.

Companion website: www.wiley.com/go/sadraey/unmanned-aerial-systems

Educational Outcomes

After study of this chapter, the reader will be able to:

1) Discuss basic fundamentals of a microcontroller
2) Explain microcontroller circuitry
3) Explain microcontroller classifications and components
4) Develop microcontroller wiring for a given requirement
5) Implement microcontroller programming
6) Develop programming in C
7) Develop Arduino programming
8) Develop Matlab programming to control a device via Arduino boards
9) Explain high-level programming languages
10) Employ Arduino boards for a set of control requirements
11) Employ open-source commercial autopilots for a flight mission
12) Understand embedded systems
13) Develop microcontroller design/development procedure
14) Conduct a design project for a control system using an Arduino Uno board.

9.1 Introduction

The autopilot is as an essential component of an Unmanned Aerial Vehicle (UAV) to execute the automatic flight control activities. Four major autopilot functions are: 1. Guidance, 2. Navigation, 3. Control, and 4. Tracking. An autopilot should be able to: 1. Measure flight parameters; 2. Process data; 3. Make decisions; 4. Create commands. All these autopilot activities may be performed via a microcontroller, so the heart of an UAV autopilot is the microcontroller. A microcontroller is essentially an integrated circuit (IC) that is programmed to a do a specific task. By reducing the size and cost compared to a design that uses a separate microprocessor, memory, and input/output devices, microcontrollers make it economical to digitally control even more devices and processes.

The majority of microcontrollers in use today are embedded in devices and machinery equipment. Microcontrollers are used in automatically controlled products and devices, such as: (a) automobile cruise control; (b) implantable medical devices; (c) remote controls; (d) office machines; (e) appliances; (f) power tools; (g) toys; (h) automatic door opener; and (i) remotely controlled airplanes. In this chapter, the application and design of microcontroller for UAV autopilot is discussed.

Microcontrollers must provide real-time response to what happens in the embedded system they are controlling. A customizable microcontroller incorporates a block of digital logic that can be personalized for additional processing capability. Microcontrollers must usually have a low-power requirement (0.5–1 W); since many devices they control are battery-operated. Microcontrollers are designed for embedded applications, in contrast to the *microprocessors* (also known as Central Processing Unit or CPU) used in personal computers or other general-purpose applications, mainly used as data processors. A microcontroller [126] can be considered a self-contained system with a processor, memory and peripherals and can be used as an embedded system. Integrating all elements on one chip saves space and leads to both lower manufacturing costs and shorter development times.

Figure 9.1 illustrates a typical microcontroller connection (inputs/outputs) in an UAV. It receives signals from devices such as gyroscope, and then sends signals to elements such as engine. Some microcontroller designers/manufacturers/architectures/vendors are Intel; Microchip; ARM; Texas Instrument; Ardupilot (e.g., see Figure 9.2); Toshiba, Philips, Atmel, Siemens, Pilots-in-Command (PIC), and Motorola.

The Raspberry Pi is a series of small single board computers developed by Raspberry Pi to promote teaching of basic computer science in schools. The concept of this the best-selling British computer to inspire children is based on a microcontroller. A major application area for microcontrollers involves *embedded systems*. In embedded systems, the control unit is integrated into the system. These units are generally used in real-time systems, where the reaction to a control event has to occur within a specified time.

This chapter presents the fundamentals of the microcontroller, and how to set up a microcontroller to perform various functions of an autopilot. The level of complexity of a microcontroller depends on the UAV mission, and the equipment to be controlled during a flight operation. For instance, when the mission range is longer than the line of

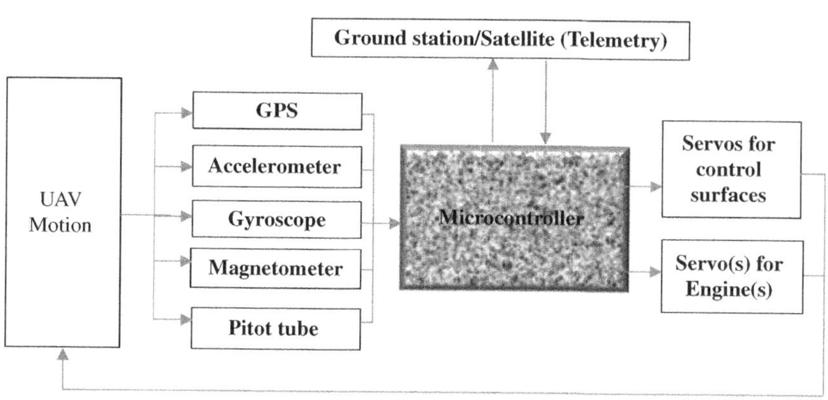

Figure 9.1 Microcontroller connections (inputs/outputs).

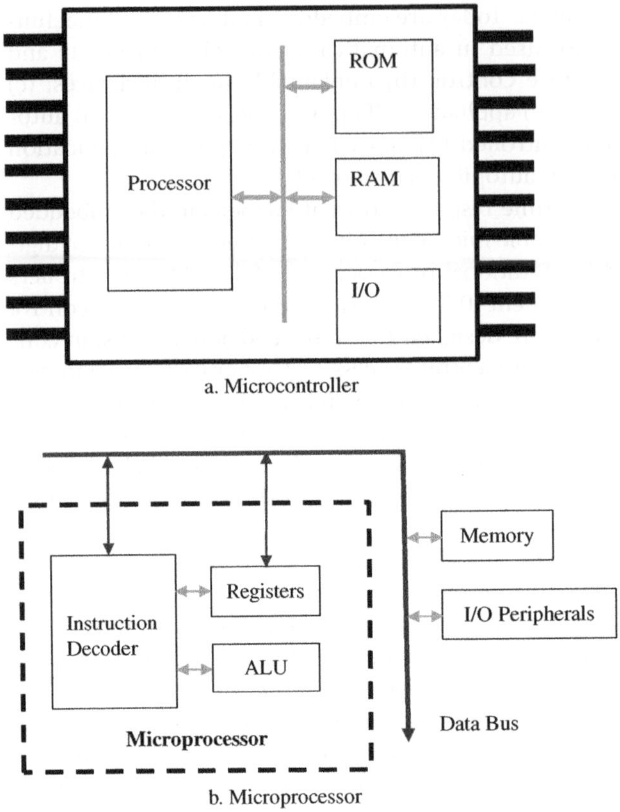

Figure 9.2 The structures of a microcontroller and a microprocessor.

sight, there is a need for sophisticated radars and powerful CPUs. As the UAV range and endurance are increased, the microcontroller requires more supporting equipment, with a higher level of complexities. Moreover, an example microcontroller, the Atmel ATmega644P, and Arduino, popular electronics ready to be employed are introduced in detail. Furthermore, a few popular ready-to-use autopilots, particularly Ardupilot, Micropilot, DJI WooKong, and PX4 PixHawk, are briefly reviewed.

Due to popularity of Arduino boards, the Arduino programming for reading UAV sensor data, and controlling UAV payloads are explained. In addition, the application of Matlab special support packages for Arduino hardware will also be introduced. Hence, the reader will learn how to employ a matlab code and Simulink model to collect a UAV sensor data and to control a UAV payload via an Arduino board.

9.2 Basic Fundamentals

9.2.1 Microcontroller Basics

A microcontroller is an electronic device with several elements to allow a UAV to monitor the flight, control its flight operations, and activate the payloads. The common features of a microcontroller are briefly described in this section. The peripherals of a microcontroller are integrated into a single chip, so, the overall weight and cost is very low. It has a CPU, in-circuit programing, and in-circuit debugging support. It is also equipped with Analog-to-Digital (A/D) converters, some include Digital-to-Analog (D/A) converters. For program, data storage, and operating parameters storage, various types of memories such as RAM[1], ROM[2], and flash memory are utilized.

Table 9.1 demonstrates the elements which send/receive commands/signals to/from microcontroller. In general, the following elements send signals to a UAV microcontroller: (a) receiver; (b) Global Positioning System (GPS); (c) pitot-tube through A/D converter; (d) accelerometer; (e) attitude gyroscope; (f) rate gyroscope; (g) magnetometer; (h) compass; (i) camera; and (j) sensors. Some of these elements send a signal (voltage or digital signal) directly to the microcontroller. However, the signals of most elements (e.g., pitot-tube) must be converted to a digital form through A/D converter before going to the microcontroller.

In contrast, the following elements receive commands/signals from the UAV microcontroller: (a) elevator servo; (b) aileron servo; (c) rudder servo; (d) lap servo; (e) engine; (f) landing gear; (g) display; (h) transmitter; (i) payload (e.g., store); and (j) sensors. These are intermediators between the microcontroller and the Ground control station (GCS). Some of these elements receive a signal (voltage or digital signal) directly from the microcontroller. However, the microcontroller signals to most elements (e.g., elevator servo) must be converted to an analog form (e.g., force) through D/A converter before coming to the element.

The receiver and transmitter usually communicatee with the GCS, so they play the role of an intermediator between microcontroller and the GCS. There are primarily four elements (Figure 9.2) within a microcontroller: (a) microprocessor; (b) ROM; (c) RAM; and (d) I/O. In contrast, there are three elements within a microprocessor: (a) instruction decoder; (b) registers; and (c) Arithmetic logic unit (ALU). The memory and input/output peripherals are physically out of the microprocessor and are communicating through the data bus.

Table 9.1 Elements which sends/receive signals/commands to/from microcontroller.

No	Devices that provide input to microcontroller → [M]	Devices that microcontroller sends commands to [M] →
1	Receiver	Elevator servo
2	GPS	Aileron servo
3	Pitot-tube	Rudder servo
4	Accelerometer	Flap servo
5	Attitude gyroscope	Engine
6	Rate gyroscope	Landing gear
7	Magnetometer	Display
8	Compass	Transmitter
9	Camera	Payload
10	Sensors	Sensors

For the purpose of system interconnect, various interfaces [127] and serial input/output such as serial ports, Inter-Integrated Circuit; I2C; (also known as Two-Wire Interface; TWI), serial peripheral interface (SPI), and controller area network are installed. Most microcontrollersalso have peripherals such as timers, event counters, Pulse Width Modulation (PWM) generators, and watchdog clock generator. Other possible items are: (a) power supply; (b) Universal Serial Bus (USB) interface; (c) pin header; (d) Liquid crystal display (LCD); (e) buttons; and (f) Potentiometers (Pot). The size and cost of a microcontroller is a function of computing needs; e.g., speed, capacity of memory, number of I/O (Input/Output) ports; power consumption, and timers. The microcontrollers are easily upgradable.

A microcontroller [128] is a small computer on a single IC containing: (a) processor core; (b) memory; (c) programmable input/output peripherals; and (d) computer code. A microcontroller is a small, low-cost computer-on-a-chip; it is often used to run dedicated code that controls one or more tasks in the operation of a device or a system. They are called embedded controllers, because the microcontroller and support circuits are often built into, or embedded in, the devices they control.

In many cases, a microcontroller communicates bi-directionally with the wireless telemetry system which relays data back to a ground station. This allows the user to receive real time data about the aircraft, and to be able to update commands or change the mission parameters on-the-fly. A remote control input is a signal that can be passed through the microcontroller to the servos and speed controllers if the user wants to manually override the flight controller with a remote control transmitter and the UAV is within range. Microcontrollers may be classified from various aspects including bits, memory, instruction set, and architecture. Figure 9.3 illustrates various types [129] of microcontrollers. The term CISC denotes Complex Instruction Set Computer, while RISC stands for Reduced Instruction Set Computer. Microcontrollers are frequently RISC.

The Harvard architecture demands that program and data are in separate memories which are accessed via separate buses. In consequence, code accesses do not conflict with data accesses which improves system performance. As a slight drawback, this architecture requires more hardware, since it needs two busses and either two memory chips or a dual-ported memory (a memory chip which allows two independent accesses at the same time).

In the Von Neumann Architecture, program and data are stored together and are accessed through the same bus. This implies that program and data accesses may conflict (resulting in the famous von Neumann bottleneck), leading to unwelcome delays.

A UAV designer has at least four options for number of bits (4, 8, 16, or 32); two options for type of memory (embedded versus external); two options for instruction set

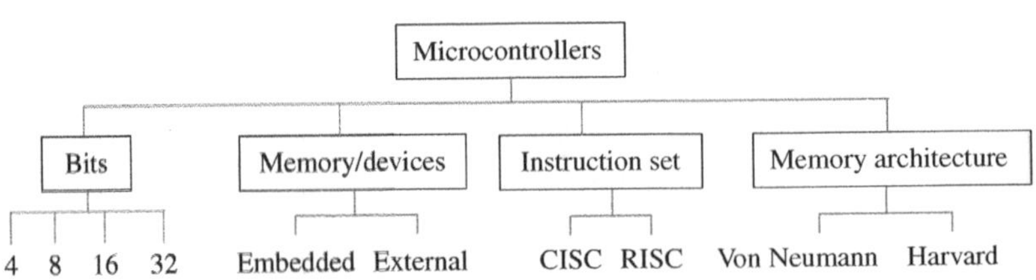

Figure 9.3 Types of microcontroller.

(CICS versus RICS); and two type of memory architecture (Von Neumann versus Harvard). The main criterion for selecting the best option is to weigh cost versus performance. The optimum selection may be determined through an optimization process. Since this chapter covers the fundamentals of the microcontroller, having at least one computer science engineer is a must in the team of UAV design/development.

9.2.2 Microcontroller Versus Microprocessor

A microcontroller (MCU) has many similarities and differences with a microprocessor (MPU). Both of them have been designed for real time application. They share many common features, while at the same time, they have various differences. Some form of microprocessors and microcontrollers look alike, such that they cannot be distinguished by visual inspection. A computer requires a processor/microprocessor to support myriad computing functions. In contrast, an autopilot requires a microcontroller to support various real-time computing/controlling functions. In both microprocessor and a microcontroller, algorithms must be executed on digital computation devices.

Microprocessor has only the processing power (such as Intel's Pentium IV, core 2 duo, and i7). The microprocessors do not have RAM, ROM, and other peripheral on the chip. Thus, RAM, ROM, and I/O are internal elements of a microcontroller. A microcontroller contains one or more CPUs (processor cores) along with memory and programmable input/output peripherals. A microcontroller is designed for embedded applications, in contrast to a microprocessor, which is mainly used in personal computers or laptops.

Table 9.2 illustrates a comparison [126] between a microprocessor and a microcontroller. Microprocessors available to aerospace generally lag the performance available for ground-based computing. Microcontrollers are an integral part of autopilots, Inertial Navigation System (INS), Vehicle management system (VMS), mission management system (MMS), and robots. Microprocessors are primarily employed in desktop computers, laptops, notepads, iPads, and cellphones. In general, microprocessors are larger and more expensive than microcontrollers.

Microcontrollers have embedded systems which usually have no keyboard, screen, disks, printers, or other recognizable I/O devices of a personal computer, and they may lack human interaction devices of any kind. Microcontrollers may not implement an external address or data bus, as they integrate RAM and non-volatile memory on the same chip as the CPU.

The datasheets for microcontrollers are frequently several times that of other devices, often running into the hundreds of pages. This is partly due to the number of devices, called peripherals, that are internal to the microcontroller, and partly due to the required listings of all the special function registers (SFR). SFRs are specific locations in the memory map that have predefined tasks. The memory map is the linking of a memory address, which is the value used to determine where the desired information is, to the physical location of that data.

9.2.3 Packaging Formats

Microcontrollers are manufactured/packaged in various forms depending upon application, functional requirements, and input/output features. In general, there are three

Table 9.2 A comparison between a microprocessor and a microcontroller.

No	Criterion	Microprocessor	Microcontroller
1	**Cost**	High	Low
2	**Energy use**	Medium to high	Very low to low
3	**Applications**	Desktop computers, Laptops, notepads, iPad, Cellphone	Appliances, digital camera, toys, robots
4	**Speed**	Very fast	Slow to fast
5	**External parts**	Many	Few
6	**Manufacturers**	Intel, ARM, AMD, IBM, Samsung, Motorola	Texas Instruments, Microchip, Fujitsu, Silicon Labs, Intel
7	**Internal Memory**	No	Yes
8	**Storage**	Hard disk (128 GB up to 2 TB)	Flash memory (32 kB –2 MB)
9	**Typical form**		

categories for microcontroller packaging: (a) through-hole mount package; (b) surface mount package; and Single chip package (SCP).

In a **through-hole mount package**, leads are used on the components that are inserted into holes (drilled in a printed circuit board; PCB), and then soldered to pads on the opposite side. This packaging form provides a strong mechanical bond compared with a surface-mount package. Most through-hole packages are in the form of dual in-line package (DIP) or dual in-line pin package (DIPP), with a regular housing and two parallel rows of electrical pins. The number of leads could vary from as low as 4 to as high as 40. The through-hole package is generally bigger, and much easier to work with. It is designed to be stuck through one side of a board and soldered to the other side.

The components in **surface mount packaging technology** (SMT) are mounted directly onto the surface of a PCB. It may have short pins or leads of various styles, flat contacts (e.g., Quad flat package; QFP), and a matrix of solder balls (e.g., Ball Grid Array; BAGs). Since this package is smaller than its through-hole counterpart, they are more widely used in mechatronic inductees. In addition, they need fewer holes to be drilled through abrasive boards. This packaging technique has a few disadvantages:

1) The manufacturing process is much more sophisticated.
2) It cannot be used directly with breadboard.
3) The solder connection may be damaged by potting components going through thermal cycling.

No:	1	2	3
Sample			
Pins	28	100	-
Name	Through-hole mount package	Surface mount package (Quad flap)	Chip-scale package (CSP)

Figure 9.4 Microcontroller Packaging.

The **chip-scale package** (CSP) is kind of IC chip, and manufactured in many forms such as flip-chip, wire-bonded, ball grid array, and leaded. A special type of CSP is the wafer level CSP. This packaging technology is more sophisticated than the other two, so it is more expensive.

Using fewer pins, the microcontroller chip can be placed in a much smaller, cheaper package. Many embedded systems need to read sensors that produce analog signals. This is the purpose of the A/D converter.

Microcontrollers usually contain from several to dozens of General Purpose Input/Output pins (GPIO). The GPIO pins are software configurable to either an input or an output state. When GPIO pins are configured to an input state, they are often used to read sensors or external signals. Configured to the output state, GPIO pins can drive external devices such as LEDs or motors, often indirectly, through external power electronics.

One of the influencing parameters in the packaging is the interfacing (due to certain I/O functions). The interface to the microcontroller is the connections to external devices, which could range from other microcontrollers, other microprocessors, or peripheral devices with which the microcontroller communicates. Figure 9.4 illustrates three samples of microcontroller packaging and their features.

9.2.4 Modules/Components

Figure 9.5 demonstrates a basic layout [130] of and a typical microcontroller connections (inputs/outputs). The main modules are processor core, Static random access memory (SRAM), EEPROM[3], time, digital I/O module, serial interface, analog module, and interrupt module. These modules typically found in a microcontroller. Typical microcontroller input and output devices include: (a) switches; (b) relays; (c) solenoids; (d) Light emitting diode (LEDs); (e) small or custom liquid-crystal displays; (f) radio frequency devices; and (g) sensors for data such as temperature, humidity, light level. A number of measurement devices such as pitot-tube, GPS, gyroscope, magnetometer, altimeter, accelerometer, and rate gyro are communicating with the microcontroller via interfaces and modules. In this section, the major modules of a microcontroller are briefly described.

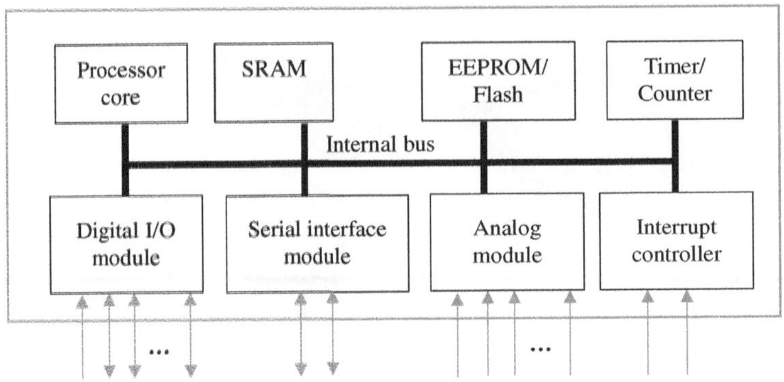

Figure 9.5 Basic layout of a microcontroller.

1) **Processor Core:** The processor core (CPU) is the main part of any microcontroller. It contains the ALU, the control unit, and the registers (stack pointer, program counter, accumulator register, register file,...). Digital equipment only use 0 and 1 to represent numbers and letters.
2) **Memory:** A memory is an element to store information. The memory is split into program memory and data memory. In larger controllers, a Direct memory access (DMA) controller handles data transfers between peripheral components and the memory.
3) **Timer/Counter:** The timer/counter is used to timestamp events, measure intervals, and count events. Most controllers have at least one timer/counter. Timers are used for a variety of tasks ranging from simple delays over measurement of periods to waveform generation. The most basic use of the timer is in its function as a counter. Timers generally allow the user to timestamp external events, to trigger interrupts after a certain number of clock cycles, and even to generate pulse-width modulated signals for motor control. Many controllers also contain PWM[4] outputs, which can be used to drive motors. Furthermore, the PWM output can, in conjunction with an external filter, be used to realize a cheap digital/analog converter. Each timer is basically a counter which is either incremented or decremented upon every clock tick.
4) **Digital I/O:** A signal is either digital (combinations of 0 and 1) or analog (such as voltage or force). The microcontroller digitizes an analog value by mapping it to one of two states, logical 0 or logical 1. The analog and digital signals are transferred differently. Parallel digital I/O ports are one of the main features of microcontrollers. The ability to directly monitor and control hardware is the main characteristic of microcontrollers. As a consequence, practically all microcontrollers have at least 1–2 digital I/O pins that can be directly connected to hardware. The number of I/O pins varies from 3 to over 90, depending on the controller family and the controller type. I/O pins are generally grouped into ports of 8 pins, which can be accessed with a single byte access.
5) **Analog I/O:** The analog I/Os are the way to transfer analog signals. Apart from a few small controllers, most microcontrollers have integrated analog/digital converters, which differ in the number of channels and their resolution (8–12 bits). The analog module also generally features an analog comparator. In some cases, the microcontroller includes digital/analog converters.

6) **Interrupt Controller:** Interrupts are useful for interrupting/stopping the normal program flow in case of (important) external or internal events. In conjunction with sleep modes, they help to conserve power.
7) **Interfaces:** Controllers generally have at least one serial interface which can be used to download the program and for communication with the user. Serial interfaces also are used to communicate with external peripheral devices; most controllers offer several and varied interfaces. Many microcontrollers also contain integrated bus controllers for the most common busses. Larger microcontrollers may also contain USB, or Ethernet interfaces.

   The basic purpose of any interface is to allow the microcontroller to communicate with other units, (e.g., other microcontrollers, peripherals, or a host computer). The implementation of such interfaces can take many forms, but basically, interfaces can be categorized according to a hand-full of properties: They can be either *serial or parallel*, synchronous or asynchronous, use a bus or point-to-point communication, be full-duplex or half duplex, and can either be based on a master–slave principle or consist of equal partners.
8) **Watchdog Timer:** Since safety-critical systems form a major application area of microcontrollers, it is important to guard against errors in the program and/or the hardware. The watchdog timer, also sometimes called COP[5],is used to monitor software execution. The basic idea behind this timer is that once it has been enabled, it starts counting down. When the count value reaches zero, a reset is triggered, thus reinitializing the controller and restarting the program. To avoid this controller reset, the software must reset the watchdog before its count reaches zero. The watchdog timer is used to reset the controller in case of software "crashes".
9) **A/D and D/A Converter:** Since processors are built to interpret and process digital data (i.e., 1s and 0s), they are not able to do anything with the analog signals that may be sent to it by a device. So, the analog to digital converter (ADC) is used to convert the incoming data (e.g., analog voltage) into representative digital value (a form that the processor can recognize). A less common feature on some microcontrollers is a digital-to-analog converter (D/A) that allows the processor to output analog signals or voltage levels. The D/A conversion is a prerequisite for some A/D converters. Microcontrollers often have little or no analog output capabilities. So, if the application requires a D/A converter, most of the time, it has to be fitted externally. A PIC[6] microcontroller has a built-in A/D converter.
10) **Debugger Unit:** Some controllers are equipped with additional hardware to allow remote debugging of the chip from a land computer. So, there is no need to install a special debugging software, which has the distinct advantage that erroneous application code cannot overwrite the debugger.

A UAV designer has various options for selecting modules. The main criterion for selecting the best option is to weigh cost versus performance. The optimum selection will be determined through an optimization process.

9.2.5 Atmel ATmega644P

In this section, characteristics of an example microcontroller, the Atmel ATmega644P [131] in the AVR family produced by Atmel is presented. The Atmel® ATmega644P is a low-power CMOS 8-bit microcontroller based on the AVR enhanced RISC

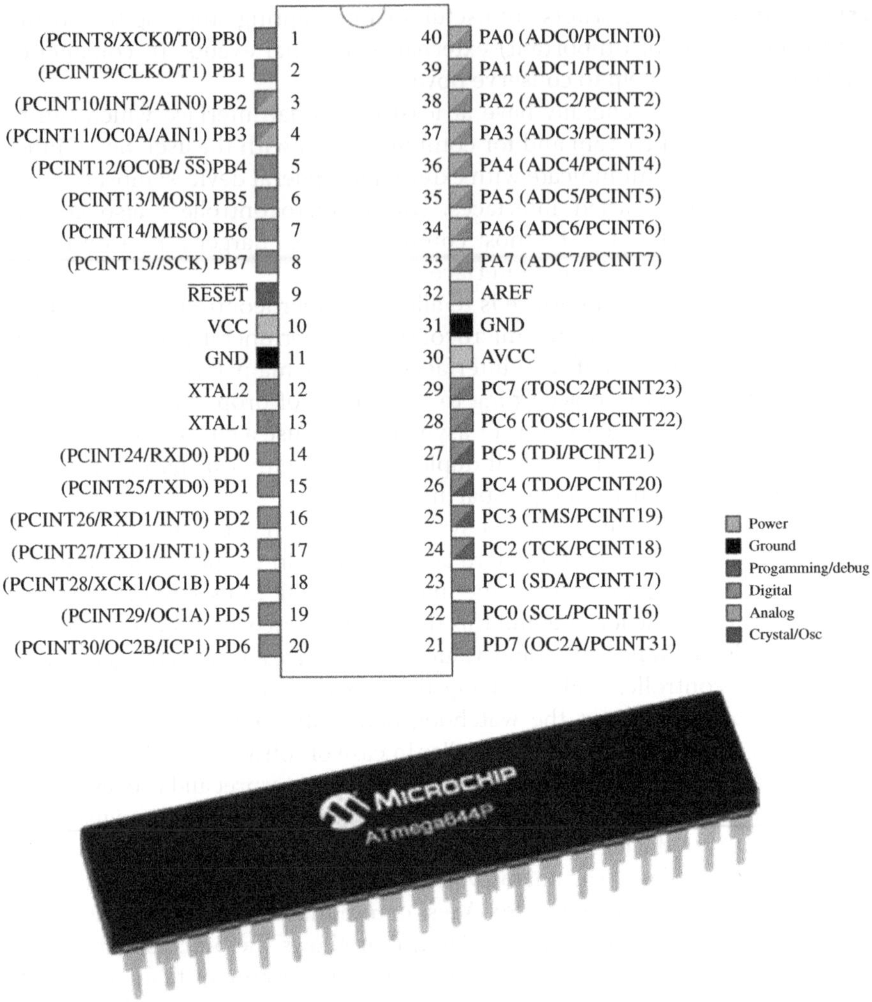

Figure 9.6 Atmel ATmega644P pinout.

architecture. By executing powerful instructions in a single clock cycle, the ATmega644P achieves throughputs close to 1 MIPS per MHz. This empowers system designer to optimize the device for power consumption versus processing speed. Power Consumption at 1 MHz with 1.8 V, at 25° C when in active Mode is 0.4 mA. In terms of I/O and Packages, it has 32 Programmable I/O Lines, 40-pin PDIP (Plastic Dual In-line Package), 44-lead TQFP (Thin Quad Flat Package), and 44-pad VQFN/QFN (Quad-flat no-leads/very thin Quad-flat no-leads). Speed grades at 2.7–5.5 V is 10 MHz.

This controller has non-volatile memory segments with 64 kb of in-system self-programmable flash program memory, 2 kb EEPROM, and 4 kb internal SRAM. Peripheral Features: (a) two 8-bit Timer/Counters with Separate Prescaler and Compare Mode; (b) one 16-bit Timer/Counter with Separate Prescaler, Compare Mode, and Capture Mode; (c) Real Time Counter with Separate Oscillator; (d) six PWM Channels; and (e) 8-channel 10-bit ADC.

Figure 9.6 illustrates the Atmel ATmega644P pinout. The pins descriptions are provided here. (a) VCC: Digital supply voltage; (b) GND: Ground; and (c) Port A [7:0]: This port serves as analog inputs to the A/D Converter. (d) REF: This is the analog reference (AREF) pin for the A/D Converter. (e) Port B [7:0], Port C [7:0], and Port D [7:0]: These are 8-bit, bi-directional I/O ports with internal pull-up resistors, individually selectable for each bit.

The ATmega644P provides the following features: 64 kb of in-system programmable flash with read-while-write capabilities, 2 kb EEPROM, 4 kb SRAM, 32 general purpose I/O lines, 32 general purpose working registers, real time counter, three flexible timer/counters with compare modes and PWM, two serial programmable USARTs, one byte-oriented two-wire serial interface (I^2C), a 8-channel 10-bit ADC with optional differential input stage with programmable gain, a programmable watchdog timer with internal oscillator, an SPI serial port. The idle mode stops the CPU while allowing the SRAM, timer/counters, SPI port, and interrupt system to continue functioning. The Power-down mode saves the register contents but freezes the oscillator, disabling all other chip functions until the next interrupt or hardware reset.

In Power-save mode, the asynchronous timer continues to run, allowing the user to maintain a timer base while the rest of the device is sleeping. The ADC Noise Reduction mode stops the CPU and all I/O modules except asynchronous timer and ADC to minimize switching noise during ADC conversions. In standby mode, the crystal/resonator oscillator is running while the rest of the device is sleeping. This allows very fast start-up combined with low power consumption. The ADC contains a *sample-and-hold* circuit, and is a 10-bit successive approximation ADC.

The device also supports 16 differential voltage input combinations. Hence, it features an ADC and a digital to analog converter. In order to provide a closed-loop control system, the microcontroller is equipped with an analog comparator (Operational Amplifier) that compares the input values on the positive pin AIN0 and negative pin AIN1. Dimensions of an ATmega644P microcontroller are 52.578 mm × 13.970 mm × 4.826 mm.

9.3 Microcontroller Circuitry

9.3.1 Microcontroller Circuit Board

Many microcontrollers have a timer/counter peripheral with an option to use an external clock signal (e.g., timer on the ATmega644p). This section will cover several types of external devices including motors, servos, and a variety of sensors. These devices are connected to a microcontroller for either providing a signal to, or controlled by, the microcontroller.

9.3.2 Electric Motor

DC Motors are one of the standard devices when it comes to controlling motion in quadcopters. They are responsible to generating thrust which is converted to lift force. There are two methods of supplying energy to and controlling a DC motor: (a) a variable DC voltage; and (b) continuous series of pulses at the maximum voltage of varying voltage. The majority of DC motors can be powered by voltage levels up to a certain level; 6–12 V is common values among RC planes. The microcontroller should be able to supply both the voltage and the current required to a desired motor. A regular motor settles into a speed/torque combination based on the load and electrical power supplied.

In general, there are two types of driver for motors: (a) linear drivers; and (b) PWM motor drivers. The linear motor drivers change the motor input voltage to change the speed. In contrast, running a motor via a PWM signal controls the speed by changing the amount of time the motor is powered for. This has the benefit of increasing the torque because full voltage, and therefore full torque, is available part of the time. PWM drivers generally utilize H-bridges.

Once the primary components (microcontroller, op-amp, and motor) are selected, the process of combining them will begin. Several resistors will also be required, their values should be calculated at some point.

9.3.3 Servo Motor

Servomotor or simply servo is a popular device in UAVs; they are mainly utilized for deflection of control surfaces. A servo is an electric motor that allows for precise control of angular or linear position, velocity, and acceleration. A servo has an error-correcting feedback signal built in, which causes it to seek the instructed speed or position. The majority of servos have a limited range of motion, commonly limited to a 180–360° or less of rotation at the output shaft, and the position they move to is set by the incoming signal.

Servos often use a PWM signal, though the duty cycle does not range from 0–100%. Rather than looking at the duty cycle, a servo employs the width (or duration) of the pulse. A pulse of 1.5 ms will cause a servo to move to its neutral position, which is the middle of its range. While the range of motion and timings vary from model to model, the relationship is linear. Servos will receive the control pulse at least every 20 ms.

Servos have three wires: (a) power (+); (b) ground (−); and (c) control. The control wire (middle wire) is usually the only one that needs be connected to the microcontroller to receive the commanded signal. However, in an Arduino Uno board, all three wires are connected to the board. The power wire is connected to either 5 or 3.3 V pins. The ground wire is connected to GND pin.

9.3.4 Sensors

There are a variety of sensors used in UAVs for different measurement purposes, such as temperature, pressure, altitude, attitude, airspeed, and position. The output of most sensors is analog, but some are generating digital signals and data. From a microcontroller's point of view, there are really only about four types; those that use (a) a serial communications protocol; (b) analog signals; (c) a one bit digital signal (on/off); and (d) a series of pulses. For instance, encoders can handle the quadrature calculations, and then communicate with a microcontroller via a serial. The signal generated by a sensor must be transferred into the microcontroller for processing. The data transfer can be done either by wire connection, or by radio waves using communication system (via Transmitter/Receiver). More details about sensors are provide in Chapter 12.

9.3.5 Potentiometer

The potentiometer (Pot) is an electro-mechanical transducer [132]; the main element is a variable resistor. It converts linear or angular position/displacement from the operator into a voltage (i.e., electric potential). So, its output resistance changes as an internal wiper moves across a resistive surface. A Pot has three leads, with the center wiper lead connected to the analog input pin. A Pot (shown in Figure 9.13) has three wires; the middle one is for the commanded signal, and the other two for electric power (+ and –). A brief technical description of this sensor (potentiometer) is provided in Chapter 11.

9.4 Embedded Systems

9.4.1 Introduction

The application of a microcontroller in UAVs falls under a broad engineering topic referred to as the embedded systems. An embedded system is a processor-based system that is built to control a function or range of functions, and it is not designed to be programmed by the end user in the same way that a PC is. Many devices such as robots, modern cars, appliances, toys, and automatic sliding doors are equipped with embedded systems. In this section, fundamentals of embedded systems are briefly described.

For any electromechanical system, the ability to offer a finer degree of control is important. A sedan car has over 50 microcontrollers controlling functions such as the engine through engine management systems, brakes with electronic anti-lock brakes, transmission with traction control and electronically controlled gearboxes, safety with airbag systems, electric windows, and air-conditioning. A modern washing machine is equipped with a microcontroller that contains the different washing programs, and this provides the power control for the various motors and pumps.

In these cases, an embedded processor controls the engine/motor and will alter the parameters and timing depending on inputs from the engine/motor such as temperature, and the accelerator position.

In an embedded system, the hardware should be designed/identified, and the hidden software that really supplies the system's' functionality must be developed. With self-contained microcontrollers, all that is visible is a circuit board with a few connections to the outside world. The software is already uploaded into the on-chip memory and is effectively impossible to access. As a result, the system is much more secure and protected.

An embedded system is primarily composed of: (a) processor; (b) memory; (c) peripherals; (d) software; and (e) algorithm. Microcontrollers can be considered as self-contained systems with a processor, memory, and peripherals so that in many cases all that is needed to use them within an embedded system is to add software.

9.4.2 Embedded Processors

The processors are constantly being developed to provide the processing heart for computer architectures. The embedded processor is designed when a processor is combined with the right software and hardware peripherals. There are many types of processors available ranging in cost, processing power and levels of integration. In embedded systems, the processor is integrated into the system. The question arises concerning how we select a processor for an embedded system.

There are essentially four basic architecture types which are usually defined as (a) 8-bit accumulator; (b) 16/32-bit CISC; (c) RISC; and (d) architectures and digital signal processors (DSP). While most of the original devices (8-bit processors) are no longer available, their architectures live on in the form of microcontrollers. An embedded processor controls the engine/motor and will alter the parameters and timing depending on inputs from the engine/motor such as temperature, and the accelerator position.

Microcontrollers are the most integrated, but they do not offer the best performance. However, the ability to pack a whole system including memory, peripherals, and processor into a single package is attractive, provided there is enough performance to perform the work required. In general, microcontroller have a lower cost and lower performance compared with microprocessors. On the other hand, microcontrollers are easier to integrate into a complex system than microprocessors.

9.4.3 Signal Flow

A microcontroller is only able to process digital signals; either sending a control signal, or receiving an external device's (e.g., sensors) signal. However, not all external devices (e.g., electric motors and electromechanical actuators) work with digital signals. Thus, such digital signals must be converted to analog signal (e.g., voltage and position) and vice versa. Two important devices for this goal are: (a) D/A converter; and (b) A/D converter.

An analog signal is sampled at a desired frequency and is converted by the A/D converter. This frequency will be first determined by the speed of this conversion. At every period, there will be a new sample to process using N instructions. Figure 9.7 exhibits the functions of D/A converter and A/D converter.

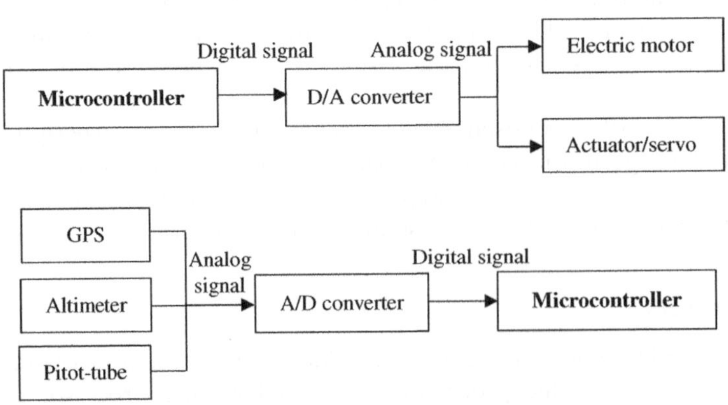

Figure 9.7 The functions of D/A converter and A/D converter.

9.5 Microcontroller Programming

9.5.1 Software Development

UAVs are software-intensive systems. A solid understanding of flight software is essential for the success of a UAV program. Many UAV development programs reach the state where a nearly completed unmanned aircraft awaits software before flight test can begin. Much of the software that was developed by the UAV designer in the past is now developed by suppliers. UAV software is becoming a commodity for flight controls, data management, and GCS software. The software development scope is underestimated, and is often not well understood. This section presents the requirements of an acceptable flight software as well as the microcontroller code. An UAV program requires a number of software to develop, including: (a) flight simulation software; (b) ground station software; (c) MMS software; and (d) microcontroller software. This section introduces only microcontroller software.

The DroneDeploy, a leading cloud software platform company for commercial UAVs, delivers UAV flight software, which enables professional mapping, 3D modeling, and reporting from any UAV on any device. The software makes the aerial data accessible and productive by transforming data collection and analysis across industries, including construction, energy, agriculture, and mining. DroneDeploy, announced[7] in January 2019, that its customers have safely conducted more than one million automated UAV flights around the world.

9.5.2 Operating System

The Operating System (OS) application programming interface (API) is a significant consideration, not only for execution, but also for ease of system development. Due to the critical nature of flight control, high reliability and real-time execution is required. A portable OS interface is a recommended. This it is widely supported, and it allows easy porting of applications programmed either to attempt for some fixed period of time to re-establish communications, to execute a fully automated operation, or to independently complete the mission.

9.5.3 Management Software

An unmanned aircraft management system (UMS) performs flight critical and unmanned aircraft operation functions. This includes interfacing to command and control communication systems, subsystems, and the autopilot. Frequently high-level flight route management occurs on the UMS ahead of the autopilot. The UMS is hosted on one or more computers. However, a MMS manages mission critical functions. These can include commanding payloads, selecting data feeds for downlink, data storage, data retrieval, data fusion, and payload downlink management. An architecture that segregates the MMS from the UMS and other flight critical systems is generally desirable for improved reliability. The MMS is hosted on one or more computers.

Large defense contractors can choose to specialize in mission system software, which runs on the MMS and the GCS. The VMS efforts involve subsystem management and integration of various avionics components provided by multiple suppliers. The contingency management algorithms usually reside on the VMS as well.

The software that estimates the unmanned aircraft state based on sensor data generally runs on a procured device. The INS contains the inertial sensors and GPS, and all of the Kalman filtering is performed on the INS. The INS provides state data to the data bus, which is routed to the data consumers. Many autopilots have built in INS as well as air data sensors and include the necessary filters to estimate the state.

Ground payload software should have the following functionalities: (a) ability to command the payload; (b) process payload state feedback; and (c) process the payload's collected data to generate usable intelligence. Ground payload software functions can be performed at: (a) MPO (MessagePlus/Open) workstation; (b) separate mission management processors; and (c) other systems.

9.5.4 Microcontroller Programing

Microcontroller programs must fit [40] in the available on-chip memory, since it would be costly to provide a system with external, expandable memory. In general, microcontroller programming procedure is as follow:

1) Write a code (in a language) for the microcontroller in an integrated development environment (IDE), a PC program.
2) Debug the code
3) Compile the code into binary code which the microcontroller can execute.
4) A programmer (a piece of hardware, not a person) is used to transfer (load) the code from the computer to the microcontroller.

Using third-party software can save nonrecurring engineering cost, but the UAV designer gives up some control. This supplier-provided software is sometimes called firmware. The UAV prime contractor can focus its attention on the UMS MMS.

Flight software should be reliable and predictable. Software that responds predictably to a set of inputs is known as deterministic. Aerospace software standards and certification methods help ensure solid design practices and software testing methods. Many government contracts require that the software developers or prime contractor achieve the certification.

Flight control software can run on an autopilot that includes the sensors or on a separate processor. The UAV prime contractor can implement the control laws themselves, though this work is increasingly performed by autopilot suppliers. Every device must be commanded, receive data, or provide data. This data exchange occurs over a real-time operating system (RTOS).

The most common type of the programmer is an ICSP (In-Circuit Serial Programmer). An example is 8051 loader. ISP programmer have two functions: (a) converts the code to binary; and (b) loads the code from the PC to the microcontroller. Common data bus types are: Ethernet; MIL-STD 1760; MIL-STD 1553; and Controller Area Network (CAN). Figure 9.8 shows an ISP output connection to an 40-lead PDIP.

9.5.5 Software Integration

Most payloads require interfaces with both the UAV and ground-station software. Payloads can be an integral part of the system, such as is common with Electro-Optic/Infra-Red (EO/IR) surveillance payloads, where the payloads are nearly always on the

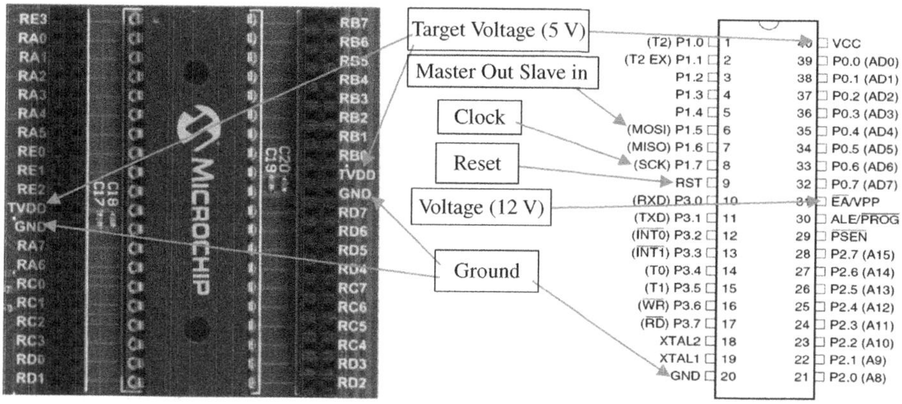

Figure 9.8 ISP output connections to a 40--lead PDIP.

UAV. The payload workstation is baseline in the GCS. Most operational UASs are required to adapt to specialized payloads that might only be required for a small portion of the missions. Specialized payloads require adaptation of the system software – both on the air and ground.

Sophisticated UAVs have a distinct MMS that interfaces to the payloads, data storage devices, and payload return link. Robust mission management architectures are segregated from flight-critical UAV systems. Dedicated on-chip hardware often includes capabilities to communicate with other devices (chips) in digital formats such as I^2C, SPI, USB, and Ethernet. UAV software interfaces typically are: (a) payload command and control; (b) receipt of unmanned aircraft state data; (c) sending payload status messages; and (d) providing the payload data output stream for storage or downlink.

The data bus is the medium by which data are distributed across the avionics architecture. The data bus routes information over an electrical circuit from the data producers and consumers. The data are prioritized such that the most critical data have a high probability of getting through. U.S. military manned aircraft typically use a 1553 data bus [127] for avionics and payload systems.

9.5.6 High-Level Programming Languages

Typical high-level programming languages are: (a) assembly; (b) C, C++; (c) Python; and (d) JavaScript. Writing a program in machine language would be tedious. Any program code written in a high-level language like C++ needs to be translated into machine code before it can be executed by a processor. This translation is done with a program aptly named Assembler.

C is an imperative procedural language [133]; it uses statements to specify actions. The most common statement is an expression statement, consisting of an expression to be evaluated, followed by a semicolon ";". By design, C provides constructs that map efficiently to typical machine instruction. There is a small, fixed number of keywords, including a full set of flow of control primitives such as "for," "if/else," and, "while". User-defined names are not distinguished from keywords. C has no "define" keyword; instead, a statement beginning with the name of a type is taken as a declaration. There is no "function" keyword; instead, a function is indicated by the parentheses of an argument list.

There are a large number of arithmetical and logical operators, such as "+" and "=." Procedures (subroutines not returning values) are a special case of function, with an un-typed return type "void". The "**void main**()" means that the function **main**() does not return a value. Complex functionality such as I/O, string manipulation, and mathematical functions are consistently delegated to library routines. Comments may appear either between the delimiters "/*" and "*/" or following "//" until the end of the line.

The basic C source character set includes the characters such as lowercase and uppercase letters; decimal digits, and graphic characters. New software development tools improve development efficiency while reducing errors. Model-based design techniques allow development in the requirements and architectural level rather than working with lines of code from the start. Auto-coding generates software code based on higher-level design.

9.5.7 Compiler

The name "compiler" is primarily used for programs that translate source code from a high-level programming language to a lower-level programming language (e.g., assembly language, machine code). Compilers and assemblers are used to convert both high-level and assembly language codes into a compact machine code for storage in the microcontroller's memory. Depending on the device, the program memory may be permanent, read-only memory that can only be programmed at the factory, or it may be field-alterable flash or erasable read-only memory. Programmable memory also reduces the lead time required for deployment of a new product.

9.5.8 Debugging

The testing and debugging of the software are very important and often time-consuming tasks, which makes up a large portion of the overall development cycle. Testing is performed with the aim to check whether the programing meets its requirements. Detected deviations from the specification may result in debugging the program code (if its cause was an implementation error), but may even instigate a complete redesign of the software in case of a design flaw.

9.6 Programming in C

9.6.1 Introduction

C language is a structured, machine independent, and high-level language. It allows the software engineer to develop programs without worrying about the hardware platforms where they are implemented. Programs written in C are efficient and fast.

9.6.2 General Structure of a C Program

A C program should be written in C editor. Then, the source program statement should be translated (i.e., compiled) into object programs which are suitable for execution by a computer. When compilation proceeds, the translated program is stored in another file with the same extension "obj". The general structure of a C program is provided below:

```
Include header file section
Global declaration section
Main()
{
Declaration part
Executable part
}
Used-defined functions
{
Statement
}
```

For description of each line and section, the interested reader is encouraged to consult with texts in C language such as [133].

9.6.3 Example Code – Detecting a Dead LED

In this section, an example code (in C language) to detect a dead LED is presented. This program will turn on an LED connected to pin 40 of the ATmega644p. In order to better understand how this code works, each line is also explained. A basic understanding of the C programming language is required for this code.

Consider that we have designed an LED control system through a circuit with a couple of resistors and one LED. We would like to know, whether the LED works or is broken. The feedback desired would be whether the LED is properly functioning. The broken LEDs can be either a short or open circuit. We would like the microcontroller to differentiate between an LED working under expected parameters (2 V forward voltage) and a broken LED. The resistance in the control interface must be increased to prevent a shorted LED from damaging the microcontroller. The microcontroller can read a digital input, seeing a digital low for a functional LED or a digital high for a broken LED. So, the microcontroller would read the results.

```
# include <avr/io.h>
# include <util / delay.h>
int main ( void ) {
unsigned int LEDstatus ;
DDRA = 0 xff;
PORTA = 0xfe;
_delay_us (1);
LEDstatus = PINB & 0x01;
return 0;
}
```

In order to better understand how this program works, each line is explained below.

```
# include <avr/io.h>
```

This line tells the compiler to include the specified header file (avr/io.h). Almost every program will include this line, as it includes additional files that define much of the memory structure such as the names of all the registers. Most microcontroller manufacturers or compilers will include such a header file to define names for all the registers in their microcontrollers. If such a file is not included, the programmer has to address the registers by memory address, rather than by name.

```
# include <util / delay.h>
```

This tells the compiler to include the delay library, which includes two functions causing the microcontroller to idle for a certain period of time.

```
int main ( void ) {
```

This is the main function. Every program written in C language must have one main function and it is where program execution starts. This line also specifies that the program does not take any information as input (*void*), but it does return an integer upon completion (*int*).

```
unsigned int LEDstatus;
```

Here, the program initializes a variable in which the program will store the information collected from the feedback.

```
DDRA = 0xff;
```

This sets the contents of the register DDRA to the hex (0x) value ff, which turns all 8 pins to output. Only one of the pins is actually used, but setting them all does not cause any issues in this case. Whenever there are several registers serving the same purpose for multiple devices, the actual register names are often generalized by replacing the distinguishing character, in this case A, with a generic character indicating that it could be any of the possibilities. This generic character is generally a lower case "n" for numerical distinguishers or "x" for alphabetic. Therefore, when some reference is made to DDRX[1], it is referring to any or all of the data direction registers rather than a specific one. This "x" means something different when preceded by a "0" and followed by a string of numbers or letters a-e. In this case the "0x" is a designator that the following value is in hexadecimal (i.e., base 16).

```
PORTA = 0xfe;
```

This sets all of the bits in the data register to high except for bit 0. As the circuit is using active-low logic, with the microcontroller acting as a sink, this turns the LED on. As a value of 0 is default here, this line could be left out. However, including this line will cause any LEDs attached to the other pins on the port to be off.

Moreover, the value "0xfe" can be replaced by any syntax that gives the same value. This syntax is for hexadecimal; other options include binary (011111110) or decimal (254). Hexadecimal is commonly used due to its compactness and easy conversion to and from binary (each character in hexadecimal corresponds to a block of four bits).

---

1 **DDRX** stands for data direction register, where X may be A, B, C or D depending upon the type of microcontroller.

```
_delay_us (1);
```

This is a call to a function in the delay library, and causes the microcontroller to idle for 1 μs. While not necessary, it allows the system to fully stabilize before the chip checks the feedback.

```
LEDstatus = PINB & 0x01;
```

There are three registers for each of the 8-bit ports on the Atmega644p: PINx, PORTx, DDRx. Here, the program is reading the inputs on port B. The rest of the line performs a bitwise AND operation on the read value, and stores it in the previously initialized variable LED status. A bitwise AND compares each bit in the two bytes (PINB and 0×01), and results in a new value that has a 1 in each bit in the byte where both the input bytes have one second.

In the program, this operation causes only the one bit of data corresponding to port B pin 0 (pin 1 on the ATmega644p) to be saved and ignores the other seven bits of the port.

```
return 0;
```

The final line of code supplies the integer requested as the return value in the function declaration. The *return* statement has the added effect of exiting the function immediately. While not necessary in this application, many compilers will throw errors or warnings, if the main function does not have a return type of "int", or if execution of the function can reach the end of the function without encountering the return command.

```
}
```

While the *return* statement tells the program to exit the function, this bracket signifies that the definition of the function is complete, and any further lines belong to some other aspect of the code.

9.6.4 Execution of a C Program

There are basically three steps in executing a C program. They are:

1) Creation of a program. A program should be written in C using an editor, then the file must be saved. The default extension is C. (e.g., filename.c). In this phase, an editor software is needed.
2) Compiling the program. C is a high-level language and it needs a compiler to convert it into an executable code so that the program can be run by a computer. The compiler translates the program statement into object program. The new file is generated and stored with the same name, but with a new extension; "obj" (e.g., filename.obj). For this phase, a C compiler (software) is needed.
3) Execution of the program. After compilation, execute the generated executable file, by typing the file name (e.g., filename) and pressing "Enter". This should result in the output. In this phase, no software is needed.

In summary, two software packages are needed prior to the execution of a C program.

9.7 Arduino

9.7.1 Arduino Overview

Arduino is an open source electronics hardware (e.g., microcontroller) and software [134, 135]. It directly interfaces with sensors (e.g., gyro and Inertial measurement unit (IMU)) and processes raw data. Arduino has full range of hardware products, including boards, modules (a smaller form-factor of classic boards), shields (elements that can be plugged onto a board to give it extra features), and kits. Arduino hardware (e.g., boards) is able to read inputs and after processing them, generates outputs.

Over the years, Arduino has become very popular, and has been the brain of thousands of industrial projects, including UAVs. The Arduino website [136] provides various resources including codes and installation procedures for Arduino software package. The Arduino software allows the user to write a program and upload it to a board. An Arduino board can be readily connected to a computer via USB cable. Popular Arduino boards are Uno, Due, Mega, Leonardo, Micro, Mini, MKR, Nano, Pro, and Fio.

To implement a UAV project with the Arduino, three main components are needed: (a) the Arduino board itself; (b) external hardware (including both shields and handmade circuits); and (c) the Arduino IDE. Since the Arduino is open source hardware, you can find various "Arduino compatible" devices available to UAV projects. Because the Arduino boards are designed to be programmable via the same IDE, one can utilize any of the modern Arduino boards for a project with zero or minor changes.

All Arduino boards have a number of key components and functions. For instance, an Arduino Uno board; has the following components: (a) Atmel microcontroller unit (often ATMega 328p); (b) USB communication interface; (c) voltage regulator and power connections; (d) I/O pins; (e) debug and power LEDs; (f) reset button; and (g) ICSP connector. The Arduino can accept between 6 V and 20 V via the direct current barrel jack connector, or into the V_{in} pin. The Arduino circuitries have built-in 5 V and 3.3 V regulators.

A graphical illustration of the wiring between an Arduino board and sensors, control surfaces, motor, battery, and communication system of an UAV is shown in Figure 9.9. The electric motor is almost the only component that directly receives the electric power via battery. Other elements will have their electric powers through Arduino board. This board (i.e., microcontroller) provides electric power to other elements; while sending control commands, or while receiving measured signals.

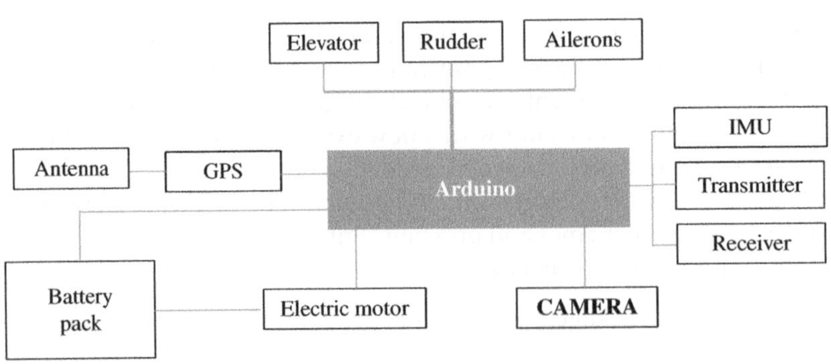

Figure 9.9 Arduino wiring in an unmanned aerial vehicle (UAV).

9.7.2 Arduino Programming

Arduino provides both hardware (board) and software; however, one must access the Arduino software, and program it in order to prepare it for a purpose. Anyone may download the Arduino software for free to a computer from the Arduino website; https://www.arduino.cc/en/Main/Software [136]. You need to have this software to write your Arduino code and upload it to an Arduino board. In addition, all Arduino codes in [134] can be downloaded at http://www.wiley.com/go/exploringarduino.

The Arduino development environment contains a text editor for writing code, a message area, a text console, a toolbar with buttons for common functions, and a series of menus. Figure 9.10 shows the Arduino software window its main menu items. When the Arduino software is used to write a code, you may directly upload it into an Arduino board without any separate compiler. In practice, the command "Verify" compiles, and checks your code for errors. Programs written using Arduino are called sketches. Additional commands are found within the five menus: File, Edit, Sketch, Tools, and Help. The toolbar button that has an arrow to the right, compiles your code and uploads it to the Arduino board.

A sketch is the name that Arduino uses for a program. It is the unit of code that is uploaded to and run on an Arduino board. When you download a header file (name.h) from Arduino libraries (i.e., open repository), you should add this file by suing "Sketch" menu. Then, insert an **#include** statement at the top of the sketch for each header file.

The Arduino programming language [137] provides the user with access to microcontroller peripherals, including A/D converters, input/output pins, communication busses, and serial interfaces. The Amtel microcontroller programs are generally written in C or Assembly and programmed via the ICSP interface using a dedicated programmer. In an Arduino board, you can program it easily via USB, without using a separate programmer and external hardware. This feature is easily implemented by an internal element, referred to ad bootloader.

Arduino programming language can be divided into three main parts: structure, values (variables and constants), and functions. Functions are employed for controlling the Arduino board and performing computations. Variables are used to introduce Arduino data types and constants. The structure of an Arduino program is very similar to a program in C (as introduced in Section 9.6). Software structure consist of two main functions: (a) setup( ); and (b) loop( ).The **setup()** is called when a process starts. It is used to initialize the variables, pin modes, start using libraries, and sets the initial

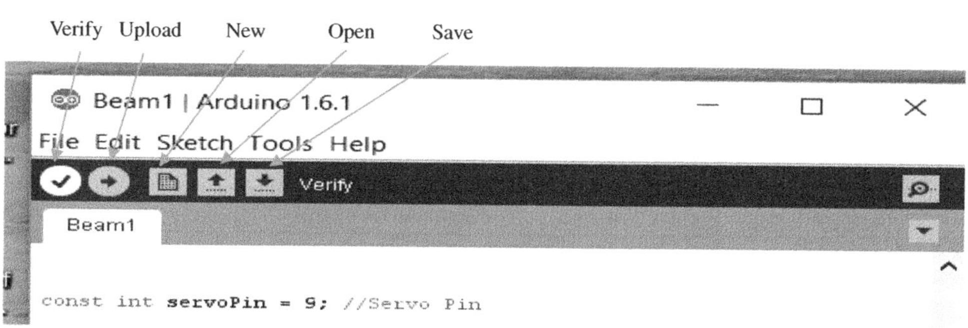

Figure 9.10 Arduino software and its main menu items.

values. As the name implies, the **loop()** loops consecutively, allowing a program to change and respond.

Tables 9.3 and 9.4 list a number of functions and syntaxes in Arduino programming. Similar to many programming platforms, the Arduino environment can be extended through the use of libraries [136]. Libraries provide extra functionality for use in sketches, for instance, working with hardware or manipulating data. Any Arduino code must be uploaded into an Arduino board via a computer and a proper cable connection.

9.7.3 Arduino Uno Board

Arduino microcontroller Uno is the flagship Arduino; it is the most popular board in the Arduino board family. Figure 9.11 demonstrates its major elements, each one/group

Table 9.3 Functions in Arduino programming.

No	Function group	Example
1	Digital I/O	digitalRead(); digitalWrite(); pinMode()
2	Analog I/O	analogRead(); analogReference(); analogWrite()
3	Time	delay(); delayMicroseconds(); micros(); millis()
4	Math	abs(); constrain(); map(); max(); min(); pow(); sq.(); sqrt()
5	Trigonometry	cos(); sin(); tan()
6	Characters	isAlpha(); isAscii(); isControl(); isDigit(); isGraph(); isPrintable()
7	Bits and Bytes	bit(); bitClear(); bitRead(); bitSet(); bitWrite(); highByte(); lowByte()
8	Communication	serial; stream
9	USB	kyboard; Mouse

Table 9.4 Elements to structure an Arduino program.

No	Syntax group	Example
1	Sketch	loop(); setup()
2	Control Structure	break; continue; do...while; else; for; goto; if...else; return; while
3	General	#define (define); #include (include); /* */ (block comment); // (single line comment); ; (semicolon); {} (curly braces)
4	Arithmetic Operators	% (remainder); * (multiplication); + (addition); – (subtraction); / (division); = (assignment operator)
5	Comparison Operators	!= (not equal to); < (less than); <= (less than or equal to); == (equal to); > (greater than); >= (greater than or equal to)
6	Boolean Operators	! (logical not); && (logical and); || (logical or)
7	Pointer Access Operators	& (reference operator); * (dereference operator)

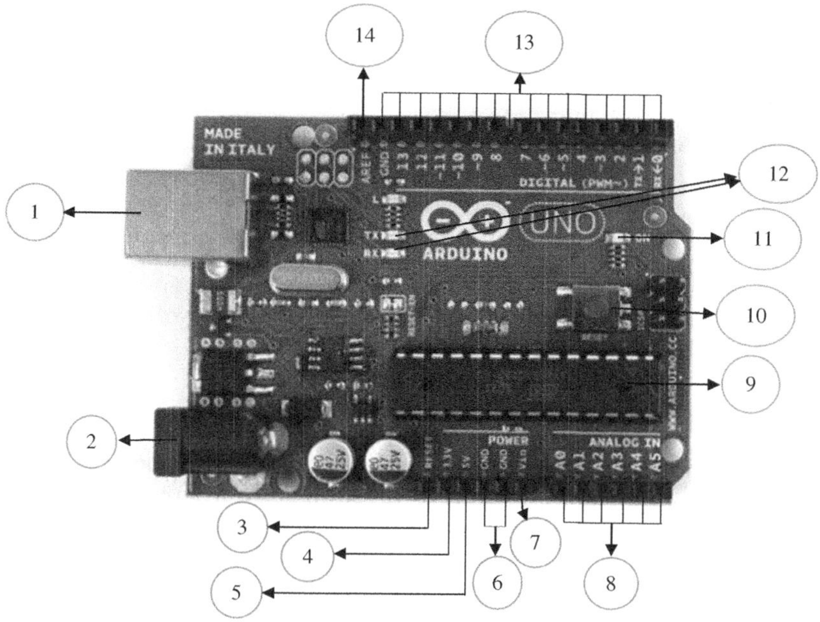

Figure 9.11 Elements of an Arduino Uno board.

is labeled with a number. The input and output pins are at the top and bottom. A 28-pin microcontroller is located in the lower center. In this section, major elements and their features are described.

- **1, 2, 7. Power in:** Arduino board can be powered by using the USB cable from a laptop/computer by connecting the USB cable to the USB connection (1). Another way is to directly connect it (2) with a power cord to the AC power supply by. Vin (7) pin also can be used to power the Arduino board from an external power source.
- **3, 10. Reset:** An Arduino UNO board can be reset to start a program from the beginning in two ways. A. by using the reset button (10) on the board. B. connect an external reset button to the pin labeled RESET (3).
- **4, 5, and 6. Power out Pins: An** Arduino board can supply electric energy to external elements (e.g., Pot). Most components (e.g., servomotor) used with an Arduino board works with 3.3 and 5 V supply. A 3.3 V power is supplied through pin (4), A 5 V power is supplied through pin (5). The ground is provided through pin (6) to close the circuit. There are a number of GND pins, any of which can be used to ground an element.
- **8. Analog pins:** There are five analog input pins; A0 through A5. External analog sensors – like the thermometer – are connected to the board via these pins, which can read the signal from a sensor, and convert it into a digital value that can be read by the microprocessor.
- **9. Microcontroller:** Each Arduino board has a microcontroller, which are usually made by the Atmel Company.
- **11. Power LED indicator:** This LED will light up when an Arduino board is plugged into a power source.

- **12. TX and RX LEDs:** The TX LED flashes while transmitting the serial data. RX LED flashes while a signal is received.
- **13. Digital I/O:** There are 14 digital input/output (I/O) pins. These pins can be configured to work as input digital pins to read logic values (0 or 1) or as digital output pins to drive different modules like LEDs, and relays. Six pins labeled "~" can be used to generate PWM.
- **14. AREF:** AREF denotes Analog Reference. It can be used to set an external reference voltage (between 0 and 5 Volts) as the upper limit for the analog input pins.

The Arduino Uno is not expensive (about $25); and it is deemed as a reliable board. The programming language used in Arduino is C, with a large number of free libraries that can be downloaded from the Arduino website.

9.7.4 Open-Loop Control of an Elevator

In this section, an experiment for using Arduino in controlling the position of a servomotor with an Arduino board is demonstrated. A general application of this experiment is the control of the UAV attitude (i.e., pitch angle) by a remote operator; when he/she deflects the control stick (here, the Pot) on the remote control box.

In a fixed-wing UAV, the deflection of a control surface such as the elevator is usually applied via a servomotor. To simplify the experiment, we replace the communication system (transmitter and receiver) by two wires. Figure 9.12 depicts the block diagram of the open-loop (i.e., no feedback) control of an elevator.

This example makes use of the Arduino servo library [136]; Knob. For the implementation, four hardware items are necessary: (a) Arduino board; (b) servomotor; (c) a Pot; and (d) hook-up wires. Both servomotor and Pot (Figure 9.13) have three wires; the red one (middle in Pot, but left/right in servomotor) is for the commanded/measured signal, and the other two are for electric power (+ and −). Hence, left/right one in Pot, and middle wire for servomotor is for electric power (e.g., 5 V).

The following is the code that must be written and uploaded into the Arduino board:

```
#include <Servo.h>
Servo myservo;
int potpin = 0;
int val;
void setup()
{
  myservo.attach(9);
}
void loop()
{
  val = analogRead(potpin);
  val = map(val, 0, 1023, 0, 180);
  myservo.write(val);
  delay(15);
}
```

Create servo object to control a servo

Connects analog pin **A0** to the **potentiometer**

Variable to read the value from the analog pin

Attaches pin **9** to the **servomotor**

Reads the value of the potentiometer (value between 0 and 1023)

Scale it to use it with the servo (value between 0 and 180)

Sets the servo position according to the scaled value

Waits for the servo to rotate

This code must be uploaded via a computer and Arduino software, using a proper cable connection. In the code, some line descriptions are added to explain the meaning of some syntaxes and commands. When the code is uploaded, you may disconnect the board from computer, and start the experiment.

Figure 9.12 Block diagram of the open-loop control of an elevator

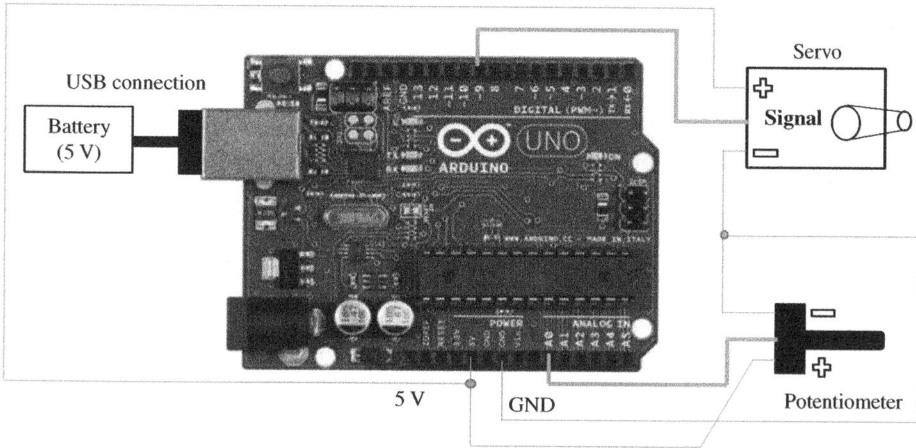

Figure 9.13 Circuit, wiring, and schematic of the servo control system.

Figure 9.13 demonstrates the circuit, wiring, and schematic of the servo control system. Servo motors have three wires: power, ground, and signal. The power wire (middle, often red) is connected to the 5 V pin on the Arduino board. The ground wire is connected to a ground pin on the board. The signal pin is connected to pin 9 on the board. Moreover, the Pot should be wired so that its two outer pins are connected to power (+5 V) and ground, and its middle pin (for signal, often red) is connected to analog input 0 on the board.

When wiring is completed, begin the experiment by pressing the start button on the board. You must see the servomotor begin to rotate and then stop at the desired angular location, which is directed by Pot. Thus, as you rotate the Pot by hand, the servomotor must rotate accordingly. As an application, the Pot is similar to a stick of a control box that an UAV operator is moving. The servomotor is similar to the one that deflects a control surface such as the elevator. The main difference is that; a transmitter sends signals from the operator's control box to the receiver of an UAV.

9.7.5 Arduino and Matlab

Matlab – with a high-level interpreted language - has a special support package for Arduino hardware. So, results from I/O instructions can be seen immediately, without compiling. Since MATLAB includes thousands of built-in math, engineering, and plotting functions, the user can quickly analyze and visualize data collected from your Arduino. The user may connect to, interactively communicate with, and control Arduino inputs and outputs from MATLAB.

The Matlab package enables the Arduino board user to perform tasks such as: (a) Acquire analog and digital sensor data from an Arduino board; (b) drive and control devices such as servomotors; (c) write scripts; (d) develop algorithms that run standalone on the Arduino; and (e) communicate with an Arduino board over a USB cable or wirelessly over Wi-Fi. For using Matlab for Arduino, you need to purchase the Arduino Engineering Kit, use add-on files, download and install hardware support packages, and use Matlab resources such as YouTube channel.

The Matlab support tools are provided for both matlab codes and Simulink models. The matlab support packages should be setup and configured. They are Matlab add-ons that will allow the user to write matlab code which is then compiled, loaded, and executed on an Arduino board. The package enables the user to read/write to IO pins on Arduino board, communicate with peripheral devices via I2C or SPI, etc. Please note that, the Matlab package does not support all hardware functional on all types of Arduino boards. The application of matlab support package in Arduino boards are a type of hardware in the loop simulation (See Section 5.9.3).

As a quick example, the following Matlab code

```
ledPin = 'D13';
deltaT_blink = 2;
port = 'COM5';
board = 'Due'
a = arduino(port, board);
```

will cause an LED on an Arduino Due to blink every 2 seconds.

For the execution, you need to follow this procedure: (a) connect an Arduino Due board to a computer using a USB cable; (b) run matlab software in computer; (c) install Matlab support package for Arduino hardware; (d) write the above Matlab script; and (e) run the code. Recall that Arduino Due already has LEDs (as part of the board) connected to digital pin 13. When the "Run" in Matlab is clicked, the computer and Matlab software will begin to communicate with Arduino Due. Hence, the matlab code will be compiled, loaded onto, and executed on the Arduino Due.

Instead of a Matlab code, a Simulink model may also be employed to control a device via an Arduino board. The reader is expected to be knowledgeable of Simulink model development. A similar procedure will be followed to control a device via an Arduino board.

9.8 Open-Source Commercial Autopilots

Open source commercial autopilots are widely used in RC planes. There are a number of open-source, open-hardware autopilot available in the market. A UAV designer has the option to select the complete autopilot from a vendor, and then focuses on its incorporation. In this section, a few popular autopilots (namely Ardupilot, Micropilot, DJI WooKong, and PX4 PixHawk) are briefly reviewed.

9.8.1 ArduPilot

The ArduPilot is an autopilot hardware for use in autonomous UAVs. It is an open source UAV platform [138], able to control autonomous multi-copters, fixed-wing aircraft, traditional helicopters, ground rovers and antenna trackers. ArduPilot is an award-winning platform for autonomous stabilization, waypoint based navigation and two-way telemetry with Xbee wireless modules. The system uses an IMU using a combination of accelerometers, gyroscopes, and magnetometers. Today's ArduPilot is almost entirely C++ and has evolved to run on a range of hardware platforms and OS.

The ArduPilot project[8] is a family of open source autopilots based on the Arduino open source computing platform. It consists of a choice of electronics and a range of free software programs for different vehicles. The autopilot can be setup, and firmware may be loaded via a point-and-click utility. The embedded programming toolkit is based on Arduino, with full mission scripting.

The ArduPilot is programmed to handle autonomous takeoff and landing, and special action commands such as video and camera controls. It is equipped with a two-way telemetry and in-flight command. Its measurement hardware includes 3-axis gyros, 3-axis accelerometers, 3-axis magnetometer, and barometric pressure sensor for altitude. Furthermore, Ardupilot has a built-in hardware failsafe processor, and can return-to-launch if radio control is lost loss. Two current active Ardupilot products are ArduPilot Mega (APM). The cost of one Ardupilot unit is about \$200–\$300. The Ardupilot APM 2.6 (sees Figure 9.14) with eight RC channels and four serial ports can convert a remote-control (RC) air vehicle into a fully autonomous UAV. Its control system employs PID control law, while the operator may tune the PID gains.

The PID parameters are mainly adjusted for stabilizing modes during initial flights and setup operation. The PID gains have default values, but there are recommended values for each undesired behavior, and also error compensation. For instance, the default value [138] of k_P for pitch control is 4.5. The PID variables for each function are modifiable in the Mission Planner software. The Pitch PID values are slaved to Roll PID values. The manufacturer recommendation is to redo stabilize PID – P settings until roll and pitch control and self-leveling work smoothly.

9.8.2 PX4 Pixhawk Autopilot

PX4 Pixhawk is an open source (open-hardware) professional Autopilot with a Unix/Linux-like programming environment. PX4 is an open source flight control software and open-hardware that can power any vehicle from flying UAVs to ground vehicles. PX4 can be customized and tailored to fit a specific need. The collaborative project is

Figure 9.14 ArduPilot AMP 2.6.

participated by industry experts and academic researchers from around the world. It enables developers create custom drone OSs with a flexible core. It originated from the PIXHAWK Project of the Computer Vision and Geometry Lab of ETH Zurich (Swiss Federal Institute of Technology) and Autonomous Systems Lab.

The Pixhawk (www.pixhawk.org) is an autopilot system designed by the PX4 project and manufactured by 3D Robotics. It features advanced processor and sensor technologies from ST Microelectronics® and a NuttX RTOS. The objective of the project was to deliver high performance, flexibility, and reliability for controlling any autonomous vehicle at low cost and high availability.

The microprocessor of PX4 Autopilot is equipped with a 32 bit STM32F427 Cortex M4 core with FPU, a 168 MHz/256 kb RAM/2 MB Flash, and a 32 bit STM32F103 failsafe co-processor. The sensors include ST Micro L3GD20H 16-bit gyroscope, ST Micro LSM303D 14-bit accelerometer/magnetometer, and MEAS MS5611 barometer. The autopilot mass is 38 g. In terms of dimensions, the length, width, and thickness are 81.5, 50, and 15.5 mm respectively.

9.8.3 Micropilot

The MicroPilot[9] Company which sells commercial autopilots – serves mini, small, and large UAV manufacturers (NASA, Raytheon, and Northrop Grumman), large-scale defense and research enterprises. MicroPilot also provides a series of autopilots, software, accessories, and customized UAV training and integration services. MicroPilot manufactures circuit board and enclosed autopilots for UAVs as well as ground control software (HORIZON), a software developer's kit (XTENDER) and a hardware in the loop simulator (trueHWIL). They also sell data links for telemetry, compasses, AGLs and ADCs.

Some sample products are: single-use autopilot (MP2x28^{xP}), and VTOL autopilot (MP2128^{HELI}). The Micro Pilot's MP2 × 28 introduced in 2004, with a mass of 28 g, (dimension of 4 cm × 10 cm) is a good fit for mini/small UAV autopilots.

Figure 9.15 MP21283X MicroPilot's triple redundant autopilot.

Table 9.5 Characteristics of MicroPilot MP21283X autopilot.

No	Parameters	Value/features
1	Altimeter	Max altitude: 12000 m
2	Sensors	• 5 g, 3-axis accelerometers • 3-axis rate gyro • Max angular rate: 150 per s
3	Supply voltage	4.2–26 V
4	Current	192 mA (at 6.5 V)
5	Dimension	10 cm in length, 4 cm in width, 1.5 cm in height
6	Mass	28 g (including GPS receiver, gyros, and all sensors)
7	GPS	Update rate: 1 Hz

In 2010 MicroPilot introduced a triple redundancy autopilot (MP21283X) for rotary- and fixed-wing UAVs. Figure 9.15 shows a MP21283X Micro Pilot's triple redundant UAV autopilot which offers such features as redundant radio modems, redundant RC override, integrated radio modems, and a backup microprocessor. It is fully integrated with 3-axis gyros/accelerometers, GPS, pressure altimeter, pressure airspeed sensors, all on a single circuit board. It is capable of GPS waypoint navigation with altitude and airspeed hold. Separate 24-bit ADC converters per gyro and a 1 kHz sample rate provides optimum performance for attitude estimation.

In its control system, it has user definable PID feedback loops. In the navigation system, the autopilot is equipped with 1000 waypoint command buffer, and changes altitude, airspeed at waypoint. It can handle 8/16/24 servos with 50 Hz servo update rate. In telemetry, it has a 100 user defined fields transmitted each second at 5 Hz telemetry update rate and 5 Hz data log update rate. Table 9.5 exhibits other characteristics of this autopilot.

9.8.4 DJI WooKong Autopilot

WooKong-Mis an autopilot system (www.dji.com) that provides self-leveling and altitude holding, which takes the stress out of flying RC multi-rotors operator. The mass of this autopilot is 118 g, and it has the following geometric dimensions: Controller: 51.2 mm × 38 mm × 15.3 mm, IMU: 41.4 mm × 31.1 mm × 27.8 mm, GPS module: 50 mm (diameter) × 50 mm. It is equipped with a receiver, 2-axis gimbal support, three autopilot modes. Three important features are go home module, auto landing capability, and intelligent orientation control.

9.9 Design Procedure

In the preceding sections, the basic fundamentals of a microcontroller and functions of its main components have been described. In this section, the design and development procedure of a microcontroller will be introduced. In general, the microcontroller development begins [139] from the requirements analysis and ends with testing and maintenance. Figure 9.16 illustrates the microcontroller design/selection/development

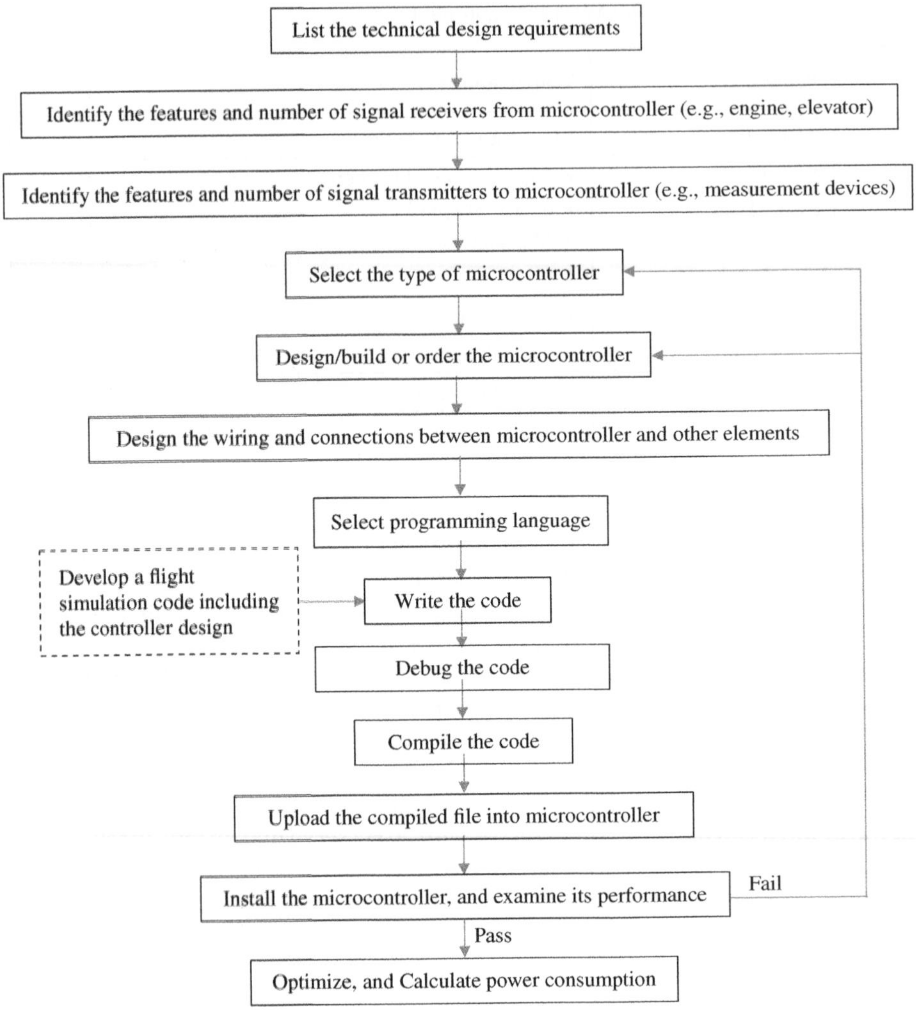

Figure 9.16 Microcontroller design/selection/development procedure.

procedure block diagram. The process has a number of feedbacks to check if the requirements are met. The iteration continues until the design requirements are met.

It should be noted that a flight simulation code including the controller design is developed in parallel. Thus, prior to the writing a code for the microcontroller, the controller design, and the flight simulation must be conducted. Thus, the flight controller must perform satisfactorily in a flight simulation, before it is included in the microcontroller code.

9.10 Design Project

In this section, the application of a microcontroller in a control system is demonstrated in the form of a design project. The design objective is to control and balance a ball on a level beam, and keep the ball at a given location. For this project, an Arduino Uno board is employed.

9.10.1 Problem Statement

A ball on a horizontal plate or a horizontal beam is naturally neutrally stable. If a disturbance is applied, the ball will move to a new location, and will not return to its initial place. The objective of this example project is to design, build, and implement a closed-loop control system to control and keep a ball on the beam at a desired location. The control objectives are to control and stabilize the ball location with a low overshoot (less than 15%), a short rise time (less than 2 seconds), and a short settling time (less than 5 seconds). The goal is to keep the ball at 20 cm from the range finder.

9.10.2 Design and Implementation

For this goal, the following equipment are needed: (a) beam balance mechanism (See Figure 9.17); (b) ball; (c) Arduino Uno board; (d) computer; (e) USB cable; and (f) wire. The sensor for this purpose is an ultrasonic range finder (lidar) for object (i.e., ball) recognition and to measure the ball distance.

For this project, three codes were written: (a) Arduino code; (b) MATLAB code for plotting; and (c) MATLAB code for determining PID gains. The Arduino code is developed to implement the control the ball on the beam, and to keep it at a desired distance (i.e., does not allow the ball to hit either ends of the beam). Both codes are provided in this section. The Arduino code is compiled and uploaded to the Arduino Uno board using the Arduino software and a USB cable. The MATLAB code is to plot the variation of the location of the ball in real time. The wiring of the Arduino Uno board is shown in Figure 9.18.

In this mechanism, one end of the beam (right) is allowed to rotate. Another end (left) was moved up and down by a servo. The command to the servo is coming from the microcontroller. The servo used in this experiment is a HS-7985MG high torque digital servo. This servo is connected by a wire to the analog pin of the board (9).

9.10.3 Arduino Code

The Arduino code for ball-beam control system is shown below:

```
#include<Servo.h>
#include<PID_v1.h>
```

Figure 9.17 Beam balance mechanism.

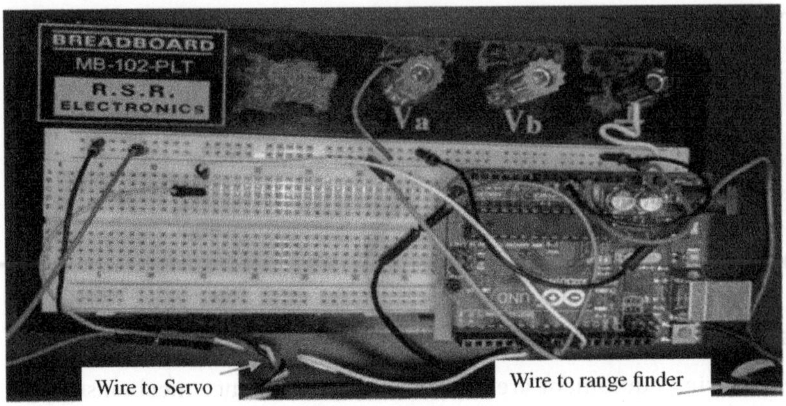

Figure 9.18 Wiring of the Arduino Uno board.

```
const int servoPin = 9; //Servo Pin
float Kp = 5; //Initial Proportional Gain
float Kd = 3; //Intitial Derivative Gain
float Ki = 3; //Initial Integral Gain
double Setpoint, Input, Output, ServoOutput;
PID myPID(&Input, &Output, &Setpoint, Kp, Ki, Kd, DIRECT);
Servo myServo;
void setup()
{
Serial.begin(9600);
myServo.attach(servoPin);
Input = readPosition();
myPID.SetMode(AUTOMATIC);
myPID.SetOutputLimits(-85,85);
}
void loop()
{
Setpoint = 20;
Input = readPosition();
myPID.Compute();
ServoOutput=99+Output;
myServo.write(ServoOutput);
}
float readPosition() {
delay(50);
const int pingPin = 7; //Trig Pin Arduino 7
const int pingPin2 = 6; //Echo Pin Arduino 6
long duration, cm;
unsigned long now = millis();
pinMode(pingPin, OUTPUT);
digitalWrite(pingPin, LOW);
delayMicroseconds(2);
```

```
digitalWrite(pingPin, HIGH);
delayMicroseconds(5);
digitalWrite(pingPin, LOW);
pinMode(pingPin, INPUT);
duration = pulseIn(pingPin2, HIGH);
cm = duration /(24.5*2);
if(cm > 50)
{cm=49;}
Serial.println(cm);
return cm;
}
```

The description of each line of the code is not shown, to only show the pure code. It can be seen on the website of this text.

9.10.4 Procedure

The experiment procedure (in four groups) in this project is presented below:

1) **Start Up**
   a) Connect the Uno board to the computer using a USB cable
   b) Turn on the computer
   c) Open MATLAB software
   d) Open Arduino software

2) **MATLAB simulation**
   a) Write a matlab code for simulation
   b) Estimate appropriate K_p, K_i, and K_d values for the system
   c) Hit run
   d) Monitor step response
   e) Adjust these values until the desired response is reached

3) **Arduino Program**
   a) Write an Arduino code
   b) Input Kp, K_i, and K_d values from matlab system simulation
   c) Plug in Arduino board on the balance to the computer
   d) Select Arduino Uno for the board
   e) Select COM port that the Arduino is located on (or assign the Arduino to a COM port)
   f) Compile (i.e., verify)
   g) Upload Arduino code – (i.e., **ON**)
   h) Press ctrl+shift +m to open serial monitor (Under Tools)
   i) Place the ball on the balance and wait for the ball to settle

4) **MATLAB program serial plot**
   a) Write a new MATLAB code for plotting
   b) Run code
   c) Stop code when a response curve is created, or steady state is reached.

BTW, step 4 can also be done with Excel software as follows:

a) Unplug the Arduino Uno board – (i.e., **OFF)**
b) Copy and paste the serial data (from window) into Excel
c) Plot the response vs time

9.10.5 MATLAB Code for Real-Time Plotting

A MATLAB code should be written for real-time plotting. The real time outputs can be created by employing this MATLAB module. This code allows the communication between MATLAB software and Arduino Uno board. Note that this MATLAB code is used for plotting, with the Arduino code running. However, the alternate Excel method is employed after Arduino is unplugged. The MATLAB code for real time plotting is shown below:

```
clear all
close all
clc
arduino=serial('COM3','BaudRate',9600); % create serial com-
munication object on port COM18
fopen(arduino); % initiate Arduino communication
CM(1)=0;
time(1)=0;
i=1;
tic;
while (toc<=100)
   CM(2)=fscanf(arduino,'%f');
   time(2)=toc;
   figure(1);
   grid on;
   axis([toc-10, toc+10, 0, 35])
   h(i)=plot(time,CM,'b','LineWidth',5);
   hold on
   CM(1)=CM(2);
   time(1)=time(2);
 if(i >= 300)
   delete(h(i-299));
   end
i=i+1;
end
fclose(arduino); % end communication with arduino
```

The MATLAB code for evaluating/determining PID gains is not provided in this section. The technique to determine PID gains for such neutrally stable system is left to the reader to investigate. This MATLAB code requires a mathematical (dynamic) model of the ball-beam mechanism. Chapter 6 provides techniques to design control systems using a PID controller.

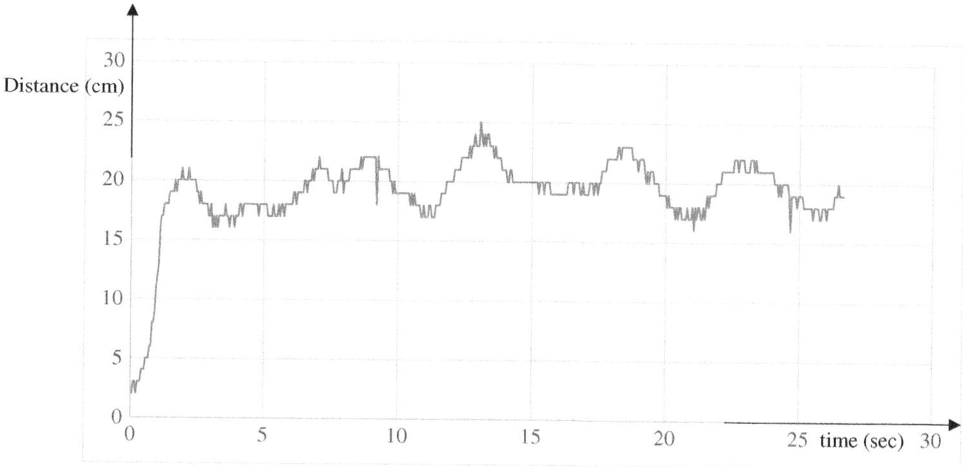

Figure 9.19 Variations of the ball location on the beam.

9.10.6 System Response and Results

The control goal is to keep the ball at 20 cm from the range finder. Figure 9.19 indicates the variations of the location of the ball on the beam. The response indicates that the overshoot and rise time requirements are met, while the settling time requirement was never met. This deficiency is expected, since the ball-beam system is naturally neutrally stable, not fully stable.

Questions

1 What is the major function of a microcontroller?

2 What are the primary functions of a microcontroller in an UAV autopilot?

3 What autopilot activities are performed via a microcontroller?

4 Name three differences between a microcontroller and a microprocessor.

5 Name at least three microcontroller designers/manufacturers/architectures/vendors.

6 What do A/D and D/A stand for?

7 Briefly describe the features of a microcontroller.

8 Describe Harvard memory architecture.

9 Name at least five main modules of a microcontroller.

10 Name at least two high-level programming languages.

11 What is the function of a complier?

12 What are the differences between flight simulation software and the microcontroller code? Briefly describe.

13 Name at least three software to develop for an UAV.

14 Describe the procedure to develop the microcontroller code.

15 What are the ground payload software functionality?

16 Describe features of the ArduPilot.

17 Describe how a code (computer program) in MATLAB (m-file) is converted and transferred to a microcontroller. In another words, write four main steps of programming procedure.

18 What is Arduino?

19 Describe the programming language in Arduino.

20 Describe the features on an Arduino Uno board.

21 What are the three wires in a Pot? Explain each one.

22 What are the three wires in a servomotor? Explain each one.

23 Draw a figure showing typical connection of a microcontroller to sensors, servos, and communication system elements.

24 List name of at least four elements/sensors to which a microcontroller sends a signal.

25 List name of at least four elements from which a microcontroller receives a signal.

26 What does CPU stand for?

27 Name two analog signals.

28 Briefly describe the features of an Atmel ATmega644P microcontroller.

29 Briefly describe the features of a DC motor as a servo.

30 Name two types of driver for DC motors.

31 What is the function of a servomotor?

32 What each of the three wires of a servos for?

33 What is the function of a Pot?

34 Name at least three software needed for developing a large UAS.

35 Briefly describe the features of an UMS.

36 Briefly describe the features of a PX4 Pixhawk autopilot.

37 Briefly describe the features of a DJI WooKong autopilot.

38 Briefly describe the features of a MicroPilot MP2128 autopilot.

39 Briefly describe the Matlab special support package for Arduino hardware.

40 Briefly describe the process to use the Matlab software for controlling a device via an Arduino board.

Problems

1 Write a code in C that converts the given temperature in Fahrenheit to Celsius using the following conversion formula:

$$C = \frac{F - 32}{1.8}$$

2 Write a code in C that converts the given temperature in Celsius to Kelvin.

3 Write a code in C that converts the given voltage (V) and current (I) of a sensor to determine its power using the following conversion formula:

$$P = IV$$

4 Write a code in C that solve the quadratic equation

$$\mathrm{a}x^2 + \mathrm{b}x + \mathrm{c} = 0$$

5 Write a code in C that solve the following equation

$$\mathrm{a}x^3 + \mathrm{b}x^2 + \mathrm{c} = 0$$

6 What is the output of the following code?

```
void main()
{
char *s="\12345s\n";
```

```
printf("%d",sizeof(s)):
}
```

7 Write a code for Arduino Uno board to sweep the shaft of a servomotor back and forth across 90°.

8 Write a code for Arduino Uno board to sweep the shaft of a servomotor back and forth across 180°.

9 Write a code for Arduino Uno board to sweep the shaft of a servomotor back and forth ±30°. This is to model one aileron.

10 Write a code for Arduino Uno board to sweep the shaft of two servomotors simultaneously back and forth ±30°. This is to model two ailerons. Hence, make sure, when one aileron is deflected upward, the other one is deflected downward, and vice versa.

11 Write a code for an Arduino Uno microcontroller to control a close/open function of a building door, while the only sensor is a camera. Assume PID gains are, $K_p = 2$, $K_i = 3$, $K_d = 0.2$. The actuator should be activated, when a person is within 2 mm from the door.

12 Write a code for an Arduino Uno microcontroller to control a close/open function of a building door, while the only sensor is a range finder. Assume PID gains are, $K_p = 24$ $K_i = 1.2$, $K_d = 0.05$. The actuator should be activated, when a person is within 2 m from the door.

13 Write an Arduino code to receive a signal from a range finder and create an analog output (voltage).

14 Write an Arduino code to receive a signal from an altimeter and create an analog output (voltage).

15 Write an Arduino code to use a range finder output to measure the distance from a UAV to an obstacle.

16 Write an Arduino code to use a rate gyro output to measure the attitude of a UAV.

17 Write an Arduino code to use an accelerometer output to measure the ground speed of a UAV.

18 Write an Arduino code to use an accelerometer output to measure the displacement of a UAV.

19 Write an Arduino code to use a pitot-tube output to measure the UAV airspeed.

20 To simulate the measurement of fuel level in a UAV fuel tank, write an Arduino code to use an ultrasound sensor to measure and illustrate the fuel level.

21 To simulate the measurement of fuel level in a UAV fuel tank, write an Arduino code to use an ultrasound sensor to measure and illustrate the fuel level.

22 Write a Matlab code to cause an LED on an Arduino Uno to blink every 1 second. In this code, the matlab support package for Arduino hardware is employed.

23 Write a Matlab code to cause an LED on an Arduino Mega to blink every 0.4 second. In this code, the matlab support package for Arduino hardware is employed.

24 Write a Matlab code to cause an LED (connected to an Arduino Uno) to flash on and off five times. In this code, the matlab support package for Arduino hardware is employed.

25 Write an Arduino code to use a thermometer output to measure the air temperature.

Design Projects

1 Design and build a control system using an Arduino Uno to control the position of a servo with an analog Pot. Instructor should provide the equipment.

2 Design and build a control system using an Arduino Uno to control the position of a servo with a digital Pot. Instructor should provide the equipment.

3 Design and build a control system using an Arduino Uno to read air pressure from a pitot tube and send the results to a computer and plot the results. Instructor should provide the equipment.

4 Design and build a control system using an Arduino Uno to calculate airspeed from a pitot tube (in a wind tunnel) output and send the results to a computer and plot the results. Instructor should provide the equipment.

5 Design and build a control system using an Arduino Uno to read temperature from a thermometer and save data. Instructor should provide the equipment.

6 Design and build a control system using an Arduino Uno to read altitude from an altimeter and save data to a laptop. Instructor should provide the equipment.

7 Design and build a control system using an Arduino Uno to read applied force from an analog force sensor and save data to a laptop. Instructor should provide the equipment.

8 Design and build a control system using an Arduino Uno to read applied force from a digital force sensor and save data to a laptop. Instructor should provide the equipment.

9 Design and build a control system using an Arduino Uno to control a group of LEDs (in series). LEDs should turn on and off each for one second. Instructor should provide the equipment.

10 Design and build a control system using an Arduino Uno to control two LEDs (in parallel). LEDs should turn on and off in turn, each for one second. Instructor should provide the equipment.

Notes

1 Random Access Memory.
2 Read Only Memory.
3 Electrically Erasable Programmable Read Only Memory.
4 Pulse Width Modulation (a type of digital signal).
5 Computer Operates Properly.
6 *Peripheral Interface Controller (also referred to as Programmable Intelligent Computer).*
7 http://suasnews.com.
8 http://ardupilot.org.
9 www.micropilot.com.

10

Launch and Recovery Systems Design

CONTENTS

Design of Unmanned Aerial Systems, First Edition. Mohammad H. Sadraey.

Companion website: www.wiley.com/go/sadraey/unmanned-aerial-systems

Educational Outcomes

After study of this chapter, the reader will be able to:

1) Discuss the current launch and recovery technologies
2) Manage a launch system design project
3) Conduct a recovery system design project
4) Conduct a Unmanned Aerial Vehicle (UAV) launcher design project
5) Explain various launch techniques
6) Explain various types of launchers
7) Explain various launcher components
8) Explain various recovery techniques
9) Discuss fundamentals of launch
10) Discuss fundamentals of recovery
11) Design elevation mechanism
12) Evaluate launch and recovery systems
13) Discuss launch/recovery systems mobility
14) Develop launch and recovery systems selection process

10.1 Introduction

Similar to other aircraft types, UAVs need means to accelerate up to a certain speed (usually slightly higher than stall speed) to be able to become airborne. All unmanned aircraft must initially take-off or be launched to become airborne. Moreover, at the end of the flight mission, they must land on an airfield, or be recovered. Launch involves transitioning the UAV from a nonflying state (e.g., stationary on the ground) to a flying state. In the case of a horizontal runway launch, this can be considered as a conventional takeoff. Various conventional and unconventional launch and recovery techniques have been applied to manned aircraft for over a century. The range of options is greater for unmanned aviation. These techniques are largely enabled by the exclusion of pilot physical constraints and the inclusion of much lower UAV weights.

In general, there are three techniques for launching a UAV: (a) conventional horizontal takeoff and landing (HTOL); (b) a catapulted or powered launch; and (c) vertical takeoff and landing (VTOL). The first method (i.e., HTOL) requires a long runway, and is primarily employed by large and medium-weight UAVs. The powered launch is mainly utilized by small UAVs, and requires a launcher. The third technique (i.e., VTOL) is applicable when the engine thrust is greater than the UAV weight. All quadcopters and helicopter UAVs are employing the VTOL technique. The conventional takeoff and landing fundamentals and analysis are presented by [9].

The vast majority of large fixed-wing UAVs employ conventional launch and recovery techniques; i.e., conventional takeoff and landing. Conventional launch and recovery require a length of horizontal flat surface (runway); and a further distance before the nearest obstacle of maximum specified height. Of all the methods for UAV launch and recovery, the VTOL one is an accurate, operationally convenient and gentle. VTOL are more appropriate for very small UAVs.

The complexity in the integration of unmanned air vehicles is more challenging when it comes to naval ships, as the technology requirements are usually higher, and space is limited. Apart from the physical aspects of the integration, it is desired to have a solution that is flexible, cost efficient and have a wide envelope of operability. As the types and sizes of UAVs increase, a need for an integrated launch and recovery system that can handle various types and sizes are being developed.

When a UAV has a conventional takeoff and landing, the launch and recovery systems is converted to the design of landing gear. The approach for designing the landing-gear for a tactical-to-large UAVs, such as the AAI RQ-7 Shadow 200 or the Northrop Grumman RQ-4 Global Hawk, is not different than for a manned aircraft. The high-altitude long endurance UAV, Global Hawk (Figure 1.2) has a high speed required for lift-off (about 100 knot). It is powered by a single turbo-fan engine whose thrust is just over 40 kN and requires a ground run of over 600 m. The landing gear design is out of scope of this book, landing gear design process is presented in Ref. [8].

The Global Hawk (Figure 1.2) representing the HTOL type, powered by a turbofan engine whose nominal thrust of just over 40 kN generates an acceleration of 0.3 m/s^2 to achieve ground roll of about 600 m. The Predator B (Figure 5.12), on the other hand, employs a turboprop engine with a variable pitch propeller, has a ground roll of about 300 m. Unlike the Hunter, the Shadow does not use an external pilot, depending instead on a launcher for takeoffs, and an automated landing system for recovery.

The launch acceleration will significantly influence the UAV structure as well as the installation of payloads, due to imposing inertia force. All of the on-board equipment, including sensitive camera, and navigation sensors must withstand this acceleration. Below 2 g, the amount of hardening and support needed for the equipment may be ignored. To withstand acceleration beyond 5 g, installations require more substantial hardening, and will add to the cost and overall weight.

Every year, new launch and recover techniques are developed by universities and industries. Recently, Aurora is developing a new a runway independent UAV launch and recovery system; called SideArm[1]; which is capable of both land and sea-based operations. It brings runway independence to a new class of large fixed wing UAVs. The SideArm system provides the versatility of rotary-wing aircraft with the payload and endurance of fixed-wing aircraft. This single apparatus launches and recovers aircraft from austere sites, and can relocate between launch and recovery events.

Typical launch methods include: (a) rail launchers; (b) rocket launch; (c) air launch; (d) hand launch; (e) tensioned line launch (i.e., catapult); (f) gun launch; and (g) ground-vehicle launch. Currently the following recovery methods are employed: (a) skid and belly recovery; (b) net recovery; (c) cable-assisted recovery; and (d) parachutes. In this chapter, the launch of micro to small UAVs are addressed, which particularly using concepts that avoids the need for a runway.

10.2 Launch Technologies and Techniques

The launch and recovery operations may be performed in various ways; here a brief description of the most common related technologies employed today are reviewed. The simplest and most cost-efficient way to launch/recover a fixed-wing UAV is conventional takeoff and landing on a runway. The power drives move launchers. The launcher positions are frequently determined by potentiometers that convert mechanical motion to electrical signals.

Typical launch methods include: (a) rail launchers; (b) rocket launch; (c) air launch; (d) hand launch; (e) tensioned line launch (i.e., catapult); (f) gun launch; and (g) ground-vehicle launch. The four most common categories of launching systems and technologies currently available are: (a) rocket assisted take-off; (b) bungee cord; (c) hydraulic launchers; and (d) pneumatic launchers. In comparison, the five most common categories of recovery systems and technologies currently available are: (a) net recovery; (b) arresting line; (c) skyhook; (d) windsock; and (e) parasail. These launch techniques are discussed in this section.

10.2.1 Rocket Assisted Launch

A rocket is small non-air breathing engine with a high thrust; the fuel may be solid or liquid. If the fuel is solid, the duration of operation is just a few seconds. In the rocket assisted launch or take-off, first, a rocket (usually with a solid fuel) is attached to the UAV in a reloadable motor system. Then, the UAV is placed on a ramp with a launch angle. The launch starts with pressing a button by an operator, to ignite the rocket. When the launch rocket is ignited, it helps in accelerating the UAV until its cruise engine takes over. The rocket is released after the launch, so it does not add to the UAV weight during the rest of the flight operation. Thus, the rocket is part of the UAV only during the launch operation. Figure 10.1 exhibits a Pioneer UAV during a rocket assisted launch.

Figure 10.1 Pioneer Unmanned aerial vehicle (UAV) during rocket assisted launch.

When released, the rocket is usually dropped into the water/ground, it will not be usable again. However, when dropped, the rocket can be collected and inspected for repair and for possible reuse. After the launch, there is an interval where the UAV is airborne, but does not have enough airspeed to produce the required aerodynamic side-force and yawing moment to directionally control the UAV. For such a case, care must be taken in operating under a cross wind. This pitfall can be minimized by adjusting the launch angle to the optimum value.

When the launch is canceled due to non-favorable weather/environment, the rocket must be removed from UAV and stowed safely. The rocket is a pyrotechnic component and needs to be housed in a special place. Furthermore, rockets requiring special care, and safety considerations are highly recommended.

The small size and small number of components needed for the launch make this technique favorable to carry onboard a ship; hence it does not require a significant alteration to the ship. Moreover, a minor installation is needed on the ship. However, the flames and the thrust generated by the rocket can be an issue, if the deck structure is sensitive to these two items. When the rocket is being fired, the flame, smoke, and explosion sound are created. These three outcomes have a negative effect, if stealth characteristics are expected to be maintained.

In this launching operation, the rocket imposes a huge structural load on the UAV, due to a high linear acceleration. This motion acceleration can be expressed in terms of gravity acceleration (g). This ratio can be referred to as the g-load or load factor:

$$n = \frac{a}{g} \tag{10.1}$$

In other words, the UAV load is expressed as a factor of the standard acceleration due to gravity ($g = 9.81\,m/s^2 = 32.17\,ft/s^2$). As the load factor increases, the load on the aircraft structure (e.g., wing and fuselage) will be increased; thus, the UAV structure gets heavier. In the UAV structural design, the highest allowable load factor is considered. Typical value for the load factor is from 1 to 5.

Major advantages of the rocket assisted launch are: (a) it is a small system; (b) the UAV requires near to zero take-off length; (c) small initial cost; and (d) it can be prepared in advance. These features make this launch system particularly suitable for a ship. The rocket assisted launch (sometimes referred to as the jet assisted takeoff [JATO] is a matured technology, and even employed by large prop-driven transport aircraft. The rocket thrust is relatively constant during launch, so you may employ Newton's second law to determine the end velocity.

10.2.2 Bungee Cord Catapult Launch

Another technology for the UAV launch operation is to use bungee cord(s) in a catapult launching system. A bungee cord is a highly elastic mechanical element that stores energy and provides it when released. The bungee cord stored energy is converted to a push force for launching the UAV. This element requires some means of transferring the load from the winch to the body of the cord. The load transfer can be done with a threaded plug or a swivel hook. The body of cord must safely handle two types of stresses: (a) shear stress; and (b) normal/tensile stress. This should be considered in the design of the winch-cord connection system with an acceptable safety factor (about 2–3).

This simple concept launch system includes the following elements: (a) a metal rail that is positioned at a launch angle; (b) a bungee cord (or multiple cords) (c) a winch to stretch the cord; and (d) release mechanism.

Due to the simplicity in the system design, and not requiring many mechanical or electrical components, it is relatively light weight, low cost, and easy to operate. This system is relatively silent, has no electrical components, and the rails could be manufactured from low signature materials. The bungee cord is probably the quietest launch technique; and is frequently limited to support UAVs of smaller sizes.

Performance of a bungee cord is similar to a mechanical spring; the released energy is equal to the applied force multiplied by the rate of change of deflection. The force (F) is proportional to cord length and its extension (x).

$$F = 2kx \tag{10.2}$$

where k is the cord spring index; and is a function of cord materials, density, and its cross-sectional area. The constant, 2, is used (Figure 10.2) due to two parts (left and right of the UAV) of the cord. When the cord has its longest extension, it will apply the maximum force; while as the extension is decreased, the force is decreased too. The value of cord spring index is a function of cord materials and its diameter.

The launch operation procedure is described here. Prior to launch, the winch stretches the bungee cord into necessary tension; then the cord is secured with a safety pin. To begin the launch, the pin is released. Then, the UAV accelerates along the ramp until it passes the rail and gets airborne. The cord is not part of the UAV and; stays on the launcher after the launch. If the cord stretching force (winch operation) is beyond the operator's force limit, a motor – to turn the winch – is required. Compared with other launch techniques, a launcher with a bungee cord requires the longest launcher ramp. The maximum achievable UAV airspeed by a bungee cord launcher is low; and hardly passes 20–30 knots. Thus, bungee cord launchers are often limited to lightweight UAVs.

The initial release of cord is usually without any dampening; thus, the launch can be a "jerky" motion in the start. This feature can cause structural/performance problems, if the UAV or its payload cannot withstand this type of acceleration. Moreover, since the deflection is decreased during the launch, the launch force is gradually decreased. Hence, the acceleration – where it is initially high – is gradually decreased too.

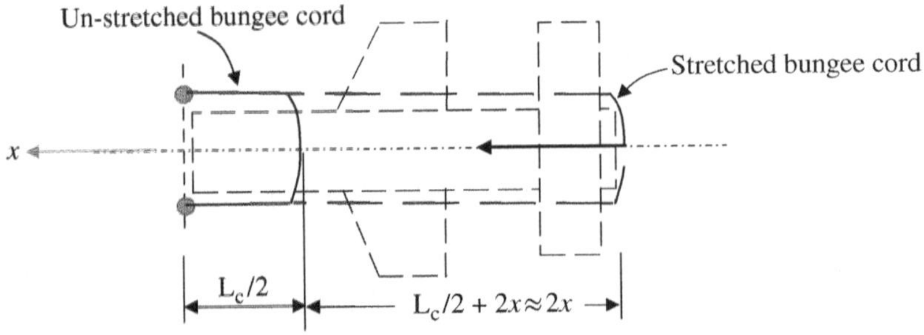

Figure 10.2 Top view of a bungee cord launcher.

In order to calculate the end velocity, employ the Newton's second law in its differential form.

$$\Sigma F = m\frac{d^2}{dt^2}x \tag{10.3}$$

where m is the UAV mass on the launcher. The contributing forces are: (a) cord force; (b) UAV weight; (c) friction between UAV and ramp; (d) lift; and (e) drag. Section 4 presents the derivation of this equation by including all contributing forces. To simplify the derivation, here, only the bungee cord force, and friction force are considered. If the short notation (i.e., $\frac{d^2}{dt^2}x = \ddot{x}$) is utilized, the governing differential equation is equivalently obtained as:

$$2kx - \mu mg\cos\Theta = m\ddot{x} \tag{10.4}$$

where θ denotes the launch angle, and μ is the friction coefficient. The equation may be reformatted as:

$$\ddot{x} - \frac{2k}{m}x = F_A \tag{10.5}$$

where F_A is the normalized applied input, $F_A = -\mu g\sin\Theta$ to this dynamic system. In terms of control theory, this second order differential equation is the standard mathematical model of an undamped oscillatory motion (with a natural frequency of $\sqrt{\frac{2k}{m}}$). However, since the UAV will be launched shortly, the system – after launch – will have an entirely different dynamics and another mathematical model. Thus, the solution to this differential equation is only valid until UAV is on the ramp and cord is pushing. Two unknowns of this second order linear differential equation is of interest: (a) velocity; $\dot{x}$; and (b) displacement, x.

$$\dot{x} = V \tag{10.6}$$

The solution of Eq. (10.5) (assuming zero initial conditions) yields the velocity of the UAV along the ramp. The transfer function of this launch system is obtained as:

$$\frac{X(s)}{F_A(s)} = \frac{1}{s^2 + \frac{2k}{m}s} \tag{10.7}$$

The direction of cord extension is opposite to the direction of velocity. The solution also illustrates the required cord extension; x (i.e., launcher length) as a function of the desired end velocity (V). Eq. (10.7) indicates that, in order to achieve a higher UAV velocity at the separation point, one must: (a) increase the cord extension (and consequently, the ramp length); (b) employ a cord with a higher spring index (k); and (c) reduce the UAV mass (m).

The general solution to a linear second order differential equation (10.5) has the following form [140]:

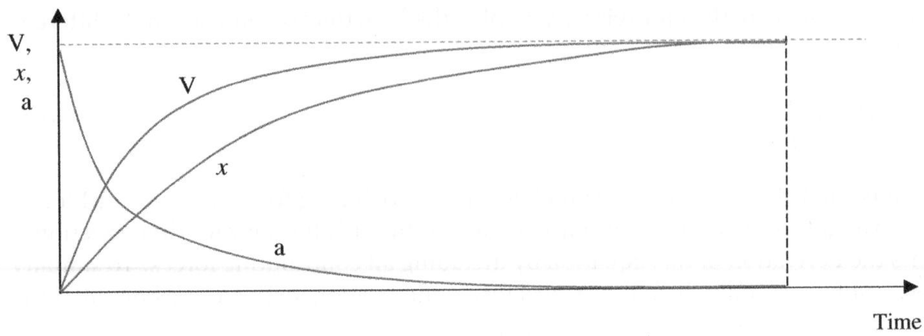

Figure 10.3 Normalized values of Unmanned aerial vehicle (UAV) displacement, velocity, and acceleration during launch.

$$x = c_1 e^{\lambda_1 t} + c_2 e^{\lambda_2 t} \tag{10.8}$$

where c_1, c_2, λ_1, and λ_2 are determined from initial conditions and equation coefficients (including k). However, this is a special case, with no damping. The complete solution (assuming zero initial conditions) of Eq. (10.4) yields the displacement [50] of the UAV along the ramp as a function of time (t):

$$x = \cos\left[\sqrt{\frac{2k}{m}}t\right] \tag{10.9}$$

This solution is valid only until the UAV is separated from the launcher, so no oscillation is involved.

Figure 10.3 demonstrates the normalized values of displacement, velocity, and acceleration during launch as a function of time. As indicated, all three parameters are nonlinear with time. The displacement, velocity will reach their maximum values, while the acceleration drops to zero.

Due to nonlinear nature of the contributing forces, the end velocity is difficult to calculate. The calculation challenge intensifies when noted that the friction at the beginning of launch is much higher than the period when the UAV is moving. The interested reader is encouraged to derive the complete nonlinear governing equation; and use numerical methods to obtain the more accurate results.

10.2.3 Pneumatic Launchers

The pneumatic launching systems are typically designed as a catapult rail launcher. The energy storage is a compressed gas, usually the air is used, due to its availability and no cost. Air is pressurized by a compressor; and is stored in a high-pressure accumulator tank. The launch is performed by applying the compressed air force on UAV through a cylinder/pipe/tube along the ramp. When the compressed air is allowed to leave the tank via a valve; it pushes and accelerates a rail-mounted cradle which holds the UAV. The applied force can be regulated by adjusting the air pressure in order to support UAVs of different mass.

An advantage of a pneumatic launcher is that there is a full control over the whole launch phase. However, a disadvantage of this technique is that, it often takes a while to

Figure 10.4 ScanEagle pneumatic launcher

pressurize the air and to refill the accumulator tank. Moreover, a compressor to pressurize the air needs to be powered (e.g., electric motor, or piston engine). Furthermore, it tends to have a high initial cost, and have a high weight. The advantages are that: (a) the pneumatic launcher is a proven working design; (b) it has a low recurring cost; and (c) adaptable to different UAVs. If this launcher is employed on the deck of a ship, the ship compressed air system can be utilized through outlets.

Many UAVs are employing pneumatic power for launching operation. A ScanEagle is launched using a pneumatic launcher (Figure 10.4), patented by Insitu, known as the "SuperWedge" launcher. Moreover, the Insitu trailer-mounted Mark 4 launcher – a self-powered by an onboard diesel (JP-5 or JP-8 fuel) generator and air compressor – is compatible with all of Insitu's unmanned aircraft. The setup of this launcher with a weight of 4,200 lb, a deployed length of 22 ft, and with two operators, takes about minutes. The RQ-7 Shadow is also launched from a trailer-mounted pneumatic catapult. Its pneumatic launcher can accelerate the 170 kg aircraft to 70 knots in 12 m.

10.2.4 Hydraulic Launchers

The hydraulic launching systems are much like the pneumatic ones. The energy storage is a compressed fluid, usually the oil is used, due to low cost, and lubricating property. The hydraulic oil is compressed by a hydraulic pump; and may be stored in a high-pressure oil tank. In hydraulic launchers, a two-compartment cylinder is employed, the

Figure 10.5 Shadow 200 hydraulic launch

compartments are separated by a piston/diaphragm that has a compressible gas (e.g., air) on one side, and compressed oil on another side. To perform a launch, a valve is quickly opened, so that the oil is pumped inside the cylinder to push the piston. Consequently, the gas is compressed on the other side. Figure 10.5 illustrates the Shadow 200 launch operation using a hydraulic launcher.

Advantages of a hydraulic launcher are: (a) it can produce relatively high launching velocities; and (b) the initial start can be adjusted to avoid very high acceleration. The hydraulic launch system is much more complex than the bungee cord or pneumatic launcher, or the rocket assisted launcher.

Thus, advantages of a hydraulic launching systems are: (a) control over the whole launch phase; (b) low recurring cost; (c) adaptable to different UAVs; and (d) repeatable launch with quick reset. In contrast, three main disadvantages are: (a) high up-front cost; (b) large deck footprint; and (c) complex mechanism.

The system can also provide energy to a winch which is attached to a rail-mounted cradle where the UAV is placed. This allows the gas in the cylinder to expand and to move the piston that pushes the oil back, to power the hydraulic motor. For this goal, the hydraulic motor turns the winch that pulls the cradle and accelerates the UAV to launch speed. The Shadow 200 can be catapult launched from a hydraulic rail launcher.

10.2.5 Air Launch

Other than ground vehicle launch, there are a number of techniques that a UAV can be launched. One launch technique is to employ another aircraft (manned or unmanned) to release a UAV. Air launch involves releasing the UAV from a host or mother platform (i.e., another aircraft) at altitude and with high initial airspeed relative to ground. The unmanned aircraft can be released from a host platform's internal bay, under wing, under the fuselage, or from a cargo door.

The mother aircraft have a station that the UAV is attached to. Typically, a launch rail will attach to a station, and that launch rail is what carries the UAV. The launch method depends greatly on the mission and characteristics of both the mother aircraft and UAV. Usually the UAV is dropped from under the host platform. The design of the UAV air launch system is challenging, and involve a lot of calculation, and may generate a risk to the mother aircraft. This technique is only used when the UAV does not have a long range to reach the target area. So, an aircraft carries the UAV to an area close to the target zone, and can then release/launch it.

10.2.6 Hand Launch

A very cheap and easy launch technique for a very small and light UAV is the human operator. This approach achieves runway independence. Very small unmanned aircraft can employ a hand-launch technique, which eliminates the need for any launch equipment. An overhead launch is the most common form for launching very small UAVs. Indeed, the human becomes the launch equipment. The individual launching the UAV is able to provide initial velocity and orientation, to enable controlled initial flight conditions.

The operator must have sufficient hand force to launch the UAV. Moreover, he/she may run while launching to increase the initial velocity. After the hand launch, the UAV will use its engine force to continue the flight. Hand launch will drive the UAV configuration; one example is eliminating a need for the landing gear. Figure 10.6 shows the launch of the small UAV RQ-11 Raven by hand.

A typical military male operator may provide a hand force of up to 100 N. Moreover, the length of an extended hand is about 70–90 cm. By application of Newton's second law and dynamics governing equations, one can determine the maximum mass (m) of an UAV that can be launched by hand.

$$F = ma \tag{10.10}$$

Moreover, the operator may run during the launch period in order to increase the initial speed. When a hand launch system is selected for a UAV, no design activity is required for the launch system. However, an analysis should be conducted to make sure that the operator can provide a safe initial velocity.

When the acceleration, a, initial velocity, V_1, and the extended hand (distance), x are known, the release speed (V_2) is obtained by:

$$V_2^2 - V_1^2 = 2ax \tag{10.11}$$

Figure 10.6 RQ-11 Raven hand launch.

The initial velocity; is usually zero.

Example 10.1
A very small UAV has a mass of 1.5 kg. A human operator with a maximum force of 90 N is arranged to launch this UAV. If his hand extends 70 cm, determine the release speed.

Solution
Newton's second law yields the linear acceleration:

$$F = ma \rightarrow a = \frac{F}{m} = \frac{90}{1.5} = 60\frac{m}{s^2} \tag{10.10}$$

Release speed:

$$V_2^2 - V_1^2 = 2ax \rightarrow V_2 = \sqrt{2ax} = \sqrt{2 \times 60 \times 0.7} \rightarrow V_2 = 9\frac{m}{s} \tag{10.11}$$

The initial velocity; V_1 is assumed to be zero.

10.3 Launcher Equipment

10.3.1 Elements

Conventionally, a relatively long runway is employed where a UAV accelerates by its own propulsion. However, when such a runway is not available, measures have to be taken (e.g., using launch equipment) to assist in the take-off. UAVs can be launched from a fixed or a moving ground vehicle. A launcher eliminates the need for conventional landing gear for the purposes of takeoff. The launcher is a mechanical devise to accelerate a fixed-wing UAV to a minimum controllable airspeed before releasing it from launcher. The ground vehicle provides the acceleration to flight speed, at which time the UAV is released. This launch technology predates conventional landing gear on flying aircraft. For UAVs which do not have a vertical takeoff capability, a launcher will be required.

In the design of a launcher, a few parameters such as launcher length, launcher weight, launch angle, and the required force and power must be calculated. Moreover, the type of source of launcher power (e.g., mechanical, spring, pneumatic, and rocket) needs to be decided. Major elements of a launcher are: (a) ramp/slipway; (b) elevation platform/mechanism; (c) mobility wheels; (d) transportation truck; (e) power source; and (f) push mechanism.

Furthermore, if the UAV is equipped with a pusher prop-driven engine, a propeller cover (element 7) is also needed. In this section, these elements are briefly described. Figure 10.7 shows a Shadow UAV and its launcher, notice that the flap is deflected. Figure 10.8 illustrates a Penguin C UAV with the launcher and ground station.

10.3.2 Ramp/Slipway

The vehicle must slide up the ramp due to the launch force. The ramp is often made up of one or two parallel rails. Rail launchers provide the UAV with a stabilizing track and launch energy to take the UAV from rest to airborne. The UAV is secured to a shuttle

Figure 10.7 Shadow Unmanned aerial vehicle (UAV) and its launcher.

Figure 10.8 Penguin C with launcher and ground station. *Source:* Courtesy of UAV factory.

that travels (slips) along the rail via guides. It usually takes the form of a ramp along which the UAV is accelerated on a trolley, propelled by a system of rubber bungees, by compressed air or by rocket, until the aircraft has reached an airspeed at which it can sustain airborne flight.

The launch method must provide a safe physical separation between the UAV and the launcher; and other hazards throughout the launch operation. Most rail launchers with pneumatic/hydraulic power unit are capable of launching UAVs with a weight range of 500–1000 lb Typical launch speed for small UAVs is about 50–70 knots.

10.3.3 Push Mechanism

Pneumatic pistons are the most common tool to providing the launch energy. A compressor is required to pressurize air in the accumulator. Launchers must have a reacting mass or be fixed to the ground for stability. The Insitu Aerosonde was car launched for a transatlantic flight, enabling a lightweight airframe without landing gear.

To minimize the length of runway/launcher, a rocket engine/booster can be employed to launch a UAV. The rocket's high thrust-to-weight ratio and modularity can enable simple installation. Rocket launch is appropriate when no runway is available and other techniques such as pneumatic rail launchers have excessive footprint. The primary forces acting on the UAV are the rocket engine thrust, the UAV weight, and the normal propulsion system thrust. Hence, the rocket engine force is addition to the conventional engine thrust. It may be used on both runway and launcher. After the launch, the rocket engine may be dropped.

There is a mechanism, called *detent* mechanism that prevents the motion of the UAV at the beginning. When the engine thrust reaches a certain value, this mechanism loses its function and the UAV is set free to move. The detent mechanism is mathematically modeled as a force. It is known at the force value that the detent mechanism loses its function. A force in an opposite direction to the UAV direction, represents the detent mechanism.

10.3.4 Elevation Platform

The pitch angle adjustment mechanism (i.e., elevation platform) is a mechanical system to adjust the launch angle. There is an optimum launch angle at which the UAV will achieve its launch speed with the lowest cost. This angle is a function of UAV mass, friction, the launch speed, and the launch power. When the UAV launch angle is adjusted, it must be locked to prevent further variation. The typical angle is about 20–30°.

The launch angle is recommended to have a value between the UAV maximum climb angle, and the climb angle corresponding to the UAV maximum rate of climb.

$$\gamma_{ROC_{max}} < \theta_L < \gamma_{max}$$

The techniques to calculate the UAV maximum climb angle, and the climb angle corresponding to the maximum rate of climb are provided in Chapter 14.

The elevation platform can be hydraulic, pneumatic, or carried out by power screw. A power screw (Figure 10.9) is a device to transfer the mechanical power by changing angular motion into linear motion, and is usually paired with a nut. Other familiar applications include the lead screws of lathes, and the screws for vises, presses, and jacks.

The torque (T_R) required for raising a load (e.g., UAV mass), while overcoming thread friction is obtained [141] by:

$$T_R = \frac{F_N d_m}{2}\left(\frac{\pi \mu d_m + l}{\pi d_m - \mu l}\right) \tag{10.12}$$

where d_m is the mean (pitch) diameter of a square-threaded power screw with single thread, and μ denotes the coefficient of friction between screw threads. The normal force is the normal component of the UAV mass (m_U) plus mass of the launcher rails (m_R).

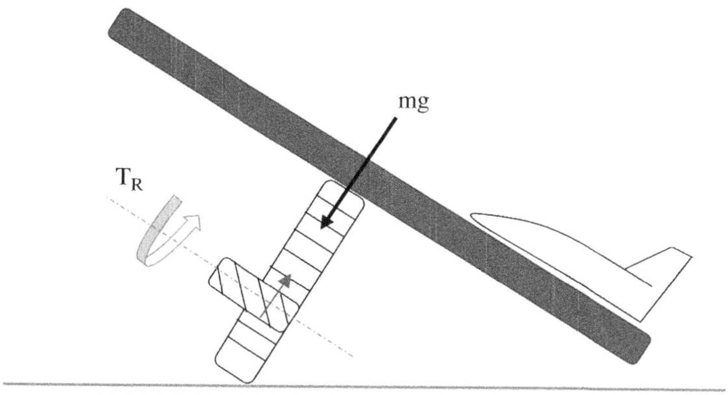

Figure 10.9 Worm-gear screw jack as used the elevation platform.

$$F_N = (m_R + m_U) g\cos\Theta \tag{10.13}$$

where θ stands for launch angle. The parameter l is referred to as the lead, and is the distance the nut moves parallel to the screw axis when the nut is given one turn. For a single thread, the lead is the same as the pitch. For a screw with n threads,

$$l = np \tag{10.14}$$

where p stands for screw pitch. A desired feature of an elevation mechanism is the self-locking, where it does not allow the UAV weight to automatically push the mechanism down. The parameter is the normal component of the sum of the UAV mass and the launcher arm. Table 10.1 illustrates [141] the friction coefficient for threaded pairs.

The self-locking is obtained whenever the coefficient of thread friction is equal to or greater than the tangent of the thread lead angle. When a positive lowering torque is obtained, the screw is said to be *self-locking*. Thus, the condition for self-locking is

$$\pi \mu d_m > l \tag{10.15}$$

The Eqs. (10.12) and (10.13) are valid for square threads where the normal thread loads are parallel to the axis of the screw. In the case of Acme or other threads, where the normal thread load is inclined to the axis, you may consult with [141].

Table 10.1 Friction coefficient for threaded pairs.

No	Screw materials	Nut materials			
		Steel	Bronze	Brass	Cast iron
1	**Steel, dry**	0.15–0.25	0.15–0.23	0.15–0.19	0.15–0.25
2	**Steel, machine oil**	0.11–0.17	0.10–0.16	0.10–0.15	0.11–0.17
3	**Bronze**	0.08–0.12	0.04–0.06	–	0.06–0.09

The mechanical power is torque multiplied by angular velocity of the screw. Hence, the power required to raise the launcher is:

$$P = T_R \omega \tag{10.16}$$

where ω represents the angular velocity of the power screw in raising the launcher rails and UAV.

Example 10.2

An elevation platform is being designed for a UAV launcher. The platform is made up of power screws with a square-thread with a pitch diameter of 30 mm, a pitch of 4 mm, and double threads. The power screws are made up of steel, and machine oil is applied. Assume the launch angle is 15°. The mass of UAV is 20 kg, and the launcher rail is 40 kg.

a) Find the torque required to raise the UAV.
b) Determine the required power, if the thread is desired to turn at the rotational speed of 100°/s.
c) This power is provided by an electric motor with the voltage of 12 V. Determine the electric current.

Solution

a) Torque: From Table 10.1, the friction coefficient for a machine oiled steel power screw is between 0.11 and 0.17. We select an average value of 0.14.
   Screw lead:

$$l = np = 2 \times 4 = 8\,mm \tag{10.14}$$

   Normal force:

$$F_N = (m_R + m_U)\, g\cos\Theta = (20 + 40) \times 9.81 \times \cos(15) = 568.3\,\text{N} \tag{10.13}$$

$$T_R = \frac{F_N d_m}{2}\left(\frac{\pi \mu d_m + l}{\pi d_m - \mu l}\right) = \frac{568.3 \times 30}{2}\left(\frac{3.14 \times 0.14 \times 30 + 8}{3.14 \times 30 - 0.14 \times 8}\right) = 1.94\,Nm \tag{10.12}$$

b) Power:

$$P = T_R \omega = 1.94 \times \frac{100}{57.3} = 3.38\,W \tag{10.16}$$

c) Electric current:

$$P = IV \Rightarrow \text{I} = \frac{P}{V} = \frac{3.38}{12} = 0.282\,A = 282\,mA \tag{3.4}$$

10.3.5 Power Supply

A launcher requires a power supply for the activation of mechanisms such as the elevation platform. The power may be provided by various sources such as electric engines, reciprocating engines, and human forces. The features of electric engines are presented in Chapter 3, and the human force limits are provided in Chapter 11.

In pneumatic or hydraulic launchers, the power is coming from compressed air or oil. However, the energy to compress the air and oil should be generated by a power supply. Chapter 11 briefly discusses the functions and characteristics of electric generators.

10.4 Fundamentals of Launch

The analysis of launch and design of launcher require an understanding of the concept of launch, and the application of governing equations. In this section, main elements, parameters, contributing forces, and governing equations of the launch operation are presented. Basically, UAV launcher is a runway-independent platform that both guides and directs the UAV. In terms of operation, it is beneficial to launch the UAV into the wind, if available, thus achieving flight speed over a shorter distance.

10.4.1 Fundamental Principles

Depending upon the mission and the weight of the UAV, there are various types of launchers. Typical launch methods include: (a) rail launchers; (b) rocket launch; (c) air launch; (d) hand launch; (e) tensioned line launch (i.e., catapult); (f) gun launch; and (g) ground-vehicle launch.

As an example, the launch system for the Phoenix UAV uses a nine-meter long hydro-pneumatic catapult and is mounted on a large truck. In contrast, the Sky Lite B UAV utilizes a bungee-powered catapult launch, while Desert Hawk is held by hand before a bungee Launch.

Main components of a launcher system (See Figure 10.10) are: (a) launching vehicle; (b) fixed platform; (c) azimuth platform; (d) elevation platform; (e) launching ramp (e.g., rail); (f) mechanisms and locks; and (g) onboard generator. Unlike missiles, there is often no rush in launching a UAV; thus, the azimuth and elevation platforms are adjusted by human operators, rather than through a closed-loop control system.

The aerial vehicle moves in the launcher for a very short time. In an ideal situation, a straight motion is desired for the UAV along the launcher. However, in practice, this is not possible due to the dynamics of the whole system. Engine thrust, flexibility properties of launcher and UAV, joint characteristics, mass unbalance and thrust misalignment have a number of effects on the dynamics of the launcher system.

In practice, it is not possible to have such an ideal case, during the design of this dynamic systems a small error in mathematical modeling must be considered. If the errors arising from the launcher system can be predicted, necessary precautions can be taken for interaction properties. The launcher components, the fixed, azimuth and elevation platforms and the launching guide are flexible bodies.

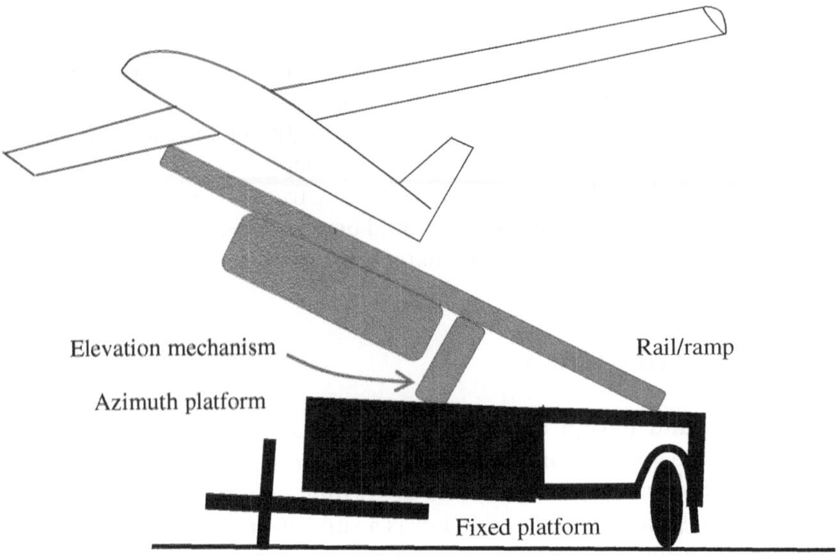

Figure 10.10 Major elements of a launcher.

10.4.2 Governing Launch Equations

The launch process is basically a linear accelerated motion, where the UAV is accelerating along the ramp. The relationship between acceleration, launch speed, and launcher length has significant implications for launcher design. The integral form of the equation is recommended because launch force is rarely constant in practice. The peak loads are often significantly higher than the average loads. From theory of dynamics [142], when a moving object with an initial velocity of V_1 accelerates to a new velocity of V_2, the distance (x) covered is governed by the following equation:

$$V_2^2 - V_1^2 = 2ax \tag{10.11}$$

where "a" represents the linear acceleration. For an UAV launcher, the initial velocity is often zero, and the distance traveled is equal to the length of launcher (L_L). So, the governing equation is expressed as:

$$V_L^2 = 2aL_L \tag{10.17}$$

In addition, in order for a launcher to create such acceleration, it must provide a sufficient launch force (F_L). From Newton's second law, along the launcher ramp, sum of the forces along x-axis creates the acceleration:

$$\Sigma F_x = ma \tag{10.18}$$

where m denotes the UAV mass. The contributing forces in this accelerated launching motion are: (a) launch force (e.g., engine thrust); F_L; (b) UAV-Ramp friction; F_f; (c) UAV weight (W); (d) UAV drag (D_L); and (e) UAV Lift (L_L). In the x direction (along the ramp), the force equation would be:

$$F_L - F_f - W\cos\theta - D = ma \tag{10.19}$$

The friction force is proportional to the UAV normal force (x-component of the weight minus lift), N. It is the product of the coefficient of friction μ with the normal force *N*, and acts to oppose the motion. Hence, the UAV-Ramp friction force (F_f) is:

$$F_f = \mu N \tag{10.20}$$

where μ is the friction coefficient between UAV legs and the ramp, and N denotes the normal force on the ramp. Table 10.1 provides [141] the friction coefficient between UAV and ramp, for various ramp materials.

Two forces are contributing to the normal force: (a) UAV weight; and (b) UAV lift, L_L. The ramp has frequently a launch (i.e., climb) angle, θ. Since, the UAV is accelerating, the lift is gradually increasing too.

$$N = W\cos\theta - L_L \tag{10.21}$$

where θ is the launcher angle (See Figure 10.11), F_f is the friction between launcher rails and the UAV (indeed, two metals), and W is the weight of UAV (W=mg).

The friction coefficient (μ) between a UAV and the launcher rails is typically about 0.05–0.12. At the beginning of the launch, the lift (L_L) and drag (D) are both zero. When the friction force is substituted from Eqs. (10.20) and (10.21) into Eq. (10.19), we obtain:

$$F_L - \mu W\cos\theta - W\cos\theta = ma \Rightarrow F_L = W\cos\theta(1+\mu) + ma \tag{10.22}$$

This is the force (F_L) that the launcher must provide to generate the desired acceleration at the beginning of the launch. The source of power to provide launcher force may be of pneumatic (e.g., ScanEagle launcher), or hydraulic, or even a spring. The engine thrust is also contributing to the launch process, it will create a portion of the required launch force.

As the launch angle is increased, the greater launcher force is required. A compromise is needed to select the launcher length, since a long launcher requires less force,

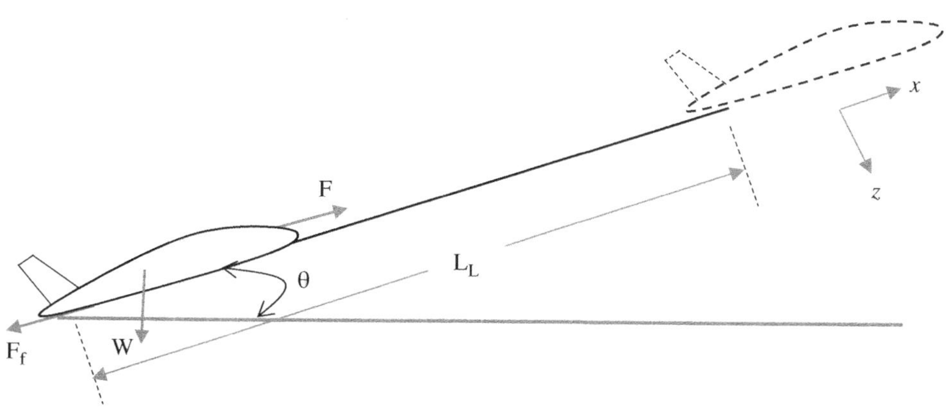

Figure 10.11 Contributing forces on a launcher.

but makes the launcher heavier and needs a larger volume for depot. For more accurate calculation, you must include UAV lift and drag forces.

The launch force is only applied until the UAV is on the ramp; when the UAV left the ramp, the only forward force is the engine thrust. It must be ensured that, the UAV has gained sufficient airspeed, before leaving the launcher.

The launch speed is recommended to be equal to the UAV velocity corresponding to the maximum rate of climb. The technique to calculate the UAV velocity corresponding to the maximum rate of climb is provided in Chapter 14. The UAV speed at the end of launcher is often about 10–30% greater than the stall speed. The UAV stall speed is a key parameter governing the energy of launch event. It is desirable to reduce the launch energy in order to lessen the UAV loads and the footprint of any supporting apparatus.

$$V_s = \sqrt{\frac{2W}{\rho S C_{L\max}}} \tag{10.23}$$

where S is the wing area, ρ is the air density (at sea level, 1.225 kg/m^3), and $C_{L\max}$ is the maximum lift coefficient (usually about 1.2–1.6). The equations in this section are employed to calculate the launcher length and required launcher force. The total launcher length should not exceed practical transportability and storage dimensions, but the rail may consist of multiple segments. Many governing equations of a normal takeoff operation is applicable to launch. Reference [9] provides details of takeoff operation and introduces its governing equations and contributing parameters.

If the UAV has a pusher prop-driven engine, it is recommended to equip the launcher with a special housing for covering the prop (i.e., propeller guard) to prevent any human-related incidence and to increase operator safety.

Right after the launch, the UAV is going to climb with an instantaneous climb angle of γ.

$$\sin(\gamma) = \frac{T - D}{W} \tag{10.24}$$

The instantaneous rate of climb (ROC) is obtained as:

$$V\sin(\gamma) = \frac{(T - D)V}{W} \tag{10.25}$$

The left-hand side of Eq. (10.25) is by definition, ROC of the UAV.

$$ROC = V\sin\gamma \tag{10.26}$$

The numerator in Eq. (10.24) (i.e., "(T-D)V"); is referred to as the excess power. In order for the UAV to climb after the launch, the excess power must be positive. Otherwise, the UAV will descend/crash after the launch operation. Please note that, during the launch, the launch force is F_L, while after the launch, it will often be decreased to engine thrust; T. In the design of the launcher, this important safety point must be considered.

It is evident that, the ROC (after launch) directly depends on the excess power; in combination with inverse of the weight of the UAV. The higher the thrust; the lower the drag; and the lower the weight, the higher will be the climb performance.

Example 10.3

A launcher is designed to launch a UAV with a mass of 500 kg. The launch angle is 20°, the launcher length is 3 m, and the friction coefficient is 0.06. The UAV is required to reach a speed of 30 knot at the end of launcher rail. Ignore the contribution of UAV lift and drag during the launch.

1. What acceleration does need to be generated?
2. What force is needed to push the UAV over the launcher?

Solution

$$V_L^2 = 2aL_L \rightarrow a = \frac{V_L^2}{2L_L} = \frac{(30 \times 0.514)^2}{2 \times 3} = 39.6\frac{m}{s^2} \quad (10.17)$$

$$F_L = W\cos\theta(1+\mu) + ma \quad (10.22)$$

$$F_L = 500 \times 9.81 \times \cos(20)[1+0.06] + 500 \times 39.63 = 25{,}300\,N$$

Please note that, part of the launch force is generated by the UAV engine.

10.4.3 Wing and Horizontal Tail Contributions

As the UAV is accelerating along the ramp, lift, and drag are increasing with velocity, while the normal force and consequently, the friction force is decreasing. The UAV lift and drag during launch are obtained by the following expressions:

$$L_L = \frac{1}{2}\rho V^2 S C_{LL} \quad (10.27)$$

$$D_L = \frac{1}{2}\rho V^2 S C_{DL} \quad (10.28)$$

where V denotes the UAV linear forward speed on the launcher, S represents the wing planform area, and ρ is the air density. Two parameters of C_{LL} and C_{DL} are lift and drag coefficients respectively. Typical values for C_{LL} is between 0.1–0.5. The launch should be continued until the lift is sufficient for lift-off.

In a UAV with a conventional configuration, the total lift during launch (L_L) is equal to the algebraic sum of wing-fuselage lift (L_{wf}), and horizontal tail lift (L_{ht}):

$$L_L = L_{wf} \pm L_{ht} \quad (10.29)$$

where the sign "±" indicates that the horizontal tail lift may be positive, or negative. This sign is a function of a number of parameters including the relation between the UAV center of gravity with respect to wing-fuselage aerodynamic center. In majority of the cases, if the UAV center of gravity is ahead of the wing-fuselage aerodynamic center, the horizontal tail will generate a negative lift (thus, $-L_{ht}$). In contrast, if the UAV center of gravity is aft of the wing-fuselage aerodynamic center, the horizontal tail will frequently generate a positive lift (thus, $+L_{ht}$). The wing-fuselage lift and the horizontal tail lift are functions of air density, airspeed, a reference area, and their angles of attack:

$$L_{wf} = \frac{1}{2}\rho V^2 S C_{Lwf} \tag{10.30}$$

$$L_{ht} = \frac{1}{2}\rho V^2 S_{ht} C_{Lht} \tag{10.31}$$

A negative sign for the horizontal tail lift in Eq. (10.29) indicates that, this force acts downward.

The wing-fuselage and tail lift coefficients are functions of their aerodynamic features (primarily wing and tail airfoil sections), and their angles of attack. Furthermore, in a UAV with a conventional configuration, the horizontal tail coefficient is controlled by elevator deflection. In the aerodynamic design process of wing and horizontal tail, their contribution to the launch motion must be considered. Therefore, the aerodynamic design of horizontal tail will significantly influence the design of the launcher.

10.4.4 UAV Longitudinal Trim

A UAV is essentially required to maintain the longitudinal trim during and after launch. This is a significant requirement which will guarantee the launch safety and success. This section explores the longitudinal trim requirements during and after launch for a UAV with a conventional configuration.

The longitudinal trim is easily maintained during launch (when the UAV is on the ramp) by using hard stops and a lock mechanism. However, at the end of launch, the UAV is free of such mechanisms, and locks; so, a provision is required to guarantee the longitudinal trim. If the UAV is not longitudinally trimmed at the end of the launch operation, the UAV will crash. A correct elevator deflection will guarantee the UAV longitudinal trim; thus, the launch success. Thus, the force is generated by upward/downward deflection of elevator.

The launch process may be designed to provide the UAV longitudinal trim in two ways: (a) open-loop control; and (b) closed-loop control. In the open-loop control mode, the elevator deflection is adjusted and fixed before the launch begins. However, in the closed-loop mode, the autopilot deflects the elevator at the end of ramp such that to suddenly increase/adjust the UAV angle of attack. In the open-loop mode, the UAV will continue to speed-up along the launcher, until the lift force (plus a vertical component of the engine thrust) exceeds weight of UAV.

Trim is one of the inevitable requirements of a safe flight (including launch operation). When an aircraft is at trim, the aircraft will not rotate about its center of gravity (cg), and aircraft will either keep moving in a desired direction or will move in a desired circular motion.

When the summation of all forces (Figure 10.12) in flight path; x-direction; and z-direction is zero; and the summation of all moments including aerodynamic pitching moment about y axis is zero, the aircraft is said to have the longitudinal trim. Since the UAV is accelerating while climbing, the sum of the forces will be in the following form:

$$\Sigma F_x = m a_x \tag{10.32}$$

$$\Sigma F_z = 0 \tag{10.33}$$

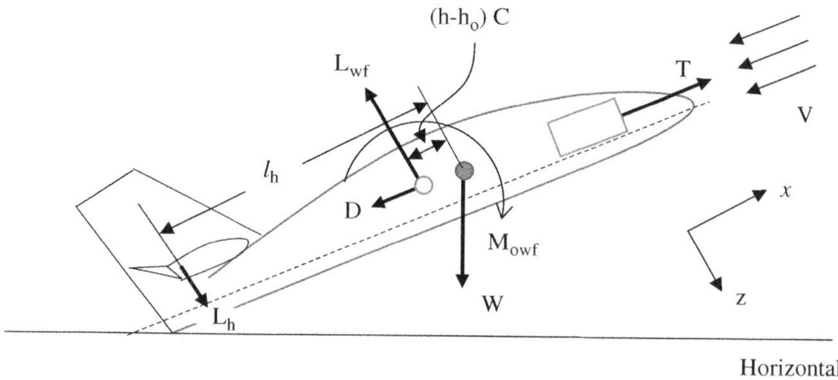

Figure 10.12 Forces and moments at the launch.

However, in order to keep the climb angle constant, sum of the moments about y-axis should be zero.

$$\Sigma M_{cg} = 0 \tag{10.34}$$

Along the x-axis (i.e., direction of flight), we can expand and write the equilibrium of the forces as:

$$T\cos(i_T + \alpha) - D - W\sin(\gamma) = ma = m\frac{dV}{dt} \tag{10.35}$$

If we ignore the engine setting angle (i_T) in our calculations, the thrust is acting parallel to the fuselage center line. In general, this is not quite true, but in conventional aircraft the effects of an inclination of the thrust vector are small enough to be neglected. Furthermore, for simplicity, we may assume the thrust line is in the direction of flight (ignoring the effect of angle of attack on engine thrust). For more simplicity we assume all forces are passing through the aircraft center of gravity and aircraft angle of attack is zero. Based on these simplifying assumptions, we have:

$$T - D - mg\sin(\gamma) = m\frac{dV}{dt} \tag{10.36}$$

Considering the above assumptions, in the z direction, the equilibrium of forces will result in:

$$L - W\cos(\gamma) = 0 \tag{10.37}$$

where γ is the climb angle. In a fixed-wing UAV, the horizontal tail (including elevator) is responsible for maintaining the longitudinal trim, and to make the summations – in Eqs. (10.36) and (10.37) – to be valid, by generating a necessary horizontal tail lift, and by contributing in the summation of moments about y-axis.

The application of the trim to moment equation (assuming thrust; T and drag; D are both passing through the UAV cg) leads to the following:

$$\Sigma M_{cg} = 0 \Rightarrow M_{owf} + M_{L_{wf}} + M_{L_h} = 0 \tag{10.38}$$

where M_{owf} denotes the wing-fuselage aerodynamic pitching moment.

$$M_{owf} = \frac{1}{2}\rho V^2 S C_{m_{owf}} \bar{C} \tag{10.39}$$

The parameter C_{mowf} represents the wing-fuselage aerodynamic pitching moment coefficient, and $\bar{C}$ (or simply C) is the wing mean aerodynamic chord. Reference [8] has further simplified the trim Eq. (10.38) to the following non-dimensional form:

$$C_{m_{owf}} + C_L(h - h_o) - \bar{V}_H C_{L_h} = 0 \tag{10.40}$$

This longitudinal trim equation provides a critical tool in the design of the horizontal tail, as well as a successful launch system. For the definition and calculations of new parameters in Eq. (10.40), consult with Ref. [8]. The horizontal tail lift coefficient is modeled [40] as:

$$C_{L_h} = C_{L_{ho}} + C_{L_{\alpha_h}} \alpha_h + C_{L_{\delta_E}} \frac{S}{S_h} \delta_E \tag{10.41}$$

where $C_{L_{\alpha_h}}$ is the tail lift curve slope, $C_{L_{\delta_E}}$ is the elevator control derivative, δ_E is the elevator deflation, and $C_{L_{ho}}$ denotes the tail lift coefficient at zero angle of attack. Most horizontal tails tend to use a symmetric airfoil section, so the parameter $C_{L_{ho}}$ is normally zero.

The worst case for satisfying the longitudinal trim at the launch separation is, when the UAV cg is located at the most forward location. For such case, the maximum elevator deflection should be sufficiently large to generate the required tail lift coefficient (C_{L_h}). If this is not feasible, the launcher ramp length (L_L) must be increased until the UAV reaches the lowest satisfying airspeed. The elevator designer must synthesize an elevator in such a way that longitudinal trim is not a limiting factor anywhere in the intended flight envelope, particularly during launch operation.

10.5 Elevation Mechanism Design

10.5.1 Elevation Mechanism Operation

In order to adjust the launch (climb) angle of a UAV, an elevation mechanism is need to move the rail up and down. The main component of an elevation mechanism is the actuator; two proper ones are hydraulic or pneumatic. In designing an actuator for a particular application, the engineer deals with this multi-variable dependency. Main unknown parameters in the actuator design for a small to medium-weight UAV are: (a) actuator's geometry; (b) required power; (c) weight; and (d) medium (oil or air).

The mechanical requirements of an actuator can be expressed in terms of force, displacement, size, mass, response time (or operating frequency), power, efficiency, cost, durability, and maintenance. Actuator can be designed (costumed) or purchased off-the-shelf (industrial). These requirements must be matched to the performance characteristics of an actuator in order to determine whether it can give the performance required for the application.

Figure 10.13 illustrates the elevation mechanism which varies the angle of the launcher ramp (θ_R). The actuator applies a force along its axis to lift the ramp for proper angle.

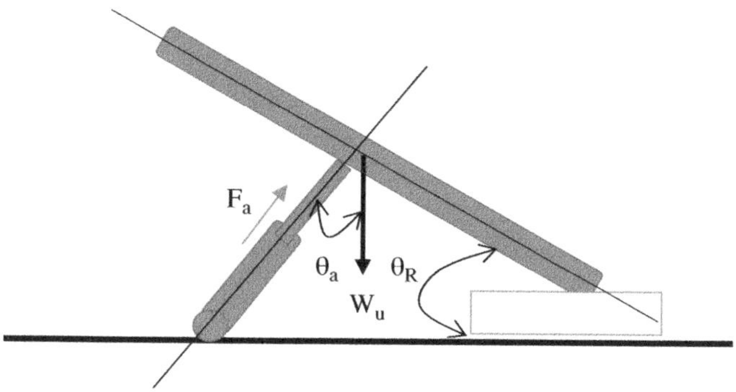

Figure 10.13 Elevation mechanism.

When the actuator piston extends its length, the actuator angle (θ_R) is increased. As the ramp angle is increased, a less force/power is required to move the ramp (including UAV). Both ends of the actuator are hinged, so they can rotate as needed.

10.5.2 Hydraulic and Pneumatic Actuators

Hydraulic and pneumatic actuators provide force and displacement via the flow of a pressurized fluid (oil or air). The actuators dynamics may be modeled with a second order differential equation:

$$F_a = m_u \ddot{x} + b_a \dot{x} + k_a x \tag{10.42}$$

where x is the actuator displacement, k_a is the elastic constant, b_a is damping coefficient, actuator force F_a and m_u represents the mass of the rail, plus mass of UAV unit along the actuator axis. The required power (P_a) to create the actuator force is:

$$P_a = F_a \dot{x} = F_a V_a \tag{10.43}$$

where V_a is the linear velocity of the actuator along the cylinder/piston (i.e., sliding speed). This velocity is a function of the angular velocity of the rail and the arm (distance between the actuator connection to rail and the rail hinge).

$$V_a = L_a \omega \tag{10.44}$$

A typical rail angular velocity (ω) is about 2–10 deg/s.

Detail analysis of elevation mechanism is beyond the scope of this book, for detail analysis and synthesis of mechanisms, you may refer to [143]. The required maximum fluid power [144] for the actuator is:

$$P_{max} = 4 C_c \sigma_{max} \varepsilon_{max} V_{max} A \tag{10.45}$$

where A is internal area of the cylinder of the actuator, σ_{max} is the maximum actuation stress, ε_{max} is actuation strain, V_{max} is the maximum sliding speeds. The V_{max} is limited to about 0.5 m/s. The coefficient C_c, referred to as the cyclic work coefficient, is an efficiency measure of the shape of the σ-ε curve in cyclic operation and is in the range of 0.4–0.5 for hydraulic/pneumatic systems.

10.6 VTOL

A vertical take-off is the most elegant mode of launch. It enables the UAV system to be operated from almost any type of terrain, requires no runway, airstrip, or cumbersome catapult equipment. The launch is independent of wind direction and the aircraft can be airborne in a short time. All quadcopters and helicopter UAVs are employing the VTOL technique. The dynamic model and control of quadcopters are presented in Chapter 5.

10.7 Recovery Technologies and Techniques

10.7.1 Fundamentals

Recovery is defined as transitioning a UAV from a flying state to a nonflying state. Recovery systems eliminate the complexities of runway landing. Recovery is the process of arresting a UAV and bringing it to rest by means of a deployable aerodynamic decelerator. The recovery is a sequence of events which includes deceleration, descent, approach, landing, and retrieval and sometimes refurbishment an aerial vehicle to enable use of the recovered vehicle, over, and over. A recovery subsystem is the combination of special components incorporated in and integrated with a Unmanned aerial system (UAS) to effect recovery of the air vehicle under predictable conditions. In the case of a conventional runway recovery, this phase is referred to as landing.

Landing involves the return of the UAV to make a controlled touch-down onto its landing gear at the threshold of runway, deceleration along the runway, followed by the UAV taxiing or being towed back to its hangar. UAV recovery (i.e., capture) can be more challenging than launch, and touch-down at the correct position and airspeed require considerable judgment For long-range UAV such as Global Hawk, a form of approach control with guidance, along a radio beam may be used. Initial positioning onto the radio beam will be performed with the help of GPS.

The recovery concept implies two objectives: (a) minimal damage; and (b) retrieval. Recovery systems eliminate a need for any runway. Recovery requires that the UAV horizontal and vertical velocity components become equal to that of the recovery platform (in most cases, it means a stationary status). Recovery of the UAV requires a safe landing, as well as the return of the UAV to its base or hangar. For UAVs weighing more than approximately 500 lb, the absorption of recovery energy by any means other than landing gear is challenging and almost impractical. A few practical recovery techniques (other than conventional landing) are: (a) parachute; (b) arrester; (c) net recovery; (d) cable-assisted recovery (suspended cables with hooks); and (d) skid and belly recovery.

A few techniques for recovery of a catapult launched UAV are: (a) guided flight into a catchment net; (b) skid or belly landing; (c) guided flight onto an arresting pole and suspended cables; and (d) deployment of a parachute – in flight – to reduce the ground impact. Suspended cables seize both the horizontal and vertical motion of the UAV relative to the recovery platform. The arresting cables on the horizontal platform only arrest horizontal motion and act in conjunction with a conventional runway landing. The dynamics have many similarities to the net recovery technique.

Skid and belly landing provide a recovery technique where the UAV contacts the ground directly without the use of conventional landing gear. Either a skid device or the UAV fuselage is the interface to the ground, providing both landing shock absorption and friction to slow the UAV to a rest. Skids or body ground interface points must be designed to contend with the friction of scraping the ground.

10.7.2 Net Recovery

Recovery is the process of arresting a state of motion and bringing to rest a UAV by means of a deployable decelerator. In the recovery purpose and concept, minimal damage and retrieval are expected. A straight forward solution for recovery of a fixed-wing UAV is to employ a net which results in a quick landing. In this technique, a net is vertically mounted that the UAV will fly into. The process is pretty rough as the deceleration is pretty quick and the initial impact to the net is considerable. For larger UAVs, the current net technology is not strong enough to absorb all the energy that an impacting UAV owes. The UAV flies into a net, and its motion relative to the net (i.e., platform) is arrested.

The net recovery technique is simple and reliable and does not involve complex components. However, for recovering a relatively large UAV, a braking system is necessary to complement the net. There is a risk of entanglement in the net that can damage the engine propeller. To solve the possible damage to UAV propellers, some UAVs utilize a pulse propeller as propulsion. This technique is commonly used for larger UAVs that are too heavy for simple belly landings. Unmanned aircraft larger than 500 lb typically avoid this technique in favor of conventional landing or parachute recovery.

The Insitu ScanEagle UAV was purposely designed to use two techniques: (a) net; and (b) hook. The vertical rope requires an airframe-mounted hook with a horizontal extension. The Insitu ScanEagle UAV with a maximum takeoff mass of 22 kg, a wing span of 3.1 m, is equipped with a 2-stroke piston engine (1.5 hp) and has a maximum speed of 80 knot, and a service ceiling of 19 500 ft.

The recovery is expected to incorporate provisions to soften the impact of landing; a problem with net recovery is that it can damage the UAV as the deceleration is relatively sudden. A goal of recovery is to disconnect the decelerator from the UAV, after being captured in the net. In a successful net recovery, capture of the UAV should not result in damage to the vehicle. A disadvantage of net recovery is that a net system is labor-intensive, time consuming, and logistically complex. The net system should be set up prior to, and disassembled after, the recovery operation. The nest is frequently very large (about 10–20 times as the UAV top area), hence, the footprint is fairly large. This large net size may create a risk to helicopter flight, if the system is on a ship.

In general, two advantages of a net recovery system are: (a) zero-length recover; and (b) simple solution. In contrast, four disadvantages are: (a) requires a lot of manning; (b) risk of damaging the UAV; (c) setup time is long; and (d) has a large deck footprint. A net recovery requires a lot of manning, which is amount of crew that is required to operate the system. The net recovery system has been successfully employed in AAI Aerosonde (Figure 10.14) and AAI RQ-2 Pioneer (Figure 10.15) UAVs.

10.7.3 Arresting Line

An arresting line as a recovery technique, commonly used in a fixed UAVs on ships and aircraft carriers. The system usually includes an arresting line, and a braking system

Figure 10.14 Net AAI Aerosonde recovery.

Figure 10.15 Net recovery of a AAI RQ-2 pioneer UAV.

[145] that are coupled together with a stanchion and a boom. The line system can be on the deck or on the side of the ship. For using this recovery system, the UAV has to be equipped with a hook in order to be able to latch on to the arresting line. Modifications can impact the effectiveness of the UAV payload such as a camera. Adding a hook to the UAV will slightly increase aerodynamic drag of the UAV.

In general, two advantages of an arresting line recovery are: (a) zero-length recovery; and (b) quick recovery. In contrast, four disadvantages are: (a) rough landing; (b) suffers from ship motions; (c) setup time is long; and (d) UAV needs a hook.

The arresting line recovery system has been successfully implemented and employed in the Insitu ScanEagle (Figure 10.16). The recovery system for a ScanEagle – called "tetherball"- on a ship employs a hook at the end of the UAV wing for recovery on a cable hung off the side of the ship from a crane which can suspend a line 10–15 ft away from the ship roughly 40 ft above sea level from an upper deck. This recovery method needs no deck space and avoids the need for UAV recovery methods using hangars, nets, or in-water retrievals involving small boats.

10.7.4 Skyhook

The skyhook recovery system is a variant of an arresting line, in whicha vertical catch cable is used. The system consists of a retractable crane and a wire. The procedure is to have the UAV fly into the wire with its wing tip. The process is pretty rough as the deceleration is pretty quick and the initial impact to the wire is considerable. From a structural point of view, the UAV will suffer from a considerable load factor, since hitting the UAV is concentrated on a small zone (i.e., wing tip). Compared with a net, the setup time of the skyhook recovery system is not very long.

A requirement for Skyhook recovery system is a large crane which in turn needs a large stowing room. When UAV is caught, it will pivot down toward the frame of the crane. The hanging UAV may turn around the crane, which may cause a risk to the carne too. The UAV has to be released later from the wire, which required skillful personnel.

In general, two advantages of a skyhook recovery system are: (a) zero-length recovery; and (b) requires a low manning. In contrast, four disadvantages are: (a) very rough recovery; (b) large stowage; (c) can suffer from ship motions; and (d) UAV is prone to structural damage. The Skyhook recovery system has been successfully implemented and employed in Insitu ScanEagle (Figure 10.16).

10.7.5 Windsock

The windsock recovery system in a theoretical concept that has a low recovery time, and a low degree of complexity. The design consists of a large conical shaped windsock (made up of strong plastic or thick fabric) mounted on top of a movable stand. The large open entry should be much larger (Figure 10.17) than the UAV wingspan, and the small closed end is near to the ground. For the recovery, the UAV should be flown into the large end of the windsock, to be decelerated by the narrowing geometry of the windsock.

Once the UAV has stopped on the smaller side, it can be readily retrieved by opening the small side of the windsock. Due to the nature of windsock, the windsock recovery of a UAV with a prop-driven tractor engine does not seem to be practical. However, a UAV

Figure 10.16 Skyhook ScanEagle recovery. *Source:* Insitu.

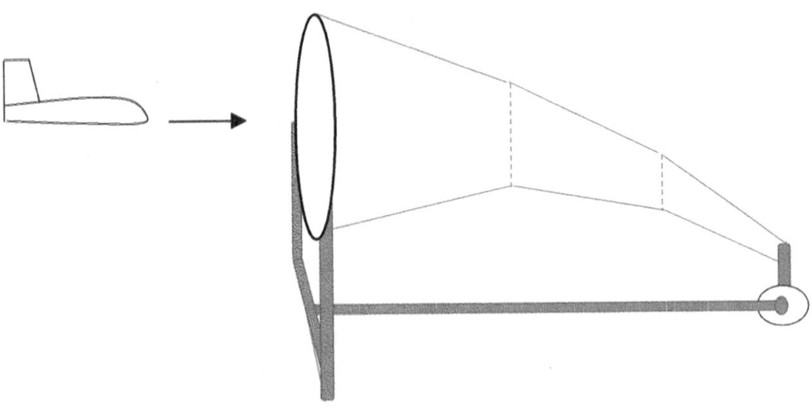

Figure 10.17 Windsock recovery system concept.

with a single pusher prop-driven engine may be successful. In the approach phase, the engine of the UAV should be turned off, and the prop should stop rotating, for a safe recovery. A prop-driven tractor engine may damage the windsock, but a careful design will minimize the risk, and will deliver a smooth recovery. This system has a large footprint, and may be employed on ground or on ship deck. The windsock is frequently cheap, and it may be replaced after each operation.

In general, a few advantages of a windsock recovery system are: (a) short-length recovery; (b) low cost; (c) low weight; (d) low degree of complexity; (e) simple design; and (f) no equipment is needed in UAV. In contrast, four disadvantages are: (a) large footprint; (b) limits types of UAV propulsion system; (c) UAV is prone to structural damage; and (d) large stowage.

10.7.6 Parachute

A parachute recovery system has no ground equipment, but the ground pilot may have an effective role in the recovery success. One approach for recovery for UAVs without a vertical flight capability, is using a parachute; which will allow a wheeled or skid-borne landing onto terrain. A parachute is installed within the UAV, carried from takeoff, and is deployed at a suitable altitude over the landing zone. As the parachute gets bigger, the impact to the ground will be slower.

However, a large parachute is heavy and adds to the weight of UAV. So, a compromise needs to be made in selecting the parachute size. The design of the parachute is based on the drag concept and terminal velocity. The landing/impact velocity will be equal to the terminal velocity. The accuracy and efficiency of the parachute recovery are heavily influenced by the wind speed and direction. The control of the landing spot will also be affected by the wind. Moreover, if the air is turbulent, the recovery system will generate a risk to the UAV, due to wake effects. The descending parachute may also be intercepted in mid-air, thus modifying the nature of the landing process. A variant of parachute is parasail, which requires a specially designed canopy wing that resembles a parachute.

Reference [146] presents the development and testing of a recovery system for the emergency flight termination of an 1800-pound air vehicle, the General Atomics Aeronautical Systems medium altitude endurance Predator A. A particularly efficient main parachute canopy was developed with very good stability and in excess of 100 ft^2 of drag area per pound of canopy.

10.8 Recovery Fundamentals

In section 10.7, various ground recovery techniques have been introduced. In this section, the fundamental principles of two recovery techniques are briefly reviewed, and the governing equation will be presented.

10.8.1 Parachute

A parachute allows a wheeled or skid-borne landing onto terrain. The landing/impact velocity will be equal to the terminal velocity (V_t). The terminal velocity relation is obtained by equating the drag and weight:

$$V_t = \sqrt{\frac{2W}{\rho S C_D}} \tag{10.46}$$

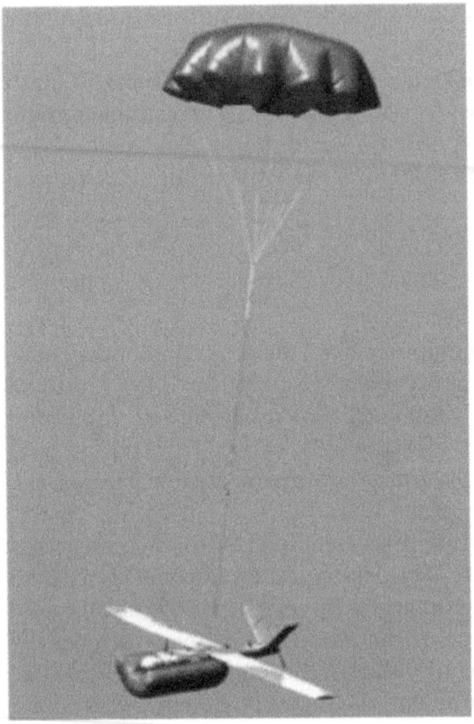

Figure 10.18 SkyLite UAV parachute recovery. *Source:* Insitu.

where W is the weight of UAV (including the parachute), S denotes the parachute projected area, and C_D is the parachute drag coefficient. Typical values for parachute drag coefficient are about 1.2–2. The drag coefficient of a personnel parachute [44] with the projected area of 500 ft^2 is about 1.3.

The landing velocity of the UAV relative to the platform is a key parameter governing the energy of recovery event. It is desirable to minimize the recovery energy in order to reduce the UAV impact force and the footprint of any supporting apparatus. Figure 10.18 illustrates the SkyLite UAV recovery using a parachute.

Example 10.4

A small UAV with a mass of 100 kg is required to land on the ground with the help of a parachute with an impact velocity of 3 m/s. What parachute projected area is needed for landing on sea level altitude? What is its radius?

Solution

At sea level, the air density is 1.225 kg/m^3. The parachute projected area is determined by using Eq. (10.46). For parachute drag coefficient, we take the average of 1.75.

$$V_t = \sqrt{\frac{2W}{\rho S C_D}} \rightarrow S = \frac{2W}{\rho C_D V_t^2} = \frac{2 \times 100 \times 9.81}{1.225 \times 1.75 \times 3^2} = 101.7\, m^2 \quad (10.46)$$

$$S = \pi R^2 \rightarrow R = \sqrt{\frac{S}{\pi}} = \sqrt{\frac{101.7}{\pi}} = 5.7\,m$$

10.8.2 Impact Recovery

Three basic forms of recovery equipment are: (a) arrester; (b) net; and (c) suspended cables with hooks. The recovery system design for these cases is based on the impulse-momentum theorem. Two techniques which are mainly utilized for micro to mini-UAVs are net and skyhook. In both cases, a large net or, alternatively, a carousel apparatus, or a hanging hook is employed to which the UAV is flown and caught. A means of absorbing the impact energy is needed, usually comprising airbags or replaceable frangible material. The arresting technique for which the impulse–momentum equation is applicable is for larger UAVs. The recovery system reaction forces that bring the UAV to rest initially should be quickly disengaged to prevent sending the UAV in rearward and upward motions.

An impulse is equal to the net force (F) on the object times the time period (t) over which this force is applied. The impulse–momentum theorem states that the impulse in an impact is equal to the change in linear momentum.

$$Ft = m(V_2 - V_1) \tag{10.47}$$

Since the final velocity is zero:

$$Ft = mV_2 \tag{10.48}$$

The energy of this impact (i.e., kinetic energy) must be absorbed by net, hook, or arrester without any harm to the UAV. The size of the net is a function of the UAV frontal area, as well as the desired reliability. The larger the net, the more reliable will be the recovery system. The net should be flexible so that the UAV get caught in a safe way. The flexibility will provide a longer time (t) for the UAV impact. The more energy absorbed for recovery, the greater the safety risk. The impact velocity and altitude for a given tension are controlled by the hook location.

10.9 Launch/Recovery Systems Mobility

10.9.1 Mobility Requirements

One of the basic requirements for the launcher/recovery system is the mobility over a wide variety of terrains. This requirement is satisfied with an appropriate roving vehicle. The roving vehicle should be effectively controlled, has acceptable maneuverability, and capable of mobility. Due to new challenges, always new concepts of vehicles for application to an operator-friendly launcher should be explored.

The primary constraining factor for conceptual design launcher/recovery system is the mission profile of the UAV. Mobility over a wide variety of terrains, effective operation control, and ability to transport the required payload are definite requirements of the mission profile. For military UAV, the battlefield environment in which the roving vehicle is expected to operate is another constraint on the vehicle design. The designer may come up with a state-of-the-art of method of mobility that can be utilized in the conceptual design as a new special purpose vehicle.

The mobility requirement formilitary use stems from the fact that the battlefield terrain is an unknown variable. The vehicle must be able to operate on as wide a variety of the possible rugged terrains as is practical. The control constraint requires that the vehicle be adaptable to the desired mode of operation.

In general, as the mobility is increased, the total weigh and cost of launcher/recovery system will be increased too. Hence, a compromise must be made to arrive at the optimal solution. In another word, the mobility is enhanced by reducing the weight of the launch and recovery systems. Moreover, in order to reduce the design cost, a regular truck is recommended as the base vehicle.

For a launcher/recovery roving vehicle, there are two options: (a) a currently available commercial wheeled vehicle; and (b) a new special purpose vehicle. Features of a conventional wheeled vehicle is presented in the next section.

10.9.2 Conventional Wheeled Vehicle

The first option to provide mobility forlauncher and recovery systems is to employ a conventional wheeled vehicle, such as a truck. This design option will reduce the design complexity, and benefits from the features of the available vehicle. However, the attachment mechanism still needs to be designed. For a small UAV, where a special purpose launching vehicle is not considered, the launcher is mounted on the ground by wheels, (as a trailer-mounted launcher).

There are various methods of improving the mobility characteristics of wheeled vehicles. Among the most obvious is incorporating four-wheel drive and using large tread tires. Another method of improving mobility is by lowering the ground contact pressure. This can be easily done by using lower pressure tires, which effectively lowers the contact pressure, and imposes the vehicle a larger footprint.

A more advanced system based on the same idea is a centrally variable tire pressure system. With this design, the tire pressures can be varied from inside the vehicle according to the type of surface encountered. Thus, high tire pressures can be used for road travel to maximize mileage, and progressively lower pressures can be used as the terrain becomes harder to traverse. Another interesting concept involving the tires is the run-flat tire. These are tires that through their structural strength, can still operate while deflated. This option is valuable in desert conditions where tires could be easily punctured.

Articulated vehicles, both wheeled and tracked, provide increased mobility over irregular, rough terrain. Obstacle negotiation can be significantly improved using two or three unit vehicles connected with articulated joints, which allows the vehicle to remain in contact with the ground for a longer time.

Another technique in which articulation can improve mobility is with controlled articulated suspension. This concept combines the good highway performance of a wheeled vehicle with the ability to maneuver in highly difficult, soft, off-road conditions. This design is accomplished with the controlled articulated suspension and individually powered wheels. An important issue with articulation is that it complicates the vehicle mechanically and with respect to control. The increased mobility that may be valuable for a short time is not necessarily worth the price paid in simplicity.

In comparing two available vehicles, the most important trade-off decision is cost versus survivability. If one vehicle with the launcher costs twice as much as the other, then the vehicle should be twice as survivable. Survivability can be evaluated through maneuverability tests. Moreover, the launch unit weight and cost reduction are achieved by using a conventional wheeled vehicle (e.g., a regular truck) which tows the launcher for transportation.

10.10 Launch and Recovery Systems Design

Design of launch and recovery systems is a challenging task; and requires a large amount of analysis and calculations. Similar to UAS design process, launch, and recovery systems design begins with conceptual design, and ends with test and evaluation. Conceptual design can be regarded in many different ways, from simple sketches from a brainstorming session to something that is almost ready for production. This phase starts with identifying needs and ends with a developed configuration. It frequently includes a feasibility study to make sure the concept is feasible with the given constraints. At this stage, a designer may come up with some solutions that do not exist entirely as proven designs, but only on a conceptual level. When choosing a technology to examine, it must be considered as plausible solution.

10.10.1 Launch and Recovery Techniques Selection

One of the major steps in the launch and recovery systems design is the selection of launch and recovery techniques. Two major sections of launch system design and recovery system design may be implemented in parallel. At any stage of the design procedure, the output is checked with the design requirements, and a feedback is taken. The process is repeated until the requirements are met.

The primary criteria for the selection of launch/recovery techniques are as follows: (a) mobility/transportation; (b) performance; (c) stability requirements; (d) structural stiffness; (e) operational requirements; (f) size (particularly length); (g) safety; (h) reliability; (i) cost; (j) UAV characteristics; (k) maintainability; (l) weight; (m) manufacturing technology; (n) operational and installation considerations; (o) power supply system; (p) maneuverability; (q) setup time; and (r) human factors. Since there are various techniques for launch and recovery systems, no details are provided for the design procedure. Each approach has a unique group of equipment, and entails a specific design procedure.

After the design requirements are developed, a list of different systems/techniques for launching/recovering a fixed-wing UAV is prepared. These techniques are evaluated and compared against each other in terms of each design requirement (e.g., Cost, Weight, size, safety issues, footprint, stowing, and impact on other systems) and how well the techniques are performing. The higher the score, the better the system performs in that criterion. Next, the results are tabulated, and the criteria are weighed against each other based on priority. Using a tradeoff study, a score (i.e., figure of merit) is assigned to each technique. The system with a higher score is the best candidate for launching/recovering process. This method is described in Chapter 1.

As an example, Figure 10.19 demonstrates the relative setup time for various recovery systems. The time it takes to set up the system and prepare it for recovery is an

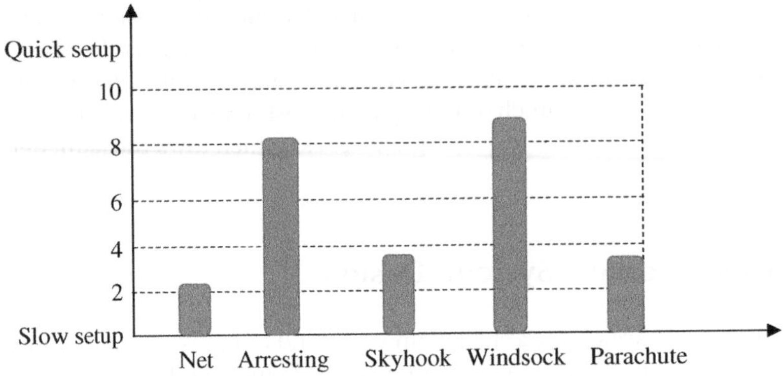

Figure 10.19 Relative setup times for various recovery systems.

important criterion for selection of the recovery system. The net recovery system has the slowest setup and the most complicated preparation. The poles must be mounted, and the net stretched up, and since it is a relatively large system it takes some time.

The skyhooks system has a slightly faster preparation process, since hydraulics can be used to enable a quick extension of the crane. However, the crane is often mounted on a trailer for stowing in a hangar. This needs to be brought out onto the ground and prepared. In comparison, the arresting line recovery is very fast, and does not need much preparation, particularly if it is a permanent system, mounted in the side of the ship.

If the system is planned to be employed on a ship or carrier, other design requirements must be included; such as impact on ship design, impact on ship operations, deck footprint, impact on helicopter operation, deployed mode, Risks of damaging the ship, and number of crew. Estimations of cost that derives from parts/components are a significant design operation. If any component/subsystem is commercially available, the overall cost will be highly reduced.

Significant design requirements for the launch and recovery system of an UAV are: (a) performance; (b) cost; (c) stability; (d) reliability; (e) safety; (f) manufacturing issues; (g) weight; (h) size; (i) marketability; and (j) handling requirements. In this chapter, different technologies or solutions for launching and recovering were presented together with positive and negative attributes. The technologies must be evaluated in terms of design drivers to be able to decide on which type of technology is chosen for further development and analysis. Figures 10.20 and 10.21 present the general design process of launch and recovery systems. The design has an iterative nature and begins with the design requirements.

10.10.2 Launch System Design

The design of launch system begins (See Figure 10.20) with development of the problem statement, and to formulate the design problem. This activity is based on the given objectives (by customer) and technical requirements (through engineering evaluations). Then, a functional analysis is performed to describe tasks or "functions" to be performed by the launch system and its major components. The functional analysis provides the baseline from which reliability requirements, maintainability requirements, human factor requirements, supportability requirements, and manufacturability requirements are identified.

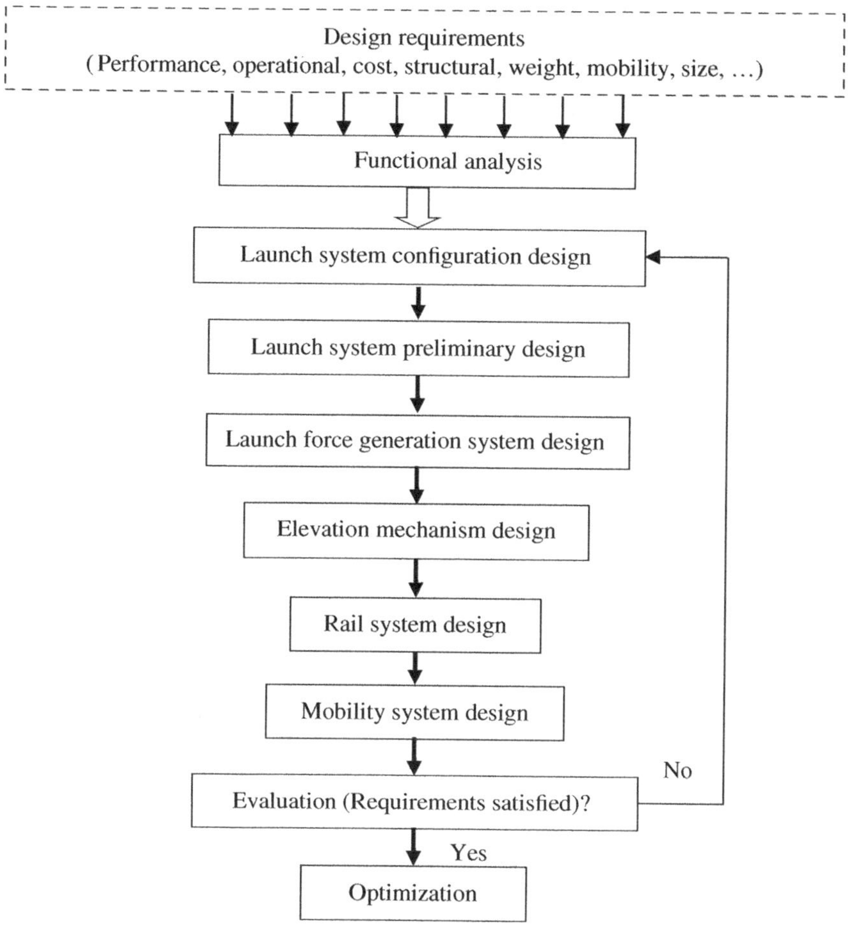

Figure 10.20 Launch system design process.

Then, a conceptual design will be performed to develop/select the configuration of the launch system. The next step is the partitioning; that is to break the launch system down into subsystems/elements/parts. A few considerable subsystems are: (a) fixed platform; (b) transportation system; (c) power supply system; (d) elevation mechanism; (e) azimuth platform; and (f) ramp system.

The power supply which contributes to the launch force generation, could be of many types including pneumatic, hydraulic, and electric systems. The design-to requirements have to be determined for each of the subsystems. The functional analysis is also performed during detailed design, since it includes breaking requirements at the system level down to the subsystem. Chapter 3 presents the fundamentals of the mechanical/power-transmission/electric systems design.

One of important design requirements for a launcher is the space needed for the operation of the launch unit. For instance, a ScanEagle launch unit takes up roughly 5 ft by 15 ft of a ship deck space, which is best located with rail access on an upper deck.

In the preliminary design phase of a launcher, primary launcher parameters such as required power, launch force, and general dimension are determined. Then, in the detail

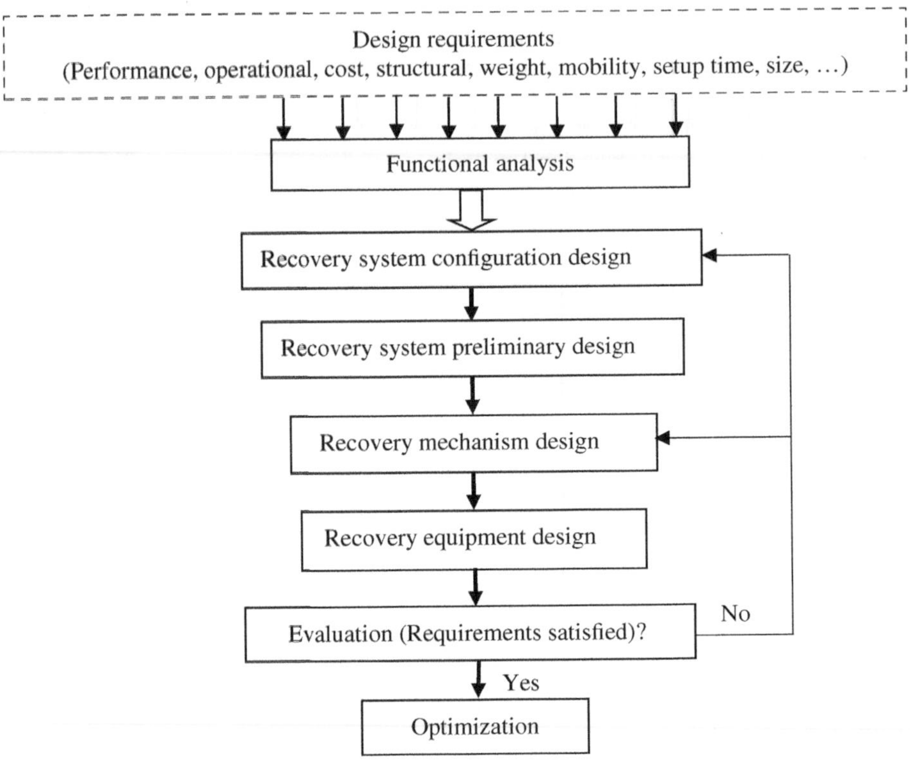

Figure 10.21 Recovery system design process.

design phase, the subsystems are designed by performing calculations, modeling, and simulation of the launch operation. Mobility design and analysis is the last step, before evaluation of the design, to determine if the design requirements are satisfied. If the answer is yes, the optimization process is performed. Otherwise, the non-satisfactory items are fed back to the appropriate phase for a redesign. The iteration is continued until all design requirements are satisfied. The primary goal of the optimization process is to minimize cost, weight, as well as the overall size, while maintaining the launcher performance. Multidisciplinary design optimization is a well-developed technique and is recommended for this purpose.

An ideal launcher is the one which is highly transportable and universal, and is suitable for several types of unmanned aerial vehicle. For instance, the KONTIO (designed by Robonic) is a transportable universal launcher and is capable to launch air vehicles of up to 140 kg with a 70 m/s exit velocity, or alternatively 500 kg at 37 m/s.

10.10.3 Recovery System Design

The design of the recovery system begins with the development of the problem statement, and to formulate (See Figure 10.21) the design problem. This activity is based on the given objectives (by customer) and technical requirements (through engineering evaluations). Then, a functional analysis is performed to describe tasks or "functions" to be performed by the recovery system and its major components. The functional analysis provides the baseline from which reliability requirements, maintainability requirements, human factor requirements, supportability requirements, and manufacturability requirements are identified.

Then, a conceptual design will be performed to develop/select the configuration of the recovery system. The design-to requirements have to be determined for each of the subsystems. The functional analysis is also performed during detailed design, since it includes breaking requirements at the system level down to the subsystem. The next step is the partitioning; that is to break the recovery system down into subsystems/elements/parts. A few considerable related subsystems are: (a) fixed platform; (b) transportation system; (c) power supply system; (d) elevation (retracing) mechanism; and (e) azimuth platform. The power supply which is contributing to the recovery handling, could be of many types including pneumatic, hydraulic, and electric systems. Chapter 3 presents fundamentals of mechanical/power-transmission/electric systems design.

In the preliminary design phase of a recovery system, primary recovery parameters such as required power, expansion mechanism, and general dimension are determined. Then, in the detail design phase, the subsystems are designed by performing calculations, modeling, and simulation of the recovery process. Mobility design and analysis is the last step, before evaluation of the design, to determine if the design requirements are satisfied. If the answer is yes, the optimization process is performed. Otherwise, the non-satisfactory items are fed back to the appropriate phase for a redesign. The iteration is continued until all design requirements are satisfied. The primary goal of the optimization process is to minimize the cost, weight, as well as the overall size, while maintaining the recovery performance. Multidisciplinary design optimization is a well-developed technique and is recommended for this purpose.

A crash detection and rescue system for quadcopters and small UAVs can be developed to provide a fast and efficient parachute safety solution to reduce the risk of harming people on the ground and safely lands the UAV when it malfunctions during flight. This system should work autonomously and be independent of the UAV autopilot, to eject the parachute in a fraction of a second to enable a safe UAV operation even over crowded area. The entire parachute can be a system–independent to deployed even with a total failure of the UAV. In an emergency, the parachute is catapulted out of its cage by an elastic cord system or dispensing pyrotechnics as the trigger mechanism. The parachute design should be light weight, while it is pendulum and wind stable (e.g., circular or rectangular).

Questions

1 List main design requirements for the launch system of an UAV.

2 List main design requirements for the recovery system of an UAV.

3 In the design of a launcher, what parameters must be determined?

4 What are the major elements of a launcher?

5 What is the function of a UAV recovery system?

6 Name a few practical techniques in the UAV recovery.

7 What technique is employed in the recovery of: a. SkyLite UAV, b. Insitu Scan Eagle, c. Global Hawk?

8 Describe the procedure of a UAV hand launch.

9 What major elements are required for the air launch of an UAV?

10 What technique is used in launching Pointer UAV?

11 What is a typical impact speed of a UAV when it is recovered by a parachute?

12 What is the maximum takeoff mass and wing span of ScanEagle UAV?

13 Describe the recovery system of ScanEagle UAV.

14 Describe the mobility requirements.

15 What are the two options in the rover vehicle of a launcher?

16 What is detent mechanism in a launcher?

17 Name three techniques of launching a UAV.

18 Briefly describe the SideArm system.

19 List seven typical launch methods.

20 List five most common categories of UAV recovery systems and technologies.

21 What are the typical values for the load factor during a launch operation?

22 What are major advantages of the rocket assisted launch?

23 Bungee Cord Catapult Launch

24 Briefly describe bungee cord catapult launch operation.

25 What are the main elements in a bungee cord catapult launch system?

26 What is the maximum achievable UAV airspeed by a bungee cord launcher?

27 List the contributing forces in a bungee cord catapult launch operation.

28 What are the design options to achieve a higher UAV velocity at the separation point in a bungee cord launcher?

29 Briefly describe the pneumatic launching system.

30 What are the advantages of a pneumatic launching system?

31 What are the disadvantages of a pneumatic launching system?

32 Briefly describe the hydraulic launching system.

33 What are the advantages of a hydraulic launching system?

34 What are the disadvantages of a hydraulic launching system?

35 What launch system is used by the RQ-7 Shadow?

36 Briefly describe the hand launch system.

37 What are typical launch speeds for small UAVs?

38 What is the function of a power screw?

39 What parameters are contributing to the torque required to raise launch rails by elevation mechanism?

40 What is the self-locking condition in an elevation mechanism?

41 What parameters do influence the power required to raise launch rails by elevation mechanism?

42 What technique is utilized in recovering the Sky Lite B UAV?

43 What are the typical values for the friction coefficient between UAV legs and the ramp of a launcher?

44 What is typical speeds for a UAV at the end of launcher compared with the stall speed?

45 What forces and moments are contributing to the longitudinal trim of a UAV after the launch?

46 What are design parameters in actuator design for an elevation mechanism of a small to medium-weight UAV?

47 Briefly describe the recovery process.

48 What objectives are implied by a UAV recovery process?

49 Briefly describe the net recovery process.

50 What is the typical size of a net as compared with the UAV top area?

51 List the advantages of a net recovery system.

52 List the disadvantages of a net recovery system.

53 List the advantages of an arresting line recovery system.

54 List the disadvantages of an arresting line recovery system.

55 Briefly describe the arresting line recovery process.

56 List the advantages of a skyhook recovery system.

57 List the disadvantages of a skyhook recovery system.

58 Briefly describe the skyhook recovery process.

59 Briefly describe the windsock recovery process.

60 Briefly describe the parachute recovery process.

61 List three basic forms of recovery equipment.

62 Discuss and compare the relative setup times for various recovery systems.

63 What are the features of an ideal UAV launcher?

Problems

1 A launcher is designed to launch a UAV with a mass of 300 kg. The launch angle is 25°, the launcher length is 2.6 m, and the friction coefficient is 0.05. The UAV is required to reach a speed of 20 knot at the end of launcher rail. What force is needed to push the UAV over the launcher? Ignore UAV drag during launch.

2 A launcher is designed to launch a UAV with a mass of 200 kg. The launch angle is 30°, the launcher length is 2.2 m, and the friction coefficient is 0.055. The UAV is required to reach a speed of 15 knot at the end of launcher rail. What force is needed to push the UAV over the launcher? Ignore UAV drag during launch.

3 A small UAV with a mass of 70 kg is required to land on the ground with the help of a parachute with an impact velocity of 4 m/s. What parachute projected area is needed for landing on sea level altitude? Ignore UAV drag during launch.

4 A small UAV with a mass of 150 kg is required to land on the ground with the help of a parachute with an impact velocity of 3 m/s. What parachute projected area is needed for landing on sea level altitude? Ignore UAV drag during launch.

5 The Shadow UAV is launched from a trailer-mounted pneumatic catapult. Its pneumatic launcher can accelerate the 170 kg aircraft to 30 knots in 12 m. Ignoring the drag, determine the applied force. Assume the launch angle of 30°, and a friction coefficient is 0.043.

6 A UAV is launched from a trailer-mounted pneumatic catapult. Its pneumatic launcher can accelerate the 120 kg aircraft to 50 knots in 9 m. Ignoring the drag, determine the applied force. Assume the launch angle of 40°, and a friction coefficient is 0.035.

7 A UAV is launched from a trailer-mounted pneumatic catapult. Its pneumatic launcher can accelerate the 50 kg aircraft to 30 knots in 6 m. Ignoring the drag, determine the applied force. Assume the launch angle of 25°, and a friction coefficient is 0.06.

8 A UAV with a mass of 5 kg is launched using bungee cord. The UAV is gaining the speed of 20 knot in 5 m with the launch angle of 25°. Determine the cord spring index. Assume a friction coefficient of 0.06.

9 A UAV with a mass of 12 kg is launched using bungee cord. The UAV is gaining the speed of 30 knot in 7 m with the launch angle of 30°. Determine the cord spring index. Assume a friction coefficient of 0.06.

10 A UAV with a mass of 180 kg is required to land on the ground with the help of a parachute. The parachute projected area is 60 m^2. Determine the impact velocity for landing on sea level altitude.

11 A very small UAV has a mass of 1 kg. A human operator with a maximum force of 80 N is arranged to launch this UAV. If his hand extends 75 cm, determine the release speed.

12 A human operator with a maximum force of 100 N is arranged to launch a very small UAV. His hand extends 70 cm, and the UAV release speed is required to be 6 m/s. Determine the maximum possible mass of the UAV that can be launched.

13 A launcher is designed to launch a UAV with a mass of 30 kg. The launch angle is 40°, and the friction coefficient is 0.05. The UAV is required to reach a speed of 25 knot at the end of launcher rail. The launch force is 700 N to push the UAV over the launcher. What should be the length of this launcher?

14 A launcher is designed to launch a UAV with a mass of 120 kg. The launch angle is 20°, and the friction coefficient is 0.06. The UAV is required to reach a speed of 45 knot at the end of launcher rail. The launch force is 10 000 N to push the UAV over the launcher. What should be the length of this launcher?

15 A launcher is designed to launch a UAV with a mass of 100 kg. The launch angle is 25°, and the friction coefficient is 0.05. The UAV is required to reach a speed of

20 knot at the end of launcher rail. The launch force is 2000 N to push the UAV over the launcher. What should be the length of this launcher?

16 The pneumatic launcher of a UAV can accelerate the 120 kg aircraft to 35 knots in 10 m. Assume the launch angle to be 40°, and the friction coefficient to be 0.08. Neglect the contribution of UAV lift and drag during the launch, determine the launch force.

17 The pneumatic launcher of a small UAV can accelerate the 60 kg aircraft to 30 knots in 6 m. Assume the launch angle to be 20°, and the friction coefficient to be 0.045. Neglect the contribution of UAV lift and drag during the launch, determine the launch force.

18 Repeat problem 16, but include the effect of aircraft drag. Assume the drag coefficient to be 0.035, and wing area to be 4.1 m^2. Hint: write a MATLAB code and use a short time-step such as 0.02 seconds.

19 Repeat problem 17, but include the effect of aircraft drag. Assume the drag coefficient to be 0.04 and wing area to be 4 m^2. Hint: write a MATLAB code and use a short time-step such as 0.02 seconds.

20 An elevation platform is being designed for a UAV launcher. The platform is made up of power screws with a square-thread with a pitch diameter of 40 mm, a pitch of 6 mm, and double threads. The power screws are made up of steel, and machine oil is applied. Assume the launch angle is 10°. The mass of UAV is 15 kg, and the launcher rail is 35 kg.

- **a** Find the torque required to raise the UAV.
- **b** Determine the required power, if the thread is desired to turn at the rotational speed of 120°/s.
- **c** This power is provided by an electric motor with the voltage of 12 V. Calculate the electric current.

21 An elevation platform is being designed for a UAV launcher. The platform is made up of power screws with a square-thread with a pitch diameter of 20 mm, a pitch of 8 mm, and double threads. The power screws are made up of steel, and machine oil is applied. Assume the launch angle is 12°. The mass of UAV is 10 kg, and the launcher rail is 20 kg.

- **a** Find the torque required to raise the UAV.
- **b** Determine the required power, if the thread is desired to turn at the rotational speed of 150°/s.
- **c** This power is provided by an electric motor with the voltage of 8 V. Calculate the electric current.

22 Consider a small UAV with a launch mass of 20 kg, a wing area of 2 m^2, and a horizontal tail area of 0.5 m^2 with the following characteristics:

$$C_{L_{ho}} = 0; C_{L_\alpha} = 5.6\,1/rad; C_{L_{\alpha_h}} = 4\,1/rad; \alpha_h = 0; C_{L_{\delta E}} = 0.4\,1/rad; AR = 8$$

$$C_{m_{owf}} = -0.09; h = 0.18; h_o = 0.24; \bar{V}_H = 0.8,$$

The launch speed is 20 knot and maximum elevator deflection is ±20°. Determine whether the UAV longitudinal trim can be maintained at the end of launch, if it is launched at a 20° launch angle. Assume sea level.

23 Consider a small UAV with a launch mass of 1000 kg, a wing area of 12 m², and a horizontal tail area of 3 m² with the following characteristics:

$$C_{L_{ho}} = 0; C_{L_\alpha} = 6\,1/rad; C_{L_{\alpha h}} = 4.7\,1/rad; \alpha_h = -1\deg; C_{L_{\delta E}} = 0.35\,1/rad; AR = 6$$

$$C_{m_{owf}} = -0.04; h = 0.12; h_o = 0.25; \bar{V}_H = 0.5,$$

The launch speed is 50 knot and maximum elevator deflection is ±25°. Determine whether the UAV longitudinal trim can be maintained at the end of launch, if it is launched at a 15° launch angle. Assume sea level.

24 A very small UAV is launched with a velocity of 20 knot at a launch angle of 18 degrees. Assume the UAV keeps the velocity and launch angle constant for climbing flight. The launch site is within a wooded area, where trees with a height of 20 m are at a distance of 50 m. The launcher is 6 m long. Is this UAV able to clear the trees safely?

25 A very small UAV is launched with a velocity of 25 knot at a launch angle of 20 degrees. Assume the UAV keeps the velocity and launch angle constant for climbing flight. The launch site is within a wooded area, where trees with a height of 22 m are at a distance of 60 m. The launcher is 5 m long. Is this UAV able to clear the trees safely?

Design Projects

1 A launcher with a pneumatic power source has been design for a fixed-wing UAV with a mass of 100 kg and conventional control surfaces. However, the UAV is crashing just after leaving the launcher. In investigating this crash, a report has concluded that the UAV speed when leaving the launcher was less than stall speed. You are required to redesign the system (either the UAV, or the launcher). Provide at least two solutions for this problem. Justify your solutions.

2 A launcher with a pneumatic power source has been design for a fixed-wing UAV with a mass of 100 kg and conventional control surfaces. However, the UAV is crashing just after leaving the launcher. In investigating this crash, a report has concluded that the elevator (at its max deflection) is unable to longitudinally trim the UAV, so the elevator control power is not sufficient at the time of leaving the launcher. You are required to redesign the system (either the UAV, or the launcher). Provide at least two solutions for each problem. Justify your solutions.

Note

1 Aurora Flight Sciences.

11

Ground Control Station

CONTENTS

Design of Unmanned Aerial Systems, First Edition. Mohammad H. Sadraey.

Companion website: www.wiley.com/go/sadraey/unmanned-aerial-systems

Educational Outcomes

After study of this chapter, the reader will be able to:

1) Discuss the current ground control station (GCS) technologies
2) Explain GCS features for several current Unmanned Aerial Vehicles (UAVs)
3) Conduct a GCS design project
4) Explain various types of GCS
5) Develop human-related design requirements
6) Discuss GCS support equipment
7) Describe GCS subsystems
8) Describe the process of sending a command from a GCS to a UAV
9) Describe the process of receiving a signal from a UAV
10) Explain mathematical model and function and a potentiometer
11) Design interior layout of a GCS for a large UAV
12) Discuss fundamentals of ergonomics
13) Design a pilot/operator console

11.1 Introduction

A significant element of an Unmanned aerial system (UAS) is the ground station, which is the man–machine interface with the UAV. Other alternative terms used in the literature are "Ground Control Station (GCS)"; "Command and Control Station"; CCS, "Mission Control Station; MCS", and "Mission Planning and Ground Control Station", MPGCS. The UAVs and their payloads are controlled and monitored by ground pilots through GCSs. The GCS is a major component of the data collection and distribution system. It receives the information collected by a UAV, processes that information, and resends it via a data link to the appropriate end user. Depending upon range and type of mission (complexity of the UAV system), smaller UAVs are controlled via visual contact (manual real-time control), while the larger ones are equipped with a communication system (engage stored on-board flight plans).

The human controller/pilot in the GCS can talk to the UAV via the communication system (up-link), in order to direct the flight or to update the flight plan. The ground station allows remote control of UAV, receives, and processes navigation data from UAV via the data-link, supports flight planning and communicates with the autopilot board on ground via interface. Direct operation of various payloads (e.g., camera, weapon) that the UAV carries may also be required. If data link is installed, ground station may support flight control without visual contact with UAV through software application. The flight plans are programmed by the operator; and uploaded through a serial connection between the UAV autopilot and the ground station. Three main functions for the ground station are mission tracking/planning, observing, and piloting.

The aircraft will return information and images to the operators via the communications down-link (See Figure 11.1), either in real-time or on command. The information will usually include data from the payloads, status information on the vehicle's subsystems, altitude and airspeed and position information. The launch and recovery of

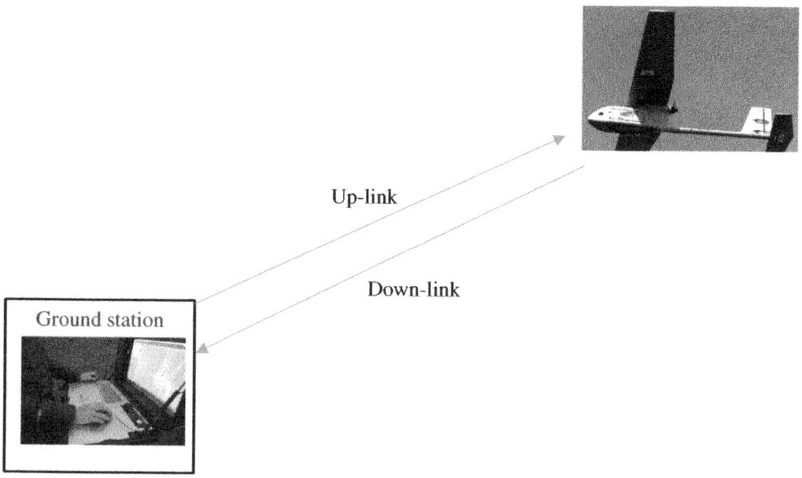

Figure 11.1 Ground control station (GCS) and the air vehicle (RQ-11 Raven Unmanned aerial vehicle [UAV]).

the vehicle may also be controlled from the station. A controller in the ground station has a place similar to a cockpit with displays, stick/wheel and buttons to control the UAV. The deflection to control surfaces and engine throttle are applied through communication signals.

Conventional remote control is responsible for autopilot engagement and disengagement as well as allows manual/remote controlled flight operations during visual contact with the UAV. The UAV autopilot is controlled by human pilot in the ground station. This chapter will introduce features of ground station, main elements, and support equipment, design requirements, and design techniques. Since one or more humans are utilizing the station, the ergonomics is briefly reviewed.

In general, the operator(s) in the ground station have three main tasks: (a) to control the UAV trajectory; (b) to employ payloads; and (c) to monitor sensors. If the UAV does not have the capability of full autonomous flight (e.g., Predator A), the controller in the ground station must engage with the vehicle in real time throughout the flight operation. However, for the case of the full autonomous UAVs, the operator(s) in the ground station will monitor the mission, and will act only if needed.

The typical operator's task includes: (a) control the UAV to orbit at a given altitude, radius, and speed around a to-be-designated terrestrial grid reference; (b) direct the UAV to hover over a selected area for surveillance; (c) descend at a controlled rate and land; and (d) descend to ground level to take an atmospheric sample and then to climb again to the operating altitude. The sensor operator in the GCS can determine what the UAV is currently looking at.

In some UAVs, the air vehicle has a fraction of the cost of the unmanned aircraft system. For instance, the cost of each RQ-7 Shadow vehicle is about $750 000 (2011 dollar), while the cost of each system including GCS, launcher, and support equipment is about $15.5 million. Moreover, the cost of each General Atomics MQ-1C Gray Eagle vehicle is about $1.5 million (2013 dollar), while the cost of each system including GCS, launcher, and support equipment is about $21.5 million.

11.2 GCS Subsystems

A GCS usually contains a number of sub-systems required to achieve its overall function. These sub-systems will depend in detail on the type (civil versus military) and mission (e.g., range, endurance, and delivery) of the UAV. In general, the following subsystems and elements may be structured into the GCS:

1) The UAV flight controls consol. The operator will connect to the UAV autopilot (i.e., control system) to control the mission; either for manual real-time control of the UAV, or to select and engage stored on-board software.
2) Sticks to control: (a) engine (for airspeed and altitude); and (b) attitudes (pitch, roll, and yaw)
3) The payload monitoring and control consol. The payload operator will monitor the status of the operators and control their operations (e.g., video camera).
4) Communication subsystem with UAV and other operators/commanders encoders, transmitter, and receiver). The communication with UAV has both uplink and downlink. This system includes the control for the status of communication elements. For example, raising or lowering radio antenna masts, steering them manually or automatically to obtain good transmission and reception of the radio waves, and changing frequencies as necessary. Section 11.5 provides more details.
5) Navigation displays for monitoring the status and flight path of the UAV. The position of the UAV will be displayed automatically on a position display.
6) Display for recording equipment, and measuring payloads
7) Terrestrial map displays.
8) Computers for data processing, calculations, mission planning, keeping the UAV flight programs, and housekeeping data storage.
9) Equipment to provide comfort to operators (e.g., heater, air conditioner, dehumidifier).
10) Support subsystem to provide electric/mechanical/hydraulic power to various equipment.
11) Office equipment (e.g., table, seat)
12) Computational software and computer programs/codes
13) Human operators/controllers

In designing a GCS, all these subsystems and elements must be designed and then integrated into one unit. In a small UAV, all subsystems/elements will be integrated into one small handheld box. However, for a large UAV, usually more than one ground vehicle is needed to locate all pilots, operators, and equipment.

11.3 Types of Ground Stations

Ground stations range from a small handheld remote controller, to a simple laptop computer, to a mobile truck, to comprehensive fixed central command stations. The mobile stations are frequently located in all-terrain vehicles and are made for close- or medium-range systems. However, the central command stations are in fixed bases and are usually employed for Medium altitude long endurance (MALE) and High altitude long endurance (HALE) systems. In general, the control station does not necessarily need to be located on the ground; they may be positioned in a submarine, ship, or another aircraft (airborne). In this section, a few types of ground stations are introduced.

11.3.1 Handheld Radio Controller

11.3.1.1 General Structure

Most small homebuilt UAVs and RC (Remotely Controlled) model airplanes are controlled via a simple handheld remote controller which weighs about 1–2 pounds. The controller has a few sidesticks/levers/buttons (see Figure 11.2) with an antenna and uses radio/infrared signal to communicate with the UAV. These types of station are utilized for visual and real time control. The operator will push/pull stick/lever to deflect any control surface, activate engine, and to change the flight path.

Figure 11.3 illustrates the process of sending a command signal from a handheld control to a flying UAV. When the joystick is deflected by a pilot on the ground, a potentiometer will generate a voltage corresponding to the deflection angle (δ_S). This voltage will be processed by the microcontroller within the box, and will generate a command analog signal to the transmitter. Then, the transmitter will generate a digital signal (e.g., radio wave) which is transmitted to the air via an antenna.

A ground pilot is fully responsible in guiding and controlling the UAV via the remote-control box. Both guidance/control decisions are made directly observing the UAV through its on-board video camera. When this GCS is adopted, the color of the UAV should not be all blue or all white, since the sky in a day is either clear (i.e., blue), or cloudy (i.e., white). It is hard for a ground pilot to detect a white or blue UAV, even if there is a line of sight. Hence, the UAV paint decision must consider this requirement.

If the UAV is at the line of sight, its antenna will receive the radio wave, and send a signal to the receiver. Then, the receiver will generate a voltage and sent it to the microcontroller within the UAV. The next step is to create a command signal (a voltage) by the microcontroller and send an analog signal (e.g., voltage) to the servo of the elevator. The

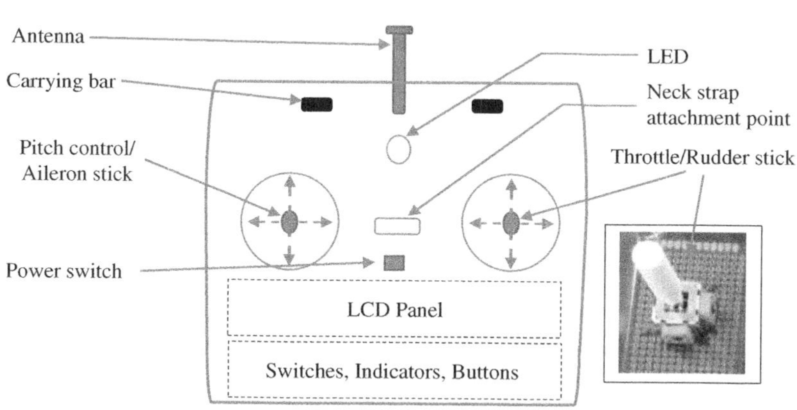

Figure 11.2 Handheld remote control of a small Unmanned aerial vehicle (UAV).

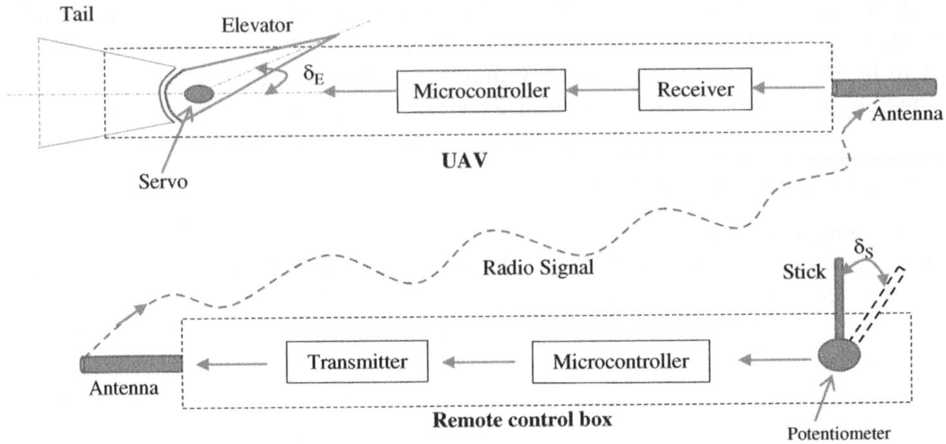

Figure 11.3 Process of sending a command from a handheld control box to a Unmanned aerial vehicles (UAV).

rotation of the servo will be a function of received voltage. As the servo is rotating, the elevator is deflected with same rate. All of this process will happen in a fraction of a second. The structure and governing equations for a potentiometer is presented in the section 11.3.1.3.

There are various commercial handheld controllers in the market for a couple of hundred dollars. A few examples are: (a) Spektrum DX8 (with 8 channels); (b) Graupner MX-20 (with 12 channels); (c) RealFlight RF8 Horizon InterLink-X (with 10 channels); and (d) Radiolink AT10II (with 12 channels). Each package usually comes with both transmitter and compatible receiver.

11.3.1.2 Stick

A remote pilot applies UAV flight control from a GCS through various tools including sticks for steering. There are primarily two types of UAV control functions: (a) engine; and (b) attitudes (pitch, roll, and yaw). These controls are implemented by pilot's hands via sticks. The motor (engine) throttle is controlled by pilot when a stick is deflected, this will change airspeed and then angle of attack, pitch angle, and consequently altitude. Moreover, three attitude angles: pitch (including angle of attack and climb angle), roll (bank), and yaw (heading) angles) are also controlled via sticks.

These four controls may be incorporated into two sticks, each have four group of deflections (i.e., forward, backward, left, and right).

For the correlation between the operator steering control stick (angle) in the GCS, and the UAV attitude angle; an estimating theory (e.g., Fuzzy logic, neural network, and adaptive control) should be employed to calculate an estimation of attitude angle.

The stick deflection in a GCS will frequently rotate a potentiometer (Figure 11.4) which in turn creates an analog signal (usually a voltage). The voltage will enter a microcontroller, which in turn sends a signal to the transmitter. The next step is to send a digital signal (i.e., command) through the antenna via radio waves to the UAV. When the signal is received by the receiver of the UAV via its antenna, it will send an analog signal (e.g., voltage) to the microcontroller. The next step is for the microcontroller to process

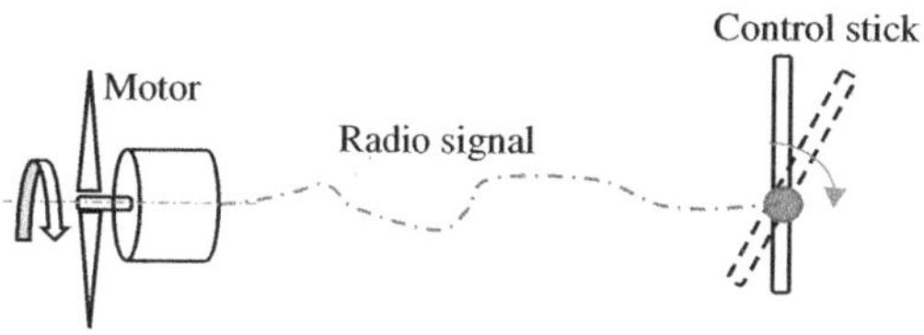

Figure 11.4 Engine control via a stick deflection.

this command and generate an analog signal (e.g., voltage) to the respective element (e.g., engine). The rotational speed of the engine (ω_E) is a function of the stick deflection (δ_s):

$$\omega_E = K_\omega \delta_{s1} \tag{11.1}$$

where K_ω is the engine-stick proportionality constant or gearing ratio. However, this constant will be influenced by a number of factors including transmitter, receiver, and microcontroller performances. Furthermore, it is also a function of engine type (e.g., electric, turboprop, piston, or turbofan). With the same token, gearing ratio for other three control surfaces are defined:

$$\delta_E = K_E \delta_{s2} \tag{11.2}$$

$$\delta_A = K_A \delta_{s3} \tag{11.3}$$

$$\delta_R = K_R \delta_{s4} \tag{11.4}$$

where K_E, K_A, K_P are the gearing ratios for elevator (δ_E), aileron (δ_A), and rudder (δ_R) deflections respectively. The subscript 1, 2, 3, and 4 are referring to various sticks and their deflections. For instance, in the case of a GCS with two sticks, subscripts 1 through 4 may refer to the aft-to-fore movement of the right stick (See Figure 11.2), the aft-to-fore of the left stick, left-to-right of the right stick, and left-to-right of the left stick respectively. In the case of a quadcopter, the functions of the stick movements are defined differently (See Section 5.5.7).

Example 11.1

When a remote pilot at a GCS deflects a stick force with the value of 30°, the piston engine of the UAV will rotate at 1500, rpm. Determine the engine-stick proportionality constant (K_ω) for this UAS.

Solution

$$\omega_E = K_\omega \delta_s \Rightarrow K_\omega = \frac{\omega_E}{\delta_s} = \frac{1500}{30} = 50\frac{rpm}{\deg} = 300\frac{\deg/\sec}{\deg} \tag{11.1}$$

In order to control UAV motor and attitudes, a GCS must have at least two sticks (trigger) with adjustable positions (fore-to-aft and side-to-side). Each aft-to-fore and left- or right-hand steering is dedicated to a UAV control variable via a channel of the transmitter. A typical transmitter of a handheld GCS has at least six channels, a dedicated frequency (e.g., 2.4 GHz), modulation, and programming for a compatible receiver. It is clear that the stick force (F_s) must be within the range of hand power of the pilot.

11.3.1.3 Potentiometer

The motions/deflections of electric motor and control surfaces are controlled by an operator via throttle and sticks. The element that converts the deflection of stick and throttle to a signal is potentiometer.

The potentiometer (or pot) is an electro-mechanical transducer [132]; the main element is a variable resistor. It converts linear or angular position/displacement from the operator into a voltage (i.e., electric potential). So, its output resistance changes as an internal wiper moves across a resistive surface. A potentiometer has three leads, with the center wiper lead connected to the analog input pin.

A potentiometer is a zero-order measurement device that can provide a feedback for a control system. The output voltage (V_{out}) as a function of input (displacement) is linearly modeled as:

$$V_{out} = \left(\frac{R_x}{R_p}\right) V_{in} \tag{11.5}$$

$$V_{out} = V_{in}\left(\frac{X_{in}}{L}\right) \quad \text{(Linear pot)} \tag{11.6}$$

$$V_{out} = V_{in}\left(\frac{\theta_{in}}{\theta_{\max}}\right) \quad \text{(Rotary pot)} \tag{11.7}$$

where R_x is the resistance between potentiometer leads, R_p is the maximum resistance of the potentiometer, X_{in} is the linear displacement of wiper, θ_{in} is the angular displacement of wiper, L is the maximum linear travel of wiper and θ_{max} is the maximum angular travel of wiper. The output voltage on the wiper line varies between 0 V (ground) and a maximum (e.g., 5 V) value (V_{in}), as the stick (knob) is turned to its maximum position. Hence, the transfer function of a potentiometer is:

$$\frac{V_{out}(s)}{X_{in}(s)} = K_L \tag{11.8}$$

$$\frac{V_{out}(s)}{\theta_{in}(s)} = K_A \tag{11.9}$$

where K_L (in volt/meter) and K_A (in volt/deg) are gains for linear and angular potentiometer respectively. Thus, the gain of a rotary potentiometer is:

$$K_A = \frac{V_{battery}}{\theta_{\max}} \tag{11.10}$$

Example 11.2

A rotary potentiometer is programmed to provide 5 V for a full 360° rotation. What is the gain for this potentiometer?

Solution

$$K_A = \frac{V_{battery}}{\theta_{\max}} = \frac{5}{360} = 0.014\frac{Volt}{\deg} = 0.796\frac{Volt}{rad} \quad (11.11)$$

Thus, its gain is 5-V/360-deg or 0.014 V/deg or 0.796 V/rad.

The ultimate choice of gain depends upon the desired linearity and accuracy of the signal and the amount of force/torque.

To interface a potentiometer to a microcontroller (See Chapter 9), the analog voltage output must be converted to digital form (using an A/D converter). A PIC[1] microcontroller has a built-in analog-to-digital converter. Figure 11.5 demonstrates a few types of typical single-turn angular potentiometer. Typical application for an angular potentiometer is to measure the angular deflection of a control surface (e.g., elevator). Most rotary potentiometers have 270° of rotation from one extreme to the other. Hence, a "single turn" potentiometer is indeed only a 3/4 turn measurement device.

Majority of the potentiometers are linear. However, the variation of the output voltage as a function of the input displacement for some potentiometers are nonlinear (e.g., logarithmic, and quadratic). When a new potentiometer is employed, you need to make sure about its governing law (sometimes referred to as taper). The governing law and also quoted tolerance of a pot could be checked by the catalog through the supplier. Otherwise, the measured signal will not be correct/accurate.

In a fixed-wing UAV, the potentiometer for the control of the UAV motor will only generate a positive voltage. However, the potentiometers for the control of the UAV control surfaces (elevator, aileron, and rudder) will generate both positive and negative voltages. This is due to the fact that control surfaces have deflections to two sides (positive and negative). For instance, the elevator and aileron have up/down deflections, while the rudder has left/right deflections.

11.3.2 Portable GCS

When the necessary equipment for monitoring and controlling a UAV is slightly larger than the size of a handheld box, the solution is to adopt a larger portable GCS. This is true for mini to small UAVs. In some cases, both UAV and GCS can be

Figure 11.5 Typical single-turn potentiometers.

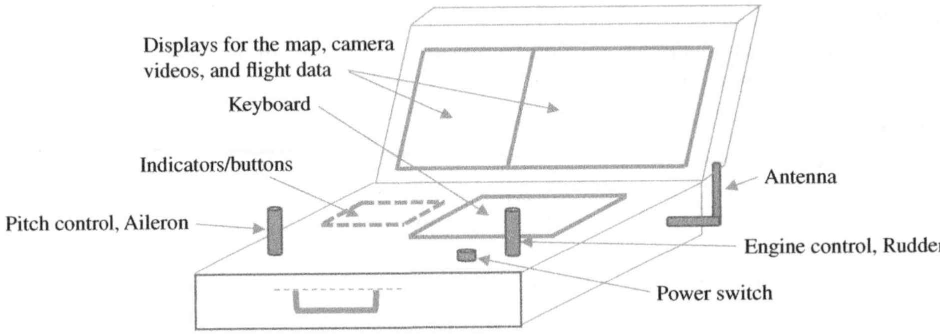

Figure 11.6 Main elements of a typical portable ground control station (GCS).

back-packed and carried in a small suitcase. The size of such GCS is comparable to the size of a laptop. Such GCS frequently incorporates the graphical user interface (GUI) and features a laptop, allowing the operator an easy access to key and frequently used features. A portable GCS requires a line-of-sight to allow the pilot to see the UAV and to send the command signals. The operator may employ real-time platform health monitoring, while looking at the display to watch the video feed from the payload in real time.

The total mass of the portable GCS should be such that (e.g., less than 3 kg) an operator can carry it for the duration of flight (e.g., 20 minutes). If the total mass is heavier than this, the GCS may be moved for a short time, but cannot be carried for a long time by a human operator. In such cases, the portable GCS is placed on a table/platform to operate. Since the operator must have a line of sight for the duration of flight, a portable GCS is suitable for a UAV with a range of less than about 500 m.

Main elements/modules (Figure 11.6) of a typical portable GCS include: (a) display for camera videos/images; (b) monitor for flight data; (c) engine control; (d) motion control joysticks; (e) timer; (f) power switch; (g) keyboard; (h) various buttons; (i) antenna; (j) payload control platform; (k) computer (processor/RAM/hard-drive); (l) external connectors; (m) interfaces; (n) battery; (o) audio subsystem; (p) transmitter; (q) receiver; and (r) battery indicator. The mission information is easily updated when the flight vehicle is re-tasked to support a changing environment. Figure 11.7 illustrates the mini UAV Desert Hawk III and its portable control station.

11.3.3 Mobile Truck

The GCS for close-range UAVs have a number of equipment and devices which do not fit in a portable box. This type of GCS provides the interfaces for controlling the flight of UAV through a customized console. The volume and weight of such equipment are such that they can be setup in back of a truck. A few advantages of a truck are: (a) the truck can transport the UAV, GCS, and the human operator to any designated area; (b) the truck can provide the GCS with electric power; (c) back of a truck can be converted to a comfortable place for the UAV controller; and (d) the truck will keep the human operators in the loop. A truck easily meets the mobility requirement of the UAV system.

Figure 11.7 Desert Hawk III mini unmanned aerial vehicle (UAV) and its portable control station.

Thus, the GCS for close-range UAVs are usually mobile and housed within an "all-terrain" vehicle. A pneumatically raised steerable radio mast is at the rear of the vehicle which carries an antenna for communication with the UAV.

The air vehicle of these systems usually will be either: 1. Ramp-launched or 2. VTOL. The requirements and capabilities of both systems will be generally similar, except for the specifics of the control during launch and recovery. Although unmanned aircraft system implies that no human operator is involved, but these systems might require the manpower equivalent to manned aircraft. There is no sensation of speed, attitude, or g's in any maneuver. Each operator will be assigned an alternative assignment task; plus, a crew member as the commander who is responsible for the inter-system communication. There is usually a flexibility of tasking and easy cooperation between crew members.

The ground vehicle is usually fitted with chassis-mounted jacks or stays which are lowered once the truck is on-site. These are necessary to stabilize the truck to prevent it from rocking under the influence of wind or the operators moving around within the vehicle. The GCS also incorporates electrical power generation and air-conditioning which is required not only for the comfort of the crew, but for the climatic control of electric equipment (e.g., monitors, computers). These pieces of equipment also have their own controls and monitoring systems provided in a separate rack. There will be a longer duty time of the operators, additional crew members may be required, especially for the more complex payloads.

There are a number of monitors and displays. For instance, a monitor carries a video image from a TV camera. Beneath the monitors are the control decks, over which are the control sticks to direct operator control of the UAV in flight. The keypad controls for inputting mission data, software updates, or way-points to the UAV in located on the control deck. Controls are provided for starting the aircraft, selecting pre-flight test data, and for activating the aircraft launch.

There will be displays showing the payload status and data (imagery/video and/or other types of data) with recording media. Thus, a major GCS section is the sub-system to recognize the type of payload installed in the UAV, and to control the operation of the payload. A display may also be included to show communication status. The display screen should be large enough, so that the operator's eye is able to take advantage of the full resolution of UAV sensors.

The operators in the GCS are not visualizing the UAV that they are controlling. All they see is an icon appears on a moving map display, and the payload imagery which is displayed on a monitor. The UAV could be aesthetically pleasing, large, or small, but the icon on the map makes no distinction. Figure 11.8 illustrates inside of the GCS (e.g., an operator, and a few displays can be seen) of a RQ-7A Shadow 200. Figure 11.9 shows NASA research pilot Mark Pestana which flies the Ikhana unmanned aircraft remotely from the ground control station at NASA Dryden. The GCSs of MQ-TB Hunter, Avenger, and MQ-1 Predator have 3, 6, and 12 screens respectively. In Predator UAV, the ground station had initially 6 displays. However, users/operators required/added more displays/monitors. Now, it has 12 displays.

When a close range UAV is launched from sea, the control station will be naturally located in a ship. The Insitu ScanEagle UAV (Figure 3.3) with 20+ hour endurance is an example where the GCS is placed onboard a navy ship. A ship, similar to a truck, is capable of providing comfort for operators, transporting GCS and UAV, and generating electric energy for GCS. However, there is a possibility that a GCS be airborne; in another aircraft (i.e., parent aircraft).

Figure 11.10 exhibits the plan view of interior layout of a typical GCS truck for a UAV. This GCS houses two UAV flight personals; one pilot, and one payload operator. It has

Figure 11.8 Inside RQ-7A Shadow 200 GCS.

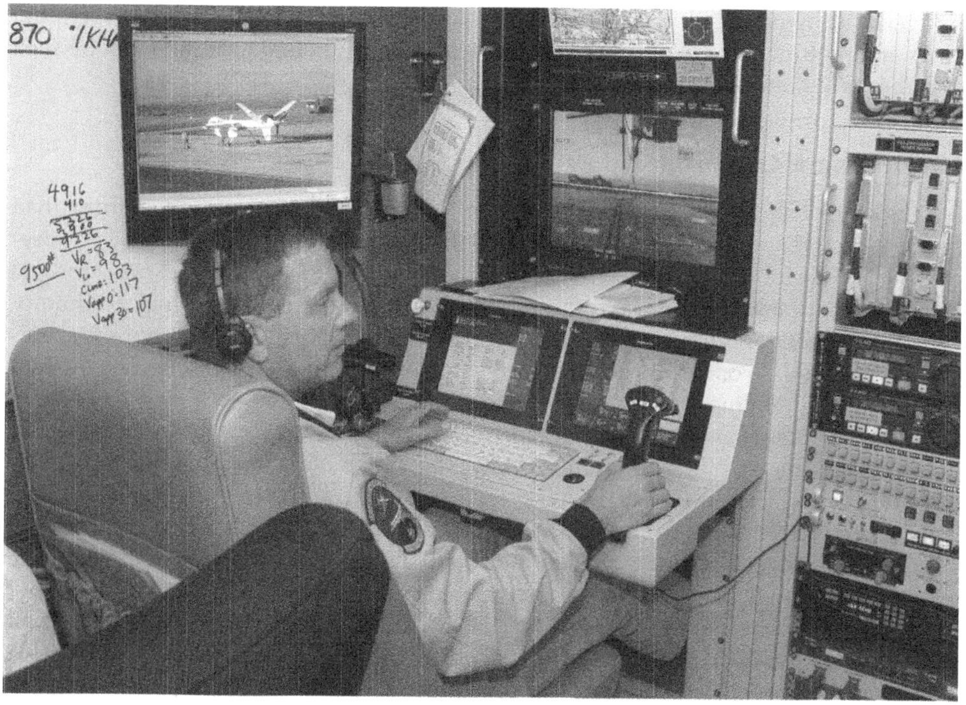

Figure 11.9 Ikhana pilot workstation.

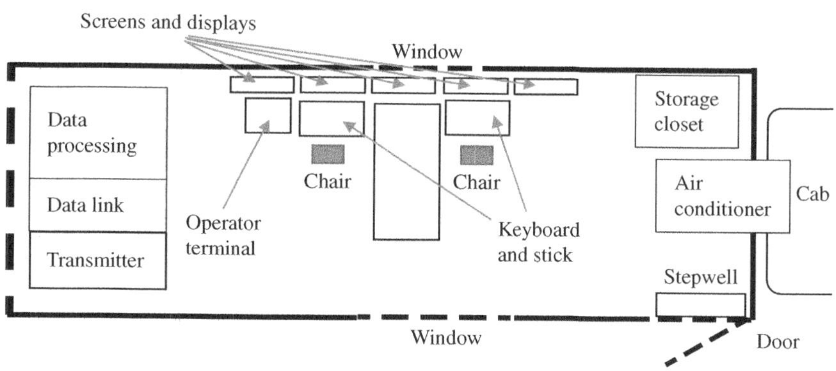

Figure 11.10 Plan view of interior layout of a ground control station (GCS) truck for unmanned aerial vehicle (UAV).

one door, two windows, and an air conditioner (and heater). Two individuals can sit on their own seats and move freely, and can look at the displays and screen. The stick of the pilot station is for deflecting control surfaces and engine throttle of the UAV. However, the stick of the payload operator station is to control the operation and attitude of payloads such as cameras, and radar.

Mechanically operated external antennas, such as weather radar antennas, feature gears and motors may fail due to erosion. Their installation and location must be such that the corrosion is minimized.

11.3.4 Central Command Station

For the case of an UAV (e.g., Global Hawk and Predator) with a long range (more than 10 000 km), and a very long endurance (more than 10 hours), the GCS must be comprehensive and be located in a fixed location. There are generally additional missions and associated payloads beyond the common types; only a fixed-base GCS may accommodate most potential types with a single GCS. A fixed GCS will provide more spacious accommodation, so operators are provided with comfort for a longer duty time. A fixed base GCS is a building with complete office and life equipment. It will accommodate several flight crew members, payload operators, and commanders. A challenging part in the design of a very long endurance UAV is the design of its fixed-based GCS.

A long-range UAV will usually take off horizontally on wheels along a runway. Provision has to be made to control the aircraft during its take-off and lift-off. This and subsequent recovery, is accomplished under direct operator control, usually with the aircraft in direct view of the operator. Additional crew members may be required, especially for the more complex payloads carried. Moreover, there are more crew involved the control of the UAV and conducting the entire mission. Most modern UAVs have a mission payload operator workstation. This position provides an operator with the ability to command payload functions and utilize the payload data. If the UAS is equipped with an EO/IR (Electro-Optic/Infra-Red) ball or perhaps a Synthetic aperture radar (SAR), these payloads are controlled and managed from the baseline workstation.

Long-ranging radio equipment is employed for wider networking. Crews will operate in shifts during the long flight-times of the UAV. Provision will therefore be made for handing over responsibility for the mission. If the aircraft carries armament, a further crew member, the Weapons Systems Operator, is required to select, monitor, release, and guide the weapons onto the target. Unmanned aerial systems such as Predator and Global Hawk have the option to launch the UAV from a GCS on an airfield relatively close to the target area, but, after launch, be controlled from a command center which may be a few thousand kilometers away. The Global Hawk system employs two ground-based GCSs. They are the launch and recovery element, known as the satellite control station, and a mission control element.

The following items will be held in a fixed GCS: operating manual, maintenance manual, consumable items, repair tools, spare parts, and special test equipment. Figure 11.11 illustrates the Global Hawk (Figure 1.2) Operations Center at National Administration for Aeronautics and Astronautics (NASA) Armstrong. In 2010, flight crew and scientists occupied the GCS for the Genesis and Rapid Intensification Processes hurricane study.

A few recent examples in restructuring and upgrading the GCSs are for the fixed-wing wing Global Hawk and the rotary-wing MQ-8 Fire Scout. In 2016, there was a plan to incorporate new payload controls for the Global Hawk UAV,[2] and to move GCSs from temporary building to permanent structures at Beale AFB (Air Force Base) in California and Grand Forks AFB in North Dakota. Another plan was to build the control station for the USN's Northrop MQ-8B Fire Scout, which recently deployed on

Figure 11.11 Global Hawk operations centeroc at National Administration for Aeronautics and Astronautics (NASA) Armstrong.

board the USS Coronado with controls that are common with the open architecture to be incorporated into the RQ-4's system.

11.3.5 Sea Control Station

When the flight operation target area is over the sea or close to a sea, a sea-based control station (or simply sea control station [SCS]) may be employed. Hence, naval operations may be conducted either using a UAV aboard a ship, or by taking over control of the UAV where initially flown/launched from a land base. Both fixed-wing UAV (e.g., Scan Eagle [Figure 3.3]), and rotary wing one (e.g., Fire-scout, Yamaha RMAX, and Sea Eagle) may be controlled by a SCS. Due to the limited length of a ship, a fixed-wing UAV should be launched and recovered from the ship. In a SCS, the required energy will be supplied by the ship's power center. There are many similarities between layout and equipment of a GCS and a SCS.

The communication system of a SCS may be incorporated with the ship communication system. This provision will result in an overall cost-effective communication system. In contrast, the UAV control may be completely or partially integrated within the ship's control center. However, it is recommended to keep them separate for the safety reasons.

11.3.6 General GCS

With the GCS of the Global Hawk unmanned system, only this UAV can be controlled, and only the data of its sensors can be processed. Similarly, the Predator GCS can only

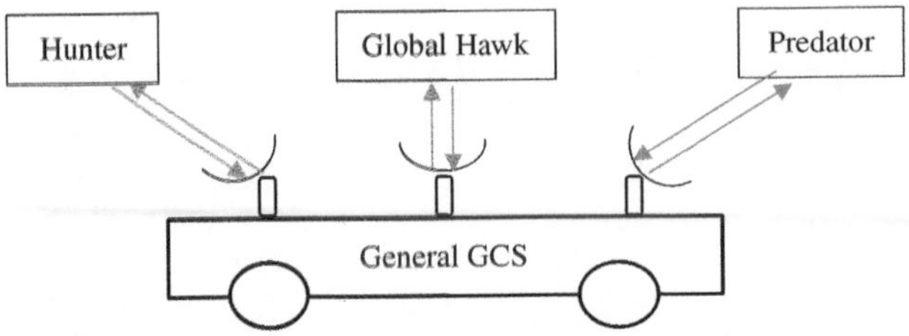

Figure 11.12 A general GCS to control various unmanned aerial vehicles (UAVs).

control one Predator UAV at a time and monitor its payloads. This is true for all current UASs, since each UAV use its own proprietary GCS systems. However, most large UAVs in the US have a common user, the government. It is desired to reduce the overall cost, and to increase the efficiency of UASs by defining a standardized, government-owned, and managed modular open ground control systems architecture applicable to government owned UASs.

The idea for a common GCS for controlling multiple types of UAVs was launched in 2008. The plan is to cover all UASs above 20 pounds maximum takeoff weight (referred to as UAS groups 2–5 by Department of Defense [DoD]) from MQ-8 Fire Scout unmanned helicopter to high altitude long-range Global Hawk (Figure 11.12).

The new GCS will be a platform for rapid integration of various databases and sharing information gathered by all types of UASs. Commercial manufacturers are developing common ground stations with standard software modules to control various unmanned aircraft. Moreover, the information gathered by various UAVs payloads. This initiative is aimed at designing better tools for visualization, data archiving, auto-tracking, or other applications to improve functionality of the ground stations. With a common GCS, command users are able to task the payload operators for raw images or video on demand.

11.4 GCS of a Number of UAVs

In this section, the general features of GCS of a number of UAVs are presented. The familiarity of a UAV designer to the current GCS technologies and their characteristics is a necessity for designing an efficient GCS. The characteristics of GCSs of the following UASs are briefly discussed: (a) Global Hawk; (b) Predator; (c) Hunter; (d) Shadow 200, and (e) Yamaha RMAX Unmanned Helicopter.

11.4.1 Global Hawk

The mission, performance, technical features, and payload of Global Hawk UAV is presented in Chapter 1. The Global Hawk is a large vehicle with a multi-element GCS, in a highly automated UAV system. A Global Hawk has three pilots in the GCS to remotely control the UAV; one for launch and recovery, one for mission control, and one as the sensor operator.

The building-based ground control station at NASA Dryden has a large payload operations area. Each GCS workstation is equipped with the following equipment: (a) map tracker display; (b) head-up display; (c) head-down display; (d) auxiliary displays; (e) common center display; (f) crew station console; (g) payload Operator Hand Controller; (h) electronically adjustable rudder/brake pedals; (i) digital flight recorders; and (j) intercom/radio system. The Global Hawk pilot/payload operator workstation is equipped with 11 displays. The plan by Northrop Grumman for the flight operations was to have five personnel in the GCS: (a) primary pilot; (b) second pilot (instructor or next shift); (c) flight operations manager; (d) payload manager; and (e) GCS systems manager.

All portions of the flight mission, including take-off and landing are pre-planned and programmed before the flight. The basic task of the crew during the flight is simply to monitor the status of the aircraft and control the payload. The pre-planning makes flying the Global Hawk simple, but the mission planning process is challenging, and requires a great deal of time and effort to accomplish.

There is a 24 in. display between the pilot and payload operator which provides space for shared data such as weather and airspace. Moreover, electronically adjustable and heated aircrew seats, and electronically adjustable rudder pedals have been provided. These pieces of equipment and their layouts are based on ergonomics standards to improve efficiency. The Raytheon-made GCS receives the high-quality imagery obtained by the air vehicle sensor suite, and then forwards the imagery to military commanders and users in the field.

Another axillary vehicle in flying the Global Hawk is a chase car as a safety measure used during takeoff and landing phases in order to give pilots (in GCS) assistance. There is a highly trained pilot in that car to act as ground-based wingman for the remote pilot talking them through runway operations. This is particularly useful since the size of the aircraft (wingspan of 131 ft) is such that there is a risk of hitting any ground obstacle with a wingtip during taxi.

The mission control ground station (made by Raytheon) includes a shelter measuring 8 ft × 8 ft × 24 ft housing the communications, command, and control, mission planning and image processing computers with four workstations for the mission control staff and officers. The Raytheon launch and recovery ground station is also housed in an 8 ft × 8 ft × 10 ft shelter equipped with two workstations and the launch and recovery mission computers.

11.4.2 Predator

The mission, performance, technical features, and payload of Predator UAV is presented in Chapter 1. The Predator UAS includes a package of four vehicles, a GCS, communications equipment, maintenance spares, and personnel. The Predator aircraft is flown from within the GCS, similarly to a manned aircraft, using a joystick and rudder pedals and a forward-looking camera that provides the pilot with a 30° field of view. The camera (infrared and night-vision cameras) is used for both takeoffs and landings.

Each Predator is operated remotely by a team of three, a pilot, a sensor operator, and a mission intelligence coordinator. The pilot is controlling the flight, while the sensor operator monitors the performance of the various sensors/payloads.

The Predators are deployed in groups of four. Each Predator is controlled by its own pilot and operator located at a GCS. The teams are actually able to switch control of the vehicle midflight. For instance, a team at an airbase in Afghanistan may control a take-off from its base, but then hand it over to a GCS in the US.

The Predator GCS is housed in a commercially available trailer or a light weight truck. The GCS incorporates five main elements: (a) a power supply; (b) an environmental control system; (c) pilot and payload operator workstation; (d) data exploitation, mission planning, and communication terminals; and (e) SAR workstation. The Predator has no onboard recording capability, so all imagery recording is located within the GCS. The ground elements for the communication link between the GCS and Predator includes: (a) encoders; (b) transmitter; (c) receiver; and (d) controls for its flight mission.

The control of the air vehicle and the sensors, data exploitation, mission planning, mission tracking, and payload monitoring, and system management are implemented via GCS. There is a plan for Predator B Reaper to redesign the GCSs with user-friendly video game-like controllers and touchscreen maps to access data without overwhelming operators. The US Air Force's new Block 50 GCS – developed by General Atomic Aeronautical Systems – for the first time controlled an MQ-9 Reaper on January 8th 2019 from flight operation facility near Palmdale, California.

11.4.3 MQ-5A Hunter

The mission, performance, technical features, and payload of MQ-5 Hunter UAV is presented in Chapter 1. The MQ-5A Hunter GCS has three control bays and an optional intelligence bay: (a) the operator control bay is used to control the flight of the UAV; (b) the observer control bay is used to control the payload functions; (c) the navigation control bay is equipped with a digital map display which traces the flight path and monitors the progress of the mission; and (d) the intelligence bay provides data processing and distribution capabilities. Two crew in the GCS, are responsible for tracking, commanding, controlling, and communicating with the air vehicle and its payloads.

An external pilot controls the Hunter takeoff and landing while standing next to the runway in visual contact with the aircraft, and remotely commanding the UAV via a control box (similar to ones used by radio-controlled airplane hobbyists). After takeoff and initial climb, control of the aircraft is transferred to another (i.e., internal) pilot, operating from a ground station. The internal pilot located within the GCS controls the Hunter by selecting the altitude, airspeed, and heading, using a set of knobs and sticks. For landing, control of the aircraft is transferred back from the GCS to the external pilot who has visual contact. Having two pilots (one internal and one external) provides more reliability and safety to the Hunter mission.

By one GCS, one air vehicle or two air vehicles in relay can be controlled. A unique mission planner system of this reconnaissance UAV provides flexible automated tactical mission planning, and also access to digital terrain elevation data, digital map data, and data from the Defense Mapping Agency.

11.4.4 Shadow 200

Two operators inside a GCS control the Shadow 200 in flight and handle its sensors and payloads. The Shadow 200 GCS is mounted on a light four-wheel truck (Figure 11.13). Each GCS has an associated ground data terminal, which takes commands generated by the GCS and modulates them into radio signals and transmits them to the Shadow. The ground data terminal also receives video imagery from the Shadow cameras, as well as telemetry from the Shadow transmitter, and sends the received data to the GCS.

Figure 11.13 GCS of a Shadow 200.

The Shadow 200 system also includes a portable GCS and portable ground data terminal, which are stripped-down versions of the basic ones designed as a backup to the two GCSs. A fielded Shadow 200 system requires 22 operators/soldiers to operate it. A trailer, towed by a support vehicle, carries the ground data terminal and houses a 10 kW generator to provide power for its associated GCS.

11.4.5 DJI Phantom

The small quadcopter DJI Phantom [147] is remotely controlled by a ground pilot via a remote controller. The Maximum transmission distance is 3.5 km, the operating frequency is 2.4 – 2.483 GHz, and the transmitter power is FCC (Federal Communications Commission) of 26 dBm. The operating current/Voltage is 7.2 A at 7.4 V with 6000 mAh Li–Po 2S battery. Video Output Port is GL300E: HDMI and GL300C: USB. The rated power of the charge is 100 W. It has a built-in display device (5.5 in. screen, 1920 × 1080, 1000 cd/m^2, Android system, 4 GB RAM+16 GB ROM).

11.4.6 Yamaha RMAX Unmanned Helicopter

The GCS of Yamaha RMAX has one internal operator (for non-LOS (Line-Of-Sight) control and observation), and one external operator (for LOS control). In the GCS, three personal computers, and three displays are provided for operators to fly the helicopter. On the GCS, a Global Positioning System (GPS), three personal computers and a communication modem [148] for the base station are used to send the correction data to the GPS on the UAV.

The flight data are viewed by the operator on a four-window split monitor. The helicopter status display provides the operator with the performance data of the vehicle during the flight, such as altitude, roll, pitch angles, speed, and rate of climb. Moreover, this display presents health data of the helicopter such as, engine speed, engine temperature, battery condition, and radio wave condition. Also, a 3-dimensional picture of the helicopter is illustrated in the display. The position of the helicopter is presented on the map on a flight control display. The image display shows images from four cameras on the helicopter, where they are viewed on a four-window split screen.

11.5 Human-Related Design Requirements

11.5.1 Number of Pilots/Operators in Ground Station

One of the most important aspects of operating an UAV is safety, which relies heavily on remote pilots. Operators in GCS directly control the UAV and its payloads with communication signals. Two of the most common UAV personnel roles are (a) air vehicle flight operator (or controller, or ground pilot); and (b) mission payload operator. Other functions that may be performed in GCS are: (a) mission commander; (b) communication operator; (c) intelligent specialist; and (d) weather observer. Most UAV losses are attributed to remote pilot/operator errors; part of which is related to ineffective interface design. Regarding the human operator, some possible deficiencies of a weak GCS design are: (a) GCS creates excessive workload; (b) GCS does not effectively present critical information; (c) GCS does not facilitate emergency procedures; and (d) uncomfortable situation inside the GCS generates operator stress.

Table 11.1 presents number of operators in GCS for a number of current UAVs. When only one operator is needed (e.g., a traditional quadcopter), it often implies that the GCS is in the hand of the operator (i.e., handheld box).

11.5.2 Ergonomics

Ergonomics (or human factors) is the science of designing user interaction with equipment and workplaces to fit the user. The field of human factors engineering uses scientific knowledge about human behavior in specifying the design and use of a human-machine system. The aim is to improve system efficiency by minimizing human error and optimize performance, comfort, and safety. The operator work environment should be comfortable; and the interface must be effective; otherwise there will be operative fatigue and probable UAV losses. Whenever an engineering device which deals with humans is planned to be designed, the ergonomic standards need to be

Table 11.1 Number of human operators in ground control station (GCS) of a number of current active unmanned aerial vehicles (UAVs).

No	UAV	Number of operators in GCS	Remarks
1	RQ-4 Global Hawk	3-5	A flight pilot, a payload/sensors operator, one for launch and recovery, and one for mission control
2	MQ-1B Predator A	3	A pilot, a payload operator, and an operation manager
3	MQ-9 Reaper/Predator B	2	A pilot and a sensor operator
4	General Atomics Avenger	2	
5	RQ-5 Hunter	3	Including one out of GCS – on the runway
6	Northrop Grumman Firebird	2	
7	Shadow 200	2	GCS on a Humvee
8	Lockheed Martin RQ-170 Sentinel	3	one for launch/recovery, one for mission control, and one as the sensors operator
9	Pioneer	2	A pilot in GCS and an external pilot for takeoff and landing
10	Quadcopter	1	With four electric engines
11	Yamaha RMAX	1	Helicopter
12	Fixed-wing RC airplane	1	Mainly used for hobby
13	Arcturus T-20	2	A pilot and a payload operator

considered. In this section, a few related data are reproduced which will be used in the design of GCS crew station.

The International Ergonomics Association defines ergonomics as follows: "Ergonomics (or human factors) is the scientific discipline concerned with the understanding of interactions among humans and other elements of a system, and the profession that applies theory, principles, data, and methods to design in order to optimize human well-being and overall system performance". References [47, 68] are helpful resources in the introduction, principles, fundamentals, and useful data for various aspects of human factors.

Because of the health risks posed by poor posture and repetitive stress, proper ergonomics are too important to ignore. Whether it is a pilot or operator, the seat must be designed such that the seater stays healthy from head to toe when he/she is sitting. The following contains ergonomic advice for healthy seating.

MIL-STD-1472G [69] establishes general human engineering criteria for design and development of military systems, equipment, and facilities. Its purpose is to present human engineering design criteria, principles, and practices to be applied in the design of systems, equipment, and facilities. The design objective should be to achieve the required performance by operator, control, and maintenance personnel, and to minimize skill and personnel requirements and training time.

Matching of working and living areas to human characteristics is one of the primary tasks of those specializing in human factors. Some of the basic characteristics of human beings are those associated with size and shape and the movements of various parts of the body. Such data is applied in several locations on board a GCS. On the remote pilot workstation and payload operator workstation, the basic data are used in the basic geometry, in the provision of adequate inside visibility, in the location and design of controls, and in seat and doors design. Furthermore, design of much equipment on board and maintenance areas take into account human dimensions to assure sufficient access to equipment and working space.

11.5.3 Features of a Human Pilot/Operator

The UAV and its GCS may be manufactured for and sold to the users in other countries in five continents. A wide difference in the physical dimensions of people can be expected between ethnic groups. Persons of African descent are relatively long-legged compared with white Caucasians. Asians, on the other hand, are smaller overall but have relatively long trunk and short legs compared with Europeans. In addition to differences between ethnic groups there are also differences between men and women within one ethnic group. In this section, a few important advices for optimizing the GCS are provided.

Proper ergonomic design is necessary to prevent repetitive strain injuries, which can develop over time and can lead to long-term disability. Human factors must be considered in deriving design requirements. Reference [47] is a helpful resource in introduction, principles, fundamentals, and useful data for various aspects of human factors. For more details, Ref. [149] is recommended which reviews human factors for command and control of unmanned air vehicles.

The human has various limitations (power, size, tolerance). Some noticeable ones are: (a) needs rest; (b) is not very accurate; (c) may get sick; (d) cannot handle a load more than 12 "g"s; and (e) Cannot work when pressure is beyond 0.75–1 atm. In contrast, a human; (a) is able to react to unexpected situations; and (b) have an intuitive feel for an aircraft.

The operator within GCS must be able to access all relative information when needed. In the meantime, they should not be overwhelmed with the high volume of data. Moreover, the temperature and humidity at the GCS must be controlled to provide maximum comfort and effectiveness. In addition, the number of simultaneous activities should be less than or equal to a normal human capability. Thus, the number of sticks, buttons, and monitors/displays should be limited. The seating environment and the seat/table relation must be designed based on the ergonomic fundamentals. The software/algorithm should be designed so that the interface is menu-driven, to allow the operator to find the necessary information in perhaps one to three mouse clicks. When there is more than one human operator inside the GCS, the tasks of each operator must be clearly defined.

A universal population survey a few years ago indicated a general increase in height over half a century of 1.3 mm a year for males and 0.9 mm for females. The most comprehensive source of such data is a three-volume edition published by the Department of Defense [76] which covers various groups. Photographic techniques have been developed to measure and collect data from a representative sample of people who are to use equipment. However, a caution necessary in applying data concerning physical dimensions is that they are slowly changing.

In the USA, the Federal Aviation Administration (FAA) regulations currently require that manned transport category aircraft be designed for operating by crew members of physical height ranging from 5 ft 2 in. (157 cm) to 6 ft 3 in. (190 cm). This criterion can also be applied to the design of GCS.

The recommendation is to accommodate 90% of the particular population, which results in excluding the top and bottom 5%. This is often referred to as the design for the 5th–95th percentile. Accommodating a very large percentage of the population could be very expensive. Since this could involve a disproportionate penalty in trying to accommodate extremes when designing a seat or equipment. Therefore, there will be some people who will be too wide to enter a GCS door or use an emergency escape hatch, or too short to see the monitors, or too tall to avoid striking their heads when entering the GCS door. The decision on how many are to be included in this disadvantaged minority is a fundamental one in the design process. Human dimensions should be employed in the seat design, and console design. The goal should be to cover the dimensions of 90% of the human pilots.

11.5.4 Console Dimensions and Limits

The GCS of a medium or large UAV often house a minimum of two personnel: (a) pilot; and (b) payload operator. Each of these two individuals has their own desk and console. For general console design guidance, see dimensions for the standing body, seated body, depth, and breadth, circumferences and surfaces, hands and feet, and head and face in MIL-HDBK-759. A console design should accommodate the desired visibility over the top of the console, operator mobility (e.g., sit, stand, or sit-stand), panel space, and volume in the area below the working surface. Moreover, the number of visual displays in a console is restricted by over-burdening; ambient light variability; operator mobility; degradation of vision by vibration, or other environmental considerations.

Incorrect monitor positioning can cause neck and eye strain, and can lead to poor seat positioning, which creates pressure on the back. The top of the monitor should be positioned just above the eye level when seated. This is the best place for the "vision cone," the most immediate field of vision, which starts at the top at one's eye level and descends at a 30° angle. When monitors are too far away, people tend to lean forward to see well. This is increasingly true as people age, since vision almost inevitably declines over time. As a rule of thumb, if a person can extend his/her arm and just touch the screen with the fingertips, then he/she is in the right position.

To keep wrists and arms at an optimum position, reducing the risk of repetitive-motion injuries, the stick/yoke and switches should be at the same level as the elbows when seated. Since not everybody has a standard size, a simple fix is an adjustable seat. Sitting properly takes 20–30% of the pressure off the lower back. The seat should be 17–19 in. deep, and it should have good lower-back support. The body should be positioned with the back against the seat and the hips open. If one finds himself/herself leaning forward to see the panel, he/she needs to move the seat forward.

Leg positioning contributes to the overall position in the seat, so make sure the legs are bent at about 90° angles at the knees. This helps alleviate pressure on the back. Movement is essential for circulation, however, so allow for subtle shifts in positioning

Table 11.2 Suggested values for the geometry of a pilot/operator console.

No	Key	Distance/Height	Dimension (cm)
1	A	Maximum total console height from standing surface	Varies
2	B	Vertical dimension of panel, including sills	Varies
3	C	Height of the console surface from standing surface	Varies
4	D	Seat height from standing surface	Varies
5	E	Minimum knee clearance	46
6	F	Foot support to sitting surface	46
7	G	Seat adjustability	15
8	H	Minimum thigh clearance at midpoint of G	19
9	I	Console surface depth including shelf	40
10	J	Minimum shelf depth	10
11	K	Eye line-to-console front distance	40

and be sure to stand, stretch and walk a few steps at least every few hours. Feet should be firmly planted on the floor. If the seat positioning required for proper wrist alignment results in the feet not reaching the floor, use some types of footrest to support the feet, such that the height of the support keeps the knees at a right angle.

Suggested values for the geometry of a pilot/operator console is shown in Table 11.2. Various distances and heights for the console are illustrated in Figure 11.14. The values of both A and B vary in function of dimensions and number of screens. Moreover, C and D are functions of human dimensions. The thickness of the shelf is assumed to be 11 in (25 mm). For other thicknesses, suitable adjustment should be applied. Since F is fixed, a heel catch must be added to the chair, if D exceeds 460 mm. Figure 11.15 illustrates the

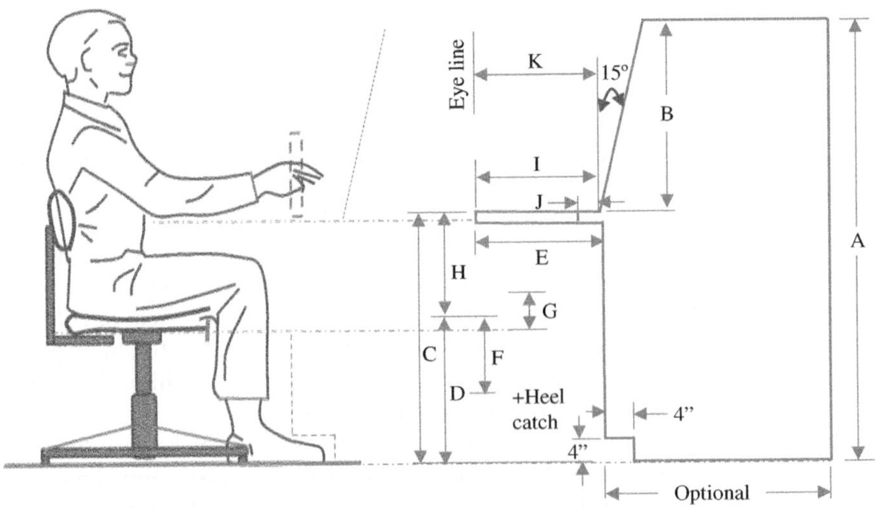

Figure 11.14 Standard console dimensions.

Figure 11.15 Reaper ground control station (GCS). *Source:* Photograph by Master Sergeant Steve Horton U.S. Air Force.

Reaper GCS. Applications of a few ergonomic recommendations are observed in this GCS.

The depth of the console is optional; however, the shorter is the better to allocate all equipment inside the GCS. Some equipment and devices may be placed inside console to save space. To illustrate various distances and heights, the human operator has been moved (separated) to the left of the console.

11.6 Support Equipment

11.6.1 Introduction

Two primary elements of an UAS are: (a) UAV; and (b) GCS. However, for a large UAV with a challenging flight mission, there are various required elements which are classified as support equipment. For instance: (a) transportation equipment; (b) electric power generator; (c) HVAC system; (d) ground data terminal; (e) satellite communication ground data terminal; (f) launch and recovery system; and (g) ground support equipment, are typical support equipment for a medium weight UAV with a significant mission. This section is dedicated to the introduction of support equipment for a UAS.

Figure 11.16 illustrates a Phoenix UAS (i.e., troop) which includes 5 Vehicles, 11 Personnel, and 1 Air Vehicle. Its ground equipment per Platoon includes: two GCSs, two ground data terminals, one satellite communication ground data terminal, one portable GCS, one portable ground data terminal, an automated takeoff and landing system, two tactical automatic landing systems, and ground support equipment.

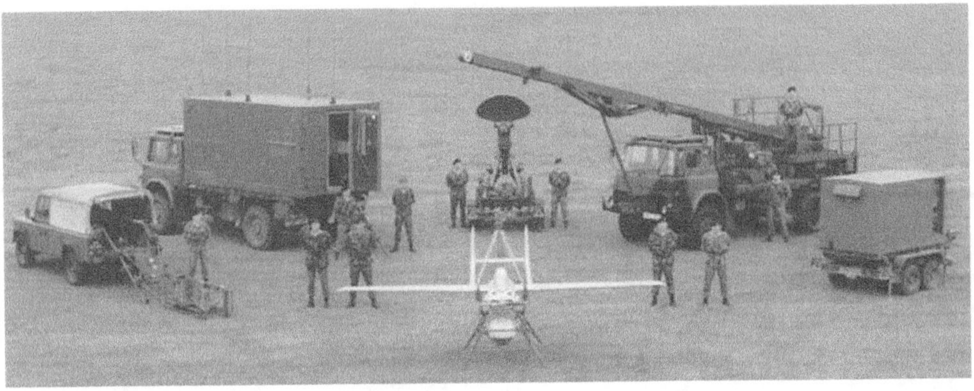

Figure 11.16 Phoenix unmanned aerial system (UAS).

11.6.2 Transportation Equipment

A GCS should be designed such that it can be prepared for transportation, if needed. A small handheld GCS is lightweight, so it can be easily carried by the pilot in command. However, a large GCS is very heavy, and needs another vehicle to move it to the center of flight mission. Transportation is the process of moving an item from point A to point B. For a large UAV such as Global Hawk, a complete transportation system is required as a facility consisting of the means and equipment necessary for the movement GCS and other support equipment. In designing a UAS with a large UAV, the designer should be aware of the transportation system challenges and incorporate them into the design process of the GCS.

The Shadow 200 UAV is part of a larger system which currently uses a number of Humvees for ground transport of all ground and air equipment. The RQ-7 Shadow 200 system has an air vehicle transport truck and a personnel carrier. A typical Shadow UAS includes three air vehicles (plus one spare air vehicle) with payloads, two GCSs, portable GCS, air vehicle transport truck, launch and recovery equipment and personnel carrier.

A Shadow 200 UAS consists of four air vehicles, three of which are transported in a special transporter. The fourth is transported in a specially designed storage container to be used as a spare. The transporter also tows the launcher. The transporter and trailer contain extra equipment to launch and recover the aircraft, such as the automatic landing system. Maintenance part and equipment for the aircraft is stored in another vehicle and its trailer.

Each GCS –mounted on a Humvee – has an associated Ground Data Terminal, which takes commands generated by the GCS and modulates them into radio waves and sends them through the transmitter to the UAV. The Ground Data Terminal receives video imagery from the UAV sensors, as well as telemetry from the aircraft, and sends this information to the GCS. The complete system can be transported to a target area in two C-130 aircraft.

The complete GCS and the launch and recovery element of the Global Hawk are transportable in a single load on the Lockheed C-5 Cargo aircraft, and in less than two loads on the Boeing C-17 aircraft.

11.6.3 Power Generator

Any GCS requires a power source as energy is consumed by various GCS equipment for the duration of the UAV flight operation. Small GCSs such as handheld or portable ones use a pack of batteries. However, a GCS on a truck or a trailer consumes a large amount of electric energy so that the truck engine is hardly able to generate adequate power. When the power required by the GCS is beyond the capacity of the truck engine, an independent power generator is necessary. A UAV designer is not expected to design such generator; it can be easily ordered from a commercial manufacturer. The primary information which is expected to be provided by a GCS designer is the amount of power required.

A power generator out of a residential area – which is usually the case for an UAV application – needs to be independent. This equipment will frequently consume fossil fuel and originally generates esmechanical power. However, a GCS needs an electric power. Hence the mechanical power of the generator must be converted to an electric power by a device, usually an alternator.

Due to the large size of such a power generator and the mobility requirement, a dedicated truck or trailer is required, too. An example is the Tenkel TK10355 system, which includes a power generator and a trailer with ancillary equipment. This power generator consumes fossil fuel and is capable of delivering 18 kW of electric power. Its four-cylinder diesel engine is of 30 kW of mechanical power at 1500 rpm, thus delivering 220 V at 50 Hz (electric power) through an alternator (a nominal power of 25 kVA). The Shadow 200 UAV trailer, towed by the GCS support vehicle, carries the GDT (Ground Data Terminal) and houses a 10 kW power generator.

For mini to small UAVs and quadcopters, a wall charger will allow the recharging of the batteries by plugging it to the power socket. For instance, the Contixo F18 quadcopter is equipped with a pack of 2100 mAh Li-polymer batteries. It takes about five hours to recharge the batteries. The flying time is approximately 18 minutes.

11.6.4 HVAC System

A remote pilot or a payload operator in a GCS must have a desired work place to have an efficient and optimal performance. The human comfort zone has been defined by ergonomics in terms of temperature, humidity, and air quality. In a comfort zone, an operator is at ease and in control of his/her environment, experiencing low levels of stress. This work situation will result in an enhanced level of concentration and focus. An important factor in the comfort zone is the temperature and humidity of the work place. Due to the change in the season, the atmospheric temperature and humidity are constantly changing. To provide a reasonable temperature for the pilot, a GCS must be equipped with a package of HVAC (Heating, Ventilation, and Air Conditioning) system to decrease/increase the temperature/humidity inside a GCS if needed. The design of the HVAC system is beyond the scope of this text, the interested reader can refer to references such as [150].

11.6.5 Other Items

The list of support equipment is long; here a list of the most essential ones is presented: (a) repair parts; (b) tools; (c) consumable items; (d) operating manual; and (e) maintenance manual. A UAS designer must incorporate the support equipment design requirements in the UAS design plan.

11.7 GCS Design Guidelines

In the preceding sections, various types of GCS have been described. In designing GCS for a UAV, one must select the type of GCS, select the measurement devices, number of operators, and then conduct some calculations and analysis. In general, the primary criteria for the design of GCS are as follows: (a) manufacturing technology; (b) required accuracy; (c) mission; (d) weather; (e) reliability; (f) life-cycle cost; (g) UAV configuration; (h) human factors; (i) maintainability; (j) UAV endurance; (k) communication system; (l) weight; (m) level of control; (n) mission control location; (o) mobility; (p) Operator work load, (q) Modularity and interface, (r) Operational availability, (s) Network infrastructure, and (t) Information displays.

Figure 11.17 illustrates a flowchart that represents the GCS design process with a number of feedbacks. In general, the design process begins with a trade-off study to establish a clear line between cost and performance (i.e., accuracy) requirements; and ends with optimization. The designer must decide about two items: (a) select the type of GCS; and (b) select the desired equipment. After conducting the calculation process, it must be checked to make sure that the design requirements are met. Otherwise, deficiency feedbacks are used in design iteration.

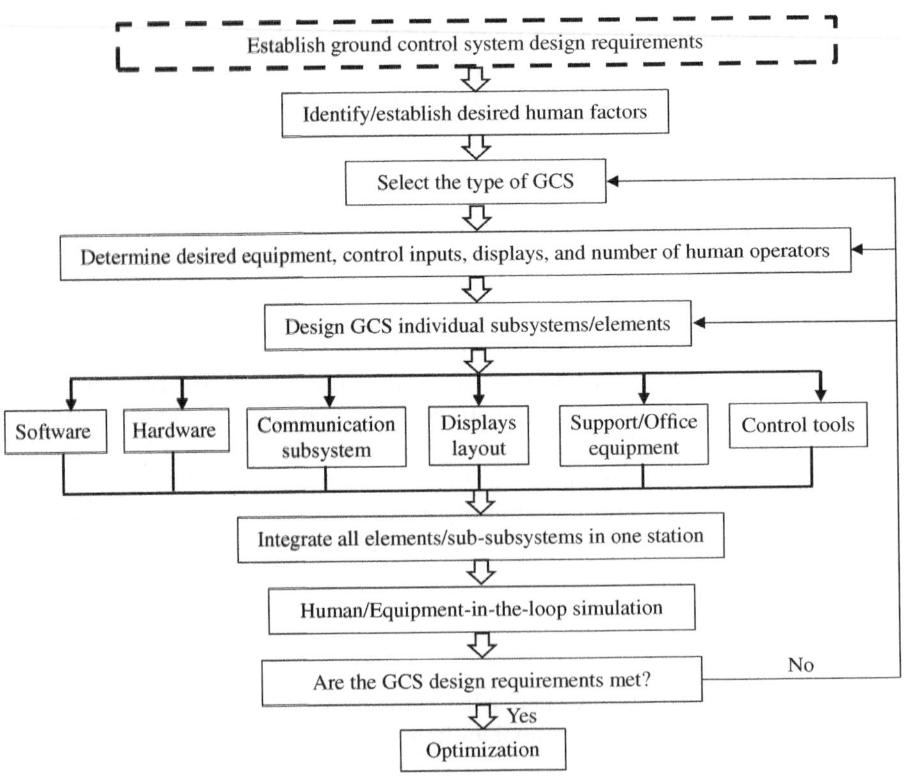

Figure 11.17 Ground control station (GCS) design process.

A crucial part of the design process [151] is to integrate the GCS with the air vehicle. If all equipment/displays are selected/purchased, the integration process must be still conducted. An increase in modularity and interface definition will help in overcoming diminishing manufacturing sources. Moreover, improving line-replacement unit access will reduce maintenance down time and will increase operational availability. The comfort level and effectiveness of the operators (ergonomic indicators) must be determined in order to make sure that the mishap and error is minimal.

Since the GCS is the arena for the interaction between human (i.e., flight operator/controller) and a machine (i.e., UAV), the fundamentals of ergonomics and human efficiency must be observed in the design layout.

In the design of a GCS, human-machine interfaces should be designed such that the workplace efficacy is achieved. The minimization of the cognitive load is highly emphasized to provide operator reaction speed and accuracy. Improved visibility of mission situations can be improved by using advanced graphical techniques, while enables the fast identification of targets. The human interface will be more efficient by employing a touch-screen and multi-screen system for fast reaction and presentation of necessary information.

There should be various displays showing the payload status and data (e.g., imagery/video) with recording media. A larger screen can be used to maximize the information presentation. It is appropriate to place the screens up and down in order to reduce the fatigue of the pilots and payload operators. Moreover, the individual convenience of the pilot/operator in terms of the configuration of screens can be implemented by the customization. Different information and flight data can be classified byusing various colors. Furthermore, a three-dimensional map is a better tool to express the details of the flight mission situation including terrain and target location.

For a rapid situation awareness for pilot and payload operator, the following GCS design directions ae recommended: (a) exhibit data through a number of symbols and geographical data; (b) use color coding for rapid recognition; (c) use simple, but effective symbols; (d) represent the target situation with ease of understanding; (e) allow the user to zoom in or zoom out of display; (f) display should be such in order to provide an enlarged view of a particular target area, while simultaneously exhibiting several display screens; (g) use 2D and 3D color representation to indicate various features of target and target area including moving targets, their dynamics, and terrain; (h) allow the user to set the style of representation to accommodate the varying situations; and (i) allow the operator to freely set in the screen in accordance with the flight mission.

Questions

1 What are the alternative terms equivalent to the GCS?

2 What are the functions of a GCS?

3 List GCS Subsystems in brief.

4 What is ergonomics?

5 What are a few outputs to operator when the GCS is not optimally designed?

6 Name the main types of GCS.

7 What is the typical weight of a handheld radio controller?

8 At what situation, does the designer need to select a fixed comprehensive GCS?

9 What are the features a mobile truck as the GCS?

10 Describe the features of a portable GCS.

11 What is the housekeeping data?

12 What factors do contribute in the range of LOS?

13 Which UAV was the first one to be controlled via satellite communications?

14 Compare a human pilot with an autopilot by providing: a. Three advantages of human pilot over an autopilot, b. Three advantages of an autopilot over a human pilot.

15 Describe the process of sending a command from a handheld control to a UAV.

16 Define transportation system.

17 How many personnel are there in the Global Hawk GCS?

18 Describe the process of a change in the rotational speed of a UAV motor from the GCS by pilot in moving a stick.

19 Define the aileron-stick gearing ratio.

20 Describe the functions of three leads for a potentiometer.

21 Define the function of a potentiometer.

22 What is the SCS?

23 List the main elements/modules of a typical portable GCS.

24 Describe the features for the GCS of the Global Hawk in brief.

25 Describe the features for the GCS of the Shadow 200 in brief.

26 Describe the features for the GCS of the Hunter UAV in brief.

27 Describe the features for the GCS of the Predator UAV in brief.

28 What are the most common UAV personnel roles in a GCS?

29 What reference does general human engineering establish the criteria for design and development of military systems, equipment, and facilities?

30 What are noticeable limitations for a typical human?

31 List typical support equipment for a large UAS.

32 List the titles of each personnel in the GCS of a RQ-4 Global Hawk.

33 List the titles of each personnel in the GCS of a MQ-1B Predator A.

34 List the titles of each personnel in the GCS of a Lockheed Martin RQ-170 Sentinel.

35 What military standard (MIL-STD) does establish the general human engineering criteria for design and development of military systems?

Problems

1 When a remote pilot at a GCS deflects a control stick with the value of 45°, the electric motor of the UAV will rotate at 2000 rpm. Determine the engine-stick proportionality constant (K_ω) for this UAS.

2 When a remote pilot at a GCS deflects a control stick with the value of 40°, the turbofan engine of the UAV will rotate at 15000 rpm. Determine the engine-stick proportionality constant (K_ω) for this UAS.

3 When a remote pilot at a GCS deflects a control stick with the value of 30°, the elevator of the UAV deflects 15°. Determine the elevator-stick gearing ratio (K_E) for this UAS.

4 When a remote pilot at a GCS deflects a control stick with the value of 40°, the rudder of the UAV deflects 30°. Determine the rudder-stick gearing ratio (K_R) for this UAS.

5 When a remote pilot at a GCS deflects a control stick with the value of −30°, the aileron of the UAV deflects −25°. Determine the aileron-stick gearing ratio (K_A) for this UAS.

6 A remote pilot at a GCS deflects an aileron control stick with the value of 45°. This control stick is connected to a potentiometer with a gain of 0.015 V/deg. The corresponding signal is transmitted to the UAV via communications system. A similar potentiometer is used in UAV for aileron control. Due to this signal, the UAV microcontroller applies 0.5 volts to the aileron potentiometer, determine the aileron deflection. Then calculate the aileron-stick gearing ratio.

7 A remote pilot at a GCS deflects an elevator control stick with the value of 25°. This control stick is connected to a potentiometer with a gain of 0.012 V/deg. The corresponding signal is transmitted to the UAV via communications system. A similar potentiometer is used in UAV for elevator control. Due to this signal, the UAV microcontroller applies 0.42 volts to the elevator potentiometer, determine the elevator deflection. Then calculate the elevator-stick gearing ratio.

8 A remote pilot at a GCS deflects a rudder control stick with the value of 30°. This control stick is connected to a potentiometer with a gain of 0.01 V/deg. The corresponding signal is transmitted to the UAV via communications system. A similar potentiometer is used in UAV for rudder control. Due to this signal, the UAV microcontroller applies 0.38 volts to the rudder potentiometer, determine the rudder deflection. Then calculate the rudder-stick gearing ratio.

9 A rotary potentiometer is programmed to provide 3.3 V for a full 360° rotation. What is the gain for this potentiometer?

10 A rotary potentiometer is programmed to provide 5 V for a 270° rotation. What is the gain for this potentiometer?

Design Problems

1 You are in the UAS conceptual design phase of a remote-controlled fixed-wing UAV for an aerial filming mission. The UAV is equipped with an electric engine and has the following features:

$$m_{TO} = 3\,kg,\ V_c = 20\,\text{knot},\ E = 20\,\text{minutes},\ R = 4\,km,\ \text{ceiling} = 500\,\text{ft}$$

The payload is just an optical day camera located under the fuselage nose. Design and layout a low cost, efficient GCS for this UAV. For this objective, develop the design requirements, and provide the following outputs: (a) GCS type; (b) approximate weight of the GCS; (c) list and overall features of required control equipment; (d) list of communications equipment; and (e) features of power supply.

2 You are in the UAS conceptual design phase of a remote-controlled quadcopter UAV for an aerial filming mission. The UAV is equipped with four electric engines and has the following features:

$$m_{TO} = 0.6\,kg,\ V_c = 10\,\text{knot},\ E = 15\,\text{minutes},\ R = 2\,km,\ \text{ceiling} = 200\,\text{ft}$$

The payload is an optical day camera and a range finder located under the fuselage nose. Design and layout a low cost, efficient GCS for this UAV. For this objective, develop the design requirements, and provide the following outputs: (a) GCS type; (b) approximate weight of the GCS; (c) list and overall features of required control equipment; (d) list of communications equipment; and (e) features of power supply.

3 You are in the UAS conceptual design phase of a remote-controlled fixed-wing UAV for a mission of aerial filming, and tracking ground moving targets. The UAV is equipped with an electric engine and has the following features:

$$m_{TO} = 5000\,kg,\ V_c = 120\,\text{knot},\ E = 5\,hr,\ R = 2000\,km, \text{ceiling} = 20000\,\text{ft}$$

The payload is an optical day camera, and an IR camera located under the fuselage nose. Design and layout a low cost, efficient GCS for this UAV. For this objective, develop the design requirements, and provide the following outputs: (a) GCS type; (b) approximate weight of the GCS; (c) list and overall features of required control equipment; (d) list of communications equipment; (e) features of power supply; and (f) plan view of interior layout of a GCS.

4 You are in the UAS conceptual design phase of a remote-controlled fixed-wing UAV for a mission of tracking hurricanes. The UAV is equipped with an electric engine and has the following features:

$$m_{TO} = 15000\,kg,\ V_c = 300\,\text{knot},\ E = 10\,\text{hour},\ R = 15000\,km, \text{ceiling} = 50000\,\text{ft}$$

The UAV integrated sensor suite includes: (a) the synthetic aperture radar (for both spot images, and wide-area search images); (b) the electro-optical and infrared sensor system; (c) an active electronically scanned array radar; and (d) ground movement target indicator. Design and layout a low cost, efficient GCS for this UAV. For this objective, develop the design requirements, and provide the following outputs: (a) GCS type; (b) approximate weight of the GCS; (c) list and overall features of required control equipment; (d) list of communications equipment; and (e) Features of power supply, and (f) Plan view of interior layout of a GCS.

Laboratory Experiments

1 Setup a circuit by using a servo, a potentiometer, and an Arduino Uno microcontroller to explore the elevator open loop control process by an operator. Use the wiring model shown in Figure 11.18, and write an Arduino code similar to the one demonstrated in Section 9.7.4. The requirement is to have −30°of elevator deflection (i.e., servo) by rotating the potentiometer for 40°. Then, write a report for this lab experiment, and include a copy of the Arduino code. The instructor will provide the necessary lab items to students. Figure 11.19 depicts the experiment.

2 Setup a circuit by using a servo, a potentiometer, and an Arduino Uno microcontroller to explore the elevator open loop control process by an operator. Use the wiring model shown in Figure 11.18, and write an Arduino code similar to the one demonstrated in Section 9.7.4. The requirement is to have −25°of elevator deflection (i.e., servo) by rotating the potentiometer for 45°. Then, write a report for this lab experiment, and include a copy of the Arduino code. The instructor will provide the necessary lab items to the students. Figure 11.19 depicts the experiment.

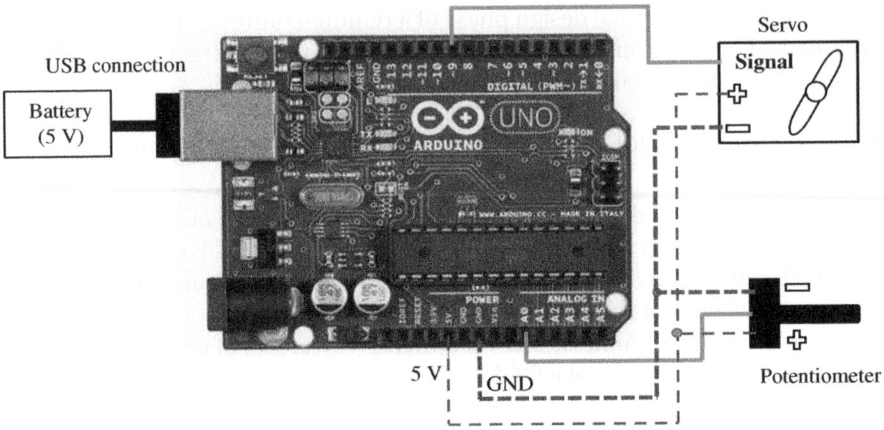

Figure 11.18 Circuit, wiring, and schematic of the servo control system.

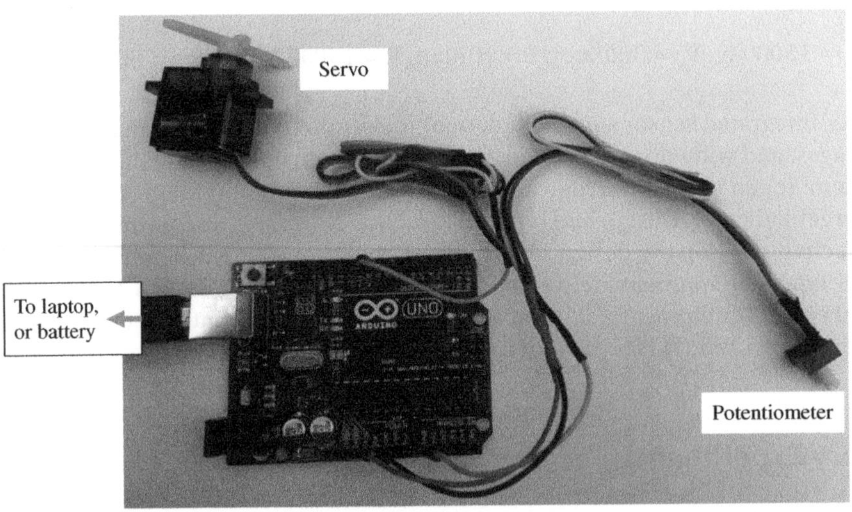

Figure 11.19 Circuit and wiring the experiment.

3 Setup a circuit by using two servos, two potentiometers, and an Arduino Uno microcontroller to explore the elevator and rudder open loop control process by an operator. Use the wiring model shown in figure 11.18 as a guide, and write an Arduino code similar to the one demonstrated in Chapter 9, Section 7.4. The requirement is to have ±25 degrees of elevator and rudder deflections (i.e., servo) by rotating the potentiometers for ±45 degrees. Then, write a report for this lab experiment, and include a copy of the Arduino code. The instructor will provide the necessary lab items to the students.

4 Setup a circuit by using two servos, one potentiometers, and an Arduino Uno microcontroller to explore the ailerons open loop control process by an operator. Use the wiring model shown in figure 11.18 as a guide, and write an Arduino code similar to the one demonstrated in Chapter 9, Section 7.4. The requirement is to have 25 degrees of both ailerons deflections (one +25 and one -25 degrees) by rotating the potentiometer for 45 degrees. Then, write a report for this lab experiment, and include a copy of the Arduino code. The instructor will provide the necessary lab items to the students.

5 Setup a circuit by using one potentiometer, one electric motor, and an Arduino Uno microcontroller to explore the motor open loop control process by an operator. Use the wiring model shown in figure 11.18 as a guide, and write an Arduino code similar to the one demonstrated in Chapter 9, Section 7.4. The requirement is to have 1000 rpm of electric motor by rotating the potentiometer for 45 degrees. Then, write a report for this lab experiment, and include a copy of the Arduino code. The instructor will provide the necessary lab items to the students.

Notes

1 Peripheral Interface Controller (also referred to as Programmable Intelligent Computer).
2 Flight Global, 24 August 2016.

12

Payloads Selection/Design

CONTENTS

Design of Unmanned Aerial Systems, First Edition. Mohammad H. Sadraey.

Companion website: www.wiley.com/go/sadraey/unmanned-aerial-systems

Educational Outcomes

After study of this chapter, the reader will be able to:

1) Discuss the current Unmanned Aerial Vehicle (UAV) payloads technologies
2) Explain various types of payloads for UAVs
3) Explain features of payloads for a number of current active UAVs
4) Conduct a payload selection project for a UAV flight mission
5) Explain features of various types of payloads
6) Explain functions and structures of various payloads
7) Identify various reconnaissance/surveillance payloads
8) Conduct an Electro-Optical\Infra-Red (EO/IR) camera selection process for a reconnaissance/surveillance mission
9) Provide payload installation guidelines
10) Know technical features of a number of commercial cameras
11) Develop payload control and management technique
12) Evaluate the application of a radar for a moving target tracking mission
13) Discuss features of UAV military payloads
14) Explain various electronic counter measure (ECM) techniques and payloads
15) Implement aerodynamic design of payloads
16) Calculate aerodynamic drag of payloads
17) Analyze payload-structure integration
18) Develop payload stabilization techniques

12.1 Introduction

A UAV is designed to carry a particular payload and to perform a given flight mission. Thus, the payload is the object/part/equipment which the basic UAV provides a platform for transportation. The payload capacity of a UAV is a measure of size, weight and energy available to perform the mission beyond the regular flight. A UAV should be able to fly safely with the payload removal.

A UAV designer is not expected to design the payload; however, the expectation is to incorporate the desired payload into the design process for s successful flight operation. Hence, a UAV designer should be aware of technical specifications and characteristics of various payloads from conventional to modern. Moreover, the weight and size of the payload will impact the design requirements, so, the design process must include review feedbacks to investigate the payload performance during the flight of the UAV. The power drives rotate the radar antennas, sensors, and camera. Sensor positions are frequently determined by potentiometers that convert mechanical motion to electrical signals.

Payload - similar to any object - has weight and volume. The UAV should have capacity (in space, and weight) to carry a payload. As the range and endurance of a UAV is increased, it can carry less payload. Table 2.1 (Chapter 2) presents characteristics of several large and small fixed-wing and rotorcraft Unmanned Aerial Systems (UAS) along with their payloads weight ratios. The payload weight along with its volume are two important factors in payload selection process.

Payloads are considered in two basic types: (a) dispensable; and (b) non-dispensable. The non-dispensable payloads will remain with the UAV until the end of flight (e.g., sensors, and camera). The dispensable payloads will be dropped/delivered/launched during flight operation (e.g., armament for the military, and crop-spray fluid or firefighting materials for civilian use). There are two groups of sensors: (a) sensor as a payload; and (b) sensors necessary for navigation/guidance systems. In this chapter, only payload sensors (not communication equipment) are addressed. Modern control theories used in servomechanisms for positioning sensors and radar antennas are discussed in Chapter 6. The sensors which are necessary for a regular flight (such as the ones used in guidance/navigation systems) have been introduced in Chapters 7 and 8.

In this chapter, the definition of payload, types of payloads, and some UAV design considerations regarding the selection, installation, and integration of payloads is presented. In addition, payloads of a number of UAVs, and the features of some popular payloads are described.

12.2 Elements of Payload

12.2.1 Payload Definition

For a manned aircraft, payload is usually passenger, cargo, and store. The term initially arose from the load carried by an aircraft the transport of which was paid for by a customer. In contrast, for a UAV, the payload is defined as the sensor/equipment installed by the customer that performs a specific task. Basically, the aircraft (manned or unmanned) should be able to fly with the payload removed. This definition, of course, should avoid any misconception. Any equipment/sensor which is necessary for a regular flight will not be assumed as a payload. For instance, communication system, the autopilot, and navigation system sensors such as accelerometers, gyros, and magnetometers are not assumed as payload. Moreover, the elements/parts which are necessary for launch and recovery are not assumed as payload. For instance, in a pizza delivery UAV, only the pizza is the payload; even, the equipment to hold and release the pizza are not part of payload.

In some references, there is a broader definition for payload, which seems ambiguous. In those references, the payload covers all equipment and sensors which are carried by the structure, including avionics and communications devices. Such a definition makes it hard to compare two UAVs in terms of their useful payloads. In this book, any equipment and store which are necessary for flight, navigation, and recovery (flight-essential items) are not considered as payload. The reader is advised to ask for clarification from manufacturer, when a UAV payload is declared.

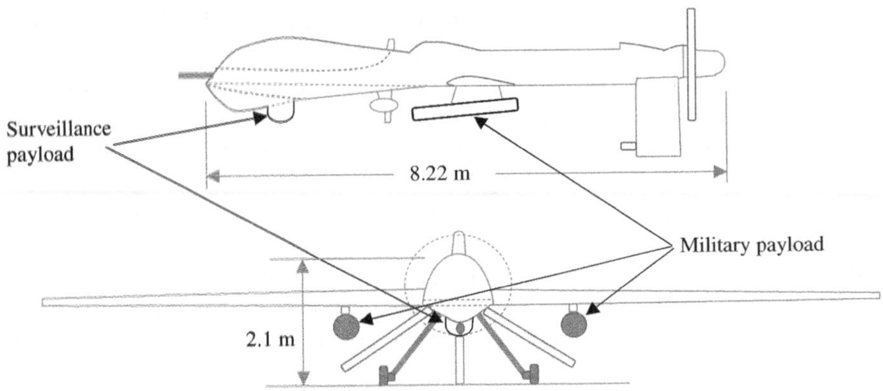

Figure 12.1 MQ-1 Predator A payloads.

12.2.2 Payloads Classifications

The primary purpose for an UAV flight operation is to carry payload(s) for a specific application. There are mainly five classes of payloads for unmanned aerial vehicles:

1) MerchandizeMerchandize/Cargo/freight
2) Reconnaissance, surveillance, and monitoring sensors
3) Scientific sensors
4) Military payload, and
5) Electronic countermeasure system.

In each class, there is a group of payloads which can be carried by a UAV for various objectives and flight missions.

Cargo/freight payloads are civil goods, merchandize, or products that are transported for commercial profit. Three primary functions of surveillance sensors are: (a) detection; (b) recognition; and (c) identification. Three types of surveillance payloads (i.e., sensors) are: (a) daytime camera; (b) IR night-vision camera; and (c) radar. Surveillance sensors (both civil and military applications) are vital in-flight safety, as well as in the success of sense and avoid system. The features of these three sensors are described in Section 12.5.

Scientific sensors are employed in scientific/research projects such as for monitoring and measuring environmental and atmospheric parameters such as temperature, humidity, and pressure. Military payloads mainly include weapons (e.g., (a) missiles; and (b) bombs) for military operations and defense purposes. An electronic countermeasure payload is a dummy air vehicle, or an electronic device, or a communication equipment aimed to deceive enemy detection systems in a battlefield.

Figure 12.1 illustrates the MQ-1 Predator A payloads: (a) military payload (missiles); and (b) surveillance payload (camera unit). The technical characteristics of some common/popular payloads are presented in the next sections.

12.3 Payloads of a Few UAVs

The technical information regarding modern payloads and payloads of the current active UAVs is a pre-requisite for a successful UAV designer. In this section, payloads of a few UAVs are briefly described: (a) RQ-4 Global Hawk; (b) RQ-7 Shadow 200;

(c) RQ-5A Hunter; (d) MQ-9 Predator B Reaper; (e) DJI Phantom Quadcopter; and (f) X-45 Unmanned combat air vehicle (UCAV).

12.3.1 RQ-4 Global Hawk

The Global Hawk aluminum fuselage contains pressurized payload and avionics compartments. The Global Hawk (Figure 1.2) integrated [152] sensor suite includes: (a) the synthetic aperture radar (SAR) (for both spot images, and wide-area search images); (b) the electro-optical and infrared sensor system; (c) an active electronically scanned array (AESA) radar (part of the Air Force airborne signals intelligence [SIGINT] platform); and (d) ground movement target indicator.

Other payloads may include hyperspectral sensors for chemical and biological agent detection, and electronic intelligence (ELINT) payload as a replacement for SIGINT platform. Moreover, a Northrop Grumman navigation system is installed on the IR/TV/SAR payload.

A 10 in reflecting telescope provides common optics for infrared and electro-optical sensors. The EO/IR sensor operates in the 0.4 to 0.8 μm visible waveband and the 3.6–5 μm infrared band. In spot collection mode the coverage is 1900 spots a day with spot size 2 km^2 to a geological accuracy of 20 m circular error of probability. The SAR and ground moving target indicator (MTI) operates at X-band with a 600 MHz bandwidth, and 3.5 kW peak power. The system can obtain images with 3 ft resolution in its wide area search mode and 1 ft resolution in its spot mode.

The RQ-4N has a Northrop Grumman AESA radar, Raytheon electro-optic/infrared sensors, L-3 communications suite and Sierra Nevada Merlin electronic support measures (ESM). High-resolution sensors, including visible and infrared electro-optical systems and SAR, can conduct surveillance over an area of 40 000 km^2 to an altitude of 65 000 ft in 24 hours.

12.3.2 MQ-9 Predator B Reaper

Since the General Atomics MQ-9 Predator B Reaper (Figure 5.12) has a much greater engine power than the original MQ-1 Predator A, it can carry 15 times more ordnance (1700, kg) payload than that of the Predator A. MQ-1 Predator has provisions to carry a variety of sensors, data links, and weapon payloads.

In terms of sensors, this UAV has streamlined payload integration capabilities, and can search for targets and observe terrain using multiple sensors including thermographic camera. The aircraft is equipped with a color nose camera (for the Ground control station (GCS) pilot forward view for flight control), a day variable-aperture television camera, a variable-aperture IR camera (for night), and a SAR for surveillance through clouds. The aircraft has options to be equipped with GA-ASI's Lynx SAR and Raytheon's MTS-B electro-optical infrared sensors, and high definition EO/IR full motion video sensor. The cameras generate full motion video, while the SAR produces still radar images. One claim is that the on-board camera is able to read a license plate from two miles away.

In terms of military payloads, the MQ-9 has provision for six stores pylons; the inner stores pylons can carry a maximum of 1500 pounds each and allow carriage of external fuel tanks. The mid-wing stores pylons can carry a maximum of 600 pounds each, while the outer stores pylons can carry a maximum of 200 pounds each.

An MQ-9 can adopt various mission kits and combinations of weapons and sensors payloads to meet combat requirements. It can carry either two AIM-114 air-to-ground hellfire missiles, or 4 AIM-92 Stinger, or 6 AGM-176 Griffin air-to-surface missiles, or the GBU-12 Paveway II laser-guided bomb, or the GBU-38 Joint Direct Attack Munition, or 500 lb laser-guided bomb units. Tests are underway to allow for the addition of the AIM-92 Stinger air-to-air missile. In 2014, it also carried European payloads such as sensors and datalinks to locate the wreckage of a crashed target.

There are future plans for improvements for operating the Reaper in contested airspace; adding readily available sensors, weapons, and threat detection and countermeasures. An example is a radar warning receiver to know when it's being targeted by an incoming missile. These provisions could increase situational awareness and enable riskier deployments, and to provide precision navigation and timing systems to fly in GPS-denied areas. In April 2017, an MQ-9 Block 5 flew with a Raytheon AN/ALR-67 Radar Warning Receiver in its payload pod to demonstrate the aircraft's ability to conduct missions in the proximity of air defenses. The Dutch MQ-9 is equipped[1] with a SAR with the Maritime search option, and also a special ground search radar with more range and electronic sensors to detect ground radar and signals.

12.3.3 RQ-7 Shadow 200

The primary payload of the RQ-7 Shadow 200 (Figure 10.5) aircraft is a group of electro-optical and infrared sensors which consists of: (a) a forward-looking infrared camera; (b) a daytime TV camera with a selectable near-infrared filter; and (c) a laser pointer.

The payloads of this UAV may include SAR, MTI, hyperspectral imagery, communications relay package, marker/illuminator, laser range finder, and target designator. The payload for each mission is selected according to the flight operation requirements.

12.3.4 RQ-5A Hunter

In the development phase, RQ-5A Hunter was used as a payload demonstration platform, so it has been proved to be capable of carrying other advanced mission payloads. The basic payload of the Hunter is the military multi-mission optic/electronic equipment, which includes television and infrared cameras to provide day/night surveillance capability. Other Hunter payloads have included a laser designator (LD) and communications devices. A communications relay payload extends Very High Frequency (VHF)/Ultra High Frequency (UHF) communications capability beyond the Line-of-sight (LOS). Electronic countermeasures payloads have included communications warning receiver, communications jammer and radar jammer. In June 2003, Northrop Grumman tested a Hunter UAV equipped with a SAR/MTI payload.

12.3.5 DJI Phantom Quadcopter

DJI Phantom – with a mass of 1.37 kg– is a small quadcopter (Figure 2.6) with four electric engines and is remotely controlled by a handheld control box. DJI Phantom 4 [147], has a three-axis stabilized camera with a 1 in. 20 Mega Pixel complementary metal oxide semiconductor (CMOS) sensor. The Phantom 4 comes with two remote controllers, one with a dedicated screen and one without. It integrates an upgraded HD video transmission system that adds 5.8 GHz transmission support and a maximum downlink video transmission range of 7 km. It upgrades its obstacle avoidance with five directional sensors. With six infrared cameras and the navigation system, the aircraft has the capability to avoid obstacles in its flight path.

It has two vision systems: (a) Forward Vision System (Field of view [FOV] of 60° [Horizontal], ±27° [Vertical]); and (b) Downward Vision System (FOV of 70° [Front and Rear],] 50° [Left and Right]). The measuring frequency is 10 Hz for forward camera, and 20 Hz for downward camera. Lens of the camera has a FOV 84° 8.8 mm/24 mm (35 mm format equivalent) and f/2.8 – f/11 auto focus at 1 m to ∞.

12.3.6 X-45 UCAV

The fuselage of X-45A (Figure 4.10) carries two internally housed weapons bays. The air vehicle can carry advanced precision guided munitions, 2000 lb bombs, or other munitions and weapons payloads. The X-45 air vehicle is also equipped with a suite of sensors including: (a) an AESA; (b) SAR; and (c) a system developed by Raytheon. All these payloads may be employed for a military operation by the ground pilot located in the GCS.

12.3.7 Yamaha RMAX

The Yamaha RMAX (Figure 1.1) has an observation platform [148], which includes a digital video camera and a digital still camera, along with three miniature cameras for navigational purposes. Images from these cameras are sent from the UAV to the GCS on a real-time basis. The helicopter is equipped with a communication modem to receive the GPS correction data and control (steering) commands from the GCS, and sends back the information regarding the position and attitude angles of the helicopter to the GCS.

Moreover, there is a camera transmitter that can move the angular position of the camera on the helicopter. Furthermore, in order to ensure the safety of the system, there is a system so that an operator can back it up by using a normal radio transmitter and the transmitter of the communication modem.

12.4 Cargo or Freight Payload

As noted earlier, cargo or freight are civil goods or product that is conveyed for commercial gains. Cargo may be stored in containers/boxes, or packed as a single item. The primary goal of carrying such payloads is mainly for delivery to a customer. The list of civilian cargo payloads for delivery by a UAV is endless; it includes items such as mails, books and pizzas, to packages, and groceries. Due to safety regulations and cost of delivery, these payloads are not still common in the market.

There is broad interest in the tech world in using UAVs to make delivery fast and efficient. In the past few years, a number of companies tried to employ UAVs for the delivery of their products. In early 2016 [153], Google delivered Chipotle burritos at Virginia Tech. It has also negotiated with Starbucks about delivering with drones. And Domino's has delivered pizza in New Zealand via a drone.

In December of 2016, Amazon [153] kicked off a private trial of its highly anticipated UAV delivery program. Amazon said it would expand to dozens of customers near its British facility in the coming months and later grow to hundreds. The trial is limited to daytime deliveries when the weather is suitable. Amazon is planning for making deliveries in 30 minutes at no extra cost. A week earlier in Cambridge, Amazon used one of its autonomous drones for its first delivery, an Amazon Fire TV and a bag of popcorn. Amazon said it took 13 minutes from the customer clicking "order", to the package being delivered.

Moreover, In March of 2017, Amazon completed its first public demonstration of a Prime Air drone delivery in the U.S, ferrying sunscreen to attendees at an Amazon-hosted conference in California. It marks the first time one of the online retailer's autonomous UAV was flown for the public in the U.S. outside of Amazon's private property. In this mission, a quadcopter glides through the air carrying a box filled with sunscreen — weighing about four pounds — under its center body. The drone then touches down on a small landing pad in a field, where it releases the payload before vertically taking off. The mission was completed fully autonomously with Amazon's own software.

More recently, the University of Maryland Medical Center successfully completed the first life-saving organs and tissue delivery by a UAV. The flight which happened on 19 April 2019 had a 2.6 miles range and took 10 minutes.

A widespread application of UAV technology to deliver packages to customers' doorsteps requires a thorough analysis of the costs and benefits of this new pick-up and delivery system. Reference [154] reviewed the basic directions of unmanned aircraft systems applications in the civil field. The concepts of creation and organization of civil unmanned aircraft systems have been discussed. Moreover, the wide range of issues concerning the use of existing capacity for design, manufacture and operation, with subsequent integration into the common air space have been addressed.

In dealing with civil cargo payloads, often, there is no design or selection process by the UAV designer. The technical characteristics of the payload are provided by the customer. However, the main job of the designer for cargos is in the accommodation, integration, platform, and load distribution of the payloads. Moreover, the designer can have some recommendations to the customer for the packaging, and transportation process of the cargo payloads. The recommendations should aim to reduce the cost for the customer, as well as increasing the efficiency of payload transportation.

12.5 Reconnaissance/Surveillance Payload

Three primary functions of reconnaissance/surveillance sensors are: (a) detection; (b) recognition; and (c) identification. Detection is defined as the activity of determining if there is an object of interest at a particular point in the target area. Recognition is the method whereby the sensor determines if the object belong to a general class such as tree, bird, human, can, or tank. Identification is defined as determining the specific identity of an object, such as enemy soldier versus own soldier, sedan car versus van, and M1 Abram tank versus T-92 tank. Aerial photo surveys carried out on behalf of utility companies and for civil engineering projects became routine as well as essential.

The three most common types of surveillance payloads (i.e., sensors) are: (a) daytime EO camera; (b) IR camera for night-vision; and (c) radar. This section provides definitions, and technical features of these three surveillance sensors.

UAVs can carry various sensor packages, including electro-optical and infrared sensors, SAR, SIGINT, and multi- and hyperspectral imagers for reconnaissance/surveillance missions. There are basically three types of surveillance missions: (a) convoy over-watch; (b) route surveillance; and (c) fixed surveillance. One potential for applying UAVs to surveillance missions is to utilize a UAV for convoy over-watch as the convoy proceeds along its route. In addition, UAVs have an effective role in fixed-site security as an integral part of larger security and surveillance systems are typically dominated by radars and fixed cameras. In military applications, a UAV may be flown in support of theater reconnaissance prior to the deployment of military forces. Moreover, UAVs are valuable for locating cargo that misses a drop zone. With current technology,river navigability using light detection and ranging (LIDAR) aboard a UAV is technically infeasible.

The reconnaissance and surveillance tasks put a premium on "real-time" sensors, or those whose results can be exploited immediately without the need for lengthy and time-consuming post-processing and annotation in order to render them intelligible. The requirement for real-time exploitation means that the options would be limited to electro-optical/infrared sensors, some varieties of SAR, and SIGINT. Other imaging systems exploiting different spectra cannot satisfy the real-time criterion.

The success for target detection, recognition, and identification based on an image is not only the function of the sensor sensitivity and resolution, but also due to the target signature. Target detection, tracking, recognition, and identification based on imaging are complicated processes and are affected by many parameters.

An important characteristic of a reconnaissance/surveillance payload (e.g., camera) is *frame rate* (i.e., frame per second, fps) which is related to the image updating and processing capability. The frame rate (i.e., frequency) is the speed at which a camera/imager updates its output image, the higher the frame rate, the faster is the image updating. A 20 Hz camera updates the image 20 times per second, and a 100 Hz camera will do that 100 times per second. A higher frame rate requires a higher electric current and power, which implies a higher cost. Faster frame rates are particularly important in applications where targets may be moving at higher speeds – such as vehicles or drones. A higher refresh rate implies a more accurate target detection. Optical cameras/TVs have two standards: (a) PAL (Phase Alternating Line); and (b) NTSC (National Television System Committee). PAL cameras use 25 fps, NTSC cameras use 30 fps. High quality cameras have much higher frame rates (e.g., 75 fps or more).

12.5.1 Electro-Optical Camera

A camera/imager is an optical instrument (i.e., imaging sensor) for recording or capturing images, which may be stored locally, and/or transmitted to another location. The images may be individual still photos or sequences of images constituting videos/movies. The camera is a remote device as it senses a subject without physical contact. Since the development of digital cameras, and low cost GPS/INS systems, the technology has advanced toward the automation of the UAV operations and the acquisition of high resolution images using off-the-shelf cameras.

Factors affecting the quality of an image/video are resolution, focus, aperture, shutter speed, white balance, metering, film speed, and autofocus point. The area of the receptor of the camera and the number of picture elements, pixels, that it contains will determine the resolution of the image.

As defined earlier, in general, there are three groups of cameras: (a) EO; (b) IR; and (c) LD. An electro-optic camera presents its output in a form that can be interpreted by an operator as an image of what the sensor is viewing. Electro-optic cameras range from simple monochrome or color single-frame cameras through color TV cameras, low-light-level television, and thermal imaging video cameras to multi-spectral cameras. Optical, or "visible light" cameras operate in the 0.4–0.7 μm wave length range. For an IR camera, the image presented represents variations in the temperature and emissivity of the object in the scene. The infrared, or heat radiation, spectrum covers longer wavelengths – in the range of about 0.7–1000 μm.

The quality of an image taken by a camera is influenced by the lighting conditions. For a camera, there is a minimum amount of light for exposure, to have a quality image. The lumen (lm) is the unit of luminous flux (i.e., illuminance), a measure of the total quantity of visible light emitted by a source. One lux is equal to one lumen per square meter (lm/m^2). Table 12.1 illustrates the illuminance levels [155] at different lighting conditions. A full moon on a clear night can shine up from 0.1 up to a single lux. For a bulb emitting 1 cd, the bulb would have a luminous intensity of 12.57 lx.

It is beneficial to differentiate three related terms in describing quality of light: (a) lumen; (b) lux; and (c) candela. The lumen measures how much light is given off by a light source, while lux deals with how bright a surface or the sky is, and candela measures the visible intensity from the light source. When the illuminance is not sufficient (e.g., at night), an IR camera must be selected to monitor a target.

One of the most significant factors in selecting a camera is image/video resolution. To describe a camera's image resolution, a standard two-dimensional matrix is typically used. The resolution of an image/video/monitor is often described by the number of pixels arranged horizontally and vertically on a picture/ screen, for example, $640 \times 480 = 307\,200$ pixels. Another method to qualify the image resolution is to use

Table 12.1 Illuminance levels at different lighting conditions.

No	Sky condition	Level of illuminance (lm/m^2)
1	Direct sunlight	$1–1.3 \times 10^5$
2	Full indirect daylight	$1–2 \times 10^4$
3	Overcast day	1000
4	Very dark day	100
5	Light bulb	10–4000
6	Twilight	10
7	Deep twilight	1
8	Night with full Moon	0.1
9	Night with quarter Moon	0.01
10	Moonless, night clear sky	0.001
11	Moonless, overcast night sky	0.0001

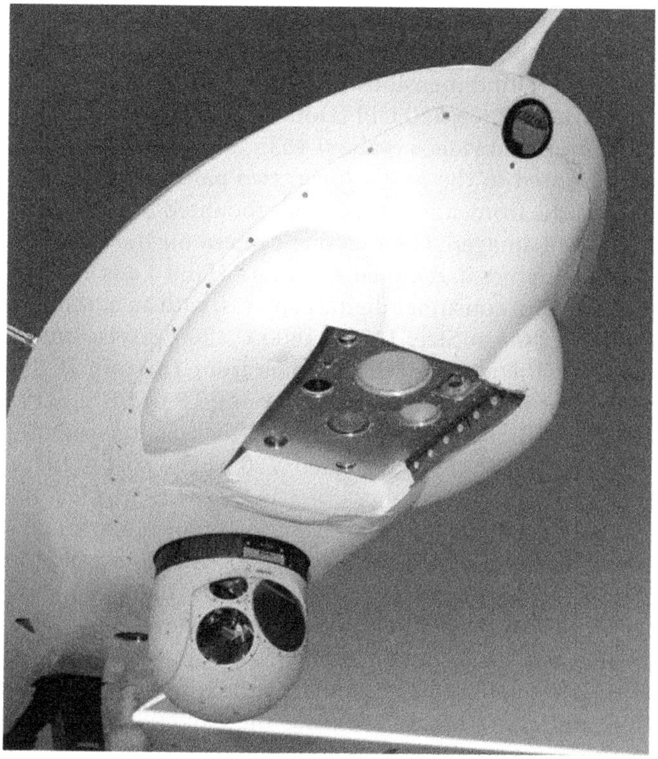

Figure 12.2 Sensors of NASA Altair UAV.

one code (e.g., 720p or 1080p). The code 720p is employed to refer to 1280 × 720 pixels, while, the code 1080p is employed to refer to 1920 × 1080 pixels. It is convenient to refer to the resolution of 720p as HD or High-Definition, while 1080p is usually known as FHD or Full HD. The resolution of 4096 × 2160 is referred to as the UHD, or Ultra HD.

Figure 12.2 depicts sensors of Altair UAV which includes an ocean color sensor, a passive microwave sounder, and an electro-optical sensor. These sensors were mounted for the NOAA-NASA flight demonstration in 2005. A small camera, the DJI Zenmuse X5 Camera with a FOV of 72° can take 4K video at up to 30 fps. The gimbal platform of the camera can tilt, roll, and pan the camera about three axes, independent of the UAV movement. This camera has no fairing, so, carrying such a camera will generate a lot of drag. This payload is only appropriate for a UAV with a very low speed.

Imaging sensors produce images of varying degrees of resolution. These degrees of resolution are characterized by the NIIRS[2] with a numerical score, typically ranges from 1 to 9, where 1 represents very-low-resolution images and 9 very-high-resolution images. For instance, at an image with a score of 5, there is enough resolution at normal size to distinguish cars and buildings, and to identify planted crops. The available real-time sensors typically generate imagery with NIIRS ratings between 5 and 6.5. With progress in miniaturization, we will probably see improvements in sensor technologies.

Some available camera classes in the market are: (a) rangefinder camera; (b) instant picture camera; (c) single-lens reflex; (d) twin-lens reflex; (e) large-format camera; (f) medium-format camera; (g) subminiature camera; (h) movie camera; (i) camcorders; (j) digital camera; (k) panoramic camera; and (l) VR (Virtual Reality) Camera.

It is worth comparing two cameras; a simple, cheap and commercial camera used in RC modelers, and an advanced one employed in the *Mars Rover*. The Wi-Fi Enabled FPV High-Definition Ominus Camera with dimensions of 73 mm × 28 mm × 14 mm, and a range of 100 mm has a Resolution: 1280 × 720 DPI (Dots Per Inch). The price tag of this camera, which takes both photos and videos is about $63.

There are nine cameras hard-mounted to the Mars Rover; two pairs of black-and-white Hazard Avoidance Cameras in the front, another two pair mounted to the rear of the rover, and the color Mars Descent Imager. There is one camera on the end of a robotic arm that is stowed in this graphic; it is called the Mars Hand Lens Imager (MAHLI). The MAHLI is similar to that of consumer digital cameras, with an autofocus ability. Some features are as follows: Image Size: Up to 1600 × 1200 pixels; Image Resolution: Possibility of 13.9 μm/pixel; Focal Length: In focus from 18.3 mm at the closest working distance to 21.3 mm at infinity; Focal Ratio: From f/9.8 to f/8.5; FOV: from 34° to 39.4°; Memory: 8 Gigabyte flash memory storage; 128 megabyte synchronous dynamic random access memory (SDRAM); HD Video: 720 p. The cost of these advanced cameras exceed millions of dollars.

A large High altitude long endurance (HALE) UAV intended for a typical reconnaissance operation to detect a target will need a wide FOV lens to focus onto a receptor with a large number of pixels. This type of camera is usually large, heavy, more complex, and requiring great electrical power for its operation. In contrast, a short-range UAV will need a lighter, low-cost, smaller, and less capable camera.

To obtain a 360° azimuth FOV, a rotatable turret should be mounted beneath the UAV. In addition, the elevation and roll gimbals, carrying the camera, should be mounted with their actuators within the turret. The Controp's medium range stabilized EO/IR cameras are 3-axis gimbal gyro-stabilized payloads for medium range (up to 3–5 km). A 3-axis gimbal provides a stable platform for the camera to capture clear image during UAV flight. Moreover, the gimbal platform can tilt, roll, and pan the camera about three axes, independent of the UAV movement.

The resolution of an image is defined as the extent of which it is possible to see the details of an image. The number of horizontal lines in an image determines the resolution of a camera. The number of vertical lines, or scanning, is constraint by the image scanning system to produce an image. For instance, PAL format utilizes 625 lines scanning at 50 Hz, and NTSC format uses only 525 lines at 60 Hz frequency. Table 12.2 shows technical features of a number of commercial EO video cameras including dimensions and weight used for small UAVs and quadcopters. It is interesting to note that the camera Tarot TL300MN FPV costs about $20, while camera DJI Zenmuse X5 has a price tag of about $2000.

The Camera DJI Zenmuse X5 used in DJI Phantom 4 Pro quadcopter (Figure 2.6) integrated with 3-axis gimbal system has the following technical characteristics: 4/3-inin. CMOS sensor, 4K video at up to 30 fps, record at 1920 × 1080 (60 fps), 16 MP photographs, 12.8 stops of dynamic range, lens with Max F1.7 aperture, FOV 72°, and a precision accuracy of 0.02 °.

It is interesting to note all HALE and MALE UAVs including the RQ-4 Global Hawk, MQ-1 Predator, and MQ-9 Reaper, are equipped with: (a) a SAR; (b) an electro-optical camera; and (c) an infrared camera. Moreover, Scan Eagle (Figure 3.3) carries an infrared camera on an inertial stabilized turret system. The RQ-7 Shadow has a gimbal-mounted, digitally stabilized, EO/IR camera – installed on a step-stare turret – that relays video in real time via a data link to the GCS.

Table 12.2 Technical features of a number of commercial cameras.

No	Model	Minimum Lux	Horizontal Lines	Current	Voltage	Mass	Dimensions
1	Micro Color	1.5	380	30 mA	7–12	0.35 oz	0.31 in × 0.31 in × 0.39 in
2	Micro B/W	0.5	380	15 mA	7–12	0.32 oz	.31 in × .31 in × .39 in
3	Sony CC-1SBHR	0.2	550	65 mA	9–14.5	10.2 oz	1.3 in × 1.3 in ×1.1 in
4	Sony CC-1XHRM	0.05	480	120 mA	9–14.5	10.2 oz.	1.4 in × 1.4 in × 1.4 in
5	KX-131G	1	330	120 mA	5	7.8 oz.	0.87 in × 1.0 in × 1.1 in
6	KX-121	5	330	120 mA	5	2.12 oz	0.87 in × 1.0 in × .59 in
7	Pencil Eraser Cam	1.5	380	0.35 mA	7–12	2.4 oz	0.32 in × 0.32 in. (w × h)
8	Miniature ProSeries	1.5	380	–	5–12	2.4 oz	2 in × 1 in ×1 in
9	Tarot TL300MN FPV	0.008	520	40 mA	12	15 g	21 mm × 26 mm × 30 mm
10	DJI Zenmuse X5	1.5	380	360–700 mA	18–22	526 g	120 mm × 135 mm ×140 mm
11	Foxeer 16:9 RCX09–459-NTSC	0.05	960	70 mA	5–40	50 g	19 mm × 13 mm × 13 mm

Figure 12.3 Epsilon 135 (EO) and Epsilon 175 (IR) gyro stabilized cameras (Courtesy of UAV factory).

Predator A has used both precision approach radar, and a video camera to assist pilot in landing on manual control. The small UAV RQ-11B Raven with a mass of 1.9 kg has the following payloads: (a) gimbaled stabilized EO/IR cameras with IR illuminator; and (b) dual forward and side-look EO stabilized cameras. The EO/IR cameras can have a continuous pan of +10 to −90° tilt.

AeroVironment RQ-11B Raven (Figure 10.6) is equipped with stabilized EO/IR cameras. The Raven is a small hand-launched remote-controlled UAV designed/manufactured by AeroVironment. It has a pusher high-wing conjugation with a mass of 1.9 kg and wing span of 1.37 m. The UAV is equipped with a 250 W electric brushless engine has a cruising speed of 30 km/h, a ceiling of 150 m, a range of 10 km and an endurance of 90 minutes. The pilot is controlling the raven from a GCS through a line of sight or the satellite communications. Figure 12.3 illustrates Epsilon 135 (EO) and Epsilon 175 (IR) gyro stabilized cameras.

Among a number of candidate cameras which satisfy the requirements such as the image resolution, and lighting conditions; one should select a camera which has the lightest weight, smallest size, lowest power consumption, and of course the lowest price. In selecting an optical camera, the UAV designer should conduct a trade-off study to determine an optimum selection.

12.5.2 Infra-Red Camera

An IR camera in a thermal sensor (imager) which operates in the Infra-Red bandwidth of the light spectrum. Infrared light extends from the nominal red edge of the visible spectrum at 700 nm to 1 mm. This range of wavelengths corresponds to a frequency range of approximately 300 GHz to 430 THz. This type of sensor is used to detect targets in low light environment. An IR sensor can distinguish between cold and warm objects, and it is beneficial in detecting human (warm) bodies. Thus, it is ideal and essential for locating objects or persons at night or in bad weather conditions, offering low visibility.

The resolution of an image or a video for an IR camera is defined in a similar manner to that of an optical camera; (i.e., number of pixels). To describe an IR camera's image resolution, a standard two-dimensional matrix is utilized. Imager format is basically the number of pixels on the detector array. Standard formats are: (a) regular resolution: 160 × 120 (19 600 pixels), and 384 × 288 (110 592 pixels); and (b) high resolution: 640 × 480 (307 200 pixels), and 1024 × 768 (786 432 pixels). Hence, in general, an infrared camera has fewer pixels than an optical camera.

Infrared cameras are measuring the differences between heat radiations of a target from surrounding environment. Many infrared cameras store images in a proprietary format that can only be read and analyzed with specialized software. However, some cameras provide a standard JPEG image format with an embedded temperature data. TV cameras and forward looking infrared (FLIR) sensors are less susceptible to detection than active devices such as radars.

The prices of IR camera have fallen dramatically over the past few years. There are lots of choices in IR camera, ranging from extremely affordable simple models to highly specialized high-resolution complex research cameras. Higher resolution infrared cameras can measure smaller targets from farther away and create sharper thermal images. Some infrared cameras allow the user to stream live image/video to computers and screen. This is especially useful for capturing a moving target for recording motorized equipment or processes in motion. Thus, it is possible to connect the camera via communication system to smartphones and GCS.

An IR camera's temperature range and sensitivity are important factor. The range provides the minimum and maximum temperatures that the camera can measure (e.g., −20 °F to 2000 °F). Human body has a normal temperature range of 97 °F (36.1 °C) to 99 °F (37.2 °C). However, manifolds and exhaust pipes on automobiles can reach 1200 ° F. The smallest temperature difference between two objects that an IR camera can distinguish is about 0.05 °C. The UAV designer should select an IR camera with a temperature range broad enough to capture all of the temperatures of the target which is typically encountered.

An EO/IR sensor turret incorporates an IR sensor and a high-resolution optical color camera in one unit. Moreover, both the IR sensor and EO color camera may be housed in the same turret, and it can be coupled with additional laser payloads.

12.5.3 Radar

12.5.3.1 Fundamentals

Radar (short for **RA**dio **D**etection **A**nd **R**anging) is an object-detection device that uses electromagnetic (e.g., radio) waves to measure the range (distance), angle, or velocity of an object. UAV radar is employed for the automatic detection and tracking of moving ground vehicles as well as flying aircraft within its range. Four main elements of a radar are: (a) transmitter; (b) receiver; (c) signal processor; and (d) antenna. Often, the transmitting antenna and the receiving antenna are the same.

There are a number of advantages and disadvantages for radars as compared with optical sensors such as IR camera. A radar provides its own source of energy; hence it does not depend on reflected light or heat emitted from the target. Thus, radars are more susceptible to detection than non-active sensors such as TV cameras and FLIR sensors. Compared with radars, under favorable conditions, optical sensors can supply high-quality images of the target for human interpretation. However, optical sensors suffer from a limited FOV and from severely reduced performance in adverse weather and smoke and dust conditions.

Long-range surveillance radars can provide rapid surveillance of large areas of foreign territory. Modern radars can rapidly scan large areas, and they are less influenced by smoke, dust, humidity, and weather. A radar sensor provides synthetic images, often includes false colors that convey information about target motion, and polarization of the returned signal. In contrast, an electro-optic camera generates clear color image/photo or video.

The wave which is sent by radar could be either (a) continuous; or (b) pulsed. A radar transmitter emits radio signals in predetermined directions. If the transmitter turns, the signal direction is turned as well. If the Earth were a perfectly flat horizontal plane, the signal would come only from the closest point, and would be a true measure of altitude. However, the Earth is not smooth, and energy is scattered back to the radar from all parts of the surface illuminated by the transmitter.

A transmitter generates the radio signal with an oscillator such as a magnetron. A receiver is an electronic device that receives radio waves and converts the information carried by them to a usable form. An antenna is an electrical device which converts electric power into radio waves and vice versa. An antenna either receives energy from an electromagnetic field or transmits (i.e., radiates) electromagnetic waves produced by a high frequency generator, or which are returned from a target. The antenna is a structure which serves as a transition between wave propagating in free space, and the fluctuating voltages in the circuit to which it is connected (i.e., transmitter). The most frequently used type of an antenna for a radar is in the form of a parabolic dish (for example, see Figure 13.6).

The radar signals that are reflected back are employed by the receiver and will be sent to the radar processor. The radar measures distance by determining the time required for a radio wave to travel to and from a target. The radar frequencies for UAVs are often from 9 to 35 GHz. The electromagnetic waves are traveling with the speed of light. Thus, when a target is located at a distance of even 100 km away from the radar, the measurement is assumed real-time. The performance of most radar is limited by the ability of radar to separate targets from clutter. If the radar beam is way larger the target dimensions, the performance can be enhanced.

A particular type of radar is the radar altimeter which is often used in aircraft during bad-weather landings. Radar altimeter is much more accurate, and more expensive than the pressure altimeters. They are an essential part of many blind-landing and navigation systems and are used over mountains to indicate terrain clearance. Special types are used in surveying for quick determination of profiles. Radar altimeters have been in use on various spacecraft, to measure the shape of the geoid and heights of waves and tides over the oceans.

A type of modern radar to provide finer spatial resolution than conventional beam-scanning radars is Synthetic Aperture Radar (SAR). The SAR is a side-looking radar system that electronically simulates an extremely large antenna or aperture, and which generates high-resolution remote sensing imagery. In this process, known as coherent detection; the phase of a return signal is compared with that of the transmitted signal. The details of radar design are beyond the scope of this book; for details see references such as [156].

Radar is an important device in a collision avoidance system, and in target identification. Once radar detects and locates a target, the sensor turret can be deployed, and, depending on range, light, and environmental conditions, either the optical or the IR camera, or both, can be directed and zoomed onto the target for detail identification. The Kalman filter (Chapter 8) has long been regarded as the optimal solution to many tracking and data prediction tasks such as tracking moving targets using radar (e.g., another aircraft, ground moving target). Search radars (20 km to 2,000 km) have high power density on the target with high antenna gain; while tracking radars have accurate angle and range measurement is required. A closed-loop control system in required to control the azimuth rotation of a radar antenna, to track a moving target when it goes past the radar. The dynamics of an antenna can be modeled with a first order transfer function; G(s) =10/(s + 10); where the time constant is 0.1 seconds.

12.5.3.2 Radar Governing Equations

A radar operates based on the concept of Doppler Effect (or the Doppler shift) which is the change in frequency or wavelength of a wave signal for an observer moving relative to its source. The signal emitted by the radar is usually in the form of a train of narrow, rectangular-shaped pulses; usually referred to as a waveform. The frequency of transmission (f_p) - number of pulses per second - is related to the time between two pulses (t_o) as

$$f_p = \frac{1}{t_o} \tag{12.1}$$

The signal travels to the target at the speed of light, hits the target, and is reflected back at the same speed. However, a portion of the emitted electromagnetic radiation is reflected back from the target. The receiving antenna receives this reflected energy and sends it to the receiver. The receiver processes this energy to detect the presence of the target, and to calculate its location, size, and its relative velocity.

The distance; or range (R) to the target is given by

$$R = \frac{c.t_r}{2} \tag{12.2}$$

where t_r is the time taken for two travels. In other words, the reflected energy is received t_r seconds at the radar after sending the pulse.

The relative motion that affects the observed frequency is only the motion in the Line-Of-Sight between the source and the receiver. When the speeds of source and the receiver relative to the medium (i.e., air) are lower than the velocity of waves in the medium, the relationship between observed frequency (f) and the emitted frequency (f_o) is given by:

$$f = \left(\frac{c \pm v_r}{c \pm v_s}\right) f_o \tag{12.3}$$

where c is the wave velocity (299 792 458 m/s), v_r is the velocity of the receiver (i.e., UAV), v_s is the velocity of the source (i.e., UAV/target), v_r is positive if the receiver is moving towards the source; v_s is positive if the source is moving away from the receiver. If the UAV is moving at an angle with respect to the target, the angular effect should be included, by multiplying both UAV and target speeds (v_r and v_s) with cos(θ). The θ is the angle between the UAV/target forward velocity and the line of sight from the UAV to the target. When the transmitted and received frequencies are measured, the target velocity is calculated by the radar processor using this equation. The frequency of any sinusoidal signal (including electromagnetic or radio waves) is expressed in terms of the wavelength (λ):

$$f = \frac{c}{\lambda} \tag{12.4}$$

Radar equation relates the returning power, P_r and transmitted power, P_t:

$$P_r = \frac{P_t G_t A_r \sigma F^4}{(4\pi)^2 R^4} \tag{12.5}$$

where G_t denote the gain of transmitting antenna, A_r the effective aperture, σ radar cross section (RCS), F the pattern propagation factor, and R the distance between UAV radar and the target. The equation indicates that the received power drops as the fourth power of the range, which means that the received power from a distant target is relatively very small.

Example 12.1
Consider a radar with a pulse repetition frequency of 500 kHz. An echo from a distant object (i.e., target) is received 12 μs after the pulse is transmitted.

a) Determine the distance of the target from radar (i.e., range).
b) Determine the time between two pulses.

Solution
a) Range

$$R = \frac{c.t_r}{2} = \frac{299{,}792{,}458 \times 12 \times 10^{-6}}{2} = 1799\,m = 1.8\,km \tag{12.2}$$

b) Time between two pulses

$$f_p = \frac{1}{t_o} \Rightarrow t_o = \frac{1}{f_p} = \frac{1}{500} = 0.002\,\text{sec} = 2\,ms \tag{12.1}$$

12.5.3.3 An Example

An airborne wide-area, nearly all-weather surveillance radar [12] for an UAV – sponsored by the Army and DARPA – was developed by Lincoln Laboratory. This radar was required to detect and track ground moving vehicles such as tanks, trucks, and low-flying helicopters within a range of 15 km. The following is a general description of this UAV radar system.

This radar system requires less than 1400 W of power and weighs 175 lb. The radar system alone weighs 110 lb; the remaining 65 lb are contributed by support equipment

Table 12.3 Lincoln Lab radar parameters.

No	Parameter	Value
1	Radar type	Coherent-pulse Doppler
2	Frequency	Ku-band
3	RF bandwidth (instantaneous)	10 MHz
4	Receiver noise figure	7 dB
5	PRF (variable)	3–10 kHz
6	Range resolution	15, 30, and 50 m
7	Linear dynamic range	40 dB minimum
8	AID quantization of I/O video	8 bits/channel
9	Antenna reflector type	Parabolic 18 in × 8 in.
10	Rotary joints	Azimuth and elevation
11	Elevation pattern	Cosecant squared
12	Azimuth beam-width	3°
13	Scan speed (variable)	0°–48° per second
14	Azimuth side-lobes	−28 dB (maximum), −35 dB (average)
15	Peak gain	30 dBi
16	Polarization	Horizontal

such as the INS, data link, and altimeter. In a free-flight configuration, the support equipment would normally be shared with the flight control system. This radar system has a 360° wide-area surveillance mode in which the radar antenna sweeps out an annular 5–15 km range swath every 18 seconds. When the UAV is flying at an operational altitude of 3 km, the 5–15 km swath corresponds to depression angles of 11° to 37°.

The antenna which weighs 14 lb and requires 75 W of power, is mechanically rotated with an 18 in by 8 in dish, a 30-azimuth bandwidth, and a cosecant-squared-weighted elevation beam between −11° and −40°. Table 12.3 presents [12] the UAV radar technical features. The weight breakdown of this radar system is provided in Chapter 2.

The radar has three basic linear FM pulse-compression waveforms and operates in a variety of modes to match different requirements for detection and tracking. For the primary wide-area surveillance mode, the radar performs a 360° scan and detects ground moving targets in a 5–15 km range swath. In this mode, a 50 m resolution waveform is utilized. In the moving-target tracking mode, the radar operates in a track-while-scan mode and a sector scan is chosen to focus on a specific area of interest.

The 50 m resolution of the primary wide-area surveillance mode is not sufficient to resolve closely spaced targets. As a result, a higher resolution (i.e., 15 m) waveform is utilized in the ground moving-target tracking mode for better separation of individual vehicles. A high pulse repetition frequency is dedicated for the helicopter-detection mode to distinguish between helicopters and ground moving targets. The high pulse repetition frequency is intended for detecting individual flashes of the helicopter main rotor blades. Table 12.4 provides characteristics of the basic waveforms for three radar modes.

Table 12.4 Characteristics of the unmanned aerial vehicle (UAV) radar waveforms.

No	Radar modes	Characteristics of waveforms
1	Ground moving target detection (Wide area surveillance)	• 360° scan every 18 seconds • 5–15 km range swath • 50 m resolution waveform • 7 kHz maximum pulse repetition frequency
2	Ground moving target tracking (specific area of interest surveillance)	• 30° scan every 5 second • 6 km range swath broken into two 3 km intervals • 15 m resolution waveform • 4.4 kHz maximum pulse repetition frequency
3	Helicopter tracking	• 25° scan every 2.5 seconds • 4 km range swath • 30 m resolution waveform • 10 kHz maximum pulse repetition frequency

12.5.3.4 A Few Applications

In this section, radar systems of four UAVs; Global Hawk, MQ-1C Gray Eagle, Hunter, and X-45 UCAV are briefly described.

The Air Force Global Hawk UAVs are equipped with Garmin GSX 70 weather radar which provides real-time weather information, and offers horizontal scanning for better visibility into the strength and intensity of convective activity, and a vertical scanning mode to analyze storm tops and gradients. This small-size, low-weight radar provides a vertical scanning mode, which provides users with the means for analyzing storm tops, gradients, and cell buildup activity at various ranges of altitudes. It also offers horizontal scan angles of up to 120°. Table 12.5 presents technical characteristics of a Garmin GSX 70 radar used in a NASA Global Hawk (Figure 1.2). Figure 12.4 demonstrates the Global express Primus LSZ-860 weather radar which detects both visible and invisible electromagnetic and electrostatic discharges indicating turbulent activity.

General Atomics MQ-1C Gray Eagle is equipped with an AN/ZPY-1 STARLite Small Tactical Radar. The radar with a mass of 29 kg occupies 1.2 ft^3, and it requires less than 750 W of electric power. This radar has a dismount moving target indicator (MTI) mode that can track a person walking on the ground from a range of 8 km.

In June 2003, Northrop Grumman tested a Hunter UAV equipped with a SAR/MTI payload. The Raytheon SAR developed for X-45 UCAV (Figure 4.10) provides a resolution of 60 cm at a target range of 80 km.

Among a number of candidate radars which satisfy the requirements such as the range, and frequency; one should select a radar which has the lightest weight, smallest size, lowest power consumption, highest efficiency, and of course the lowest price. In selecting a radar, the UAV designer should conduct a trade-off study to determine an optimum selection.

Table 12.5 Garmin GSX 70 radar used in Global Hawk.

No	Parameter	Values
1	Output frequency range	X-Band (9.3 to 9.5 GHz)
2	Receiver gain adjustment	+12 to −64 dB
3	RX sensitivity	−126 dBm MDS
4	Manual gain	+12 to −64 dB
5	Antenna sizes and beam-width	18″ Antenna: 5.3°
6	Size	• Diameter at base; Depth: 6.3″ • Antenna: 18″ diameter flat-plate
7	Unit weight (mass)	12.7 lbs (4.31 kg)
8	Power requirements	• <3 A at 14 V and • <1.5 A at 28 V
9	Transmitter power (nominal)	40 W

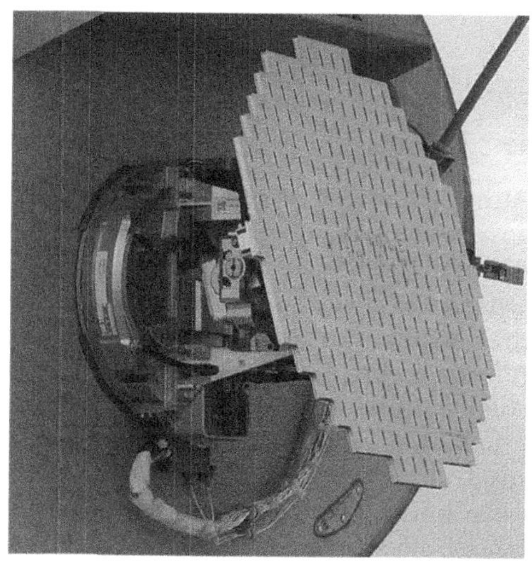

Figure 12.4 Global express weather radar.

Example 12.2

A UAV flying at an altitude of 1 km is directly behind and closing at a horizontal range of 5 km from a ground moving truck. The UAV with a speed of 120 knot parallel to the road, is tracking the truck with a radar with a frequency 10 GHz. Doppler shift between the echo and source frequency is 2948 Hz. Calculate the truck velocity.

Solution

The radar wave has a speed of light (i.e., 3×10^8 m/s). We use equation 12.3 to calculate the truck speed (v_r):

$$f = \left(\frac{c + v_r}{c + v_s}\right) f_0 \tag{12.3}$$

Since the radar signal is echoed back from truck, the wave frequency is shifted twice. Thus, we need to use equation 12.3 two times. The easier approach is to divide the frequency shift by two and use it once. Hence, the frequency shift at the truck is 2948/2 = 1474 Hz. So, the frequency at the truck will be 10.000001474 GHz. Moreover, the line of sight is

$$\theta = tan^{-1}\left(\frac{H}{D}\right) = tan^{-1}\left(\frac{1}{5}\right) = 11.3\ deg$$

In the equation 12.3, the components of both UAV and truck velocities along the line of sight need to be employed. Hence, by plugging these values, we obtain:

$$10.000001474 = \left(\frac{3\times10^{8}\times v_r \times \cos(11.3)}{3\times10^{8} - 120\times0.514\times\cos(11.3)}\right)\times 10 \tag{12.3}$$

where 0.514 is used to convert knot to m/s. Furthermore, the velocity of UAV (v_S) is negative, since the source (UAV) is moving towards the receiver (truck). The solution of this equation yields a value of -16.67 m/s or -60 km/hr for v_r:

$$v_r = -60 \text{ km/hr}$$

Therefore, the truck is moving away from UAV with a velocity of 60 km/hr.

12.5.4 Lidar

Although images – provided by a camera – contain rich information about the environment, their real-time processing to retrieve useful cues suffers from a significant disadvantage. Image processing has a high computational requirement, and is highly sensitive to environmental factors, such as light and, textures. One solution to resolve these issues, is to employ acoustic signatures instead of images of a target. A Lidar can provide accurate information about a target or the environment structure with lower computational requirement.

Lidar is a similar device to radar to make use of other parts of the electromagnetic spectrum, which uses ultraviolet, visible, or near infrared light from lasers rather than radio waves. Lidar is an optical distance measurement sensor that measures the distance from an observer to a target by illuminating the target with pulsed laser light and measuring the reflected pulses with a sensor. Differences in laser return times and wavelengths can then be used to make digital three-dimensional representations of an object. This laser radiation device is a kind of range finder, and it can be utilized to measure altitude.

Two weaknesses of a lidar are its higher weight and greater power consumption (compared with optical cameras) which limit its application to UAVs with a relatively heavy payload (e.g., the unmanned helicopter Yamaha RMAX [Figure 1.1]). Recent advances in Lidar miniaturization has allowed its use on quadrotors as the main perception sensor. Most Lidar-based perception techniques for UAVs are based on building environment's maps.

12.5.5 Range Finder

A range finder (or rangefinder) is a device that measures the distance from an observer (e.g., UAV) to a target. The rangefinder emits laser[3] (invisible light) beams at a target.

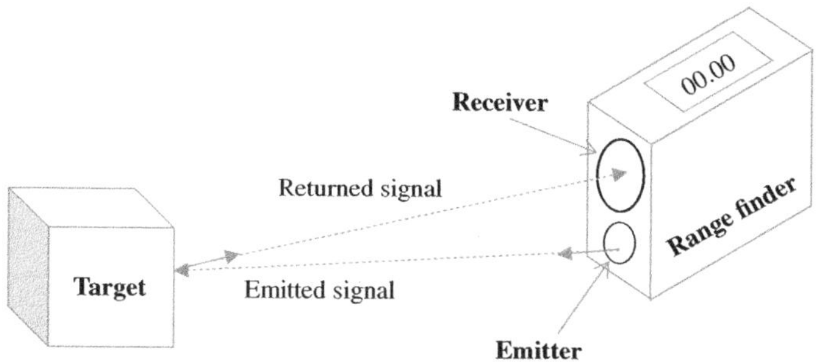

Figure 12.5 Range finder mechanism.

Table 12.6 Features of a number of commercial range finders.

No	Model	Dimension (cm × cm × cm)	Mass (kg)	Voltage (V)	Current (mA)	Min Range (m)	Max Range (m)
1	BOD 63M-LA04-S115	3.5 × 7 × 9	0.26	24	200	1	6
2	Opti Logic	8 × 8 × 3	0.22	8	14.4	91	550
3	Nikon Coolshot 80 VR	10 × 5 × 8	0.22	3	–	8	915
4	Micro-Laser	4 × 4 × 10	0.57	1.5	–	25	1500

Those beams will bounce off distant target and received by the receiver (Figure 12.5) of the range finder. The rangefinder's clock measures the total time it took from when the beams left the observer, until they returned. Hence, a range finder has four main elements: (a) signal generator; (b) emitter; (c) receiver; and (d) processor.

Lasers are continuous wave and repetitively pulsed lasers with wavelengths between 0.18 μm wavelengths between 1.5 μm. Since the travel speed of laser is known (i.e., 299 792 458 m/s), a processor calculates the range (distance the signal traveled) by multiplying this period by the speed of light. Then, this value (i.e., range) is displayed the distance to the user. There are a number of factors that influence the performance of a rangefinder, including target dimensions, distance from the target, and atmospheric conditions, such as humidity, cloud, or rain.

A popular commercial range finder is Lidar-Lite. If the goal is to measure the distance to target, a camera fitted with a rangefinder does the same function. A rangefinder camera is a split-image rangefinder by employing a focusing mechanism that allows the operator/photographer to measure the subject distance and takes photos that are in sharp focus. A range finder is a vital tool in the collision avoidance system, sense-and-avoid system, or anti-collision system. Table 12.6 presents features of a number of commercial range finders including dimensions, mass, power consumption and range limits.

Table 12.7 presents the optical features of the range finder BOD 63M-LA04-S115. This range finder uses a laser signal with the frequency of 2050 kHz, and wavelength of 660 nm.

Table 12.7 Optical features of the range finder BOD 63 M-LA04-S115.

No	Parameter	Value/type	Unit
1	Light type	Laser red light	
2	Principle of optical operation	Light time-of-flight	
3	Beam characteristic	Collimated	
4	Ambient light max	10 000	Lux
5	Average power; P_{omax}	1	mW
6	Light spot size at 6 m	Ø 10	mm
7	Pulse duration; t_{max}	0.007	μs
8	Pulse frequency	2050	kHz
9	Pulse power Pp_{max}	70	mW
10	Wave length	660	nm

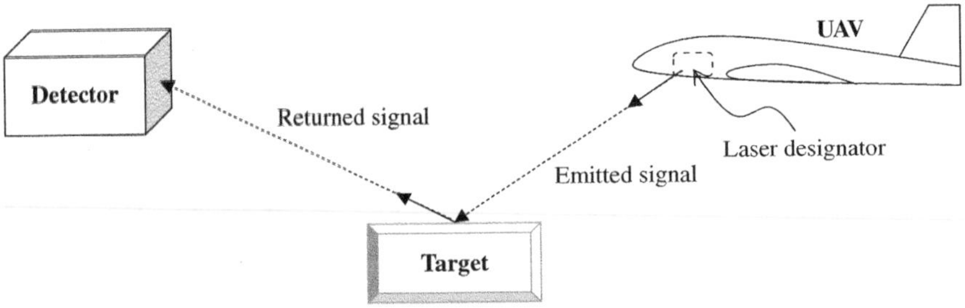

Figure 12.6 Laser designator (LD) mechanism.

12.5.6 Laser Designator

As the name implies, a LD is a laser light source which is employed to designate a target. A laser beam is invisible, and usually, LD transmits a series of coded pulses of laser light. Thus, when a target is marked by a designator, a detector/seeker (Figure 12.6) can sense these signals that bounced off the target to another observer. In military application, a line of sight must exist between the designator and the target, as well as between the target and the tracker or laser guided weapon (i.e., another observer).

Hence, a LD has three main elements: (a) signal generator; (b) transmitter; and (c) laser signal to pulse convertor. The Predator UAV (Figure 8.5) carries a multi-spectral targeting system, which integrates an infrared sensor, daylight TV camera, image-intensified TV camera, laser illuminator, and LD for intelligence, surveillance, and reconnaissance missions.

The Northrop Grumman has combined a light weight LD and a range finder– AN/PED-1 LLDR – in one unit; that can recognize targets in day/night/obscurant conditions, range to the target at an eye-safe wavelength, and calculate grid coordinates with its own GPS/Elevation/Azimuth capability.

12.5.7 Radar Warning Receiver

Radar warning receiver is employed to detect the radio emissions of a threatening radar system; and then to issue a warning when this radar signal is a threat (such as an incoming missile's radar). The next step is for the UAV to either deceive or evade the missile by its ECM system, or destroy it by its defense systems. This warning system is frequently installed at the rear of the aircraft (e.g., inside the horizontal tail), so it can look aft. To improve survivability, the Global Hawk UAV is fitted with a Raytheon developed self-protection suite which includes an AN/APR-49 Radar warning receiver.

12.6 Scientific Payloads

Autonomous flight and payload control on a UAV platform show great promise for future science missions. A few recent NASA science mission concepts are: (a) climate variability and change (to observe solar radiation in a vertical profile); (b) carbon cycle, ecosystems and biogeochemistry (for greenhouse gas (CO_2) flux measurements); (c) water and energy (to measure cloud properties in a vertical profile); (d) atmospheric chemistry and composition (for tropospheric ozone and pollution tracking); (e) Earth surface and interior structure (to observe polar ice sheet dynamics); and (f) weather (to observe hurricane cyclogenesis, evolution and landfall).

These science mission concepts require new UAV payload (scientist sensors) development and autonomy levels. For instance, a NASA Global Hawk UAV, with a Cloud Physics Lidar (CPL), collects atmospheric data in airborne science missions. The CPL is a compact elastic airborne lidar system that provides multi-wavelength measurements of cirrus clouds, and aerosols with high temporal and spatial resolution. It uses a laser operating at 1064, 532, and 355 nm with a repetition rate of 5 kHz. In this section, scientific payloads classifications and a list of UAV sensors are provided. As an example, the technical specifications of a thermal sensor are described in detail in Section 12.6.2.

12.6.1 Classifications

There is a long list for non-reconnaissance/surveillance, non-military payloads which may be utilized by UAVs for various scientific missions. Some significant ones are spectrometers, radiation detectors, environmental sensors, and atmospheric sensors. A brief introduction to these sensors is presented here.

- A spectrometer analyzes the spectrum of light to identify the chemical elements of the target. The technology may also be used in other extreme conditions such as nuclear reactors and the sea floor. For instance, Mars Rover is equipped with a rock-zapping laser and telescope. It produces more than a million watts of power focused for five one-billionths of a second to hit targets up to seven meters away. The result is a glowing ionized gas which it observes with the telescope. The light is then analyzed by spectrometer.
- A radiation detector monitors high-energy atomic and sub atomic particles coming from the sun, from distant supernovae and other sources to determine whether conditions are favorable to life and preserving evidence of life. Naturally occurring radiation is harmful to microbes and human life. The Earth atmosphere has a protective magnetic field against radiation.

- An environmental sensor records environmental information such as carbon monoxide, water vapor, aerosols, ozone, gases composing air, air viscosity, chemicals in the air, and ultraviolet radiation. The wind speed and direction, temperatures, and humidity are measured by electronic devices. The atmospheric pressure is measured by a barometer. Some other atmospheric phenomena include: (a) rain; (b) ice; (c) snow; (d) wind; (e) gust; (f) turbulence; (g) hurricane; (h) tornado; (i) thunderstorm; and (j) lightning.
- An atmospheric sensor records atmospheric information such as wind speed, wind direction, air pressure, relative humidity, air temperature, ground temperature. The wind speed and direction, temperatures, and humidity are measured by electronic devices. The atmospheric pressure is measured by a barometer.

The scientific payloads are manufactured in various configurations, sizes, and weights. For details characteristics, refer to the manufacturers' websites, or design your own. From 2006, NASA acquired [157] a General Atomics MQ-9 Predator (Named Ikhana) to support Earth science missions and advanced aeronautical technology development. Ikhana (Figure 13.7) flew over several of the major Southern California wildfires in 2007, capturing thermal-infrared imagery to aid firefighters battling the remaining uncontrolled blazes.

The flights took Ikhana over the major blazes in the in the San Bernardino National Forest. Ikhana also imaged the Santiago Fire in the Cleveland National Forest in Orange County on Friday, Oct. 26. Ikhana carried the Autonomous Modular Scanner that incorporates thermal-infrared imaging sensors and real-time data communications equipment. The sensor is capable of peering through thick smoke and haze to record hot spots and the progression of wildfires over a lengthy period. The data is downlinked to NASA Ames where it is overlaid on Google Earth maps, then transmitted in near-real time to the Interagency Fire Center in Boise, Idaho, and made available to fire incident commanders to assist them in allocating their fire-fighting resources.

Moreover, in 2009, NASA equipped a Global Hawk (Figure 12.7) with a number of scientific sensors such as weather sensors, and lightning sensors, and Ku-band radar for research missions and to demonstrate new technologies. The bulbous nose houses

Figure 12.7 NASA Global Hawk. *Source:* Courtesy of NASA, Carla Thomas.

communications and sensor payloads on Earth science missions. Some science applications included: (a) measurements of the ozone layer; (b) cross-Pacific transport of air pollutants and aerosols; and (c) monitoring the development of Atlantic basin Hurricanes. A UAV with a lightning sensor would track air to ground lightning strikes to identify potential fire targets, prioritize potential targets based on risk to people and property, guide flight trajectory, and investigate and report fires.

Recently developed scientific sensors developed by NASA and NOAA are: (a) Airborne compact atmospheric mapper (ACAM); (b) Cloud physics lidar (CPL); (c) Focused cavity aerosol spectrometer (FCAS); (d) Meteorological measurement system (MMS); (e) Microwave temperature profiler (MTP); (f) High-definition video (HDV) system; (g) Nuclei-mode aerosol size spectrometer (NMASS); (h) Dual-beam UV photometer for accurate Ozone measurements (O_3); (i) Chromatograph for atmospheric trace species (UCATS); (j) Ultra-high sensitivity aerosol spectrometer (UHSAS); and (k) Laser hygrometer (ULH) for in-situ high-accuracy atmospheric water vapor measurements. All of these sensors are employed as scientific payloads in a NASA Global Hawk for various Earth science flight missions.

The InSitu ScanEagle small unmanned aircraft system (sUAS) is outfitted with special sensors consisting five modular cartridges and four pass-through particle collectors, enabling multiple measurements. It is controlled by satellite and has the capability of traveling 250 nautical miles and then loitering over the area for an hour.

12.6.2 Temperature Sensor

The list of UAV sensors as scientific payloads is very long. In this section, the technical specifications of a thermal sensor – as an example - is presented. The TMP37 temperature sensor (from Analog Devices[4]) is a silicon based temperature sensor (i.e., thermometer) with three pins, the ground pin, the analog voltage pin, and the positive voltage pin with a range of 2.7–5 V as shown in Figure 12.8. The change in temperature around the sensor provides the proportional voltage across the diode of the sensor. The amplification of the generated voltage provides an analog signal.

Figure 12.8 TMP37 temperature sensor.

The TMP37 is a low voltage, low cost, precision centigrade temperature sensor. It provides a voltage output that is linearly proportional to the Celsius temperature. The low output impedance of the TMP37 and its linear output and precise calibration simplify interfacing to temperature control circuitry and ADCs. The TMP37 is available in 3-lead TO-92, 8-lead SOIC_N, and 5-lead SOT-23 surface-mount package.

This device is intended for single-supply operation from 2.7 V to 5.5 V maximum. The supply current runs well below 50 μA, providing very low self-heating – less than 0.1 °C in still air. In addition, a shutdown function is provided to cut the supply current to less than 0.5 μA. Table 12.8 summarized the features of a TMP37 temperature sensor.

Table 12.8 Features of a TMP37 temperature sensor.

No	Parameter	Value
1	Temperature range	−40 °C to +125 °C
2	Voltage	2.7 V to 5.5 V
3	Current	Less than 50 μA
4	Linearity	±0.5 °C
5	Scale factor	20 mV/°C
6	Accuracy	±2 °C
7	Dimension	3 mm × 3 mm × 1.45 mm

The TMP37 is intended for applications over the range of 5 °C to 100 °C and provides an output scale factor of 20 mV/°C. The TMP37 provides a 500 mV output at 25 °C. The TMP37 does not require any external calibration to provide typical accuracies of ±1 °C at +25 °C and ± 2 °C, over the −40 °C to +125 °C temperature range.

12.7 Military Payloads

In general, there are two groups of military missions: (a) delivery of lethal warhead to a target; and (b) military operations with no lethal warhead. The first group involves operations such as strike, interception, and attack (e.g., improvised explosive device, and landmine detection and destruction). The latter includes missions such as military reconnaissance and surveillance, electronic intelligence (ELINT), and target designation by laser. Please note that guided weapons (e.g. cruise missiles) are not covered in this book. Payloads relating to the latter group have been discussed in Section 12.5. In this section, only weapon payloads are addressed. The UAV which utilizes such payloads are in the category of UCAV.

The weapon payloads primarily include stores and munitions such as: (a) missile; (b) rocket; and (c) bomb. These weapons are frequently attached underneath the UAV, and the time of need, they are released or dropped, or launched. The number of weapons required for a mission can be traded off against longer range, lighter weight, smaller size, and lower cost. If the radius of action is shorter than the designated range, more payload can be carried with less fuel.

An important requirement for an armed UAV is that it should be capable of taking off with a useful load of weapons. There are a number of factors affecting the capability of a UAV to carry a lethal payload including the payload weight. In general, the Army missiles are less heavy than the Air Force missiles. For this reason, the Predator UAV was originally equipped with Hellfire missiles; simply because the Air Force air-to-ground missiles are too heavy for the Predator to carry.

There are variety of weapon payloads that can be carried by a combat UAV in terms of weight, size, configuration, type of warhead, and the mission. In this section, as an example only one missile is introduced. For more information about any other aerial weapons, you may consult with Department of Defense (DOD) publications.

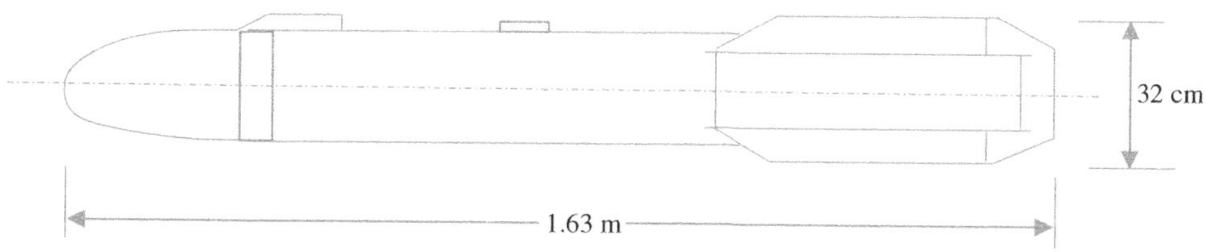

Figure 12.9 AGM-114 Hellfire missile.

The AGM-114 Hellfire (Figure 12.9). is an air-to-surface missile initially developed for anti-armor application, but later models were developed for precision strikes against other target types. The basic Hellfire has a length of 163 cm, a mass of 45 kg, and is equipped with a solid fuel rocket motor, and a semi-active laser homing guidance system. It can target tank, armored vehicles, and even individuals. The range of this missile is 8 km. a Predator UAV can carry up to two Hellfire missiles; and have been used in a number of targeted killings of high-profile individuals.

Air-to-surface missiles are sophisticated and very expensive weapon payloads, and currently controlled (i.e., launched) from a GCS. The design of a UAV to carry such payload is a very challenging task. There are mechanical and electrical interfaces, extreme care must be taken to guarantee the safety of UAV over the launch period.

General Atomics MQ-1 Predator has two hard-points under the wing. It has provisions to carry either two AIM-114 hellfire missiles, or 4 AIM-92 Stinger, or 6 AGM-176 Griffin air-to-surface missiles.

The Predator can carry two Hellfire missiles, at a maximum speed of 135 mph and has a range of 770 miles. The Hellfire missiles organizationally do not belong to Air Force; they are utilized, since they are light (14.2 kg) and small (53 cm long) enough to be attached and carried by a Predator. The air force missiles are relatively heavy (86 kg) and big (2.89 m long).

12.8 Electronic Counter Measure Payloads

A military UAV should be able to have a self-defense system, otherwise it will crash, or will be shut down by enemy missiles. There are a number of ways that a UAV may become unable to conduct its mission, or to follow the ground pilot's command. For instance, when a UAV loses the communication link with its GCS, its control will be disrupted. Another threat to a UAV flight safety is jamming. Jamming attacks on UAVs are a subset of denial of service (DoS) attacks in which malicious nodes block legitimate communication by causing intentional interference in data link.

When survivability becomes vital, the survivability of a UAV can be increased with a number of countermeasures. Techniques to nullify DoS attacks, are grouped in a method called ECMs. An ECM is a dummy object, or an electronic device, or a communication operation (e.g., anti-jamming) aimed to deceive or fool enemy detection systems in a battlefield. The deceive operation may include: (a) make an object (e.g., UAV) to disappear to an eye of a radar of an incoming missile; (b) make one object as many separate targets to the enemy; and (c) make an object to seem moving about randomly. Most military UASs use ECM to protect their UAVs against an attack.

Basic self-protecting (defensive) ECM strategies are: (a) radar interference; (b) target modifications; and (c) changing the electrical properties of air. Two ECM interference techniques are: (a) jamming; and (b) deception of homing missiles.

Jamming is defined as the disruption of existing signal by decreasing the signal-to-noise ratio at enemy missile side through the deliberate transmission of interfering signals. Jamming is accomplished by a UAV platform to transmit signals on the radar frequency to produce a noise level sufficient to hide echoed signals of the UAV. The jammer's continuous noise transmissions provide a clear direction to the enemy radar, but no range information. Jamming as a defense is to counteract eavesdropping attacks.

A UAV communication system can be jammed in various ways using different types of jammers. In such case, jamming is an attack, not defense. Jamming makes use of intentional radio interferences to harm communications [158] by keeping transmitter/receiver busy. This causes a transmitter to back-off whenever it senses busy wireless medium, or corrupted signal at receivers.

Deception employs a transponder to mimic the radar echo with a delay to indicate incorrect range. Transponders increase return echo strength to make a small decoy object appear to be a large target. One way is to use cheap radar decoys to trick the enemy into firing missiles at the decoys. Another deception technique is the dispersal of small metal (usually aluminum) strips called chaff/flare to change the electromagnetic properties of air in order to provide confusing radar signals. The chaff is such that it generates an RCS comparable to that of the UAV. An alternative technique for deception is to employ inexpensive escort UAVs to attack enemy search radars. The decoy and chaff can only be employed once in a flight, while the jamming can be continuously used. Hence, decoy and chaff are limited ECM resources.

The Global Hawk has three levels of defense operations against probable incoming missiles. This vehicle is equipped with a modular self-defense system which includes: (a) an AN/ALR 89 radar warning receiver; (b) an on-board jamming system; and (c) an ALE 50 towed decoy system (Figure 12.10).

ECM platforms vary widely in size, weight, cost, electric power, and countermeasure capability. Selection and design of electronic countermeasure payloads for a UAV depends upon potential threats, and incoming signal. Further discussion is beyond the scope of this text, the interested reader may refer to references such as [158, 159] for more details.

The integrated defensive electronic countermeasures (IDECM), suite; developed by ITT Electronic Systems and BAE Systems [159]; incorporates the ALQ-214 receiver/

Figure 12.10 Towed decoy system for Global Hawk.

processor/jammer and the ALE-55 fiber-optic towed decoy. It also includes the ALR-67 radar warning receiver and the ALE-47 chaff/flare dispenser. The individual systems are integrated to increase aircraft survivability. The IDECM suite will include: (a) IR/EO missile approach warning system; (b) ALR-67(V) radar warning receiver; (c) ALE-47(V) countermeasures dispenser; (d) three ALE-50(V) towed decoy (GFE); and (e) ALQ-214 jamming techniques generator. The unit cost – with a total volume of 0.06 m^3 and a total mass of 73.8 kg– including both ALQ-214 and ALE-55 is $1.7 million.

The IDECM provides three layers of defense: 1. Suppression, which denies, delays, and degrades adversary acquisition and tracking. 2. Deception, which misleads guided weapons away from the carrying aircraft if a track solution is achieved and a launch occurs. 3. End-game, makes the decoy the preferred target, seducing adversary missiles that leak through the first and second layers of defense.

12.9 Payload Installation

When the payload for a UAV is selected, the installation is the next step which may even affect the UAV configuration. The payload will require at least a space and power allocation. The payload installation has various aspects, four of which are briefly addressed in this section. (a) payload wiring; (b) payload location; (c) payload aerodynamics (if mounted externally); (d) payload-structure integration; and (e) payload stabilization.

It is recommended to employ open-architecture avionics system without altering the airframe itself. This initiative will move away from heavily integrated payloads, sensors, and communications systems, which leverages highly modular components with standard interfaces.

12.9.1 Payload Wiring

All payloads often consume electric energy, so they are connected to a power source (VCC) such as a battery pack; this requires a wire connection (wire 1). In addition, every payload is controlled by the microcontroller via a wire (wire 2, or Com Port). Moreover, the output of all payloads is sent to the microcontroller for processing and further application, and this is implemented through wire 2. Finally, to close the circuit, all circuits are connected back to the power source with the ground (GND) wire (wire 3). This is the case for almost all types of payloads from small to large ones. In this section, the wiring of a camera – as an example – is briefly described.

Cameras often have three wires: (a) one for video signal (usually white); (b) one for electrical input (usually red); and (c) one for ground (usually black). The camera is stabilized with the help of three gimballed gyros (Figure 12.11). which are mechanically rotated. The rotation mechanism is controlled from microcontroller based on the pre-program. The camera images/videos are sent to the microcontroller, and after processing, they are sent to transmitter for transmitting the data to GCS as a radio signal.

A 3-axis gimbal provides a stable platform for the camera to capture quality image during UAV flight maneuvers. The gimbal platform can rotate the camera about three axes, independent of the UAV attitudes. A gimbal head rotates a camera around its center of gravity (CG). Wiring/circuit for a camera has more elements, the details are

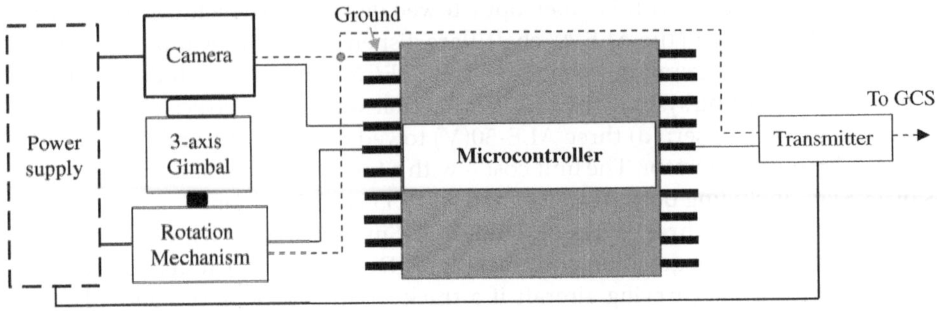

Figure 12.11 Wiring for a camera in a unmanned aerial vehicle (UAV).

not shown here for clarity. For instance, the connections of elements with the power supply are controlled by the microcontroller. These connections are not exhibited in Figure 12.11.

Compared to traditional cameras, miniaturized cameras (e.g., JPEG Camera module) with serial communication interface have been recently introduced. The camera is capable of capturing high-resolution images, sending them as JPEG through the serial port. Images will be saved on a SD-card in JPEG format file. This camera has a four-pin connector where two pins are for the power supply (+5V and GND) and two for the serial port (RX and TX). To connect the camera to Arduino, connect the red wire (RX) to Arduino Digital 3 pin (data into the module); the brown (TX) to Digital 2 pin (data from the module); the gray to the + 5V signal, and the violet to ground (GND). To save camera images, you need to connect an external storage (such as SD card) using two more wires. One wire is connected to the second Ground pin, and the second wire to the CVBS pin.

12.9.2 Payload Location

Generally, there are two locations for a payload: (a) inside UAV (i.e., buried); and (b) outside of UAV (usually podded/faired). If the payload has a large size, and it does not have an aerodynamic shape (e.g., radar), it is recommended to be mounted inside the UAV.

If a payload is considered to be installed inside a UAV (e.g., a radar), space management and CG adjustment are required. In contrast, if the payload is installed out of the UAV (e.g., below the fuselage), lofting technique is employed to reduce payload drag. When a payload is installed inside the body, the designer must make sure that the payload performance is not negatively affected. For instance, the design of radome (i.e., its material, thickness, and curvature) will influence the performance of a radar.

Since a radar needs to have a signal connection to the target, the radar radome should not be made of metallic materials. This is also true for a satellite communication antenna. An antenna radome made of composite material can handle surface structural stress, while allowing the signal to pass through for both directions.

Two important factors affecting the payload location are the payload mass and dimensions, which is subsequently driven by the needs of the operational tasks. The type and performance of the payloads can range from: (a) relatively simple sub-systems consisting of an un-stabilized video camera with a fixed lens having a mass as little as 200 g; to (b) a video system with a greater range capability, employing a longer focal length lens with zoom facility, gyro-stabilized and with pan and tilt function with a mass of probably 3–4 kg, to 3. a high-power radar having a mass, with its power supplies, of possibly up to 1000, kg.

The distribution of aircraft weight will influence the airworthiness and performance via two aircraft parameters: (a) aircraft CG; and (b) aircraft mass moment of inertia. Aircraft must be stable, controllable, and safe for all allowable aircraft cg locations during flight envelope. One of the primary concerns during aircraft design process related to the payload location is the UAV weight distribution. The distribution of UAV weight (sometimes it is referred to as weight and balance) will greatly influence airworthiness as well as aircraft performance. The aircraft CG is the cornerstone for aircraft stability, controllability, and trim analysis.

If the payload is expendable (e.g., weapon), the location of the payload will influence the cg range too. When the military payload is delivered, the UAV cg will move to a new location. As a rule of thumb, the best aircraft cg location is to be around wing-fuselage aerodynamic center. The distance between the most forward and most aft CG limits is called the CG range or limit along x axis. In general, the location of a payload must be such that the entire UAV cg is within the acceptable range (both initially, and after the payload is released). For technique to distribute weight, and to determine the UAV range of CG, refer to Chapter 11 of Ref. [8].

12.9.3 Payload Aerodynamics

Most payloads do not originally have a streamlined shape, and they are usually installed outside of the UAV body (e.g., below fuselage). For such cases, techniques (e.g., lofting) should be employed to reduce payload drag. This is the case for almost all weapon payloads and cameras. In designing a cover (i.e., Fairing) for payload, both aerodynamics and overall weight need to be considered. The fairing is a partial cover (Figure 12.12) that has a streamlined shape such as airfoil. A well-designed fairing can reduce the drag of a covered component as much as 90%. Moreover, a fairing will reduce the flow interferences with lifting surfaces such as wing and tail. Thus, the wing and tail will be aerodynamically more efficient with faired payloads.

A short fairing is light, but it may create a high drag. In contrast, a long fairing may decrease drag, but it will have a higher weight. Hence, a trade-off is necessary between weight and size. As a general rule, the thickness-to-length of an effective fairing should be less than 20%. The length is measured along the flight path.

Keep the aerodynamic drag of the UAV as low as possible commensurate with the practical installation and operation of the payload, and radio antennae. Where the payload is connected to the fuselage, some forms of filleting (e.g., fillet) are required to avoid airflow separation and turbulence. The exact shape may be determined by wind

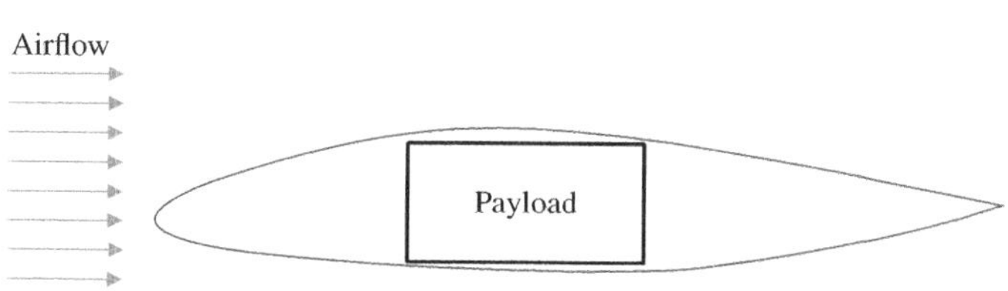

Figure 12.12 Payload fairing (side-view, and top-view).

Figure 12.13 Fairing for two external antennas of a Predator B Reaper.

tunnel experiments. Figure 10.8 demonstrates the aerodynamic fairing for Penguin C stabilized EO/IR Cameras. Figure 12.13 shows the fairing for two antennas (e.g., data link LOS) of a Predator B Reaper.

The aerodynamic design of the air vehicle – including some guidelines in payload installation – is presented in Chapter 4. For fundamentals of aerodynamics, please refer to references such as [14, 15].

The best place for a radar to track ground moving targets (e.g., truck) is under the forward of fuselage nose. In contrast, the best location for a satellite communication dish antenna is inside forward of top of the fuselage. The dish antenna is usually large (20 to 60 in. in diameter), and it needs to rotate at least 120°. For a fuselage nose when it houses a large radar dish, or a massive satellite communication dish antenna, there are at least five requirements: (a) allow the antenna to rotate freely; (b) low drag; (c) low distortion; (d) low cost; and (e) light. These requirements will make the fuselage nose parabolic, bulbous, or whale-shaped.

A bulbous nose allows a massive satellite dish to rotate freely; and to connect to and track the communications satellites. The whale-shaped forward fuselage of Global Hawk covers a parabolic antenna for a high-bandwidth datalink. In is interesting to note that, large commercial aircraft all have GPS, but they do not use that kind of large dish to communicate with satellites. Using satellite communications, the operators in GCS

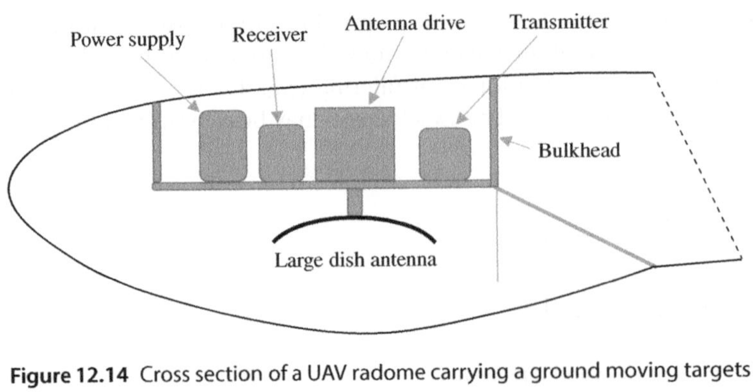

Figure 12.14 Cross section of a UAV radome carrying a ground moving targets radar.

will receive the reconnaissance data in real time. Figure 12.14 illustrates the cross section of the low-drag, low-distortion radome of a UAV carrying a ground moving targets radar. Such a nose allows a massive radar dish to rotate freely; and to efficiently track the ground moving targets, while generating a low drag.

Techniques to determine UAV drag (including fairing drag) are presented in references such as [9]. The aerodynamic drag of a payload (e.g., camera) is a function of the following factors: (a) surface angle of attack; (b) geometry (e.g., frontal area; S_f); (c) fairing; (d) airspeed (V); (e) air density (ρ); and (f) Reynolds number of the airflow. The drag is determined by:

$$D_P = \frac{1}{2}\rho V^2 S_f C_D \tag{12.6}$$

where C_D is the payload drag coefficient. A number of symmetrical shapes having drag values [160] at low speed are illustrated in Table 12.9. The drag coefficient values in this table are based on the frontal area. In this table, the flow is coming from left to the right.

Table 12.9 Drag coefficient values for various geometries and shapes.

a. Two dimensional bodies (L: length along flow, D: length perpendicular to the flow)				
No	**Body**	**Status**	**Shape**	**C_D**
1	Square rod	Sharp corner	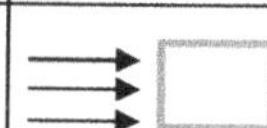	2.2
		Round corner		1.2
2	Circular rod	Laminar flow	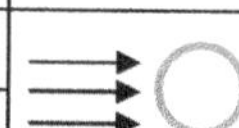	1.2
		Turbulent flow		0.3
3	Equilateral triangular rod	Sharp edge face		1.5
		Flat face	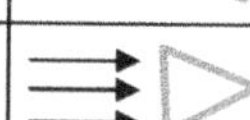	2
4	Rectangular rod	Sharp corner 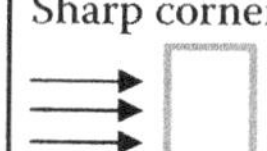	L/D = 0.1	1.9
			L/D = 0.5	2.5
			L/D = 3	1.3
		Round front edge 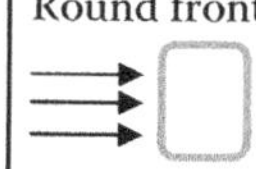	L/D = 0.5	1.2
			L/D = 1	0.9
			L/D = 4	0.7
5	Elliptical rod 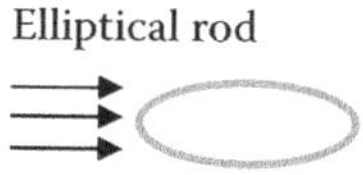	Laminar flow	L/D = 2	0.6
			L/D = 8	0.25
		Turbulent flow	L/D = 2	0.2
			L/D = 8	0.1

(Continued)

Table 12.9 (Continued)

a. Two dimensional bodies (L: length along flow, D: length perpendicular to the flow)				
No	**Body**	**Status**	**Shape**	**C_D**
6	Symmetrical shell	Concave face	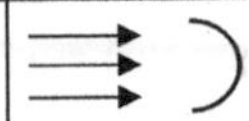	2.3
		Convex face	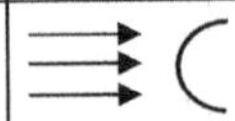	1.2
7	Semicircular rod	Concave face	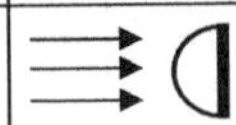	1.2
		Flat face	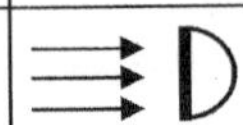	1.7

b. Three dimensional bodies (L: length, D; diameter)				
No	**Body**	**Laminar/turbulent**	**Status**	**C_D**
1	Cube	Re > 10 000		1.05
2	Thin circular disk	Re > 10 000		1.1
3	Cone (θ = 30°)	Re > 10 000		0.5
4	Sphere	Laminar Re ≤ 2×10^5		0.5
		Turbulent Re ≥ 2×10^6		0.2
5	Ellipsoid	Laminar Re ≤ 2×10^5		0.3–0.5
		Turbulent Re ≥ 2×10^6		0.1–0.2
6	Hemisphere	Re > 10 000	Concave face	0.4
		Re > 10 000	Flat face	1.2
7	Rectangular plate	Re > 10 000	Normal to the flow	1.1–1.3
8	Vertical cylinder	Re ≤ 2×10^5	L/D = 1	0.6
			L/D = ∞	1.2
9	Horizontal cylinder	Re > 10 000	L/D = 0.5	1.1
			L/D = 8	1
10	Parachute	Laminar flow		1.3

Techniques to determine the drag of major UAV components such as wing and fuselage are provided in another text [9] written by the author.

Example 12.3

A small quadcopter has a camera installed under the body. The camera has a cubical shape with round corners, a frontal area of $5 \times 10\,\text{cm}^2$, and has no fairing. Determine the drag for the camera. The quadcopter is cruising at sea level with a speed of 20 m/s. Assume the Reynolds number is greater than 10 000.

Solution

According to Table 12.9, the drag coefficient of the cubical camera is 1.05. At sea level, the air density is $1.225\,kg/m^3$. The frontal area of the camera is $5 \times 10\,cm^2 = 0.005\,m^2$.

$$D_P = \frac{1}{2}\rho V^2 S_f C_D = \frac{1}{2} \times 1.225 \times 20^2 \times 0.05 \times 0.1 \times 1.05 = 1.28\,N \quad (12.6)$$

12.9.4 Payload-Structure Integration

Whether the payload in inside or outside of the UAV, the structural integrity of the UAV must be maintained such that the various flight loads (e.g., gust, aerodynamic, and weight) and stresses (e.g., normal, shear, bending) are handled safely by the structure. The load is significantly higher in high-g turn and hard landing. When the payload is installed inside the UAV, a special mounting should be designed to allow the payload to perform properly.

If the payload is installed out of the UAV (e.g., below fuselage), provisions are required for mounting payloads on so-called "hard-points" under the wing or fuselage. The structure must be designed to provide mounting points capable of supporting the launch rails or bomb racks. Installation of a payload outside of structure requires a cutout by reinforcing the neighboring structural elements (e.g., stiffener, frame). When a missile is launched, the reaction force is passed from rails to the structure. Thus, the structural elements near the rails must be reinforced. The UAV structural design is outside the scope of this book, but the reader may refer to references such as [16].

In a stealth UAV, one concern for the installation of a payload is the radio/radar signature. There are basically three methods of minimizing the reflection of pulses back to a receptor: (a) to manufacture appropriate areas of the UAV from radar-translucent material such as Kevlar or glass composite as used in radomes which house radar scanners; (b) to cover the external surfaces of the aircraft with RAM (radar absorptive materials); and (c) to shape the aircraft externally to reflect radar pulses in a direction away from the transmitter.

Payload integration and operation of an instrumented Yamaha RMAX unmanned helicopter (Figure 1.1) is implemented by [21]. The hardware components that make up the baseline flight avionics include general purpose processing capabilities and sensing, and add approximately 35 lb to the basic airframe, leading to a total weight of approximately 157 lb. These components have been packaged into four exchangeable modules: (a) two computer modules; (b) the GPS module; (c) the data link module (wireless Ethernet, wireless serial, Ethernet switch); and (d) the inertial measurement unit (IMU) module.

Figure 12.15 demonstrates a sample payload configuration in a UAV with a pusher engine for a surveillance mission. All equipment, components, and payloads must be located within fuselage to allow an efficient flight operation. The integration of payloads with the flight control system, guidance system, and other internal elements is a significant challenge, since several requirements are conflicting.

The Global Hawk aluminum fuselage contains pressurized payload and avionics compartments. Figure 12.16 illustrates the payload and instrument layout in a NASA Global Hawk. Please note that some of the sensors should be placed in a pressurized space, in order to perform normally and efficiently. For instance, a cloud physics lidar (CPL) can perform in an unpressurized space, while Dual-beam UV photometer for accurate Ozone measurements (O_3) needs to be placed in a pressurized compartment.

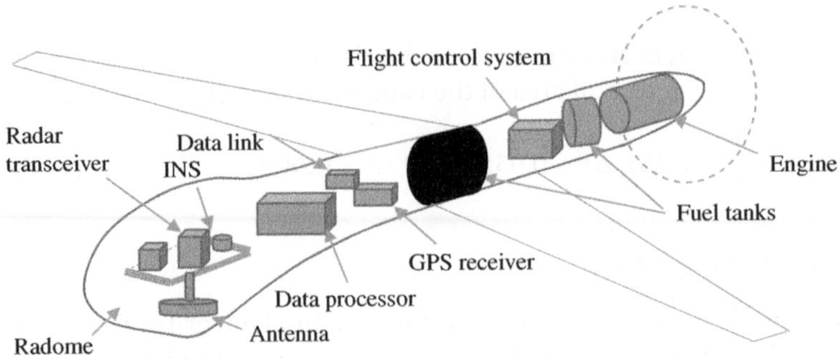

Figure 12.15 Payload configuration in a UAV with a pusher engine.

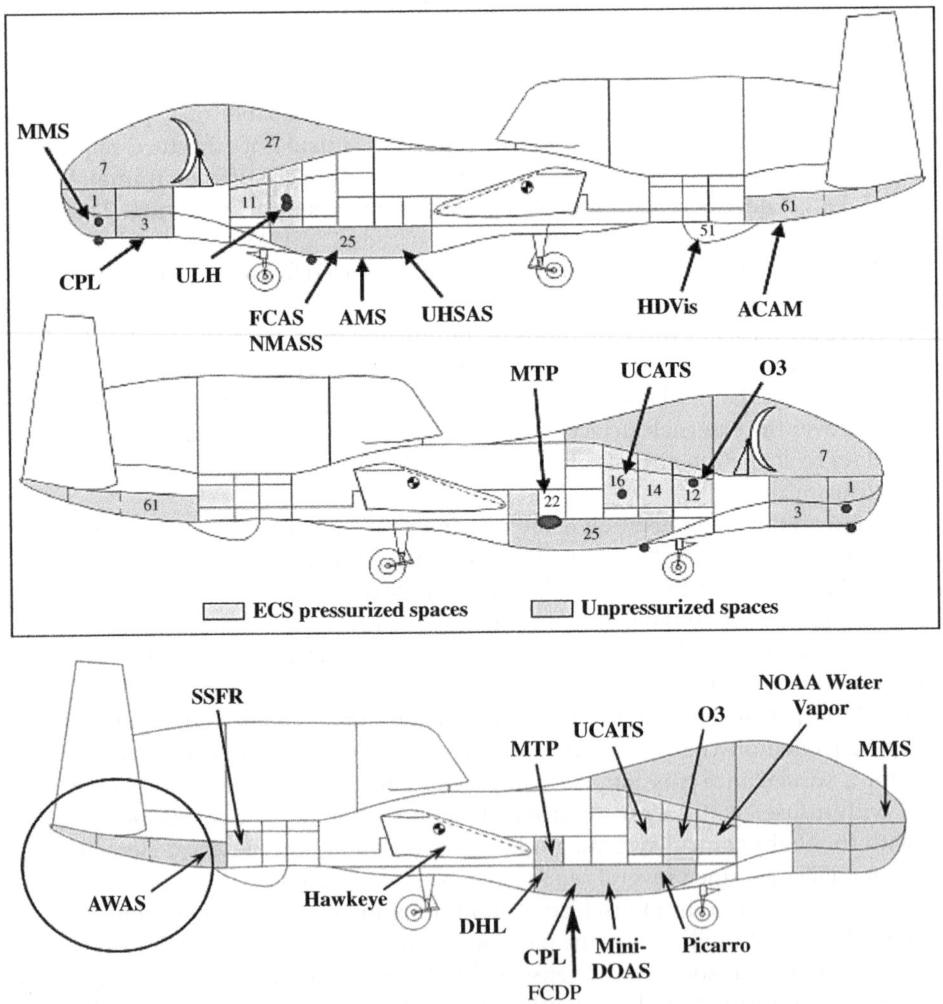

Figure 12.16 Instrument layout in a NASA Global Hawk (*Source:* NASA, https://espo.nasa.gov/attrex-epo/image/ATTREX_Global_Hawk_Instrument_Layout).

12.9.5 Payload Stabilization

Most payloads (e.g., cameras) and sensors (e.g., thermal imaging sensor, laser range finder, and illuminators) need a stable location, in order to have an acceptable photographic performance. Moreover, in order to obtain a quality image, the line-of-light must be directed to angular space at a desired angle. These two requirements are achieved by a stabilization system, which is a closed-loop system in which the motion of the base is sensed and fed back. Camera stabilization systems often use gyroscopes to sense disturbances. A camera stabilizer (i.e., camera–stabilizing mount), is a device designed to hold a camera in a way that prevents or compensates for undesired camera movement, such as vibration.

Rotary elements (e.g., propellers), and reciprocating elements as in piston engines are two frequent sources of vibration in UAVs. Moreover, atmospheric disturbances such wind gusts and clear air turbulences impose undesired motions to cameras and sensors.

Single-axis gimbal heads are used in order to allow a balanced movement for cameras for easy and smooth manipulation while tracking moving subjects. The 3-axis gimbals are designed to give the camera the independence of UAV structure without camera vibration. Powered by three electric motors, the gimbals have the ability to keep the camera level on all axes as the UAV vibrates or shakes. An IMU feels the feedback and responds to movement and utilizes its three separate motors to stabilize the camera.

With the application of an algorithm, the stabilizer is able to notice the difference between deliberate movements such as desired rotation from unwanted shake. This allows the camera to not feeling any unexpected movement.

A popular and effective element for mounting cameras is the use of gimbals which can provide stabilization. Gyro-stabilized gimbals which house multiple sensors are utilized for airborne surveillance applications such as pipe/power-line/bridge inspection, airborne law enforcement, mapping, surveying, and surveillance.

A controller sends commands to actuators to rotate the gimbals. So, any motion of the base is nullified by the gimbals, hence the camera remains still. For instance, if the base is pitched up 5°, a gimbal is pitched down by −5°.

A camera may be nose-mounted or in turret-mounted configuration. Stabilization of the sight-line is necessary to off-set the effects of both UAV movements and displacements of the aircraft due to gusty/turbulent air. To make the lens FOV narrower, the sight-line must be held more precisely. The three-axis (x, y, and z) stability is the highest possible way to provide the sight-line stability.

There are two basic techniques of maintaining a stabilized line-of-light: (a) the camera assembly is mounted on a stabilized platform with attitude and rate sensors; and (b) the camera assembly is mounted on gimbals remote from the attitude sensors. In the first technique, the camera with its gimbals is mounted on a platform which is gyro-stabilized. Fast-reacting actuators are ready to correct any movement away from the horizontal. Thus, the sight-line keeps its direction of view relative to the horizontal base. In the second technique, the whole camera is contained within an exchangeable payload module with the gimbals' base fixed to the structure of that module to provide a stable picture. In general, the second method offers a better maintainability, while it is less expensive. The aircraft attitude data is coming from the navigation system and used to drive the gimbal actuators.

12.10 Payload Control and Management

Weapon payloads have mechanical and electrical interfaces; extreme care must be taken to guarantee the safety of the UAV at the time of launch. An external payload (e.g., missile, camera) has an electrical connection to the UAV via a connector; referred to as umbilical. The connector can be plugged into the payload and will come free when the payload is released/launched. The configuration of the connector assembly must be such that it provides a safe separation for both payload and UAV. Various data is exchanged through an umbilical. A few examples are: (a) imagery from camera/seeker; (b) control signals; and (c) launch signals.

In addition to maintaining control and stability of the aircraft, it is just as important to achieve that for the payload. Control of the UAV is needed to get the vehicle over the target area, but will be useless, unless the payload is properly controlled. The payload control is achieved using a system which is either part of the aircraft flight control system or by using a separate module. For a camera, the output of the control process, is to bring the sight-line accurately on to the target and keep it there. The gyro-stabilization of the sight-line is a function of the camera control system.

The integration of the payload control and stability systems and flighty control system is a challenging task. In an integrated system, the same set of gyros will support the control and stabilization of both aircraft and payload. The integrated system manages two sets of coordinate axes, those of the aircraft and those of the payload, even though the latter is fixed within the aircraft. Thus, the payload sight-line is at the target while the UAV is flying in another direction.

Another aspect of the payload control is the data management, processing, and transmission. Images and/or data from the various payloads must be transmitted to the GCS. The output from sensors within the payload will require processing and converting to radio signals. Different sensors will require different amounts of radio bandwidth and at different data rates. Thus, it is necessary that as much processing of the signals as is practical is carried out within the UAV in order to reduce signal loads on the communication system. In general, the basic fundamentals of payload control are very similar to flight control system. Both have feedback, controller, and measurement devices. The control theory and control system design are covered in Chapter 6, so they are nor repeated here.

12.11 Payload Selection/Design Guidelines

Selection of the payload has a significant impact on the UAV configuration, since it is tied with UAV mission and performance. A UAV designer must be aware of the ambiguity about payload capacity, and so use a consistent definition. Sometimes, the payload in-flight function may interfere with the UAV flight management. This interference should be accurately planned and managed in order not to compromise the flight safety.

New challenging flight missions demand new, modern, and sophisticated payloads, which in turn require new platform development. However, certain payloads may also require access to UAV system data, such as position, attitude, or airspeed. Thus, a mechanism must exist to provide UAV data to the payload in a manner such that the failure of the payload cannot impact the safety of the UAV's own systems. In some cases, a UAV is required to be capable of a range of different roles, and of carrying various payloads and different capability in the mission. Thus, a compromise will be necessary in designing payload compartment.

In the preceding sections, features of various payloads and techniques to control, handle, and manage have been described. Moreover, the installation issues including drag reduction, payload-structure integration, and optimum payload location were addressed. In selection/design process of payload(s) for a UAV, one must select the type of payloads, select the payload devices, and then conduct some calculations and performance/stability analysis.

In general, the primary factors affecting the selection/design process of payload are as follows: (a) manufacturing technology; (b) accuracy and precision; (c) size; (d) volume; (e) reliability; (f) life-cycle cost; (g) UAV configuration; (h) stealth requirements; (i) maintainability; (j) energy consumption; (k) integration; (l) installation; (m) power consumption; (n) interference with flight safety; (o) compatibility with control system; (p) weight; (q) radio/radar signature; (r) environmental factors; and (s) aerodynamic considerations. Figure 12.17 illustrates a flowchart that represents the payload (s) selection/design process.

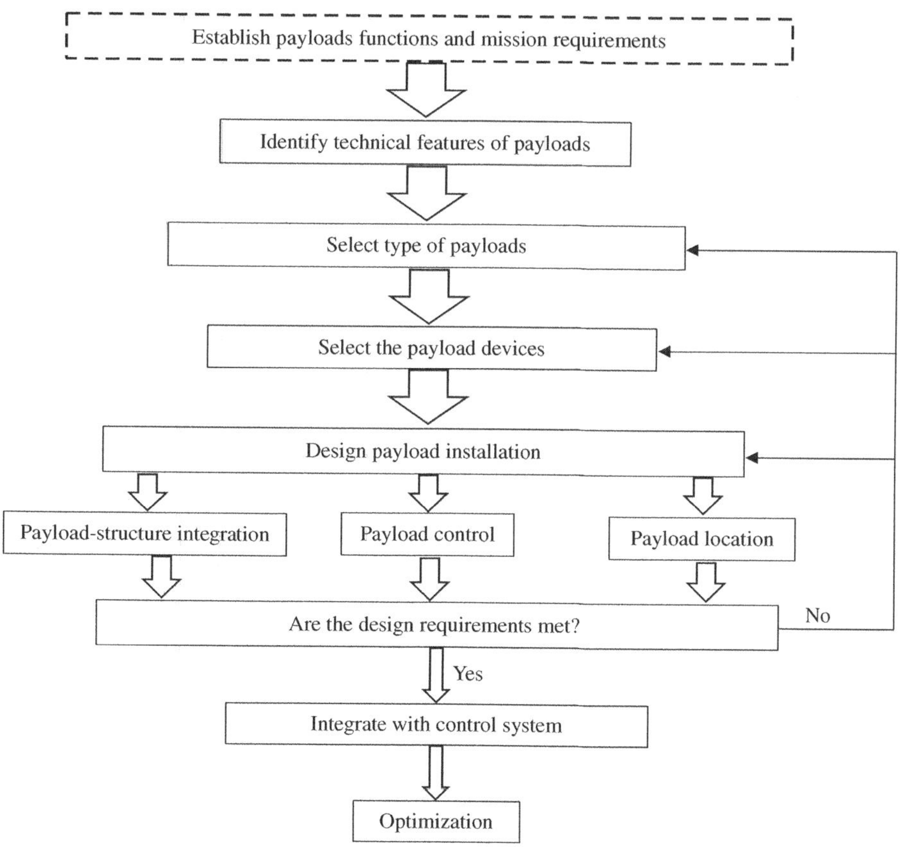

Figure 12.17 Payload selection/design process.

For a given UAV, the range and endurance can be increased by decreasing the payload weight. Hence, one way to improve the performance of a UAV is to replace the classical payloads with modern payloads. For instance, light weight cameras and very light Micro Electro-Mechanical Systems (MEMS) electronic devices may be employed. An easy technique to increase the range/endurance of a military UAV is to replace part of the weapon on the wing hard-points with external fuel tanks.

In general, the design process begins with a trade-off study to establish a clear line between cost and performance (i.e., accuracy) requirements; and ends with optimization. The designer must decide about two items: (a) select type of payload; and (b) select the payload devices. Most payload devices may be ordered/purchased from off-the-shelf commercial market. After conducting the calculation process, it must be checked to make sure that the design requirements are met.

A very crucial part of the design process is to integrate the payloads with structure and control system (i.e., a consistent autopilot). If the complete payloads (e.g., camera) is selected/purchased, the integration process must be still conducted. This includes activities such as structural installation, management, interfaces, and electric power requirements. More details are outside the scope of this text.

A vision system (e.g., camera) is the most basic payload for majority of UAVs (from a small quadcopter to large Global Hawk). Selecting the right camera for an aerial imaging is a challenging task, since it will influence the UAV flight performance too. Camera technical parameters to be considered are: (a) pixel size; (b) focal point; (c) megapixel; (d) sensitivity; (e) sensor size; (f) shutter type; (g) dynamic range; and (h) bit depth. A few of these parameters are discussed here in brief.

Resolution is an important objective in an aerial digital imaging/filming. The ground sampling distance (GSD) is a measure of sampling limitation to image resolution. GSD is determined by the UAV's altitude (H), the focal length (f) of the camera lens, and the pixel size (px):

$$GSD = \frac{H.px}{f} \tag{12.7}$$

Hence, to have a higher image resolution, select a camera with a higher pixel size, and a shorter focal length. The shorter the focal length (e.g., 20 mm), results in the wider the angle of view and the greater the area captured. A longer focal length is necessary for objects further away (i.e., high altitude flight). Similarly, a shorter focal length captures more of the scene in front of camera. In an image with a 2-meter GSD, adjacent pixels image locations are 2 meters apart on the ground. However, while smaller pixels provide greater ground resolution, it will limit the speed of the UAV. The UAV speed is a function of on the pixel size, height above the ground, and the lens' focal length.

It is interesting to note that the GSD is not a function of the megapixel. The pixel size is determined by camera sensor. Indeed, the pixel size dictates the image resolution in an aerial imaging, not the number of megapixels. A low megapixel, but high pixel size will translate to a less UAV flight endurance, less fuel/power consumption, and lower overall operating costs. The sensor size determines the field of view.

Sensitivity and noise are parameters that influence both image resolution and the UAV's maximum speed. The more sensitive the camera, the less exposure time is needed, and the faster the UAV can fly. With a lower read noise, more gain can be used, while maintaining a quality image. Thus, it allows a shorter exposure time, which implies a faster allowable UAV speed. A higher dynamic range is very beneficial for imaging shaded areas on the ground, while maintaining visibility in well-lit areas. Required bit depth is determined by the application. If the images are being viewed with a human operator eyes, 8-bits of data per color channel are sufficient.

Questions

1. Define payload.
2. Provide at least five types of payloads that a UAV may carry for various missions.
3. What factors are affecting the quality of an image/video in a camera?
4. What provisions are necessary in designing payload compartment?
5. What are the primary factors affecting the payload selection/design process?
6. Name four scientific payloads.
7. Define the Doppler effect.
8. What are the typical radar frequencies used for UAVs?
9. What can be concluded from radar equation?
10. What is the typical outcome of the Doppler effect?
11. What factors do influence the resolution of a camera?
12. Describe the SAR.
13. What is the speed of a radio wave?
14. Describe receiver, transmitter, and antenna in brief.
15. What is LIDAR?
16. Provide three examples for the weapon payloads.
17. Name four aspects of the payload installation.
18. What is the functions of a payload umbilical?

19 Describe the overall payload selection/design process.

20 Define ECM?

21 What are the basic self-protecting ECM strategies?

22 Describe jamming of homing missiles.

23 What is chaff in ECM?

24 What is lumen?

25 How image resolution is measure?

26 What is the illuminance in an indirect daylight?

27 What is the illuminance in a night with full moon?

28 How the resolution of an image/monitor is described?

29 What are the frame rate of cameras with 1. PAL and 2. NTSC systems?

30 What can a 3-axis gimbal platform provide to a UAV camera?

31 What is the function of a range finder? Draw a figure to describe it.

32 What is the main elements of a range finder?

33 What is the function of a LD? Draw a figure to describe it.

34 What is the main elements of a LD?

35 What is the function of a radar warning receiver?

36 What is the function of the payload AN/APR-49?

37 Provide one advantage and one disadvantage of radar compared with an optical sensor.

38 What are three basic deception techniques for a UAV against an incoming missile?

39 What are the three layers of defense provided by the IDECM?

40 What is the role of chaff/flare in ECM?

41 Describe how gyros are used in stabilizing cameras.

42 What weather activities are typically detected by a weather radar?

43 List a number of recently developed scientific sensors by NASA for UAV applications (mention at least five items).

44 What is the type of output signal of a TMP37 temperature sensor; analog or digital?

45 What are the potential applications for a UAV with a lightning sensor which tracks air to ground lightning strikes?

46 What is the application of a laser hygrometer developed by NASA?

47 Describe features of a cloud physics lidar developed by NASA.

48 List technical specifications of a TMP37 temperature sensor.

49 What NASA Global Hawk sensors should be placed in a pressurized space, in order to perform normally and efficiently? Write at least three sensors.

50 A small camera has three wires for the connection to a microcontroller (e.g., Arduino board). Explain the connection destination for each wire.

Problems

1 A sound wave is moving with the frequency of 800 Hz. Assume the sound travels at 290 m/s, determine the wavelength.

2 Consider a radar with a pulse repetition frequency 1200 kHz. An echo from a distant target is received 15 μs after the pulse is transmitted.
   a What is the distance of the target from the radar (i.e., range)?
   b Determine the time between two pulses.

3 Consider a radar with a pulse repetition frequency 2000 kHz. An echo from a distant target is received 15 μs after the pulse is transmitted.
   a What is the distance of the target from the radar (i.e., range)?
   b Determine the time between two pulses.

4 An antenna of a UAV transmitter radiates a radio signal with the frequency of 20 GHz. Assume the wave travels with a speed of 299 792 458 m/s, determine the wavelength.

5 A UAV flying at an altitude of 2 km is directly behind and closing at a horizontal range of 10 km from an aircraft carrier. The UAV with a speed of 100 m/s parallel to

the ground, is tracking the ship with a radar with a frequency 12 GHz. The ship cruises (away) at 5 m/s. If the speed of radar wave is 3×10^8 m/s, calculate: a. the radar frequency detected by the aircraft carrier, b. the echo frequency detected by the UAV, c. the Doppler shift between the echo and source frequency, d. wave length.

6 A UAV flying at an altitude of 3 km is directly behind and closing at a horizontal range of 10 km from a carrier. The UAV with a speed of 50 m/s parallel to the ground, is tracking the ship with a radar with a frequency 20 GHz. The ship cruises (away) at 7 m/s. If the speed of radar wave is 3×10^8 m/s, calculate: a. the radar frequency detected by the aircraft carrier, b. the echo frequency detected by the UAV, c. the Doppler shift between the echo and source frequency, d. wave length.

7 A fixed sound source emits a sound with a frequency of 1 MHz on a day when the speed of sound in air is 340 m/s, and there is no wind. What is the frequency you will receive if you move toward the source at 20 m/s?

8 A moving sound source emits a sound with a frequency of 2 MHz on a day when the speed of sound in air is 330 m/s, and there is no wind. What is the frequency you will receive if the source moves toward you at 30 m/s? Assume you are stationary.

9 A small quadcopter has a camera installed under the body. The camera has a cubical shape with round corners, a frontal area of $3 \times 5\,\text{cm}^2$, and has no fairing. Determine the drag for the camera. The quadcopter is cruising at sea level with a speed of 15 m/s. Assume the Reynolds number is greater than 10 000.

10 A fixed-wing small UAV has a camera installed under the fuselage. The camera has a frontal area of $2 \times 3\,\text{cm}^2$, and it is covered with an ellipsoid fairing. Determine the drag for the camera. The UAV is cruising at sea level with a speed of 25 knot. Assume the Reynolds number is greater than 2×10^6.

11 A small UAV has an antenna installed above the fuselage. The antenna has a thin circular disk shape, and a frontal area of $1 \times 10\,\text{cm}^2$. Determine the drag for the antenna. The UAV is cruising at 12000, ft with an airspeed of 120 knot. Assume the Reynolds number is greater than 10 000.

12 A large UAV has an EO/IR camera unit installed under the fuselage. The camera has a frontal area of $10 \times 15\,\text{cm}^2$, and it is covered with a fairing that looks as a hemisphere with a concave face. Determine the drag for the camera unit. The UAV is cruising at 10000 ft with an airspeed of 90 knot. Assume the Reynolds number is greater than 10 000 ft.

13 A UAV flying at an altitude of 500 m is directly behind and closing at a horizontal range of 3 km from a ground moving truck. The UAV with a speed of 150 knot parallel to the road, is tracking the truck with a radar with a frequency 15 GHz. Doppler shift between the echo and source frequency is 6383 Hz. Calculate the truck velocity.

14 A UAV flying at an altitude of 300 m is directly behind and closing at a horizontal range of 1 km from a ground moving truck. The UAV with a speed of 90 knot parallel to the road, is tracking the truck with a radar with a frequency 20 GHz. Doppler shift between the echo and source frequency is 4497 Hz. Calculate the truck velocity.

15 A UAV flying with a speed of 130 knot is directly behind a manned aircraft at a horizontal range of 2 km. The UAV is tracking the manned aircraft with a radar at a frequency 25 GHz. Doppler shift between the echo and source frequency is 12012 Hz. Calculate the velocity of the manned aircraft.

Design Problems

16 You are required to select a payload (camera) for a small quadcopter to take image/video during hover and cruising flight. The vehicle is planned to fly only during daylight. The payload requirements are:

(a) cost of payload: less than $50; (b) UAV cruise velocity: 30 knot; (c) allowable payload mass: 30 g; (d) allowable frontal area of payload: $50 \times 50\,mm^2$; (e) required power: less than 2 W; (f) Voltage: less than 3 Volts; (g) FOV: more than 60°; (h) image/video resolution: 1080p; (i) endurance: more than 10 minutes; and (j) data could be stored in internal hard or SD card.

Analyze the given requirements, and nominate at least one camera to satisfy the desired requirements. You may use the manufactures' websites to search for various commercial cameras. Camera should be compatible to be mounted on a 3-axis gimbal gyro-stabilized turret. Provide the model and technical features of the selected camera.

17 You are required to select a payload (camera) for a fixed-wing UAV to take image/video during cruising flight. The vehicle is planned to fly only during daylight. The payload requirements are:

(a) cost of payload: less than $200; (b) UAV cruise velocity: 50 knot; (c) allowable payload mass: 100 g; (d) allowable frontal area of payload: $100 \times 100\,mm^2$; (e) required power: less than 10 W; (f) voltage: less than 5 Volts; (g) FOV: more than 45°, (h) image/video resolution: FHD: 4096 × 2160; (i) endurance: more than 15 minutes; and (j) data could be stored in internal hard or SD card.

Analyze the given requirements, and nominate at least one camera to satisfy the desired requirements. You may use the manufacturers' websites to search for various commercial cameras. Camera should be compatible to be mounted on a 3-axis gimbal gyro-stabilized turret. Provide the model and technical features of the selected camera.

18 You are required to select a payload (camera) for an unmanned helicopter to take image/video during hover and cruising flight. The vehicle is planned to fly only during daylight. The payload requirements are:

(a) cost of payload: less than $500; (b) UAV cruise velocity: 25 knot; (c) allowable payload mass: 800 g; (d) allowable frontal area of payload: $120 \times 150\,mm^2$; (e) required power: less than 15 W; (f) Voltage: less than 12 Volts; (g) FOV: more than 70°; (h) image/video resolution: UHD: 4096×2160; (i) endurance: more than 30 minutes; and (j) data could be stored in internal hard or SD card.

Analyze the given requirements, and nominate at least one camera to satisfy the desired requirements. You may use the manufacturers' websites to search for various commercial cameras. Camera should be compatible to be mounted on a 3-axis gimbal gyro-stabilized turret. Provide the model and technical features of the selected camera.

19 You are required to select a payload (camera) for a fixed-wing UAV to take image/video during cruising flight. The vehicle is planned to fly during day and night. The payload requirements are:

(a) cost of payload: less than \$1000; (b) UAV cruise velocity: 70 knot; (c) allowable payload mass: 1000 g; (d) allowable frontal area of payload: $200 \times 100\,\text{mm}^2$; (e) required power: less than 60 W; (f) Voltage: less than 20 Volts; (g) FOV: more than 60°; (h) image/video resolution: day: FHD, night: 384×288 (110 592 pixels); (i) endurance: more than 30 minutes; and (j) data could be stored in internal hard or SD card.

Analyze the given requirements, and nominate at least a set of two cameras to satisfy the desired requirements. You may use the manufacturers' websites to search for various commercial cameras. Camera should be compatible to be mounted on a 3-axis gimbal gyro-stabilized turret. Provide the model and technical features of the selected cameras.

20 A mini quadcopter UAV has a mission to travel inside a busy urban area during daytime, so it is desired to have sense-and-avoid system. The UAV is desired to have a full autonomous capability, and no ground pilot controller, with a range of 5 km, and endurance of one hour. What are the minimum payloads for such UAV? Just list the names and types.

21 A small quadcopter UAV has a mission to travel inside an open area during daytime. The UAV is controlled by a ground pilot via a handheld remote controller via line of sight. The only mission is to take video and send the videos live to the GCS. The desired range and endurance are 2 km and 20 minutes respectively. What are the minimum payloads for such UAV? Just list the names and types.

22 A military UAV has a mission to travel inside an unknown area during day and night times. It is controlled by a pilot and a payload operator via a GCS. The desired range and endurance are 5000 km and 20 hours respectively. One of the flight missions for the UAV is to identify specific targets and designate them. What are the minimum payloads for this UAV for such mission? Just list the names and types.

23 A military UAV has a mission to travel inside an unknown area during day and night times. It is controlled by a pilot and a payload operator via a GCS. The desired range and endurance are 10 000 km and 30 hours respectively. The flight mission for the UAV is to identify and attack specific fixed ground targets with military stores. What are the minimum payloads for this UAV for such mission? Just list the names and types.

24 A large military UAV has a mission to travel inside an unknown area during day and night times. It is controlled by a pilot and a payload operator via a GCS. The desired range and endurance are 12 000 km and 25 hours respectively. The flight mission for the UAV is to identify and attack specific moving ground targets with military stores. Moreover, the UAV should be able to deceive potential incoming enemy missiles. What are the minimum payloads for this UAV for such mission? Just list the names and types.

25 In a quadcopter project for aerial filming, you are required to design the circuit to connect a small EO camera (with three wires) with one capture mode to an Arduino Uno board. The Arduino should start the camera 30 seconds after the takeoff and stop filming after two minutes. Draw the wire connection between camera and the board. Then, write an Arduino code to control the camera output signal by Arduino board.

Notes

1 General Atomics Readies Drone for European Skies, https://www.military.com, 25 Jun 2015.
2 National Imagery Interpretability Rating Scale.
3 Light amplification by stimulated emission of radiation.
4 www.analog.com.

13

Communications System Design

CONTENTS

Educational Outcomes

After study of this chapter, the reader will be able to:

1) Discuss the current Unmanned Aerial Vehicle (UAV) communications system technologies
2) Discuss communication process
3) Explain features of communications system for a number of current active UAVs
4) Explain function and elements of a UAV communications system
5) Provide communications system installation guidelines
6) Evaluate communications-system-structure integration
7) Explain features of a transmitter

Design of Unmanned Aerial Systems, First Edition. Mohammad H. Sadraey.

Companion website: www.wiley.com/go/sadraey/unmanned-aerial-systems

8) Explain features of a receiver
9) Describe telemetry operation
10) Explain the command transmission process from a ground station to a UAV
11) Explain the data transmission process from a UAV to a ground station
12) Know antenna technologies
13) Know commonly used frequency bands in communications
14) Discuss encryption process
15) Conduct a communications system design process

13.1 Fundamentals

An important subsystem of an unmanned aerial system (UAS) is the telecommunication (or simply communication) system which provides a connection between the UAV and the ground station for the purpose of transmitting and receiving digital information. Through this subsystem, the data of air vehicle and of the Ground control station (GCS) is linked together (i.e., data link); thus, an uplink and downlink. Another important function of the communication is in the avoidance of mid-air collision between a UAV and other flying aircraft. In this section, various aspects of the communication system are discussed.

The UAVs are more dependent on communications systems than manned aircraft. This is especially true for a UAV that collects large volumes of images/data with multi-sensors such as EO/IR camera, high-resolution spot mode SAR, wide area search SAR, and Full Motion Video (FMV) for ISR. The communication relay is a traditional military application for Navy UASs (particularly HALE UAVs). Instead of storing the UAV sensor data for post-mission analysis, the communications system can transmit the data and intelligence products to the ground station to allow real-time control and to provide live operation.

A telecommunication system requires [161] three main elements: (a) transmitter, for conversion of data into digital or analog signal, and transmitting; (b) receiver, for receiving and decoding the signal into data; and (c) medium. The communication signal needs a medium to travel through. In general, the wireless communication between the GCS and vehicle may be achieved by three different mediums: (a) radio wave; (b) laser beam; and (c) infrared light.

A digital signal consists of information encoded as a set of discrete packets (i.e., 0 and 1), while in an analog signal, the information is encoded in the form of a continuous wave (e.g., a cosine wave). Radio waves can be categorized into three groups: (a) narrow-band radio waves which transmits the signal using bandwidth of 25 kHz or lower; (b) spread-spectrum radio waves which transmits the signal spread over wide band of frequencies; and (c) ultra-wide band radio waves which transmits the signal over even wider frequency than spread spectrum radio making it difficult for detection.

The GCS serves two major purposes: (a) converts the pilot command to a signal and sends it to the active antenna for transmitting it to UAV using uplink radio transmission; and (b) it receives UAV positional and health parameters using downlink. These two objectives are realized through an effective communication system. There are mainly two groups of communication equipment: (a) one onboard the UAV; (b) one inside the GCS.

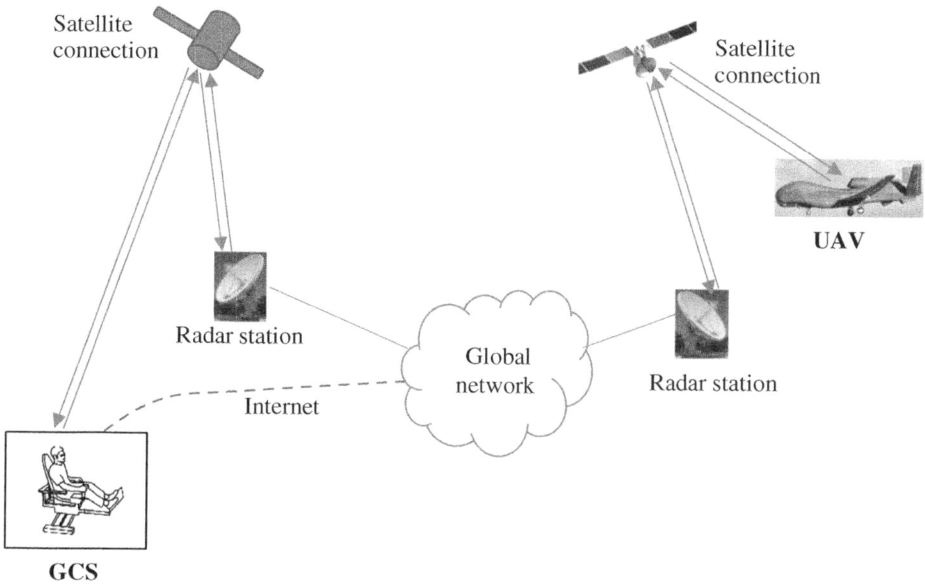

Figure 13.1 Data link network diagram.

The principal issues of communication technologies are flexibility, adaptability, security, and cognitive controllability of the bandwidth, frequency, and information/data flows. The communication system should be able to transmit data; in both ways; at an adequate rate, reliably, and securely.

Figure 13.1 depicts the global data link network diagram for an UASUAS which includes satellites (e.g., Inmarsat), radar stations (e.g., Gateway), internet network, and broadband terminals. The satellite and the radar station for a single case may be the same.

Flight data are usually received by GCS with a frequency of 20–2000 Hz; they should have a consistent frame rate compatible with the vehicle autopilot onboard sensors. A signal speed of 100 Hz is equivalent to a time step of 0.01 seconds, which implies the data of a signal is recorded 100 times per second. SBG IG-500N, a miniature Global positioning (GPS) enhanced inertial navigation system has an update rate of 100 Hz.

Three major sources of positioning system and communications using satellite are: (a) GPS; (b) Inmarsat; and (c) Iridium. The GPS system is provided by the United States government, which can selectively deny access to a user or degrade the service at any time. For instance, an access denial happened to the Indian military in 1999 during the Kargil War.

Inmarsat is a British satellite communications company, offering telephone and data services to users worldwide, via portable terminals. The service is provided by communication with ground stations through 13 geostationary satellites which orbit the earth at 35 786 km altitude. However, Inmarsat 5-F3 is a commercial communications satellite built by the Boeing Company for operation by the British company Inmarsat to provide global mobile satellite services.

The Iridium satellite constellation provides L-band voice and data communications coverage. The Iridium's coverage is provided by over 66 satellites, which obit at a height of approximately 485 miles. The ScanEagle UAV is equipped with an Iridium data link for beyond-LOS communications. Inmarsat's satellites orbit the Equator, while Iridium's orbit the Poles.

The air traffic control (ATC) and command-and-control (C2) of NASA Global Hawk from Dryden Flight Research Center is applied in two distinct regions: (a) line-of-sight (LOS); and (b) beyond line-of-sight (BLOS). The communications link used for LOS are through UHF/VHF links. The primary communications links used for BLOS are two Iridium Satcom links. However, an Inmarsat Satcom link provides a backup communications capability.

The NASA Global Hawk payload communications architecture is independent of the communications links utilized to control the aircraft. Four dedicated Iridium SatCom communication links are used for continuous narrow band communications between the ground station and the UAV payloads. Moreover, two additional Iridium links are used to monitor power consumption by individual payloads, and to control features such as lasers and dropsonde. The use of the Iridium system provides a complete global coverage, including the Polar regions.

GPS satellites fly in medium Earth orbit at an altitude of approximately 20 200 km. Each GPS satellite circles the Earth twice a day. Each GPS satellite continuously transmits a radio signal containing the current time and data about its position. In an UAV operation as a navigator, four, or more GPS satellites must be visible to obtain an accurate result. As of early 2015, augmented FAA-grade GPS receivers have been provided a horizontal accuracy of better than 3.5 m.

The original communications system for Predator UAV was slow. Pilots experienced a delay of several seconds between moving their sticks and the UAV's response. One of the modifications was to include the better satellite communications and data links to improve the efficiency of flight operation. The satellite link provides communications between the GCS and the aircraft when it is beyond line-of-sight (LOS).

When the modified Predator UAV is out of sight of the GCS, a satellite data link is employed by the operators. Video signals received in the Predator GCS are resent directly to operational users via a land commercial broadcast system, or to global users via satellite communication. External communications of a Predator are via High Frequency (HF) /Ultra high frequency (UHF)/Very High Frequency (VHF) (data), GCS cellular/landline telephones, and hardwire connectivity with a satellite communication terminal. Other SatCom (Satellite Communications) systems may be used to link the GCS to an intelligence architecture.

13.2 Data Link

An important function of the communication system is providing the data link between various elements of an UAS. The data link is the means of connecting one location to another one (e.g., UAV to GCS) for the purpose of transmitting and receiving digital information. Through the communication system, the data of air vehicle and of the GCS is linked together (i.e., data link); thus, an uplink and downlink. Thus, the communication between a GCS (in ground) with an UAV (in the air) has both uplink and downlink. If data link is installed, the ground station may support autopilot without LOS with UAV through software application. For instance, the Predator UAS includes a package of four vehicles, a GCS, and a data link suite.

The data link of both UAV and GCS typically consists of a transmitter and a receiver, an antenna, and a modem to link these parts with the sensors. The GCS serves two major purposes: (a) converts the pilot command to a signal and sends it to the active antenna for transmitting it to UAV using uplink radio transmission; and (b) it receives UAV positional/health parameters and payload data using downlink. The combination of a receiver and a transmitter is a *transceiver* which is the device by which receiver and transmitter functions will be simultaneously implemented.

Part of the data link is *telemetry* which is a communication process by which UAV flight measurement data are collected and automatically transmitted to the receiver of a GCS. Since there is no physical connection (e.g., wire) between a GCS and an UAV, communication process is referred to as the *wireless telemetry*. The telemetry data link is from the UAV to GCS containing all its internal health parameters. The onboard telemetry data link is transmitted by radio link via an onboard telemetry antenna. If the UAV is equipped with the range data link (a subset of the telemetry data link), the GCS can provide the positional information of the UAV with respect to GCS reference.

The UAV data received by the GCS are live processed by the data processing unit to generate a master information referred to as the cue data. The cue data serves as a guiding information for the range sensors, if they have lost the track of the UAV due to system failure.

In small UAVs and quadcopters, copper cables are utilized as a means for sending signals within a UAV and within a GCS. A major progress in a high-capacity data link is replacing copper cables with optical fibers that are quite reliable and provide a number of benefits. Optical fibers offer less weight, smaller diameter, lower cost, and immunity to water and corrosion. They are also completely free from electromagnetic interferences. Moreover, the optical fibers are immune to interference and contamination from generators, transmitters, antennas, and other noise sources. This not only eliminates unwanted radiation, but also prevents unauthorized interception of sensitive information by cable sniffers.

All these benefits of optical fibers are vitally significant to a UAV and a GCS, where various subsystems, sensors, components, and antennas are crowded together in close proximity. Thus, the UAV and the GCS data links will be more effective and reliable by employing optical fibers.

The data link between a UAV and a GCS may be provided by a LOS, or by internet connections using satellites. When the Predator drone is out of sight of the GCS, the operators utilize a satellite data link. The Guardian UAV is employing the BST100 satellite data link, with a component (box) on the UAV and another component on the GCS. The Predator UAV can be programmed to autonomously return to ground base if the datalink is lost.

Designing UAS wireless data links is highly challenging due to long Distance between UAV and ground station, high-speed flight, and limited availability of radio frequency spectrum. These challenges and the spectrum affect the performance of the data link. UAVs and commercial manned aircraft will share the same non-segregated airspace, and would need to be aware of each other's presence. UAS data link design requirements are: (a) high-availability; (b) networked and non-networked controllers; (c) preemption; (d) chaining; and (e) compatibility with manned aircraft datalinks.

13.3 Transmitter

Both the air vehicle and the GCS (Figure 13.2) need a transmitter to transmit the signal/data/command to the other side. The GCS primarily sends the command to the air vehicle, while the air vehicle is sending the status of its flight, plus the output of its measurement sensors to the GCS.

A transmitter is an electronic device [162] which produces radio waves through an antenna. The transmitter initially generates an alternating electric current which is then is converted to radio waves by an antenna. When excited by this current, the antenna radiates radio waves. Antennas are required by any radio wave transmitter to couple its electrical connection (e.g., copper wire or optical fiber) to the electromagnetic field. The optical fibers are immune to interference and contamination from generators, transmitters, antennas, and other noise sources.

In a transmitter, electromagnetic waves are generated by time-varying electric currents, flowing through a metal conductor, antenna. One of the functions of a transmitter is to add the desired information (data) to a radio wave by a process called modulation. There are mainly three modulation techniques: (a) amplitude modulation; (b) frequency modulation; and (c) phase modulation.

Electromagnetic (EM) waves, composed of oscillating magnetic and electric fields, are traveling with the speed of light (about 3×10^8 m/s or 3×10^5 km/s). Unlike mechanical waves (e.g., sound wave), EM waves do not need any media to travel. Hence, EM waves can be transmitted through vacuum or space. The relationship between wavelength (λ), frequency (f), and velocity of an EM or radio wave (c) is governed by the following equation:

$$\lambda f = c \tag{13.1}$$

Since the velocity of light (c) is constant; as the frequency of the wave is increased, the wavelength is decreased. The design of a transmitter (and its antenna) is directly a function of radio frequency. The design of transmitter is beyond the scope of this text, the interested reader can refer to references such as [162–164].

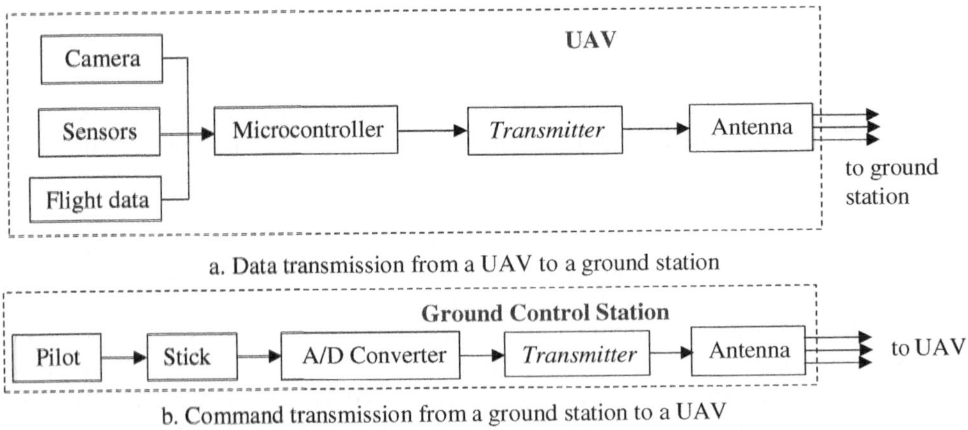

Figure 13.2 Transmitter in two locations.

Example 13.1

A maritime variant of Guardian UAV is employing the Inmarsat communication system. The UAV – flying at 15000 m – is transmitting a data through its Inmarsat antenna to its GCS. How long does it take for this signal to reach the GCS? Ignore the distance between GCS and the UAV.

Solution

Inmarsat satellites orbit the Earth at an altitude of approximately 35 786 km altitude, while the UAV is flying at an altitude of 15 km from sea level. Moreover, the speed of the radio wave is about of 3×10^5 km/s. The distance between GCS and the Inmarsat satellite is 35 786 km, while the distance between the UAV and the satellites is 35 786–15 or 35 771 km. Thus:

$$t_u = \frac{D_{UAV}}{c} = \frac{35771}{3 \times 10^5} = 0.1192 \sec \tag{8.4}$$

$$t_d = \frac{D_{GCS}}{c} = \frac{35786}{3 \times 10^5} = 0.1193 \sec \tag{8.4}$$

Thus, the total duration is:

$$T = t_u + t_d = 0.1192 + 0.1193 = 0.2385 \sec$$

The signal strength of EM waves decreases the farther they travel from the transmitter, hence a signal can only be received within a limited range of its transmitter. The transmitter range depends on the power of the transmitter, the sensitivity of the receiver, and atmospheric and internal noise (e.g., engine vibrations).

In a handheld GCS, each aft-to-fore and left- or right-hand steering is dedicated to a UAV control variable via a channel of the transmitter. A typical transmitter of a handheld GCS has at least six channels, a dedicated communication frequency (e.g., 2.4 GHz), modulation, and programming for a compatible receiver.

The size, weight, and cost of a transmitter is directly a function of its desired electric output power. The transmitter power output is a function of the amount of energy that a transmitter produces and radiates at its output (i.e., transmitting antenna). The antenna energy is primarily a function of the desired range (to the UAV or a target), radio frequency, transmitter location, radiation resistance, and losses. The radiated power is measured by unmodulated radio frequency as:

$$P_t = I_t V_t \cos(\phi) \tag{13.2}$$

where ϕ is the phase between oscillating voltage (V_t) and current (I_t).

13.4 Receiver

Both the air vehicle and the GCS (Figure 13.3) need a receiver to receive the signal/data from the other side. The air vehicle primarily receives the command from the GCS, while the GCS is receiving the status of the UAV flight, plus the output of the sensors of the air vehicle. Basically, a receiver can be regarded as the opposite of a transmitter. It

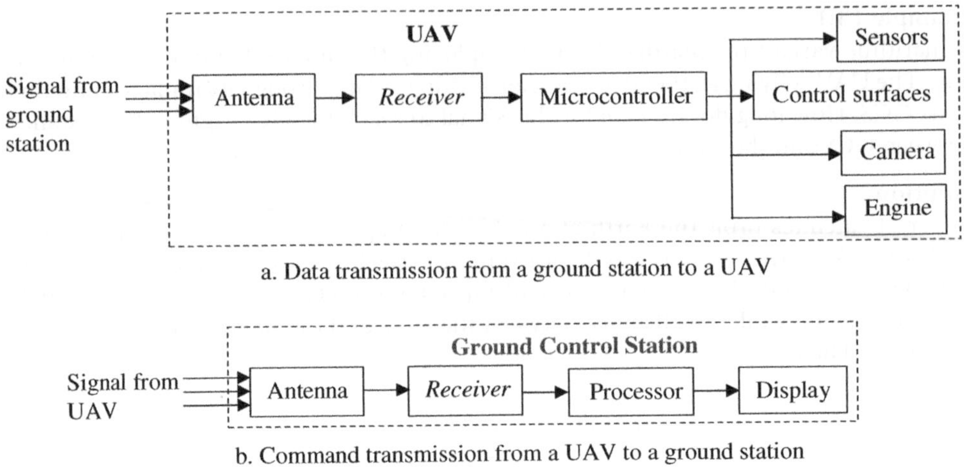

Figure 13.3 Receiver in two locations.

uses an antenna to capture radio waves, processes those waves to extract only those waves that are vibrating at the desired frequency, extracts the signals that were added to those waves, amplifies the signals, and send them to the user (e.g., microcontroller).

A receiver is an electronic device that receives radio waves and converts the information carried by them to a usable form (e.g., digital data or image). It is connected to an antenna to convert the energy of the incoming radio wave into an Alternating Current (AC) voltage which is applied to the receiver's input. A receiver extracts the desired information from a signal through demodulation.

Receivers perform three basic functions on an incoming signal from the antenna: (a) filtering; (b) amplification; and (c) demodulation. Antennas are required by any radio wave receiver to couple its electrical connection to the electromagnetic field. The receiver uses filters to separate the desired frequency signal from all the other signals picked up by the antenna. The demodulation is a process through which the data is extracted from the radio waves. The data will be then sent to the microcontroller for further processing and implementation.

The signal strength of a radio signal from a receiver's antenna is mainly a function of three factors: (a) distance to the transmitter; (b) transmitter power; and (c) propagation conditions along the path of the radio waves. In a modern receiver, the signal is sampled and digitized, and the filtering and detection functions are performed by digital signal processing.

A typical receiver working with a handheld GCS has at least six channels, a dedicated frequency (e.g., 2.4 GHz), demodulation, and programming for a compatible transmitter. In some UAVs, a VOR (Very High Frequency (VHF) Omni-Directional Range) receiver is employed to determine the UAV direction to the beacon, for air navigation. This short-range navigational instrument uses the VHF signal from VOR navigational beacons between 108 and 117.95 MHz.

The power of a receiver is a function of various factors including radio frequency, path loss, signal data, type of antenna, and signal strength. The size, weight, and cost of a receiver is directly a function of its desired electric power. The receiver power is a function of the amount of energy needed for filtering, amplification, and demodulation. The design of a receiver is beyond the scope of this text, the interested reader can refer to references such as [162–164].

13.5 Antenna

Both transmitter and receiver are equipped each with an antenna; which is employed for both broadcasting and receiving signals. An antenna is the interface between radio waves propagating through space, and electric currents moving in metal conductors, used with in receivers/transmitters. An antenna is an array of conductors, electrically connected to the receiver or transmitter. For the theory and design of antenna, you may consult with references such as [73]. In this section, the fundamentals of the antenna are briefly reviewed.

The manned transport aircraft Boeing 787 has more than 20 antennas protruding from its fuselage. They include: (a) GPS receiver; (b) SatCom (satellite communications); (c) marker beacons; (d) weather radar; (e) UHF DME (distance-measuring equipment); (f) transponder-based radio navigation (UHF radio signals); (g) VOR (omni-directional radar-ranging); (h) ELT (emergency location transmitter); (i) TCS (Terminal Cellular System); (j) HF; (k) VHF; (l) CWLU (Crew wireless LAN unit); (m) ADF (automatic direction-finder); (n) RA (Radio altimeter); (o) TCA (traffic collision-avoidance system); (p) ATC (air traffic control); (q) ILS (instrument landing system) Glideslope and Localizer; (r) LAN (local area network); and (s) ELT (Emergency locator transmitter). Similarly, a large High altitude long endurance (HALE) UAV such as Global Hawk is required to have multiple antennas for various transmitting and receiving functions.

Antennas are required by any radio wave receiver or transmitter to couple its electrical connection to the electromagnetic field. In reception, the receiver antenna intercepts some of the power of an incoming electromagnetic wave in the air in order to produce an electric current at its terminals. In transmission, a transmitter supplies an electric current to the antenna's terminals, and the antenna radiates the energy from the current as electromagnetic waves to the air. In general, there are four types of antennae (Figure 13.4) applicable to UAV systems: (a) monopole; (b) dipole (most common, Yagi-Uda antenna); (c) array; and (d) aperture.

- A mono-pole antenna consists of a single conductor such as a metal rod, mounted, either vertical or parallel to the ground, or on the UAV body. The most common form of a monopole antenna is the quarter-wave which is one-quarter of a wavelength long. A few other commonly used monopole antennas are whip, rubber ducky, umbrella, mast radiator, and ground plane. This antenna is frequently used in the RC model airplanes where the air vehicle in within LOS of the operator. However, it is not safe in regard to jamming and electronic warfare. This antenna can be employed on both UAV and its GCS.
- A basic dipole antenna consists of two conductors; usually metal rods or wires; arranged symmetrically, with one side of the balanced feedline is attached to transmitter or receiver. The most common type of dipole antenna is the half-wave dipoles which are used alone as omnidirectional antennas. This antenna is a building block of many other more complicated directional antennas. **Yagi-Uda** is one of the most common

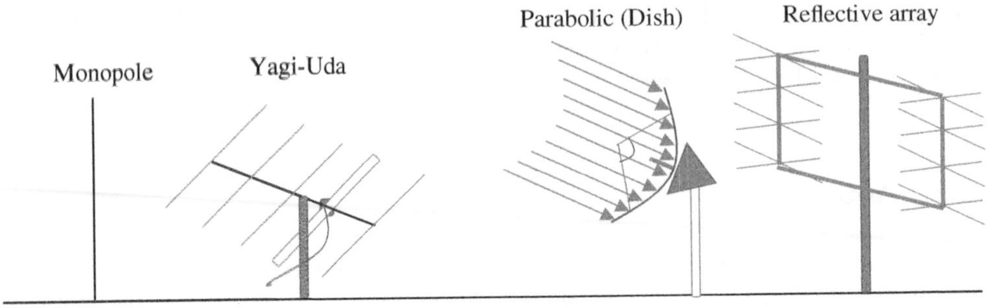

Tracking antenna (courtesy of UAV factory)

Figure 13.4 Antenna types.

directional antennas at HF, VHF, and UHF frequencies. It consists of multiple half wave dipole elements in a line, with a single driven element and multiple parasite elements which serve to create a uni-directional antenna. A few other commonly used dipole antennas are turnstile, patch, and long periodic dipole array. This antenna cannot be frequently employed on a UAV due to its size and drag, but it may be utilized in the GCS.

- Array antennae consists of multiple identical driven elements (e.g., single dipole) working as a single antenna. A few examples are collinear, **reflective array**, phased array, curtain array, and batwing. This antenna cannot be frequently employed on a UAV due to its size and drag, but it may be utilized in the GCS.
- Aperture antennas are the main type of directional antennas used at microwave frequencies and beyond. They consist of a small dipole inside a three-dimensional large structure with an aperture to emit the radio waves. There are mainly four types of aperture antenna: (a) parabolic (the most commonly used); (b) horn; (c) slot; and (d) dielectric resonator. A parabolic utilizes a large (compared to the wavelength) dish-shaped metal parabolic reflector with a feed antenna at the focus. It can have some of the highest gains of any antenna type; up to 60 dB. The **dish** antenna has many applications including radars, point-to-point data links, satellite communication, and radio telescopes. The most common form of aperture antenna is shaped like a dish and is popularly called a dish antenna. The main advantage of a parabolic antenna is that it has high directivity. This antenna can be employed on both UAV and its GCS.

It is always desirable to have the radio antennae mounted on the GCS to minimize conductor lengths. Meanwhile, it is necessary to situate the antennae in a position to obtain good transmission and reception of the radio waves. In a hilly terrain area, this may mean on a crest. For some military operations, unless the GCS is small and readily concealed, its position may become obvious to hostile forces and so vulnerable. A solution would be to mount the antennas on a small, mobile, and less-detectable platform which can be located in a suitable position.

The Predator UAV is equipped with a couple of antennas. Its GCS also has a few antennas. The primary satellite link provides over-the-horizon communications for the UAV. The ground data terminal antenna provides LOS communications for takeoff and landing.

13.6 Radio Frequency

An important step in the design of the communication system (either transmitter or receiver) is the selection of radio frequency. The communication between the GCS and a UAV and between the UAV and GCS may be achieved by two different media: (a) electromagnetic (radio) waves; and (b) laser beam. The communication by radio wave between the UAV and the GCS can be direct or via satellites or other means of radio relay. Typical usable radio frequencies for communication range from 3 MHz to 300 GHz (below the infrared spectrum of electromagnetic waves). Frequencies in the range 3 Hz (extremely low frequency[ELF]) to 3 GHz (UHF) are effective radio frequencies as they are refracted in the lower atmosphere to curve to some degree.

The selection of an operating frequency requires a compromise. A lower frequency offers a better and more reliable propagation, while having reduced data-rate ability. In contrast, a higher frequency is capable of carrying high data rates, but this requires direct LOS and higher power to propagate the signal. For instance, a high-resolution TV camera will produce a data rate of order 75 megabytes per second.

There are two techniques to track a UAV is by radio communication: (a) fit the UAV with a transponder which will receive, amplify, and return a signal from the control station; and (b) UAV down-link should transmit a suitable pulsed signal. The successful operation of the communication system depends upon the integration of the various components of the system to supply adequate energy to achieve the required range. Four contributing factors are: (a) transmitter power output; (b) receiver sensitivity; (c) antenna gain; and (d) path loss.

These communications for UAV are done primarily through the use of RF applications, usually, satellite communication links in UAV are used LOS mode. The most common frequency bands of this type of links are Ku, K, S, L, C, and X bands. The Ku band has been historically used for high speed links. Due to its short wavelengths and high frequency, this band suffers from more propagation losses. Yet, it is also able to trespass most obstacles thus conveying great deals of data. K band possesses a large frequency range which conveys large amounts of data. However, it requires powerful transmitters and it is sensitive to environmental interferences.

The S and L bands do not allow data links with transmission speeds above 500 kbs. Their large wavelength signals are able to penetrate into terrestrial infrastructures and the transmitter require less power than in K band. The C band requires a relatively large transmission and reception antenna. The X band has been reserved for military

Table 13.1 Commonly used frequency bands in communications.

No	Band	Frequency	Applications
1	HF	3–30 MHz	Over-The-Horizon radar surveillance
2	VHF	30–300 MHz	Very long range surveillance
3	UHF	300–1000 MHz	Very long range surveillance
4	L	1–2 GHz	• Long range surveillance • Enroute traffic control
5	S	2–4 GHz	• Moderate range surveillance • Terminal traffic control • Long range weather
6	C	4–8 GHz	• Long range tracking • Airborne weather detection
7	X	8–12 GHz	• Short range tracking • Missile guidance • Mapping • Marine radar • Airborne intercept
8	K_u	12–18 GHz	• High resolution mapping • Satellite altimetry
9	K	18–27 GHz	Little use (water vapor absorption)
10	K_a	27–40 GHz	• Very high resolution mapping • Airport surveillance

application. In the past, the common radio frequency was 72 MHz, but now, it is 2.4 GHz. The L band (1–2 GHz) seems to be the preferred route for safety of flight in UAVs. Table 13.1 shows the commonly used frequency bands [165] for communication systems.

If the GCS communication system uses a frequency higher than that of the UAV camera, the images are displayed/updated on the pilot's screen with the same frame rate. Thus, the frequency of the GCS communication system should be at least equal to the frequency of the surveillance system (e.g., camera).

Figure 13.5 illustrates the GCS Command, Control, and Communications (C3) model. Technologies and operating procedures related to C3 of the UAVs are divided into two categories: (a) RF line of sight; and (b) RF beyond line of sight. Under each category UAV technical issues may be divided into two approaches: (a) Command and Control (C2); and (b) ATC. Under Command and Control, both full autonomy, and remote control are applicable.

Table 13.2 demonstrate typical radio frequencies for various applications for a large UAV such as Global Hawk. There is a military recommendation to employ 14/15 GHz band for line of sight links, while to use 20/30 GHz beyond line of sight links. The Predator data link suite consists of: (a) UHF and VHF radio relay links; (b) a C-band LOS data link; and (c) Ku-band satellite data links. In Predator UAV, the GCS is using C-band and Ku-band data links for LOS and non-LOS communications with the UAV, respectively.

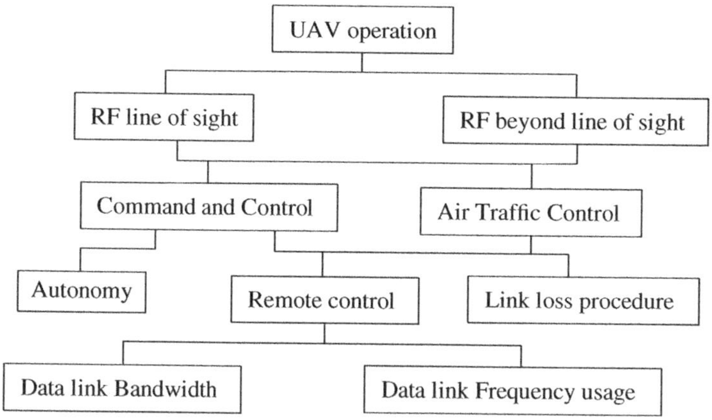

Figure 13.5 Command, Control, and Communications (C3) model.

Table 13.2 Typical radio frequencies for a large unmanned aerial vehicle (UAV).

No	Category	Application	Frequency
1	Communications links	Ku Satcom Uplink	14 GHz
2		Ku Satcom Downlink	11.7 GHz
3		UHF Satcom	200–400 MHz
4		Flight test telemetry	2387 MHz
5	NAVAIDS	Radar Altimeter	4300 MHz
6		IFF	1030 MHz
7		IFF	1090 MHz
8		GPS L1	1575 MHz
9		GPS L2	1227 MHz
10		Differential GPS	113 MHz
11	Payloads	EO/IR Sensors	8–10 GHz

The Garmin GSX 70 radar used in a NASA Global Hawk is employing Output frequency range of 9.3–9.5 GHz (X-Band). The mission control center has data up- and down-links to the Global Hawk vehicle directly and via the Ku satellite and the UHF satellite systems. The launch and recovery station has both up- and down-data communications links to the Global Hawk vehicle and to the UHF communications satellite.

Example 13.2
A large UAV is employing the satellite communication system with a radio frequency of 15 GHz. Calculate the wavelength of the signal for such communication.

Solution

The radio wave is an electromagnetic wave with the velocity of 3×10^8 m/s.

$$\lambda f = c \Rightarrow \lambda = \frac{c}{f} = \frac{3 \times 10^8}{15 \times 10^9} = 0.02\,m = 2\,cm \tag{13.1}$$

The signal wavelength is 2 cm.

13.7 Encryption

There have always been major concerns on UAVs security. Data link spoofing, hijacking, and jamming are major security [77] issues facing the UAV communication system. UAVs are vulnerable [166] to cyber-attacks, as they communicate via wireless networks. UAVs carry sensitive information related to individuals' privacy and this causes critical problems if the data is exploited by hackers. Securing the necessary bandwidth and electromagnetic spectrum for civilian UAS communication between the UAV and the GCS is a serious challenge.

There are a lot of attacks [167] which are related to wireless networks on UAVs, such as Telnet/FTP attack, Denial of Service (DoS), ARP (Address Resolution Protocol) spoofing, ad-hoc network, and man in the middle attack. A security researcher was able to take control of a Parrot AR Drone by de-authenticating the UAV and commanding it through his script [167] that was implemented on Raspberry pi. The U.S. Air Force has long tasked Northrop Grumman to upgrade the abilities of the Global Hawk to resist computer hackers' attempts to commandeer the vehicle while in flight. Thus, the Global Hawk's maritime SATCOM links is upgraded to prevent computer hackers from breaking through existing data security to take unauthorized control of Global Hawk reconnaissance UAVs.

Many UASs, including the Air Force's RQ-4B Global Hawk, the Navy's RQ-4N BAMS (Broad Area Maritime Surveillance) UAS, and Predator variants, depend on the availability of satellite communications for: (a) receiving command and control; and (b) sending sensor data, to the ground control stations. Similar to the case for GPS, satellite communications are vulnerable to noise jamming. As the radiated power of the noise jammer is increased, the UAV achievable communication data rate for a fixed quality of service will be significantly decreased. Providing immune communications allows military commanders to leverage some of the advantages that the UAS has, particularly in dirty, dull, and dangerous missions.

There were many techniques to solve security problems and cyber-attacks on UAVs. One solution to protect the UAV's information is a second channel security system. If the security second channel is compromised, the self-destroy function can be activated. However, a standard solution is encryption, which is the process of encoding a data or information in such a way that only authorized parties can access it, but not the people who are not authorized. Encryption is one of the most important methods for providing data security, especially for end-to-end protection of data communicated across data link. Cryptography is vital to keeping information safe, in an era when the world becomes more and more challenging.

Encryption [168] is the process of taking a data signal, a readable document, or image and changing that information to an extent that it is no longer readable. Encryption converts any type of data from a readable form to an encoded version that can only be decoded by another entity if they have access to a decryption key. The objective of encryption is to protect and hide the contents of an information source from improper disclosure. Hence, the primary purpose of encryption is to protect the confidentiality of digital data or signal transmitted via a UAV communication system. Modern encryption algorithms play a key role in the security assurance of communications as they provide confidentiality.

It is suitable to clarify the difference between encoding and encryption. An encoder is a device, circuit, transducer, software program, or algorithm that transforms information from one format or code to another, for the purposes of standardization, speed, or compression. Different types of technologies (e.g., mechanical, magnetic, resistive, and optical) are employed in encoders to create a signal. Thus, encoding is for maintaining data usability and uses schemes that are publicly available, while encryption is for maintaining data confidentiality and thus the ability to reverse the transformation codes are limited to certain individuals. Encryption is the process of translation of data into a secret code.

13.8 Communications Systems of a Few UAVs

General Atomics RQ-1 Predator UAV (Figure 8.5) was the first UAV to be controlled via satellite communications, and the first UAV to provide voice radio communications. Large modern UAVs (e.g., MQ-9 Reaper [Figure 5.12], and NASA Altair [Figure 13.6]) are equipped with satellite antenna. The GCS for RQ-1A Predator uses C- and Ku-Band data links for LOS and non-LOS communication with the UAV, respectively. Range of the non-LOS link, and therefore the effective operational radius of the aircraft, is about 740 km. Non-LOS communications in the first three *Predator* systems was handled via a Trojan SPIRIT SATCOM link.

The up-link transmits command and control from the operators to the UAV (or UAVs in multiple operations), and the down-link returns payload data, and the flight status and images from the UAV to the GCS and to any other satellite receiving stations. The UAV status data (e.g., location, airspeed, and attitudes, weather conditions) is often known as housekeeping data. Loss of communication during flight operations will likely result in the UAV crash.

The range of communication system is a function of the height of the communication antenna and the flight altitude of the air vehicle. These two factors determine the availability of LOS between UAV and GCS. As the UAV altitude increases, the LOS range will be increased. Figure 13.7 illustrates a number of antennas of NASA's Ikhana Predator B, which is employed to support Earth science missions and advanced aeronautical technology development.

Another variant, the Maritime variant of a Guardian UAV has six antennas: (a) Ku SATCOM[1] antenna; (b) C-band LOS antenna; (c) Inmarsat[2] antenna; (d) ARC-210 Radio antenna; (e) AIS antenna; and (f) Wulfsburg antenna. BTW, it has a SeaVue surveillance radar, and a group of EO/IR sensors. The design of antenna is beyond the scope of this text, the interested reader can refer to references such as [162].

Figure 13.6 NASA Altair satellite communications antenna.

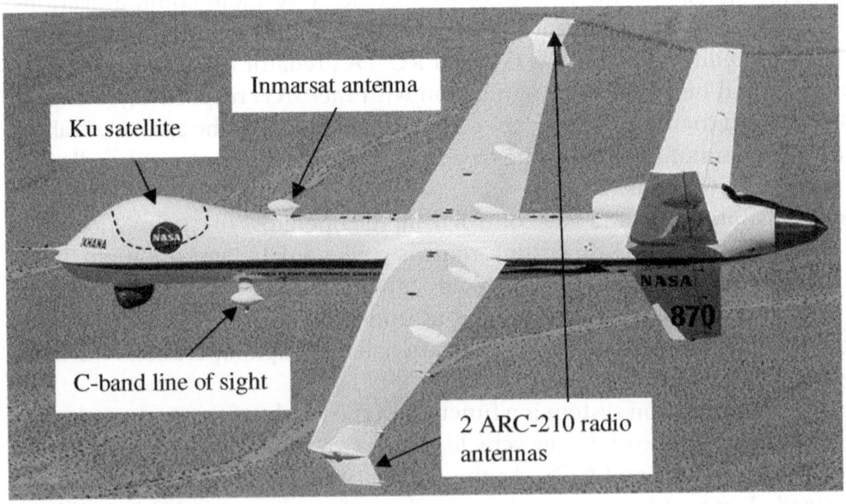

Figure 13.7 Antennas of NASA's Ikhana Predator B.

The Global Hawk UAV is equipped with a number of antennas: (a) 1.2 m ku-band SatCom antenna; (b) UHF antenna; (c) LOS antenna; (d) Synthetic aperture radar (SAR) combined antenna; and (e) dual band common data link antenna. Global Hawk has wide band satellite data links and line of sight data links. The top front surface of the fuselage houses the 48 in Ku-band wideband satellite communications antenna. Data is transferred by: (a) Ku-band satellite communications; (b) X-band LOS

Table 13.3 Weight and electric power of a few components of Global Hawk communications system.

No	Component	Weight (lb)	Power (W)	Remarks
1	Ku-band SatCom antenna	46	35	Gimbaled 8 in dish, Ku band
2	Ultra high frequency (UHF) antenna	8	–	–
3	LNA/Diplexer	3	7	Low noise amplifier
4	UHF PA	14	700	Power amplifier
5	UHF RX/TX	13	150	Transmitter/receiver
6	HPA	56	33	High power amplifier
7	CAMA	85	310	Modem assembly

links; and (c) both Satcom and line of sight links at UHF-band. The SAR and ground moving target indicator operates at X-band with a 600 MHz bandwidth. Table 13.3 illustrates the weight and electric power of a few components of Global Hawk communications system.

Large UASs (e.g., Global Hawk and Predator) are usually encrypting their communicated data to protect the UAVs against cyber-attacks. By using the satellite communication and the encrypted data link, Global Hawk UAVs can safely fly anywhere in the world. In MQ-5A Hunter, the communications uplink and downlink channels employ fixed coded frame format. There is also an optional spread spectrum modem on the main uplink channel to present anti-jam capability.

13.9 Installation

When the communications system elements for a UAV and the GCS are selected, the installation is the next step which may even affect the UAV configuration. The transmitter/receiver will require at least a space and power allocation, while the antennas for the UAV also require a kind of fairing. The installation of transmitter/receiver/antennas in the UAV is much more challenging than that of the GCS, since the space is limited, and design requirements are more conflicting.

The transmitter/receiver installation on a UAV has various aspects which a designer should be concern of. Four main aspects are: (a) wiring; (b) locating; (c) aerodynamics (if mounted externally); and (d) transmitter/receiver/antennas-structure integration. Fundamentals of installation of transmitter, receiver, and antennas are very similar to those of the installation of payload. Please study Section 12.9 for more details and guide lines.

13.10 Communications System Design

In general, the primary criteria for the design of communications system are: (a) current available technology; (b) required accuracy; (c) mission complexity; (d) weather; (e) reliability; (f) cost; (g) UAV configuration; (h) compatibility; (i) maintainability; (j) UAV range; (k) UAV service ceiling; (l) weight; (m) mobility; (n) UAV drag; (o) cybersecurity; (p) payloads; (q) volume and type of data; and (r) see-and-avoid system.

If the UAV is equipped with the range data link (a subset of the telemetry data link), the GCS should provide the positional information of the UAV with respect to GCS reference. The GCS serves two major purposes: (a) converts the pilot command to a signal and sends it to the active antenna for transmitting it to UAV using uplink radio transmission; and (b) it receives UAV positional and health parameters using downlink. These two objectives should be realized through an effective communications system.

The key challenges in the design of UAV communications system are the large distances that they need to cover and the high speed of aircrafts. These along with the limited availability of radio frequency spectrum affect the performance of the data link. Designing UAV wireless data link is much more challenging than other wireless links. The key challenges are long distance, high speed, and spectrum. New data links need to be developed for UAVs and for commercial manned flight, since they will share the same air space and would need to be aware of each other's presence.

The communications system design process is demonstrated in Figure 13.8. The design of communications systems of both UAV and ground control systems are performed in parallel. In both elements, transmitter, receiver, antenna, power supply, and the processor are designed in detail. The frequency of communications should be carefully selected such that both UAV and GCS can function effectively in the worst-case scenario.

In designing communications system, growing bandwidth and multiple band requirements should be considered. Frequency bands are divided into three categories: (a) exclusive federal; (b) exclusive civil; and (c) shared federal/civil. Both civil and military UAS spectrum needs are constantly growing. Potential mitigation techniques are: (a) data compression; (b) bandwidth efficient modulation techniques; and (c) smart pull.

If the contents of the communicated information need to be protected and hidden from improper disclosure, then encryption should be implemented. The encryption will protect the confidentiality of digital data or signal transmitted via a UAV communication system as a security assurance.

13.11 Bi-directional Communications Using Arduino Boards

Arduino is an open source electronics hardware (e.g., microcontroller) and software (e.g., codes). For more descriptions, refer to section 7 of Chapter 9. In this section, the bi-directional wireless communications using Arduino boards is presented.

13.11.1 Communications Modules

The Arduino has designed and fabricated a variety of microcontrollers/boards, and a number of modules for wireless bi-directional communications. A few example modules are: (a) NRF24L01 radio transceiver; (b) ESP8266 for Wi-Fi application; (c) HC-05 for Bluetooth; and (d) CC2541 for Bluetooth. These transceiver modules are Arduino compatible, and should be used in a pair (each with one Arduino board); one as the transmitter, and one as the receiver. To establish a connection and receive data with a laptop computer rather than a second Arduino, a Bluetooth module (e.g., HC-05) should be utilized.

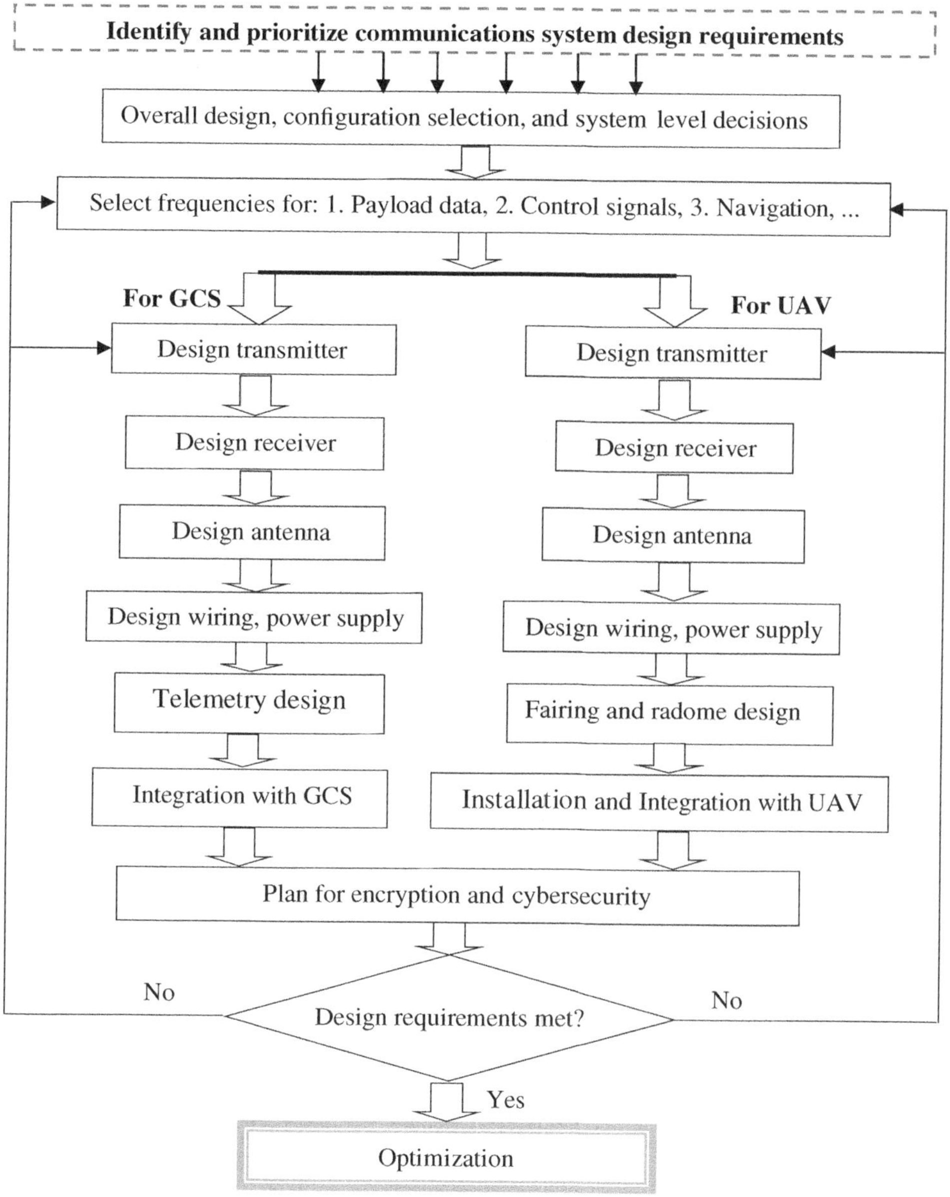

Figure 13.8 Communications system design process.

13.11.2 NRF24L01 Module

The NRF24L01 module (Figure 13.9) employs the 2.4 GHz frequency band, and operates with baud rates from 250 kbps up to 2 Mbps. If used in an open space and with lower baud rate, the range can reach up to 100 meters. The operating voltage of the module is from 1.9 V to 3.6 V, and the power consumption is about 12 mA during

(a)

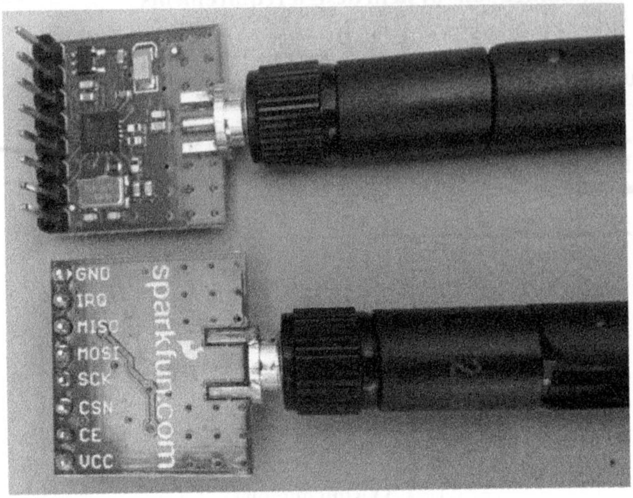

(b)

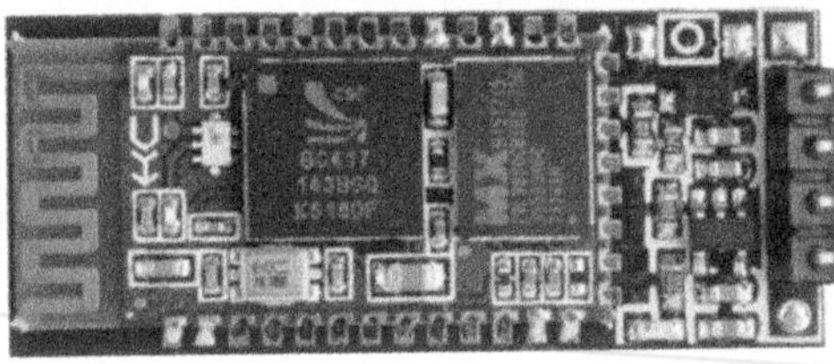

Figure 13.9 NRF24L01 and HC-05 modules. (a) Two NRF24L01 modules with antennas (front and back views), (b) HC-05 module and its pins.

Table 13.4 Technical specifications of a NRF24L01 module

No	Parameter	Value
1	Frequency	2.4 GHz
2	Baud rate	250 kbps – 2 Mbps
3	Current	12 mA
4	Voltage	1.9 V – 3.6 V
5	Power	22.8 mW – 43.2 mW
6	Range	100 m
7	Dimensions (without antenna, and not including header pins)	29 mm × 15.2 mm × 4.6 mm

transmission. Table 13.4 provides technical specifications of an NRF24L01 module. An NRF24L01 module creates and reads digital signals, hence; all pins of the module should be connected to digital pins of the Arduino Uno board.

NRF24L01 module has eight pins: 1. GND, 2. IRQ, 3. MISO, 4. MOSI, 5. SCK, 6. CSN, 7. CE, 8. VCC. The VCC and GND pins are for power (voltage) and ground respectively. If you use a breakout board, the VCC lead can go to the 5 V Arduino pin. Three pins

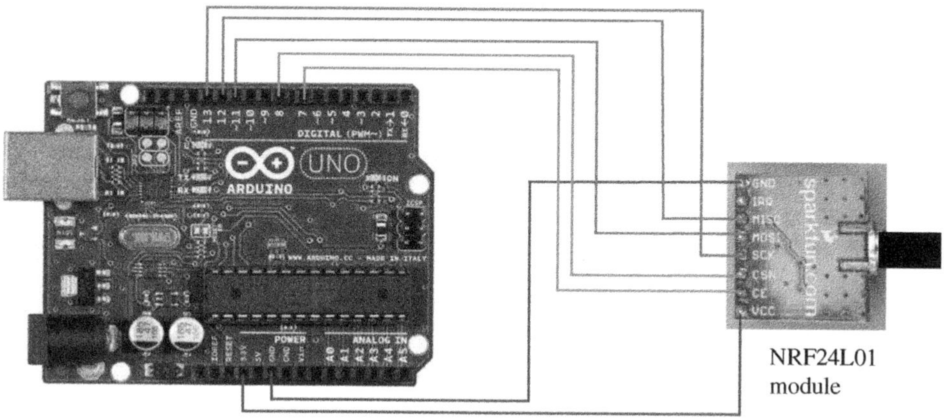

Figure 13.10 Typical pin connections of a NRF24L01 module with an Arduino Uno.

Table 13.5 Connections of pins of an NRF24L01 module to an Arduino Uno

Board	**Pins**							
NRF24L01 module	VCC	GND	CSN	CE	SCK	MOSI	MISO	IRQ
Arduino Uno	3.3 V	GND	8	7	13	11	12	10

(i.e., MOSI, MISO, and SCK) are for the SPI communication and they need to be connected to the SPI pins (Figure 13.10) of the Arduino (i.e., 11, 12, and 13). The pins CSN and CE can be connected to any digital pin (e.g., 8 and 7 (for digital signals), or A1, A2 (for analog signals)) of the Arduino board. They are used for setting the module in standby or active mode, as well as for switching between transmit or command mode. The IRQ pin is an interrupt pin, and does not have to be used. Table 13.5 summarizes the connection of pins of a NRF24L01 module to an Arduino Uno.

The NRF24L01module may employ 125 different channels simultaneously, which provides the possibility of having a network of 125 independent modems. Each channel can have up to 6 addresses; or each unit can communicate with up to 6 other units at the same time.

Once the NRF24L01 modules are connected to the Arduino boards, two separate Arduino codes should be written and uploaded; one to the transmitter, and one to the receiver. In the following, two Arduino codes [131] for the wireless communications to send the message of "Hello World", via NRF24L01 modules are provided.

Transmitter Code:

```
#include <SPI.h>
#include <nRF24L01.h>
#include <RF24.h>
RF24 radio(7, 8);
const byte address[6] = "00001";
void setup()
```

```
{
  radio.begin();
  radio.openWritingPipe(address);
  radio.setPALevel(RF24_PA_MIN);
  radio.stopListening();
}
void loop()
{
  const char text[] = "Hello World";
  radio.write(&amp;text, sizeof(text));
  delay(1000);
}
--------------------------------
```

Receiver Code:

```
#include <SPI.h>
#include <nRF24L01.h>
#include <RF24.h>
RF24 radio(7, 8);
const byte address[6] = "00001";
void setup()
{
  Serial.begin(9600);
  radio.begin();
  radio.openReadingPipe(0, address);
  radio.setPALevel(RF24_PA_MIN);
  radio.startListening();
}
void loop()
{
  if (radio.available())
{
    char text[32] = "";
    radio.read(&amp;text, sizeof(text));
    Serial.println(text);
  }
}
----------------------------
```

Since an Arduino Uno board has six analog pins, you may connect up to six analog input devices (e.g., Potentiometers) simultaneously, to send up to six commands at once. Moreover, an Arduino Uno board has 14 digital pins, of which 6 pins are utilized by an NRF24L01 communication module. Hence, up to 8 output signals may be simultaneously communicated. For instance, if you would like to drive six servos of s1, s2, s3, s4, s5, s6; in the loop of the code, insert the following functions: s1.attach(2); s2.attach(3); s3.attach(4); s4.attach(5); s5.attach(6); s6.attach(7);.

13.11.3 Bluetooth Module

Bluetooth is a wireless technology for exchanging data between fixed and mobile devices over short distances. This technology is using short wavelength UHF radio wave and operates at frequency range of 2.4 to 2.485 GHz. A Bluetooth module in an UAV can be paired to a master Bluetooth device such as a computer, laptop, smart phone, and tablet. This is of particular interest, when a pilot in the ground control station or a payload operator needs to have a live communication with the UAV to monitor the flight data or to see the images/videos of the camera.

The HC-06 module (Figure 13.9) is a class 2 slave Bluetooth transceiver designed for wireless serial communication. Its operating frequency is 2.4 GHz (i.e., industrial, scientific and medical), and the supply voltage is 3.3 Volt DC, with a 50 mA current. The module has a dimension of 36.5 mm by 16 mm, with a mass 4 grams.

This module is set with serial interface, which is easy to use. It has four pins: 1. VCC (positive pole for connection to 5 Volts of a power source), 2. GND (for connection to ground), 3. TXD (transmitting terminal), 4. RXD (receiving terminal). In connecting to an Arduino Uno board, the HC-06 TXD pin is connected to the digital pin of D0 (i.e., Reception pin RX0), and HC-06 RXD pin is connected to the digital pin of D1 (i.e., transmission pin TX0). The default baud rate is 9600 bps.

Configuring the HC-06 module is needed to verify that it is working, hooked up correctly, and to modify its parameters such as its name, PIN code, and communication speed (i.e., baud rate). Once the module is configured, you can pair the module HC-06 to any Bluetooth device. An Arduino code should be developed and uploaded to modify the parameters using the serial monitor. Do not connect the Bluetooth module to the Arduino board, when the code is uploading.

A code similar to the following code [131] should be compiled and uploaded before building the circuit. Through this code, on the display, it can be seen that, a "Hi" is transmitted, and a "Hello" is returned. Moreover, when the signal (i.e., Hi) is sent, the LED on Arduino board will blink. In the first line of the code, the pin 13 built-in LED light is defined.

```
#define ledPin 13
int val;
void setup()
{
  Serial.begin(9600);
  pinMode(ledPin, OUTPUT);
}
void loop()
{
  if(Serial.available())
  {
    val = Serial.read();
    if(val =='Hi')
    {
      digitalWrite(ledPin, HIGH);
      delay(500);
```

```
        digitalWrite(ledPin, LOW);
        delay(500);
        Serial.println("Hello");
    }
}
```

This code may be revised to transmit any other signal such as the UAV trajectory or the payload measurement data to a ground control station.

13.11.4 An Application

In this section, an application of the NRF24L01 communication module in the form of an example is presented. The example indicates the open-loop remote control of an UAV control surface (e.g., elevator) via a ground control station. The Arduino code development and wiring connections are fully described.

Example 13.3

Using two Arduino Uno boards and two NRF24L01 modules, design a communications system to remotely drive the position of a servo with an analog potentiometer. The servo is representing a joystick of a ground control station, while a servomotor is representing the drive for any control surface of an UAV.

Solution

For this experiment, the following parts are needed: 1. Two Arduino Uno boards, 2. Two NRF24L01 modules, 3. Two breadboards, 4. Two LEDs, 5. Two 220 ohm resistors, and 6. Generic jumper wires. The breadboards, LEDs, and resistors are optional. The breadboards allow for a 5 V connection and easier wiring. The LEDs are blinking to indicate that a signal is received or transmitted. The communication process constantly repeats, and both Arduino boards are constantly sending and receiving data. The physical build of the two circuits are very similar (almost identical).

- **Wiring diagram**

  Servo motors have three wires: ground, power (central), and signal (often orange). In the **receiver** side, the signal lead of the servomotor is arbitrarily connected to the digital pin 6 of the Arduino Uno board. You may change 6 to whatever digital pin your servo is connected to. The power wire (middle, often red) is connected to the 5 V pin on the Arduino board. The ground wire is connected to a ground pin on the board.

  The potentiometer should be wired so that its two outer pins are connected to power (+5 V) and ground, and its middle pin (for signal, often red) is connected to analog input 0 on the board. In the transmitter side, since an analog potentiometer is employed, the signal pin (i.e., center lead) of the potentiometer is arbitrarily connected to the analog input pin 0 of the Arduino Uno board. You may change 0 to whatever analog pin your potentiometer is connected to. Since a potentiometer has an analog resistor, there is no polarity. Hence, any side pin may be connected to the power, and the remaining side to the ground.

  The CE and CSN pins of both NRF24L01 modules are arbitrarily connected to pins 9 and 10 of the Arduino Uno boards. The connection of other pins of a NRF24L01 module to an Arduino Uno is shown in Table 13.6. Figure 13.11 demonstrates the circuit, wiring, and schematic of the open-loop remote control system of a servomotor.

Table 13.6 Connection of pins of NRF24L01 module, potentiometer, and servo to Uno board

Board	Pins									
NRF24L01 module	VCC	GND	CSN	CE	SCK	MOSI	MISO	IRQ		
Potentiometer	+	-								middle pin
Servomotor	+	-							Orange wire	
Arduino Uno	5 V	GND	10	9	13	11	12	8	D6	A0

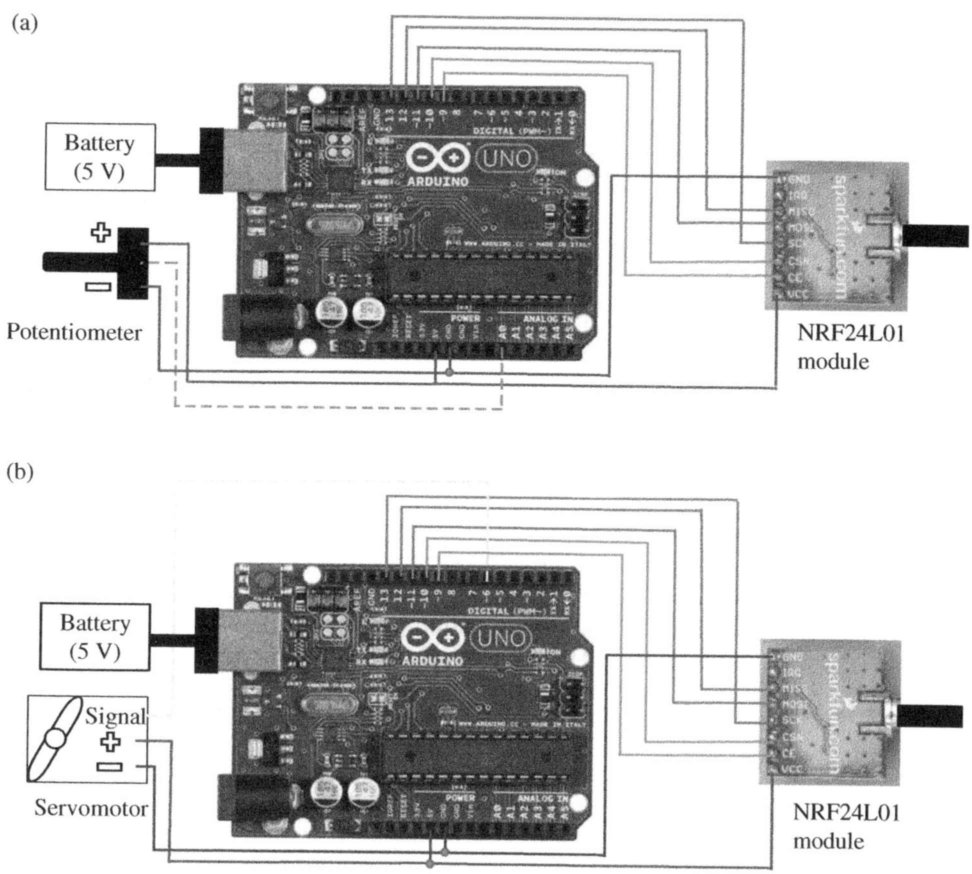

Figure 13.11 Wiring of NRF24L01 module, potentiometer, and servo to Uno board. (a) Transmitter (NRF24L01 module, potentiometer, and Uno board), (b) Receiver (NRF24L01 module, servo, and Uno board.)

Table 13.6 demonstrates the connection of pins of a NRF24L01 module, a potentiometer and a servo to an Arduino Uno board.

- **Transmitter and Receiver Codes**

In the following, two Arduino codes [131] for the wireless communications are developed. This is to send a potentiometer command to drive a servomotor via

NRF24L01 modules. There are some codes differences for the reading and writing pipes for the respective boards.

In the transmitter side, the center lead of the potentiometer is arbitrarily connected to the analog input pin 0 (i.e., function "PotVal analogRead(A0)"). In the receiver side, the center lead of the servomotor is arbitrarily connected to the digital pin 6 (i.e., function "S1.attach(6)").

In both Uno boards, we need to include the basic SPI and the RF24 libraries to create an RF24 object. The two NRF24L01 module arguments are the CSN and CE pins; we arbitrarily choose to connect them to pins 9 and 10 of the Uno (i.e., "RF24 radio (9, 10)"). The function "const byte addresses[][6] = {}" creates an array of arrays of six bytes.

For a bi-directional communication, we need to create two pipes or addresses. A byte array to represent the address is used, hence; through the so called pipe, two modules will communicate. In the setup section, we need to define both pipes, and note that the writing address at the first Arduino needs to be the reading address at the second Arduino, and vice versa. The reading address at the first Arduino needs to be the writing address at the second Arduino. The value of these addresses can be changed to any 5 letter string; and this enables to choose to which receiver we talk to.

At the transmitter code, we write:

- radio.openWritingPipe(addresses[1]); // 00001
- radio.openReadingPipe(1, addresses[0]); // 00002

At the receiver code, we write:

- radio.openWritingPipe(addresses[0]); // 00002
- radio.openReadingPipe(1, addresses[1]); // 00001

In the setup section, we need to initialize the transmitter object and using the radio.openWritingPipe() function, we set the address of the receiver to which we will send data. On the other side, at the receiver, using the radio.setReadingPipe() function, we set the same address to enable the communication between two modules.

Then, using the radio.setPALevel() function, we need to set the Power Amplifier level. In our case, we will set it to minimum as our modules are identical. If a higher level is employed, it is recommended to use a bypass capacitor across GND, so that, they have a stable voltage while operating. At the transmitter, we set the first Arduino as receiver, and with an empty "while" loop, we may wait for the second Arduino to send the signal.

Next, we use the radio.stopListening() function to set one module as the transmitter (in transmitter code). In the receiver code, we use the radio.startListening() function to set the second module as the receiver.

In the loop section, we first assign a delay, by adding 10 milli-seconds delay (i.e., delay(10)). We read and map the value of potentiometer (i.e., joystick) deflection from 0 to 180. The function "angle = map()" allows us to scale the potentiometer value of 0 to 1023 to the servo value range of 0 to 180 degrees. At the transmitter, we need to create an array of characters to which we assign the transmitting signal. Using the radio.write() function, we will send the command signal to the receiver.

- delay(10);
- int angleValue = map(PotVal, 0, 1023, 0, 180);
- radio.write(&angleValue, sizeof(angleValue));

By using the "&" before the variable name (i.e., angle value), we set an indication of the variable that stores the data which we want to be sent. Using the second argument, we set the number of bytes that we want to take from that variable. In this case, the sizeof() function gets all bytes of the "angle value".

At the receiver side, in the loop section, using the radio.read() function, we read the incoming signal, and store the data into the "servo angle" variable.

While there is data available, we will read it, save it to the "servoAngle" variable, and then use that value to rotate the servo motor. By using the servo.write() function, it returns immediately, and the servo will subsequently move to the new position. Since we only have one servo, we may use S1 instead of servo (i.e., "Servo S1;").

- delay(10);
- radio.read(&srevoAngle, sizeof(srevoAngle));
- S1.write(srevoAngle);

By the way, you may add LEDs that will light up when a signal is communicated (transmitted or received). However, to have the correct voltage for the LEDS, a resistor must be connected in series with each LED. A 220 ohm resistor is typical for small LEDs. In this case, connect an analog Arduino pin (e.g., 2) to an LED long lead (anode), and the short lead (cathode) to the resistor. Then connect other end of the resistor to GND. If you utilize an Arduino Uno R3, it has already embedded LED and resistor.

When the pot is deflected, or, when the servo is rotating, the LED will light up. So, the process constantly repeats, and both Arduino boards are constantly sending and receiving signals.

Once both codes are uploaded, run the serial monitor at the receiver. You will notice that the command signal is saved each 10 milliseconds.

Transmitter Code:

```
#include <SPI.h>
#include <nRF24L01.h>
#include <RF24.h>
#define PotVal analogRead(A0)
RF24 radio(9, 10);
const byte addresses[][6] = {"00001", "00002"};
void setup()
{
        radio.begin();
        radio.openWritingPipe(addresses[1]);
        radio.openReadingPipe(1, addresses[0]);
        radio.setPALevel(RF24_PA_MIN);
        radio.stopListening();
}
void loop()
{
        delay(10);
        int angleValue = map(PotVal, 0, 1023, 0, 180);
        radio.write(&amp;angleValue, sizeof(angleValue));
}
```

Receiver Code:

```
#include <SPI.h>
#include <nRF24L01.h>
#include <RF24.h>
#include <Servo.h>
RF24 radio(9, 10);
const byte addresses[][6] = {"00001", "00002"};
int srevoAngle = 90;
Servo S1;
void setup()
{
        S1.attach(6);
        radio.begin();
        radio.openWritingPipe(addresses[0]);
        radio.openReadingPipe(1, addresses[1]);
        radio.setPALevel(RF24_PA_MIN);
        radio.startListening();
}
void loop()
{
        delay(10);
        radio.read(&amp;srevoAngle, sizeof(srevoAngle));
        S1.write(srevoAngle);
}
```

Questions

1 What is the range of frequencies for UAV communication?

2 What are the main elements of the communication system in the GCS?

3 What are the primary functions of the UAV communication system?

4 List applicable types of antenna to UAVs.

5 What is the typical frequencies for Inmarsat communication system?

6 What is the typical frequencies for GPS signals?

7 What factors paly role in the design of a transmitter?

8 What factors do affect the design of a receiver?

9 Define encryption.

10 What is SatCom?

11 What is encoder?

12 Discuss the differences between encoding and encryption.

13 What are the different media for communication between the GCS and a flight vehicle?

14 Define telemetry.

15 Define transceiver.

16 Define uplink and downlink.

17 The signal strength of a radio signal from a receiver's antenna is mainly a function of three factors. What are they?

18 What are two major sources of positioning system and communications using satellite?

19 What are the advantage of optic fiber over copper wire in connecting electronic systems?

20 List Antennas of maritime variant of a Guardian UAV.

21 What is typical radio frequencies for EO/IR sensors for a large UAV?

22 What are the three modulation techniques?

23 What are main aspects of transmitter/receiver installation on a UAV which a designer should be concern of?

24 What is the range of frequencies for Ku bands?

25 What are the range of frequencies for VHF and UHF bands?

26 Which types of antennas are recommended for use in UAVs?

27 What is radome?

28 What does C2 stand for?

29 What does C3 stand for?

30 Describe elements for the Predator data link suite?

31 Which one is more challenging: a. installation of an antenna in a GCS, b. installation of an antenna in a UAV? Why?

32 List antennas for the Maritime variant of a Guardian UAV.

33 List antennas for the Global Hawk UAV.

34 What is frequency range for the Garmin GSX 70 radar used in a NASA Global Hawk?

35 How much power is required by Ku-band antenna of Global Hawk communications system?

36 How much power is required by LNA (low noise amplifier)/Diplexer of Global Hawk communications system?

37 How much is the weight of UHF RX/TX (Transmitter/receiver) of Global Hawk communications system?

38 How much power is required by UHF power amplifier of Global Hawk communications system?

39 What is the diameter of the Ku-band Satcom antenna of Global Hawk?

40 What does BAMS stand for?

Problems

1 A large UAV is employing the satellite communication system with a radio frequency of 18 GHz. Calculate the wavelength of the signal for such communication.

2 A large UAV is employing the satellite communication system with a radio frequency of 30 GHz. Calculate the wavelength of the signal for such communication.

3 A large UAV is employing the satellite communication system with a radio frequency of 25 GHz. Calculate the wavelength of the signal for such communication.

4 A large UAV is employing the Inmarsat communication system. The UAV – flying at 10 000 m – is transmitting a data through its Inmarsat antenna to its GCS. How long does it take for this signal to reach the GCS? Ignore the distance between GCS and the UAV.

5 A reconnaissance UAV is employing the Inmarsat communication system. The UAV – flying at 20 000 m – is transmitting a data through its Inmarsat antenna to its GCS. How long does it take for this signal to reach the GCS? Ignore the distance between GCS and the UAV.

6 A reconnaissance UAV is employing the Inmarsat communication system. The UAV – flying at 18 000 m – is transmitting a data through its Inmarsat antenna to its

GCS. How long does it take for this signal to reach the GCS? Ignore the distance between GCS and the UAV.

7 A UAV transmitter antenna is consuming 65 mA at 3.5 volts. The phase angle between voltage and current is 75 degrees. How much power is radiated by this antenna?

8 A GCS transmitter antenna is consuming 0.6 A at 120 volts. The phase angle between voltage and current is 30 degrees. How much power is radiated by this antenna?

9 A GCS transmitter antenna is consuming 1.5 kW of power with 120 volts. The phase angle between voltage and current is 45 degrees. How much current is flowing through this antenna?

10 A UAV transmitter antenna is consuming 20 W of power with 24 volts. The phase angle between voltage and current is 0 degrees. How much current is flowing through this antenna?

Laboratory Experiments

1 Use a servo, a potentiometer, two Arduino Uno microcontrollers, a transmitter and a receiver to explore the elevator open loop control process by an operator. In this lab, setup two circuits by using the wiring model (similar to the one in Figure 13.11), and write two Arduino codes similar to the one demonstrated in example 13.3. The requirement is to have −20° of elevator deflection (i.e., servo) by rotating the potentiometer for 45°. Then, write a report for this lab experiment, and include a copy of the Arduino codes. The instructor will provide the necessary lab items to the students.

2 Use a servo, a potentiometer, two Arduino Uno microcontrollers, a transmitter and a receiver to explore the rudder open loop control process by an operator. In this lab, setup two circuits by using the wiring model (similar to the one in Figure 13.11), and write two Arduino codes similar to the one demonstrated in example 13.3. The requirement is to have 30° of rudder deflection (i.e., servo) by rotating the potentiometer for 45°. Then, write a report for this lab experiment, and include a copy of the Arduino codes. The instructor will provide the necessary lab items to the students.

3 Use a servo, a potentiometer, two Arduino Uno microcontrollers, a transmitter and a receiver to explore the ailerons open loop control process by an operator. In this lab, setup two circuits by using the wiring model (similar to the one in Figure 13.11), and write two Arduino codes similar to the one demonstrated in example 13.3. The requirement is to have two 20° of aileron deflection (one +20° and one −20°) by rotating the potentiometer for 60°. Then, write a report for this lab experiment, and include a copy of the Arduino codes. The instructor will provide the necessary lab items to the students.

4 Use a servo, a potentiometer, two Arduino Uno microcontrollers, a transmitter and a receiver to explore the motor open loop control process by an operator. In this lab, setup two circuits by using the wiring model (similar to the one in Figure 13.11), and write two Arduino codes similar to the one demonstrated in example 13.3. The requirement is to have full engine throttle (i.e., servo) by rotating the potentiometer for 30°. Then, write a report for this lab experiment, and include a copy of the Arduino codes. The full engine throttle is implemented by a 4 VV from the microcontroller. The instructor will provide the necessary lab items to the students.

5 Use two servos, two potentiometers, two Arduino Uno microcontrollers, a transmitter and a receiver to explore the elevator and rudder open loop control process by an operator. In this lab, setup two circuits by using the wiring model (similar to the one in Figure 13.11), and write two Arduino codes similar to the one demonstrated in example 13.3. The requirement is to have ±30° of elevator and rudder deflections (i.e., servos) by rotating each potentiometer for ±45°. One potentiometer should control the elevator, and another one should control the rudder. Then, write a report for this lab experiment, and include a copy of the Arduino codes. The instructor will provide the necessary lab items to the students.

Design Projects

1 Design and build a control system using two Arduino Uno boards, a transmitter and a receiver to remotely control the position of a servo with an analog potentiometer. Instructor will provide necessary parts and equipment.

2 Design and build a control system using two Arduino Uno boards, a transmitter and a receiver to remotely control the position of a servo with a digital potentiometer. Instructor will provide necessary parts and equipment.

3 Design and build a control system using two Arduino Uno boards, a transmitter and a receiver to remotely read air pressure from a pitot-tube and send the results to a computer and plot the results. Instructor will provide necessary parts and equipment.

4 Design and build a control system using two Arduino Uno boards, a transmitter and a receiver to remotely read airspeed from a pitot-tube (in a wind tunnel) and send the results to a computer and plot the results. Instructor will provide necessary parts and equipment.

5 Design and build a control system using two Arduino Uno boards, a transmitter and a receiver to remotely read temperature from a thermometer and save data. Instructor will provide necessary parts and equipment.

6 Design and build a control system using an Arduino Uno, a transmitter and a receiver to read altitude from an altimeter and save data to a laptop. Instructor will provide necessary parts and equipment.

7 Design and build a control system using an Arduino Uno, a transmitter and a receiver to transmit the images taken by a digital camera to a ground station. Instructor will provide necessary parts and equipment.

8 Design and build a control system using an Arduino Uno, a transmitter and a receiver to transmit the UAV position data - measured by a GPS - to a ground station. Instructor will provide necessary parts and equipment.

9 Design and build a control system using an Arduino Uno, a transmitter and a receiver to transmit the UAV speed - measured by a GPS - to a ground station. Instructor will provide necessary parts and equipment.

10 Design and build a control system using an Arduino Uno, a transmitter and a receiver to transmit the UAV distance from an obstacle - measured by a range finder - to a ground station. Instructor will provide necessary parts and equipment.

Notes

1 Satellite Communications.
2 Inmarsat PLC is a British satellite telecommunications company, offering global mobile services with ground stations through 13 geostationary communications satellite.

14

Design Analysis and Feedbacks

CONTENTS

Design of Unmanned Aerial Systems, First Edition. Mohammad H. Sadraey.

Companion website: www.wiley.com/go/sadraey/unmanned-aerial-systems

Educational Outcomes

After study of this chapter, the reader will be able to:

1) Discuss the techniques to analyze a Unmanned Aerial Vehicle UAV configuration
2) Conduct a UAV design analysis process
3) Conduct a UAV stability analysis process
4) Conduct a UAV performance analysis process
5) Conduct a UAV controllability analysis process
6) Plan for a UAV cost analysis process
7) Implement the weight and balance for a UAV
8) Evaluate a UAV design program and provide feedbacks

14.1 Introduction

An essential technical activity within a UAV design project is the design analysis and evaluation. UAVs are remotely-piloted or self-piloted air vehicles that can carry payloads for a given flight mission. A design analysis is providing a formalized check of the proposed system design with respect to specification requirements. In each major design phase (conceptual, preliminary, and detail), an analysis should be conducted to review the design and to ensure that the design is acceptable at that point before proceeding with the next stage.

A civilian UAV is designed to perform a desired mission at a lower cost than a manned aircraft equivalent. The cost is not a priority for a military UAV designer, and the number one priority for a military UAV designer is UAV performance (e.g., maneuverability). However, the design requirements for a military UAV are much more challenging (e.g., stealth, combat capability).

The Final design review is usually scheduled after the completion of the detail design phase and prior to entering the production phase. It accomplishes two major tasks: (a) discovering and resolving any design problems and the subsystem/component level; and then (b) verifying and documenting the system capabilities for certification or customer acceptance. The UAV design project has an iterative operation in its nature, and it is open-ended. Feedbacks of each analysis report should be used to refining the design, and to improve the final configuration and specifications to satisfy the requirements.

There are mainly two approaches to handle the design activities and to establish design groups: (a) design groups based on aircraft components; and (b) design groups based on expertise. There are various advantages and disadvantages for each of the two planning approaches in terms of ease of management, speed of communication, efficiency, and similarity of tasks. However, if the project is large such as the design of a large UAV, both groupings could be applied simultaneously.

If the approach of groups based on expertise is selected, the chief designer must establish (Figure 14.1) the following teams: (a) structural analysis; (b) aerodynamic analysis; (c) flight stability analysis; (d) controllability analysis; (e) UAV performance analysis; (f) mission success analysis; and (g) cost analysis. In this chapter, a few design analyses are briefly presented: (a) flight stability analysis; (b) flight performance analysis; (c) controllability; and (d) cost analysis.

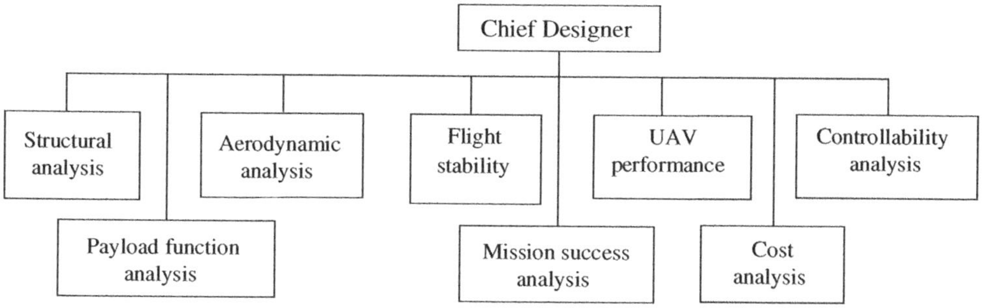

Figure 14.1 Unmanned Aerial Vehicle (UAV) analysis teams.

14.2 Design Feedbacks

A UAV is composed of various components such as wing, fuselage, tail, engine, and landing gear, plus payload, fuel and autopilot. Two features of each component have a significant impact on stability and control of a UAV. They are: (a) mass; and (b) moment of inertia. Any physical component has a fixed center of gravity (cg). The aircraft cg and mass moment of inertia calculations provide two feedbacks in the overall design process as depicted in Figure 14.2. The aircraft cg location, aircraft cg range and aircraft mass moment of inertia about three axes of x, y, and z are three required groups of information to analyze aircraft stability and control.

Hence, in the first place, mass, center of gravity, and mass moment of inertia of each component should be determined. Section 14.3.1 presents the technique to determine the UAV center of gravity. The technique for the calculation of mass moment of inertia of each component has been provided by Chapter 10 of [169].

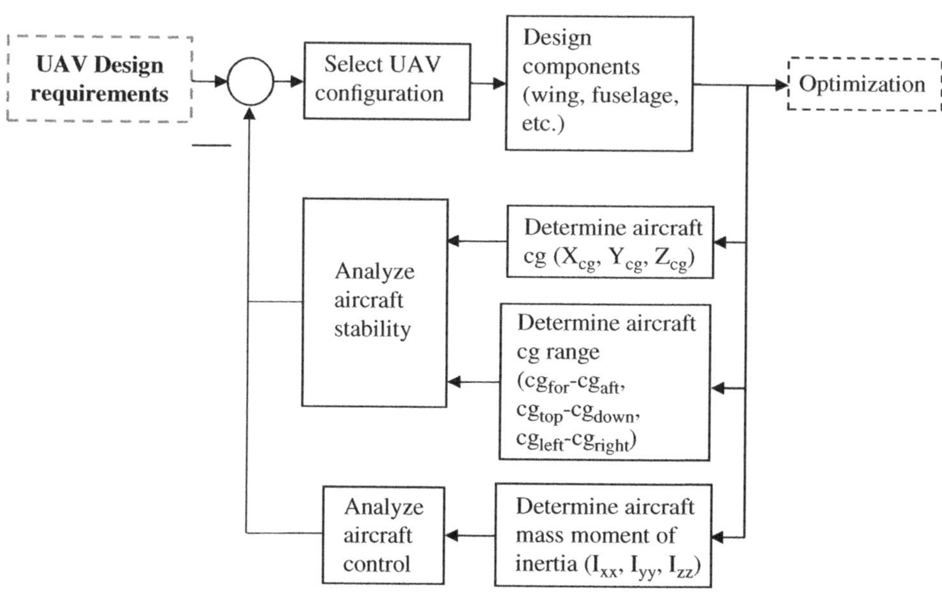

Figure 14.2 Feedbacks provided by cg and moment of inertia calculations.

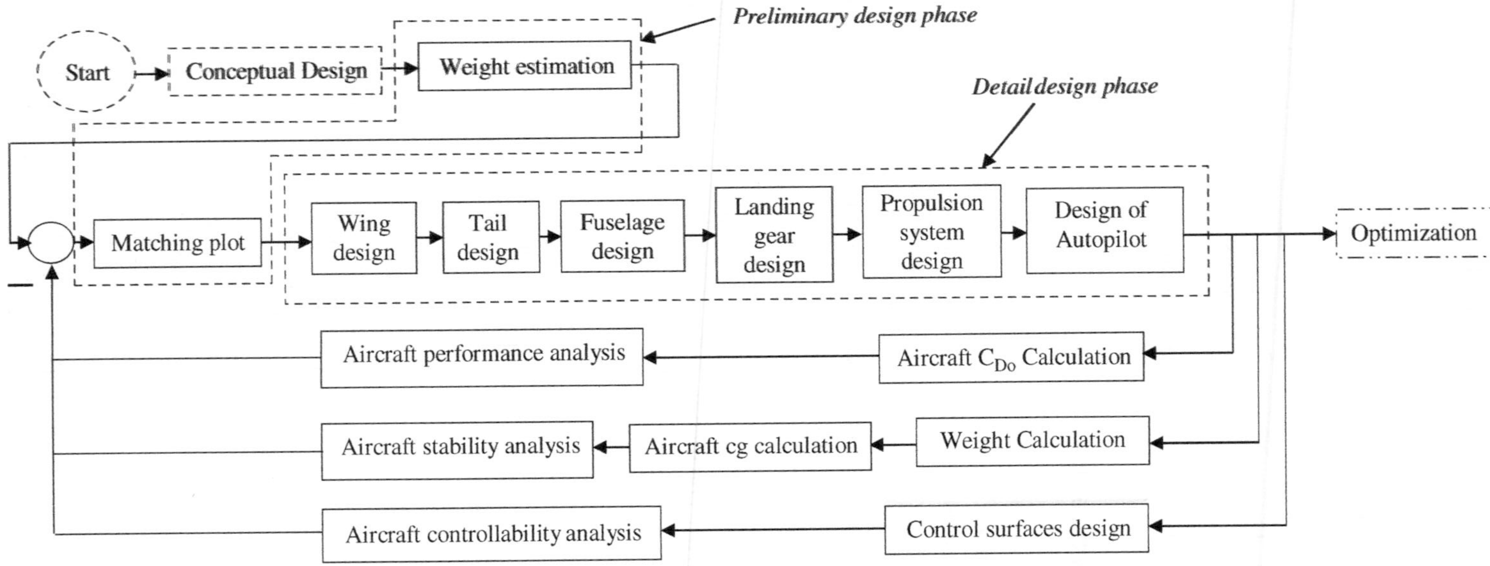

Figure 14.3 Three major feedbacks in the Unmanned Aerial Vehicle (UAV) design process. Notes: (i) Loop 1: UAV performance analysis (ii) Loop 2: UAV stability analysis (iii) Loop 3: UAV controllability analysis

Three major feedbacks (i.e., loops) in the entire design process are illustrated in Figure 14.3. The aircraft weight calculation is in the second loop which provides a feedback on aircraft stability analysis. This feedback initially uses the weight of components to balance the aircraft by allocating them on the right place to submit the desired center of gravity location. As the Figure 14.3 shows, the calculated aircraft weight is employed to modify all components that were previously designed based on an estimated aircraft weight. The relationship between three design phases (conceptual, preliminary, and detail) is also briefly depicted.

14.3 Weight and Balance

14.3.1 UAV Center of Gravity

An aircraft is made up of several components such as wing, fuselage, tail, engine, and landing gear, plus payload, fuel and autopilot. Each component has a unique mass (weight), unique center of gravity, and thus contributes to the aircraft overall center of gravity. In order to determine aircraft cg, a coordinate axis system must be defined. Moreover, for each axis, a reference line should be designated.

There are mainly four coordinate axis systems; namely (a) Earth fixed axis system; (b) body axis system; (c) wind axis system; and (d) stability axis system. The body, wind, and stability axes systems move and rotate with the aircraft motion and are centered at the aircraft cg. Each of these four axis systems has applications and advantages; here, a body axis system as depicted in Figure 14.4 is selected.

In a body axis system, an orthogonal axis is defined where x axis is along fuselage center line; y is defined using right-hand rule. Thus, y-axis is to the right, and z is downward. The coordinate of the aircraft cg along x axis is represented by X_{cg}, along y axis by Y_{cg}; and along z axis by Z_{cg}. The aircraft cg locations (X_{cg}, Y_{cg}, and Z_{cg}) with n components are determined using the following formulas:

$$X_{cg} = \frac{\sum_{i=1}^{n} m_i x_{cg_i}}{\sum_{i=1}^{n} m_i} \tag{14.1}$$

$$Y_{cg} = \frac{\sum_{i=1}^{n} m_i y_{cg_i}}{\sum_{i=1}^{n} m_i} \tag{14.2}$$

$$Z_{cg} = \frac{\sum_{i=1}^{n} m_i z_{cg_i}}{\sum_{i=1}^{n} m_i} \tag{14.3}$$

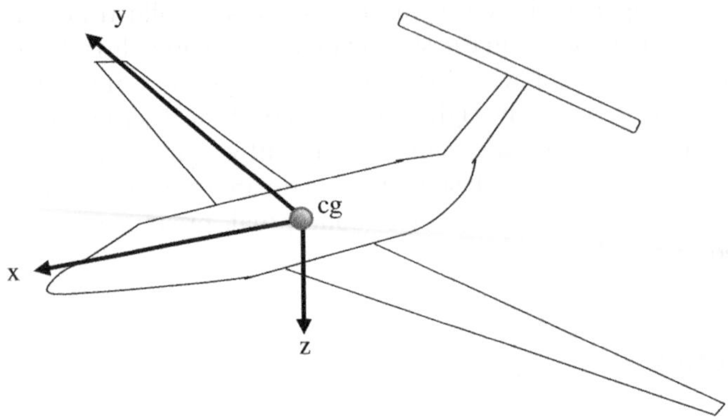

Figure 14.4 Definition of body-axis coordinate system.

where m_i denotes the mass of each aircraft component, and x_{cgi}, y_{cgi}, and z_{cgi} are the coordinates of each individual component. The coordinates are measured with respect to particular reference lines. The selection of reference lines is arbitrary, and it does not affect the final result. However, it is recommended to select a vertical line passing through the foremost point of the aircraft (e.g., fuselage nose) as the reference line for x coordinate. The reference line for y coordinate is recommended to be the fuselage centerline; i.e., xz plane. The reference line for z coordinate is recommended to be the ground level (i.e., the contact surface between the landing wheels with the ground).

When an aircraft carries the maximum allowable payload, and maximum fuel, the summation of the weights of all components is equal to maximum take-off weight.

$$\Sigma W_i = W_{TO} \tag{14.4}$$

Otherwise; the summation is equal to the weight of the components which are present in that particular configuration and condition:

$$\Sigma W_i = W_w + W_F + W_{HT} + W_{VT} + W_E + W_{LG} + W_{PL} + W_{fuel} + W_A + \ldots \tag{14.5}$$

where W_w, W_F, W_{HT}, W_{VT}, W_E, W_{LG}, W_{PL}, W_{fuel}, W_A represent wing, fuselage, horizontal tail, vertical tail, engine, landing gear, payload, fuel, and autopilot respectively. Thus,. Eqs. (14.1) through (14.3) are revised into the following forms:

$$X_{cg} = \frac{\begin{array}{c}\Sigma W_w x_w + W_F x_F + W_{HT} x_{HT} + W_{VT} x_{VT} + W_E x_E \\ + W_{LG} x_{LG} + W_{PL} x_{PL} + W_{fuel} x_{fuel} + W_A x_A + \ldots\end{array}}{\Sigma W_i} \tag{14.6}$$

$$Y_{cg} = \frac{\begin{array}{c}\Sigma W_w y_w + W_F y_F + W_{HT} y_{HT} + W_{VT} y_{VT} + W_E y_E \\ + W_{LG} y_{LG} + W_{PL} y_{PL} + W_{fuel} y_{fuel} + W_A y_A + \ldots\end{array}}{\Sigma W_i} \tag{14.7}$$

$$Z_{cg} = \frac{\begin{array}{c}\Sigma W_w z_w + W_F z_F + W_{HT} z_{HT} + W_{VT} z_{VT} + W_E z_E \\ + W_{LG} z_{LG} + W_{PL} z_{PL} + W_{fuel} z_{fuel} + W_A z_A + \ldots\end{array}}{\Sigma W_i} \tag{14.8}$$

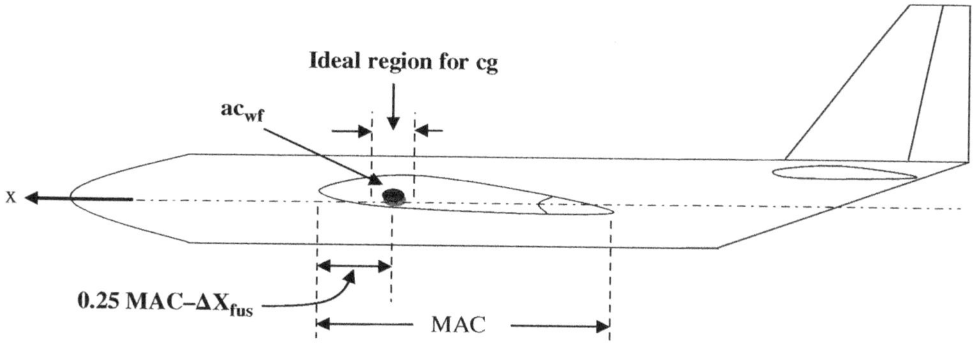

Figure 14.5 Ideal region for the cg location along x-axis.

These equations must be modified based on the components of an aircraft. For instance, if an aircraft has a special system (such as automatic flight control system; or autopilot in an unmanned aerial vehicle); this system must be included in these equations.

Another topic which is directly related to the aircraft weight distribution is to decide about whether to widen the cg range or shorten it along x axis. The answer to this challenge by two groups of aircraft-related people (i.e., designer, customer) is almost opposite. In this specific issue, the interest or wish of the designer does not match with that of the customer. Generally speaking, an aircraft customer often seeks an aircraft with a wider cg range, while the designer will try to shorten the cg range.

The decision about how wide the cg range needs to be must be made after consultation with the marketing department. In summary, a load-relating customer would rather purchase an aircraft with a large cg range, while an aircraft designer tries to limit the cg range to the shortest distance.

Figure 14.5 provides the ideal region for the cg location along x–axis. For controllability purposes, it is recommended to locate UAV loads such that the total cg (along x-axis) is located close to the wing-fuselage aerodynamic center (ac_{wf}). This will provide less trim drag, stable UAV, and less control power requirement. The aerodynamic center for a wing-alone is often located at the quarter chord mean aerodynamic chord (25% MAC).

14.3.2 Weight Distribution

The load balancing is often used to distribute the autopilot, weight of battery, cargo, and fuel throughout an aircraft, so as to keep the aircraft's cg close to its wing-fuselage aerodynamic center to minimize elevator deflection for longitudinal trim. In military UAV, it is common to have a weapon installation master as a part of the Ground control station (GCS) team; their responsibilities include installing payload, ensuring the weapon is properly placed and secured to prevent its shifting.

In large aircraft, multiple fuel tanks and pumps are often used, so that as fuel is consumed, the remaining fuel can be positioned to keep the aircraft balanced, and to reduce stability problems associated with the free surface effect. There are various ways to maintain the aircraft cg which allowable limits during flight operations. An in-flight technique is used to transfer fuel among various fuel tanks.

In general, design requirements that influence the decision on the aircraft weight/load distribution (i.e., cg location and limits) are: (a) controllability requirements; (b) stability requirements; (c) flying qualities requirements; and (d) operational requirements and constraints. However, various aircraft have different missions and requirements, so they do not have the same priority and similar constraints.

In weight distribution, the locations of some items and components such as autopilot, tail, and some payloads (e.g., camera) are self-evident. Other components such as wing, fuel tanks, engines, and other payloads (e.g., radar) can be shifted around, to a certain extent to achieve the desired cg location. A table similar to Table 14.1 should be prepared and completed, to make sure the UAV masses are correctly placed; and at any flight condition, the UAV can be balanced.

For a civil/commercial aircraft, the static longitudinal stability criterion requires that the center of gravity is never allowed to be behind the aircraft neutral point or aircraft aerodynamic center (X_{np}). In terms of non-dimensional derivative, the rate of change of pitching moment with respect to angle of attack must be negative for an aircraft to be statically longitudinally stable. This derivative is determined by:

$$C_{m_\alpha} = C_{L_\alpha}\left(\bar{X}_{cg} - \bar{X}_{np}\right) \tag{14.9}$$

Table 14.1 Typical weight and balance table.

No	Components	Weight/Force	Moment arm	Moment
1	Wing			
2	Fuselage			
3	Tail			
4	Engine			
5	Fuel/battery			
6	Autopilot			
7	Navigation system			
8	Communications system			
9	Payloads (e.g., camera and radar)			
10	Landing gear			
11	Maximum wing-body lift			
12	Minimum wing-body lift			
13	Minimum wing-body aerodynamic pitching moment			
14	Maximum wing-body aerodynamic pitching moment			
15	Engine thrust			
16	UAV drag			
17	**Maximum positive horizontal tail lift**			
18	**Maximum negative horizontal tail lift**			
Sum				

This criterion dictates that for a statically longitudinally stable aircraft, the most aft location of the cg must be forward of the aircraft neutral point. Equation (14.9) indicates that the location of aircraft cg will directly influence the longitudinal static stability.

14.4 Stability Analysis

14.4.1 Fundamentals

An important UAV design requirement is flight stability. When the UAV is designed, its stability should be analyzed, to make sure, the design, configuration, components, and payload installation are all acceptable.

Stability is defined as the tendency of an aircraft to oppose a disturbance (e.g., gust) and return to its initial steady state trim condition if disturbed. Stability is often divided into two branches:

1) Static stability
2) Dynamic stability

Static stability is defined as the initial tendency of an aircraft, without autopilot assistance/interference, to develop forces and moments to oppose an instantaneous perturbation of a motion variable from a steady state flight condition. Dynamic stability is defined as the tendency of an aircraft, without autopilot assistance/interference, to return to the initial steady state trim condition after a disturbance disturbs the trim values. When an aircraft is dynamically stable, it definitely has static stability. However, if an aircraft is statically stable, there is no guarantee that it has dynamic stability.

Figure 14.4 provides an aircraft body coordinate system, plus three rotational motions of roll, pitch, and yaw. The convention is that the clockwise rotation about any axis, when you look from UAV cg, is assumed as the positive rotation. An aircraft motion (flight) has six Degrees-Of-Freedom (6 DOF), due to two types of freedom (one linear and one rotational) about each of the three axes of x, y and z. Therefore, stability is measured about these three axes:

a) Lateral stability
b) Longitudinal stability
c) Directional stability

Lateral stability is defined as the stability of any rotational motion about x axis (i.e., roll) and any corresponding linear motion along yz plane (i.e., side motion). Longitudinal stability is defined as the stability of any rotational motion about y axis (i.e., pitch) and any linear motion along xz plane (i.e., forward and aft, and up and down). Directional stability is defined as the stability of any rotational motion about z-axis (e.g., yaw) and any corresponding linear motion along xy plane (e.g., sideslip).

The requirements for aircraft static and dynamic stability (longitudinal, lateral, and directional) are different. When the aircraft derivative Cm_α is negative, the aircraft is referred to as statically longitudinally stable. Table 14.2 shows the summary of static longitudinal and lateral-directional stability requirements.

Table 14.2 Static longitudinal and lateral-directional stability requirements.

No	Requirements	Stability derivatives	Symbol	Typical value (1/rad)
1	Static longitudinal stability	Rate of change of pitching moment coefficient with respect to angle of attack	C_{m_α}	−0.3 to −1.5
2	Static directional stability	Rate of change of yawing moment coefficient with respect to sideslip angle	C_{n_β}	+0.05 to +0.4
3	Static lateral stability	Rate of change of rolling moment coefficient with respect to sideslip angle	C_{l_β}	−0.1 to −2

14.4.2 Static Longitudinal Stability

The static longitudinal stability is examined through the sign of the longitudinal stability derivative Cm_α or the location of the aircraft neutral point. The static longitudinal stability is satisfied provided that the following condition is met.

$$C_{m_\alpha} \prec 0 \tag{14.10}$$

When the derivative Cm_α is negative or when the neutral point is behind the aircraft cg, the aircraft is said to be statically longitudinally stable. For an aircraft with a fixed aft tail, the aircraft static longitudinal stability derivative is determined [70] as:

$$C_{m_\alpha} = C_{L_{\alpha wf}}\left(h - h_o\right) - C_{L_{\alpha h}}\eta_h \frac{S_h}{S}\left(\frac{l}{\overline{C}} - h\right)\left(1 - \frac{d\varepsilon}{d\alpha}\right) \tag{14.11}$$

where S_h and S represent horizontal and wing planform areas respectively. Parameters h and h_o are the non-dimensional locations of the aircraft cg and wing-fuselage aerodynamic center respectively. Moreover, l and C denote horizontal tail moment arm and wing aerodynamic chord.

Among major aircraft components, the horizontal tail has the largest contribution to the aircraft longitudinal stability. The reason is that the horizontal tail is able to generate the counter pitching moment in order to restore the longitudinal trim position.

Three basic ways to improve static longitudinal stability to: (a) increase the horizontal tail area; (b) increase the horizontal tail moment arm; and (c) relocate/redistribute the components, payloads, and fuel tanks in order to move the UAV cg to be further ahead of the aircraft neutral point.

14.4.3 Dynamic Longitudinal Stability

The dynamic longitudinal stability analysis is performed after all aircraft components are designed and the roots (λ) of the longitudinal characteristic equation are calculated. A general form of the aircraft longitudinal characteristic equation looks like the following:

$$A_1\lambda^4 + B_1\lambda^3 + C_1\lambda^2 + D_1\lambda + E_1 = 0 \tag{14.12}$$

where coefficients A_1, B_1, C_1, D_1, and E_1 are functions of the several stability derivatives such as Cm_α and Cm_q. An aircraft is dynamically longitudinally stable, if the real parts of all roots of longitudinal characteristic equation are negative. Another way to analyze dynamic longitudinal stability is to make sure that longitudinal modes (i.e., short period and long period [Phugoid]) are damped.

If you are using a state-space dynamic model of the UAV,

$$\begin{aligned} \dot{x} &= Ax + Bu \\ y &= Cx + Du \end{aligned} \tag{14.13}$$

The UAV is dynamically longitudinally stable, if the real parts of all eigenvalues of state matrix (A) are negative. The matrix A may be calculated from:

$$A = \frac{A^*}{E} \tag{14.14}$$

where A* and E are introduced in Eq. (5.73) in Chapter 5.

The reader is encouraged to consult with [70] to see how to derive the aircraft longitudinal characteristic equation. The longitudinal stability derivatives cannot be determined unless all aircraft components including wing and fuselage have been designed. This is the reason to resort to a simplifying criterion that could be a base for the horizontal tail preliminary design. When the horizontal tail volume coefficient ($\bar{V}_H$) is in ballpark number, we are 90% confident that the longitudinal stability requirements have been satisfied. When other aircraft components such as fuselage and wing have been designed, the horizontal tail design will be revised and optimized in the longitudinal stability analysis process.

For the dynamic longitudinal stability, the aircraft response to a longitudinal disturbance is studied. The response of a longitudinally dynamically stable aircraft to a vertical gust is an oscillatory one and comprises of two modes: (a) short period mode; (b) long period mode.

14.4.4 Static Lateral-Directional Stability

An aircraft is said to be statically laterally stable, when the aircraft derivative C_{l_β} (known as dihedral effect) is negative. When the aircraft derivative Cn_β is positive, the aircraft is said to be statically directionally stable.

When a laterally-directionally dynamically stable aircraft experiences a lateral-directional disturbance (i.e., a horizontal gust hits the vertical tail), the aircraft will oppose the disturbance, and eventually returns to the initial trim point. There are static and dynamic lateral-directional stability requirements. The static lateral-directional stability is satisfied provided the following two conditions are met.

$$C_{l_\beta} \prec 0 \tag{14.15}$$

$$C_{n_\beta} \succ 0 \tag{14.16}$$

The dynamic lateral-directional stability is investigated through the response of an aircraft to a lateral-directional disturbance. In a conventional aircraft, the response of the aircraft involves a second-order oscillatory mode (often called Dutch roll), plus two first order modes (spiral and roll). This section also examines some aspects of the lateral-directional handling qualities of an aircraft in terms of these three modes.

The vertical tail area must be large enough to satisfy directional stability, control, and trim requirements. Two most effective parameters in the directional stability are Cn_β and Cn_r. The major contributor to the static directional stability derivative (Cn_β) is the vertical tail [40]:

$$C_{n_\beta} \approx C_{n_{\beta V}} = K_f C_{L_{\alpha V}} \left(1 - \frac{d\sigma}{d\beta}\right) \eta_V \frac{l_{Vt} S_V}{bS} \tag{14.17}$$

where $C_{L_{\alpha V}}$ denotes the vertical tail lift curve slope, $\frac{d\sigma}{d\beta}$ is the vertical tail sidewash gradient, and η_V is the dynamic pressure ratio at vertical tail. The parameter K_{f1} represents the contribution of fuselage to aircraft Cn_β and depends strongly on the shape of the fuselage and its projected side area. The fuselage contribution to directional static stability tends to be strongly negative. The typical value of K_f for a conventional aircraft is about 0.65–0.85. The value of C_{n_β} for a statically directionally stable aircraft is positive. A higher value for Cn_β implies a more directionally statically stable aircraft. A typical value for Cn_β is about 0.2–0.4 1/rad.

Table 14.2 shows the summary of static longitudinal and lateral-directional stability requirements. On the other hand, the vertical tail has the largest contribution to the aircraft directional stability. The vertical tail is able to generate the counter yawing moment in order to restore the directional trim position.

14.4.5 Dynamic Lateral-Directional Stability

When all modes and oscillations (including (a) Dutch-roll; (b) spiral; and (c) roll subsidence) are damped, an aircraft is said to be dynamically laterally-directionally stable.

Both horizontal tail and vertical tail has significant contributions to the aircraft lateral stability, since both are capable of generating counter rolling moment in order to restore the lateral trim position.

The primary function of the vertical tail is to maintain the aircraft directional stability. The static lateral-directional stability requirements were discussed in Section 14.4.4. In summary, the stability derivatives Cn_β must be positive (to satisfy the static directional stability requirements), but the stability derivatives Cn_r must be negative (to satisfy the dynamic directional stability requirements).

Two major contributors to the value of these stability derivatives are vertical tail area (S_V) and vertical tail moment arm (l_V). If vertical tail area is large enough and vertical tail moment arm is long enough, the directional stability requirements could be easily satisfied. The directional stability analysis is performed after all aircraft components are designed and the roots (λ) of the lateral-directional characteristic equation are calculated. A general form of the aircraft lateral-directional characteristic equation looks like the following:

$$A_2\lambda^4 + B_2\lambda^3 + C_2\lambda^2 + D_2\lambda + E_2 = 0 \tag{14.18}$$

where coefficients A_2, B_2, C_2, D_2, and E_2 are functions of the several stability derivatives such as Cn_β and Cn_r. An aircraft is dynamically directionally stable, if the real parts of all roots of lateral-directional characteristic equation are negative. Another way to analyze dynamic directional stability is to make sure that directional modes (i.e., Dutch roll, and spiral) are damped.

The reader is encouraged to consult with [70] to see how to derive the aircraft lateral-directional characteristic equation. The directional stability derivatives cannot be determined unless all aircraft components including wing and fuselage have been designed. Hence, we have to resort to some other simplifying criterion that could be a base for the vertical tail preliminary design. Similar to horizontal tail volume coefficient, a new parameter that is referred to as vertical tail volume coefficient (V_V) is defined. If the value of this parameter is in ballpark number, we are 90% sure that the directional stability requirements have been satisfied. When other aircraft components have been designed, the vertical tail design will be revised and optimized in the directional stability analysis process.

1) **Roll subsidence**
   The roll subsidence mode is a part of aircraft response to any lateral-direction disturbance. In addition, when the trim bank angle (even in a level cruising flight where the bank angle is zero) is disturbed, a laterally-directionally dynamically stable aircraft will return to its initial bank angle through a rolling motion mode.

   As the name implies, the roll mode is a rolling motion about x-axis that involves a change in bank angle. At any rate, the roll mode or roll subsidence mode is a first order response which is characterized by a time constant. For an aircraft to feature acceptable lateral-directional handling qualities, the roll time constant (T_R) of the roll subsidence mode is required to be less than the specified values given in given references.

2) **Spiral Mode**
   Another contributing mode of response to a lateral-directional disturbance is the spiral mode. A spiral mode is a first order response which is characterized by a time constant. As the name implies, the spiral mode is a yawing motion about z-axis which involves a change in yaw angle. In UAVs with a conventional configuration, the spiral mode is often unstable. Thus, there are no specific requirements for spiral stability in any aircraft. Therefore, spiral mode is often allowed to be lightly unstable, but limits are placed on the minimum time for the mode to double the amplitude (i.e., allowable divergence of the mode).

3) **Dutch Roll Oscillation**
   The Dutch roll mode is a second order mode of response to a lateral-directional disturbance and consists mainly of simultaneous side-slipping and yawing. The rolling motion is also present in most Dutch-rolls; however, it has a relatively negligible participation. Although the Dutch roll mode has a very little useful part to play in the lateral-directional dynamic stability of an aircraft, it does possess significant nuisance value. The Dutch roll (i.e., a second order) response is characterized by a damping ratio (ζ_d) and a frequency of oscillation (ω_d).

14.4.6 Typical Values for Stability Derivatives

Tables 14.3 and 14.4 illustrate typical values for the longitudinal, and lateral, and directional stability derivatives for a fixed-wing UAV with a conventional configuration.

These values can be employed for checking the calculation results.

Table 14.3 Longitudinal stability derivatives.

Derivatives	Typical values (1/rad)		
	All types of UAVs (from UCAV to large UAV)		Small fixed-wing UAV (at cruise)
	Minimum	Maximum	
C_{D_u}	−0.01	+0.3	0
C_{L_u}	−0.2	+0.6	0
C_{m_u}	−0.4	+0.6	0
C_{D_α}	+0	+2.0	0.1
C_{L_α}	+1.0	+8.0	4–5
C_{m_α}	+0	−20	−0.5 to −1.5
$C_{D_{\dot{\alpha}}}$	~	~	0
$C_{L_{\dot{\alpha}}}$	−5.0	+15.0	2
$C_{m_{\dot{\alpha}}}$	+0	−20.0	−3 to −6
C_{D_q}	~	~	0
C_{L_q}	+0	+30.0	2 to 6
C_{m_q}	0	−90	−10 to −15

Table 14.4 Lateral-directional stability derivatives.

Derivatives	Typical values (1/rad)		
	All types of UAVs (from UCAV to large UAV)		Small fixed-wing UAV (at cruise)
	Minimum	Maximum	
C_{y_β}	−0.1	−2.0	−0.4
C_{l_β}	+0.1	−0.4	−0.1
C_{n_β}	0	+0.4	0.05 to 0.1
C_{y_p}	−0.3	+0.8	−0.4
C_{l_p}	−0.1	−0.8	−0.5
C_{n_p}	−0.5	+0.1	−0.04
C_{y_r}	0	+1.2	0.2
C_{l_r}	0	+0.6	0.1
C_{n_r}	0	−1.0	−0.1

14.5 Controllability Analysis

The fundamentals of aircraft controllability have been presented in Sections 6.5 and 6.6 in Chapter 6. In this section, the applications and analysis techniques are provided.

14.5.1 Longitudinal Control

As discussed in Chapter 6, the longitudinal control is defined as any rotational motion control in the x-z plane (e.g., pitch about y-axis, plunging, climbing, cruising, pulling up, and descending). Any change in lift, drag and pitching moment have the major influence on this motion. The pitch control is assumed as a longitudinal control. Two major longitudinal control inputs are the elevator (δ_E), and engine throttle (δ_T).

The control surfaces must be designed such that aircraft possesses acceptable flying qualities anywhere inside the operational flight envelope; and allowable cg range, and allowable aircraft weight. The operational flight envelopes define the boundaries in terms of speed, altitude and load factor within which the aircraft must be capable of operating in order to accomplish the desired mission. A typical operational flight envelope for a HALE UAV is shown in Figure 14.6.

An aircraft must be longitudinally controllable, as well as maneuverable within the flight envelope (Figure 14.7). In a conventional aircraft, the longitudinal control is primarily applied though the deflection of elevator (δ_E), and engine throttle setting (δ_T). There are two groups of requirements in the aircraft longitudinal controllability: (a) required elevator servo force; and (b) aircraft response to the servo input.

The aircraft response in the longitudinal control is frequently expressed in terms of pitch angle (θ) and pitch rate (q). However, the forward speed and angle of attack would be varied as well. The most critical flight condition for pitch control is when the aircraft is flying at a low speed. Two flight operations which feature a very low speed are take-off

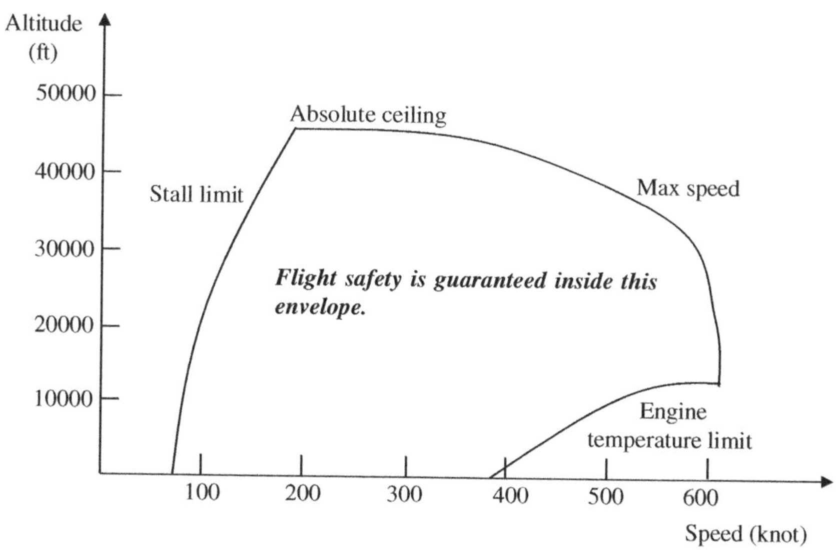

Figure 14.6 A typical operational flight envelope.

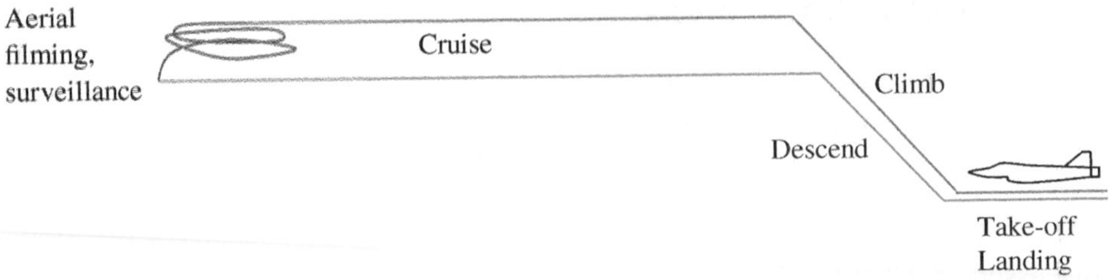

Figure 14.7 Main operations in a typical flight.

and landing. Take-off control is much harder than the landing control due to the safety considerations. A take-off operation is usually divided into three sections: (a) ground section; (b) rotation or transition; and (c) climb. The longitudinal control in a take-off is mainly applied during the rotation section which the nose is pitched up by rotating the aircraft about main gear.

The longitudinal control handling qualities requirements during take-off operation is stated as follows: in an aircraft with a tricycle landing gear, the pitch rate should have a value such that the take-off rotation does not take longer than a specified length of time. Since the take-off rotation dynamics is governed by Newton's second law, the take-off rotation time may be readily expressed in terms of the aircraft angular acceleration ($\ddot{\theta}$) about the main gear rotation point. For instance, in a large UAV, the acceptable value for the take-off rotation time is three to five seconds. The equivalent value for the angular rotation rate to achieve such requirement is 5–7 °/s. This requirement must be satisfied when the aircraft center of gravity is located at the most forward location.

The longitudinal controllability criterion requires the center of gravity is not allowed to be forward of a specific location such that the aircraft is longitudinally controllable. In majority of the cases, it is translated to take-off rotation requirements. This criterion dictates the most forward location of the cg. For an aircraft with a tricycle landing gear, the forward position of the cg must be at the location such that elevator is able to rotate the aircraft about the main gear and lift the nose when aircraft has obtained 80% of its take-off speed. The initial angular acceleration about the main gear (rotation point) should have a value of 6-8 °/s^2 [170]. Then a constant angular velocity of 2–3 °/s should be maintained such that the takeoff rotation process does not take more than 3–4 seconds.

Furthermore, in a conventional aircraft with a tricycle landing gear, when the aircraft is on the ground, the aircraft cg should not be located aft of the main gear. Otherwise, the aircraft will tip on its tail and the rear fuselage/tail will be damaged.

14.5.2 Lateral Control

As discussed in Chapter 6, the lateral control is defined as the rotational motion control about x-axis (e.g., roll about x-axis). Any change in lift distribution and rolling moment have the major influence on this motion. The rolling control is assumed as a lateral control. The primary lateral control input is the aileron (δ_A) deflection. However, the rudder (δ_R) deflection has an indirect influence on this motion too.

Roll or lateral control requirements govern the aircraft response to the aileron deflection; thus, the requirements are employed in the design of aileron. It is customary to specify roll power in terms of the change of bank angle achieved in a given time in response to a step function in roll command. Thus, the aircraft must exhibit a minimum bank angle within a certain specified time in response to aileron deflection.

The notation "60 degrees in 1.3 seconds" indicates the maximum time it should take from an initial bank angle (say 0°) to reach a bank angle which is 60° different than the initial one, following the full deflection of aileron. It may also be interpreted as the maximum time it should take from a bank angle of −30° to +30°. For class IV aircraft, for level 1, the yaw control should be free. For other aircraft and levels, it is permissible to use the yaw control to reduce any sideslip which tends to retard roll rate. Such yaw control is not permitted to induce sideslip which enhances the roll rate.

14.5.3 Directional Control

As discussed in Chapter 6, the directional control is defined as the rotational motion control about z-axis and any motion along y-axis (e.g., yaw about z-axis, side-slipping, and skidding). Any change in side-force and yawing moment have the major influence on this control. The yaw control is assumed as a directional control. A level turn is a combination of lateral and directional motions. The primary directional control input is the rudder deflection (δ_R). However, the aileron deflection (δ_A) has an indirect influence on this motion too.

In a conventional aircraft, directional control is usually maintained by the use of aerodynamic controls (e.g., rudder) alone at all airspeeds. There are a number of cases that directional control must be achievable within a specified limits and constraints. In this section, most important ones are presented. Directional control characteristics shall enable the rudder servo to balance yawing moments and control yaw and sideslip. Sensitivity to yaw control forces shall be sufficiently high that directional control and force requirements can be met, and satisfactory coordination can be achieved without unduly high forces, yet sufficiently low that occasional improperly coordinated control inputs will not seriously degrade the flying qualities.

In a multi-engine aircraft, at all speeds above 1.4 Vs with asymmetric loss of thrust from the most critical factor while the other engine(s) develop normal rated thrust, the airplane with yaw control may be balanced directionally in steady straight flight. The trim settings shall be those required for wings-level straight flight prior to the failure.

When an aircraft in is directional trim with symmetric power/thrust, the trim change of propeller-driven airplanes with speed shall be such that wings-level straight flight can be maintained over a speed range of ±30% of the trim speed or ±100 knots equivalent airspeed, whichever is less (except where limited by boundaries of the Service Flight Envelope) with yaw-control-device (i.e., rudder). In the case of one-engine-inoperative (asymmetric thrust), it shall be possible to maintain a straight flight path throughout the Operational Flight Envelope with yaw-control-device.

14.5.4 Typical Values for Control Derivatives

Table 14.5 illustrates typical values for the longitudinal, lateral, and directional control derivatives for a fixed-wing UAV with a conventional configuration.

Table 14.5 Typical values for longitudinal, lateral and directional control derivatives.

	Typical values (1/rad)		
	All types of UAVs (from UCAV to large UAV)		
Derivatives	Minimum	Maximum	Small fixed-wing UAV (at cruise)
$C_{D_{\delta_e}}$	~	~	0
$C_{L_{\delta_e}}$	+0	+0.6	0.4
$C_{m_{\delta_e}}$	+0	-4.0	−1 to −2
$C_{y_{\delta_a}}$	~	~	0
$C_{l_{\delta_a}}$	0	+0.4	0.2
$C_{n_{\delta_a}}$	-0.08	+0.08	-0.03
$C_{y_{\delta_r}}$	0	+0.5	0.2
$C_{l_{\delta_r}}$	-0.04	+0.04	0.01
$C_{n_{\delta_r}}$	0	-0.15	-0.05

14.6 Flight Performance Analysis

Another design analysis is the UAV flight mechanics evaluation which mainly concentrates on performance parameters such as maximum speed, rang, and endurance. Figure 14.7 illustrates general operations in a typical UAV flight mission. Take-off, climb, cruise, descent and landing are operations necessary to have a conventional flight mission. Different UAVs in performing their missions such as aerial filming often require maneuvers and multiple turns. The flight performance governing equations in this section are reproduced from [9], for the derivations and assumptions, you may refer to the reference.

14.6.1 Maximum Speed

The maximum speed (V_{max}) for a UAV with a jet engine (e.g., turbofan) is determined by solving the following equation:

$$\frac{1}{2}\rho V_{\max}{}^{2} S C_{D_o} + \left(\frac{2KW^2}{\rho S}\right)\frac{1}{V_{\max}{}^{2}} - T_{\max} = 0 \tag{14.19}$$

where S represents wing area, W, aircraft weight, C_{Do}, aircraft zero lift drag coefficient, K, induced drag correction factor, and ρ is air density at given altitude. The calculation of the aircraft zero lift drag coefficient is presented in [9].

Figure 14.8 NASA Helios Unmanned Aerial Vehicle (UAV). *Source:* https://upload.wikimedia.org/wikipedia/commons/4/4c/Helios_in_flight.jpg.

The only unknown in Eq. (14.19) is the maximum speed. This equation is an algebraic nonlinear equation with the order of 4. When solving this equation, we will have four solutions, only one solution is acceptable that is frequently the highest.

The maximum speed (V_{max}) for a UAV with a prop-driven engine (e.g., turboprop) is determined [9] by the following equation:

$$P_{max_{SL}}\eta_P\left(\frac{\rho}{\rho_o}\right)^{0.9} = \frac{1}{2}\rho V_{max}^3 S C_{D_o} + \frac{2K(mg)^2}{\rho V_{max} S} \quad \text{(Turboprop engine)} \tag{14.20}$$

$$P_{max_{SL}}\eta_P\left(\frac{\rho}{\rho_o}\right)^{1.2} = \frac{1}{2}\rho V_{max}^3 S C_{D_o} + \frac{2K(mg)^2}{\rho V_{max} S} \quad \text{(Piston/electric prop engine)} \tag{14.21}$$

where $Pmax_{SL}$ is the maximum engine power at sea level, and η_p is the prop efficiency.

Figure 14.8 illustrates the NASA Helios UAV that was developed for the Environmental Research Aircraft and Sensor Technology (ERAST) program. After Pathfinder and Centurion, the Helios was the fourth aircraft developed as part of an evolutionary series of solar-powered UAV. They were built to develop the technologies that would allow long-term, high-altitude aircraft to serve as atmospheric (or pseudo) satellite. The Helios HP03 with a wing span of 75.3 m and a mass of 1052 kg was powered by 10 electric engines each rated at 2 hp. On June 26, 2003, the Helios broke up and fell into the Pacific Ocean. The investigation report identified a two-part root cause of the accident: (a) lack of adequate analysis methods; and (b) low margins of safety aircraft configuration.

14.6.2 Maximum Range

The maximum range (R_{max}) for a UAV with a jet engine is determined [9] by the following equation:

$$R_{max} = \frac{0.866 V_{max_R} (L/D)_{max}}{C} \ln\left(\frac{1}{1-\left(\frac{W_f}{W_1}\right)}\right) \tag{14.22}$$

where C is the specific fuel consumption, W_1 is the initial UAV weight, and W_f is the total fuel weight. The calculation of the fuel weight is presented in Chapter 2. The maximum lift-to-drag ratio is a function of the aircraft zero lift drag coefficient (C_{Do}), and the induced drag correction factor (K) as:

$$\left(\frac{C_L}{C_D}\right)_{max} = \frac{1}{2\sqrt{KC_{D_o}}} \tag{14.23}$$

The maximum range often is obtained when flight program is at constant-airspeed, and constant-lift coefficient flight mode. This maximum range speed for a jet aircraft is obtained by:

$$V_{max_R} = \sqrt{\frac{2mg}{\rho S\sqrt{\frac{C_{D_o}}{3K}}}} = \left(\frac{2mg}{\rho S}\right)^{\frac{1}{2}}\left(\frac{3K}{C_{D_o}}\right)^{\frac{1}{4}} \tag{14.24}$$

The maximum range (R_{max}) for a UAV with a prop-driven engine is determined by the following equation:

$$R_{max} = \frac{\eta_P (L/D)_{max}}{C} \ln\left(\frac{1}{1-G}\right) \tag{14.25}$$

This equation is not directly a function of altitude. The parameter G is fuel weight ratio; $\frac{W_f}{W_1}$.

Eq. (2.33) can be used to determine the range of a UAV with electric engine(s) when flying with a cruising velocity. For calculation of the maximum range for a UAV with a prop-driven electric engine, consult with [171].

14.6.3 Maximum Endurance

The maximum endurance (E_{max}) for a UAV with a jet engine is determined by the following equation:

$$E_{max} = \frac{(L/D)_{max}}{C} \ln\left[\frac{1}{1-G}\right] \quad (14.26)$$

The maximum endurance for a UAV with a prop-driven non-electric engine is determined by the following equation:

$$E_{max} = \frac{0.866\eta_P (L/D)_{max}}{CV_{min_P}} \ln\left(\frac{1}{1-G}\right) \quad (14.27)$$

The minimum power speed is calculated by:

$$V_{min_P} = \sqrt{\frac{2mg}{\rho S\sqrt{\frac{3C_{D_o}}{K}}}} \quad (14.28)$$

The maximum endurance (in hour, h) for a UAV with a prop-driven electric engine is determined by the following equation:

$$E_{max} = \frac{E_D m_B}{P_{min}} \quad (14.29)$$

where P_{min} denotes the minimum required power (in Watt, W) for an endurance mission, E_D for energy density (in Wh/kg) of the batteries, and m_B stands for the mass of battery pack. Table 3.5 in Chapter 3 provides energy density of some rechargeable batteries.

14.6.4 Climb Performance

The climb performance of a UAV is analyzed by evaluating the fastest climb and the steepest climb. These two climb performance features are presented in this section. Both the fastest climb, and the steepest climb are significant in the success of a UAV flight mission particularly in a see-and-avoid maneuver, and a formation flight.

14.6.4.1 Fastest Climb

A climbing flight with the maximum rate of climb (ROC_{max}) is referred to as the fastest climb. Techniques to analyze the fastest climb for a UAV with jet engine(s), and a UAV with prop-driven engine(s) are different.

The fastest climb for a UAV with jet engine(s) is achieved when the aircraft climbs with a speed corresponding to the maximum rate of climb (V_{ROCmax}), and employs the climb angle that submits the maximum rate of climb (γ_{ROCmax}).

$$ROC_{max} = V_{ROC_{max}} \sin(\gamma_{ROC_{max}}) \quad (14.30)$$

Two parameters; V_{ROCmax} and γ_{ROCmax} are determine separately. The speed for maximum rate of climb is found by solving the following expression for an acceptable solution:

$$V_{ROC_{max}} = \sqrt{\frac{T_{max}}{3\rho C_{D_o} S}\left[1+\sqrt{1+\frac{3}{\left[\left(\frac{L}{D}\right)_{max}\frac{T_{max}}{W}\right]^2}}\right]} \tag{14.31}$$

The climb angle that submits the maximum rate of climb is obtained by:

$$\gamma_{ROC_{max}} = -\sin^{-1}\left[\frac{0.25}{KW}\left(\sqrt{-8K\rho T_{max} S V_{ROC_{max}}^2 + \rho^2 S^2 V_{ROC_{max}}^4 \left(1+4KC_{D_o}\right)} - \rho S V_{ROC_{max}}^2\right)\right] \tag{14.32}$$

where T_{max} denotes the maximum engine thrust, and W is the UAV weight.

However, the fastest climb or the maximum rate of climb (ROC_{max}) for a UAV with prop-driven engine(s) is achieved (Eq. 14.30) when an aircraft climbs with a speed for maximum rate of climb (V_{ROCmax}) and employs the climb angle that submits the maximum rate of climb (γ_{ROCmax}).

The aircraft speed for maximum rate of climb for a propeller-driven aircraft is obtained by:

$$V_{ROC_{max}} = \sqrt{\frac{2mg}{\rho S\sqrt{\frac{3C_{D_o}}{K}}}} \tag{14.33}$$

Note that this speed should be greater than the stall speed.

The angle corresponding to the maximum rate of climb for a UAV with prop-driven engine(s) is obtained by:

$$\gamma_{ROC_{max}} = -\sin^{-1}\left[\frac{0.25}{KW}\left(\sqrt{16K^2W^2 + \rho^2 S^2 V_{ROC_{max}}^4 \left(1+4KC_{D_o}\right) - 8P_{max} K S \rho \eta_P V_{ROC_{max}}} - \rho S V_{ROC_{max}}^2\right)\right] \tag{14.34}$$

where P_{max} is the maximum engine power.

14.6.4.2 Steepest Climb

A climbing flight with the maximum climb angle is referred to as the steepest climb. Techniques to analyze the steepest climb for a UAV with jet engine(s), and a UAV with prop-driven engine(s) are different.

The maximum climb angle for a UAV with jet engine(s) is obtained by:

$$\gamma_{max} = \sin^{-1}\left(\frac{1}{W}\left[T_{max} - 2W\sqrt{KC_{D_o}}\right]\right) \tag{14.35}$$

The maximum climb angle for a UAV with prop-driven engine(s) is obtained by:

$$\gamma_{\max} = \sin^{-1}\left[\frac{P_{\max}\eta_P}{WV_{\gamma_{\max}}} - \frac{1}{2W}\rho V_{\gamma_{\max}}^2 SC_{D_o} - \frac{2KW}{\rho V_{\gamma_{\max}}^2 S}\right] \tag{14.36}$$

where the speed for maximum climb angle for a propeller-driven airplane is determined by:

$$V_{\gamma_{\max}} = \frac{4W^2K}{P_{\max}\rho\eta_P S} \tag{14.37}$$

Note that this speed should be greater than the stall speed. For analyzing other topic of interest to a climbing flight such as the most economic climb, time to climb, and the gliding flight, you may consult with [9].

14.6.5 Takeoff Performance

For a fixed-wing UAV without a launcher that should take off by its landing gear using a runway, the takeoff analysis is required. The primary parameter of interest in takeoff operation analysis is the takeoff run.

The takeoff run is divided into three segments or sections: (a) ground roll (S_G); (b) rotation or Transition section (S_R); and (c) airborne (initial climb) section (S_A). The takeoff run (S_{TO}) is the summation of these three sections:

$$S_{TO} = S_G + S_R + S_A \tag{14.38}$$

To determine the ground roll, the following equation is utilized:

$$S_G = \frac{m}{\rho S\left(C_{D_{TO}} - \mu C_{L_{TO}}\right)} \ln\left[\frac{\frac{T_{TO}}{mg} - \mu}{\frac{T_{TO}}{mg} - \mu - \frac{k_{LO}^2\left(C_{D_{TO}} - \mu C_{L_{TO}}\right)}{C_{L\max}}}\right] \tag{14.39}$$

where μ is the static rolling friction coefficient. The takeoff drag coefficient ($C_{D_{TO}}$) is:

$$C_{D_{TO}} = C_{D_{oTO}} + KC_{L_{LTO}}^2 \tag{14.40}$$

The takeoff lift coefficient ($C_{L_{TO}}$) is a function of cruise lift coefficient and flap deflection as:

$$C_{L_{TO}} = C_{L_C} + \Delta C_{L_{flapTO}} \tag{14.41}$$

The ΔC_{DoTO} for a small RC fixed-wing UAV is about 0.005–0.01; and for a large UAV (e.g., Global hawk) is about 0.01–0.015.

The takeoff engine thrust (T_{TO}) for a UAV with jet engine(s), is almost 90% of the maximum engine thrust. However, in a prop-driven aircraft, where propeller efficiency is a function of aircraft speed; the takeoff engine thrust (T_{TO}) for a fixed-pitch propeller is estimated from the following equation:

$$T_{TO} = \frac{0.5P_{max}}{V_R} \tag{14.42}$$

The rotation distance (S_R or S_T) is simply modeled by the following linear equation:

$$S_R = T_R V_R \tag{14.43}$$

where T_R is the duration of rotation, as is about 0.3–1 seconds for a small UAV, and 1–2 seconds for a large UAV.

The last term in Eq. (14.38) is the airborne section (S_A) of the takeoff run. This length is calculated by.

$$S_A = \sqrt{\left(S_A'\right)^2 - h_o^2} \tag{14.44}$$

where the parameter S_A' is determined from

$$S_A' = \frac{mg}{T - D}\left[\frac{V_2^2 - V_{LO}^2}{2g} + h_o\right] \tag{14.45}$$

The obstacle height (h_o) is governed by airworthiness standards. For micro to small UAVs, this height is recommended to be about 3 ft. However, for medium to large UAVs, the height is is recommended to be 50 ft.

14.6.6 Turn Performance

Turn is an important maneuverability requirement, and influences various maneuvers, particularly when needed for a sense-and avoid objective. Two aspects of turn performance are: (a) fastest turn; and (b) tightest turn. In this section, only the tightest turn is presented. For a full turn performance analysis, consult with [9].

When an aircraft turns with the lowest possible radius (R_{min}), the turn is called tightest turn (so, subscript tt). We are looking for a set of parameters (e.g., load factor, n; bank angle, ϕ; and airspeed, V) that yields the minimum turn radius. For a small RC fixed-wing UAV with propeller-driven engine(s) or jet engine(s), the turn radius for the tightest turn (R_{tt}), or the minimum (shortest) turn radius (R_{min}) is obtained by:

$$R_{min} = \frac{V_{tt}^2}{g\sqrt{n_{tt}^2 - 1}} \tag{14.46}$$

The velocity for the tightest turn is often the corner speed (V^*), and is defined as:

$$V^* = \left[\frac{2n_{max}W}{\rho S C_{L_{max}}}\right]^{\frac{1}{2}} \tag{14.47}$$

The corner speed is the speed when the aircraft is turning with the maximum possible bank angle while having the maximum lift coefficient. It is often located at the top left corner of the n-V diagram.

Figure 14.9 NASA highly maneuverable aircraft technology (HiMAT) Unmanned Aerial Vehicle (UAV).

The bank angle corresponding to the tightest turn is written as:

$$\varphi_{tt} = \cos^{-1}\left(\frac{1}{n_{tt}}\right) \tag{14.48}$$

The value of load factor for tightest turn, n_{tt} or (or n_{max}, the maximum load factor) will be decided by the structural limits of the UAV.

Figure 14.9 illustrates a NASA highly maneuverable aircraft technology (HiMAT) UAV. From mid-1979 to January 1983, two remotely piloted, experimental Highly Maneuverable Aircraft Technology vehicles were used at the NASA[1] Dryden Flight Research Center, to develop high-performance fighter technologies. Each aircraft was approximately half the size of an F-16 and had nearly twice the fighter's turning capability. The aircraft were flown 26 times during the 3 1/2 years history of the project.

HiMAT research at Dryden was conducted jointly by NASA and the Air Force Flight Dynamics Laboratory. The planes were controlled from a ground control station, to employ the experimental technologies for high-risk maneuverability tests.

The HiMAT plane's rear-mounted swept wings, digital flight control system, and forward controllable canard made the plane's turn radius twice as tight as that of conventional fighter planes. At near the speed of sound and at an altitude of 25 000 ft, the HiMAT vehicle could sustain an 8-g turn. For comparison, at the same altitude, an F-16's maximum sustained turning capability is about 4.5-g.

An interesting aspect of the HiMAT project was the flight test maneuver autopilot, which was based on a design developed by Teledyne Ryan Aeronautical. An important contribution of the HiMAT project to flight technology was the use of new composite materials in structural design.

Example 14.1

Consider a mini fixed-wing UAV with the following characteristics:

$$m = 700\,g, S = 0.2\,m^2, C_{Lmax} = 1.2, b = 1\,m, n_{max} = 1.5$$

The airplane which is employing a prop-driven electric motor is cruising at sea level altitude. UAV is equipped with a sense-and-avoid system. The range-finder is suddenly identifying an obstacle at 8 m away in front of the UAV. Is this UAV able to avoid a collision with this obstacle by turning away? Assume the aircraft is flying with the corner speed.

Solution

We need to find the shortest turn radius, which is a function of corner speed.

$$V^* = V_{tt} = \left[\frac{2n_{max}W}{\rho SC_{L_{max}}}\right]^{\frac{1}{2}} = \left[\frac{2\times1.5\times0.7\times9.81}{1.225\times0.2\times1.2}\right]^{\frac{1}{2}} = 8.37\, m/s = 30.1\, km/hr \quad (14.47)$$

The turn radius for the tightest turn (R_{tt}) is:

$$R_{min} = \frac{V_{tt}^2}{g\sqrt{n_{tt}^2 - 1}} = \frac{8.37^2}{9.81\sqrt{1.5^2 - 1}} = 6.39\, m \quad (14.46)$$

The turn radius is about 1.4 m shorter than the distance to the obstacle. If the UAV has a very quick reaction, yes, it is able to avoid the collision. Even if we had the semi-span to the radius (1/2 + 6.39 = 6.89 m), there is still a safe distance between UAV and the obstacle (i.e., 8–6.89 = 1.11 m). Figure 14.10 exhibits the geometry of UAV flight direction and the obstacle.

14.6.7 Absolute Ceiling

Absolute ceiling is defined as the highest altitude that an aircraft can safely have a straight level flight. It is the highest altitude that an aircraft can reach by its own engine and have sustained flight. The calculation of the absolute ceiling is presented in two sections: a. UAV with a jet engine (turbofan or turbojet), b. UAV with propeller-driven (piston prop or turboprop) engine(s).

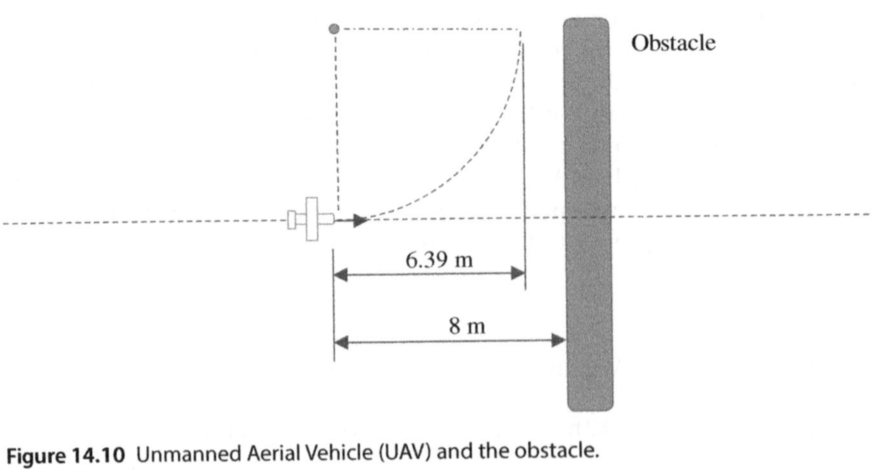

Figure 14.10 Unmanned Aerial Vehicle (UAV) and the obstacle.

14.6.7.1 UAV with Jet Engine(s)

For a UAV with jet engine(s), the air density (ρ) corresponding to the absolute ceiling is determined from one of the following equations:

$$\rho_{ac_1} = \rho_o \left(\frac{2m_{ac} g \sqrt{K C_{D_o}}}{T_{\max_{SL}}} \right)^{\frac{1}{1.2}} \tag{14.49}$$

$$\rho_{ac_2} = \frac{2m_{ac} g \sqrt{K C_{D_o}} (\rho_o)^{1.2}}{T_{\max_{SL}} (\rho_{11000})^{0.2}} \tag{14.50}$$

where ρ_{11000} denotes the air density at 11000 m (i.e., 0.365 kg/m^3). Eq. 14.49 is used when the absolute ceiling lies within the first layer of atmosphere (below 11,000 m), while Eq. 14.50 is employed, when the absolute ceiling lies within the second layer of atmosphere (higher than 11,000 m). The parameter m_{ac} denotes the UAV mass at the ceiling altitude. When the air density (ρ_{ac}) is calculated, refer to atmospheric tables to determine the corresponding altitude.

14.6.7.2 UAV with Propeller-driven Engine(s)

For a UAV with piston engine(s), the air density (ρ) corresponding to the absolute ceiling is determined by:

$$\rho_{ac} = \left(\frac{2\rho_o^{2.7} V_{\min P_E}{}^3 S C_{D_o}}{\eta_P P_{\max_{SL}}} \right)^{\frac{1}{1.7}} \tag{14.51}$$

The equivalent minimum power airspeed ($V_{\min_{P_E}}$) is determined from equation 14.27 when the sea level air density (ρ_o) is employed.

For a UAV with turboprop engine(s), the air density (ρ) corresponding to the absolute ceiling is determined from one of the following equations:

$$\rho_{ac_1} = \left[\frac{2\rho_o^{2.4} V_{\min P_E}{}^3 S C_{D_o}}{\eta_P P_{\max_{SL}}} \right]^{\frac{1}{1.4}} \tag{14.52}$$

$$\rho_{ac_2} = \left[\frac{2\rho_o^{2.4} \rho_{11000} V_{\min P_E}{}^3 S C_{D_o}}{\eta_P P_{\max_{SL}}} \right]^{\frac{1}{2.4}} \tag{14.53}$$

where $Pmax_{SL}$ represents the engine(s) maximum power at sea level. When the air density (ρ_{ac}) is determined, refer to the atmospheric tables to find its corresponding altitude.

14.7 Cost Analysis

A significant step in a UAV design project is the cost analysis, which forecasts the UAV life-cycle cost. This initiative is conducted to make sure that the entire project is

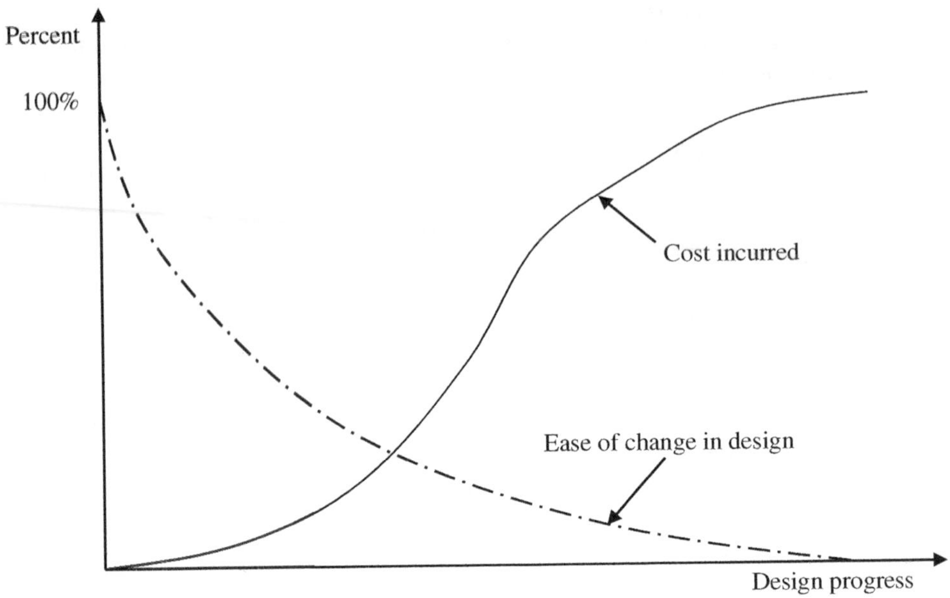

Figure 14.11 Status of cost and ease of change in design during design progress.

financially supportable, and also, the UAV is competitive. The cost of the UAV design is about 1–10% of the total life cycle cost; however, this 1–10% determines the other 90–99%. Furthermore, the design cost is about 20% of the production (acquisition) cost. Thus, any necessary investment on the design team members is worth it. Most UAV manufacturers do not make any profit in the first couple of years of production, in the hope that in future, they make money.

Figure 14.11 conceptually illustrates [172] the status of cost and ease of change in design during the design progress. It is estimated that about 70% of the projected life-cycle cost for a given product can be committed based on engineering design and management decisions during the early stages of design proves. As the design progresses, a change in the design is getting harder and harder. Therefore, the impact of a decision at the early stage of a design program is more profound that a decision at the later stages. Hence, it is crucial to be highly confident about any decision a designer makes at the conceptual design phase. A model for life-cycle cost analysis for small unmanned aircraft systems is provided by [5].

Spare part requirements are a function of multiple factors, including the number of systems, the amount of use, part reliability, and the amount of time that must be covered by the spares. The amount of use of a small unmanned aircraft system (sUAS) part is tied to flight hours or number of evolutions for the launch and recovery mechanisms. The maintenance cost per flight hour is based on the consumption of spare parts.

A cost list should be prepared to indicate the cost of each component/element that comprise an UAS, with the exception of spare parts. It includes items such as aerial vehicle, ground control station, and launch and recovery system. For instance, Table 14.6 indicates [5] the cost of sUAS equipment deployed aboard one coast guard cutter. Ref. [172] provides a cost model to estimate the life cycle cost of an engineering program.

Every year, US DOD publishes the program acquisition cost by weapon system. Table 14.7 indicates [173] the acquisition cost of some UASs for 2017. The UASs list includes Predator/Gray Eagle, Reaper, Global Hawk/Triton, Shadow, Raven, Puma, and Blackjack. Tables such as 14.6 and 14.7 provide guidelines for predicting design and development cost for future UASs.

Table 14.6 Cost of small unmanned aircraft system (sUAS) equipment deployed aboard one coast guard cutter.

No	Procure Ops Test System	Cost
1	Airframe Unit Cost – Baseline	$ 133 671
2	Avionics – Baseline	$ 149 047
3	Propulsion – Baseline	$ 40 601
4	EO/IR sensor – Baseline	$ 268 268
5	Communications relay	$ 131 301
6	AIS	$ 7851
7	GCS – Ship	$ 392 054
8	Launcher – Ship	$ 236 050
9	Recovery – Ship	$ 236 050
10	Peculiar support equipment per system	$ 69 624
11	EO/IR surface detect system	$ 75 000
12	Remove video terminal for boarding team (1)	$ 10 000
13	Initial spares	$ 174 952
14	Standard work stations	$ 16 800
One ship system with 1 aircraft		$ 1 936 473

Table 14.7 Acquisition cost of some small unmanned aircraft systems (sUAS) for FY 2017.

No	Aircraft and Related Systems		Cost ($ in million)
1	MQ-1B/MQ-1C	Predator/Gray Eagle	308.1
2	MQ-9	Reaper	1053.1
3	RQ-4/MQ-4C	Global Hawk/Triton/NATO AGS	1213.6
4	RQ-7/RQ-11/ RQ-20/RQ-21	Shadow, Raven, Puma, and Blackjack	522.4

Questions

1 Draw a figure to show the UAV analysis groups.

2 What are major UAV design analysis initiatives?

3 What feedbacks can be provided by cg and moment of inertia calculations?

4 What are three major feedbacks in the UAV design process?

5 Describe the technique to determine UAV center of gravity.

6 Where is the recommended UAV cg location along x-axis?

7 What is the relation between X_{cg} and X_{np} for a longitudinally statically stable UAV?

8 Define static stability.

9 Define dynamic stability.

10 Define lateral stability.

11 Define longitudinal stability.

12 Define directional stability.

13 What is the criterion for static longitudinal stability?

14 What is the criterion for static lateral stability?

15 What is the criterion for static directional stability?

16 What is typical value for the stability derivative C_{m_α} ?

17 What is typical value for the stability derivative C_{n_β} ?

18 What is typical value for the stability derivative C_{l_β} ?

19 Explain how dynamic longitudinal stability analysis is performed if the longitudinal transfer function is given.

20 Explain how dynamic lateral-directional stability analysis is performed if the lateral-directional transfer function is given.

21 Explain how dynamic longitudinal stability analysis is performed if the longitudinal state-space dynamic model is given.

22 Explain how dynamic lateral-directional stability analysis is performed if the lateral-directional state-space dynamic model is given.

23 Define longitudinal control.

24 Define lateral control.

25 Define directional control.

26 List a number of UAV flight performance parameters.

27 What are two topics of interest in climb performance analysis?

28 What are two topics of interest in turn performance analysis?

29 What is the primary parameter of interest in takeoff operation analysis?

30 What is the corner speed?

31 Briefly Describe the NASA HiMAT research program.

32 What is the typical cost of the UAV design as compared with the total life cycle cost?

33 What is the DOD projected cost for the acquisition of Predator/Gray Eagle UAVs in 2017?

34 What is the DOD projected cost for the acquisition of Reaper UAVs in 2017?

35 What are the names and units of the following stability derivatives: C_{l_p}, C_{m_q}, C_{n_r}?

36 What design parameters will influence the maximum endurance of a UAV with an electric engine?

37 What design parameters will influence the maximum range of a UAV with an electric engine?

38 What design parameters will influence the absolute ceiling of a UAV with a piston-prop engine?

39 What design parameters will influence the rate of climb of a UAV with a turbofan engine?

40 What design parameters will influence the maximum speed of a UAV with an electric engine?

Problems

1 Consider a mini fixed-wing UAV with the following characteristics:

$$m_{TO} = 1.2\,kg,\ S = 0.3\,m^2,\ C_{Lmax} = 1.4,\ b = 1.2\,m,\ n_{max} = 1.4$$

The airplane which is employing a prop-driven electric motor is cruising at sea level altitude. UAV is equipped with a sense-and-avoid system. The range-finder is suddenly identifying an obstacle at 8 m away in front of the UAV. Is this UAV able to avoid a collision with this obstacle by turning away? Assume the aircraft is flying with the corner speed.

2 Consider a small fixed-wing UAV with the following characteristics:

$$m_{TO} = 12\,kg,\ S = 1.6\,m^2,\ C_{Lmax} = 1.6,\ b = 3\,m,\ n_{max} = 2$$

The airplane which is employing a prop-driven electric motor is cruising at sea level altitude. UAV is equipped with a sense-and-avoid system. The range-finder is suddenly identifying an obstacle at 30 m away in front of the UAV. Is this UAV able to avoid a collision with this obstacle by turning away? Assume the aircraft is flying with the corner speed.

3 Consider a small fixed-wing UAV with the following characteristics:

$$m_{TO} = 1000\ kg,\ V_s = 40\ knot,\ C_{Lmax} = 1.7,\ b = 15\,m,\ n_{max} = 2.5$$

The airplane which is employing a prop-driven piston engine is cruising at sea level altitude. UAV is equipped with a sense-and-avoid system. The range-finder is suddenly identifying an obstacle at 100 m away in front of the UAV. Is this UAV able to avoid a collision with this obstacle by turning away? Assume the aircraft is flying with the corner speed.

4 Consider a large UAV with the following characteristics:

$$m_{TO} = 15\,000\,kg,\ V_s = 80\ knot,\ C_{Lmax} = 1.8,\ AR = 12,\ n_{max} = 2.7$$

The airplane which is employing a jet engine is cruising at sea level altitude. UAV is equipped with a sense-and-avoid system. The range-finder is suddenly identifying an obstacle at 400 m away in front of the UAV. Is this UAV able to avoid a collision with this obstacle by turning away? Assume the aircraft is flying with the corner speed.

5 A surveillance UAV with the following characteristics is searching for a target:

$$T_{max} = 64\ kN,\ S = 41\,m^2,\ m_{TO} = 17\,000\,kg,\ C_{Lmax} = 2.2,\ K = 0.09,\ C_{Do} = 0.025,$$
$$m_f = 5000\ kg,\ C = 0.84\,lb/hr/lb$$

How long this aircraft can search for its target?

6 A UCAV with a jet engine has the following characteristics:

$$T_{maxSL} = 300\ kN,\ S = 217\,m^2,\ m_{TO} = 63\,000\,kg,\ C_{Lmax} = 2.1,\ K = 0.043,$$
$$C_{Do} = 0.021$$

Is this UCAV able to fly with Mach 1.7 at 20000 ft altitude?

7 Consider a UAV with a jet engine with the following features:

T_{max} = 27 kN, S = 56 m^2, m_{TO} = 11 000 kg, C_{Lmax} = 1.7, K = 0.05, C_{Do} = 0.018, C = 0.7 lb/hr/lb, m_f = 2,000 kg

Determine the Maximum range.

8 Determine the maximum speed for a small UAV with a piston prop-driven engine and following features:

m_{TO} = 1.5 kg, S = 1.6 m^2, K = 0.05, C_{Do} = 0.025, P = 0.6 hp, η_P = 0.82, C_{Lmax} = 1.3

9 Determine the maximum speed for a small UAV with a piston prop-driven engine and following features:

m_{TO} = 1500 kg, S = 18 m^2, K = 0.04, C_{Do} = 0.022, P_{max} = 300 hp, η_P = 0.82, C_{Lmax} = 1.6

10 Determine the maximum range for a small UAV with a piston prop engine and following features:

m_{TO} = 2 kg, S = 0.5 m^2, K = 0.06, C_{Do} = 0.042, η_P = 0.6, m_f = 0.3 kg, C_{Lmax} = 1.6, C = 0.6 lb/hr/hp

11 Determine the maximum endurance for a UAV with a turboprop engine and following features:

m_{TO} = 1200 kg, S = 17 m^2, K = 0.03, C_{Do} = 0.024, P_{max} = 200 hp, η_P = 0.75, m_f = 400 kg, C_{Lmax} = 1.6, C = 0.5 lb/hr/hp

12 Determine the takeoff ground roll for a small UAV with an electric engine and following features:

m_{TO} = 1.2 kg, S = 0.3 m^2, K = 0.03, C_{Do} = 0.028, P_{max} = 0.2 hp, η_P = 0.75, V_s = 12 knot, C_{LC} = 0.3

The UAV – which has a fixed landing gear, and no flap – is taking off on a concert runway at 2000 ft altitude.

13 Consider a UAV with an electric engine and the following features:

m_{TO} = 10 kg, S = 2 m^2, K = 0.03, C_{Do} = 0.028, P_{max} = 1.2 hp, η_P = 0.75, V_s = 15 knot

Determine the maximum climb angle.

14 Consider a UAV with an electric engine and the following features:

$$m_{TO} = 20\,\text{kg},\ S = 2\,\text{m}^2,\ K = 0.05,\ C_{Do} = 0.032,\ P_{max} = 2\,\text{hp},\ \eta_P = 0.73,\ V_s = 20\,\text{knot}$$

Determine the maximum rate of climb.

15 A large UAV has the following bank-angle-to-aileron-deflection transfer function for a cruising flight:

$$\frac{\varphi(s)}{\delta_A(s)} = \frac{120s^2 + 35s + 150}{810s^4 + 550s^3 + 670s^2 + 410s + 16}$$

Determine if the UAV is dynamically laterally-directionally stable? Why?

16 A small UAV has the following lateral-directional characteristic equation for a cruising flight:

$$207s^4 + 3045s^3 + 7426s^2 + 38587s - 337.9 = 0$$

Identify the DutchD roll, roll, and spiral modes of lateral-directional dynamics; and then determine damping ratio, natural frequency, and period of the DutchD roll mode, and time constants of roll mode and spiral mode. Is the aircraft dynamically lateral-directionally stable? Why?

17 A small UAV has the following longitudinal characteristic equation for a cruising flight:

$$125s^4 + 308s^3 + 354s^2 + 30s + 29 = 0$$

Determine the features of the longitudinal modes (i.e., damping ratio, natural frequency and period). Is the aircraft dynamically longitudinally stable? Why?

18 A large UAV has the following angle-of-attack-to-elevator-deflection transfer function:

$$\frac{\alpha(s)}{\delta_E(s)} = \frac{-18s^3 - 1055s^2 - 85s - 2}{880s^4 + 890s^3 + 1600s^2 + 121s + 1.6}$$

Is the aircraft dynamically longitudinally stable? Why?

19 A surveillance UAV has the following longitudinal state space model for a cruising flight:

$$\begin{bmatrix} \dot{u} \\ \dot{\alpha} \\ \dot{q} \\ \dot{\theta} \end{bmatrix} = \begin{bmatrix} -0.1 & 7 & 0 & -32.2 \\ -0.03 & -2.71 & 0.99 & 0 \\ 0 & -103.9 & -1.42 & 0 \\ 0 & 0 & 1 & 1 \end{bmatrix} \begin{bmatrix} u \\ \alpha \\ q \\ \theta \end{bmatrix} + \begin{bmatrix} 0 \\ 0.22 \\ -15.83 \\ 0 \end{bmatrix} \delta_e$$

- Is the aircraft dynamically longitudinally stable? Why?
- If yes, determine the features of the longitudinal modes (i.e., damping ratio, natural frequency, and period).

20 A search and rescue UAV has the following longitudinal state space model for a cruising flight:

$$\begin{bmatrix} \dot{u} \\ \dot{\alpha} \\ \dot{q} \\ \dot{\theta} \end{bmatrix} = \begin{bmatrix} -0.07 & 11.34 & 0 & -32.08 \\ -0.0022 & -0.6063 & 0.9866 & -0.0165 \\ 0.0007 & -1.7581 & -1.1160 & 0.0049 \\ 0 & 0 & 1 & 0 \end{bmatrix} \begin{bmatrix} u \\ \alpha \\ q \\ \theta \end{bmatrix} + \begin{bmatrix} 0 \\ -0.0458 \\ -2.8663 \\ 0 \end{bmatrix} [\delta_e]$$

- Is the aircraft dynamically longitudinally stable? Why?
- If yes, determine the features of the longitudinal modes (i.e., damping ratio, natural frequency, and period).

21 A small UAV has the following lateral-directional state space model for a cruising flight:

$$\begin{bmatrix} \dot{\beta} \\ \dot{p} \\ \dot{\phi} \\ \dot{r} \end{bmatrix} = \begin{bmatrix} -0.084 & 0 & 0.1886 & -0.9954 \\ -1.6754 & -0.3742 & 0 & 0.435 \\ 0 & 1 & 0 & 0 \\ 0.8686 & -0.071 & 0 & -0.1517 \end{bmatrix} \begin{bmatrix} \beta \\ p \\ \phi \\ r \end{bmatrix} + \begin{bmatrix} 0 & 0.0161 \\ 1.437 & 0.1403 \\ 0 & 0 \\ -0.2965 & -0.4231 \end{bmatrix} \begin{bmatrix} \delta_a \\ \delta_r \end{bmatrix}$$

Determine the features of the lateral-directional modes (i.e., roll, spiral, and dutch roll). Is the aircraft dynamically laterally-directionally stable? Why?

22 For the UAV given in problem 20, plot the variations of pitch angle, and airspeed to a $-2.3°$ elevator deflection. If the initial airspeed and pitch angle are 160 knots, and $3°$ respectively, what are the new values for these two flight parameters? The state-space equation is based on the ft/sec as the unit of airspeed, and radian as the unit of angles and deflection.

23 For the UAV given in problem 18, plot the variations of angle of attack to a $-1.5°$ elevator deflection input. If the initial angle of attack is $4°$, what is the new value for this flight parameter?

24 Consider a UAV with a turbofan engine and the following features:

$T_{maxSL} = 30$ kN, $S = 50$ m^2, $m_{TO} = 15{,}000$ kg, $C_{Lmax} = 2.1$, $K = 0.04$, $C_{Do} = 0.022$

Determine the absolute ceiling. Assume the UAV mass at the ceiling is 95% of the maximum takeoff mass.

25 Consider a UAV with a piston-prop engine and the following features:

$P_{maxSL} = 85$ kW, $S = 12$ m^2, $m_{TO} = 1{,}000$ kg, $C_{Lmax} = 1.8$, $K = 0.036$, $C_{Do} = 0.024$, $\eta_p = 0.74$

Determine the absolute ceiling. Assume the UAV mass at the ceiling is 94% of the maximum takeoff mass.

26 Consider a UAV with a turbo-prop engine and the following features:

$P_{maxSL} = 130$ kW, $S = 15$ m^2, $m_{TO} = 1{,}300$ kg, $C_{Lmax} = 1.9$, $K = 0.033$, $C_{Do} = 0.023$, $\eta_p = 0.72$

Determine the absolute ceiling. Assume the UAV mass at the ceiling is 93% of the maximum takeoff mass.

27 The UAV in problem 12 has a pack of Li-Po batteries with a mass of 0.15 kg. Determine the UAV maximum endurance. Hint: Calculate minimum required power first.

28 The UAV in problem 13 has a pack of Nickel–cadmium batteries with a mass of 0.8 kg. Determine the UAV maximum endurance. Hint: Calculate minimum required power first.

29 Determine the maximum speed of the UAV in problem 14 at 5,000 ft.

30 Determine the maximum speed of the UAV in problem 13 at sea level.

Note

1 http://NASA.gov.

References

1 Fitzpatric A. (2018). Drones are everywhere, get used to it., *Time magazine*, 11 June 2018

2 Federal Aviation Regulations, Part 107 (2016). *Operation and Certification of Small Unmanned Aircraft Systems*. Washington DC: Federal Aviation Administration, Department of Transportation.

3 Peters, J.E., Seong, S., Bower, A. et al. (2012). *Unmanned Aircraft Systems for Logistics Applications*. RAND.

4 Austin, R. (2010). *Unmanned Aircraft Systems: UAVs Design, Development and Deployment*. Wiley.

5 Erdman, T.J. and Mitchum, M. (2013). *Life-Cycle Cost Analysis for Small Unmanned Aircraft Systems*, NPS-CE-13-096, Naval Postgraduate School.

6 Daily Launch. (2018). American Institute of Aeronautics and Astronautics, Reston, VA, Oct 30, 2018

7 Swan, S. (2006). X-45A Unmanned Combat Vehicle on Display, *Aerotech News and Review*, Nov 17 2006

8 Sadraey, M. (2012). *Aircraft Design; A Systems Engineering Approach*. Wiley.

9 Sadraey, M. (2017). *Aircraft Performance Analysis: An Engineering Approach*. CRC Press.

10 Su, Y., Liahng, H., and Wu, J. (2008). "Multilevel Peukert Equations Based Residual Capacity Estimation Method for Lead-Acid Batteries," *IEEE International Conference on Sustainable Energy Technologies*, IEEE Publication, Piscataway, NJ, 2008, pp. 101–105.

11 Simpson, C. (1995). *Characteristics of Rechargeable Batteries*. Santa Clara, CA: National Semiconductor.

12 Schwartz, C.E., Bryant, T.G., Cosgrove, J.H. et al. (1990). A radar for unmanned air vehicles. *The Lincoln Laboratory Journal* 3 (1): 119–143.

13 Up: A Chinese firm has taken the lead in a promising market. *The Economist*, 11 April 2015.

14 Anderson, J.D. (2016). *Fundamentals of Aerodynamics*, 6e. McGraw-Hill.

15 Shevell, R.S. (1989). *Fundamentals of Flight*, 2e. Prentice Hall.

16 Megson, T. (2012). *Aircraft Structures for Engineering Students*, 5e. Elsevier.

17 Paul, K. (1956). *Stresses in Aircraft and Shell Structures*. McGraw-Hill.

18 Farokhi, S. (2014). *Aircraft Propulsion*, 2e. Wiley.

Design of Unmanned Aerial Systems, First Edition. Mohammad H. Sadraey.

Companion website: www.wiley.com/go/sadraey/unmanned-aerial-systems

19 Mattingly, J.D. *Elements of Gas Turbine Propulsion*, 2e. AIAA.
20 Dorf, R.C. and Svoboda, J.A. (2013). *Introduction to Electric Circuits*, 9e. Wiley.
21 Johnson, E.N. and Schrage, D.P. (January 2004). System integration and operation of a research unmanned aerial vehicle. *Journal of Aerospace Computing, Information, and Communication* 1: 5–14.
22 Linden, D. and Reddy, T.B. (2002). *Handbook of Batteries*, 3e. New York: McGraw-Hill.
23 Associate Administrator for Commercial Space Transportation (2011). *Flight Safety Analysis Handbook*. Federal Aviation Administration.
24 ARP, SAE. 4761 (1996). *Guidelines and methods for conducting the safety assessment process on civil airborne systems and equipment.*
25 Roskam, J. (2007). *Lessons Learned in Aircraft Design*. DAR Corporation.
26 Roskam, J. (2006). *Roskam's Airplane War Stories*. DAR Corporation.
27 Boyne, Walter J. (2009). How the Predator Grew Teeth. *Air Force Magazine*, July 2009
28 National Transportation Safety Board (2011). *Table 10: Aviation Accidents, Fatalities and Rates, 1990 Through 2009*. Washington, DC: U.S. General Aviation.
29 Sommer, G., Smith, G.K., Birkler, J.L. et al. (1997). *The Global Hawk Unmanned Aerial Vehicle Acquisition Process*, a summary of phase 1, DARPA.
30 Hodge N. (2011). U.S. Says Drone Cargo Plane Collide Over Afghanistan, *Wall Street Journal*, 17 August 2011
31 Tvaryanas, A. P. (2004). USAF UAV mishap epidemiology, 1997–2003, *Human Factors of Uninhabited Aerial Vehicles First Annual Workshop*, Phoenix, AZ, May 24–25, 2004.
32 Schaefer, R. (February 2003). *Unmanned Aerial Vehicle Reliability Study*. Washington, DC: Office of the Secretary of Defense.
33 Houghton, E.L. and Carpenter, P.W. (2003). *Aerodynamics for Engineering Students*, 5e. Elsevier.
34 Katz, J. and Plotkin, A. (2006). *Low-Speed Aerodynamics*, 2e. Cambridge University Press.
35 Eppler, R. (1990). *Airfoil Design and Data*. Berlin: Springer-Verlag.
36 Abbott, I.H., Von Doenhoff, A.F., and Stivers, L.S. Jr. (1945). *Summary Of Airfoil Data, National Advisory Committee For Aeronautics*. Report No. 824.
37 Abbott, I.H. and Von Donehoff, A.F. (1959). *Theory of Wing Sections*. Dover.
38 Selig, M.S., Guglielmo, J.J., Broeren, A.P., and Giguere, P. (1995). *Summary of Low-Speed Airfoil Data*. Virginia Beach, Virginia: SoarTech Publications.
39 Etkin, B. and Reid, L.D. (1996). *Dynamics of Flight, Stability and Control*, 3e. Wiley.
40 Roskam, J. (2007). *Airplane Flight Dynamics and Automatic Flight Controls*. DARCO.
41 Martins, J.R.A. and Lambe, A.B. (2013). Multidisciplinary design optimization: a survey of architectures. *AIAA Journal*, 51 (9): 2049–2075.
42 Liebeck, R.H. (January–February 2004). Design of the blended wing body subsonic transport. *AIAA Journal of Aircraft* 41 (1): 10–25.
43 Sobester, A. and Forrester, A.I. (2014). *Aircraft Aerodynamic Design: Geometry and Optimization*, 1e. Wiley.
44 Leishman, J.G. (2000). *Principles of Helicopter Aerodynamics*. New York, NY: Cambridge University Press.
45 Clough, B.T. (2005). Unmanned aerial vehicles: autonomous control challenges, a researcher's perspective. *Journal of Aerospace Computing, Information, and Communication* 2 (8): 1542–9423.

46 Beard, R.W., Kingston, D., Quigley, M. et al. (2005). Autonomous vehicle technologies for small fixed-wing UAVs. *Journal of Aerospace Computing, Information, and Communication* 2 (1): 1542–9423.
47 Salyendy, G. (2006). *Handbook of Human Factors and Ergonomics*, 3e. Wiley.
48 Sadraey, M. (2009). *Robust Nonlinear Controller Design for Complete UAV Mission*. VDM Verlag Dr. Müller.
49 Dorf, R.C. and Bishop, R.H. (2017). *Modern Control Systems*, 13e. Pearson.
50 Ogata, K. (2010). *Modern Control Engineering*, 5e. Prentice Hall.
51 James Stewart, (2015). *Calculus*, 8e, Brooks Cole.
52 Hoak, D.E., Ellison, D.E. et al. (1978). *USAF Stability and Control DATCOM*. Wright-Patterson AFB, Ohio: Flight Control Division, Air Force Flight Dynamics Laboratory.
53 Nelson, R. (1989). *Flight Stability and Automatic Control*. McGraw Hill.
54 Stevens, B.L., Lewis, F.L., and Johnson, E.N. (2016). *Aircraft Control and Simulation*, 3e. Wiley.
55 Sadraey, M and Colgren, R. (2006). Derivation of coupling stability derivatives and coupling transfer function. *6th AIAA Aviation Technology, Integration and Operations Conference*, Wichita, Kansas, Sept 2006
56 Butcher, J.C. (2016). *Numerical Methods for Ordinary Differential Equations*, 3e. Wiley.
57 The Mathworks, Inc., (2005). MATLAB, Simulink, www.mathworks.com.
58 Buschek, H. (Jan-Feb 1999). Full envelope missile autopilot design using gain scheduled robust control. *Journal of Guidance, Control, and Dynamics* 22 (1): 46–59.
59 McFarland, M. B. and Hoque, S. (2000). "Robustness of a Nonlinear Missile Autopilot Designed using Dynamic Inversion," *AIAA Guidance, Navigation, and Control Conference*, 14–17 August 2000.
60 Wang Q., and Stengel, Robert F. (1999). Robust Nonlinear Control of a Hypersonic Aircraft. *Guidance, Navigation, and Control Conference*, AIAA-99-4000.
61 Menon P. K., Iragavarapu V. R., and Ohlmeyer E. J. (1997). Nonlinear Missile Autopilot Design using Time-scale Separation. *Guidance, Navigation, and Control Conference*, AIAA-97-3765.
62 Snell, S.A., Enns, D.F., and Garrad, W.L. (July-August 1992). Nonlinear inversion flight control for a supermaneuverable aircraft. *Journal of Guidance, Control, and Dynamics* 15 (4): 976–984.
63 Mahony, R., Kumar, V., and Corke, P. (2012). Multirotor aerial vehicles: modeling, estimation, and control of quadrotor. *IEEE Robotics and Automation Magazine* 19 (3): 20–32.
64 Wingo, B. (2016). *Mathematical Modeling of Quadcopter Dynamics*. Rose-Hulman Undergraduate Research Publications, Rose-Hulman Institute of Technology.
65 Zhicheng H., Isabelle F., and Arturo Z. (2013). Modeling and Decentralized Control for the Multiple UAVs Formation based on Lyapunov design and redesign, *2nd IFAC Workshop on Research, Education and Development of Unmanned Aerial Systems*, 20–22 November 2013. Compiegne, France
66 Chapra, S.C. (2017). *Applied Numerical Methods with MATLAB for Engineers and Scientists*, 4e. McGraw-Hill.
67 Holmberg, Jessica A., King, Daniel, J., and Leonard, John R. (2008). Flying Qualities Specifications and Design Standards for Unmanned Air Vehicles, *AIAA Atmospheric Flight Mechanics Conference and Exhibit*, 18–21 August 2008, Honolulu, Hawaii
68 Bridger, R.S. (2008). *Introduction to Ergonomics*, 3e. CRC Press.

69 MIL-STD-1472G (2012). Department of Defense Design Criteria Standard: Human Engineering, 2012

70 Federal Aviation Regulations, Part 23, Airworthiness Standards: Normal, Utility, Aerobatic, and Commuter Category Airplanes. Washington, DC: Federal Aviation Administration, Department of Transportation

71 Federal Aviation Regulations, Part 25, Airworthiness Standards: Transport Category Airplanes. Washington, DC: Federal Aviation Administration, Department of Transportation.

72 Gieras, J.F. (2002). *Permanent Magnet Motor Technology: Design and Applications*, 2e. CRC Press.

73 Dell, T.W. (2015). *Hydraulic Systems for Mobile Equipment*. Goodheart-Willcox.

74 Beater, P. (2007). *Pneumatic Drives: System Design, Modelling and Control*. Springer.

75 MIL-STD-1797 (1997). *Flying Qualities of Piloted Aircraft*. Washington, DC: Department of Defense.

76 MIL-F-8785C (1980). *Military Specification: Flying Qualities of Piloted Airplanes*. Washington, DC: Department of Defense.

77 Watterson, Kent B. (1983). Bank-to-turn cruise missile terminal guidance and control law comparison, Master's Thesis, Naval postgraduate school, Monterey, CA, 1983

78 Anderson, B.D.O. and Moore, J.B. (2007). *Optimal Control: Linear Quadratic Methods*. Dover.

79 Kendoul, F. (2012). Survey of advances in guidance, navigation, and control of unmanned rotorcraft systems. *Journal of Field Robotics* 29 (2): 315–378.

80 Doyle, J.C. and Glover, K. (1989). State-space solution to standard H_2 and H_∞ control problems. *IEEE Transactions on Automatic Control* 34 (8): 831–847.

81 Phillips, C.L., Nagle, T., and Chakrabortty, A. (2014). *Digital Control System Analysis & Design*, 4e. Pearson.

82 Anonymous, Literature Review on Detect, Sense, and Avoid Technology for Unmanned Aircraft Systems. (2009). DOT/FAA/AR-08/41, National Technical Information Service.

83 Dalamagkidis, K., Valavanis, K.P., and Piegl, L.A. (2009). *On Integrating Unmanned Aircraft Systems into the National Airspace System: Issues*, Challenges, Operational Restrictions, Certification and Recommendations, International Series on Intelligent Systems, Control, and Automation: Science and Engineering, vol. 26. Springer-Verlag.

84 Blakelock, J.H. (1991). *Automatic Control of Aircraft and Missiles*, 2e. Wiley.

85 Heath, R.L. and Bryant, J. (2000). *Human Communication: Theory and Research: Concepts, Contexts, and Challenges*, 2e. Lawrence Erlbaum.

86 Wagner, S., Prinz, R., Bierstedt, C. et al. (2004). Accessible multimedia: status-quo, trends and visions. *IT – Information Technology*: 346–352.

87 Pastrick, H.L., Seltzer, S.M., and Warren, M.E. (1981). Guidance laws for short-range tactical missiles. *Journal of Guidance, Control, and Dynamics* 4 (2): 98–108.

88 Gu, D. (Feb. 2011). A game theory approach to target tracking in sensor networks. *Transactions on Systems, Man, and Cybernetics – Part B: Cybernetics* 41 (1): 2.

89 Di Cairano, S., Bemporad, A., and Caldelli, A. (2007). Moving Target Detection and Tracking in Wireless Sensor Networks, *Proceedings of the European Control Conference*, Kos, Greece, 2–5 July 2007

90 Uijlings, J.R., Van De Sande, K.E., Gevers, T., and Smeulders, A.W. (2013). Selective search for object recognition. *International Journal of Computer Vision* 104 (2): 154–171.

91 Davies, R. (2017). *Computer Vision: Principles, Algorithms, Applications, Learning*, 5e. Academic Press.

92 Shneydor, N.A. (2008). *Missile Guidance and Pursuit: Kinematics, Dynamics and Control*. Cambridge: Woodhead Publishing Limited, UK.

93 Zarchan, P. (2013). *Tactical and Strategic Missile Guidance*, 6e. Reston, VA: American Institute of Aeronautics and Astronautics.

94 Siouris, G.M. (2004). *Missile Guidance and Control Systems*. Springer.

95 Meenakshisundaram, V.S., Gundappa, V.K., and Kanth, B.S. (2010). Vector field guidance for path planning of MAVs in three dimensions for variable altitude maneuvers. *International Journal of Micro Air Vehicles* 2: 255–265.

96 Cichella, V., Choe, R., Mehdi, S. Bilal et al. (2013). A 3D Path-Following Approach for a Multirotor UAV on SO(3), *2nd IFAC Workshop on Research, Education and Development of Unmanned Aerial Systems*, November 20–22, 2013. Compiegne, France.

97 Degen, S.C., Alvarez, L.M., Ford, J.J. and Walker, R.A. (2009). Tensor Field Guidance for Time-Based Waypoint, Arrival of UAVs by 4D Trajectory Generation, *IEEE Proceedings of Aerospace Conference*, USA.

98 Capello, E., Guglieri, G., and Quagliotti, F. (2013). A Waypoint-Based Guidance Algorithm for mini UAVs, *2nd IFAC Workshop on Research, Education and Development of Unmanned Aerial Systems*, November 20–22, 2013. Compiegne, France

99 Prats, X., Delgado, L., Ramirez, J. et al. (May 2012). Requirements, issues, and challenges for sense and avoid in unmanned aircraft systems. *Journal of Aircraft* 49 (3): 677–687.

100 FAA (2007). Unmanned Aircraft Operations in the National Airspace System, 72 Fed. Reg. 6689, 6690, 2007

101 Angelov, P. and Angelov, P. (2012). *Sense and Avoid in UAS: Research and Applications*. Wiley.

102 Unmanned aircraft system airspace integration plan, Version 2.0 ed, in Series Unmanned Aircraft System Airspace Integration Plan, Department of Defense, 2011

103 US Code of Federal Regulations. Title *14 Aeronautics and Space*, Part 91 General operating and flight rules, Section 225, Automatic Dependent Surveillance – Broadcast (ADS-B).

104 Theodore, C., Rowley, D., Hubbard, D. et al. (2006). Flight trials of a rotorcraft unmanned aerial vehicle landing autonomously at unprepared sites, *Proceedings of the 62nd Annual Forum of the American Helicopter Society*, Phoenix, AZ, May 2006

105 Airborne Collision Avoidance System Manual, Doc 9863, 1st ed. (2006). International Civil Aviation Organization.

106 Minimum operational performance standards for traffic alert and collision avoidance system II, version 7.1, DO-185B, RTCA, Inc., June 2008.

107 Han, S.C. and Bang H. (2004). Proportional navigation-based optimal collision avoidance for UAVs. *Second International Conference on Autonomous Robots and Agents*, pp. 76–81, Massey University, New Zealand.

108 Ota, T. Nagati and Lee, D. (1998). Aircraft collision avoidance trajectory generation, in *Proceedings of the AIAA Guidance, Navigation, and Control Conference*, Boston, MA, pp. 828–837, August 1998.

109 Kim, Hae-Jun and Ahn, Hyo-Sung (2016). Realization of Swarm Formation Flying and Optimal Trajectory Generation For Multi-Drone Performance Show, *Proceedings of*

the 2016 IEEE/SICE International Symposium on System Integration, Japan, 13–15 December 2016.

110 Marques, P. and Da Ronch, A. (2017). *Advanced UAV Aerodynamics, Flight Stability and Control: Novel Concepts, Theory and Applications*. Wiley.

111 Kluever, C.A. (November–December 2004). Unpowered approach and landing guidance using trajectory planning. *Journal of Guidance, Control, and Dynamics* 27 (6): 967–974.

112 Ben-Asher, J.Z. and Yaesh, I. (1998). *Advances in Missile Guidance Theory*. Reston, VA: American Institute of Aeronautics and Astronautics.

113 Palumbo, N.F., Blauwkamp, R.A., and Lloyd, J.M. (2010). Basic principles of homing guidance. *Johns Hopkins APL Technical Digest* 29 (1): 25–40.

114 Jamal, S.Z. (2012). Tightly coupled GPS/INS airborne navigation system. *IEEE Aerospace and Electronic Systems Magazine* 27 (4): 39–42.

115 Kortunov, V.I., Dybska, I.Y., Proskura, G.A., and Kravchuk, A.S. (Jan. 2009). Integrated mini INS based on MEMS sensors for UAV control. *IEEE Aerospace and Electronic Systems Magazine* 24: 41–43.

116 Merrigan, M., Swift, E., Wong, R., and Saffel, J.; "A Refinement to the World Geodetic System 1984 Reference Frame"; Proceedings of the ION-GPS-2002; Portland, Oregon; September 2002

117 Grewal, M.S., Andrews, A.P., and Bartone, C.G. (2013). *Global Navigation Satellite Systems, Inertial Navigation, and Integration*, 3e. Wiley.

118 Zarchan, P. (2015). *Fundamentals of Kalman Filtering*, 4e. Reston, VA: American Institute of Aeronautics and Astronautics.

119 Grewal, M.S. and Andrews, A.P. (2008). *Kalman Filtering: Theory and Practice Using MATLAB*, 3e. Wiley.

120 El-Rabbany, A. (2006). *Introduction to GPS: The Global Positioning System*, 2e. Artech House Publishers.

121 Santos, J.P.M.D. (2012). *Robotic Navigation in Reduced Visibility Scenarios: State of The Art*, Technical report, Faculty of Science and Technology, University of Coimbra.

122 Fang, Z., Yang, S., Jain, S. et al. (2015). Robust Autonomous Flight in Constrained and Visually Degraded Environments, in *Field and Service Robotics*, Toronto, Canada, 24–26 June 2015.

123 Motion Design (2018). Supplement to Tech Briefs magazine, Page 13, *SAE International*, Vol 42, No. 12 Dec. 2018

124 Rauw, M.O. (2001). *FDC 1.2 – A Simulink Toolbox for Flight Dynamics and Control Analysis*, 2. Zeist, The Netherlands: Dutchroll.

125 Hibbeler, R.C. (2015). *Engineering Mechanics: Dynamics*, 14e. Pearson.

126 Cady, F.M. (1997). *Microcontrollers and Microcomputers*. Oxford University Press.

127 Ball, S. (2001). *Analog Interfacing to Embedded Microprocessors*. Newnes.

128 Berger, A.S. (2002). *Embedded Systems Design*. CMP Books.

129 Miller, G.H. (2004). *Microcomputer Engineering*, 3e. Pearson Prentice Hall.

130 Wilmshurst, T. (2009). *Designing Embedded Systems with PIC Microcontrollers, Principles and Applications*, 2e. Newnes.

131 ATmega644P microcontroller Datasheet Complete. (2016). Atmel.

132 Alciatore, D.G. (2011). *Introduction to Mechatronics and Measurement Systems*, 4e. McGraw-Hill.

133 Graham, D.L. (2016). *C Programming Language: A Step by Step Beginner's Guide to Learn C Programming in 7 Days*. CreateSpace Independent Publishing Platform.

134 Blum, J. (2013). *Exploring Arduino*, 1e. Wiley.

135 Geddes, M. (2016). *Arduino Project Handbook: 25 Practical Projects to Get You Started*, 1e. No Starch Press.

136 www.arduino.cc

137 Monk, S. (2016). *Programming Arduino: Getting Started with Sketches*, 2e. McGraw-Hill.

138 ArduPilot 2.6 manual (2003). ArduPilot Development Team, http://ardupilot.org,

139 Lavagno, L., Martin, G., and Scheffer, L. (2006). *Electronic Design Automation for Integrated Circuits Handbook*. CRC Press.

140 Bronson, R. (2003). *Outline of Theory and Problems of Differential Equations, Schaum's Series*, 2e. McGraw-Hill.

141 Budynas, R.G. and Nisbett, J.K. (2015). *Shigley's Mechanical Engineering Design*, 10e. McGraw-Hill.

142 Tenenaum, R.A. (2004). *Fundamentals of Applied Dynamics*. Springer-Verlag.

143 Erdman, A.G., Sandor, G.N., and Kota, S. (2001). *Mechanism Design: Analysis and Synthesis*, 4e. Pearson.

144 Gomis-Bellmunt, O. and Campanile, L.F. (2010). *Design Rules for Actuators in Active Mechanical Systems*. London: Springer.

145 Adamski, M.D., Root, R.G., Jr., and Watts, A.M. (2007). *UAV recovery system*, U.S.A. Patent US7219856 B2.

146 Butler C Manley, Loney, Troy. (1995). Design, Development and Testing of a Recovery System for the Predator UAV. *13th AIAA Aerodynamic Decelerator Systems Technology Conference*, 15–19 May 1995, Clearwater Beach, FL

147 DJI (2017). DJI Phantom Instruction Manual (https://www.dji.com/phantom-4-adv/info#specs).

148 Sato, A. (2003). *The RMAX Helicopter UAV, Aeronautic Operations*. Japan: Yamaha Motor Co.

149 Tso K., Tharp G. K., Tai A. T., et al. (2003) A Human Factors Testbed for Command and Control of Unmanned Air Vehicles. *22nd Digital Avionics Systems Conference*, Oct 2003, Indianapolis, IN.

150 Larsen Angel, W. (2011). *HVAC Design Sourcebook*. McGraw-Hill.

151 Kwon, Y., Heo, J., Jeong, S. et al. (2016). Analysis of design directions for ground control station. *Journal of Computer and Communications* 4: 1–7.

152 Jackson P., Jane's All the World's Aircraft, Jane's information group, Various years

153 CNN, Washington. 14 December 2016, http://money.cnn.com.

154 Kharchenko, V. and Prusov, D. (2012). Analysis of unmanned aircraft systems application in the civil field. *Transport* 27 (3): 335–343. https://doi.org/10.3846/16484142.2012.721395.

155 Ostroverkhova, O. (2013). *Handbook of Organic Materials for Optical and Electronic Devices: Properties and Applications*. Woodhead Publishing.

156 Richards, M.A., Scheer, J.A., and Hilm, W.A. (2010). *Principles of Modern Radar: Basic Principles*. SciTech Publishing.

157 Daily Launch. American Institute of Aeronautics and Astronautics, Reston, VA, June 13, 2018

158 Bellardo, J. and Savage S. (2003). Denial-of-service attacks: Real vulnerabilities and practical solutions, *Proceedings of the 12th Conference on USENIX Security Symposium*, 802. 11, pp 15–28, 2003

159 Electronic Warfare Forecast. (2007) www.forecastinternational.com, Archived report, February 2007.

160 Cengel, Y.A. and Cimbala, J.M. (2013). *Fluid Mechanics: Fundamentals and Applications*, 3e. McGraw-Hill.

161 Toolev, M. and Wyatt, D. (2007). *Aircraft Communications and Navigation Systems: Principles, Maintenance and Operation*. Routledge.

162 Ellingson, S.W. (2016). *Radio Systems Engineering*, 1e. Cambridge University Press.

163 Tooley, M. and Wyatt, D. (2017). *Aircraft Communications and Navigation Systems*, 2e. Routledge.

164 Sen, A.K. and Bhattacha, A.B. (2018). *Radar Systems and Radio Aids to Navigation*. Mercury Learning & Information.

165 Frenzel, L. (2015). *Principles of Electronic Communication Systems*, 4e. McGraw-Hill.

166 Rodday, N.M., Schmidt, R.D.O., and Pras, A. (2016). Exploring security vulnerabilities of unmanned aerial vehicles, *Network Operations and Management Symposium*, 2016 IEEE/IFIP, 2016, pp. 993–994

167 Yoon, K., Park, D., Yim, Y. et al. (2017). Security Authentication System using Encrypted Channel on UAV Network, *First IEEE International Conference on Robotic Computing*, Taiwan, 10–12 April, 2017

168 Ferguson, N. (2010). *Cryptography Engineering: Design Principles and Practical Applications*. Wiley.

169 ArduPilot 2.x manual (2003) ArduPilot Development Team, http://ardupilot.org.

170 Mclean, D. (1990). *Automatic Flight Control Systems*. Prentice-Hall.

171 Traub, L.W. (March–April 2011). Range and endurance estimates for battery-powered aircraft. *Journal of Aircraft* 48 (2): 703–707.

172 Blanchard, B.S. and Fabrycky, W.J. (2006). *Systems Engineering and Analysis*, 4e. Prentice Hall.

173 FY 2018 Program acquisition cost by weapon system (2017). US Department of Defense, Ref. ID: E-7DE12B0.

Index

Design of Unmanned Aerial Systems, First Edition. Mohammad H. Sadraey.

Companion website: www.wiley.com/go/sadraey/unmanned-aerial-systems

d

e

f

t

u

v

w

Printed and bound by CPI Group (UK) Ltd, Croydon, CR0 4YY
17/09/2026
14964271-0003